YEARS AGO (MILLIONS)	
135×10^6	**Cretaceous Period Begins**
	First fossil angiosperms
110×10^6	Earliest known pines
100×10^6	
	Spread of mammals
65×10^6	First extant genera of angiosperms
	Beginning of modern forests
63×10^6	**Tertiary Period Begins**
	Earliest apes and monkeys
25×10^6	Rise of herbaceous angiosperms
10×10^6	
3×10^6	***Australopithecus***
2×10^6	First stone tools
1×10^6	

YEARS AGO (TENS OF THOUSANDS)	
100×10^4	***Homo erectus*** First glacial period begins. Specialization of Herbs
90×10^4	
80×10^4 – 40×10^4	Sequence of glacial & inter-glacial periods in the north; alternating wet & dry periods in the south, throughout this time
60×10^4	
40×10^4	

YEARS AGO (TENS OF THOUSANDS)		
	First use of fire by humans	
30×10^4		
20×10^4	Neanderthal man	
2×10^4	Modern man ***(Homo sapiens)***	
1×10^4	8000 BC	End of glacial periods
		Origins of agriculture
0.9×10^4	7000 BC	Near East
0.7×10^4	5000 BC	Meso-America
0.6×10^4	4000 BC	Europe

PLANTS **HUMANS**

continued on inside back cover

Plant Biology and Its
Relation to
Human Affairs

"The greatest service which can be rendered any country is to add a useful plant to its culture."

Thomas Jefferson (from Volume I of his Writings).

Leonardo da Vinci (1452-1519) was one of the few men equally successful in pure and applied science and in the arts. Although less famous for his work on plants than for that in other fields, he was as interested in them as in all living things, and his work and life present a striking example of "science in its relation to human affairs". Our cover shows a page from one of his many notebooks, with careful drawings of two plants and their description in his well-known "mirror writing", which reads from right to left.

BOTANY

Plant Biology and Its Relation to Human Affairs

Jean H. Langenheim
Kenneth V. Thimann
University of California, Santa Cruz

John Wiley & Sons, New York • Chichester • Brisbane • Toronto • Singapore

Production Supervisor—Nina R. West
Designer—Sue Taube
Picture Editor—Kathy Bendo
Picture Research—Terri Leigh Stratford
Illustrations—John Balbalis

Cover drawing from Leonardo da Vinci's Studies in Nature.
From The Royal Library at Windsor Castle.
By gracious permission of Her Royal Majesty Queen Elizabeth II.

Library of Congress Cataloging in Publication Data

Langenheim, Jean H.
Botany: plant biology and its relation to human affairs.

Includes index.
1. Botany. 2. Botany, Economic. I. Thimann,
Kenneth Vivian, 1904- II. Title.
QK47.L37 581.6'1 81-7466

ISBN 0-471-85880-3 AACR2

Printed in the United States of America

10 9 8 7 6 5 4 3

PREFACE

The developing interest of undergraduates in the world's problems — the demands for food, fiber, energy, the protection of the environment and the need to understand the interrelations between the human population and the plant world — has made desirable an alternative kind of botany course. Although this book covers the same basic principles of plant science as those generally presented in introductory texts, there is a change in focus, a focus on plants that are closely interwoven with the fortunes of mankind. Angiosperms and gymnosperms are of necessity in the center of the stage, but other groups (traditionally considered as plants) play lesser parts. The subjects discussed in this book include not only the hundred or so special plants selected over the centuries to furnish us the necessities of life, but also the diversity of plants making up ecosystems of which many have been modified by human activities. The teaching of botany has for too long been separated from its human aspects, and this book is a step toward reestablishing this relationship. In other words, it offers a balanced perspective toward what have euphemistically been called "pure" and "applied" approaches. This concept is discussed in more detail in Chapter 1, "What This Book Is About," so that students may understand the overall design.

The plan of the book allows it to be used either for a general course in plant science, with a stress on the flowering plants, or for courses often called "economic botany" in which a truly solid scientific content is desired. For the latter, it is not essential that the topics be covered with our preferred sequence. Some of the sections dealing with general principles could be used mainly for those students who require a deeper understanding of the structure and function of plants and the dynamics of populations and communities. In many of the topics, historical aspects prove enlightening, and it is hoped that the student will gain a sense of excitement from our current understanding of plants and its steady growth. Perhaps a few may be moved to participate as future researchers in developing such understanding even further.

A number of experts have given thoughtful reviews of chapters or sections of the book when in draft form; they include Professors Elso Barghoorn, Harry Beevers, H. L. Mason, Lincoln Taiz, Ralph Wetmore, and Adrienne Zihlman. Helpful comments were contributed also by students, graduate and undergraduate, who either took the course or have been teaching assistants in it over the years that it has been presented at Santa Cruz. In expressing our thanks to all these interested collaborators, we make the usual proviso that the responsibility for everything in the book rests with us alone. Typing of the many drafts was patiently undertaken by Gladine Clayton, Susan Curtis, Elinor Gossen, Dottie Hollinger, Kay House, and Joyce Motz. Their labors are greatly appreciated. The publishers also rendered most valuable assistance in many ways.

Santa Cruz, California.

Jean H. Langenheim
Kenneth V. Thimann

CONTENTS

BOTANY

Plant Biology and Its Relation to Human Affairs

PART 1

GENERAL AND HISTORICAL BACKGROUND

CHAPTER 1

What This Book Is About

The aim of this book is to show how plants live and grow and how they influence our lives and our ideas. It is also designed to show you the opposite side of the coin, namely how human beings have modified the types of plants on earth and the patterns of vegetation. Changes in the natural vegetation and ecosystems began with the use of fire, but vastly increased with the development of agriculture and the related growth of the human population and its technology. Agriculture, in turn, has had a tremendous impact on the rise of civilization. A reliable supply of food from cultivated plants gave our ancestors a settled life, from which arose some division of labor and an opportunity to engage in thoughtful and creative pursuits.

In presenting plants and their relation to human activities, we first follow the evolution of plants and of our human species; then we trace the gradual growth of our understanding of plants and our relationships with them. We see them as individual inhabitants of soil, as providers of our food, in populations, and as components of vegetation. Many kinds of plants populate the earth and their diversity can be bewildering. Plants range in size from microscopic single cells to giant trees, and in lifetimes from an hour to many centuries. A list of the major groups of organisms traditionally treated as plants is shown in Table 1-1. A more formal classification of the organisms used in this book, together with a discussion of the view that some are sufficiently distinctive to be included in other kingdoms, is given in Chapter 13.

The bacteria were the first living organisms, and they are the smallest and most abundant ones in the world. They occur in all habitats, and because their metabolism is very versatile they can live in environments that support no other form of life. For these reasons they play a major role in decomposition, fermentation, and the cycling of organic matter. The algae are a large group of plants, nearly all aquatic, that have developed in several evolutionary directions. They represent the photosynthetic line of development (foreshadowed by a few rather special and perhaps primitive photosynthetic bacteria), and among those the green algae were the precursors of the land plants. In marine environments the algae play an ecological role comparable to that of the green plants on land. A few of them serve us for food and other products, but their role as food for the animals whose homes are in fresh water or the oceans is of far greater significance.

The various types of fungi have ramified in directions quite other than those of the green plants. One of their roles in the context of this book is that they, together with the bacteria, are the decomposing agents of the biosphere, and their activities in thus recycling all organic mat-

TABLE 1-1

Brief Survey of the Groups Traditionally Regarded as Plants

I. Plants mostly without supporting or conducting tissue. Plant body often relatively small; some are nonphotosynthetic.

BACTERIA The smallest of all organisms, unicellular. Mostly multiplying by fission, some forming long filaments. Some photosynthetic but the majority not. They have no true nucleus. They grow in or on living organisms or dead matter. The blue-green algae are usually placed here, too, and called Cyanobacteria because, although all are photosynthetic, these are like the bacteria in having no true nucleus.

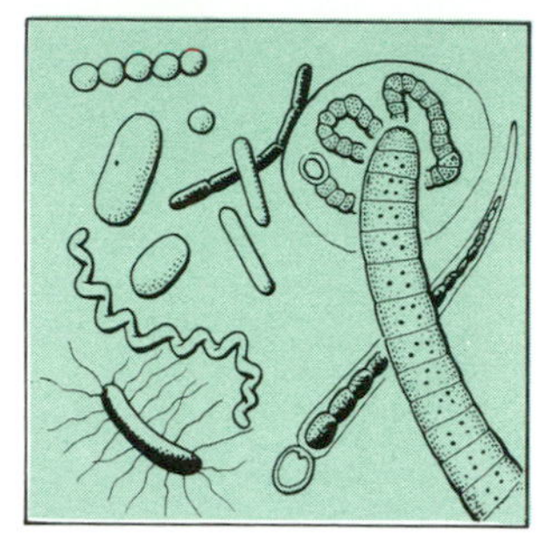

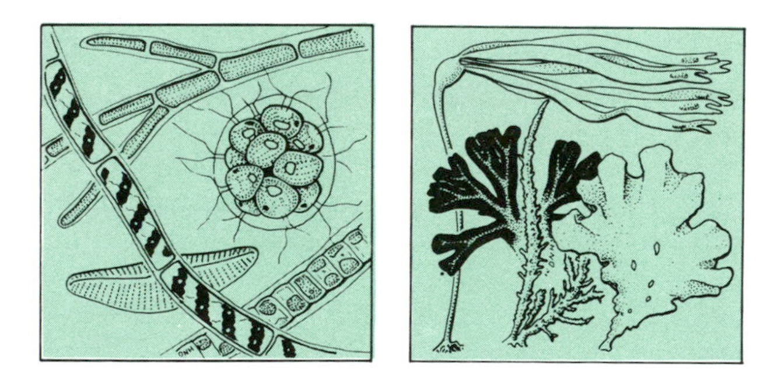

ALGAE Unicellular or multicellular plants. Almost wholly aquatic. Microscopic forms in fresh water, soil, and the open ocean; large seaweeds mainly near the seashore (Some of these have conducting tissue). A large and varied group usually classified by their pigments into green, brown, red, golden, and so on. These, and all the organisms below, have true nuclei.

FUNGI Nonphotosynthetic, growing on living organisms or dead matter. From single cells (e.g., yeasts) through microscopic forms with spreading filaments (hyphae) to large multicellular mushrooms and bracket fungi.

TABLE 1-1 continued

LICHENS Composed of an alga and fungus growing in mutual relationship. Usually found on tree trunks, rocks, or soil. Crustlike, leaflike, or tiny shrublike.

LIVERWORTS AND MOSSES Small photosynthetic, moisture-loving plants. The liverworts are creeping or prostrate; the mosses with short, erect leafy stems, bearing stalked spore capsules.

II. Plants with supporting or conducting tissue. Plant body relatively large; photosynthetic (with a few exceptions).

A. Plants reproducing only by spores.

FERNS Leaves mostly large and commonly divided, bearing spores in minute sporangia on the lower surface.

CLUB MOSSES Trailing or creeping plants with scalelike leaves. Sporangia usually grouped into slender clublike or conelike structures at the ends of the branches.

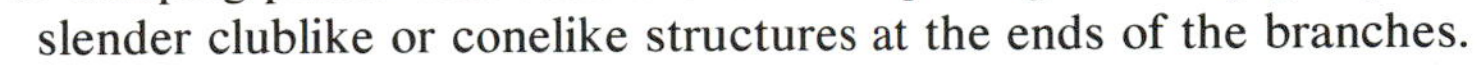

TABLE 1-1 continued

HORSETAILS Upright plants with hollow, jointed stems and minute leaves. Sporangia in conelike structures.

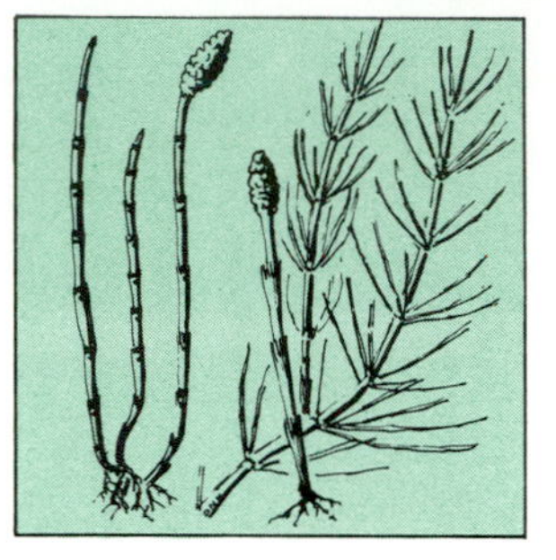

B. Plants forming seeds.

GYMNOSPERMS Mostly evergreen trees and shrubs. Primitive forms still extant include the shrubby cycads and ginkgo trees. Conifers (such as pines, spruces, redwoods, hemlocks, and Douglas firs) are the abundant forms today. All these except a few (e.g., ginkgos) bear their seeds in cones, unenclosed.

ANGIOSPERMS (FLOWERING PLANTS) Plants bearing flowers; seed contained within a fruit. Subdivided into two groups: monocotyledons, with one seed leaf, and dicotyledons, with two. The groups differ with regard to the structure of the leaf, stem, and flower (See Table 4-1).

ter are every bit as necessary as those of the producers. As decomposers, however, they often come into conflict with human interests; for example, fungi make no distinction between a fallen tree in the forest and a telephone pole. Their other major role is that the fungi are responsible for some of the most destructive diseases of our cultivated plants and forests.

The simplest of all land plants, the mosses and liverworts, play a crucial role in plant succession and in the colonization of new habitats. The ferns of today and their allies (club mosses and horsetails) had giant ancestors that populated the vast coal-forming forests that once covered the earth. They early on developed conducting and supporting tissue in the plant body that led to the subsequent dominance of the conifers and then of the flowering plants. A crucial adjustment to this dominance, however, was the development of the seed. The conifers are primarily evergreen trees that bear their seeds in cones and have remained relatively close to their primeval forms. The flowering plants, on the other hand, have given rise to a wide variety of growth habits, including the herbaceous form, and consequently have adapted to a much wider range of environments. The plants we use for food, fiber, refreshment, healing and for many other functions (including aesthetic satisfaction) are for the most part flowering plants, and hence the flowering plants are the focus of this book.

The endpapers present a scheme for the evolution of plants and mankind in geologic time and for the development of plant science in historic time. The biologist deals with a range of time values that is difficult to grasp at first. In the early stages of the earth's development, for which information is scarce, we must deal in thousands of millions of years. By about the time when land plants appear on the scene we can begin to deal in millions of years. At that time more than four-fifths of the earth's development had been completed. When *Homo sapiens,* the species to which we belong, appeared, we can deal in thousands of years. Agriculture began about 10,000 years ago, which is only about 400 human generations. Historical time, that is, the time for which records of human activities are available, is conveniently measured in centuries, whereas the growth of modern plant science took place over only a few decades. Many plants live within the cycle of a single year. Growth is commonly measured over days or hours, while the response of plants to light involves reactions timed in microseconds or even nanoseconds (billionths of seconds).

Most of the plant science with which this book deals took place within the last 150 years, although sophisticated agricultural techniques have been in use for 2000 or 3000 years, along with the first steps in the understanding of organisms. Reference to the endpapers will show that, after scientists struggled to classify the great diversity of plants and worked out how to distinguish one plant species from another, most of our major steps forward have been taken in little more than the past century.

After dealing with the long history of plant evolution and the much shorter history of plants' interrelations with the human species, our book takes up first the individual green plant and its role in world processes; its use of light to produce all the kinds of organic matter that make up the biosphere; its uptake of water and nutrients from the soil; its structure from seed to flowering; its growth and its inheritance. Next we discuss how plant populations have become diversified and how all these multifarious plant forms can be classified. Following this, we explore the ways in which ecological processes and ecosystems work, the formation of different types of soil, and our use of the land.

This fundamental information is a prerequisite to the understanding of the plants that people utilized and of our mutual relationships with them. We thus take up in turn the chief groups of plants and their products, showing something of how these products are formed and how they are processed for human use. The cereals naturally come first, since they provide the largest part of human energy and offer an outstanding example of the outcome of selection and breeding over the centuries. Other plants notable for their carbohydrate production logically follow. Description of specific groups of plants and their products is often accompanied by treatment of the principles underlying the formation and use of these materials. Thus, study of sugar plants and of fruits leads to a discussion of fermentation and thus of the yeasts; discussion of fiber plants entails treatment of cell walls and their constituent polysaccharides, as well as of textiles, and presentation of the edible roots and tubers leads to some understanding of the mechanisms of storage and conversion of food reserves. Fruits are discussed in relation to the variety of reproductive structures that plants have evolved and the physiology of ripening. In describing wood products we take up both the development of secondary tissues and the ways forests can be most effectively utilized.

Under the topics concerned with nitrogen, we deal with protein formation and destruction, the nitrogen cycle, and the special role of legumes in it. There follows a discussion of medicinal alkaloids and their part in history, and of caffeine and its role in beverages and hence in social life. Also, because the beverage plants are a leading commodity in some tropical economies, we have discussed here some agricultural practices used for many tropical crops. Next we deal with the lipids, including fats and oils, and the terpenoids, and here we consider not only their biosynthesis within the tissues and the use of fats as food but the importance of the terpenoid-producing plants in providing perfumes, spices, resins, and rubber. Historical, anthropological, and even biochemical approaches thus become a part of the broad interpretation of the plant sciences, as indeed they must be.

In the last part of the book we try to draw together much of the information from the preceding parts into a discussion and analysis of world plant usage, its scope, its problems, and its future. Included here are such matters as how gardens have been centers of knowledge about nature, the impact of plant diseases and pests on our supplies of food and fiber, and our all-important food supply in the context of human nutrition and the growth of population. Last comes a historical perspective of our modification of the natural ecosystems, and the development of models of ecosystems for their optimal usage and preservation. Throughout we have tried to compare the natural scene with the results of human activity.

CHAPTER 2

The Origin of Life and the Evolution of Plants

So far as we know now, there is no evidence of living organisms on other planets of our solar system, although, since astronomers tell us there are thousands of solar systems in existence, there is likely to be life of *some* sort in *some* of them. But we see from experience on the moon that rather special conditions must be needed for living organisms to develop. Perhaps the most important single condition is the presence of liquid water. Water is such a wonderfully effective solvent, dissolving sugars, amino acids, proteins, and a great many salts, that its special ability thus to bring biological agents into contact with each other is of the greatest value in making life possible. No other liquid is such a broadly acting solvent. Once biological chemicals are thus in contact with one another, of course, they can interact. Even though some reactions occur on the surfaces of solids rather than in true solution, still the watery solution was needed to bring the materials to the active surface. Water is liquid only between certain temperatures and pressures, so this sets the possible range for life. Conditions on other planets in our solar system are unsuitable for liquid water, so we cannot expect to find living organisms there. Thus life, if it exists elsewhere, must be found on other solar systems in the universe. Or it must be of some form we do not now perceive.

Water also has other very valuable properties that are not common in other liquids; for example, it has a high heat capacity; that is, it takes a relatively large amount of heat to raise its temperature, so that the temperatures of lakes and oceans are relatively stable. This stability is increased by the fact that ice floats, so that plants and animals can overwinter in the lower layers. The importance of this for the fauna and flora of lakes in the temperate zone needs no emphasis. Water is exceptional in this respect, for in most materials the solid is denser than the melted form. Also clear water is relatively transparent, so that plants can receive light down to depths of 50 meters or so; and thus can populate many bodies of water.

Besides the importance of liquid water, there are other considerations that further limit the conditions under which life might develop. Naturally we do not know them all, nor indeed do we really know the course of evolution of life on earth. Much of how we visualize the origin of life began as a sort of scientific speculation in the fertile brains of J. B. S. Haldane in England and A. I. Oparin in Russia (1938). Over the years, however, more and more facts have fitted in with their general concepts and have amplified the picture, till it is now rather generally accepted, though it may be modified as new facts are discovered. It goes as follows.

1. THE EARLIEST STAGES: FORMATION OF ORGANIC COMPOUNDS

In the earliest days, when the earth was, as the Book of Genesis says, "without form and void," the atmosphere contained little or no free oxygen. The oxygen that came with the earth's formation had largely been fixed by combining with iron, carbon, and other elements, so that the atmosphere was mostly nitrogen and carbon dioxide (CO_2), while the rocks contained fixed oxygen in such forms as silica, iron oxide, and calcium sulfate. Ultraviolet light from the sun then reached the earth in far higher intensity than it does now. The ultraviolet wavelengths, those shorter than 0.4 microns or 400 nanometers (400×10^{-9} meters),* are of higher energy than visible light; that is, the units or *quanta* of which they are composed are larger than the quanta in the visible part of the spectrum. (The reason for this is that the energy of a quantum of light is inversely proportional to its wavlength, $e \alpha 1/\lambda$ (see Chapter 5). In the early days, therefore, the earth was bombarded with ultraviolet light of high intensity. This strong light activated the small molecules it struck, making them chemically reactive so that they combined with one another, and over the years the resulting products underwent further chemical reactions, finally giving rise to relatively elaborate compounds that we could consider "organic" because we generally know them now as products of living organisms.

Attempts have been made to test this view of the origin of organic compounds by studying mixtures of simple gases that are known to occur elsewhere in the universe, such as ammonia, methane, hydrogen, carbon dioxide, and water, and analyzing the products. They have been exposed to X rays, ultraviolet light, or the electric spark, the last giving the best results so far. Simple compounds usually of biological origin such as formaldehyde, glycine, alanine, and lactic and aspartic acids have been identified as products in this type of experiment. They are formed by the interaction of CO_2 with the hydrogen of the other compounds, that is, by the combination of CO_2 with hydrogen, whereby part of the oxygen is converted to water, H_2O. Recently astronomers have found traces of such simple organic compounds free in interstellar space. Some of the reactions would be prevented by oxygen because it would combine at once with unstable intermediate compounds in these processes; thus, these reactions could have occurred only in the primeval absence of free oxygen. In this way a series of small reservoirs of compounds of "biological" type would have been built up on earth, and the continuing influx of high-energy quanta led gradually to the formation of larger and larger molecules.

2. THE DEVELOPMENT OF PRIMITIVE ORGANISMS

At some stage a sort of *membrane* could have been formed, doubtless made of oily materials that do not mix with water, and this would have enclosed a limited volume of solution and separated it from its surroundings. Within this enclosed part, it is assumed that processes energized by light, that is, photochemical reactions, continued until the products included some larger and more complex molecules, such as proteins. Chemical reactions not needing light, but activated instead probably by heavy metals like copper and iron, would begin to take place to an increasing extent alongside the photochemical reactions. The membrane-enclosed body had thus become a sort of "protobiont," more concentrated than the external solution and continually undergoing chemical changes or "metabolism." After a time the growing mass might well split into two and thus the "protobiont," or reaction system, would have mul-

* 1 millimeter (mm) = 1/1000 meter (m) or 10^{-3} meter.
1 micron (μ) = 1/1000 millimeter or 10^{-6} meter.
1 millimicron (mμ) or 1 nanometer (nm) = 1/1000 micron or 10^{-9} meter.

tiplied. At what point the term "living" might be appropriately applied is hard to say, but probably the most critical property of a living organism is the existence of *interrelation* and *control* among the various metabolic processes.

Metabolism itself, or the exchange of atoms and groups of atoms between the chemical materials that compose an organism, is made possible, or *catalyzed,* by enzymes. These are proteins, existing in wide variety, whose effect is to speed up reactions that would otherwise occur far too slowly to support life. Each *catalyzes* one specific reaction-type. We meet a number of enzymes as we consider the processes that take place in living plants. A limited amount of metabolism can even go on in plant extracts, which can hardly be considered living, and indeed some enzymes can still be active for a long time after separation from the living plant. But the activities of enzymes in extracts are generally independent of one another, while in the living plant the activities of the different enzymes are usually coordinated. Thus it is, as just stated, *interrelation* between enzymatic reactions (or groups of such reactions) that is really characteristic of *organisms.*

Once an enzyme molecule has been formed it will tend to promote or catalyze those reactions for which it is fitted, and this process can then go unchecked unless there is some scheme to regulate it in the organism. So far as we know now, such regulation in simple cells is achieved by a "feedback" mechanism. Thus, when the concentration of any one compound gets high enough, the process producing it becomes inhibited, so that its concentration levels off. This can happen either (a) by direct participation of the product in the reaction, thus slowing down the enzyme action producing it, or (b) by the product's decreasing the formation of the enzyme itself. The latter method is probably more common, at least in single-celled organisms, and is achieved through the *nucleic acids,* since it is these substances that control the formation of the enzyme proteins. Thus it will only be when the complexity of the chemical reactions has reached the stage of producing nucleic acids, with all the results that that implies, that we would be justified in applying to the material enclosed by its membrane the term "cell."

3. THE COMING OF FREE OXYGEN AND ITS RESULTS

Somewhere in the early stages of evolution there was a crucially important step forward in metabolism. As we have seen, these primitive organisms would have been using the energy of ultraviolet light for their first chemical syntheses and then using the energy liberated by chemical reactions among these products for further changes. They would still have been living without free oxygen. (Life without oxygen is termed *anaerobic.*) We know today of organisms using the energy of light for growth and living without free oxygen—the purple bacteria. Some other types of organisms have succeeded in living by merely breaking down the complex molecules and applying the chemical energy thus released to the synthesis of their own substances, that is, living by *fermentation,* as do anaerobic bacteria today.

Those organisms that continued to use light energy would have been obtaining their carbon compounds by reducing the CO_2 of the air with hydrogen, this hydrogen coming from methane (CH_4), ammonia (NH_3), hydrogen sulfide (H_2S), or other easily available sources. For example, $CO_2 + 2H_2S \rightarrow (CH_2O)$ (= organic compounds) $+ H_2O + 2S$. This meant that organisms would have been limited to locations where pockets of such hydrogen-rich gases occurred. However, it is postulated that around a billion and a half years after the earth's formation, one of these primitive types "discovered" a more powerful method of using light energy; it used the hydrogen of a much more stable molecule, namely water, to reduce CO_2. Since the occurrence of water was worldwide, this discovery enabled such organisms to spread everywhere. More importantly, it meant that the oxygen of the water decomposed was *set free* into the atmosphere thus: $CO_2 + 2H_2O \rightarrow (CH_2O)$ (= organic compounds) + $H_2O + O_2$. As these new organisms spread, the CO_2 level of the atmosphere would have decreased and the oxygen level steadily increased. This had two fundamental and quite different effects.

1. It enabled the development of a different type of organism altogether—one that was able to make organic compounds combine with oxygen. The special value of this *oxidation* is that it yields chemical energy. This new organism thus had a *source of energy alternative to that from light*. Such an organism, deriving energy for chemical synthesis from oxidations, would have been the forerunner of the colorless, *aerobic* organisms of today, including the animals.
2. Under the influence of the high energy of the solar ultraviolet light, some of the oxygen (O_2) became converted to ozone, O_3. This substance in turn strongly *absorbs* ultraviolet light. The result was that the lower layers of the atmosphere became shielded from the effects of most of the solar ultraviolet, so that the earlier, primitive reactions no longer took place—at least not near the earth's surface. Instead the very delicate, easily decomposed molecules that make life possible were able to accumulate in larger amounts, and they could be further modified by using the chemical energy of breakdown of other compounds. The reason they could not accumulate earlier is that too much ultraviolet light, especially at the shorter wavelengths, is lethal to virtually all organisms now on earth, because it decomposes the material of nuclei. With the absorption of the shorter wavelengths by ozone, and with the presence of oxygen, atmospheric conditions became more like those we know today. The longer wavelengths of ultraviolet do reach the earth in moderate intensities, and, though invisible to us, are used by some organisms, for example, by insects for vision and by plants for orientation towards a light source.

4. TRUE ORGANISMS AS WE KNOW THEM

At this point in evolution the picture becomes supported by fossil evidence. The age of the earth is about 4.6 billion (4.6×10^9) years, a figure supported in various ways by geologists and astronomers. In the oldest rocks that contain traces of carbonaceous matter, dated at 3.4 to 3.5×10^9 years old, fossil forms identifiable as bacteria have been seen, their shapes like those of the bacteria we have today (cf. Fig. 2-1*A*). Along with these are some that resemble existing blue-green algae (Fig. 2-1*B* and 2-1*C*). In formations not quite so ancient, there are curious linear deposits ("stromatolites") that are believed to be masses of blue-green algae. The periods when these occurred appear on the "clock" that represents the order of biological developments on a human time scale (Fig. 2-2). It is notable that in essentially the first 6 hours of the 24 hours represented there is no evidence for biological developments at all.

Among the bacteria there is a small group that is photosynthetic, but its photosynthesis does not evolve oxygen. Organisms like these could well have pioneered the photosynthetic mode of life under the anaerobic conditions that first prevailed. The blue-green algae are generally felt to be the next most primitive of photosynthetic organisms since: (a) they contain no true nucleus, (b) their cell walls are chemically like those of bacteria rather than like those of the green plants, and (c) like many bacteria, some of them have the ability to convert free nitrogen gas to organic compounds, which other kinds of algae and higher plants cannot do. Some authors therefore prefer to class them with the photosynthetic bacteria under the name *Cyanobacteria*. They are blue-green because they contain, besides green chlorophyll, the blue pigment *phycocyanin*. Blue-greens apparently belonging to all but one of the groups that we know today have been recognized in ancient rocks and stromatolites around 1×10^9 years old (Fig. 2-3).

In one respect the photosynthetic bacteria

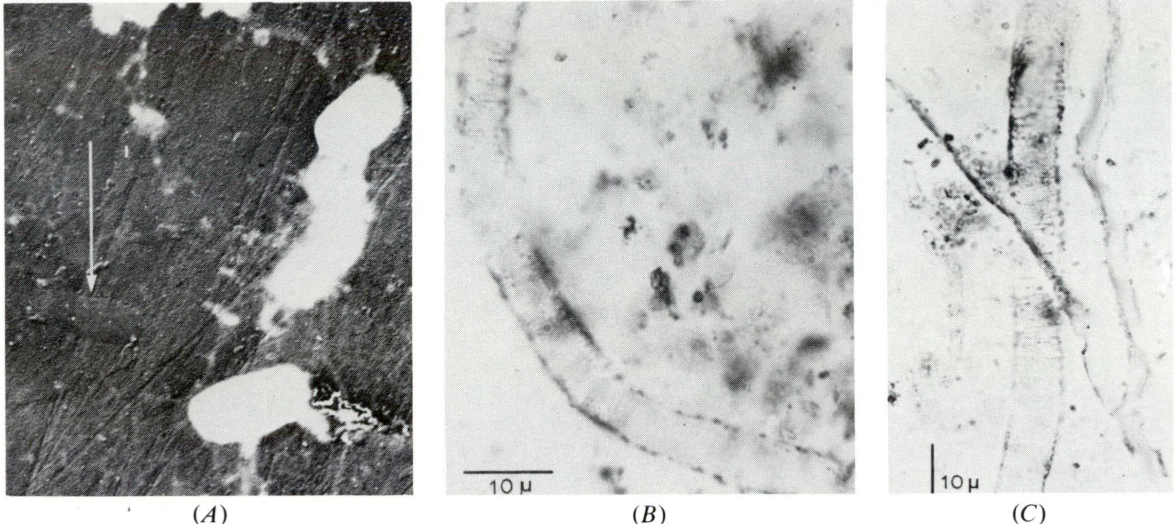

Figure 2-1 *A*. Rod-shaped bacteria in the Gun-flint Chert dated as 2.0×10^9 (= 2.0 billion) years ago. The long, thin lines are polishing scratches. Magnified 19,000 times. *B*. and *C*. Filaments of algae or possibly fungi from the same Gun-flint Chert. Both are 9 microns in diameter and show septa or cell walls. Note the scales; in *A* the magnification is about ten times higher, in *B* and *C* the dark bar is 10 microns.

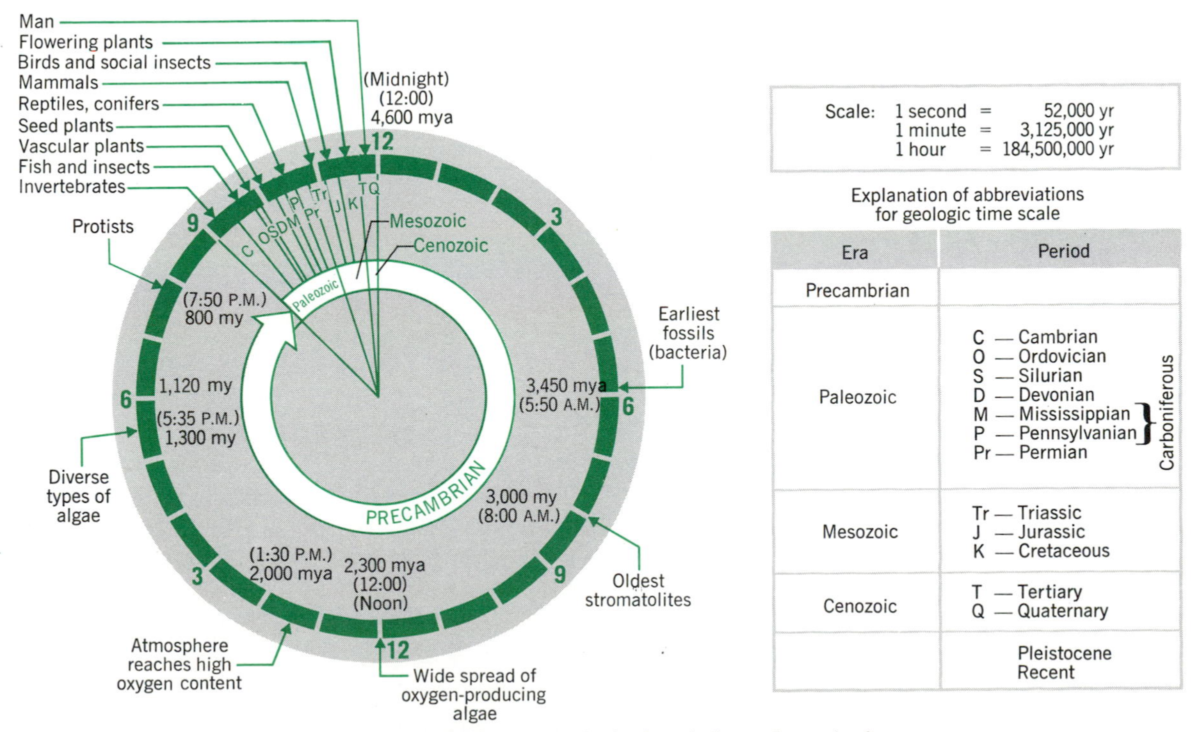

Era	Period
Precambrian	
Paleozoic	C — Cambrian O — Ordovician S — Silurian D — Devonian M — Mississippian } Carboniferous P — Pennsylvanian } Carboniferous Pr — Permian
Mesozoic	Tr — Triassic J — Jurassic K — Cretaceous
Cenozoic	T — Tertiary Q — Quaternary
	Pleistocene Recent

Figure 2-2 Significant developments of life on earth depicted through geologic time as on a clock (mya = millions of years ago).

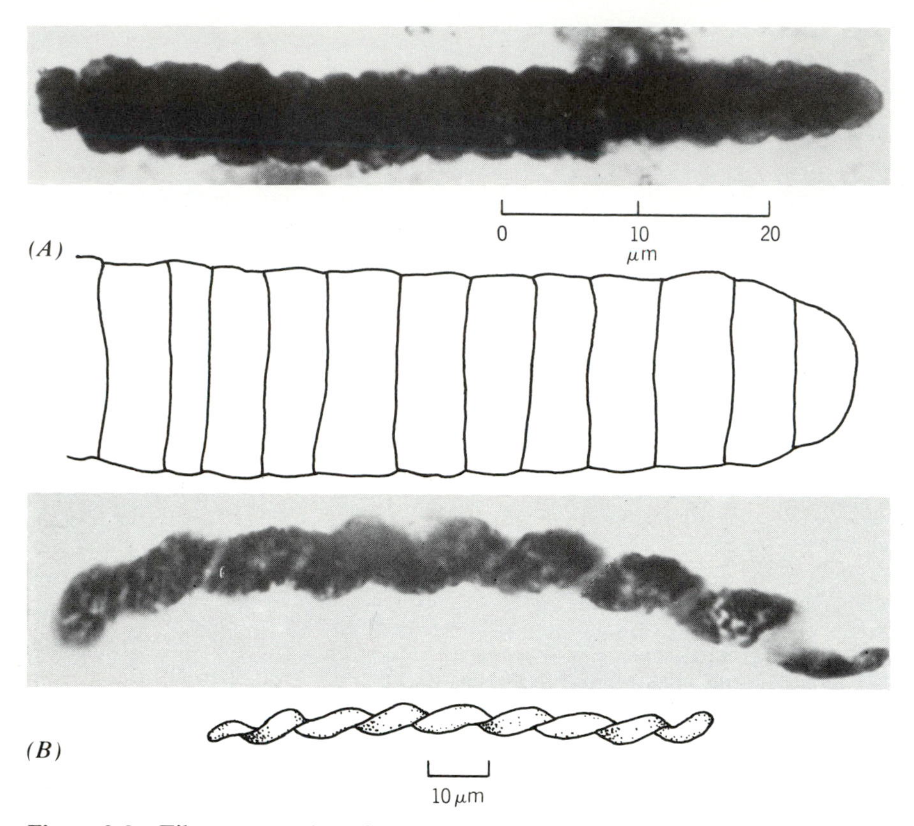

Figure 2-3 Filamentous algae from the late pre-Cambrian Bitter Springs formation of central Australia, dated at 1.0 billion years ago, and two modern forms. *A, Oscillatoria. B, Spirulina.*

and the blue-green algae that we now know are not truly primitive, for they use *visible* rather than ultraviolet light as their source of energy. This was made possible by the appearance of *chlorophyll* and its associated pigments, which (as will be seen in Chapter 5) comprise an elaborate system for generating organic matter by the aid of visible light energy. When the molecules of chlorophyll absorb visible light they become *activated,* and this starts a chain of processes that ends by converting CO_2 to organic compounds. The great decrease in the intensity of ultraviolet light at the earth's surface, of course, powerfully favored the development of this use of visible light, and all later plants have followed it.

In these early organisms there was a crucially important development of nucleic acids. Once these were "discovered" they were so effective that the system has carried through to the present day. There are two kinds of nucleic acids, one that carries the pattern of specific types of molecules from one generation to the next (called DNA) and one that essentially "carries out the bidding" of DNA by synthesizing the enzymes necessary to build that particular pattern (called RNA). DNA stands for Deoxyribo-Nucleic Acid, that is, a nucleic acid containing the sugar *deoxyribose,* whereas RNA stands for Ribo-Nucleic Acid, which contains the related sugar *ribose.* The ways in which these materials function are taken up in Chapters 9 and 10.

The next step in evolution was to develop a *true nucleus,* in which the DNA and some RNA are combined with proteins, sequestered

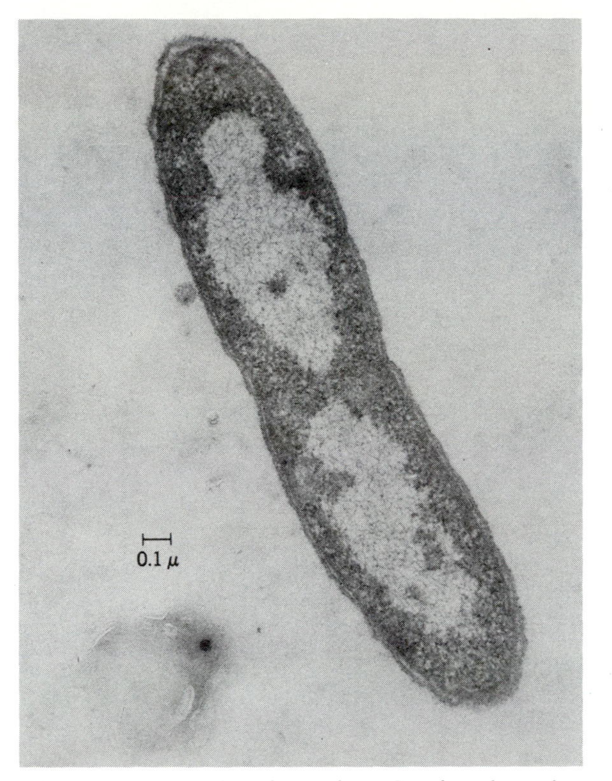

Figure 2-4 A modern bacterium. Section through a dividing cell of *Pseudomonas*. Each of the two daughter cells is about 1.7 μ long. The nucleic acid is seen as fine fibers in the light-colored areas of the cell. The cell wall is finely layered. The transverse wall has not yet formed. The small dark line at right is 0.1 μ.

inside a double membrane, and, at the time when the cells divide, assembled as organized *chromosomes* (see Fig. 4-3). The bacteria and blue-green algae have no such nucleus but their DNA is a mass of fine fibers in the middle of the cell (Fig. 2-4). The true nucleus contains most of the DNA (but see Chapter 4), whereas most of the three forms of RNA are found in the cytoplasm. Not only are the true nuclei surrounded by a double membrane, but they also contain an elaborate device for the equal distribution of the nucleic acid between daughter cells during cell division (see Chapter 4).

Organisms with a true nucleus are called *eucaryotic;* those without (the bacteria and blue-green algae) are *procaryotic* (*eu* from Greek "true"; *caryon* from Greek for a kernel, used for the nucleus; *pro* also from Greek "before"). The number of chromosomes in the true nucleus is highly characteristic for each species. Many fungi, and some unicellular algae, have two nuclei per cell. Among the animals, the single-celled forms (protozoa) often contain several nuclei, each within a membrane, and the nuclei are commonly of two different types. This suggests that perhaps the single nucleus of today's advanced organisms resulted only after a number of evolutionary "trials."

This development of a single nucleus, along with other evolutionary inventions, such as the differentiation of cells for special purposes, made possible the advent of multicellular organisms. Some of these were using light; others were using the energy of oxidations for their syntheses. Those using light, that is, carrying out photosynthesis, have given us the green plants; those using oxidation energy, together with a few still living as *anaerobes,* have given rise to colorless groups, that is, the animal kingdom and the fungi.

5. THE MODERN ALGAE

The step from the blue-green algae or Cyanobacteria to the other algae (Chapter 4 and Fig. 2-6) is a big one for two reasons. First, it involves the appearance of a nucleus and the organized control of cell division that is implied by that. The eucaryotic algae were evidently the beginning of the whole development of eucaryotic organisms. The second big change is that the all-important pigments in the eucaryotic algae became concentrated into *chloroplasts* instead of being distributed around the cell as in the blue-greens. Each chloroplast is enclosed within a double membrane (Fig. 2-5). The change is important, too, in another way, because it has recently been discovered that each chloroplast contains *DNA of its own.* This remarkable new fact may explain why chloroplasts are sometimes seen to divide, and why

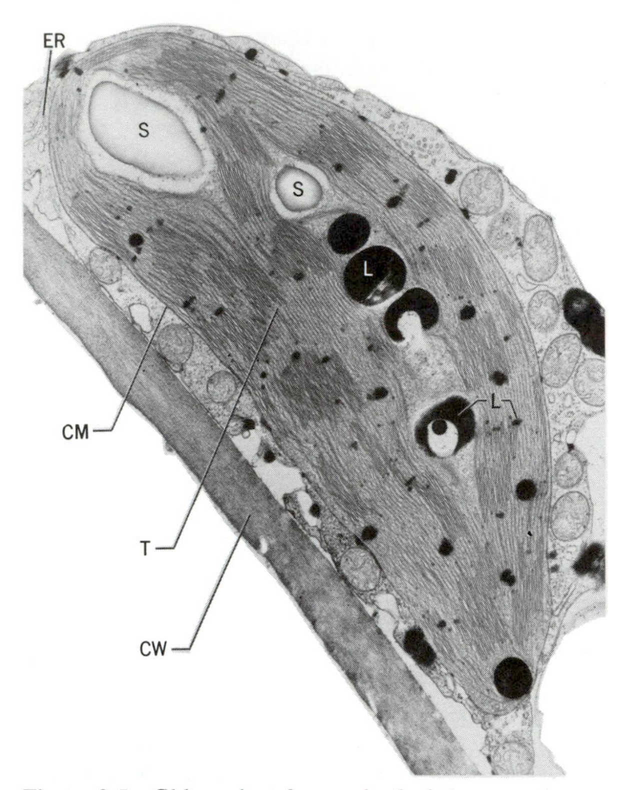

Figure 2-5 Chloroplast from a leaf of the succulent plant *Crassula*. Magnified 12,800 times. *CM*, chloroplast outer membrane; *T*, thylakoids; *L*, lipid granules (heavily stained); *ER*, endoplasmic reticulum; *S*, starch grains.

chloroplasts can have for some characters a heredity separate from that carried by the nucleus. Some biologists have drawn the conclusion that the chloroplast represents the evolutionary residue of a primitive single-celled alga that has somehow taken up permanent residence inside each cell of a colorless plant, and thus become *symbiotic* with it. A few such symbiotic partnerships are known; there are blue-green algae in some liverworts, and the little invertebrate animals that make coral harbor single-celled green algae. Lichens are associations of fungi with algae, but the algae are not inside the fungal cells. The whole idea is intriguing, but extremely hard to prove. A reason to be skeptical is that in the course of evolution there have been three changes of similar type—assembly of the chloroplast, of the nucleus, and of the mitochondrion (see Chapter 4). All contain DNA, and all are enclosed within double membranes. This suggests a general evolutionary trend.

By whatever means the nucleated, chloroplast-bearing types of algae did develop, it has been a most successful enterprise, since the algae have branched out into what are now recognized as ten different categories, such as the reds, browns, several types of yellow forms, the siliceous-shelled diatoms, and the all-important green algae. It was doubtless the green algae that migrated into fresh water and eventually up on to the land, as described below, to inaugurate the colonization of the land by the green plants. Most of the brown, red, and green algae are multicellular.

The multicellular types were probably at first colonies of single cells that adhered together after dividing; some single-celled algae indeed do that today (Fig. 2-6). One group of colorless organisms, the slime molds, continually goes through the stage of single cells, which first multiply and then aggregate together, to form a flat plant body (see Fig. 30-3). A perhaps comparable behavior is seen in one of the simplest groups of animals, the sponges. These, although multicellular, can still be artificially broken up into single cells and thereafter will reassemble spontaneously into a new multicellular animal.

6. THE MIGRATION TO LAND

Whereas single cells, spread out in relatively thin layers, were more efficient than multicellular ones in utilizing the sun's rays in the sea, multicellular plant bodies were more successful along the shore and on land. There is good reason for this, for the multicellular structure has certain advantages. Terrestrial habitats offered plants abundant light unfiltered by turbulent water, carbon dioxide and oxygen circulating more freely in the atmosphere than in water, and unoccupied space. On the other hand, there were difficult problems, and radical

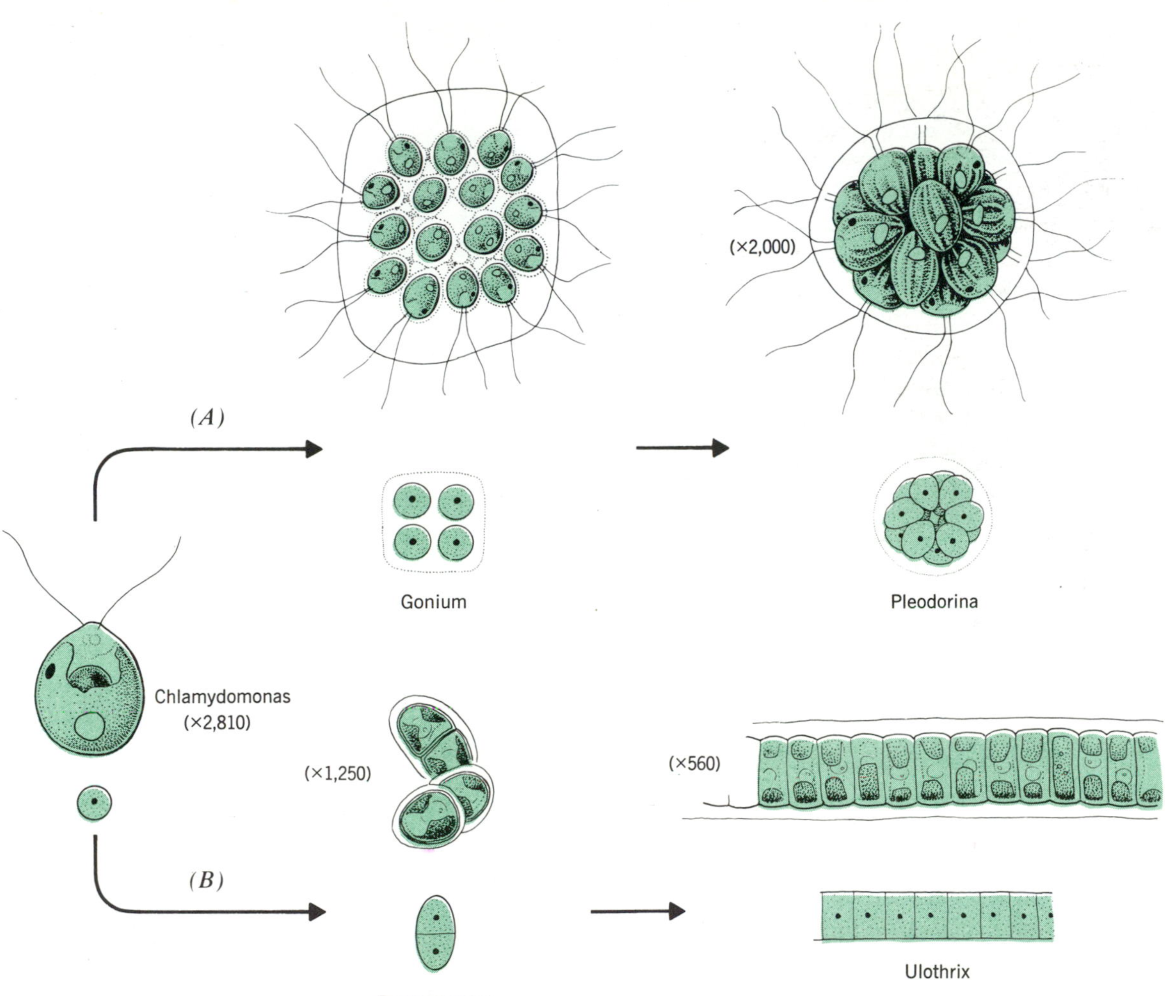

Figure 2-6 The colonial states seen as a pathway towards the multicellular form. Arrows show the possible direction of development. *A*. Each cell retains its own flagella and can swim away. But as the size of the colony grows it becomes more nearly like a multicellular organism. *B*. In other types the free movement is lost and the colony becomes truly multicellular. (Adapted from Jensen & Salisbury, 1972)

changes were necessary before land plants could prosper. The most critical problem, perhaps, was the availability of water for both vegetative and reproductive processes. In the ocean, both water and mineral elements are in direct contact with the plant body. But land plants have to bring water and minerals up from the ground to the growing parts of the plant. Correspondingly, if there were parts in the ground, the plant had to conduct products of photosynthesis down to these portions to keep them fed. We can see why a conducting system became of great importance. Only one evolutionary line of plants has been able to meet this challenge of the supply system; this they have done by developing an efficient conducting or *vascular* system that transports water and solutes from one part of the plant to another, both upwards and downwards. For this reason they are called *vascular plants*.

The fossil record offers limited evidence about one of the most important events in the history of the plant kingdom—the forms of the transitional plants that migrated from the ocean to estuaries or tidal swamps and finally to dry land. It is generally assumed that the red and brown algae, before there was vegetation on land, had become about as specialized as they are today, whereas the green algae were, and still are, more primitive in body form and methods of reproduction. They also show greater plasticity in their adaptation to various environments, a particularly important factor in the migration to land. Thus they are generally assumed to be the immediate progenitors of land plants. For the development of the typically two-component form of the vascular plant, namely a shoot above ground and a root below, an attractive theory runs like this: the green algae that gradually migrated landward developed two types of filaments—prostrate ones that spread out and, growing out of these, erect ones that gave greater exposure to light. The cells of the erect filament became divided, forming a multicellular, three-dimensional shoot (Fig. 2-7 and 2-9). The origin of the root is obscure, but it may have originated from a subterranean branch of plants like those in Figure 2-7 and 2-9. The root acquired a protective cap and retains today the type of vascular cylinder characteristic of the stem of many ancient plants (Fig. 2-7). This vascular cylinder connects directly to a vascular conducting system in the shoot.

The leaf was an elaboration that probably originated in several ways. In some of the early plants (e.g., the group to which the club mosses belong—see Table 1-1), the leaf was thought to have arisen from the stem as a superficial outgrowth that eventually developed vascular strands (Fig. 2-8A). In these cases the leaves have remained small and are called *microphylls*, meaning small leaves (Greek *phyllos*, a leaf). In more advanced forms, such as the ferns and flowering plants, the leaf is considered to represent a flattened and expanded portion of a

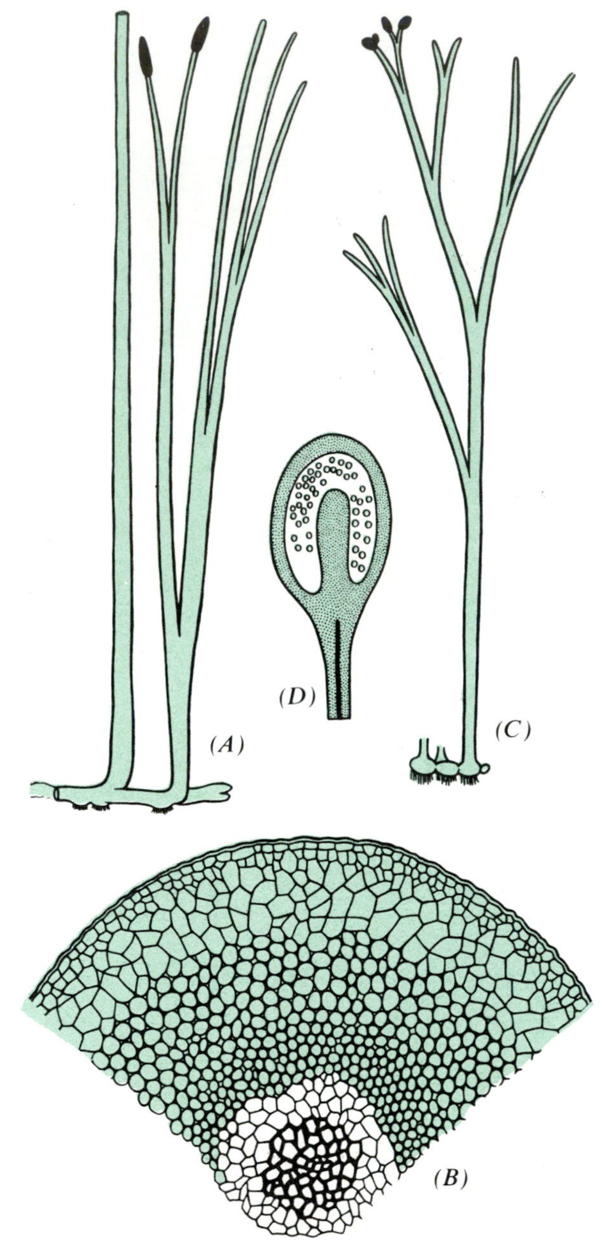

Figure 2-7 *A, B. Rhynia gwynne-vaughani. A.* Restoration. *B.* Transverse section of stem showing small vascular cylinder and broad cortex. *C, D. Horneophyton lignieri. C.* Restoration. *D.* Diagrammatic longitudinal section of sporangium. *Rhynia gwynne-vaughani* was a small plant (20 cm in height) that grew in dense aggregations with *Horneophyton.* The cellular detail was preserved in chert beds of Devonian age near Rhynie, Scotland. (From Andrews, 1961).

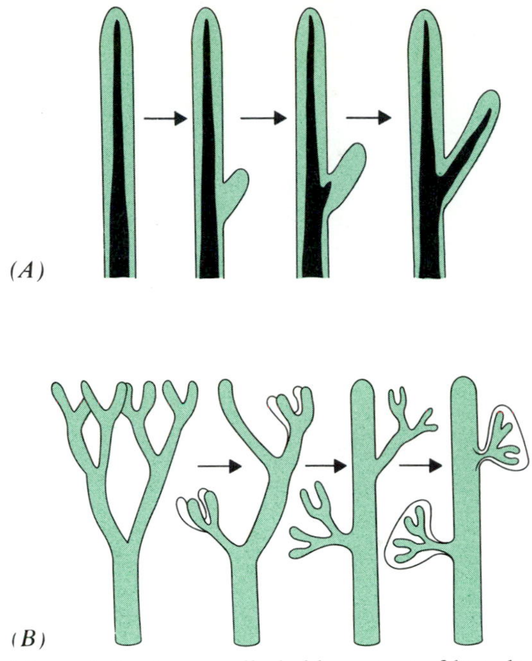

Figure 2-8 A generally held concept of how leaves evolved. *A*, microphylls; *B*, megaphylls. (After Raven, Evert, Curtis, 1976)

branch: it has become large and complex (Fig. 2-8B) and is called a *megaphyll* (large leaf).

Another modification suitable to the land habitat was that the upper portion of the plant body became protected by a waxy substance (*cutin*), excreted from the epidermal cells, that restricted water loss; however, specific cellular openings (*stomata*) developed in the epidermis permitting exchange of carbon dioxide and water vapor with the atmosphere. Modifications also occurred in the reproductive structures, and, with improving vascular systems, the plants began to increase in size. Hypothetical and fossil plants arranged to illustrate increases in complexity associated with migration to land are illustrated in Figure 2-9.

The advent of plants thus able to grow under terrestrial conditions appears to have occurred at about 9:55 P.M. in the clock symbolizing geologic history (Fig. 2-2). At that time the invertebrates, fish, and insects were already present. Also at that time not just one type of vascular plant appeared but numerous types, representing several plant groups that include club mosses, horsetails, and pre-ferns. Figure 2-10 shows an imaginary landscape several million years later, which by then included tree-like descendants of the first invaders.

7. REPRODUCTION CHANGES AND THE GROWING IMPORTANCE OF DIPLOIDY

Development of the reproductive processes was also taking place during these early evolutionary stages. This story is quite a long one, essentially in four parts.

a. HAPLOIDS AND DIPLOIDS

Hereditary information is carried in molecules of DNA (*genes*) in both procaryotic and eucaryotic organisms (see Chapter 11 for a detailed discussion). In the procaryotes the circular molecules of DNA are essentially free in the cytoplasm but in eucaryotes they are organized into larger units (*chromosomes*) that contain proteins in addition to DNA. Cells with one set of chromosomes are called *haploid* ($1n$); those with two sets are *diploid* ($2n$). Procaryotic organisms (like bacteria and blue-green algae) are also called haploid because they contain only one complement of DNA. They multiply primarily by dividing into two—simple or cell division—as in Figure 2-4. In this process the amounts of DNA (and RNA) are first doubled; then half the amount goes into each daughter cell. This same doubling and equal division (*mitosis*) takes place each time the cell divides (see Chapter 4, section 2 for a detailed discussion). But in some bacteria today two haploid cells combine most of their nucleic acids by means of a small tube and thus a "diploid" cell is temporarily formed; however, this quickly divides and distributes its nucleic acids between the two daughter cells, so that they are again haploid. This is a type of genetic recombination, perhaps a forerunner of, but not true, sexual reproduction.

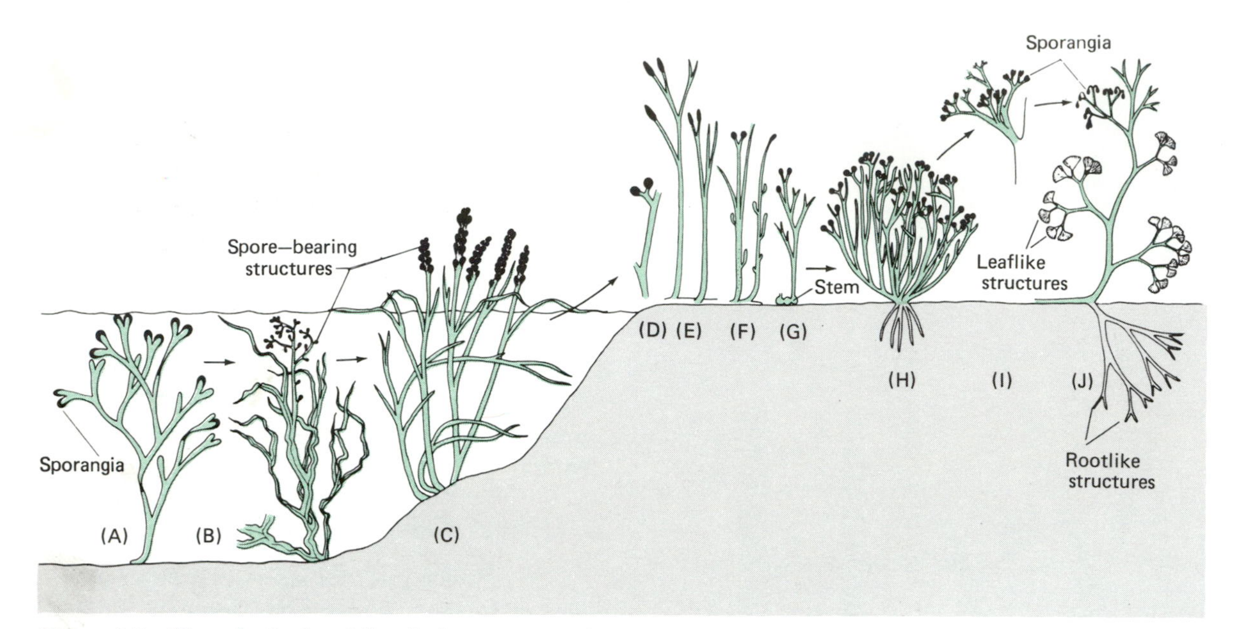

Figure 2-9 Hypothetical and fossil plants arranged to represent the increases in complexity thought to have occured in migration of the first plants to land in the Silurian and early Devonian geologic periods. All the plants were small (much less than 1 m tall) with some types (*A* to *C*) growing totally or partially submerged and the remainder growing on land. *A*. A hypothetical aquatic plant with flattened two-parted branches and spore-bearing structures (sporangia) located on the tips of the branches. *B*. *Taenocrada,* an aquatic plant that displayed the hypothetical two-parted, flattened branches and sporangia. *C*. *Zosterophyllum,* another aquatic plant with sporangia that occurred laterally on emergent branches above the water. *D*. *Cooksonia* had a simple branched axis with sporangia at the tips. *E*. *Rhynia major* had a prostrate stem from which upright branches arose. *F*. *Rhynia gwynne-vaughani* had erect branches that bore lateral appendages. *G*. *Horneophyton lignieri* had upright branches that also arose from a prostrate stem. *H*. *Hicklingia* had a compact, erect stem with upright branches, and perhaps some rootlike branches that entered the ground. *I*. *Trimerophyton* had complex branching of both sporangia-bearing and sterile branches. *J*. A hypothetical early land plant illustrating how the shoot might have become modified into rootlike structures that grow in the ground and leaflike structures and special sporangia-bearing branches on a stem that grew above ground. (From Neushul, 1973).

b. SEXUAL REPRODUCTION AND ALTERNATING GENERATIONS

One of the major distinctions between procaryotic and eucaryotic organisms is that most eucaryotes reproduce sexually whereas procaryotes do not. Sexual reproduction involves a regular alternation between a special type of nuclear division, *meiosis,* (see Chapter 4, section 2b for a full discussion) in which the number of chromosomes is reduced to half, to form haploid ($1n$) cells, and then the diploid ($2n$) number of chromosomes is reestablished by *fertilization,* in which two haploid cells (*gametes*) fuse to form a diploid *zygote*. Once sexual reproduction was established among unicellular eucaryotes, the stage was set for the persistence of organisms in the diploid state. In the evolution of diploidy, two kinds of life cycles can be envisaged. In one, the zygote divides at once by

Figure 2-10 The first land plants probably occurred in moist lowland areas about 395 million years ago. Several million years later they were accompanied by treelike descendants as shown in this imaginary mid-Devonian landscape. *A, B,* and *C* are plants previously described in Figure 2-9. *D* and *E* are early treelike forms represented today by clubmosses; *F* is a fernlike plant, and *H* is a seed fern; *G* is a relative of our modern horsetails. (From Neushul, 1973).

meiosis thus restoring the haploid condition. Here the zygote is the only diploid cell. But if (possibly by accident) the zygote divided by *mitosis* instead of by meiosis, it would then produce a whole organism composed of diploid cells. This important step probably took place in several evolutionary lines. In a few organisms, the haploid stage was lost completely; in the majority, however, meiosis was not completely suppressed but only delayed. In animals, the process results in the production of two different kinds of gametes—eggs and sperms. When the gametes fuse, the diploid ($2n$) condition is immediately restored. Thus gametes are the only haploid cells in higher animals (some insects do have haploid individuals).

In almost all plants, on the other hand, there is an additional stage, the *spore*. Spores, as shown in Figure 2-11, are haploid cells that divided again to produce a haploid organ or plant, the *gametophyte,* (*phyton* is Greek for plant; hence, the plant that produces the gametes). When two gametes fuse to produce a zygote, the zygote grows up into a diploid plant, called the *sporophyte,* that produces the spores.

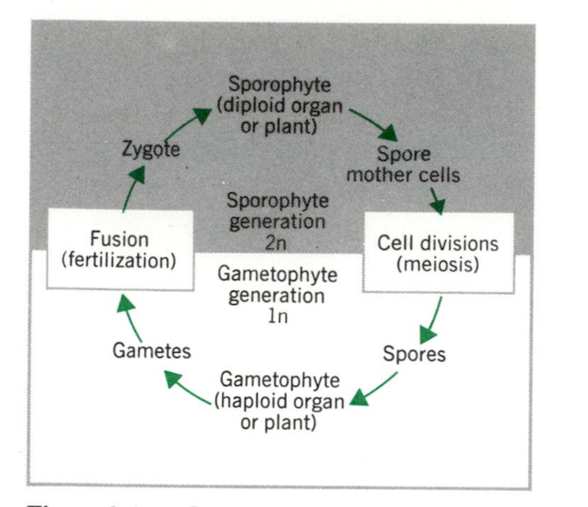

Figure 2-11 Generalized life cycle illustrating alternation of haploid and diploid generations in plants. In some plants sporophytes and gametophytes are essentially the same size (e.g., some brown algae); in others the gametophyte is dominant (e.g., mosses) and in others, such as the flowering plants, the sporophyte is dominant.

As a result, there is a continuous alternation of $1n$ and $2n$ generations.

Among many algae, some green, most of the reds, and a few browns, the haploid and diploid plants, (the gametophytes and sporophytes) have the same external appearance. Life cycles like this are thought to have occurred first in evolution. Later on, the gametophyte and sporophyte became notably different from one another. In some brown algae the generations are of similarly large and complex structure, but different in form. In other brown algae, as in all land plants, one generation is clearly more prominent and long-lasting than the other. The gametophyte (haploid) is the longer-lived phase among the mosses and liverworts, whereas among the vascular plants the sporophyte is both larger and longer lived. Thus there has been a long-continued evolutionary trend toward reduction in size and complexity of the haploid generation and a parallel increase in size and elaboration of the diploid generation. The same trend holds even more throughout the animal kingdom, the diploid state being virtually universal and the haploid gametes short-lived.

c. MALE AND FEMALE GAMETOPHYTES

The first vascular land plants produced spores that were borne in special organs (*sporangia*) located at the ends of the branches, as seen in Figure 2-9; the greater size of the diploid plant, among other things, permitted more effective spore dispersal. Later in evolution the spores lost mobility and developed a waxy cuticle that protected them from water loss and thus allowed their dispersal by wind. These haploid spores were of only one kind. Our modern seed plants, however, which form most of the subject of this book, produce *two types* of spores, in two different sporangia. This variation, which has arisen several times during the evolution of vascular plants, occurred relatively early—about 350 million years ago. The two types of spores are called *microspores* and *megaspores*. Microspores give rise to the male gametophyte and megaspores to the female gametophyte.

These two types of gametophytes develop in different ways. In the angiosperms (flowering plants), they are both small, the female gametophyte consisting of a group of only eight nuclei, one of which becomes the female gamete—the egg. From the male gametophyte, the *pollen grain*, a special outgrowth called the pollen tube brings its gametes (*sperms*) directly to the egg, inside the female gametophyte. Fertilization then produces the zygote. In most primitive land plants, although the shoot grows in the air, the gametes can function and fuse only in water. The pollen tube, by directly bringing the sperms to the egg, avoids the need for having water available for fertilization, and thus flowering plants can reproduce in many dry habitats where other plants are successful only under special conditions (see section 11 of Chapter 4 for a full discussion).

d. SEEDS AND SEED PLANTS

One of the most dramatic innovations of the vascular plants was the development of the seed. Where the female gametophyte remains within parental tissue, that is, within the megasporangium, it becomes protected by a two-layered envelope, the *integuments,* and the entire structure, called the *ovule,* develops after fertilization into a seed.

Seeds probably evolved at different times and in different evolutionary lines. Figure 2-12 shows actual seedlike structures in some fossil plants ("seed ferns") (compare Figure 4-13) and suggests how the integuments may have gradually come to enclose the megasporangium. Some of the seed-producing lines became extinct, but the conifers and flowering plants now dominate the vegetation of the earth. One of the main reasons why they were so successful in populating many environments is probably that the seeds have special protective features. In the first place the seed is protected both from physical conditions and from some predators by remaining inside the parental tissue. In the second place, it is provided with nutrients by a special vascular supply system. And yet, in spite of these favorable conditions, its development goes rapidly only for a few days or weeks and then comes completely to a stop. With the formation of the seed coat from the integuments, it can remain dormant, sometimes for years, until conditions are favorable for germination.

8. GROWTH IN SIZE: THE DEVELOPMENT OF SECONDARY TISSUES

With increase in size and specialization, new kinds of tissues, such as the *cambium,* developed in the plant body. The cambium, a thin sheet of cells developing in connection with the vascular system (*xylem* and *phloem*), divides and enlarges in the direction at right angles to the long axis of the shoot, which results in thickening of the stem (Chapter 4). The tissue thus produced from the cells of the cambium is considered *secondary* (Chapter 22), that

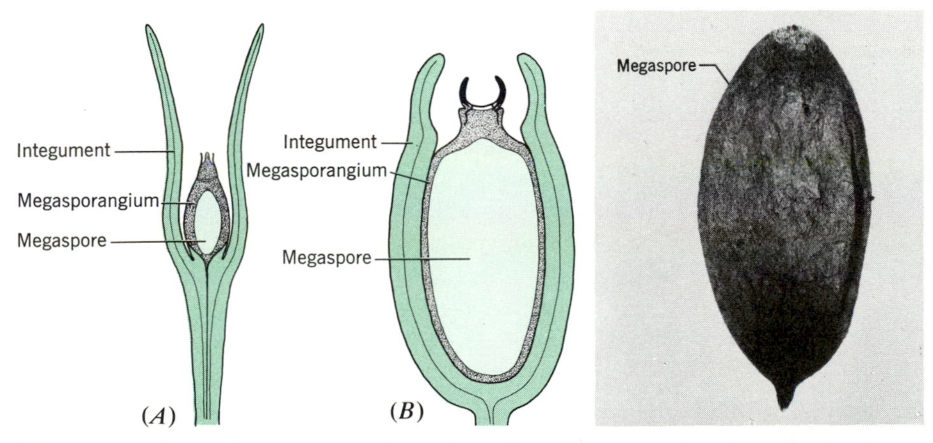

Figure 2-12 Stages in evolution of seeds as shown by several ancient plants. *A.* In *Genomosperma* (Greek *genomein,* "to become" and *sperma,* "seed"), eight fingerlike processes arise at the base of the megasproangium. *B.* In *Eurystoma angulare,* these processes are fused. *C.* A megaspore from *Archeosperma arnoldei* that was borne in a "cupule" (megasporangium) and branched stalk similar to that of *Genomosperma.* The most significant events in the evolution of the seed are the retention of the megagametophyte within the megasporangium of the parent sporophyte and enclosure of the megasporangium in protective integuments. Since *Archeosperma* was a Devonian plant, these fossils demonstrate that by about 350 million years ago seeds had essentially evolved in those groups.

formed from the cells of the apex being *primary*. Some modern plants, such as ferns and grasses, have survived with primary tissue alone, and the thickness of their stems results from combining (either by interweaving or by growing together) a large number of individual primary systems. But in most of the larger land plants, secondary tissue provides much better support and more efficient conduction of materials than that available from primary tissues. A cambium seems to have developed in a number of evolutionary lines by the equivalent of 10:15 P.M., but many of these lines are now extinct, and cambial activity today is largely restricted to seed plants. It is difficult to visualize what a modern landscape would look like if cambium had not evolved, for without cambial tissue there would be no trees, as we know them today, and we have depended upon forest products from the beginning of our history.

9. EARLY PLANT COMMUNITIES AND ECOSYSTEMS

Once plants had successfully occupied the land, plant-eating animals also appeared on the land and complex communities or *ecosystems* then came into being. The typical landscape of 395 million years ago (about 10:00 P.M. on the "clock") was probably sparsely covered with small, simple vascular plants growing primarily in lowland areas (Fig. 2-10). Their descendants colonized the higher ground to produce the first trees. About 40 million years later the land became covered with spectacular forests of woody trees having large, fernlike leaves. During the Carboniferous period or coal age, which lasted from 345 to 280 million years ago, swamps similar to those found today in Florida or the Amazon basin were widespread over most of the continents (Fig. 2-13). Amphibia, reptiles that resembled blunt-nosed alligators, and numerous kinds of insects (such as dragonflies, large spiders, and cockroaches) were abundant in these forests. Primitive trees, closely related to our modern herbaceous club mosses, grew to 30 m high in lush abundance together with giant tree-like horsetails, seed-bearing ferns as well as the true ferns, and ancestors of our present-day conifers. The trees grew on a mulch of organic material and in death made their own further contributions to this. Bacterial oxidation was limited by lack of sufficient oxygen in the stagnant water (as still happens today) and the fallen, preserved vegetation ultimately created vast coal deposits. Often the primitive plants have been sufficiently preserved in these environments so that their cellular structure may still be studied in detail. We are still harvesting these trees as fossils, in the form of coal and lignites to use as fuel (Fig. 2-13).

About this time regional uplifts of the continents gradually began, which put the warm swamp plants at a disadvantage and favored species adjusted to a drier and cooler climate. It was during this time (the Permian period, 280 million years ago – 10:30 P.M. on the "clock") that conifers became widespread. Conifers such as shown in Figure 2-14 were dominant, with cycads and ferns forming an open undergrowth beneath the trees. These conifers then rapidly evolved into modern types, which became the dominant components of the first of our modern forest ecosystems. Varied types of reptiles, the dinosaurs, arose (around 230 million years ago) and the earliest mammals appeared around 30 million years later.

About 180 million years ago there developed what may have been the greatest cosmopolitan flora in geologic history. It included ferns, conifers, ginkgos, and cycads and it fed the herbivorous dinosaurs, which became very large (Fig. 2-14). This widespread flora and the reptiles and early mammals were supported by a relatively uniform warm, moist climate. Birds and social insects also appeared during that time. A little later, in the Cretaceous period, the dominance of the conifers gradually gave way to flowering plants in most parts of the world.

Figure 2-13 During the Coal Ages (10 : 15 on the "clock") plants flourished in an environment particularly favorable for plant growth. Apparently relatively uniform vegetation occupied vast swamps, including coniferlike trees with strap-shaped leaves (*A* and *F*), fernlike plants (*B*), arborescent relatives of the club-mosses with prominent leaf scars (*C* and *E*), giant relatives of modern horsetails (*D*) and seed ferns (*G*). (From Neushul, 1973).

Since there is no evidence for dramatic climatic changes at this time, the great rise of the flowering plants probably involved complex biological interactions of the plants and animals, particularly with insects.

With the development and subsequent diversification of the flower came also the rise and diversification of the long-tongued insects such as bees, wasps, butterflies, and moths. Since in many cases flowers are their only source of nutrition, the evolution of these insects appears to be a direct result of the evolution of the flowering plants on which they feed. In compensation, the cross-pollination of plants by insects has enabled the flowering plant to transcend its rooted condition and attain a type of mobility. Although conifers and numerous flowering plants are pollinated passively by the wind, a

Figure 2-14 The Coal Ages were followed by drier conditions less favorable for plant growth and their preservation in the fossil record. The flora of the Mesozoic Era (230 to 165 mya) was dominated by gymnosperms – many of them conifers. In the early part of the Mesozoic Era much of the landscape was arid, to which many gymnosperms were well adapted (although there were moist areas as shown here). *A* and *B* are conifers and *C* is a remnant of the coniferlike trees so common in the Coal Ages; *D* is related to the early Ginkgos; *E* is a cycadlike plant and *F* is a fern; *G* is a columnar palmlike tree. (From Neushul, 1973).

more specific and effective pollination system was introduced by the insects. Many insects are highly selective in their choice of flowers, being attracted by certain forms, colors, and odors. In fact there has been in many cases a co-evolution of insects and flowers. Attraction of insects to flowers did, nonetheless, raise additional problems of protecting the seed from predator insects. Enclosing the seed in the ovary, a major distinction between the *gymnosperms* (naked seed) and *angiosperms* (seed enclosed), may have been one of the evolutionary adaptations to protect the seed from being eaten by a pollinator.

Certainly the origin of this most important group of plants has been, as Darwin put it, "an abominable mystery." Our first undoubted fossil remains of flowering plants are from about 125 million years ago (early Cretaceous period, 11:30 P.M. in Fig. 2-2). They consist of pollen similar to, but distinguishable from, the spores of ferns and the pollen of gymnosperms. By 120 million years ago, however, fossil pollen and leaves display the diversification of the angiosperms into a large number of modern genera and families. This was a time, as new geological information has dramatically told us, that South America, Antarctica, India, and Australia were linked in a great supercontinent called Gondwanaland (Fig. 2-15). Within the central regions of West Gondwanaland, formed by what are now the continents of South America and Africa, the kind of subhumid tropical habitats occurred that may have favored early angiosperm evolution. Africa and South America began to separate but did not come completely apart in the tropical regions until about 90 million years ago. By that time, the angiosperms became abundant in the fossil record on a worldwide basis. The milder world climate also may have contributed to the expansion and success of the flowering plants. By about 80 million years ago they had established a dominance that has been maintained into the present-day.

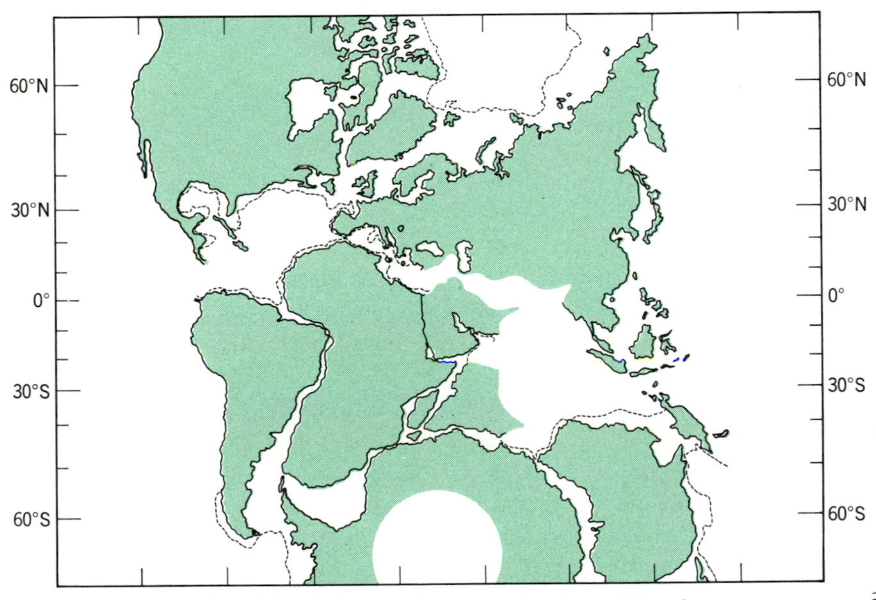

Figure 2-15 Relationship of the continents at the time of the first appearance of flowering plants in the fossil record about 125 million years ago. About 15 million years before this South America was directly connected with Africa, Madagascar, and India, and via Antarctica with Australia—this vast supercontinent is called Gondwanaland. Africa and South America began to separate but did not come apart until about 90 million years ago. At about the same time India began to move northward and collided with Asia about 40 million years ago; Australia began to separate from Antarctica about 55 million years ago, their separation taking about 15 million years.

10. RECENT DEVELOPMENTS INTRODUCTORY TO THE ARRIVAL OF MANKIND

By about 65 million years ago (at the end of the Cretaceous period) much of the world was occupied by a rich and well-developed angiosperm flora, including such familiar plants as magnolia and its relatives, figs, breadfruit, palms, oaks, willow, ash and maple. Many plants we now consider tropical or subtropical were found far north of their present distribution. Probably the warmest interval in recent earth history occurred about 50 million years ago when subtropical vegetation occurred in Alaska and Greenland and warm temperate vegetation in Antarctica. At this time there was a dramatic and widespread extinction of the giant reptiles; this happened on land, in the air, and in the sea and has not been adequately explained. In any event, with the demise of the giant reptiles, the mammals, which had been spreading during the Cretaceous, became dominant on land from then on. It was also about this time that mammalian primates (the group to which mankind belongs) appeared—with the multistoried rain forests presenting an array of aerial habitats for their evolution. The primates are divided into two groups. The *prosimians* (lower primates) include lemurs, tree shrews, and a large number of fossil forms that descended from insectivores. The *anthropoids* (higher primates) include monkeys, apes, and humans. The monkeys arose more than 60 million years ago, first in the New World and then in Africa, and subsequently there was radiation of the baboons, apes, and manlike creatures (*hominids*) (Fig. 2-16). Although the primates were descended from insectivores, they early on evolved the ability to obtain and consume a wide variety of food, from plant materials to animal proteins. In addition, they were characterized by their grasping feet and hands and well-developed vision. By about 25 million years ago distinctively *herbaceous* (nonwoody) flowering plants arose, and along with climatic cooling and drying the tropical and temperate forests contracted and grasslands began to spread. In Africa *savannas* developed, which are characterized by a mosaic of grassland, with scattered low trees or island patches of trees, and forests along the rivers. The rapidly evolving primate stock, for example the baboons, came to the ground in these savannas, as indicated by numerous fossils found in such habitats. Hominids began to diverge from the evolutionary pathway, that apes had pioneered 4 to 6 million years ago. In contrast to the baboons, the early hominids (including *Australopithecus*, precursor of *Homo*, "Man") learned to walk on two legs, thus freeing the hands for foraging for food and for carrying and using tools (Fig. 2-16 and Chapter 3).

Many important changes affecting the evolution both of our human ancestors and of plants took place during the 1½-million year interval of the Pleistocene epoch, which ended some 10,000 years ago. In northerly latitudes there were several major advances of ice, known as "glacial periods," alternating with warmer "interglacial periods"—the last being the one in which we live. Vast areas in temperate latitudes were denuded of vegetation by invasion of glaciers, but when the ice masses melted, plants could colonize the area again. Although many kinds of plants and animals became extinct during this interval, it was also a time for rapid evolution of herbaceous flowering plants. In tropical latitudes various wet/dry oscillations resulted in similarly rapid plant and animal evolution, thus increasing the variety of food resources for the hominids. Evolution of humans is discussed in the following chapter but here we note that our own species *Homo sapiens* appears to have evolved about 200,000 years ago. At first, the human impact on the biosphere probably was similar to that of most other mammals, but it then increased with the

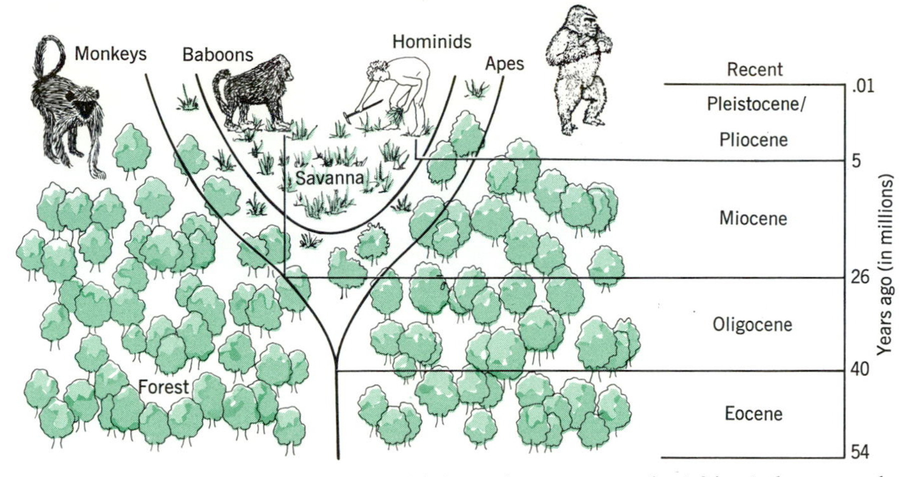

Figure 2-16 In the radiation of the higher primate groups in Africa baboons and hominids adapted to life in the areas adjacent to the tropical forests in which trees became scattered in grassland.

development of hunting technology and, with the origin of agriculture some 10,000 years ago, the stage was set both for the expansion of human population and for the specialization of technology. This in turn resulted in a progressive control of the environment. Thus we developed the ability to shape both the natural ecosystems and, to some extent, organisms in them, to our own ends both consciously and unconsciously. Hence indeed plants and people have become inextricably related in the last few thousand years.

CHAPTER 3

Our Predecessors: Their Development of Agriculture and of the Plant Sciences

In the previous chapter we glimpsed some of the major changes in the evolution of plants as individuals, as populations, and as constituents of ecosystems through approximately 3.5 billion years; we saw in perspective how very recent human appearance is upon this evolutionary scene, and we must now trace how our many interactions with plants have developed.

1. OUR PREDECESSORS

The development of the human species and its immediate ancestors is a picture that has been put together from fossil evidence found throughout Africa, Asia, and Europe during the previous century (Fig. 3-1). This work on the puzzle of human origin is still being painstakingly assembled piece by piece and, because of the many gaps, some conclusions remain controversial. However, certain key generalizations appear relatively clear, and these will set the stage for our understanding of human relationships with plants.

Fossil hominids, *Australopithecus,* became abundant about 3.5 million years ago, radiating into two or more lineages by 2 million years ago (Fig. 3-1). Also at that time the first evidence of stone tools appears. Man's erect posture, freed hands, and bipedal (two-legged) locomotion, the qualities that make an individual look human, appeared long before the increased brain capacity that distinguishes our species. By 3 million years ago *Australopithecus* had this human posture, combined with a brain not much larger

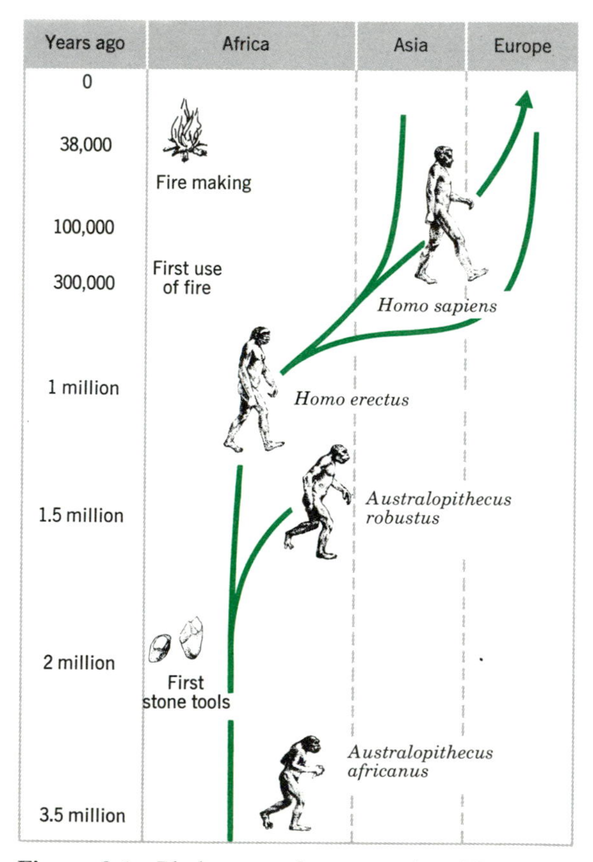

Figure 3-1 Phylogeny of our species (*Homo sapiens*) and our immediate ancestors.

than a chimpanzee's, large grinding teeth, and small canines. It has recently been proposed that bipedal carrying of gathered plant foods for sharing was the basis of divergence of the australopithecines from the ancestral apes. Plants were undoubtedly a major food source for these early hominids as they moved from the forests into the patchy, mosaic savanna vegetation. Bipedal locomotion probably developed to enable the hominids to cover the large home range necessary for obtaining widely dispersed resources on the savanna. The large, worn, and chipped teeth that have been found suggest that gritty plant foods were an important part of the diet, and markings on the skull and jaw indicate prominent muscle attachments for large chewing muscles developed from tearing tough foods. Gritty and tough foods could have included seeds, roots, fruits, and generally fibrous plant foods. Anthropologists supporting this view indicate that meat consumption probably increased over that of the ancestral apes—first by more frequent predation of small animals, then by butchering scavenged meat, killing animals at close range, and, finally, much later, pursuing and killing animals with specialized tools. They point out that hunting with tools is a "high risk, low return activity" that probably became common only after gathering was fully developed. Plant gathering offered a secure nutritional base, allowing hunters to spend time and energy for more uncertain success. This view stresses the gathering of plant foods rather than hunting to get large amounts of meat as a critical innovation in human evolution. Furthermore the omnivorous diet of chimpanzees (over 90% being plant foods) finds a parallel among living human gathering-hunters, for whom plant foods may comprise 50 to 90% of the diet.

Between 2 and 1 million years ago the early hominids were evolving in response to changes in the African grasslands. By 1 million B.P. (years Before Present) the robust australopithecine (*A. robustus*), with heavy skull, rugged face, and enormous teeth, had become extinct; the reason is unknown. Meanwhile, the physically less robust species (*A. africanus*) had evolved into *Homo erectus* ("erect man"), the precursor of modern humans (Fig. 3-1). Fossils from several African sites have documented this transition from *Australopithecus* to *Homo* as occurring between 1.8 and 1.2 million years ago.

2. MIGRATION THROUGH THE OLD WORLD

By 1 million years ago *H. erectus* was well established throughout Africa and relatively soon afterwards in the Near East and Java (Fig. 3-2). Between 1 million and 500,000 years ago there are records of *H. erectus* in northern Asia and Europe. There in cold Eurasia around 350,000 to 300,000 years ago, archaeologists have found the first evidence for use of fire. At this time hominids apparently could not make fire but collected it from natural sources.

Homo erectus had a larger cranial capacity than earlier African hominids, a thicker skull, smaller cheek teeth (implying a less tough diet), and increased stature and body size. Most of what we know about these predecessors during this period is based on tools, which are often associated with animal bones. These people apparently lived in rock or cave shelters and open sites along the ocean, lake, or river fronts. The remains point to their having eaten antelope, deer, wild boar, rhinoceros, elephant, baboon, and other animals. Here plant food may have been less abundant year round than in the tropical savannas, but herd animals were continually available. It is in this environment that our ancestors may have further developed their technology for obtaining more meat than previously. They apparently used a mixture of large and small tools manufactured with varied degrees of skill. Although *H. erectus* is often characterized as a "big game hunter" the large herd animals characteristic of northern Asia, Africa, and Europe are rare in southern Asia

Figure 3-2 *Homo erectus* foraging in the wilderness of Java perhaps half a million years ago.

and he presumably relied more on small animals, insects, and plant foods there. The variability of these sites during the middle of the Ice Ages implies increasing flexibility in adaptation to different environments—especially in the colder climates. These gatherers and hunters probably lived in small groups that were relatively mobile and widely dispersed. Their ways of obtaining food probably required relatively large territories and also imply the evolution of a high level of community organization and cooperation—itself an important innovation.

Between about 200,000 and 100,000 years ago the transition to *H. sapiens* ("knowing man"—the species to which we belong) occurred and the evolutionary tempo accelerated over the slow changes characterizing the previous 500,000 years. This acceleration was apparent in the hominid's physical characteristics, his habitations, and his tool technology.

By 100,000 years ago the human brain was nearly as large as today, increasing the capacity to process social and environmental information and communicate it with others—thus laying the neurological basis for the skills that enabled the development of "culture" as we now know it. There are more fossils after 100,000 years, attributable partially to the beginnings of deliberate burials, indicating the development of rituals, communication, and the like. Hunting was a dramatic part of the western European culture, and these new people with more technological and organizational skills recorded their activities in the first cave art around 35,000 B.P.

The early fossils of *H. sapiens* include two varieties commonly known as Neanderthal man and Cro-Magnon man; Cro-Magnon man is physically indistinguishable from us. Specimens of Neanderthal man abound from about 80,000 to 40,000 B.P. Then about 40,000 to 30,000 B.P. Neanderthal man became extinct, for reasons still unknown, and modern man appeared; within the course of 10,000 to 20,000 years this variety spread throughout the world (Fig. 3-3). It was also about this time that humans added fire making to their cultural heritage. The universal use of controlled fire for cooking was an impressive evolutionary adaptation, for it opened the possibility of expanding the human

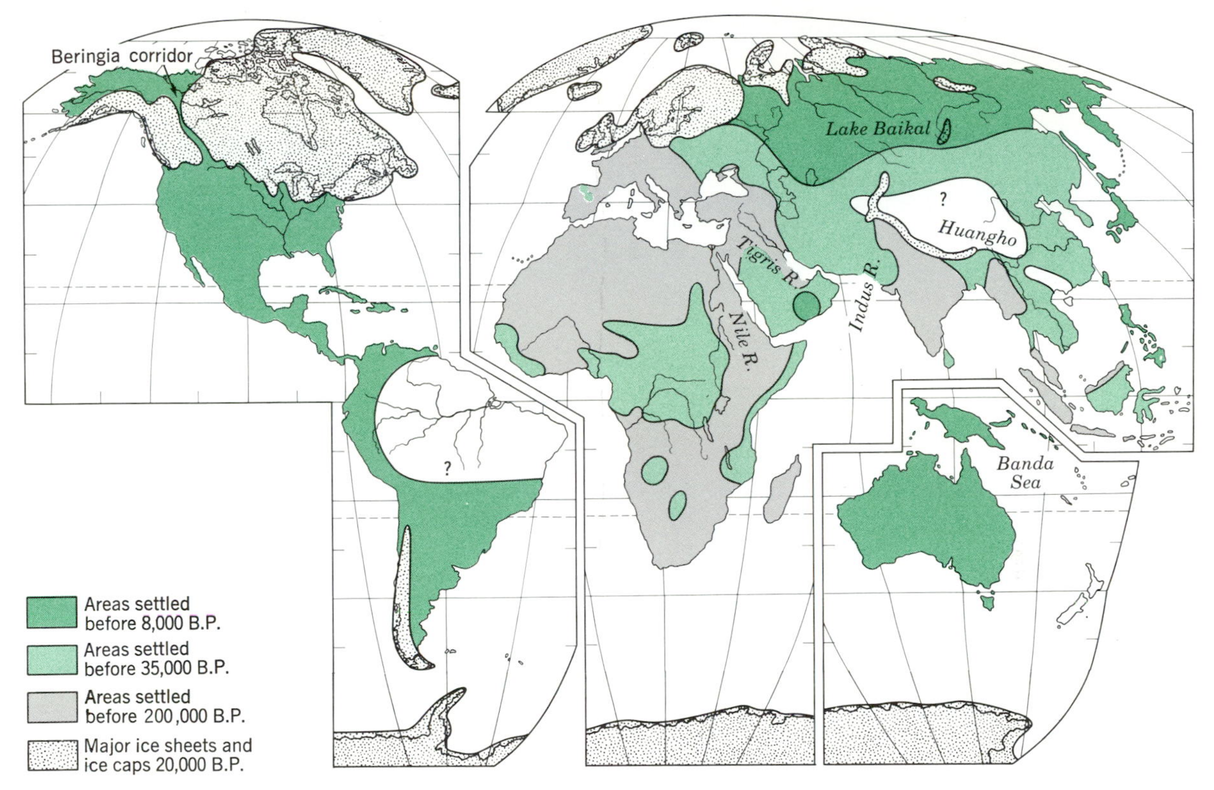

Figure 3-3 Areas settled by humans before 200,000 B.P., and before 35,000 B.P., and expansion of the inhabited world during the course of the Pleistocene Epoch (1½-million-year interval that ended about 10,000 years ago). For the two blank areas we have insufficient data. (Adapted from Butzer, 1977)

food supply by making some plant foods more digestible (e.g., cereals) and by removing toxic chemicals from others (e.g., manioc). Indeed the general adoption of cooking is coincident with the rapid expansion of our species.

3. MIGRATION TO THE NEW WORLD

Of particular interest between 35,000 and 10,000 B.P. is the first colonization of Australia and the Americas (Fig. 3-4). This ability of *Homo sapiens* to colonize new areas during the Late Pleistocene shows how adaptable the human race had become. It is a curious and still unexplained circumstance that no other hominid has been found in the New World. These first human settlers had to migrate across the harsh environment of northeastern Siberia to Alaska via the Bering Strait, which was a land bridge during certain periods of the Pleistocene. There are only three dates possible when sufficient areas of that land bridge were exposed for this crossing (75,000 to 50,000 B.P., 32,000 to 28,000 B.P., and 22,000 to 15,000 B.P.). Opinion is divided regarding the time of the first migration: good evidence points to 20,000 but a date prior to 27,000 is also conjectured. These migrants then slowly penetrated southward. Beginning in 12,500 B.P., however, they moved more rapidly across North America and all the way down the mountains of the Americas to Tierra del Fuego, a journey that took less than four millennia. Their rapid movement across almost the full range of available world environments is one of the most impressive feats of

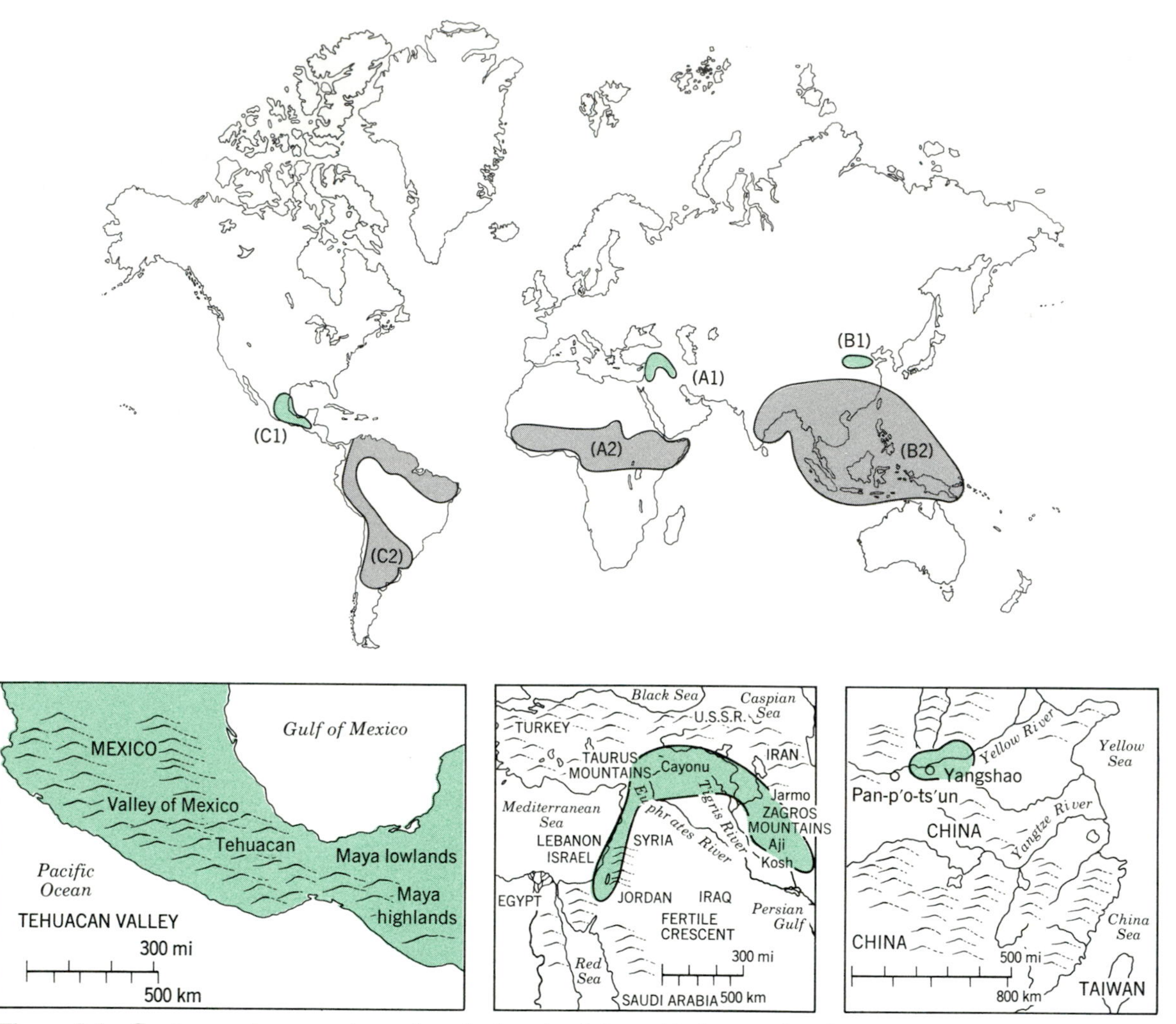

Figure 3-4 Centers and noncenters of agricultural origins with insets showing location of some of the principal archeological sites. *A* 1. Near East center. *A* 2. African noncenter. *B* 1. North Chinese center. *B* 2 Southeast Asian and South Pacific noncenter. *C* 1. Mesoamerican center. *C* 2. South American noncenter. (larger map, from Harlan, 1971)

Pleistocene humans. It strongly supports organizational skills, initiative, flexibility, and viable subsistence models.

It has also been suggested that, as a result of human hunting, a "Pleistocene overkill" occurred between 15,000 and 12,000 B.P. when many of the large mammals represented in the Rancho La Brea tar deposits in California (e.g., camels, horses, bison, mammoths, saber-toothed cats) became extinct. The human invaders are thought to have brought perhaps 50,000 years' experience in hunting similar animals (i.e., the ecological equivalents) from Asia and Africa. The hapless New World mammals would never have been exposed to any hominid-type predator and thus were easily killed by the technologically advanced hunters with projectile points. Careful studies of these events show that animal extinction proceeded at a relatively steady rate during most of the Tertiary period (Chapter 2) but during the late Pleistocene (approximately 50,000 years ago) there were sudden surges of extinctions proceeding at a higher rate and appearing to coincide either with migrations of technologically advanced humans or with a combination of environmental stress and human activity. In any event, these types of human interactions with hunted animals constitute, along with fire, some of the first major impacts upon ecosystems.

4. THE ORIGIN OF AGRICULTURE

Thus by 12,000 B.P. we had become an ecologically dominant species that could migrate throughout ecosystems of the world and influence their structure. With the development of agriculture around 10,000 B.P. we undoubtedly assumed an ecological role without parallel in the history of the planet. Through the cultivation of plants we developed a more dependable source of food, which led to ever-increasing growth of population and the first real attempts to control environment, including both the raw materials and the land itself. The sedentary existence generally associated with cultivation practices also led to the development of villages. The increased efficiency in food production allowed more people freedom from this endeavor and probably led to the development of civilization in several different locations.

In recent years archeological work has greatly increased our knowledge about the origin of agriculture. Previously archeologists had emphasized the study of spectacular artifacts from temples and tombs, but more interest now centers on how our ancestors lived and how they managed their environment. One of the most critical problems in studying what people ate is that food is perishable. Therefore, plant remains are best preserved in arid regions. Even under dry conditions most of the remains are limited to seeds, along with some flower bracts (reduced and modified leaves), stalks, leaves, and fruit rinds. This means that our ideas regarding the early stages of agriculture tend to be restricted to seed crops growing in dry areas. Some ethnobotanists and anthropologists have wondered if agriculture began with root crops in the humid tropics, but unfortunately evidence for this tantalizing suggestion has not been preserved.

After the identification of the ancient plant material, the next step is to decide whether it is from wild or cultivated plants. Sometimes domestication has been accompanied by changes in form that can be recognized. Among the cereals, of all the adaptations that separate wild from cultivated plants, the "nonshattering trait" of cultivated plants (i.e., the ability of mature seed to remain on the plant) is the most conspicuous. Most of the seeds that do not shatter are harvested; most of the seeds that shatter escape harvest. Shattering as a means of seed dispersal is, of course, important in the life of the wild plant. Also, in beans the parchment layer and the fibrous thickening of the sutures of the pods (Chapter 23) have become reduced, producing fruits that often do not fall. Again, in chili peppers the fruits of the domesticated forms do not drop at maturity.

As information about early agricultural practices has accumulated, several questions have arisen such as: (1) Where were the different centers of the origin of agriculture located and what were the possible connections between these areas? (2) What led to the development of agriculture, that is, why did people bother to change their mode of subsistence? (3) When did the society in each center come to depend upon agricultural practice? (4) How have crop plants changed and spread since they were first taken into cultivation?

Although there are many variations on the theme, three primary and three minor agricultural centers or "hearths" are frequently discussed. The primary centers were (1) the Near East, that is, the hill country of southwestern Asia, (2) Mesoamerica and the Andean highlands, (3) southeastern Asia, the margins of the Bay of Bengal, and the Malay Peninsula. The three minor centers were (1) North China, (2) Ethiopia, and (3) West Africa. These hearths actually are adaptations of the eight centers of origin of cultivated plants recognized by the great Russian plant breeder, N. I. Vavilov, who built upon the earlier work of the Swiss plant geographer Alphonse de Candolle. Vavilov, in his extensive studies on the origin of cultivated crops, came to the conclusion that the best guide to the center of domestication of any plant is the area where there is greatest variation among the types of that plant. This diversity in primary centers was considered a consequence of ancient cultivation and thus was distinguished from that in the secondary centers, which resulted from natural processes such as hybridization and selection. Later critics of Vavilov's work have suggested that ecological diversity, farming practice, human migration patterns, and/or breeding of the crop could also have produced localized centers of diversity—in which case diversity would not be a result of the length of time for which the crop was cultivated.

Relatively recently a slightly different idea has been proposed. It assumes that agriculture arose in three different regions, each having a relatively small area that was a center of diversity. However, associated with the center was a large area of domestication dispersed over 5000 to 10,000 kilometers (Fig. 3-4), which has been called a "noncenter." Centers and noncenters form systems in which evidence suggests some stimulation and feedback in terms of technology and materials within the system. One system includes a definable Near East center and a noncenter in Africa; another system includes a North Chinese center and noncenters in Southeast Asia and the South Pacific; the third encompasses a Mesoamerican center and South American noncenter (Fig. 3-4). Our discussion will generally follow this plan.

5. AGRICULTURE IN THE NEAR EAST AND AFRICA

a. ORIGIN IN THE NEAR EAST AND MOVEMENT INTO EUROPE

The earliest archeological evidence for the origin of agriculture comes from the Near East, particularly sites in Iraq, Iran, and Turkey (Fig. 3-4 and Figs. 3-5*A*, *B*, and *C*). A large and impressive body of data has been accumulated over the past 20 years, not from the fertile river valleys of Mesopotamia (important as early centers of civilization) but rather from the nearby semiarid mountain areas (Fig. 3-5*A*). Sickles and grinding stones appear in excavations of a hillside village near Jarmo, Iraq, from before 8000 B.C. (10,000 B.P.) when wild grains were being collected (Fig. 3-5*B*, Table 3-1). By 8750 B.P. there were cultivated wheat and barley seeds as well as bones of goats and sheep there. It is thought that incipient domestication may have occurred even before this documented record. In later deposits evidence of other plants (peas, lentils, vetch, grapes, olives, dates, pears, cherries, etc.) and other animals (pigs and dogs) become more abundant. This early domestication of both plants and animals in the Near East was possible because of overlap between the natural habitats of western Asiatic herd animals and the native distribution of wheat and barley on which the plant culture came to depend. If the distribution of wild wheat and barley 10,000 to 11,000 years ago was as it is today, the cradle of western agriculture was in a winter rainfall belt in the oak woodland and adjacent grassland zone of the Tagros Mountain foothills surrounding Mesopotamia (Fig. 3-4). The fossil pollen in this area indicates that prior to 11,000 B.P. few oaks grew there. The steppe vegetation was dominated by sagebrush and associated plants much like that of the cool, arid areas now occurring in Nevada. Then the percentage of oak, pistachio, and plantain increased, indicating a warmer and more

(A)

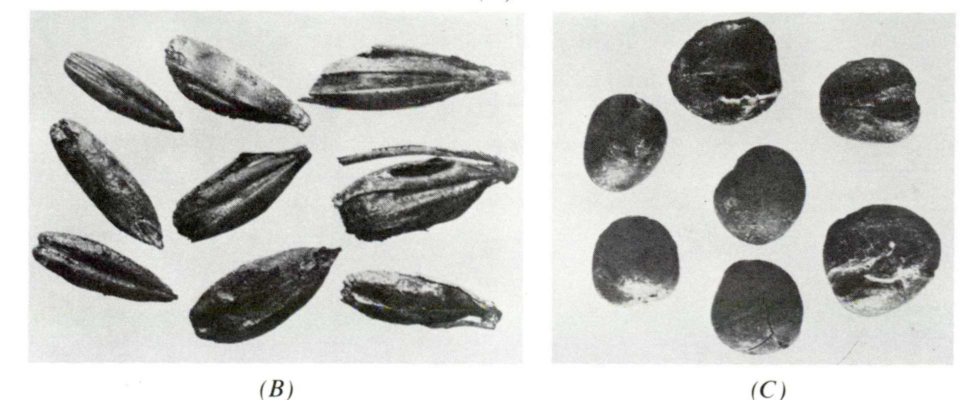

(B) (C)

Figure 3-5 *A*. Jarmo in Iraqi Kurdistan is the site of the earliest village-farming community yet discovered. This photograph of an upper level of excavation shows the foundation of a house and paving stones. The site was occupied for perhaps 300 years, about 8750 B.P. *B*. Carbonized barley kernels from Jarmo, enlarged four times, are from two-row grain (see Chapter 16). The internodes attached to the kernels indicate tough spikes of cultivated barley. *C*. Peas from the same site.

moist habitat. By 7500 B.P. the oak frequency had reached that found today. Since the distribution of wild wheat today is coincident with that of oak, it seems that rising temperatures about 11,000 years ago brought both oaks and wheat to the foothills from protected areas in which they had survived during the cold intervals of the Pleistocene period. Some investigators think that this timing was critical in making plants available for agriculture in the Near East.

Why did humans shift from a subsistence based on hunting and gathering from the wild to

TABLE 3-1
Chronology of Agriculture in Three Centers and Their Noncenters from Archeology.

B.P. (Years Before Present)	*Near East–Africa*	*North China–Southeast Asia*	*Mesoamerica–South America*
10,000	Sickles and grinding stones from villages (Iraq)		
9,000		Cultivated (?) peas and beans (Thailand)	Beginning of shift from gathering to cultivation
8,765	Cultivated wheat and barley (Iraq)		
8,500	Cultivated peas and lentils		
8,000			Cultivated beans and lima beans (intermontane Peru)
7,000	Move to Tigris and Euphrates river valleys; productive irrigated agriculture	Rice (Thailand)	10 to 15% of used plants now cultivated (Mexico); primitive maize, squash, chili peppers, beans, avocado, and amaranth
6,500	Dispersal of agriculture to central and coastal Europe		
6,000	Urban civilization in Tigris–Euphrates area; reports of plant cultivation in the Sahara	Yang-Shao farming culture (China)	
5,500			30% used plants cultivated (added other beans, other fruits, etc.); development of villages (Mexico)
5,000			Beans, lima beans, squash, cotton cultivated (coastal Peru)
4,500	Dispersal of agriculture into western and northern Europe		
4,000			Maize (Peru)
3,500			Full-fledged agriculture (Mexico)
2,900			Irrigation (Mexico)
2,500			Highly productive agriculture (Mexico)

one based upon domestication of plants and animals? In classical mythologies of all civilizations, agriculture is regarded as of divine origin, although it arrived in different ways from different deities and under various circumstances. V. Gordon Childe, the archeologist who called the invention of agriculture the "Neolithic Revolution," proposed what has come to be known as the "propinquity theory." Childe thought that man and herd animals were brought into contact, in the Near East and in North Africa, where there was water during dry periods. In the 1950s many investigators still thought that humans went through a three-stage development: hunter, then herder, then cultivator. If humans were herders, the disturbance of soil and vegetation at camp sites, together with manuring, could have encouraged weedy plants to grow. This idea, which dates back to Greco-Roman times, led to the assumption that it would be a short step from gathering these weeds to growing them on purpose. Unfortunately, the evidence does not bear out this propinquity theory.

The most extensively developed model for agricultural origins has been that cultivation was a discovery. Charles Darwin expressed ideas on this subject that included; (1) people must be sedentary before they can cultivate plants, (2) useful plants are most likely to be discovered in manure refuse heaps and then planted there, and (3) "a wise old savage" is required to start the process. These concepts of Darwin have provided the basis for several amended theories, the most prominent one being that of the geographer Carl Sauer. Sauer thought that Southeast Asia was the oldest hearth of agriculture because it had a mild climate, varied terrain rich in edible plants, and fresh water resources. In this kind of site people could settle down and start cultivating without the pressures of periodic scarcity. The plant geneticist Edgar Anderson added some genetic aspects. He saw weeds as potential domesticated plants and thought that an increase in hybridization with disturbed habitats could result in increased variation and new genetic combinations from which useful selections could be made. Anderson also thought that agriculture began in the tropics on dump heaps and that vegetative propagation predominated at the beginning. However, in Mesoamerica good archeological evidence indicates that the people remained nomadic long after they were purposely growing plants for food, and thus sedentary life is not essential to the development of agriculture. Furthermore, in the Near East agriculture developed neither in the tropics nor by people dependent upon water resources. What then were the circumstances motivating humans to domesticate plants? A much-cited model in current literature, based on proposals by Lewis Binford and Kent Flannery, attempts to integrate ethnographic and archeological data. Explicit in this new model is the recognition that gatherers are sophisticated, applied botanists who know their plants and how to exploit them. They are prepared to grow plants if and when they think it is worth the effort. Furthermore, it has been discovered that the difference between intensive gathering and cultivation is minimal. It is thought that population pressures pushed people into less and less favorable habitats at the margins of the optimum wild food zone, where these people attempted to duplicate by cultivation the dense stands from which they had gathered wild grain. Thus this type of domestication probably has taken place independently and even simultaneously in a number of areas over the world.

In the Near East around 7000 years ago people moved down from the dry upland areas into the Tigris–Euphrates River valleys; here seasonal flooding provided an annual nutrient supply in the sediments, which led to a more productive agriculture. This steady supply of food probably had considerable influence on the development of an urban civilization by 6000 B.P. (Table 3-1). Even before 6500 B.P. and the beginnings of a highly developed civilization, migration was proceeding into the adjacent parts of Europe (the Balkan Peninsula) and thence northward along the river systems into woodland areas with good soils. This continuous dispersal may have been due to chronic overpopulation and the primitive practice of shifting agriculture (or it may have been simply the result of human enterprise). By 4500 B.P. agriculture had moved into western and northern Europe on to the acidic soils in the mixed coniferous and hardwood forests. This resulted in a shift from cultivation of wheat and barley to rye and oats which, because they mature earlier, could compete better in such cool climates (Chapter 16). The early human transformations of natural ecosystems to agricultural and urbanized landscapes are recorded by the pollen deposited in bogs of Europe, particularly in

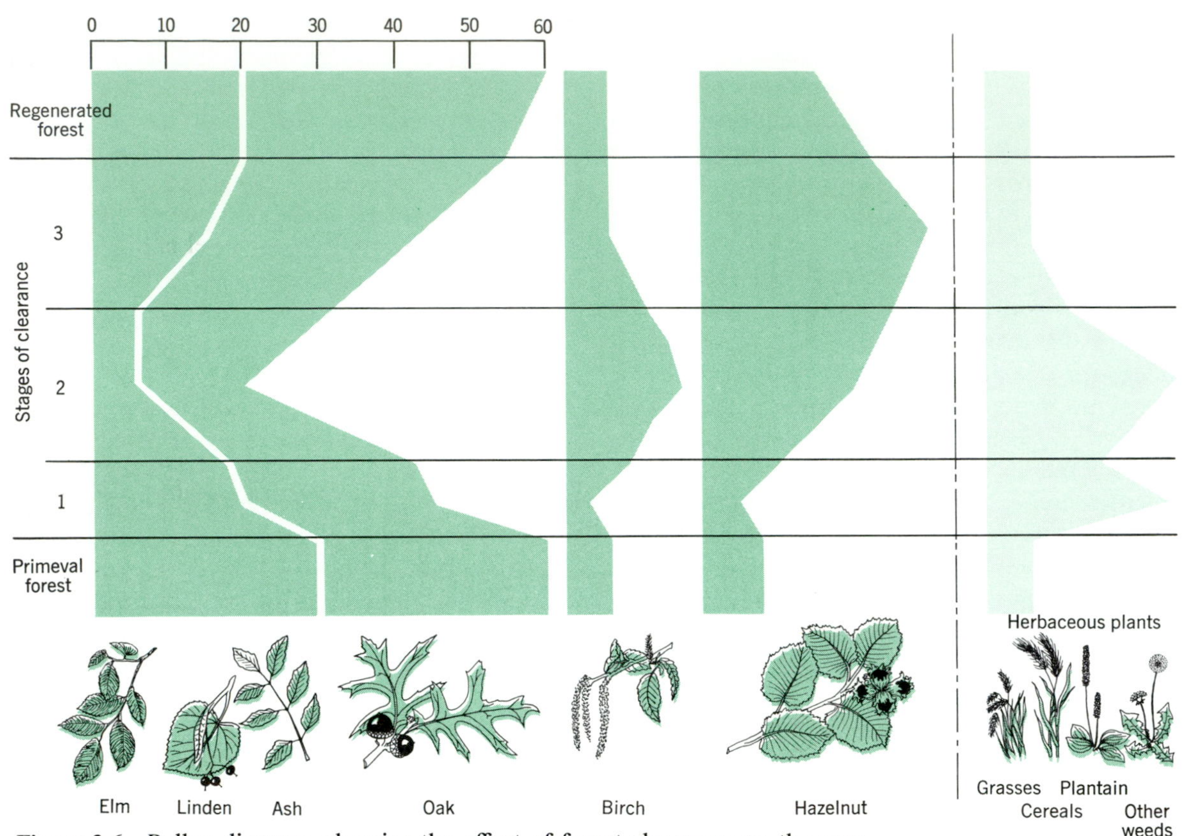

Figure 3-6 Pollen diagram, showing the effect of forest clearance on the vegetation of Denmark between about 4500 B.P. and 4300 B.P., is based on many samples of pollen taken by boring down into bogs. The width of each colored area on the diagram represents the percentage of pollen from one species in comparison to that from all others. The scale of the proportions is given at the upper left. Apparently as forests grew dense in Denmark toward the end of the last Ice Age, most people made their living by fishing and seal hunting. At this time in the primeval forest (colored areas below the lowest horizontal line) the distribution of pollen was 30% elm, linden, and ash, 30% oak, 5% birch, and 10% hazelnut. In samples dated about 1000 years later the forest pollen rapidly declines and in its place there is a sharp rise in pollen of herbaceous plants and emergence of cereals and new weeds. The distribution of herb pollen during three stages of clearance (1, 2, and 3) is shown at the right of the break in the horizontal lines. Shortly after clearance, a new growth of the species that typically follow forest clearance—willow, aspen, birch—springs up. The presence of birch suggests that fire was used to help clear the forest, for on fertile soil birch succeeds a mixed oak forest only after burning. Meanwhile, the herbaceous ground flora undergoes changes, with pasture plants dominating. Possibly cattle grazed in these grassy meadows bordered by scrub forests of birch and hazelnut. Then in the third phase, forest dominated by oak takes over, indicating that the people moved on to clear a new area. According to the pollen record, some of their settlements can hardly have lasted more than 50 years. (Adapted from Iversen, 1970)

Denmark, where the pollen left in the many bog lakes gives us a well-documented record of forest clearing, crop planting, and livestock grazing (Figure 3-6).

b. DEVELOPMENTS IN AFRICA

Little ancient plant material is available from Africa to support the origins of plant cultivation; most of our information comes from the domesticated plants we know today. However, grinding equipment dated at 14,000 B.P. has been found on terraces of the Nile. This is an earlier date than any grinding equipment found in the Near East. Could hunter–gatherers have learned to harvest and grind wild grass seeds in volume first in the Nile Valley and then transferred the techniques to the Near East where wild wheat and barley were more abundant and these plants were later cultivated? Could the idea of plant cultivation have later returned to Africa where unique cultivated strains ("cultivars") were developed? These are questions that anthropologists and ethnobotanists are asking today. Unfortunately, the idea of primitive cultivation in Africa comes only from reports of cultures existing over wide areas in the Sahara around 6000 B.P. when the rainfall was greater than today and the area supported dense Mediterranean vegetation. However, quite a number of cultivated plants probably originated in Africa and there supported a widespread development of sedentary agriculture. Most important of these crops is sorghum, one of the major cereals of the world (Chapter 16); there were also various millets and yams.

6. AGRICULTURE IN THE FAR EAST

Much archeological work supports the idea of a center for cultivation of plants in Northern China and outward dispersal from this area. The earliest farming culture known in China, at Yang-shao in the Yellow River Valley, is dated at 6000 B.P. (Fig. 3-4). The size of the villages and apparent social development, however, suggest that agriculture had evolved some time before that. Nonetheless, archeological remains are meager to support a *center* of origin in Southeast Asia and the South Pacific. Since much of this area is in the moist tropics, food products decompose rapidly and leave no record. The number of domesticated plants is impressive but varies over an area of some 10,000 square kilometers and over a period of several thousand years. Peas and beans, perhaps cultivated, have been found in Thailand from 9000 years ago (Table 3-1). We have no idea when the Asian staple, rice, was first cultivated but there is now evidence for rice in Thailand 7000 years ago (Table 3-1). It is thought that sugarcane was domesticated in New Guinea, many citrus types in South China and Indochina, and the coconut anywhere or perhaps everywhere along the shores of this vast region (Fig. 3-8). Bananas probably had a complex history involving selection of seedless plants that were produced vegetatively in Indochina, Thailand, and Malaya. Later a polyploid strain was produced and spread through other parts of the South Pacific (Chapter 17).

7. AGRICULTURE IN THE NEW WORLD

Our record for the New World—Mesoamerica (see Fig. 3-8) and the Andean region—is much better than that for Asia because again remains have been preserved in arid regions. Agriculture seems to have developed in four different areas in the Americas but apparently developed quite independently from the Near East. Good archeological data are available from the arid highlands of Mexico and from the coasts and valleys between the high mountains in Peru (Fig. 3-4). In the mesquite desert region of

Tamaulipas, Mexico, there is evidence for a shift from plant collecting toward cultivation about 9000 B.P. (Table 3-1). In both areas the first unequivocal domesticated plants appear about the same time (8000 to 7500 B.P.) but the species cultivated in these two areas are often different and appear in a different sequence (Table 3-1). One of the most complete archeological records of New World agriculture occurs at Tehuacán, Puebla, Mexico. The development here was slow, compared to that in the Near East, with a long period of "incipient cultivation" (Table 3-1). Between 9000 and 7000 B.P. plants were primarily gathered, but around 7000, with the onset of a warm period during postglacial times, 10 to 15% of the plants used by these people were being cultivated. They included primitive maize, squash, chili pepper, amaranth (a grain), and avocado. By 5500 B.P., 30% of the plants used were probably cultivated. By 3500 the people of Tehuacán were full-fledged agriculturalists, with hybridized maize (Chapter 16) and various species of squash, beans, chili peppers, and fruit, as well as fiber crops such as cotton. The dog, which was an important food in Mexico, also appeared in the archeological record about this time. Later the increase in population led to a need for expanded usage of lands and for the increase in yield provided by irrigation, and by 2500 there was evidence of productive, irrigated agriculture. The inhabitants of Tehuacán were also eating turkey at the beginning of the Christian era. About this time plant remains of South American origin appear—this fact shows that the Tehuacán people were then in contact with people from South America. A third complex of crops was domesticated in the humid lowland tropics, but it is difficult to determine how early agriculture began here because conditions do not favor preservation of identifiable botanical remains. Lowland domesticates that could survive in semiarid conditions (sweet potato, manioc, plantain, pineapple, and guava) spread to Mesoamerica and the Andes, while crops that could adapt to high rainfall (cotton and maize) spread into the lowland rainforests. In eastern North America a fourth complex of cultivated plants may have developed, but this was almost completely supplanted by the northward spread of the Mesoamerican domesticates.

Climatically these areas in Mexico were similar to the Near East locations, but the diversity of useful plant species was even greater (80 species of agriculturally important plants); however, quite different from the Near East was the absence of herd animals. Thus the diet in Mesoamerica was largely vegetables.

Productive agriculture in the New World led to the development of a series of cultures culminating in the Aztec in Mexico, the Mayan in Central America, and the Inca in the Andean region. These groups adapted their agricultural practices to a diversity of environmental conditions. For example, the Aztecs developed a unique form of land reclamation and agriculture known as the *chinampa* system. This system, practiced on the margins of lakes in the Valley of Mexico, was one of the most intensive and productive methods of farming that has ever been devised. Chinampas are long narrow strips of land surrounded on at least three sides by water (Fig. 3-7). Properly maintained, they can produce several crops a year and will remain fertile for centuries without having to lie fallow. This chinampa agriculture may have enabled the Aztecs of the Valley of Mexico to control most of this country for 1500 years before the arrival of the Spaniards. The Mayans used the corn stalk as a support for twining beans and thereby developed a balanced cropping program and balanced diet that still sustains the Indians of this tropical region. The agriculture of the Incas, on the other hand, was uniquely adapted to their high mountainous home. Their economy was based on the potato, which produces prolifically at high altitudes and could be preserved by dehydration. Stone terraces were developed as well as long-distance irrigation systems. More people were maintained at a higher standard of living in this forbidding habitat 500 years ago than at present.

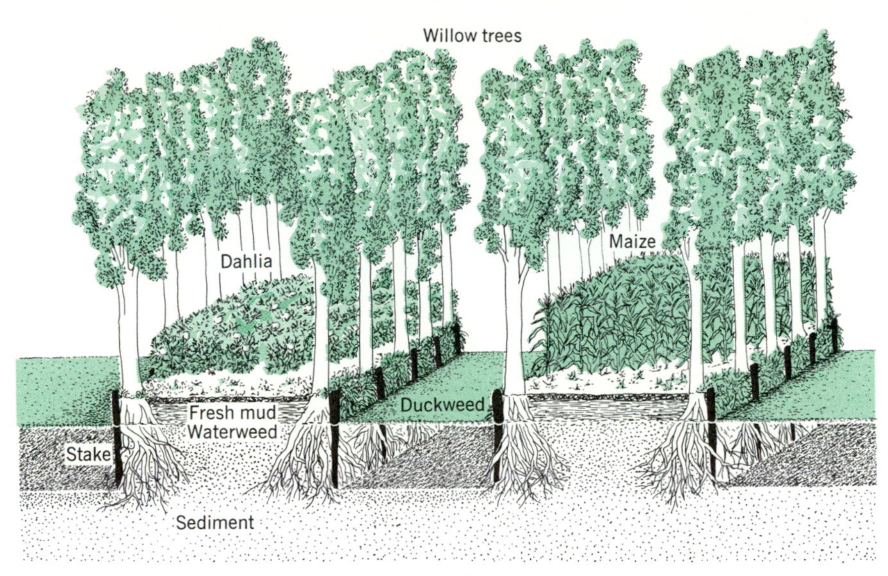

Figure 3-7 Cross section diagram of chinampas and canals, developed by the Aztecs in Mexico, gives an idea of their construction. Fresh mud from bottom of canals and weeds for compost beneath the mud are used to fertilize the chinampas. Trees and stakes keep their sides firmly in place. (Adapted from Coe, 1970)

Although many of the Spanish explorers set out on their expeditions in search of spices and gold and silver, they and their successors came back not only with these but many "strange and wonderful" plants that were to transform the European diet—corn, potato, tomato, squash, pumpkin, peanuts, kidney and lima beans, cranberries, avocado, pineapple, cacao, chili, Brazil nut, pecan, cashew, rubber, tobacco, and others (Fig. 3-8).

8. THE GROWTH OF TECHNOLOGY IN AGRICULTURE

The ancient cultures of Mesopotamia (aptly called the "cradle of civilization")—Sumeria, Babylon, and Assyria—developed an increasingly complex agriculture, their ruins indicating irrigated terraces, parks, and gardens. By 2700 B.P. an Assyrian Herbal had been compiled that contained the names of 900 plants. The Old Testament frequently refers to agriculture during this time.

Particularly well-preserved records of ancient agricultural development occur in the desert sands of Egypt. The agricultural kingdom of the great Egyptian civilization, to which our culture owes so much, flourished on the soil fertility supplied by flooding of the lower Nile. By 5500 B.P. the Egyptians had a centralized government and by 4800 B.P. had developed a sufficiently sophisticated technology to build the pyramids. They were masters at developing techniques of drainage and irrigation and also worked out refinements of the first hoe and plow for land preparation. They developed various refinements in the baking, wine-making, and food storage industries. They created the first pharmacopoeia and set up spice, perfume, and cosmetic industries. Along the Nile there were great formal gardens of ornamental plants (see Chapter 27). Orchards included cultivation of dates, figs, lemons, and pomegranates, and vegetable gardens had such familiar items as artichokes, garlic, leeks, onions, lettuce, endive, radish, mint, and various melons.

In Greece, partially because the Greek agricultural base was not sufficient to support the

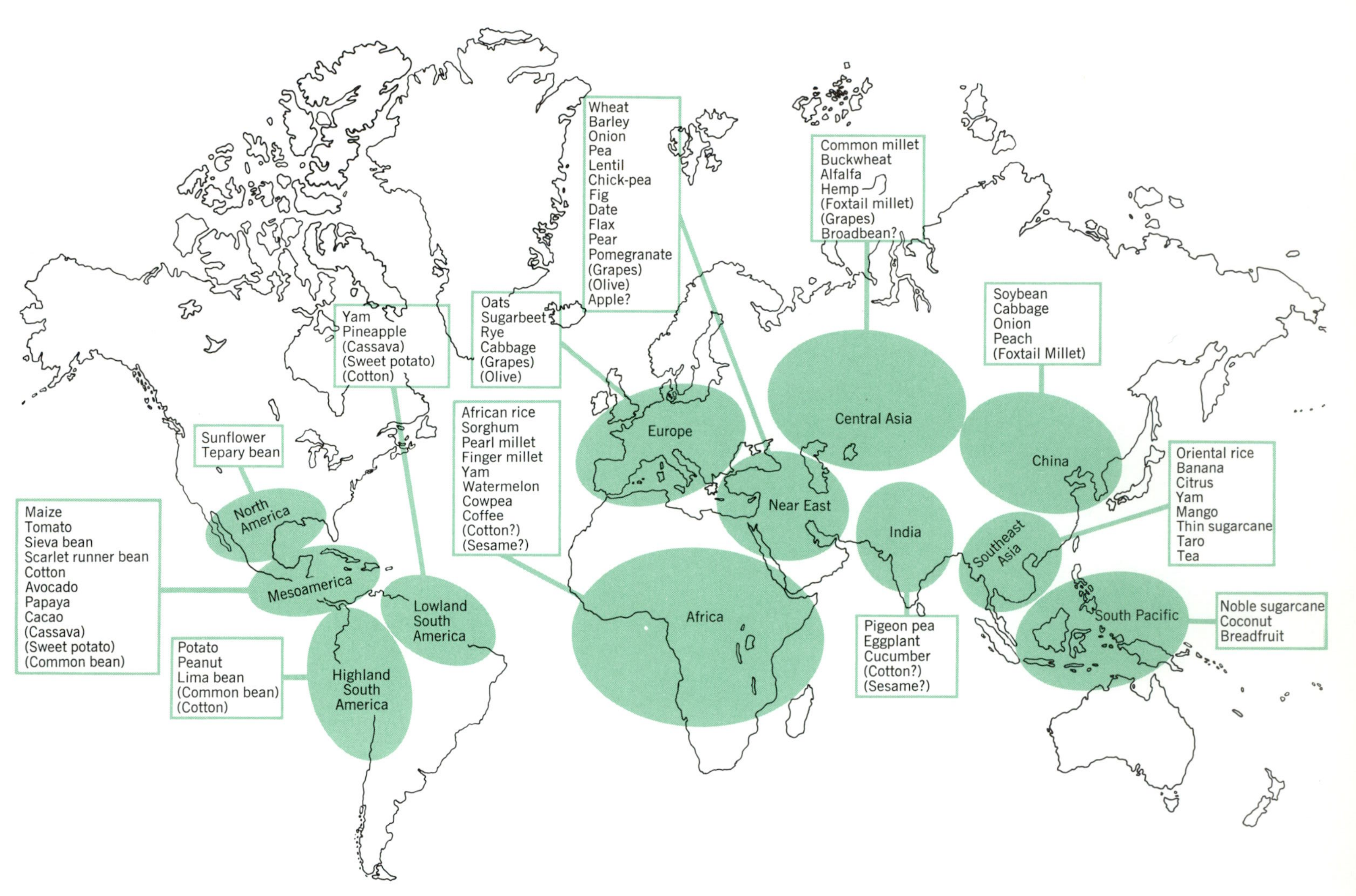

Maize
Tomato
Sieva bean
Scarlet runner bean
Cotton
Avocado
Papaya
Cacao
(Cassava)
(Sweet potato)
(Common bean)
Sunflower
Tepary bean
North America
Mesoamerica
Potato
Peanut
Lima bean
(Common bean)
(Cotton)
Highland South America
Lowland South America
Yam
Pineapple
(Cassava)
(Sweet potato)
(Cotton)
Oats
Sugarbeet
Rye
Cabbage
(Grapes)
(Olive)
African rice
Sorghum
Pearl millet
Finger millet
Yam
Watermelon
Cowpea
Coffee
(Cotton?)
(Sesame?)
Wheat
Barley
Onion
Pea
Lentil
Chick-pea
Fig
Date
Flax
Pear
Pomegranate
(Grapes)
(Olive)
Apple?
Europe
Africa
Near East
Common millet
Buckwheat
Alfalfa
Hemp
(Foxtail millet)
(Grapes)
Broadbean?
Central Asia
India
Pigeon pea
Eggplant
Cucumber
(Cotton?)
(Sesame?)
Soybean
Cabbage
Onion
Peach
(Foxtail Millet)
China
Southeast Asia
Oriental rice
Banana
Citrus
Yam
Mango
Thin sugarcane
Taro
Tea
South Pacific
Noble sugarcane
Coconut
Breadfruit

Figure 3-8 Areas (with boundaries generalized) are indicated where plants were domesticated. The name appears in each area, except for wheat, whose different genera and species were domesticated independently in different areas. Where the same species may have been domesticated independently in two areas, the name is enclosed in parentheses in each area; among examples are grapes, cassava and cotton. Where the area of domestication is doubtful a question mark follows the plant name. (From Harlan, 1976).

increasing population, their civilization was absorbed by the Roman Empire. In contrast to the Greeks, the Romans were interested in the practical aspects of agriculture. The Romans can be credited with relatively few original ideas, but they greatly improved upon what they learned from others. The combination of empiricism and a genius for political management produced the largest contiguous empire the world has ever seen, and it lasted intact almost 600 years. The rise and fall of the political fortunes of imperial Rome were paralleled by similar trends in agriculture; which was "cause" and which "effect" is not apparent. With the decline of Rome and Western civilizations between 2500 B.P. and 1400 B.P. (500 A.D.) the Eastern civilizations attained their peak.

During this decline in the West, ideas regarding improvement of agricultural techniques survived in monastic gardens where plants provided food, wine, decorations, and medicinals (Chapter 28). The job of gardener (*hortularium*) became a regular monastic office. Fruit and vegetable varieties were preserved and improved. There were few new botanical writings during this period, most of them being compilations and many traceable to Pliny's *Natural History*.

9. FROM TECHNOLOGY TO SCIENCE

Theophrastus of Eresos (2732 to 2288 B.P.), student and disciple of Aristotle, appears to have been among the first to have become interested in plants for their own sake, that is, scientifically. His two treatises, *History of Plants* and *Causes of Plants,* have earned him the title of "Father of Botany" and influenced botanical thinking until the seventeenth century. Theophrastus treated over 500 kinds of plants, both wild and cultivated, recognizing the characteristics that later led taxonomists to distinguish between major categories in plant classification (e.g., between angiosperms and gymnosperms and between monocotyledons and dicotyledons). He dealt with such diverse topics as morphology, classification, propagation, geographic botany, pharmacology, horticulture, forestry, and even plant flavors and odors.

The accumulation of information through observation and experiment began to reach significant proportions in the sixteenth century, maturing first in the discoveries of astronomy and physics. Renewed interest in botanical studies accompanied the revival of learning and was indicated by the appearance of the herbals —books with illustrated lists of plants and their characteristics, particularly to be used for medicine. This coincided with the European invention of printing from movable type, which greatly accelerated the spread of ideas. The growing interest in drugs led many professors of medicine to study plants and to classify them according to their medical uses and effects. These developments are discussed in Chapters 13 and 28.

Since our present botanical concepts and theories grew out of earlier, and especially medieval ones, it is often helpful to present them in the framework of their historical development. For this reason we shall present a number of the topics in subsequent chapters along the lines of the way they grew and were gradually clarified.

PART 2

THE INDIVIDUAL PLANT

CHAPTER 4

How the Flowering Plant Is Constructed

The fifteenth and sixteenth century botanists made fine drawings of plants to grace the pages of their herbals, but they seem to have had little curiosity about the internal structure of the plants that they grew and used as food or medicine. It was the invention of the microscope in the seventeenth century that stimulated the study of plant anatomy. The magnification provided was low at first, but after about 1650 microscopes began to reveal the details of the fine structure of plants and animals. Antony van Leeuwenhoek of The Netherlands constructed microscopes powerful enough to discover bacteria (1676); he observed the different cells in wood, the detailed structure of seeds, and the cellular structure of stems. Robert Hooke, an Englishman, had earlier (1655) with a lower-powered microscope observed the cellular structure of cork and applied the name "cell." Somewhat later Marcello Malpighi, an Italian, though mainly concerned with animal anatomy, observed the conducting vessels in wood and theorized about their action, and Nehemiah Grew, in England, studied and pictured a large number of plant structures, In the eighteenth century the classification of plants began to be understood, and Stephen Hales with his *Vegetable Staticks* (1727) and Duhamel de Monceau with his *Physics of Trees* (1758) began to combine the study of structure with that of function, and thus to work out how plants absorb and conduct their nutrients.

Although the outward appearance of flowering plants is immensely varied, their basic structures conform in general to a common plan, one that they share with the other vascular plants. As we saw in Chapter 2, the plant body represents the sporophyte (diploid) generation. It is in essence an axis (Fig. 4-1), one end becoming the shoot and the other the root. Leaves develop in a continuing series as small outgrowths (*primordia*) in the bud at the apex of the stem, and as the shoot extends the leaves grow out from the *nodes* (with *internodes* in between). In dicotyledons the leaves are usually supported by a stemlike *petiole* and in many monocotyledons by a more or less cylindrical *leaf sheath*. Buds form at the bases of the leaves in the angle (*axil*) between leaf and stem, and can grow out into branches that duplicate the structure of the main shoot. Lateral roots develop out of the main root in a somewhat different way, though the resulting *root system* comes to resemble the branched shoot in size and extent. In the reproductive stage the flowers develop either at the apex or in the axils of the leaves. The two great groups of flowering plants—monocotyledons and dicotyledons (Chapter 1) differ not only in the number of cotyledons but also in the structure of the stem,

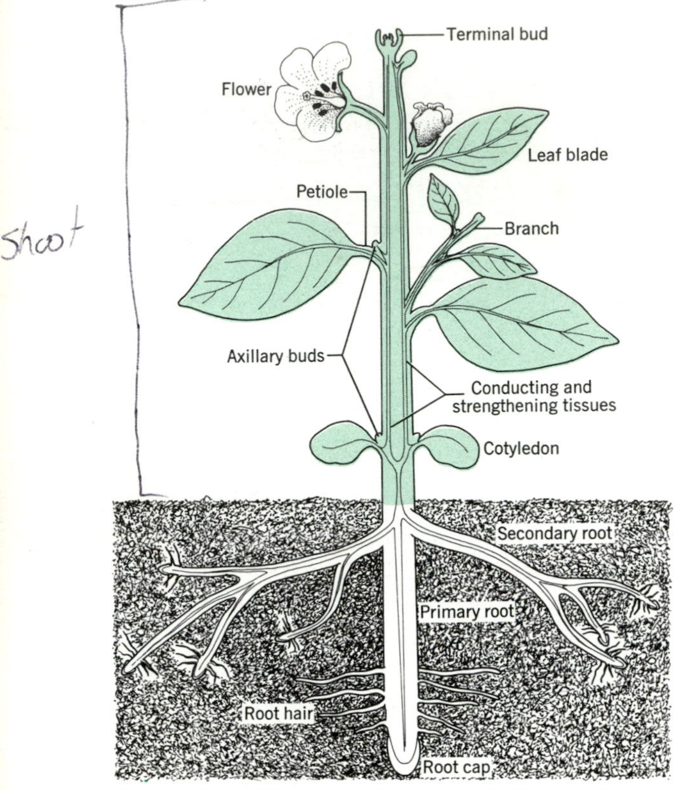

Figure 4-1 Diagram of a typical dicotyledonous flowering plant showing the major parts.

the form of the leaves, and the number of parts of the flower.

The plant body is made up of the organs—root, stem, leaves, and the parts of the flower—later on, fruits and seeds. Organs are composed of tissues and these in turn of cells.

1. THE PLANT CELL IN GENERAL

Figure 4-2 presents a diagrammatic plant cell. It is unspecialized—a *parenchyma* cell. Around the outside is the cell wall, whose composition and usefulness, especially in textiles, lumber, paper, and other materials, will be detailed in Chapter 21. In brief, there are two kinds of cell walls: the *primary wall,* formed while cells are still enlarging, and the *secondary wall,* laid down within the primary wall, generally only after all enlargement has ceased. Inside the wall, and usually pressed tightly against it, is the cell membrane or *plasmalemma.* Since membranes usually act to separate more-or-less watery constituents from one another, this one, like all other membranes, is of necessity oily in composition. It is a double layer, composed of fatty substances (lipids) and proteins, and is probably fluid.

Inside the membrane is the cytoplasm, an aqueous medium in which are suspended a number of *organelles.* Much of the cytoplasm is occupied by a system of folded membranes called the *endoplasmic reticulum* (*endo* means inner and *reticulum* means a net, which it is not really, though early electron microscopists thought of it as one). Along many of these folded membranes, tiny *ribosomes* are often attached in long lines as shown at the top of the drawing. These bodies are the seat of protein synthesis (see Chapter 9). The endoplasmic reticulum also makes connections from one cell to another, by way of tiny holes in the cell walls called *plasmodesmata,* through which the ER extends. Other groups of membranes, much smaller and closely appressed to one another, make up the Golgi bodies or *dictyosomes.* These curious structures continuously bud off from their sides little globules, probably containing carbohydrate, that migrate through the membrane and join the cell wall. Apparently this is the main way the cell wall is added to during growth.

If the cell is in the leaf or other green tissue, it may contain green *chloroplasts,* which contain long series of parallel membranes. The chlorophyll of these bodies catches the light and converts it to chemical energy (see the next chapter). Chloroplasts often store much of the end-product of this process in the form of starch.

Nearly all cells also contain organelles of another kind, the *mitochondria,* which also consist largely of folded membranes, and on these are located the enzymes of the cytochrome system that carry out oxidations

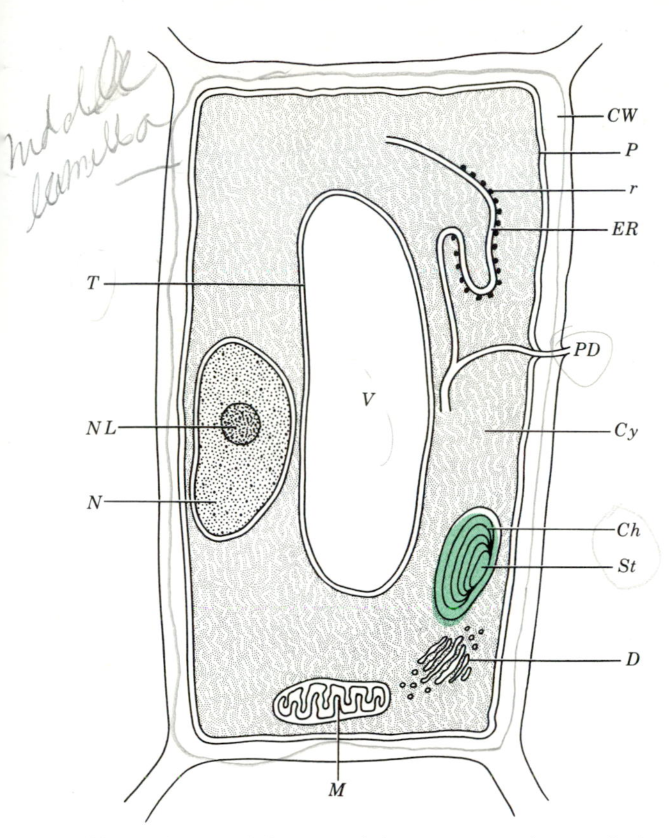

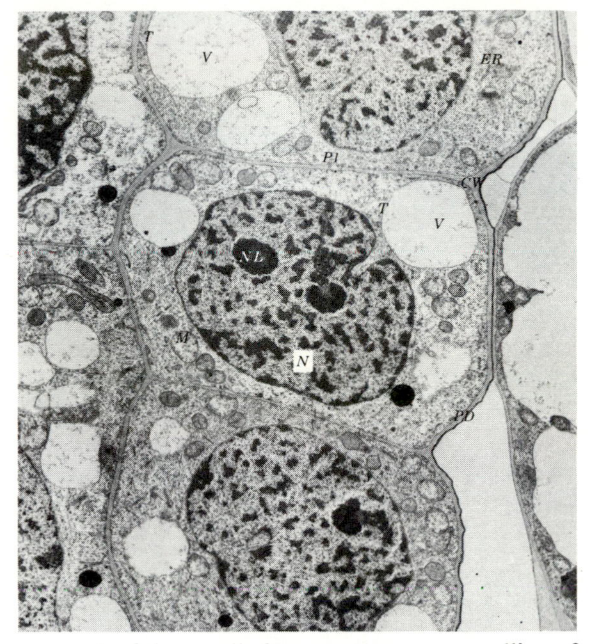

Figure 4-2 *A*. Diagram of the organelles in a typical plant cell: *Ch*, chloroplast (compare Fig. 2-5); *CW*, cell wall, *Cy*, cytoplasm; *D*, dictyosome (Golgi body); *ER*, endoplasmic reticulum; *M*, mitochondrion (compare Fig. 6-3); *N*, nucleus; *NL*, nucleolus; *P*, plasmalemma; *r*, ribosome; *St*, starch; *T*, tonoplast; *V*.vacuole; *PD* Plasmodesma, *B*. A recent electron micrograph of three cells from a very young seedling of rye. All details in Fig. 4-2*A* cannot be made out, but the cell wall, *CW*, the nucleus, *N*, with nucleolus *NL*, plasmalemma, *P*, tonoplast, *T*, and several vacuoles, *V*, are clear, Magnification about 4,000 ×.

(Chapter 6 & Fig. 6-3). Since the energy of oxidations and the energy of light are alternative and parallel ways for supplying the cell with energy, the mitochondria and the chloroplasts have certain similarities. Both have their internal structure made up of tightly packed membranes to which enzyme systems are attached, and both have an outer, double membrane that sets them off from the cytoplasm. In the mitochondria at least, the internal membranes are formed by infolding of the inner of the two surrounding membranes.

Another double membrane encloses the *nucleus,* with its *nucleolus*. The nucleus carries the hereditary material, DNA, which controls the synthesis of all the characteristic materials of the cell, and, when the nucleus divides, that hereditary material is equally distributed by an ingenious means into the two daughter nuclei (see below). In this way each newly formed nucleus exercises the same control over its surrounding cell material. The nucleus also contains RNA, which is excreted into the cytoplasm, as well as special proteins. The nuclear membrane opens to the cytoplasm with a number of circular pores, which doubtless allow communication between the nuclear material and the cytoplasm. Other kinds of RNA, operating in protein synthesis, occur in the ribosomes and the cytoplasm.

Lastly, in all of the larger plant cells there is a *vacuole,* consisting usually of a dilute solution of salts and a few organic compounds, together with a number of enzymes of the kind that break down large molecules (e.g., proteins and fats) to their smaller constituents. The vacuole is enclosed in a double membrane, called the *tonoplast*, and when the cell enlarges this vacuole accounts for most of the increase in volume. In rapidly dividing cells the vacuole may become subdivided into many small, usually narrow, vacuoles, but as the cell enlarges these fuse into the large central lake as shown in Figure 4-2*A*.

2. CELL DIVISION

When the cell is to divide, it is the nucleus that divides first, and only after it has separated into

Figure 4-3 Mitosis. Photos and diagrams of mitosis in the endosperm of the blood-lily *(Haemanthus)*. Photos *A* through *C* correspond to diagrams *A* through *D*. Photos *D* through *F* roughly correspond to diagrams *E* through *H*. In photo *A*, the nuclear membrane can be seen; in photo *B*, it has dissolved. In photo *F*, the cell plate has formed. Duration, about 6 hours. Photos, magnification about 4000×.

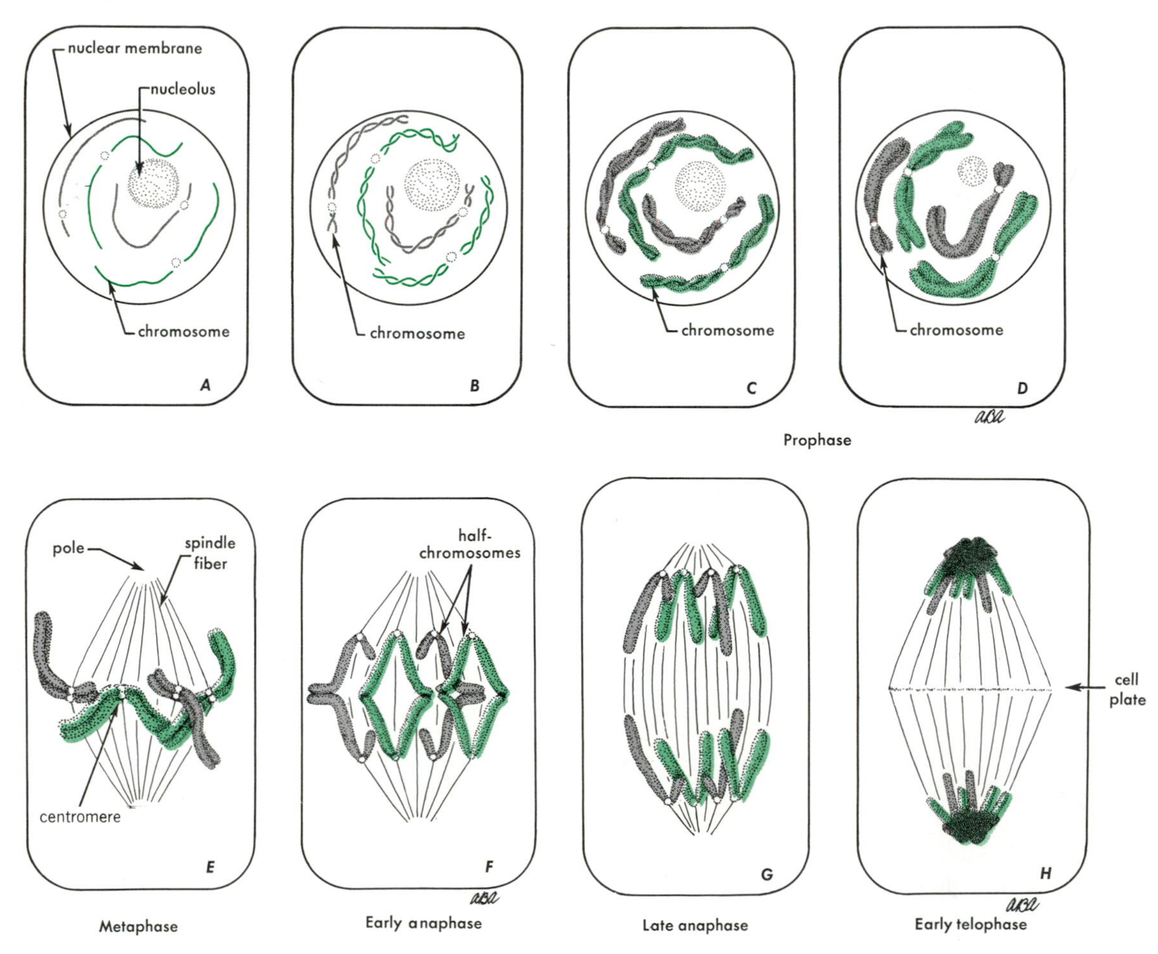

two halves does the rest of the cell also divide into halves. The division of the nucleus is a remarkable process. As we saw in Chapter 2, in ordinary cell division, more DNA is first synthesized, to produce twice the initial amount; then, by the mechanism described below, that material is divided equally between the two daughter cells. In this way each daughter cell has the same amount of DNA as the original cell. This system, *mitosis*, is described first. Another type of cell division, *meiosis*, occurring in reproductive cells, is discussed later.

a. MITOSIS

After the doubling of the amount of DNA by chemical reactions, the visible part of the process comprises four phases as follows:

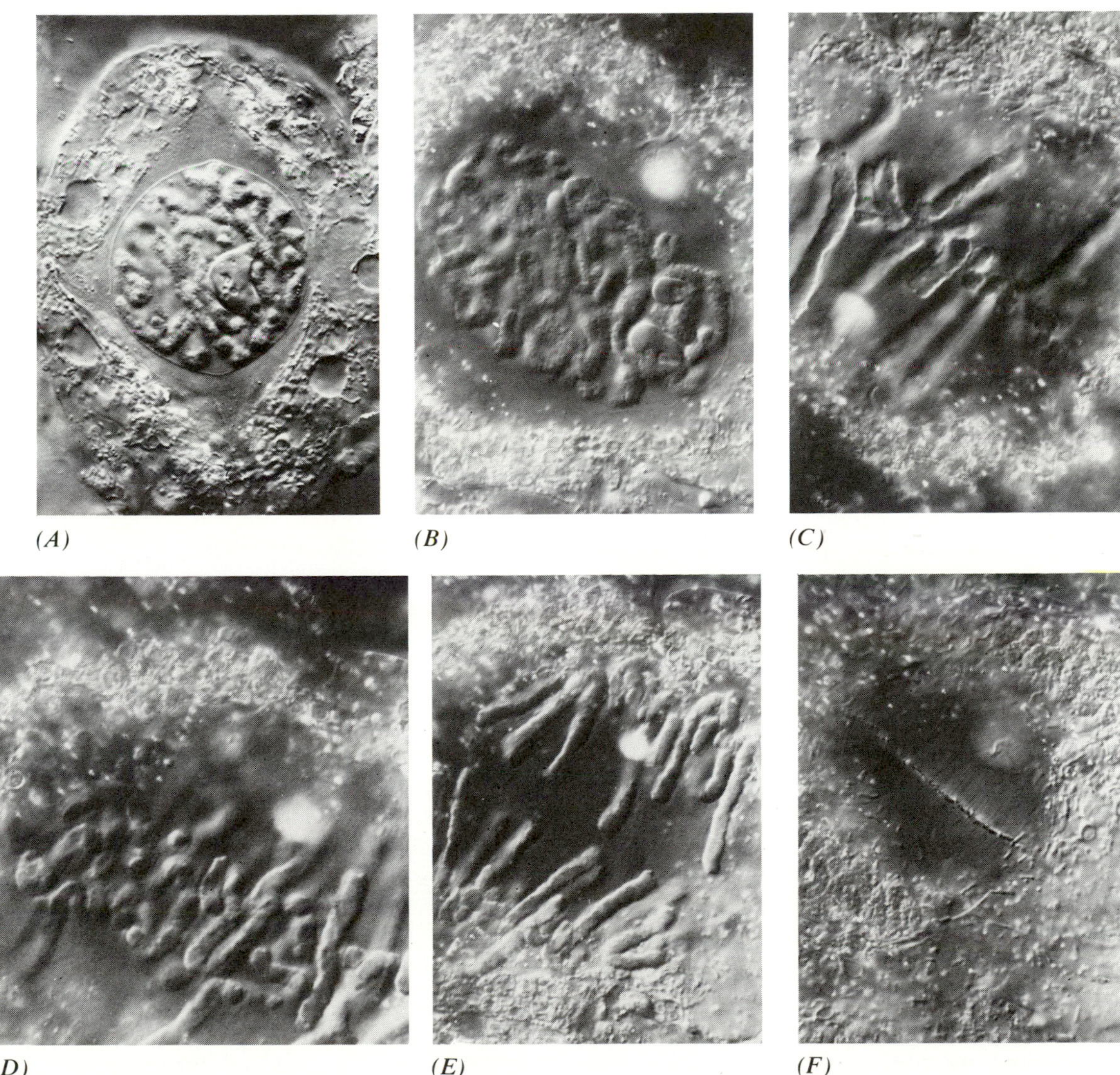

(A) *(B)* *(C)*

(D) *(E)* *(F)*

Prophase. The nuclear material condenses into a series of fine threads (Fig. 4-3*A* to 4-3*D*), which gradually shorten and thicken by coiling upon themselves, twisting and untwisting meanwhile. These are the *chromosomes*. (The name means *colored bodies* because they become highly colored when treated with certain dyes). The nuclear membrane is dissolved and the nucleolus disappears, or at least loses its ability to become colored by dyes.

Metaphase. Two "division centers" appear at opposite ends or sides of the cell (depending on the plane in which the cell is going to divide) and from each of these a bundle of very fine fibers grows out towards the center of the cell, forming a *spindle* (Fig. 4-3*E*). Each fiber is made up of still finer *microtubules* closely aggregated together. As the spindle fibers from each division center meet, the chromosomes, by now very much shorter and thicker, attach themselves to the spindle fibers by means of a small, more weakly staining zone, the *centromere*, which is generally located near the middle of each chromosome. The chromosomes thus become spread out across the central region of the cell in a "metaphase plate" (Fig. 4-3*E* and photo *D*). This is the stage at which the number of chromosomes can most easily be counted.

Anaphase. Each chromosome can now be seen (with the ordinary microscope) to be divided lengthwise into two halves (*chromatids*). Actually, this division took place in the DNA-synthesizing phase before prophase started, and in favorable cases the fine threads of chromosomes in the early prophase can be seen to be double. The spindle fibers now begin to shorten, pulling the halves of the chromosomes apart (Fig. 4-3*F* and *G*) and drawing them toward the two ends of the cell.

Telophase. The chromosome halves (chromatids) are now in two tight groups. The remains of the spindle fibers break up into large numbers of microtubules and these are joined, across the center of the cell, by some tiny vesicles. These vesicles begin to fuse with the central regions of the microtubules, beginning in the center of the cell and gradually spreading outwards toward the walls. This forms a *cell plate*, which effectively divides the cytoplasm, with the included organelles, into two parts (Fig. 4-3*H* and 4-4*F*) and is the beginning of the cross-wall. Membrane material is deposited in the region of the vesicles and joins up to form the two new plasmalemmas, while the microtubules become the sites of the plasmodesmata in the new cell wall. At the same time the two tight knots of chromosomes lose their condensed appearance and reconstitute the original appearance of the nucleus, now forming the two daughter nuclei, each with a nucleolus. The nuclear membrane and the nucleolus reappear.

In this way the hereditary material, or *genes*, situated in the chromosomes in the chemical form of DNA in combination with protein (chromatin), becomes equally divided between the two daughter nuclei, and each new nucleus finds itself in one half of the original cell, hence also with one half of all the organelles.

b. MEIOSIS

The early emergence of sexuality in the evolution of both plants and animals was discussed in Chapter 2. For the hereditary material of two gametes to be able to fuse and thus bring together characteristics from two different parents, the amount of hereditary material in each gamete must first be halved. This process is *meiosis*. When cell division takes place by meiosis, each daughter cell receives only *half* the original number of chromosomes and is thus *haploid*. After fusion the diploid content of DNA is reestablished. Meiosis begins in apparently the same way as mitosis, the hereditary material of chromatin first becoming visible as fine threads throughout the nucleus (Fig. 4-4*A*) and then becoming long thin chromosomes (Fig. 4-4*B* and 4-4*C*). The fact that the normal nucleus is diploid means, of course, that there are two of each kind of chromosome, one of each coming originally from each of the parents. (The diploid number of chromosomes is nearly always an even number; e.g., corn has 24, humans 46.) Now, instead of identical *halves* assembling as in mitosis, these similar (*homologous*) *whole* chromosomes line up together in tightly associated pairs (Fig. 4-4*D*). This process is called *synapsis,* and indeed the association is so close that occasionally the two chromosomes interchange segments, in the manner described in Chapter 11, and such *crossing-over* plays an important part in heredity. As the pairs form, it can be seen that each chromosome has in fact already divided lengthwise; that is, the DNA has been doubled, but the two halves remain joined (except at the tips in some plants (Fig. 4-4*F*) and thus the pair actually contains four similar bodies and is a *tetrad.* Figure 4-4 shows only seven pairs, but most plants contain many pairs of chromosomes.

At this stage the nuclear membrane and the nucleolus disappear as in mitosis; then the spindle begins to form. The paired chromosomes move to the equatorial area of the cell and attach to the spindle by their centromeres. The picture at this point differs a little from the metaphase plate of mitosis, in that the centromeres appear to repel each other and lie towards the poles.

Anaphase now takes place, fibers attaching to each whole chromosome to separate the pairs, which are drawn towards the poles (Fig. 4-4*E*). The *telophase* that follows is usually not completed, for although the nuclear membrane forms, the nucleoli do not reappear and often the cell wall does not form between the two halves of the cell (Fig. 4-4*G*). Instead, in each half of the cell a spindle forms again, this time in a direction at right angles to the original spindle, and a second nuclear division takes place (Fig. 4-4*G* and 4-4*H*). The chromosomes were already divided lengthwise, and now these *halves* separate as in mitosis. The number of chromosomes, and therefore the amount of DNA, in each of the resulting four groups is now only *half* the original (Fig. 4-4*I*). Finally the cell plates form, the nuclear membranes are reconstituted, and all four nucleoli reappear.

In the majority of cases these four haploid cells each undergo one further normal mitosis. In animals, as well as in many fungi and algae, the resulting eight cells would form into eight gametes. In flowering plants the haploid cells first form spores, which then give rise to small haploid *gametophytes* (pollen grain and embryo sac); these in turn produce gametes. With either arrangement the gametes eventually fuse in pairs to form a zygote and reestablish the diploid condition.

The basic significance of meiosis is that the chromosomes that pair up are homologous but not identical. They carry genetic material (i.e., genes) that has comparable functions. For instance, one gene may produce tallness and the corresponding one dwarfness, or one may produce green fruit and the other yellow. Thus these individual characters become separated in meiosis and then, when the gametes fuse, are recombined in *different groupings.* These most fundamental consequences will be treated in detail in Chapters 11 and 12.

3. CELL MODIFICATIONS FOR TRANSPORT

Many of the cells formed in a growing apex do not change much thereafter and are called *parenchyma* cells. Others first lengthen and then divide lengthwise to form *procambial initials,* which give rise to special transporting tissues. Since it was the development of transport systems that made land plants possible (Chapter 3), differentiation into transporting structures is one of the most prominent modifications. Two types of transport tissues occur.

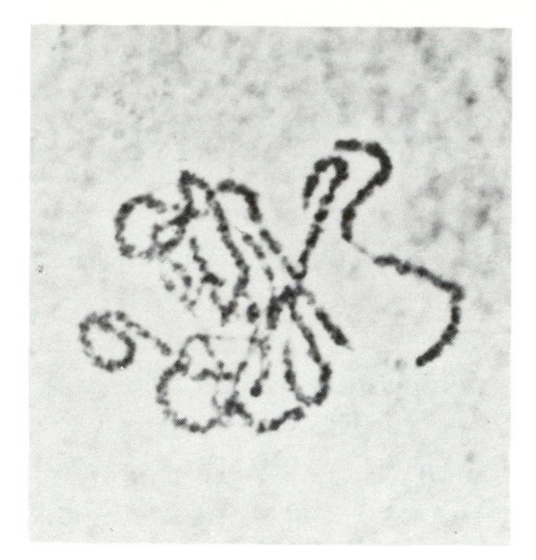

(*A*)

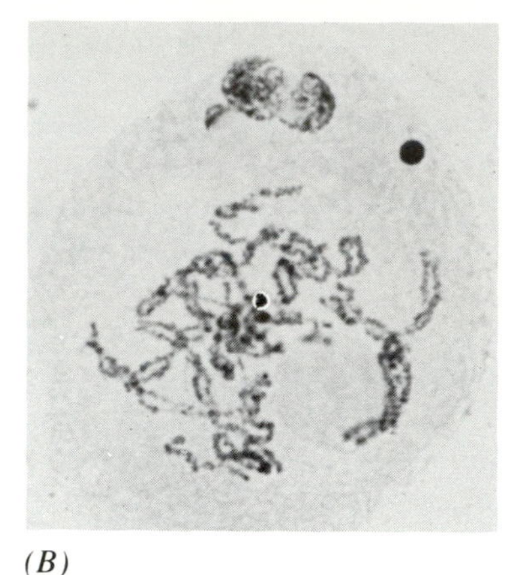

(*B*)

(*C*)

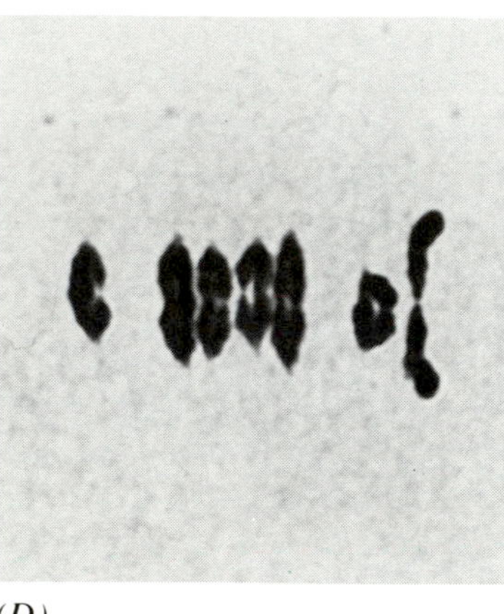

(*D*)

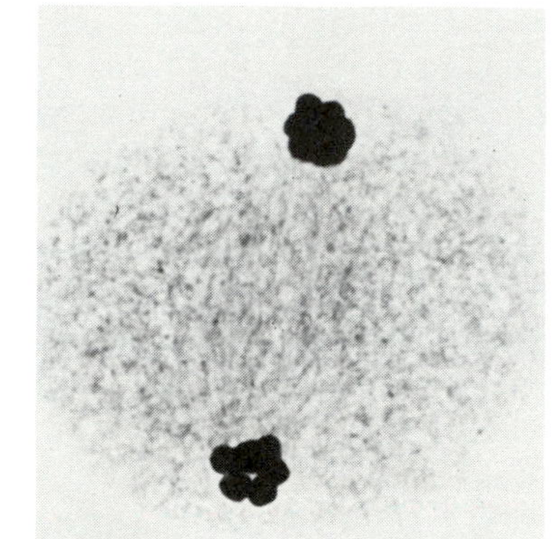

(*E*)

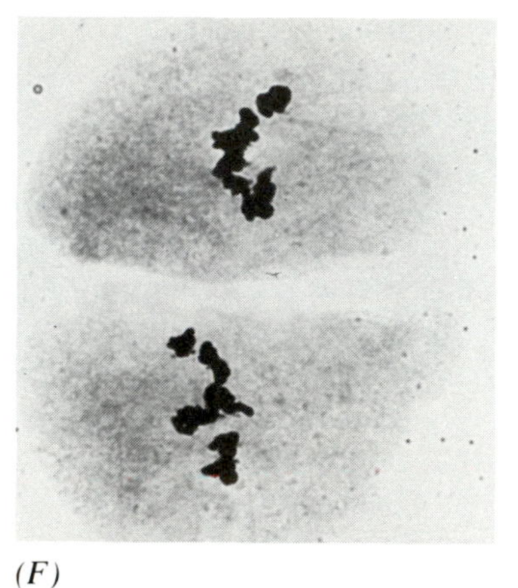

(*F*)

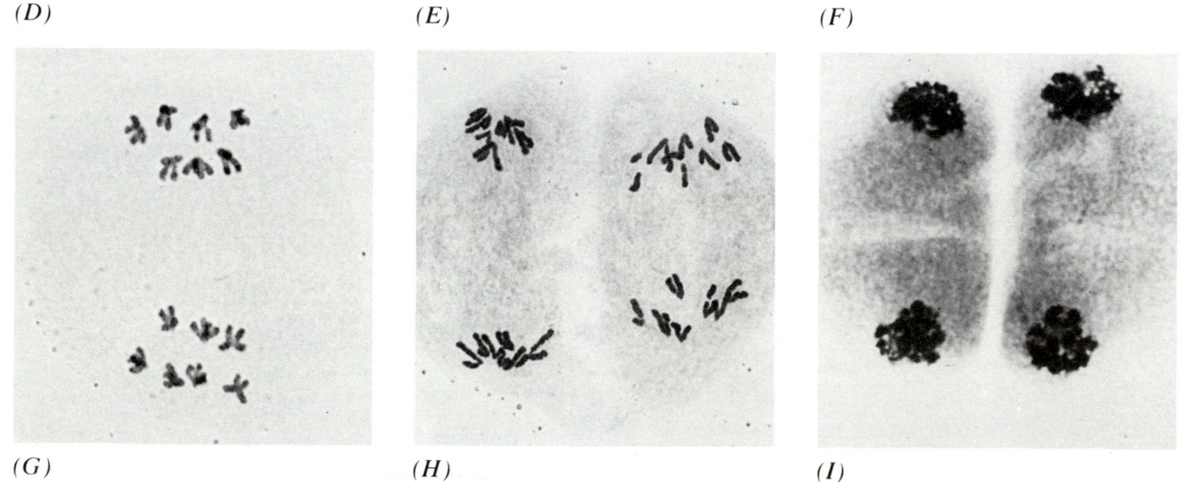

(*G*) (*H*) (*I*)

Figure 4-4 Meiosis. Stained preparations from anther of the grass *Lolium perenne* (rye-grass). (This plant is pictured in Fig. 16-15). *A* The chromosomes assembled and beginning to thicken. *B* Homologous chromosomes join in pairs. *C* Chromosomes thickened; 7 pairs. *D* The metaphase plate of the first division (with undivided chromosomes). *E* Anaphase of first division (with undivided chromosomes). No cell plate forms, and the two nuclei proceed to the second division. *F* Metaphase of the second division (similar to an ordinary mitotic division). Equatorial view. *G* Metaphase of the second division, seen in polar view. showing the two metaphase plates. *H* Anaphase of the second division. *I* Final telophase, with four haploid nuclei.

a. XYLEM CELLS

As we shall see in Chapter 7, the water supply is often under tension, the loss of water from the leaves causing a tension in the stem and thus tending to draw up water from the roots. Correspondingly, the cells of the water-conducting system in the *xylem* are mechanically strengthened to prevent their collapsing inward under the tension. Figure 4-5 gives a visual impression of the depth and strength of these reinforced tubes.

In cells that are to become xylem transport units, the volume of cytoplasm and the number of organelles both decrease. At the same time additional cell wall (*secondary wall*) is laid down, forming thick ridges on the inner side of the wall (Figs. 4-5 and 4-6). In the first xylem cells to be formed, constituting the *protoxylem*, these ridges are in the form of *spirals* or *rings* (Fig. 4-6*D*). The spirals get drawn out (like a spring when it is pulled) and the rings are drawn further from each other. Later, when elongation has ceased, cells with spiral thickening continue to be laid down with much tighter spirals, but others are formed in which the whole inner surface is thickened in a complex pattern of ridges and gaps—*scalariform* ("ladderlike") and *reticulate* ("netlike") thickening. These changes are probably hormonally controlled (Chapter 10). Lastly, the end walls of the individual cells or *vessel elements* become either coarsely perforated or else dissolve away completely (Fig. 4-6*B*) to form the long tubes or *vessels* characteristic of xylem in the flowering plants. Even these vessels are seldom continuous from top to bottom of the plant but have an occasional cross-wall; the longest are probably those in oak trees, which may extend for many meters. By the time the secondary thickening is complete, little or no cytoplasm remains, and the xylem units are dead. In the absence of cytoplasm, and therefore of the cytoplasmic membranes, the sap can flow freely through the vessels, as in a system of pipes.

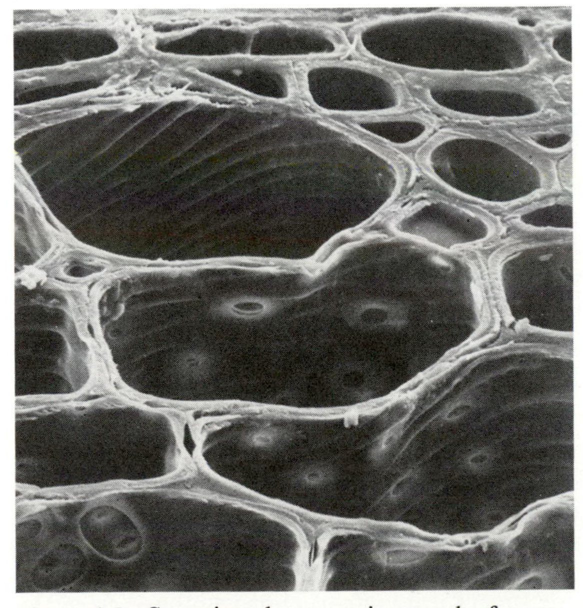

Figure 4-5 Scanning electron micrograph of a cross section of the wood of a New Zealand tree (*Pseudopanax*). (×540). Circular pits can be seen on the vessel walls.

Not all of the xylem consists of vessels, however; many of the cells grow to a few millimeters in length and have pointed ends that do not dissolve; these are *tracheids* (Fig. 4-6*A*). Their role in conduction depends on the presence in their walls of *bordered pits* (Fig. 4-6*C*,*E*), which are areas without thickening, composed simply of the permeable primary wall. The "borders" are the overlapping secondary wall. A little disc of thickened wall remains in the center of the pit too. The xylem sap can flow freely through these pits, since they give communication from one tracheid to another. In pine trees and other gymnosperms, the conducting function of the xylem is done only by tracheids, and there are no vessels. Since the gymnosperms developed earlier than the flowering plants, the tracheids doubtless represent a more primitive type of conduction system than the vessels; they give supporting strength as well as conduction.

The material that forms the secondary thickening is rich in *lignin*, which is chemically quite different from the carbohydrates of the

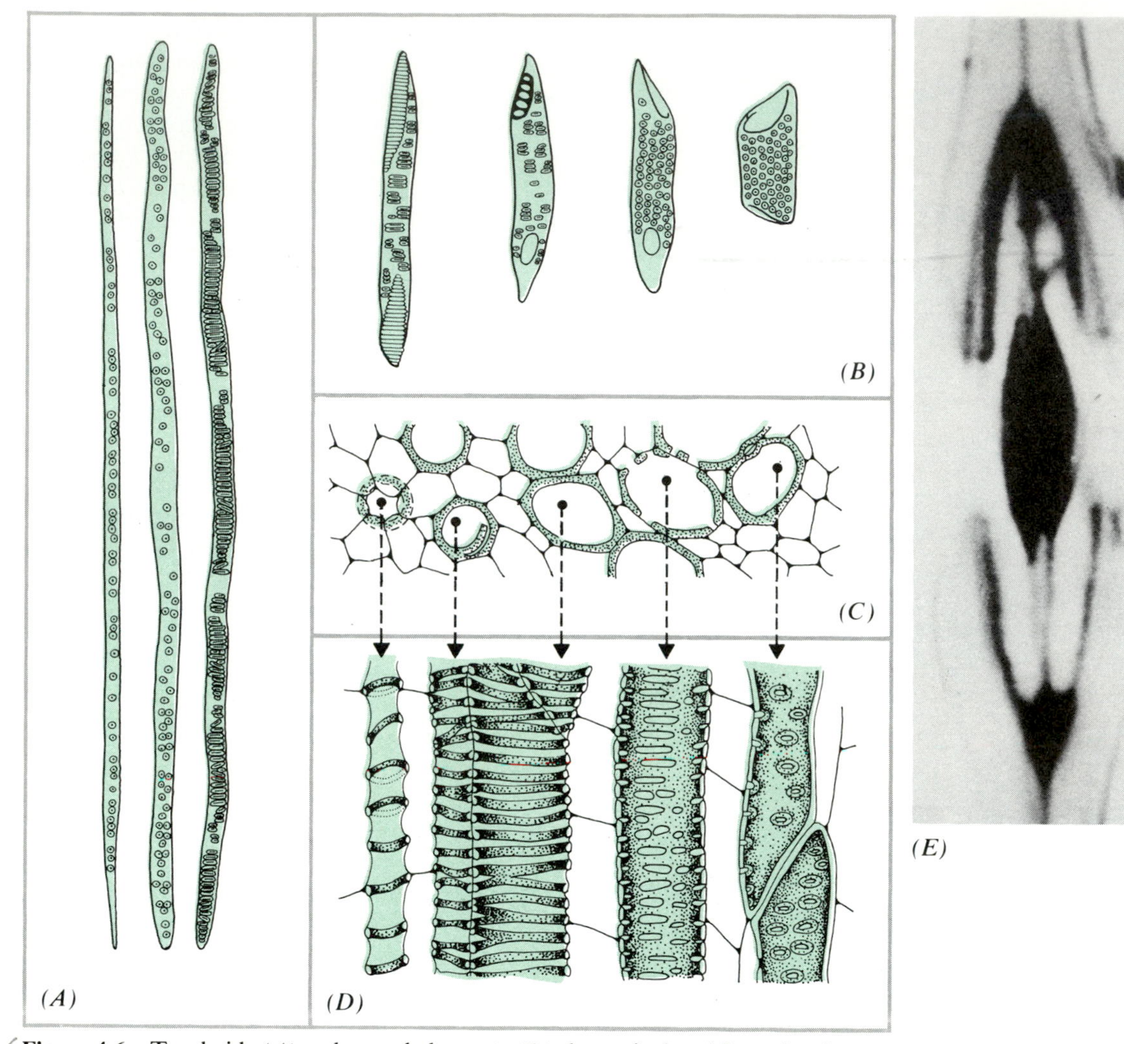

Figure 4.6 Tracheids *(A)* and vessel elements *(B)* shown isolated from the tissue. *(C)*. cross section. *(D)*. in longitudinal section. *(E)*. High magnification view of bordered pit in tracheid of pine × *ca* 3000. *(A-D* from Jensen and Salisbury, 1972)

primary wall (Chapters 6 and 21). As a result lignin takes different stains and is readily recognized under the microscope. Lignin was important in the development of land plants because it helps them have rigidity.

Besides vessels and tracheids, there are groups of living parenchyma cells at right angles to the long axis, sometimes with a few tracheids in the same orientation (Fig. 22-2). These *rays* give much of the "figure" to lumber (see Chapter 22). The xylem usually has thick-walled *fiber* cells as well (Chapter 21).

b. PHLOEM CELLS

A different type of transport system brings the sugary products of photosynthesis from the leaves to all the other parts of the plant. This also comprises long cells with very little residual cytoplasm (though probably alive), but, un-

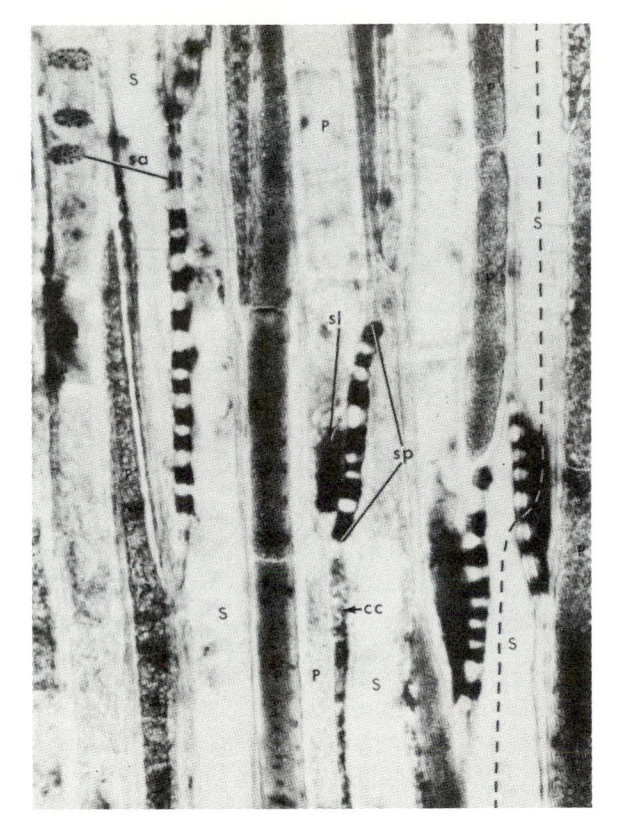

Figure 4.7 Phloem of a grape vine. *s,* sieve element; *sa,* sieve area; *sp,* sieve plate; *sl,* slime. Additional cells are *p,* parenchyma; *cc,* companion cell. The dashed line at right shows the path of soluble carbohydrate moving down the stem from one sieve element to another. (×300)

like those of the xylem elements, their walls are thin. It is the cross-walls that are characteristic here; they are usually very oblique (Fig. 4-7) and are perforated with fine pores, forming a sort of sieve. There are comparable *sieve areas* on the side-walls. These *sieve tube elements* may contain a few organelles, but basically they allow the phloem sap to move freely, under a slight *positive* pressure. But if the stem is wounded, a gum or slime is at once secreted that blocks the *sieve plate* and thus prevents the sap from pouring out. Some biologists have compared this phenomenon to the blood-coagulating system of animals; although the latter has a different biochemical basis, still the effect is the same—to save excessive loss of a valuable fluid. The flow of the sap through the phloem is not entirely passive because the continued respiration of the tissue is necessary to maintain it. But whether the respirational energy is somehow used in energizing the flow or simply in keeping the sieve tube elements healthy and functional is not known.

Alongside each sieve element is a small elongated *companion cell* (sometimes several), rich in cytoplasm, and thus sharply different from the almost cytoplasm-free sieve element. This small cell is probably the means whereby carbohydrates are loaded into the sieve tubes or unloaded from them (Chapter 6). The companion cells contain a number of enzymes active in carbohydrate metabolism.

Besides the sieve tubes, phloem also contains long fiber cells, which, in flax and some other plants, furnish valuable textile threads.

4. CELL MODIFICATIONS FOR SUPPORT

The shoot of the land plant, growing upwards into the air, is subject to the lateral forces of wind and rain, as well as often being weighted down by heavy flowers or fruits. To some extent the presence of a large number of parenchyma cells full of liquid pressing against one other gives it rigidity as can be seen if we try to bend a potato. But such organs are massive, while stems are relatively thin. The xylem transport cells with their heavily thickened secondary walls contribute further rigidity, but in addition there are cells whose function seems to be only that of providing support. In some stems, and often in petioles, a cylindrical layer of living cells forms, with unevenly thickened walls. These *collenchyma* cells have only primary wall and they can continue to elongate, laying down additional wall material as they do so. The edible petioles of celery have very obvious collenchyma, which form the "strings" on the outer surface. Still more rigid are the *sclerenchyma* cells, usually in the form of long *fibers,* pointed at each end and so heavily thickened with secondary wall that almost no hollow space or *lumen,* and usually no cytoplasm, remains. This wall material is impregnated with lignin, which makes the fibers very stiff, especially since the tapering end of each of the long cells overlaps with the next fiber cell for a considerable distance. Schlerenchyma fibers often occur in wood, in phloem (the gardener's "bast" is fibrous phloem), or even among the parenchyma cells of the cortex. Less organized into groups are individual *sclereids,* which may have almost any shape but are equally thick-walled, lignified, and rigid. They often occur in fruit, forming the "stone cells," gritty to the teeth, of pears and guavas; they are also found in a number of leaves.

The horsetails, *Equisetum,* form very little lignin and, perhaps to provide the missing support, secrete a thin layer of silica, in the form of opal, on the outside of their stems. Some grasses lay down opal in their stems too.

5. CELL MODIFICATIONS FOR SURFACES

One of the evolutionary modifications for land plants was the deposition of *cutin,* a waxy material that decreases the evaporation of water through the cell wall. The continuous layer of cutin (the *cuticle*) covers almost every epidermis; even flower petals have a thin cuticle. But since some evaporation of water is necessary for the life of the plant, the cuticle is perforated with a pattern of small holes, the *stómata* (the singular is *stoma*), whose apertures can be opened or closed according to the conditions (Chapter 7). This gives the plant considerable control over its loss of water. The cuticle is also penetrated by numerous very fine pores, through which a kind of wax is excreted from the epidermal cells. This cuticle forms complex patterns of fine threads over the leaf surface as shown in Figure 4-8.

Cells of the epidermis (those on the outermost surface) are usually flattened, rectangular in cross section, and without chloroplasts. They also have a tendency to produce long thin outgrowths or *hairs*. Sometimes a hair is just a single cell, part of the epidermal cell itself; sometimes it consists of a series of cells (Fig. 4-8). The form and arrangement of the hairs, singly or in groups of characteristic shape, may be used to establish relationships between plants in classification schemes. Some hairs act

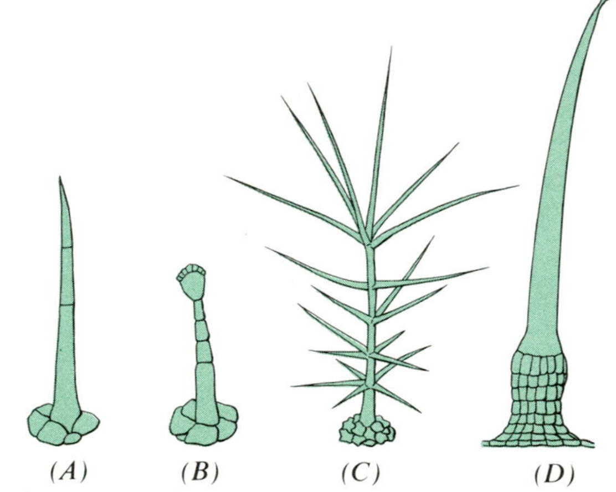

Figure 4-8 Above hairs of the lower surface of cucumber *(Cucumis sativus)* leaf. They are multicellular and on many plants such hairs are living ×120. Below, types of hairs *A*. Simple hairs of geranium *(Pelargonium)*. *B*. Glandular hairs of geranium. *C*. Branched hair of mullein *(Verbascum)*. *D*. Stinging hairs of nettle *(Urtica)*. (From Wilson, Loomis and Steeves, 1971)

Figure 4-9 Root of a radish *(Raphanus sativus)* seedling. The root cap and most of the elongating region are devoid of hairs, but large numbers of hairs, some of them very long, develop just behind this region, and they explore a large volume of soil (×10).

as glands, forming sticky, acrid, or even poisonous secretions.

The epidermal cells of roots are not in contact with the air, and they form only a very thin cuticle, but they do produce large numbers of hairs (Fig. 4-9). These root hairs grow out at a little distance behind the root tip, and new root hairs are continually formed as the roots elongate. The long, thin-walled unicellular hairs present a very large surface to the soil and are responsible for most of the uptake of water and dissolved salts from the soil (see Chapter 7). As new hairs are formed the old ones die off. In the fibrous roots of a four-month-old rye plant, it has been calculated that at least *100 million* root hairs are formed and die *every day*. They collapse and die also if the root is exposed to dry air, even for a few minutes, and this is one of the difficulties in transplanting seedlings.

6. BASIC STRUCTURE OF THE STEM

In higher plants, most of the tissues are laid down by small groups of rapidly dividing cells called *meristems* (Greek *merisu,* to divide up; hence *meristos,* divided or divisible). In these, cell divisions go on rapidly and the resulting daughter cells are differentiated to produce conducting systems, outer surfaces, and so on, thus forming the body of the plant. Some cells remain minimally differentiated also, to form the parenchyma of pith or cortex. Some just go on dividing, thereby maintaining the meristem tissue.

The typical shoot meristem (Fig. 4-10) is dome-shaped. The top of the dome is covered by a single layer of cells, the *tunica,* which do not often divide, and beneath this are the active meristematic cells. The smallest cells are packed with cytoplasm, but as they divide—predominantly in the direction of the axis—the lower daughter cells begin to form small vacuoles, which soon fuse together to form a central vacuole like that in Fig. 4-2. An important principle of the meristem's action is that the two daughter cells produced by mitosis are not exactly alike; the one closer to the apex retains its meristematic nature, whereas the one further from the apex has more tendency to lose that quality and to elongate instead. As a result, the basal end of the meristem becomes elongated and the meristem proper is continually pushed upwards.

A second major development takes place on the flanks of the meristem. Here there is renewed cell division activity and, as the resulting group of cells enlarges, these cells tend to be pushed out laterally. These groups of cells then begin to elongate upwards, to form flat structures that, as they continue to grow, overarch the apex. These are the young leaves. They may be formed opposite one another as in Figure 4-10, or in more complex arrangements, spiraling around the stem. As they enlarge, xylem and phloem become differentiated, first in the apical meristem tissue below the leaf and then spreading upwards into the young leaf itself. The differentiated tissue makes contact with the main conducting tissues coming up to the apex (Fig. 4-10) so that each leaf becomes connected to the whole supply system of the plant—a remarkable process of integration of plant structure. At the base of each leaf a fragment of the meristem remains, and this later develops into a lateral bud. We shall see something of how all this is controlled in Chapter 10.

The overall structure of the flowering plant

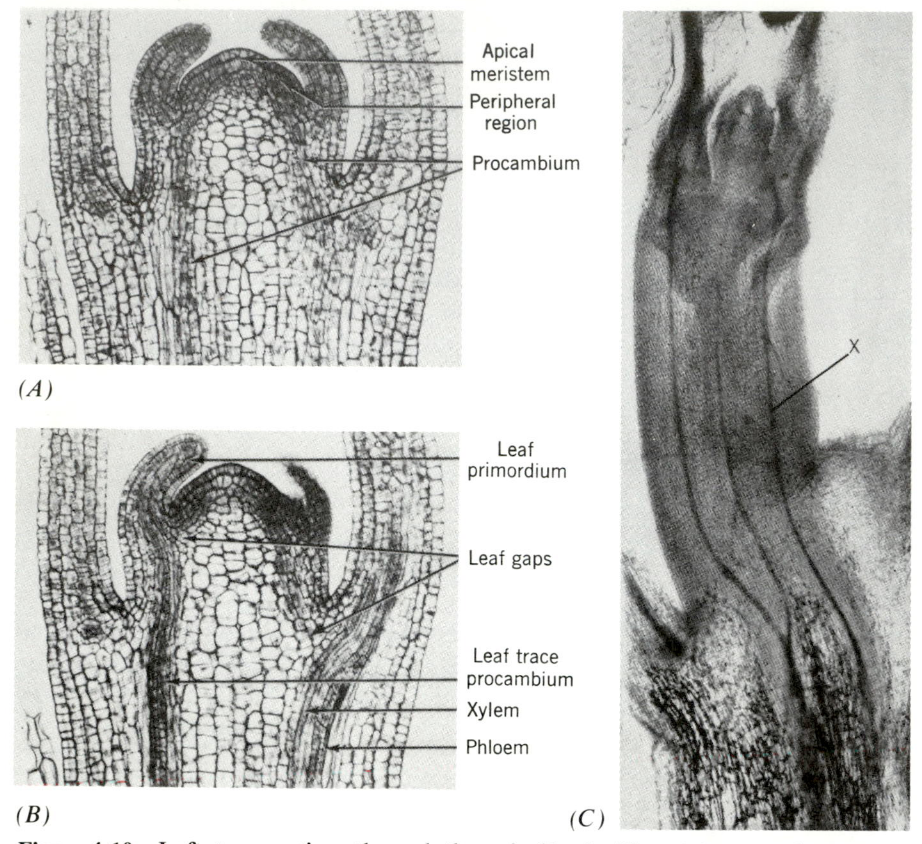

Figure 4-10 Left, two sections through the apical bud of flax *(Linum)*. *A* is a section through the center showing how the conducting tissue from the apex develops down the stem. Section *B* is a little off center and shows how the conducting tissue of the growing leaf merges into the xylem of the stem. *C*., fresh section through a developing lateral bud of a pea seedling that was decapitated a week earlier. The three xylem strands (one marked *X*) from the bud apex and from its two growing leaves are clearly seen. (×30).

body falls into two general types, those with one cotyledon in the seedling stage and those with two: monocotyledons and dicotyledons (Table 1-1). In these two types (usually abbreviated to *monocots* and *dicots*) the difference in cotyledons is only one of the many characteristics that have led to our separation of these two groups as the major subdivisions of the flowering plants in the classification system (compare Table 13-1). The main differences between the monocots and dicots are summarized in Table 4-1.

If we cut a cross section one or 2 mm below the apex we see the xylem and phloem units becoming grouped together into *vascular bundles* (Fig. 4-11). Each bundle is a self-contained conducting system, the xylem for water and solutes being drawn up into the leaves and the phloem for soluble sugars made in the leaves and being carried to the rest of the plant. This part of the plant is sometimes called the *primary body,* to distinguish it from the later-formed tissues now to be described.

As we go further down the stem, we see

TABLE 4-1

Some of the Main Differences between the Monocots and Dicots

Characteristics	*Dicots*	*Monocots*
Cotyledons	Two	One
Flower parts	Usually in 4s or 5s	Usually in 3s
Venation of leaves	Usually netlike	Usually parallel
Arrangement of primary vascular bundles in stem	In a ring (Fig. 4-11,*D*)	Scattered (Fig. 4-11*A* & *B*)
Vascular cambium	Usually present	Absent

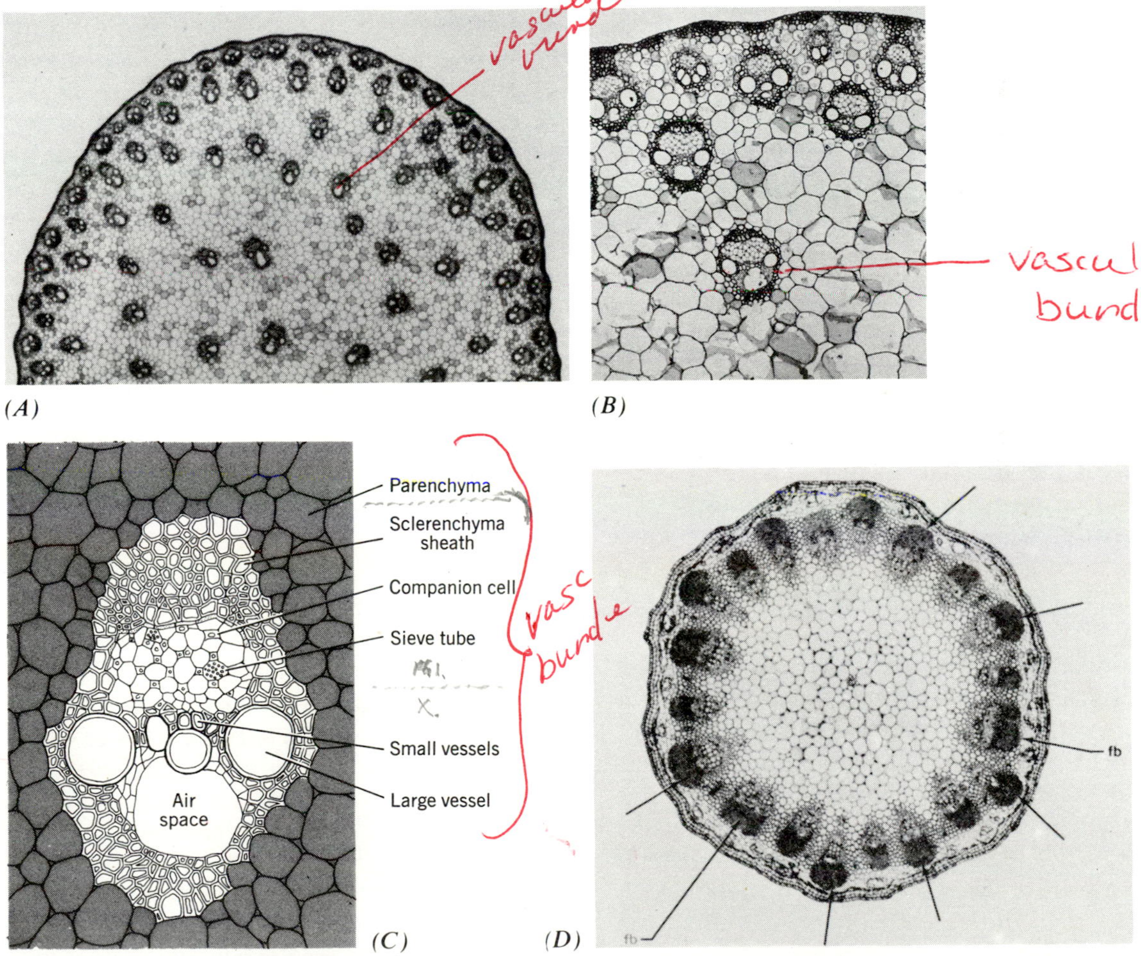

Figure 4-11 The separated vascular bundles in the stem of a monocotyledon (corn, *Zea mays.*). Three magnifications are approximately *A*, ×30; *B*, ×220; and *C*, ×550, respectively. (from Wilson, Loomis, Steeves, 1971). *D* The concentric bundles in the stem of a dicotyledon *(Helianthus)* for comparison. (From Esau, 1977). The bundles marked fb or with arrows are partly fused together.

that, in each vascular bundle between the xylem and the phloem, thin-walled cells appear that divide in the direction perpendicular to the axis, that is, along the radii, producing new cell walls that are *tangential* or parallel to the outside circumference of the stem (Figs. 22-1, 22-2, and 22-3). The region of such cells quickly spreads out *between* the vascular bundles to form a complete ring. This ring, the *cambium,* as was noted in section 8 of Chapter 2, is the origin of the *secondary body* of the stem, for as its cells continue to divide tangentially they undergo a special pattern of differentiation; on the outside of the ring the cells become phloem and on the inside they become xylem (see Chapter 22 for details of secondary development). In this way the amounts of xylem and phloem are increased as we get further from the apex. Usually some parenchyma cells remain undifferentiated in the center, forming *pith* (Fig. 4-11*D*). In woody plants, the xylem becomes an almost solid woody mass, with closely appressed cell walls impregnated with lignin. Just outside it the phloem forms a continuous ring, too, which constitutes the inner part of the bark.

What we have seen up to now holds for essentially all plants belonging to the *dicotyledons*. The stem of *monocotyledons* has a different pattern. The main difference is that most of the vascular bundles serving the individual leaves do not become joined into a single structure, but each remains intact and separated from the others by undifferentiated parenchyma as in Figure 4-11, which shows the stem of corn, a typical grass. There are some small connecting groups of xylem at certain points, but basically the bundles remain separate all the way to the ground. They branch occasionally, giving rise to the vascular strands in the leaves. In the stem of a woody monocotyledon like one of the palms, there may be hundreds of these vascular bundles, of two sizes. On the outside of the stem, instead of bark, the epidermal cells form a thick cuticle, often interpenetrated with wax, which prevents loss of water from inside and also sheds the rain outside.

7. BASIC STRUCTURE OF THE LEAF

Because the leaf has the prime function of intercepting light for photosynthesis, it seems natural that most leaves are thin, flat structures covering a considerable area. Structure and function are closely integrated.

The leaf needs to be supplied with water from the roots, and it exports its photosynthetic products to the rest of the plant; its vascular connections are therefore important, and leaves normally have prominent *vascular bundles* or *veins,* usually most marked on the lower (or "dorsal") side. The gamut of leaf shapes is shown in Figure 4-12. Simplest is the linear leaf of the typical lily, with parallel veins; this form is adopted by a majority of the monocotyledons. In the grasses, however, the leaf changes shape in the middle and its lower half is a cylindrical hollow sheath that surrounds the stem (Fig. 16-1). The simplest form in the dicotyledons is that of the first leaf shown, with two parts, the blade and the petiole. Through the petiole run the vascular bundles that will ramify to form the veins of the leaf. Some dicotyledonous leaves have no petiole, that is, are *sessile*. More complex forms are those of the *compound* leaves, which comprise a number of *leaflets,* growing either directly out of the petiole or out of a prolongation of the petiole called the *rachis*. The leaflets growing directly out are in the form of the palm of a hand; hence the leaf is called *palmate;* in the leaflets growing from the rachis the form is that of a feather; hence the leaf is termed *pinnate* (Latin for feather is *pinna*). In Figure 4-12 *F* and *G, H* and *I,* these two forms are contrasted with modifications in which the blade remains essentially simple but the veins are in palmate or pinnate form. Many intermediate types are known in which the blade is partially palmate or pinnate, being subdivided into lobes. Within a single family the leaf-blade can take several different forms. For example, within the legume family, peas and beans have pinnate leaves with little flat outgrowths at the base of the petiole

(A)

(B)

(C)

(D)

(E)

(F)

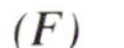

(G)

(H)

(I) *(J)*

(K)

Figure 4-12 Examples of leaf shapes. Simple dicot leaves: – Edge entire: A, *Magnolia tripetala* (umbrella tree), B, *Catalpa bignonoides* (Eastern catalpa), C, *Abies alba* (white fir). Edge serrate: D, *Populus deltoides* (poplar), E, *Betula populifolia* (white birch). Form modified by venation: F, *Quercus alba* (white oak) "pinnatifid", G, *Ricinus communis* (castor bean) "palmatifid". Compound dicot leaves: – H, *Juglans regia* (Persian walnut) "pinnate", I, *Aesculus californica* (California buckeye) "palmate". Monocot leaves: – J, *Bambusa*, sp., (bamboo), K, *Commelina erecta* (slender dayflower).

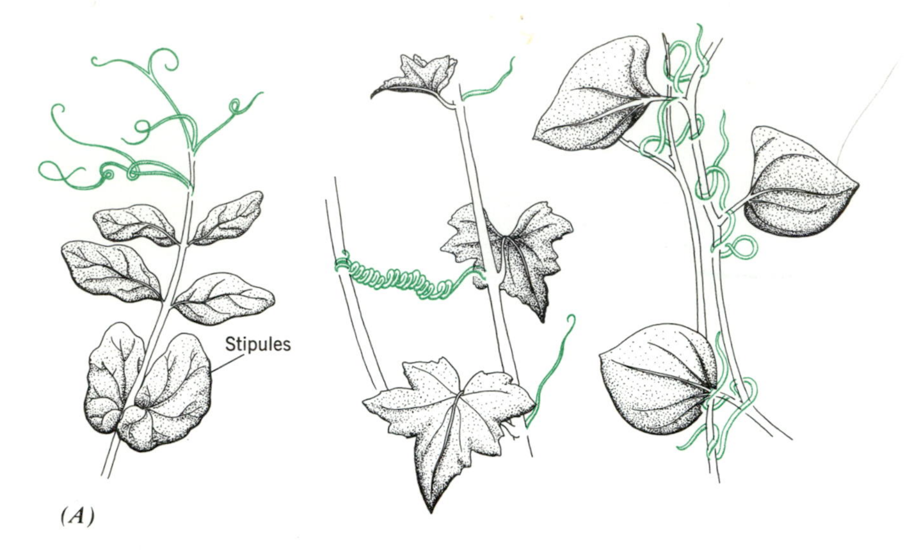

Figure 4-13 Tendrils. *A*. Left, compound leaf of pea *(Pisum)* showing tendrils instead of leaflets; center, young shoot of Bryony *(Bryonia dioica)* showing two unstimulated tendrils, opposite leaves, and one tendril stimulated to coil by attachment; right, shoot of greenbriar *(Smilax)*, in which the tendrils grow out of the petiole. *B*. Photograph of tendrils of squash *(Cucurbita)* in the stimulated (coiled) state. In this plant the tendrils replace whole leaves. *(A,* left and right from Wilson, Loomis and Steeves, 1971; center from Jost, 1913)

called *stipules,* whereas lupine leaves are palmate and without stipules. In climbing plants some of the leaflets may be replaced by *tendrils,* which grow to twine around any object they touch and thus support the stem (Fig. 4-13).

The leaflets of a compound leaf are distinguished from true leaves in that a leaf bears a small bud in its *axil* (Fig. 4-1); a leaflet does not. Other green leaflike forms include (a) the *stipules* (mentioned above), (b) *scales,* which are minimally expanded leaves enclosing a bud, and (c) *bracts,* which grow out just beneath a flower or an inflorescence.

A three-dimensional view of a typical leaf is shown in Fig. 4-14. The epidermis is covered with a cuticle, penetrated at intervals by *stomata*. There are usually more stomata on the lower side than on the upper. The chloroplast-containing cells beneath the epidermis are of two types, *palisade* cells, long and perpendicular to the surface, and *spongy parenchyma,* with marked intercellular spaces occupying the lower part. As noted above, some epidermal cells grow out into hairs, which probably serve to slow down the rate of air movement and thus decrease the water loss from the leaf; they may

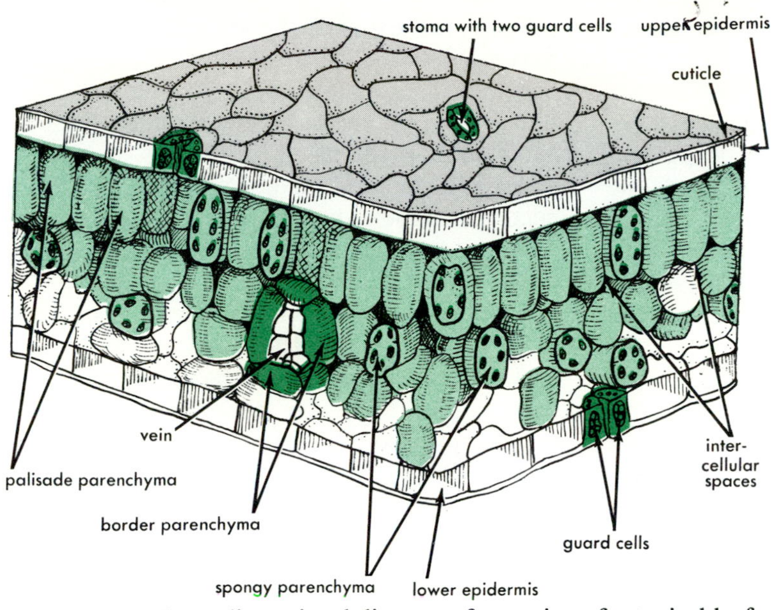

Figure 4-14 Three-dimensional diagram of a section of a typical leaf.

also—by increasing the surface area—act like the fins of a radiator to cool the leaf when in the hot sun. In any event, dense hairs are common on the leaves of plants of dry habitats.

Some leaves have areas in which the cells contain no chlorophyll and thus appear yellow. Such *variegated* leaves are grown for their appearance, but the value of variegation to the plant is unknown. *Mosaic* diseases are named for the variegation they cause. Both mineral deficiency and senescence result either in generalized yellowing or in yellowing in specific areas (see Chapters 5 and 29).

8. BASIC STRUCTURE OF THE ROOT

Like the shoot apex, the root apex is also a meristem, but the pattern is entirely different. It differs in these two important ways: (a) lateral organs are not formed as part of the meristem, but arise much further back, (b) the domed meristem not only forms layers of cells behind it, but also forms a small mass of cells, the *root cap,* in front of it (Fig. 10-1). These cells are short-lived, and as new ones are formed from the meristem the old ones, in a dying condition, are sloughed off into the soil. In this way the root produces a little trail of organic matter as it makes its way through the soil and the soil close around the root develops a special population of bacteria, living on the organic matter of the sloughed-off cells; this region of the soil around each root is called the *rhizosphere.* The root cap seems to have a special function in regard to the orientation of the root to gravity (see Chapter 10).

In one respect the basic pattern of the root meristem is like that of the shoot meristem; the cells divide unequally, one daughter cell remaining meristematic and the other often elongating. The elongation thus pushes the meristem forward. There is in the center of the meristem a small group of perhaps a dozen cells that do not divide at all—the "quiescent center." Some people think that these represent the management, sending out hormonal messages that stimulate the other cells to divide, but we do not know this to be so.

The divisions of the meristem produce the usual layers of cells – epidermis, cortex, and the primary body consisting of xylem and phloem as in the shoot. But some millimeters behind the meristem three new differentiations begin. First, (compare section 5 above), certain cells of the epidermis grow out into long, thin, single-celled *root hairs* (Fig. 4-9). Second, while the xylem (in dicots) does coalesce into a solid concentric body, the phloem elements converge into groups that lie between lobes or "arms" of xylem (Fig. 4-15*A*). Outside these develops a ring of peculiar cells with a corky belt ("casparian strip") around the middle of each. These cells, the *endodermis*, separate the vascular bundle in the center from the cortex outside. Third, a few of the ring of cells directly next to this endodermis, called the *pericycle*, begin to divide tangentially. This happens only at certain limited spots, adjacent to where the xylem lobes come to the outer edge, and results in little local masses of cells. Each of these becomes an individual meristem and, as in other meristems, some of the daughter cells continue dividing while others elongate, and the elongation pushes the new meristem out *laterally* (Fig. 4-16). This new meristem becomes a lateral root, and as it emerges through the cortex it is seen as a full-fledged root meristem, producing a root cap in front and layers of typical root tissues behind. It usually continues to grow at something less than a right angle to the main root, and soon forms root hairs and even lateral roots of its own.

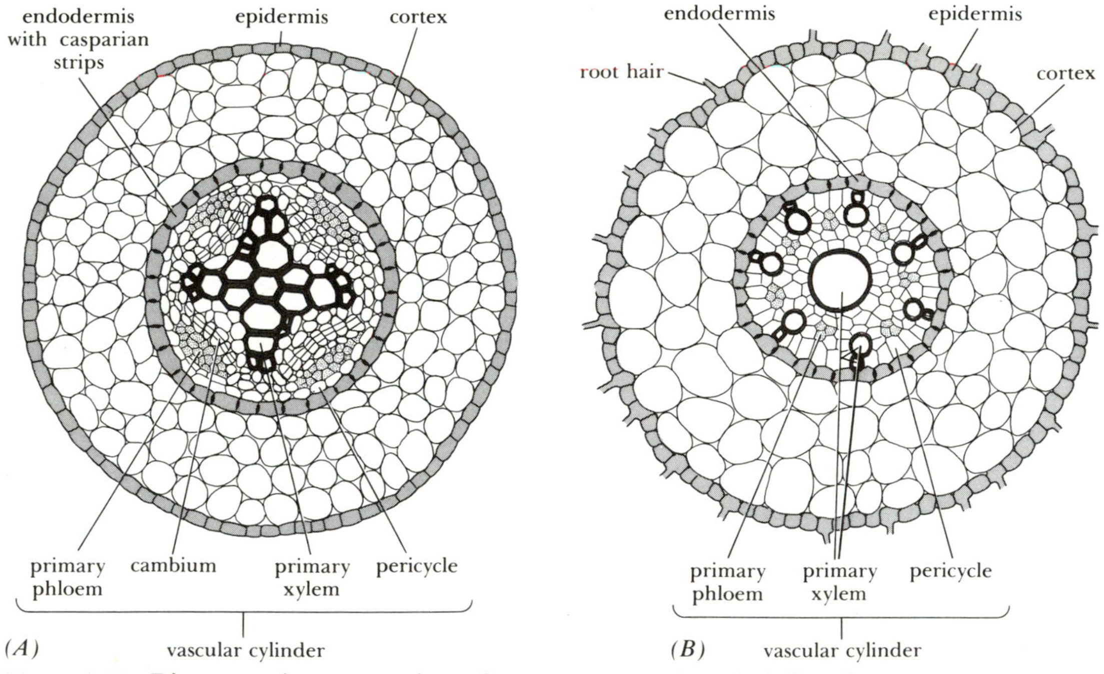

Figure 4-15 Diagrammatic cross sections of young roots. *A*. A typical dicotyledon. *B*. A monocotyledon. Note the alternating arrangement of the primary xylem and phloem (from Wilson, Loomis, Steeves, 1971).

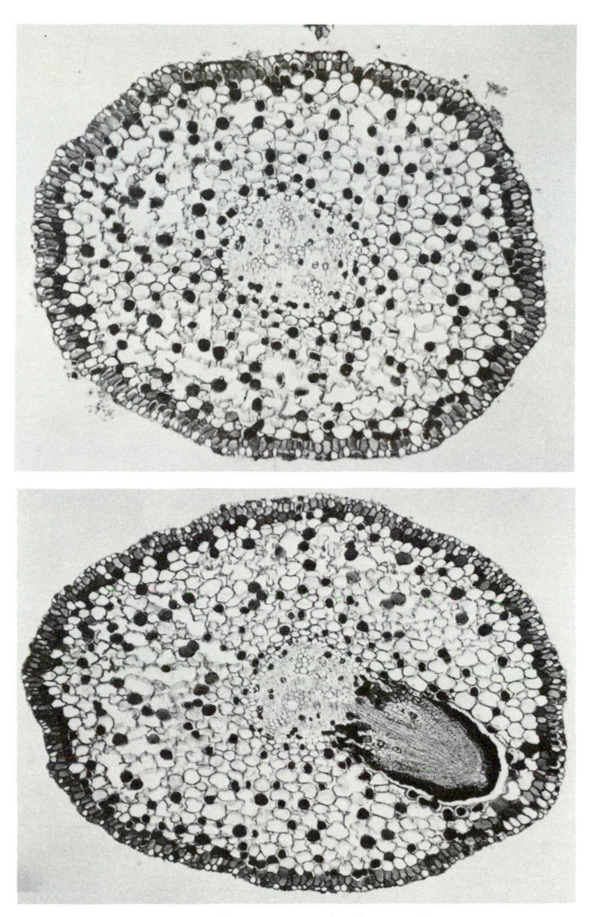

Figure 4-16 Development of a lateral root on a root of willow *(Salix)*. The initiating divisions were in the pericycle (compare Fig. 4-15*A*). Above, just before initiation; below, root initial moving through the cortex.

9. STRUCTURE OF THE FLOWER

The single characteristic that sets angiosperms (flowering plants) apart from all other groups is the flower. The essential steps of sexual reproduction, (meiosis and fertilization) take place in the flower; the reproductive cells, the haploid gametes themselves (sperm and egg), are essentially the same as those found in other plants. The flower, although following a general plan, has been modified into a tremendous diversity of forms.

a. THE GENERAL PLAN

The flower is essentially a greatly modified shoot, and some of its parts may accordingly represent modified leaves. The flower is borne on a specialized stalk, the *pedicel,* whose apex, where the floral parts are attached, is commonly enlarged into a receptacle (Fig. 4-17). The region of the receptacle in the young flower is similar to the shoot apex of a vegetative stem, but instead of producing leaves it typically gives rise to four types of *floral organs,* always arranged in the same order (Fig. 4-17). The first of these (beginning at the base of the flower) are the *sepals,* collectively forming the *calyx.* Next come the *petals,* collectively the *corolla.* The calyx and the corolla constitute the *perianth.* The sepals are usually green, somewhat leaflike, and thin; they enclose the other flower parts in the bud. The petals are commonly colored (other than green); however, in some plants, like the anemone and clematis, the petals are small or absent and it is the sepals that are brightly colored. In dogwood and poinsettia, both sepals and petals are inconspicuous, and the brightly colored structures, incorrectly called petals, are really bracts that surround a group of small flowers.

The next two of the four types of organs are specifically for reproduction. The *stamens* form a ring or *whorl,* the *androecium,* lying inside the corolla. Each stamen has a slender stalk or *filament,* at the top of which is an *anther,* the pollen-bearing organ. The *pistil,* located in the center of the flower, is commonly flask-shaped with a swollen part, the *ovary,* connected by a fine stalklike *style* to a terminal expanded portion, the *stigma,* which receives the *pollen,* and can be rough, smooth, sticky, branched, or feathery. (The name "pistil" derives from the fancied resemblance of some pistils to the pestle used in pharmacy to pound

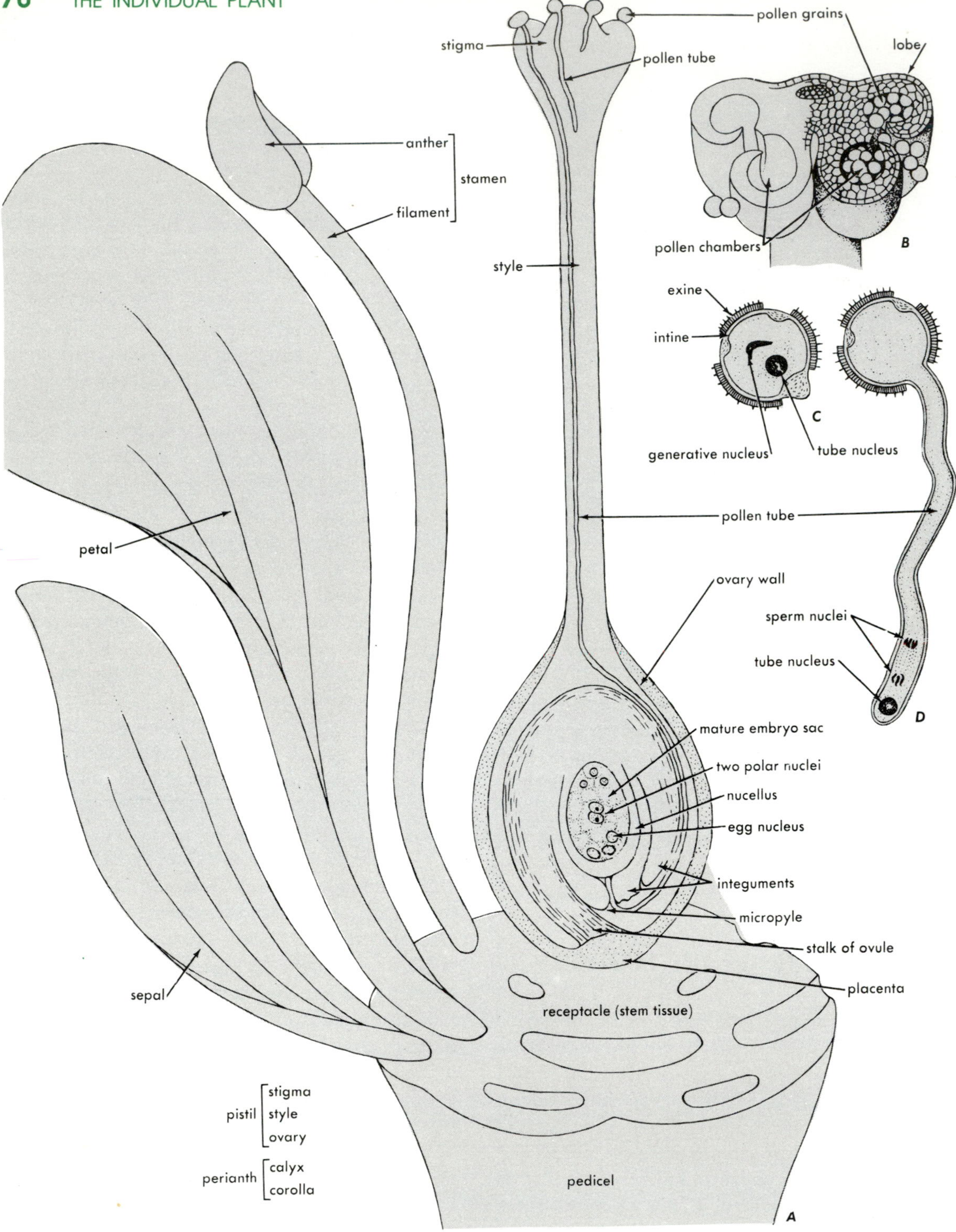

Figure 4-17 Diagram of a typical complete flower displaying all of the basic floral structures.

up plant parts in a mortar.) The pistil is composed of either a single or several united *carpels;* the pistil, or in some flowers the whorl of pistils, is known as the *gynoecium.*

Each carpel encloses within the fold one or more *ovules,* which later develop into seeds. The pea, for instance, has an ovary comprising one carpel with a row of ovules (to become peas) attached to its fused margins. The region of attachment is called the *placenta,* analogous to the organ connecting the mammalian embryo to maternal tissue. A flower may have many simple pistils (carpels), as in the magnolia, buttercup, or strawberry. But in the lily the pistil is compound containing three carpels with ovules in each.

Although it is relatively easy to envision the perianth parts as modified leaves, the question arises as to why stamens and carpels are considered modified leaves. One reason is that the early development stages of these parts closely resemble those of leaves. The second concerns the shape of spore-bearing leaves in such flowers as *Magnolia* (Fig. 13-3) and *Degeneria* where the stamen is very leaflike in appearance. Also the folded-open carpels of primitive angiosperms are leaflike in appearance. Perhaps even more striking is the case of *Sterculia platanifolia* (Fig. 4-18) where five simple pistils are united only by their stigmas. When these stigmas mature, they open to show five leaflike carpels that bear seeds along their margins.

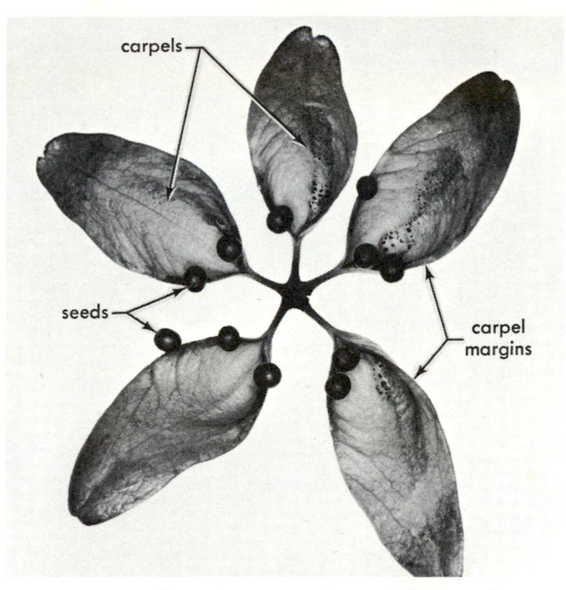

Figure 4-18 Carpels suggestive of foliage leaves. In *Sterculia platanifolia,* dehiscence of the mature ovary shows seed attached to margins of the five leaflike carpels (×5/8).

b. MODIFICATIONS OF THE FLOWER

Although all flowers follow the same general plan, they exhibit many modifications in design, some of which perhaps increase their chances of being pollinated. They often reveal evolutionary relationships among plants, enabling botanists to compare structures and thus to assign plants to related groups. Major ways in which the flower plan can vary include (1) number and arrangement of floral parts, (2) fusion of parts, (3) omission of some of the parts, (4) symmetry of the flower.

One of the distinguishing characters between the monocotyledons and dicotyledons is the number of floral parts in a whorl. In the dicotyledon the number is usually four or five or multiples of these numbers, whereas in the monocotyledons it is commonly three or multiples of three. A flower with all four sets of floral organs is said to be *complete* (Fig. 4-17); an *incomplete* flower has one or more of the four whorls lacking (Fig. 4-19*A* and 4-19*B*). For example, the flowers of grasses and poplars have no perianth. In the chenopod family, which includes the beet and spinach, the flowers have greenish sepals but no petals, but in clematis and a number of other plants that also have no petals, the sepals may be brightly colored.

Unisexual flowers with either carpels or stamens, but not both, are quite common; examples are squash, corn, oak, willow, asparagus, and date palm. These flowers are called *imperfect* (Fig. 4-19*C* and 4-19*D*); bisexual flowers are *perfect.* When (imperfect) staminate and pistillate flowers occur on the same individual plant, as they do in corn or squash and many

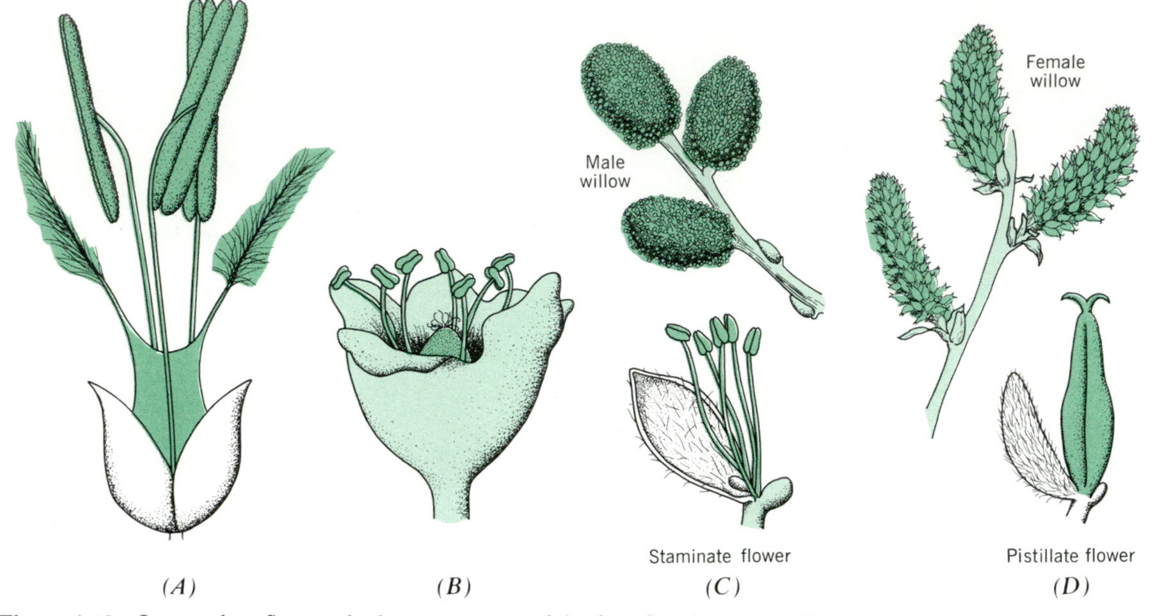

Figure 4-19 Incomplete flowers lack one or more of the four floral parts. *A*. Grass flowers lack a perianth. *B*. Rhubarb flowers have an undifferentiated perianth. An imperfect flower is an incomplete flower that lacks either stamens or pistils. The willow *(Salix)* is an example of imperfect flowers that are borne on different plants (i.e., *dioecious,* "living in two houses"). *C*. Staminate. *D*. Pistillate.

Figure 4-20 Floral symmetry. *A,* Radially symmetrical (actinomorphic) flower of columbine *(Aquilegia)*. *B*. Bilaterally symmetrical (zygomorphic) flower of a mint *(Salvia)*.

other species, the plant is said to be *monoecious* (*oikos* in Greek means "house"; hence, "having one house"). Thus in corn the staminate flowers are borne on the branched tassel at the top of the plant while pistillate flowers are borne lower down, attached to the axis of the ear or cob and surrounded by the husk (see Fig. 16-8). A single pistillate flower of corn consists of an ovary which is attached to a thread-like silk 8-20 cm long. The silk bears numerous hairs which catch the pollen grains. Each thread of corn silk serves as a combined style and stigma. When the staminate and pistillate flowers are borne on separate plants, as in asparagus, spinach, and willow (Fig. 4-19*C* and 4-19*D*) the plant is called *dioecious* ("in two houses"). In a commercial asparagus field, about half the individual plants bear only staminate flowers and the other half pistillate flowers. The young shoots of both are edible, but only pistillate plants bear fruit. In the date palm, some individuals are staminate and others pistillate, but since the edible fruit (date) is produced only by pistillate palms, there is no fruit production unless staminate and pistillate plants grow near enough together for pollen transfer. In commercial plantations of date palms, most of the individuals are pistillate, and only a few staminate palms are scattered throughout to ensure a supply of pollen (Chapter 26).

In many flowers, all parts of each whorl are alike in size and shape, and the flower could be divided into two similar parts in more than one longitudinal plane. Examples are the rose, tomato, raspberry, stonecrop, lily, and many others (Fig. 4-20*A*). Such flowers are *radially symmetrical* and are termed *regular* or *actinomorphic* (literally, "ray-shaped"). The flowers of many other species are *bilaterally* symmetrical because they are capable of division into two identical parts along only one longitudinal plane. Bilaterally symmetrical flowers are called *irregular* or *zygomorphic* ("yoke-shaped"). In such flowers, for example, mints, peas, or orchids (Fig. 4-20*B*), the corolla is commonly of more specialized form than other floral parts.

In the buttercup, all parts of the flower are separate and distinct; that is, each sepal, petal, stamen, and carpel is attached at the base to the receptacle (Fig. 4-20*A*). In many flowers, however, members of one or more whorls are to some degree united with one another, or are attached (coalesced) to members of other whorls. Union of the sepals along their edges occurs in the flowers of mints, snapdragons, peas, and many other plants (and is called *synsepaly*): if petals are attached to one another the corolla is said to be *sympetalous* (the opposite is *apopetalous*). When the corolla is sympetalous (Fig. 4-21) it may have the form of a funnel (petunia), a tube (fuchsia), or a bell (harebell). Stamens and carpels sometimes also coalesce.

Union of members of two different whorls (*adnation*) also occurs frequently; for example, in snapdragon and honeysuckle, the stamens are attached to the corolla rather than to the receptacle; thus the stamens are said to be adnate to the corolla. In the buttercup and magnolia, the receptacle is convex and the floral parts are arranged one above another, with the sepals lowest and then followed by petals, stamens, and carpels (Fig. 4-17). The gynoecium, therefore, is located on the receptacle *above* the points of origin of the perianth parts and stamens. An ovary in this position is said to be *superior*. In the daffodil and sunflower the ovary occurs below the points of attachment of the perianth parts and the stamens and thus is said to be *inferior*. In plants with an inferior ovary, the basal portions of the three outer whorls have fused to form a tube or *hypanthium*. The ovary in the daffodil (Fig. 4-21*C*) and the evening primrose is completely adnate with the hypanthium.

In flowers with a superior ovary, sepals, petals, and stamens arise from the lower portion of the receptacle, *below* the point of origin of the ovary; they are therefore said to be *hypogynous* (i.e., "under gynoecium," Fig. 4-21*A*). If the perianth parts arise *around the ovary* from the rim of the receptacle (Fig. 4-21*B*) they are *perigynous* ("around gynoecium"). But if the perianth parts arise *above* the point of origin of the ovary (Fig. 4-21*C*), they are *epigynous* ("above gynoecium").

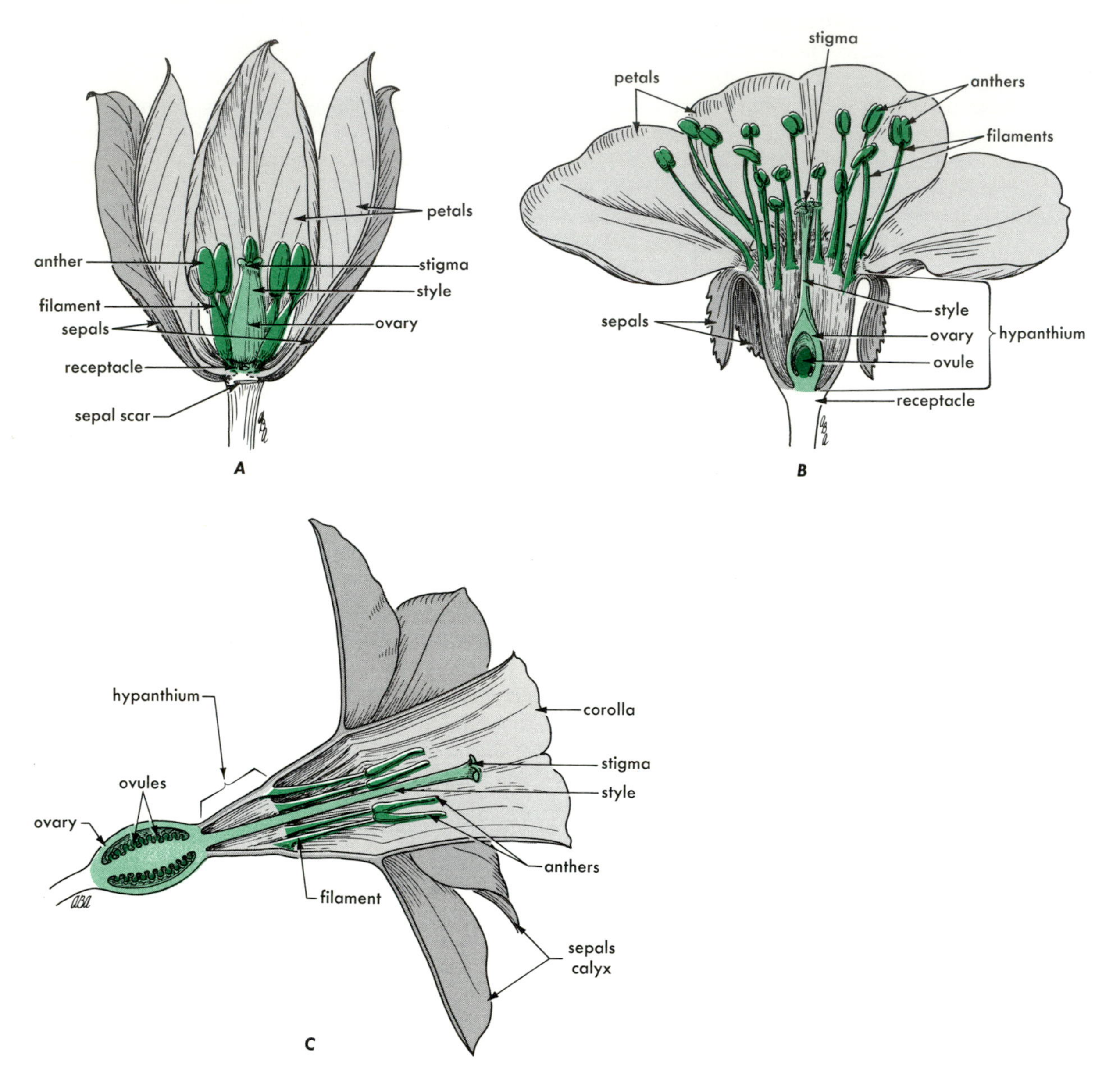

Figure 4-21 Diagrams showing elevation and fusion of floral parts. *A*. Hypogyny in the tulip *(Tulipa)*. Sepals, petals, and stamens all arise below the pistil; hence the ovary is *superior*. *B*. Perigyny in a cherry *(Prunus)* flower. The sepals, petals, and stamens all arise from the rim of the hypanthium and surround the lower part of the pistil. *C*. Epigyny in the daffodil *(Narcissus)*. All four sets of floral parts have a common origin; the fusion of the receptacle with the ovary wall results in the sepals', petals', and stamens', apparently arising from the top of the ovary—thus the ovary is *inferior*. Note sympetaly and synsepaly in this flower.

The position of the floral parts relative to the ovary or gynoecium is significant in establishing relationships between plants (Chapter 13) and also helps in interpreting many kinds of fruits (Chapter 18).

c. INFLORESCENCES

Flowers are sometimes borne singly, as in buttercups or tulips, but commonly they occur in a cluster or group on a floral axis or on branches of such an axis. Such a cluster is called an *inflorescence* (Fig. 4-22). It merely has bracts below it. The arrangement of flowers in an inflorescence varies so greatly that categories of inflorescences have been recognized and given names. Since any particular kind may be constant for a species, genus, or even family, the kind of inflorescence is useful in identifying plants. The grouping of flowers into an inflorescence may well have been of selective advantage to many insect-pollinated plants, especially if the flowers are small, for they are thereby made more conspicuous to the pollinator.

In the *raceme* the stalked flowers are arranged along a single floral axis, as in the blackberry; the *spike* is like the raceme but its flowers are sessile, as in the weedy plantain (Fig. 4-22). The *catkin* resembles the spike but the flowers are unisexual (also without a perianth) and the staminate inflorescence is shed as a unit after flowering. Catkins are only borne on woody plants such as willows, oaks, and birches and generally are among the earliest of spring flowers. The inflorescence of the grasses is either a simple raceme or spike or, often, a branching cluster of racemes known as a *panicle* (Fig. 16-1 and 16-2). In the *umbel* the internodes of the floral axis are so short that the flower stalks appear to arise from a common point, like the ribs of an umbrella. The umbel is commonly compound, as in the carrot or parsley plant. The *head* consists of a large number of stalkless flowers that are closely associated on a very short or vertically flattened floral branch.

10. THE LIFE CYCLE OF THE FLOWERING PLANT

Up to this point we have presented the structural plan of the sporophyte diploid plant, and the types of cells (gametes) that participate in the reproductive process. It remains now to describe the alternation of sporophyte ($2n$) and gametophyte ($1n$) generations that constitutes the life cycle of the flowering plant.

We can start with the diploid plant body (Fig. 4-23). When the plant flowers, two kinds of spores are produced. Within each anther, numerous specialized spores divide by meiosis to produce four haploid *microspores* (Fig. 4-23 and 4-24). The conversion of these to pollen grains (Fig. 4-24) results from a mitotic division of each microscopic nucleus, which produces two adjacent cells, usually separated by a thin wall; one of these becomes the *tube or vegetative cell* and the other the *generative cell*. The two-celled structure is the pollen grain. It germinates, when it falls on a receptive stigma, to produce a long, thin tube (Fig. 4-23, right) that grows rapidly down the style until it reaches and enters the ovule. Either before or during germination the generative cell divides to form two sperms or male gametes, which are transported down the tube to the egg cell within the ovule. There is a special device for its entry called the filiform apparatus (Fig. 4-25*R* and *S*). The germinated pollen grain, with the tube nucleus and two sperms, constitutes the mature male gametophyte (or gamete-producing "plant").

Meanwhile, within the ovule four *megaspores* develop from a single specialized cell in the sporangium (Fig. 4-23 and 4-25). Three of these disintegrate, while the fourth undergoes meiosis and mitosis to form the female gametophyte (see Fig. 4-25, which is a magnification of the same stages in Fig. 4-23). At maturity this gametophyte has seven cells and eight nuclei and is known as the *embryo sac*. Only three of these eight nuclei participate in fertilization, one which becomes the egg cell and two others "polar nuclei"; the rest disap-

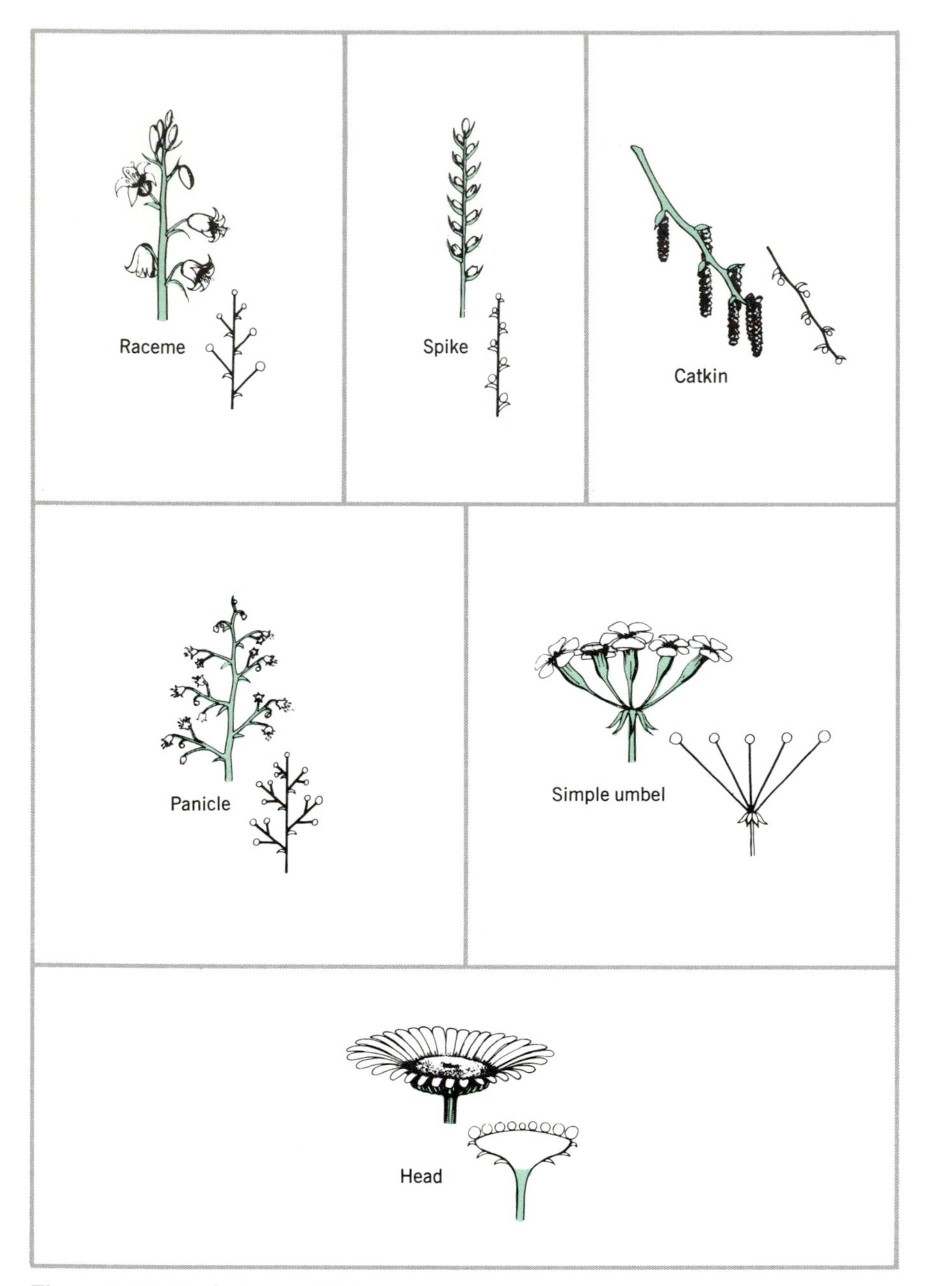

Figure 4-22 Basic types of inflorescences.

pear. When the pollen tube enters the ovary the tip of the tube breaks open, and one sperm nucleus fuses with the egg to produce the zygote (Fig. 4-25). In the majority of flowering plants, the second sperm nucleus fuses with the *two* polar nuclei to produce a $3n$ nucleus, which forms the *endosperm*. This process of *double fertilization* is known only among the flowering

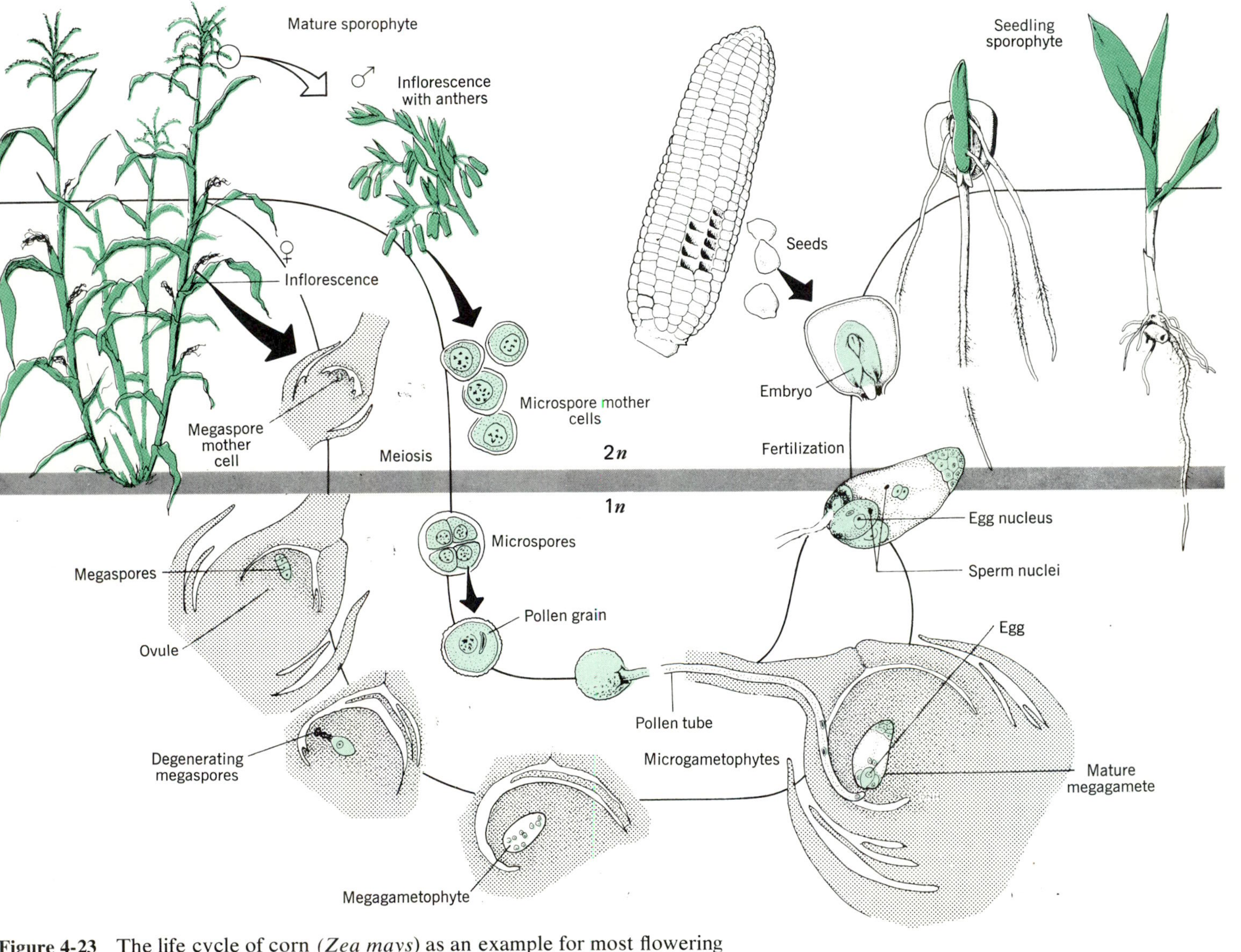

Figure 4-23 The life cycle of corn (*Zea mays*) as an example for most flowering plants (Adapted from Jensen & Salisbury, 1972).

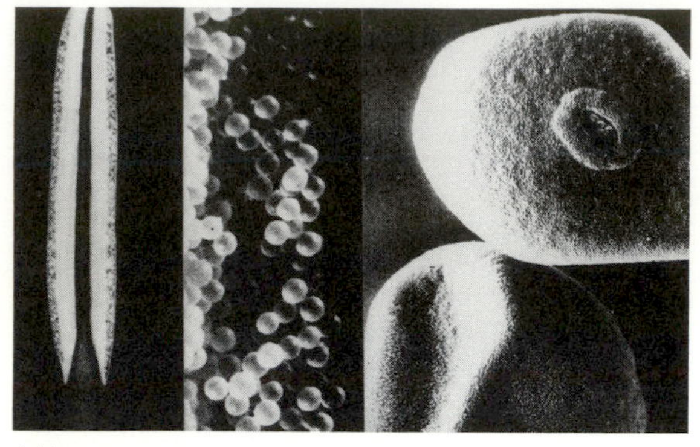

(A)

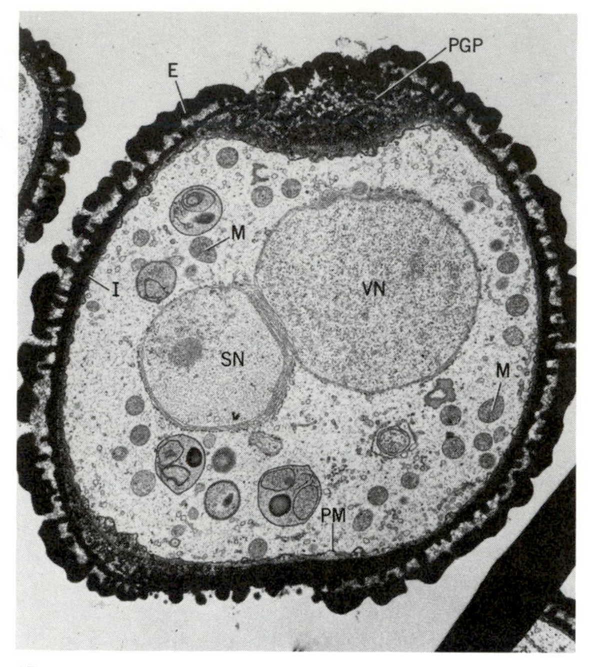

(B)

Figure 4-24 Male gametophyte. *A*. Left to right, another of *Crocus* (×18), pollen grains from the same (×24), pollen grains of wheat *(Triticum speltum)*, highly magnified (×5700). *B*. Section through pollen grain of African violet *(Saintpaulia)* (×7,200). E, exine; I, intine; M, mitochondrion; PGP, pore; PM plasma membrane; SN, sperm nucleus; VN, vegetative or tube nucleus.

plants, although it is not universal among them. The zygote develops into an embryo within the embryo sac, and the integuments immediately surrounding the ovule develop into a seed coat. In many plants, the endosperm later divides into a mass of small cells, but in some families (e.g., *legumes*) the endosperm becomes absorbed into the cotyledons. In either case, the endosperm furnishes the nutrient for the first stages of growth of the seed. Later, when the seed is shed from the ovary, it germinates to produce the *seedling,* which bears a small shoot, the *plumule,* and a single root, the *radicle*. It grows into a mature plant (as described in Chapter 10), thus completing the alternation of generations.

Figure 4-25 A higher magnification of the events in the development of the ovule and its fertilization. Structurally the ovule is relatively complex, consisting of a stalk, the *funiculus* (Fig. 4-25*K*), which contains the megasporangium enclosed by integuments. Early in the development of the ovule, the megaspore mother cell divides meiotically to form four haploid megaspores (Fig. 4-25*L*). Three of these usually disintegrate, the one farthest from the micropyle surviving and developing into the megagametophyte (Fig. 4-25*M*). The functional megaspore divides again. At the end of its third mitotic division, the eight resulting nuclei (Fig. 4-25*P*) become arranged with three at the micropylar end, three at the opposite end, and two that migrate to the center of the central cell (the "polar nuclei"). The three nuclei at the micropylar end of the megagametophyte form cell walls and with the eggs these make up the egg apparatus (Fig. 4-25*Q*). At the opposite end, cell walls form around the antipodal nuclei, and these cells later disappear. The central cell remains binucleate. *R* and *S* show in detail the events in the process of fertilization, in which the antipodal cells no longer play a role. The sperm nuclei are seen entering, one moving to the egg nucleus and the other to the two polar nuclei. *R*. Entrance of pollen tube. *S*. Discharge of pollen tube (From Jensen, 1978).

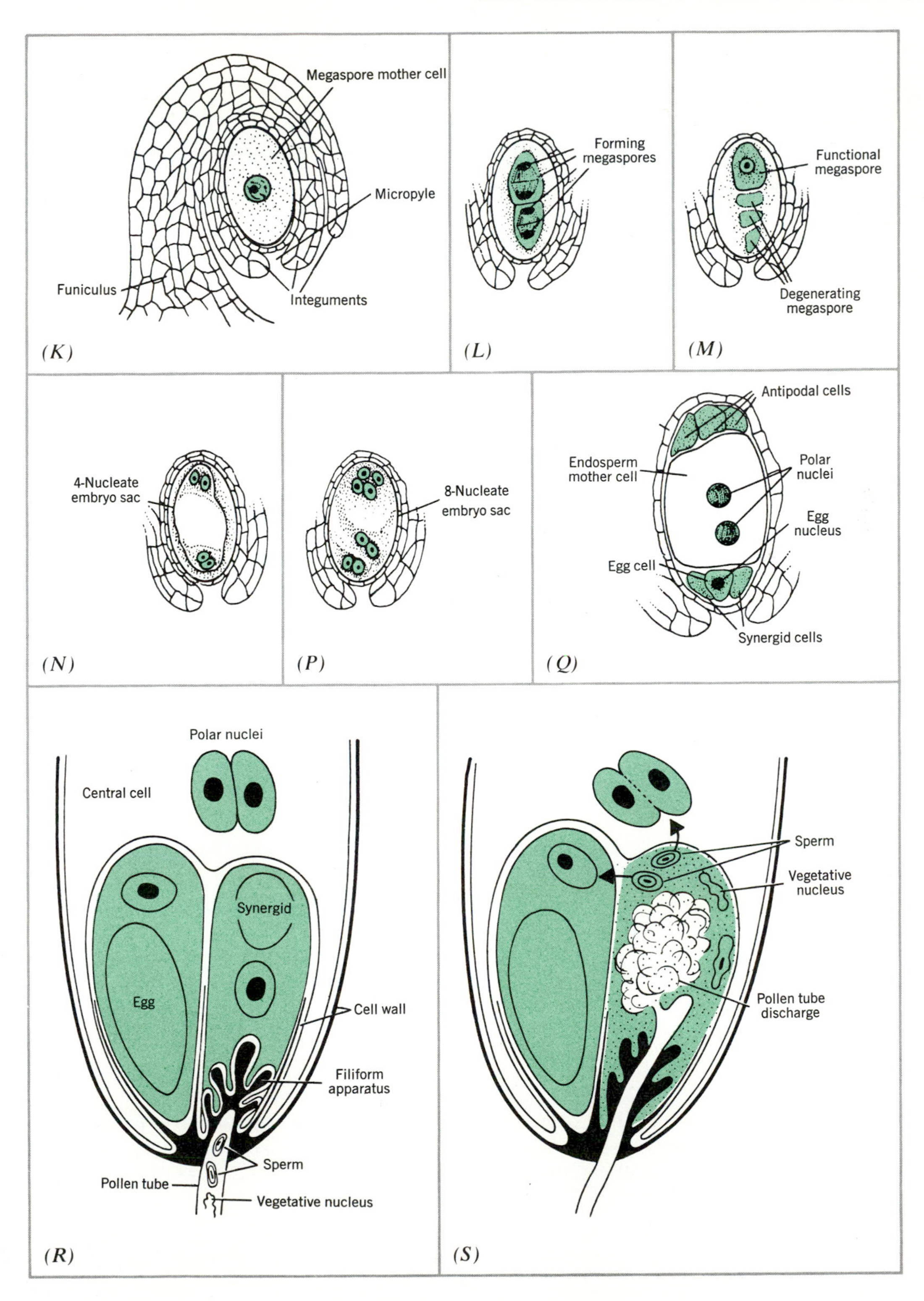
Megaspore mother cell
Micropyle
Funiculus
Integuments
(K)
Forming megaspores
(L)
Functional megaspore
Degenerating megaspore
(M)
4-Nucleate embryo sac
(N)
8-Nucleate embryo sac
(P)
Antipodal cells
Endosperm mother cell
Polar nuclei
Egg nucleus
Egg cell
Synergid cells
(Q)
Polar nuclei
Central cell
Egg
Synergid
Cell wall
Filiform apparatus
Sperm
Pollen tube
Vegetative nucleus
(R)
Sperm
Vegetative nucleus
Pollen tube discharge
(S)

CHAPTER 5

Photosynthesis: The Key Process

Life is possible on earth because of two conditions—the presence of oxygen in the air and the presence of carbon compounds on the earth's surface. Both of these are the result of photosynthesis. The coming of photosynthesis, perhaps within a billion years of the earth's formation (Chapter 2), was thus an event of unparalleled importance; it made possible the whole world of living organisms, the biosphere, and none of us would be here without it.

In this chapter we first trace how the puzzle of photosynthesis was unravelled, then how the light-absorbing pigments work, partly to harvest the light and partly to convert it to chemical energy, then (in some detail) how the oxygen and the carbon compounds are formed, and finally how photosynthesis works out in the whole plant.

1. THE FIRST STEPS IN UNDERSTANDING

The understanding of photosynthesis has been a long time developing and indeed is not complete after 200 years of study. Part of the difficulty in the beginning was due to the fact that the participation of light in a chemical process was a novel idea, but the greater part is due to the unexpected ramifications of the photosynthetic reactions—chemical, physical and biological.

The story starts in 1771 when Joseph Priestley found that a candle burning under a bell-jar went out, although some four-fifths of the original volume of air was still present. He deduced that the flame "vitiated" the air and went on to show that the breathing of a mouse in a similar limited space also vitiated the air, since the mouse died in it after a while, and then a candle would no longer burn in it. A mint plant tested in the same way, however, showed no signs of dying, although plants and animals were then thought to have the same needs. The climax came when the mouse and the plant were installed together; the mouse now survived, and hence Priestley concluded that *plants can purify the vitiated air*. Subsequently, Dr. Jan Ingenhousz, during a leave of absence from the court of Maria Theresa (where he was physician), found that the "purified air" was only produced when the plants were *in bright light*. Green algae on the bottom of a water tank gave off bubbles of "pure air" when sunlight fell on them, and so did leaves immersed in water (as actually Charles Bonnet had observed back in 1754, thinking the bubbles due to the extrusion of air held within the leaves).

Thus the process was *dependent on light*. Then in 1782 the Rev. Jean Senebier showed that what was involved was the conversion of "fixed air" (now known to be carbon dioxide) into "pure air" (oxygen). Boiled water, from which the gases had thus been expelled, produced no bubbles in sunlight, nor did leaves that had been squeezed or scratched; also, confirming Ingenhousz's experiments, he showed it was the light and not the warmth of sunlight that was effective. Leaves that had lost their greenness by growing in darkness, or by aging or drying, gave no bubbles, although fragments of green leaves were still effective. At about the same time the purely chemical studies of Antoine Lavoisier in 1775 to 1778 made it evident that "fixed air" is a compound of carbon with oxygen. This concept opened the door to a true understanding of what goes on in photosynthesis and made possible the quantitative demonstration by Théodore de Saussure in 1804 that plants *gain in weight* by converting carbon dioxide (CO_2) to oxygen. Since this gain was actually shown to be in the carbon content of the plants, he concluded that the process is:

$$CO_2 \xrightarrow{\text{Light}} (C) + O_2$$

Although his experiments were done with an artificial gas mixture enriched in CO_2, he showed that growth of plants in light in ordinary air also entails a gain in carbon, while in darkness, on the other hand, plants lose some carbon (i.e. they respire). Thus photosynthesis is not just an odd property of green tissue; it is their *means of growth*.

De Saussure made another important discovery. He found that the gain in dry weight of the plants in light was even greater than that due to the carbon alone; since no material other than water was present, he reasoned that water must enter into the *dry* weight, and wrote, "Plants appropriate the oxygen and the hydrogen of water, making it lose the liquid state. This assimilation is only marked when they incorporate carbon at the same time." Thus the true reaction should be written:

$$CO_2 + H_2O \xrightarrow{\text{Light}} (CH_2O) + O_2$$

where (CH_2O) represents the composition of the plant's dry weight. This is photosynthesis.

2. THE LEAF PIGMENTS

Up to this time the problem of just how light entered the process, and what absorbed it, had not been a central one. The green pigment was extracted (crudely) in 1818 by Pelletier and Caventou in Paris (they coined the name "chlorophyll"), but it was not until 1864 that Sir George Stokes, in Cambridge, separated the green extract into two green and two yellow pigments. One of these, the color of carrots, was named *carotene*. The second yellow pigment, *xanthophyll*, was shown to be a mixture, differing in different plants. This resulted from the discovery of the Russian Mikhail Tswett (1906), followed by many others, that all the pigments could be neatly separated by being adsorbed on a column of powdered chalk or other fine powder. This discovery was the beginning of chromatography, which has turned out to be a widely used analytical tool. Later Richard Willstätter, Hans Fischer, and collaborators (1906 to 1930) purified and worked out the structure of the two chlorophylls, *a* and *b*, the two yellow pigments, carotene and the xanthophylls, and also the brown pigment fucoxanthin, which masks the presence of chlorophyll in the brown algae. These yellow and brown pigments are called *carotenoids*. They absorb light only in the blue region of the spectrum.

But this highly successful chemical work did not explain how light entered in, and in some degree we still do not know it today. Only gradually did the idea gain acceptance that chlorophyll somehow absorbed the energy of the light and passed it on to CO_2, which in consequence underwent chemical changes. The concept that energy is *needed* for the conversion of CO_2 to plant material developed out of J. R. Mayer's studies (1842) on the conservation of energy and the conversion of work into heat. Later on, several workers found that the effectiveness of light of different wavelengths in causing photosynthesis was roughly proportional to the degrees to which they were *absorbed* by the chlorophyll, for it is only when light is absorbed that its energy can be passed on to the absorbing substance. Chlorophyll appears green because it absorbs red and blue light, thus reflecting or transmitting the yellow and green wavelengths; photosynthesis, therefore, takes place in green plants primarily in the red and blue parts of the spectrum, that is, 660 to 770 nm and 440 to 460 nm* (Fig. 5-1).

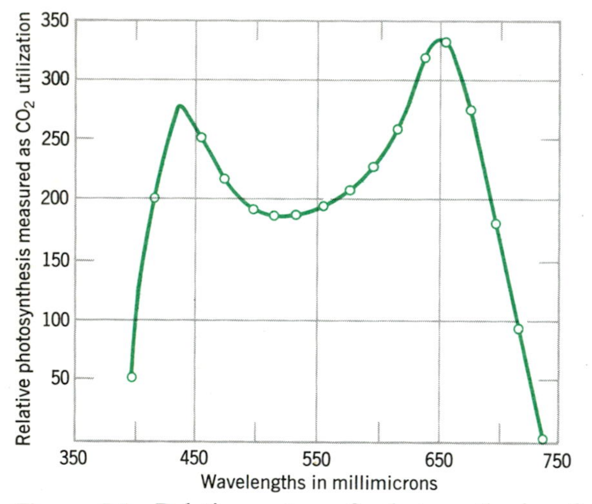

Figure 5-1 Relative rates of photosynthesis of young wheat plants at different wavelengths of equal light intensity. The peaks lie at about 445 and 660 nm (nanometer) (From Hoover, 1937).

*m = meter.
mm = millimeter = 1/1000 meter = 10^{-3} meter.
μm or μ = micrometer or micron = 1/1000 millimeter = 10^{-6} meter.
nm (or mμ) = nanometer or millimicron = 1/1000 micron = 10^{-9} meter.

3. A SKELETON OUTLINE OF THE WHOLE PHOTOSYNTHETIC PROCESS

In the following sections what is now known about the inner workings of photosynthesis is set out in some detail. To keep the many ramifications in mind, we present here a skeleton of the photosynthetic process. It has nine steps:

A. Light falls on the chlorophyll molecules and imparts its energy to them, exciting them.

B. Most of these excited molecules transfer their energy to neighbor molecules, which pass it on to others, until a select few receive enough energy to be able each to give off an electron (designated *e*).

C. Some of the emitted electrons join up, through a little chain of intermediate carriers, with hydrogen ions (H^+) from water (HOH) to reduce an organic acid $C_3H_6O_4$ forming the product $C_3H_6O_3$.

D. The other part of the emitted electrons give up their energy to phosphate molecules to produce the special energy-carrying molecule ATP. This compound then makes all the chemical reactions of the sequence occur.

E. The $C_3H_6O_3$ formed in step C condenses to make glucose, $C_6H_{12}O_6$, and some of it condenses further to make the large aggregate molecule called starch. This is the *end product* (in most leaves).

F. The glucose molecules that do not form starch undergo a cycle of reactions and finally combine with a molecule of carbon dioxide of the air to form the organic acid that is reduced in step C.

G. The chlorophyll molecules that lost electrons in step B now have their missing electrons replaced in a cycling system which also makes ATP as in (D) above. Again a little chain of intermediates takes part.

H. This second group of chlorophylls is in turn excited by absorbing light energy, which makes them somehow able to take electrons from water.

J. The water molecules lost their hydrogen ions in step C and have now lost electrons; nothing is left but the oxygen, which is given off:

$$H_2O \rightarrow 2\ H^+ + 2\ e + \tfrac{1}{2}\ O_2$$

Nearly all these constituent reactions are pretty well understood; only step *H*, the second photoreaction, remains mysterious. Some of the steps actually comprise many smaller steps or reactions. In the following sections, steps *A* and *B* will be discussed first, then the chemical steps *C*, *E*, and *F*, then the light reactions *G* and *H*, and with *D*, the whole sequence will be assembled. In the drawing of the process, shown below, the items underlined are the initial and final products. The letters correspond to the lettered stages in the outline.

4. LIGHT ENERGY AND ITS TRANSFER (Steps A and B of the Outline)

That the green and yellow pigments were concentrated in small units in the leaf cells, the *chloroplasts*, was noted in the nineteenth century, but it was the electron microscope in the 1950s that showed the full significance of these organelles. For the electron microscope revealed that, not only are the pigments concentrated into chloroplasts, they are even concentrated *within* the chloroplast, forming dense layers of "thylakoids" (see Fig. 2-5). Only in blue-green algae, (sometimes called Cyanobacteria) are the pigments apparently free in the cell, and even there they are concentrated in the same kind of dense layers, mainly around the periphery (see Fig. 5-2). Evidently the pigment molecules must be *close together* for photosynthesis. And indeed a careful study of these algae, as well as of the red algae that contain similar pigments, shows why. These cells con-

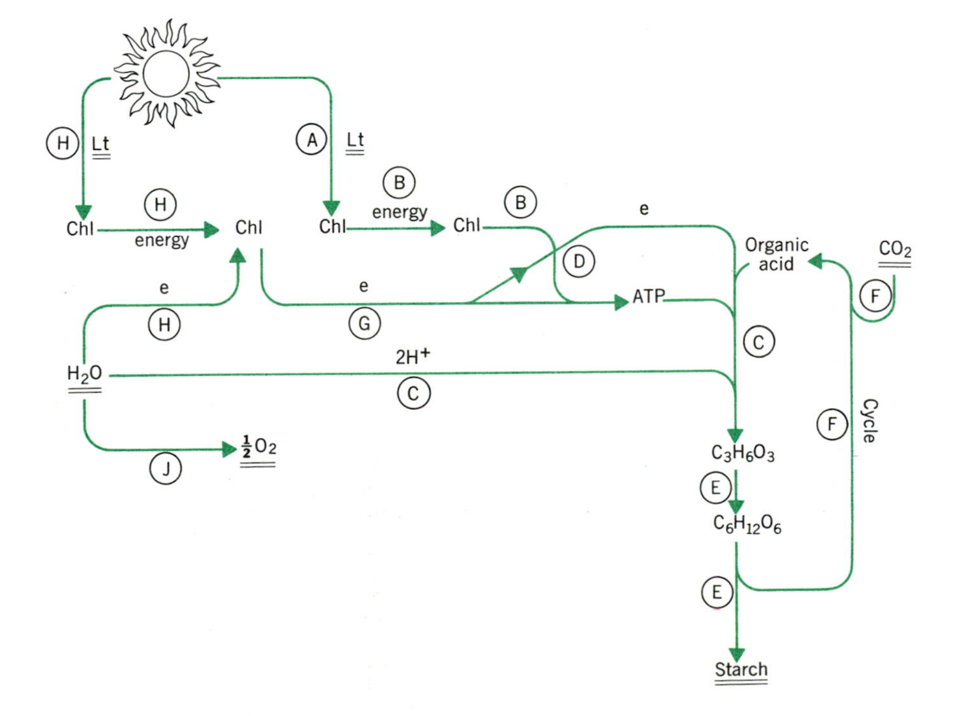

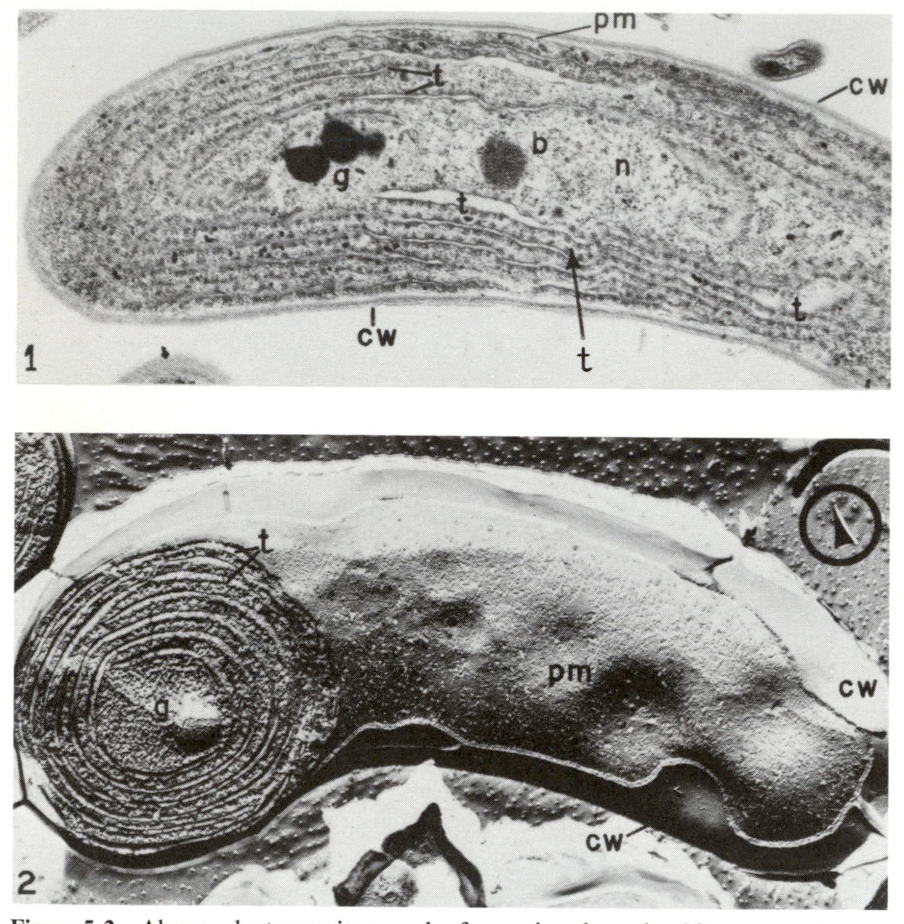

Figure 5-2 Above, electron micrograph of a section through a blue-green alga or Cyanobacterium *(Synechococcus)*. Beneath the cell wall (cw) and membrane (pm) are the parallel photosynthetic membranes or *thylakoids* (t) (compare Fig. 2-5, where they are inside the chloroplast). The particles lining the thylakoids contain the phycocyanin. In the center is the nuclear material (n). The granules g and b are not critical (×2800). Below, a frozen-etched cell of the same alga showing a transverse section of the thylakoids (×2000).

tain, besides the chlorophylls, a blue pigment called *phycocyanin,* or the red pigment *phycoerythrin* or sometimes both. And in these plants, the yellow and yellow-green wavelengths absorbed by the blue or red pigments have a special function. What they do is to absorb the light energy and *transfer* it to the chlorophyll. Such transfer could only take place between closely appressed molecules.

This transfer of energy was shown directly by measuring the efficiency of light of different wavelengths in causing photosynthesis. In higher plants the efficiency is high in the red (around 680 nm) and in the violet (around 440 nm) where chlorophyll absorbs, but low in the yellow and green; it is also low in the blue-green where the carotenoids absorb light (Fig. 5-1). But the accessory pigment of blue-green algae, phycocyanin, absorbs in the yellow light of 560 to 640 nm, and in these cells the efficiency in the yellow remains just as high as at the wavelengths where the chlorophyll absorbs (Fig. 5-3). Similarly the accessory red pigment of red algae, phycoerythrin, absorbs best in the yellow-green, and correspondingly photosynthetic efficiency in these algae is maximal in the yellow-green part of the spectrum. It follows that, whereas light absorbed by the carotenoids is evidently not effective for photosynthesis, light of the wavelengths absorbed by phycocyanin and phycoerythrin *is* effective. Incidentally, since thick layers of water look blue-green to our eyes (and therefore transmit the blue-green but absorb the red) green algae are confined to rather shallow waters, but red algae can grow down to depths of 50 feet or more, using the blue-green wavelengths that seawater transmits.

Thus the blue-green and red pigments are *light-harvesting* pigments, becoming activated by absorbing the light and then transferring that activation to chlorophyll. In both blue-greens and reds the pigments are localized in tiny bodies, the *phycobilisomes,* closely attached to the chlorophyll-bearing membranes. Why are these pigments not present in the evolutionarily more advanced forms, the green algae and higher plants? The answer is twofold. In the first place the phycocyanin of the blue-greens is not *essential* for their photosynthesis; cells from which phycocyanin has been chemically removed can still carry out some photosynthesis. In the second place (and more important), in these higher plants, which do not possess the red and blue pigments, the function of light-harvesting is done in a subtler way. In the chloroplasts of higher plants, there are several slightly different combinations of chlorophyll *a* with protein, and these combinations absorb light (in the red region) at slightly different wavelengths, between about 670 and 700 nm. When each absorbs a quantum of light it is set into more active vibration. Most of them—more than 90%—transfer this vibration to their neighbor molecules (because they are so closely in contact) and thus add further to the excitation those molecules have undergone by absorbing light. In this way the activity induced by the light absorption is intensified in some of them, and this interaction continues until a few—a very few—are sufficiently activated to give off an electron. These few chlorophyll-proteins are the ones that absorb at the two longest wavelengths, 680 and 700 nm, for the energy of a quantum of light decreases with increasing wavelength (i.e., the longer the wavelength, the smaller the quanta).* It follows that energy can be passed only in the direction toward longer wavelengths—otherwise there would need to be a gain in energy. Only 1 in a 1000 or so of the chlorophyll-proteins can reach this extreme state where they give off an electron. They are referred to as P_{680} and P_{700}. All the others act simply as light-harvesters (Fig. 5-4).

Another aspect of the differences in light absorption came out of careful determinations of the efficiency of photosynthesis. Efficiency means the number of oxygen molecules produced *per quantum* of light absorbed (just as efficiency in working of a machine is the amount of work it does *per calorie* of energy expended). Robert Emerson, at the University of Illinois, in making such efficiency measurements at different wavelengths in the red region, found that at wavelengths longer than 685 nm the efficiency dropped off sharply. The light absorption also decreases, of course, but the efficiency is calculated per quantum of light absorbed. Thus from about 660 to 680 nm it took 10

---

*This was Max Planck's great generalization, noted earlier in Chapter 2. The energy of a quantum, E, is given by $E = h\nu$, where h is a constant ($= 6 \times 10^{-27}$) and ν is the frequency of the light waves, which is inversely proportional to the wavelength. Hence the energy of a quantum decreases as the wavelength increases. Correspondingly, the quanta of really short wavelength, the ultraviolet, pack so much energy that they often distort molecules, and this may be lethal to organisms.

† Note that in this and following equations the hydrogen ions have a positive charge (they are H^+) and the electrons have a negative charge. The sum of + and − charges on each side of the equation must be the same.

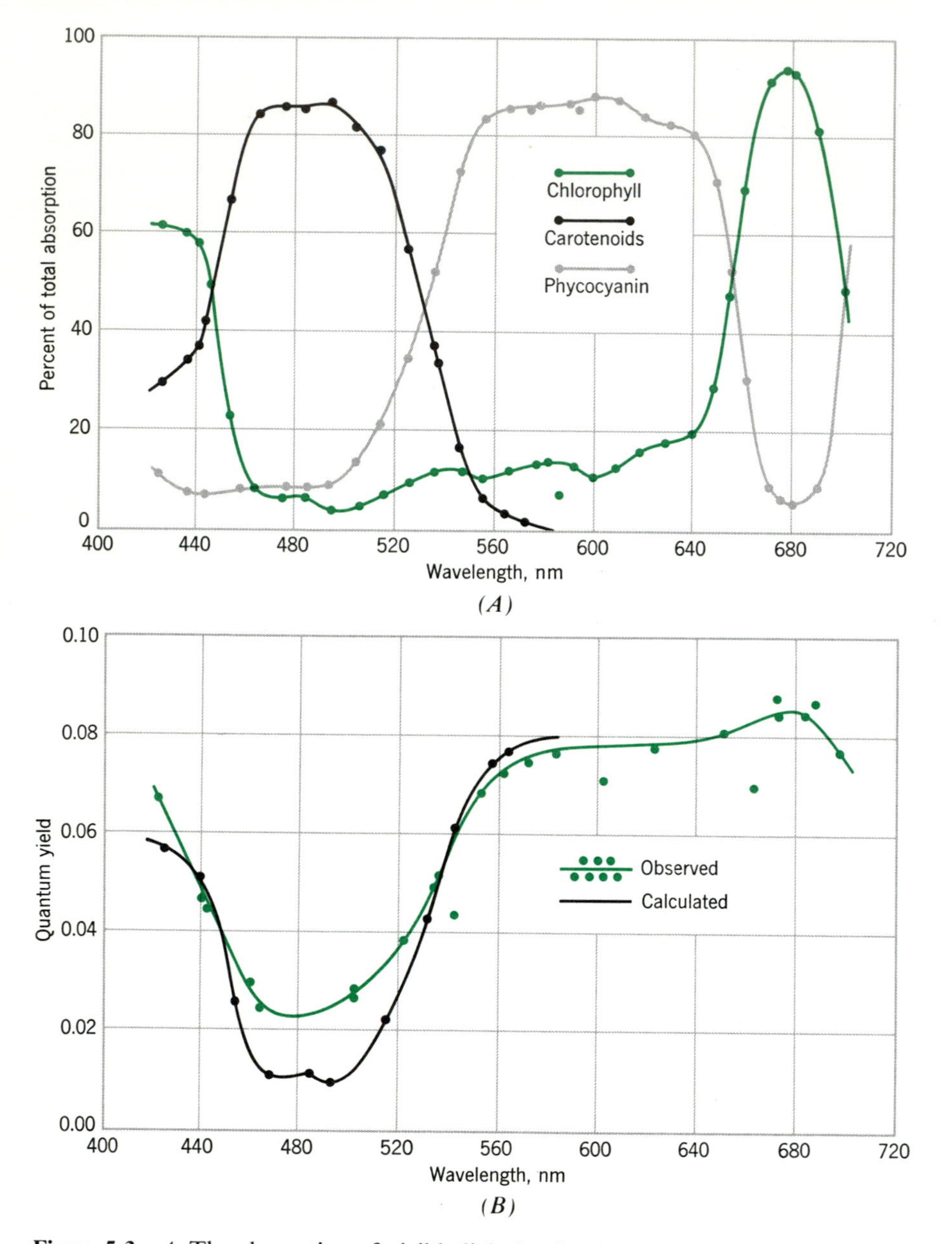

Figure 5-3 *A*. The absorption of visible light by the three pigments of blue-green algae. The chlorophyll (color line) shows peaks at 680 nm in the red and around 430 nm, and the phycocanin falls in between at 560 to 640 nm. *B*. Efficiency of photosynthesis measured as yield of oxygen per quantum of light absorbed. The color line is experimental; the black line is calculated from Figure 5-3*A* above by assuming that light harvested by phycocyanin is almost as effective as that directly absorbed by the chlorophyll, whereas light absorbed by the carotenoids is ineffective. Note particularly that from 560 to 640 nm chlorophyll has almost no absorption, yet the efficiency is almost as high as at the chlorophyll peak (From Emerson & Lewis, 1942).

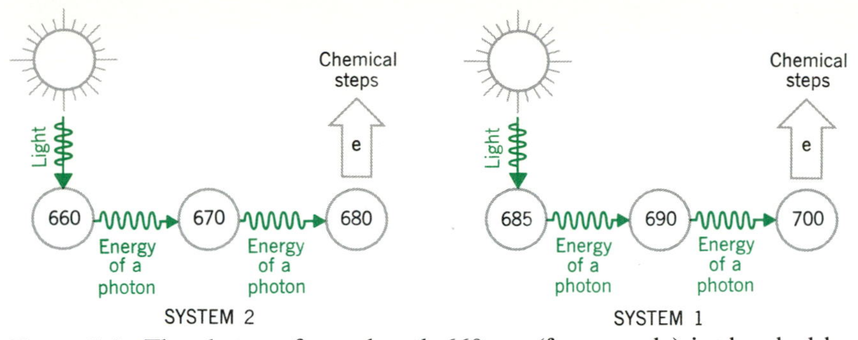

Figure 5-4 The photon of wavelength 660 nm (for example) is absorbed by a chlorophyll-protein molecule that absorbs at 660 and is energized thereby. The activated molecule passes its energy to a chlorophyll-protein molecule that absorbs at a slightly longer wavelength (shown here as 670 nm) and this again passes its energy to one that absorbs at 680, called P_{680}. This is sufficiently energized that it gives off the energy as an electron, *e*. Similarly the wavelengths shorter than 700 nm, pass energy to P_{700}. The fate of the two emitted electrons is described in Section 6.

quanta per molecule of O_2 produced, but at 690 nm it took about 24 and at 700 nm the value rose to nearly 100. Yet if a small amount of light of a somewhat shorter wavelength (e.g., in the orange red at 640 to 660 nm) fell on the cells, the efficiency was restored. Emerson's deduction from this was that two wavelengths, a shorter and a longer, collaborate to bring about efficient photosynthesis, or in other words there must be two light processes that interact. This is a different story from mere harvesting and has led to great efforts by many researchers, worldwide, to establish what these two reactions are. But in order to understand them, we must first describe the chemical basis of photosynthesis.

5. THE ORIGIN OF THE OXYGEN (Steps C and J of the Outline)

Insight into the nature of the reaction between carbon dioxide and water came first from a study of the purple sulfur bacteria by Cornelis van Niel. These microorganisms, which contain chlorophyll overlaid with a purple pigment (Fig. 5-5), grow in light in sulfur springs, ditches, and other places where hydrogen sulfide (H_2S) is being generated by decomposing organic matter. Growing them in the laboratory had been a puzzle, since, unlike algae, they produced no

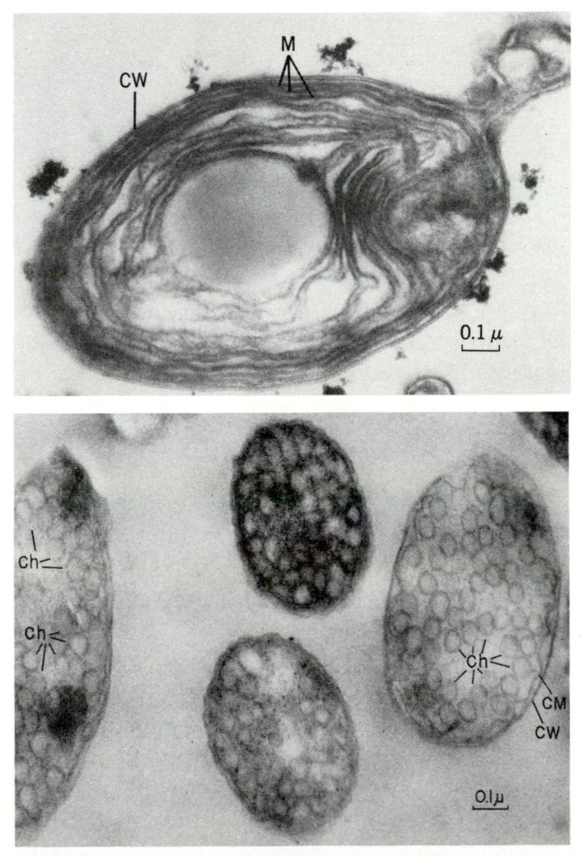

Figure 5-5 Two purple bacteria. Above, *Rhodomicrobium,* with its photosynthetic system located in a series of parallel membranes M, like those of Figure 5-2. Below, *Rhodospirillum* with a photosynthetic system in small oval chromatophores, Ch. CM, cell membrane; CW, cell wall.

free oxygen. In fact, van Niel found they were completely *anaerobic;* that is, they die in oxygen. Nevertheless since they could be grown with H_2S but without organic matter, evidently getting their carbon from CO_2, it was clear they must be photosynthetic. The puzzle was solved by finding that these bacteria use up H_2S, depositing sulfur granules:

$$CO_2 + 2H_2S \rightarrow (CH_2O)^* + H_2O + \underline{\underline{2S}}$$

where (CH_2O) represents cells formed. This equation is like that for green plant photosynthesis, for if we add H_2O to both sides of the green plant equation on page 81 we get a perfect parallel:

$$CO_2 + 2H_2O \rightarrow (CH_2O)^* + H_2O + \underline{\underline{O_2}}$$

The important implication of this is that, just as the sulfur comes from H_2S, the oxygen must come from H_2O, water. Thus photosynthesis *uses the hydrogen of water* **to reduce** CO_2 to organic matter, while the oxygen of the water is liberated. This deduction is contrary to the earlier idea (which seemed like a logical interpretation of the equation on page 81) that the oxygen came from CO_2. However, it was soon confirmed by H. A. Barker's group at University of California, Berkeley, making use of the fact that, while most atoms of oxygen have the atomic weight 16, a very small number are heavier, with atomic weight 18. Physicists had found a way to concentrate this in the form of "heavy water," $H_2{}^{18}O$. When algae were set to photosynthesize in this "heavy water," the gas evolved was rich in ${}^{18}O_2$.

*Note here that in all these formulations the plant material is of the overall composition CH_2O. Compounds of this formula are *carbohydrates* (the name means *carbon and water*). As we shall see in the next chapter, the largest part of plant material does have this composition. Glucose, which is $C_6H_{12}O_6$, equivalent to $(CH_2O)_6$, is a common example; starch and cellulose have a similar composition.

An independent approach came from experiments by Robert Hill at Cambridge, England, who had observed that isolated chloroplasts in light could give off a little oxygen even if *no* CO_2 was supplied; all that was needed was a ferric salt (ferric oxalate), which became converted to ferrous, that is, reduced. The ferric ion has three positive charges, the ferrous ion two, so that the conversion means the acceptance of one electron by each ferric ion. Not only does this show that Fe^{3+} can substitute for CO_2 and be reduced (i.e., accept electrons) instead of CO_2, but it confirms that the oxygen of photosynthesis cannot come from CO_2 and therefore must come from the water.

6. THE PATH OF CARBON DIOXIDE (Steps C, E, and F of the Outline)

If oxygen comes off, hydrogen must be left behind. But the Hill reaction makes clear that the primary effect of light must be *to liberate electrons,* since the ferric ion is reduced to ferrous:

$$Fe^{3+} + e \rightarrow Fe^{2+}$$

In normal photosynthesis these electrons, instead of reacting with iron, *could* have reacted together with the residual hydrogen ions, to *hydrogenate* (= *reduce*) certain organic compounds:

$$X + 2H^+ + 2e \rightarrow XH_2$$

Here X represents a reducible organic compound, that is, one that can accept additional hydrogen atoms. The hydrogen is set free as ions because water dissociates into H^+ and OH^- (see equation 6 on p. 91).

The reduction of X by hydrogen ions plus electrons is a typical organic *reduction.* Hundreds of such reactions are known; they can be brought about either chemically by reducing agents or biochemically by dehydrogenase enzymes. For instance, acetaldehyde is reduced to ethyl alcohol.†

$$CH_3CHO + 2H^+ + 2e \rightarrow CH_3CH_2OH$$

A comparable reaction is the reduction of an acid to an aldehyde:

$$R.COOH + 2H^+ + 2e \rightarrow R.CHO + H_2O$$

(Actually acids are hard to reduce directly and are only readily reduced when combined with a phosphate group, as we shall see below.)

The path of carbon in photosynthesis has been traced in detail by using the fact that, just as with oxygen, there are several kinds (isotopes) of carbon. Ordinary carbon atoms have atomic weight 12, but a few have atomic weight 14; these are *radioactive* and so are easily detected. When carbon dioxide containing $^{14}CO_2$ is used in photosynthesis, then after only a few seconds the radioactive ^{14}C is not found in the usual sugars and starches but is found in phosphorylated compounds, especially in the 3-carbon compound *phosphoglyceraldehyde,* $C_3H_5O_3.H_2PO_3$. Since this is an aldehyde, it was deduced that it must have been formed by hydrogenation of the related acid, phosphoglyceric acid, just as shown above. The full formulae, with the radioactive carbon shown as C*, are:

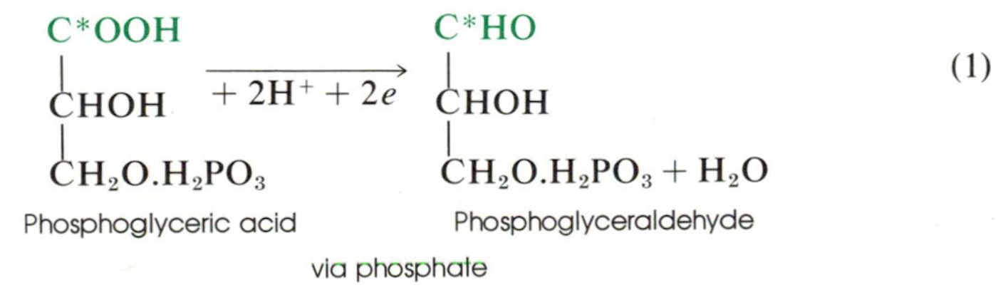

This is the *prime* chemical step in photosynthesis, the reduction of an acid group to an aldehyde. For the acid group must have been formed from CO_2, as shown by the radioactive carbon in the formula. (It is a detail, but important chemically, that the acid group is not reduced directly but is first converted to a phosphorylated form

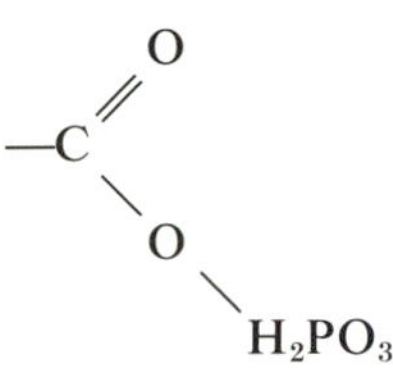

and this then accepts two (H) to form

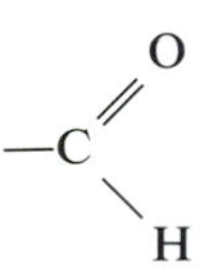

and set free H_3PO_4.)

Thus it is not CO_2, strictly speaking, that is reduced in photosynthesis; it is CO_2 that has become incorporated as the acid group, —COOH, of an organic acid. How did the applied CO_2 get incorporated into phosphoglyceric acid? The answer is that it combines with a 5-carbon compound of carbohydrate type, namely ribulose bis-phosphate, which then splits into two identical molecules of phosphoglyceric acid:

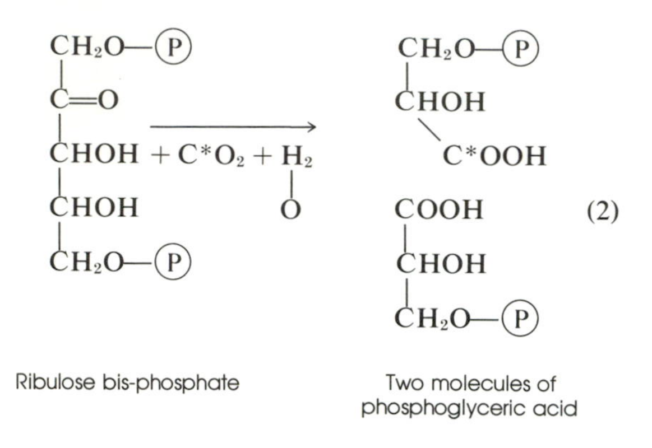

Ribulose bis-phosphate

Two molecules of phosphoglyceric acid

Here, for simplicity, (P) represents the phosphate group, $—H_2PO_3$.

The ribulose bis-phosphate arises by a long series of chemical events worked out by Melvin Calvin and his colleagues and called the Calvin cycle (Fig. 5-6). The details of this are not necessary for the understanding of photosynthesis. Briefly, five molecules of the 6-carbon sugar glucose become converted to six molecules of the 5-carbon sugar ribulose. All the compounds taking part are phosphorylated.

The main point of the process can be seen from the overall formulae, for phosphoglyceraldehyde is $C_3H_5O_3.H_2PO_3$, which means that free glyceraldehyde would be $C_3H_6O_3$, in which the atoms are in the proportion of carbon and water. This, as we saw above, is the definition of a carbohydrate; thus the immediate product of photosynthesis is a *phosphorylated carbohydrate* and by relatively straightforward reactions can give rise to all the carbohydrates of which the plant body is composed.

Two such molecules condense to the 6-carbon compound glucose phosphate (Fig. 5-6). Twelve C_3 compounds make six C_6 compounds. Of these six glucose phosphate molecules, five go into the cycle (Fig. 5-6) becoming six C_5 compounds, and one is deposited in the cell as net profit. Then the cycle starts again.

7. PUTTING THE WHOLE PROCESS TOGETHER (Stages D, G, and H of the Outline)

Basically the action of light is thus to split water into protons, electrons, and oxygen, making possible all the chemistry that follows. To complete our understanding we need three additional points—the role of intermediaries, the exact function of the second photosystem, and the source of the active phosphate groups.

a. INTERMEDIARIES

The reduction of the acid to the aldehyde by electrons and protons is achieved via a little chain of intermediates: the electrons liberated from chlorophyll are first passed through a brownish iron-containing pigment, *ferredoxin*. In turn this substance reduces a yellow enzyme called *flavoprotein* (FP) and that reduces the hydrogen-transferring compound which is called (for short) NADP. NADP is a compound that can be reduced and reoxidized with great ease and thus acts as a carrier in many oxidations and reductions:

$$NADP^+ + H^+ + 2e \rightarrow NADPH \quad (3)$$

then

$$NADPH + H^+ + RCOOH \rightarrow RCHO + H_2O + NADP^+ \quad (4)$$

The other major intermediary, and a central one, is Plastoquinone (PQ), of which there are 10 to 14 molecules for every one of P_{680} or P_{700}. This key substance is a phenolic compound that is easily oxidized and reduced, much as shown for simple phenols in section F (d) in Chapter 6.

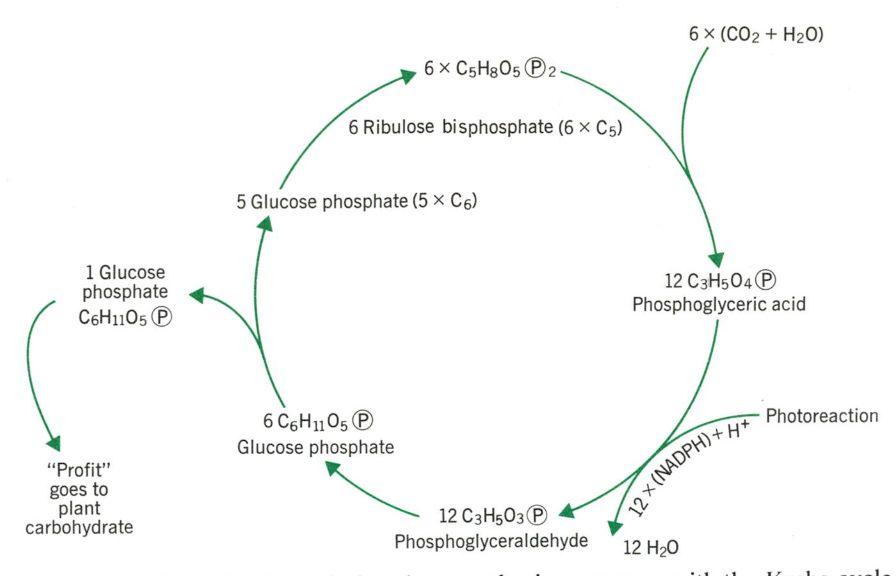

Figure 5-6 The Calvin cycle for photosynthesis compares with the Krebs cycle (Fig. 6-7) for respiration. Here the 5-carbon compound ribulose bis-phosphate combines with CO_2 and splits into two equal 3-carbon portions, and these undergo a series of changes to form the 6-carbon compound glucose-6-phosphate. Of these, five molecules go back into the cycle and one is deposited in the leaf as "profit." Ⓟ stands for the phosphate group.

b. PHOSPHORYLATION BY LIGHT

We said above that the phosphate group activates all the compounds taking part in the chemical sequence. Ribulose itself would not combine with CO_2, nor would glyceric acid be reduced to glyceraldehyde. How this phosphorylation occurs was shown by Daniel Arnon of the University of California Berkeley, when he discovered *photo-phosphorylation*. In this reaction, light absorbed by chlorophyll energizes an enzyme that combines free phosphate with adenosine diphosphate to make the highly reactive triphosphate, ATP. This then transfers phosphate to the constituents (written here as X) of the chemical sequence:

$$\begin{aligned} &\text{ADP} + \text{Ⓟ} \xrightarrow{\text{Light}} \text{ATP} \\ &\text{ATP} + \text{X} \longrightarrow \text{X Ⓟ} + \text{ADP} \end{aligned} \quad (5)$$

c. INTEGRATION

To put the whole process together we now return to the two light reactions mentioned above. Careful study of the thylakoids in the chloroplast shows that some parts (the inner surfaces of the stacks, see Fig. 2-5) are rich in photosystem I and others (mainly the outer surfaces) have system II. In the past it was thought that the two systems interact, but this separation in space requires that they can act independently. The latest scheme is that P_{680} takes the protons (H^+) from water, releasing oxygen, while the negatively charged hydroxyl ion restores the electron to P_{680}:

$$\begin{aligned} &H_2O \rightleftharpoons H^+ + OH^- \\ &OH^- + 2P_{680}^+ \rightarrow \tfrac{1}{2}O_2\ (\text{gas}) + 2P_{680} + H^+ \end{aligned} \quad (6)$$

The protons, together with other electrons from light-energized P_{680}, then cycle through plastoquinone (PQ in Fig. 5-7) alternately oxidizing and reducing it. This has the result of giving off the protons to the inner surface of the thylakoid. A special carrier (CF in Fig. 5-7) then brings them through to the outer side, causing a gain in energy sufficient to make ATP from ADP and phosphate. Once on the outside, these protons combine with electrons drawn from water, also by P_{680}, to reduce ferredoxin (Fd) and thence NADP as described above.

In system I, electrons from P_{700} cycle through ferredoxin and PQ as before but do not take protons from water; instead, the oxidation and reduction of PQ brings protons from outside the thylakoid to the inside. On their return through the thylakoid membrane, they make ATP as before. It appears important for energy exchange that ferredoxin and NADP are on the outside, where they are in contact with the chemical constituents of the process. Thus both systems offer separate paths for electrons and protons. In system II, electrons and protons, ultimately from water, reduce NADP and also make ATP, whereas system I electrons only make ATP. In the withdrawal of electrons from water, manganese and chloride are somehow involved.

Time will tell whether this concept explains all the intricacies of photosynthesis. But new understandings are developing steadily from a worldwide group of eager investigators.

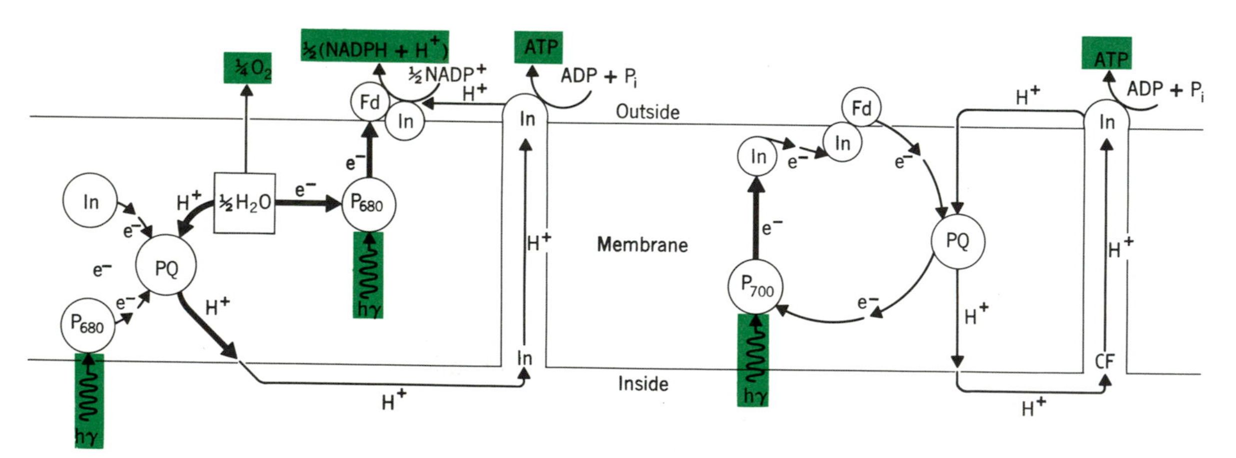

Figure 5-7 The latest integration of the light reactions of photosynthesis, with System 2 (oxygen-releasing) separated from System I on the thylakoid membrane.

At left, P_{680}, energized by light (Fig. 5-4) sends electrons via a receptor to plastoquinone, PQ, which also draws protons from water, releasing oxygen. In the reoxidation of PQ the electrons recycle, but the protons emerge on the inner side of the membrane. When the protons return to the outside, the energy gained causes synthesis of ATP. They then join with other electrons from P_{680} to reduce ferredoxin, Fd, thence NADP, and so to the chemical sequence.

At right, electrons from light-energized P_{700} go directly via acceptors to reduce ferredoxin, and recycle to P_{700} via PQ. Their reduction of PQ draws protons from the outside of the thylakoid, and releases them to the inner side; their return through the thylakoid again generates energy to form ATP. This part is "anoxygenic" and was earlier called the photophosphorylation cycle. (The compounds marked In are different intermediaries). (From Arnon, 1981).

8. A MODIFICATION: THE C_4 SYSTEM

In the leaves of some members of the grass family, Gramineae (or Poaceae) (Chapter 16) the chloroplasts are of two different types: those of the parenchyma, which are not very different from those illustrated in Chapter 2, and those of the larger cells surrounding the vascular bundles. These "bundle-sheath" cells, which in most plants contain no chloroplasts, do in plants of a special type contain chloroplasts that are even larger and more uniform in structure than in parenchyma cells (Fig. 5-8). These differences have suggested some difference in function. The difference turns out to be this: these leaves show unusually high activity of an enzyme that controls the reaction between oxalacetic acid, $C_4H_4O_5$, phosphopyruric acid, and CO_2:

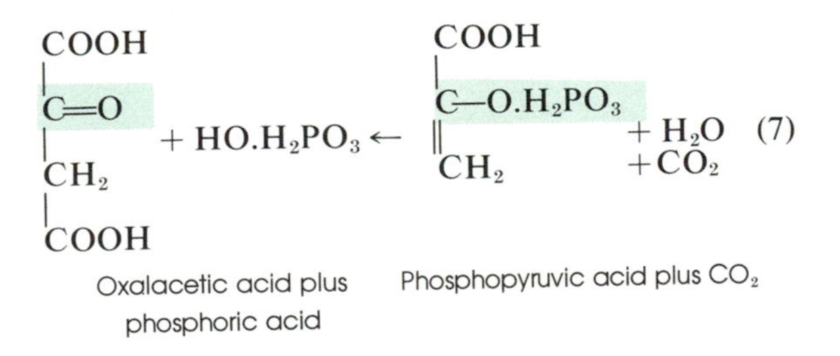

When this reaction goes from *right* to *left*, CO_2 is taken up by the 3-carbon acid and "fixed" into a carboxyl group, just as it is by ribulose bis-phosphate in the system described above. Reaction (7) for the uptake of CO_2 is the one mainly operative, for example, in the ordinary cells of the sugarcane leaf. The resulting acid, which contains four carbon atoms (one of which is that of the fixed CO_2), diffuses or is transported to the cells surrounding ("sheathing") the vascular bundles (Fig. 5-8). Here the oxalacetate is decarboxylated and the extra carboxyl group is given up to the bundle-sheath chloroplasts of special structure; it is then taken up by ribulose bis-phosphate as in the normal system and so converted to the C_3 compound phosphoglyceric acid, and thence to C_6 sugars.

This so-called "C_4 system" (because oxalacetic is a C_4 acid), is active in many tropical and semitropical grasses such as sugarcane, corn, and sorghum, as well as in some cacti, whereas the C_3 system operates in most grasses and other plants of temperate regions. Because of its frequent occurrence in the tropics, especially the drier regions, the C_4 type is believed to be a dual adaptation to (1) high temperatures, where the reaction between CO_2 and ribulose bisphosphate is known to be less rapid, and (2) low water supply, for C_4 plants tend to be more efficient in water usage. It is clear that most C_4 plants grow better in hot climates and out-yield their C_3 relatives there. Indeed the high productivity of corn and sugarcane had been recognized long before the C_4 variation was discovered. The process, however, does put heavy demands on the transport system, for the C_3 and C_4 acids have to go back and forth from cell to cell in the leaf. The C_3 plants do better on the whole in cool climates.

We see that photosynthesis in its entirety is an incredibly closely integrated series of individual chemical and physical processes. Of course we see it as it has now evolved; primitive forms may have been much simpler. Note that Nature tends to operate *by small steps;* this is just as evident in the light reactions as in the many chemical, enzymatic processes. While its evolution and refinement are certainly a triumph of Nature, we must also say that its extensive understanding, even to this imperfect point, is a triumph of our own persistence and ingenuity.

9. PHOTOSYNTHESIS UNDER NATURAL CONDITIONS

a. THE PRINCIPLE OF LIMITING FACTORS

It was F. F. Blackman in 1905 who put forward the principle that "when a process is

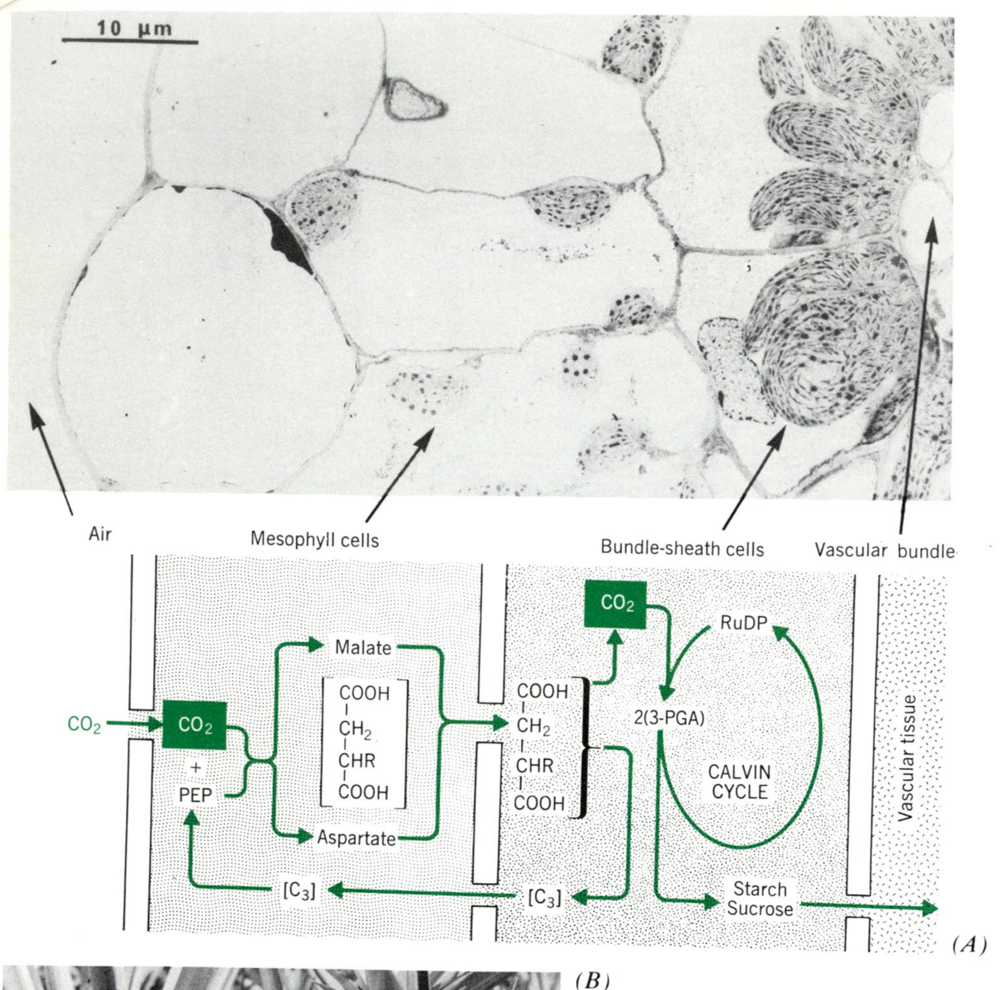

Figure 5-8 *A*. Photosynthesis by the C_4 pathway, in chemistry and in cellular location. The vascular bundle at the extreme right is surrounded by a sheath of cells with large chloroplasts that take in the 4-carbon compounds malate and aspartate and split off CO_2 to be used in normal photosynthesis. The mesophyll cells use their small chloroplasts to fix CO_2 into the COOH group of the malate or aspartate by combining it with phospho-pyruvate, PEP. (From Hatch, 1976). *B*. The pineapple, one of a group of modified C_4 plants ("CAM plants") that make the C_4 acid only at night and convert to the C_3 cycle in daytime (cf. Fig. 6-4).

conditioned as to its rapidity by a number of separate factors the rate of the overall process is limited by the pace of the slowest factor." The idea is relatively simple, and the same principle holds for fertilizers: in a soil deficient in iron the plants will be yellow, and this condition cannot be removed by supplying more phosphate or more magnesium, only by supplying more iron (see Chapter 8). Iron is "the limiting factor." In photosynthesis there are mainly three environmental limiting factors: light, carbon dioxide, and temperature.

When we follow the rate of photosynthesis at differing light intensities, we find that the rate increases steadily as the light intensity increases, till it reaches a maximum (*A* in Fig. 5-9*A*); at this point something else is limiting the response. If we now increase the supply of CO_2 at that same light intensity the rate will increase, to *B*. However, increase of CO_2 at very low light intensity (between 0 and *D*) will not help because the rate is limited by the light. At the higher CO_2 level, then, the relation between photosynthesis and light intensity will follow the upper line, fusing completely with the first curve at its lower end, and approaching light saturation at a higher light intensity *C*. Correspondingly, if we choose a given light intensity L_1 to start with, and follow photosynthetic rate at different CO_2 levels, we find the rate increasing with CO_2 to a level L_1 where light is limiting, and only by increasing the light to L_2 at this CO_2 level can the rate be increased (Fig. 5-9*B*). At still higher light intensities the curves L_3 and L_4 result. In each case the range of proportionality is extended by raising the second factor; the range over which photosynthetic rate is proportional to CO_2 is greater at high light intensity than at low.

Since photosynthesis as a whole involves many chemical reactions and these are sensitive to temperature, it follows that at high light intensities, where light is not limiting, temperature usually is. Thus the rate of photosynthesis often increases with increasing temperature in bright light. Some plants do not show much temperature dependence, wheat, for instance, giving very little change between 15° and 30°. Bright sunlight supplies warmth as well as light.

On sunny days the level of CO_2 close to the plants is often so depleted that supplying extra CO_2 will promote photosynthesis and hence growth. It is difficult and expensive to do this in the field, but in greenhouses, where the CO_2 will not simply escape into the vast reservoir of the air, and in the laboratory, moderate increases of CO_2 are often effective (see below). Many plants, including wheat, can increase their photosynthetic rate with increase in CO_2 from the normal 350 parts per million to 1000 or more parts per million. Those who worry that pollution increases the CO_2 content of the air may take comfort, at least in this respect, from the principle of limiting factors (see Chapter 14 for further discussion).

b. SUN AND SHADE PLANTS

Some plants grow best in full sun, others in shade. Some species are restricted to one or the

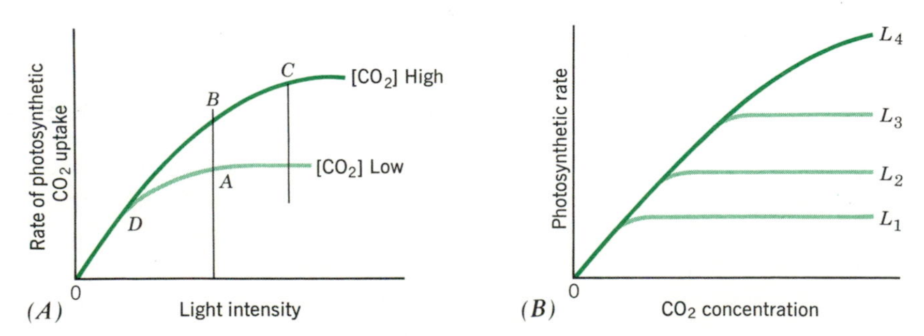

Figure 5-9 Schematic plots of photosynthesis against light intensity (A) and CO_2 concentration (B).

other contrasting light habitats, but sometimes individuals within species can become physiologically adapted to sun or shade. [Other examples of such ecological types (*ecotypes*) are discussed in Chapter 14]. Typically the leaves of shade plants are darker green and tend to be relatively large and thin. Their chlorophyll content is high and the ratio chlorophyll : carotenoid (which averages about 3) is higher than in sun plants. The lower light intensity obviously favors leaves in which every possible effort is made to catch what light there is. In some shade plants we also find mechanisms for bringing leaves out from under one another, by curvature of their petioles, so as to present maximum light-catching surface. An example is the "leaf mosaic" developed by ivy plants (Fig. 5-10). The leaves of shade plants often lose water too rapidly when in full sun and "burn" or lose chlorophyll; this happens because in taking in CO_2 they let water vapor out, as shown in Chapter 7. The leaves of sun plants, on the other hand, accustomed to tremendous light intensity, can "afford" to close off their gas exchange when water becomes deficient. The differences in response to light between these two types are exemplified in Fig. 5-11, which shows two ecotypes of one species, *Solidago virgaurea,* the European goldenrod. The leaves of the shade plant respond effectively to low intensities but are easily light-saturated, while those of the related sun plant are less efficient at low intensities but go on to photosynthesize at high rates in full sunlight.

Figure 5-10 Ivy leaves against a shaded wall gradually curve their petioles to bring the leaf blades out from under one another into full light. Thus they fit together into a mosaic that effectively covers the surface.

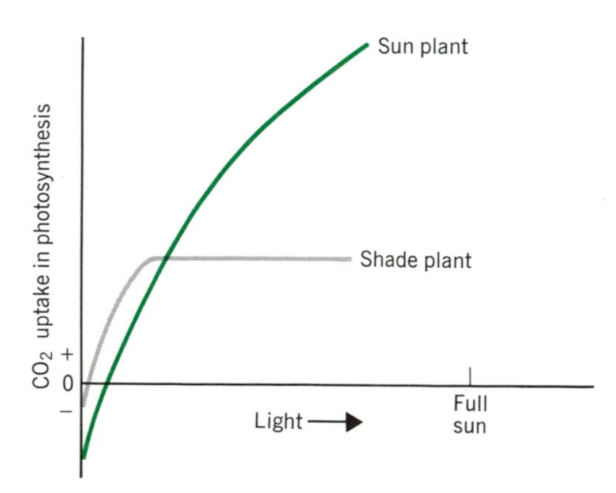

Figure 5-11 Photosynthetic rates of two types of goldenrod *(Solidago)* show how the shade plant responds to low light intensities but becomes saturated at medium intensities, while the sun plant, though slightly less effective at low light intensities, is able to increase its rate up to full sunlight. (Modified from Björkman and Holmgren, 1963)

There are several internal differences between the leaves of sun and shade plants. One is that the enzyme that combines ribulose bisphosphate with carbon dioxide is present at lower levels in the shade plants, and may even be a limiting factor there. In contrast to true shade plants, the genetic constitution of most sun plants enables them to adjust to, or "tolerate" the shade. Naturally the majority of agricultural plants are of the sun type, while herbs of the forest floor are usually of the shade type. In temperate woods they include ferns and mosses, and among the flowering plants are primroses, bluebells, hepaticas, violets, and anemones. In tropical rain forests some woody plants growing in the lower layers under the canopy are truly shade plants adjusted only to these light intensities (which are, in fact, the lowest known in terrestrial habitats). However tree seedlings are really sun plants that grow very slowly in the shade, but when openings occur they shoot up into the high light of the upper canopy of the forest.

c. THE ABSORPTION AND TRANSMISSION OF LIGHT BY LEAVES

The thickness and chlorophyll content of leaves vary widely, of course, not only with the species of plant but with the nutrition – nitrogen and iron being the most potent materials in controlling greenness. Sun and shade also cause marked variation. A rough guide is that an average leaf absorbs about 60% of the red and blue light falling on it. This means that a second leaf underneath it receives only 40% of the incident light and, since it absorbs 60% of that, a third leaf still lower down will receive only 16% of the incident light and a fourth leaf only 6.4%, which is the equivalent of heavy shade. Some plants indeed can survive with only one-tenth of this light intensity, but in most plants only four or five leaves, one above the other, will be likely to survive for long (Fig. 14-1). The floor of tropical forests with tightly closed canopy of leaves may have relatively few plants, despite favorable temperature and moisture (Fig. 14-2). Rapid growth of numerous plants in every opening only emphasizes the contrast. Similarly the lowest leaves on a tall plant, or the lowest branches on a pine or oak tree, gradually die out for lack of light.

The movement of the sun around the sky, and the perturbation of leaves by the wind, both help to provide some light to the lower levels and alleviate the strictness of the four-to-five-leaf rule above. In grasses, too, the leaves, or at least their more basal parts, are nearly vertical, so that light penetration is favored. The familiar stiff vertical leaves of the iris offer an even more marked example. Some plant breeders working on corn are aiming to develop varieties that hold their leaves at modified angles to the vertical, thus hoping to improve the grain yield by intercepting a little more sunlight.

d. EFFICIENCY OF UTILIZATION OF SUNLIGHT

The solar energy that is locked up in the plant body as leaves, fruits, stems and roots can be recovered and measured by burning the tissue to carbon dioxide and water and measuring the heat evolved. This is done in specially designed calorimeters, which are heavy, tightly sealed vessels immersed in water. The plant material, together with sufficient oxygen to burn it all, is inserted and then ignited electrically; the heat given out is measured by the rise in temperature of the water outside. The energy of the sunlight that made the plant can be measured by photometers, summed up over the whole growing period, and then compared with the amount of energy in the plant material. Thus we can see what percentage of the energy that the plant absorbed from the sun was actually used in photosynthesis.

However, during the whole growing season the plant was not only storing light energy, it was also using some up, by oxidation or *respiration*. This process keeps the plant alive in the dark, and it also goes on continuously in the light. It is essentially the reverse of photosynthesis, converting carbohydrates to water and carbon dioxide, and forming ATP in the process. The amount of photosynthetic energy that has been spent in this way can be estimated by measuring the plant's rate of respiration in the dark and again summing over the whole growing period. Respiration in the light can be neglected because most measurements are of the *net* photosynthesis.

The first attempt at such a balance sheet, made for a corn field in Ohio by E. N. Transeau (in 1926), ran like this:

Total dry weight of corn plants:	6000 kg
Less inorganic matter (ash):	300 kg
Organic matter (largely cellulose and starch):	5700 kg

Equivalent of this in glucose (see Chapter 6):	6700 kg
Organic matter lost as respiration during the season, expressed as glucose (estimated):	2000 kg
Total sugar formed (expressed as glucose):	8700 kg
Energy needed to synthesize this (from known heat of combustion at 3.8 cal/g):	33 million calories
Total solar energy falling on the area during the season (visible rays only):	2040 million calories

Percent of available energy used in synthesis:

$$\frac{33 \text{ million}}{2040 \text{ million}} \times 100 = \ldots\ldots\ldots\ldots\ 1.6\%$$

The figure of 1.6% seems surprisingly low. Other plants have given similar or even lower values, however. With sugarcane, which is not only a C_4 plant but also a perennial—and so presumably has to use less energy in root formation—the value lies between 2 and 3%, which evidently represents the best that can be expected. In a 1973 study on corn in Italy, extended over three growing seasons, values of 2.3, 2.1, and 2.4% were obtained. The corn was a modern high-yielding hybrid. These values were reached only when the full requirement of water was supplied, and they decreased towards 1.0% when the plants received only a quarter of their maximum requirements.

Why is the utilization so inefficient? In the first place the system itself has a certain amount of inevitable loss. Figure 5.3*B* shows that it takes 12 to 14 quanta of absorbed light to produce one O_2 molecule and from this one can calculate a *theoretical* efficiency of under 30%. Secondly, at least with an agricultural crop, the growing season starts with a period when the leaves do not cover the ground, so that some of the sunlight falls uselessly* on the earth. Even later on, there is a long period during which the necessary four leaf thicknesses (see section c page 97) are not yet present, so that some light passes through unutilized. Thirdly, part of the visible spectrum is not effective in photosynthesis, as we saw above. Also during cool weather the temperature is often below optimum and is thus a limiting factor in temperate climates. Lastly, in full sunlight the light is commonly *not the limiting factor,* for if seedlings are raised under artificial light of *very* low intensity, efficiencies of energy utilization of up to 19% can be obtained.

Correcting the observed 2% for all these factors gives us at least 11%, which comes a good deal nearer to the theoretical 30%.

Evidently the fraction of the solar energy converted to organic matter is small, and it will vary with the weather, the growth rate of the plants, and the specific reactions of their leaves to heat and water stress. The limitation by water stress is particularly critical; this is discussed in Chapter 7. In agriculture the yield varies also with the crop, and the particular cultivar of that crop that the farmer selects. He has to optimize the yield by the proper use of the soil, fertilizers, water, and the control of pests. Under natural conditions the energy yield depends somewhat on the plant community or ecosystem and is considered as "Primary Productivity" (see Chapter 14).

Even if the yield is low, it is all we have. Integrated over time and space, it represents the overall ability of the sun and the earth to support the animal kingdom, to provide for human needs, and to build the wealth of nations.

* Not quite uselessly, for it contributes heat to the earth. An additional part of sunlight is in the infrared, which supplies only heat—part to the earth and part to the atmosphere, forming winds and clouds. This part was not included in the above calculations.

CHAPTER 6

The Many Routes from Photosynthesis to All Other Plant Products

Since the primary products of photosynthesis are compounds like glucose phosphate (i.e., phosphorylated carbohydrates) we must now consider how such a limited group of compounds becomes translated into all compounds of which plants are composed. Much of the special area of nitrogen metabolism, that is, the formation and breakdown of proteins, nuclear components, and alkaloids, will be left to Chapters 9 and 26. We deal here first with the formation of the larger carbohydrates—sucrose, starches, and cellulose—then with the process of oxidation and its products, and lastly with how plants synthesize other materials very different from carbohydrates. There are an extraordinary array of organic compounds, of tremendously varied structure, making up virtually the entire biosphere. They are shown in Figure 6-6. Three principal types of change are involved.

1. THE MAIN SYNTHETIC PROCESSES

a. THE CHAIN OF EVENTS LEADING TO SUCROSE, A DISACCHARIDE

The plant body consists essentially of *polysaccharides,* multiple sugar molecules, linked together by loss of water. The simplest example of the linking process (and the commonest plant sugar) is a *disaccharide, sucrose,* in which two simple sugars, glucose and fructose, are linked together. Since both these simple sugars have the same overall formula, $C_6H_{12}O_6$ ("hexoses") differing only in their internal arrangement, the reaction can be written in its simplest terms as:

$$\underset{\text{Glucose}}{C_6H_{12}O_6} + \underset{\text{fructose}}{C_6H_{12}O_6} \rightarrow \underset{\text{sucrose}}{C_{12}H_{22}O_{11}} + \underset{\text{water}}{H_2O}$$

The actual process is more complicated, but the complexity is worth following, because so many plant processes follow similar patterns.

Before we get into that we shall make a little diversion to present the *nucleotides.* These crop up continually, not only in metabolism but in heredity, because they are the stuff of DNA and RNA. DNA contains the four bases adenine, guanine, cytosine, and thymine; RNA contains the first three plus uracil instead of thymine. In DNA each base is linked to the sugar deoxyribose, $C_5H_{10}O_4$. (DNA stands for deoxyribonucleic acid). RNA has the sugar ribose, $C_5H_{10}O_5$. The nucleotides also contain one to three phosphate groups each (see Chapter 11).

In glucose metabolism, uracil triphosphate, UTP, is the first nucleotide we meet. In it the base uracil, $C_4H_4N_2O_2$, is linked to ribose and three phosphate groups in line, through the loss of H_2O. Its structure is as follows:

$$\text{Uracil} - \text{Ribose} - (\text{P}) \sim (\text{P}) \sim (\text{P})$$

ATP is correspondingly

$$\text{Adenine} - \text{Ribose} - (\text{P}) \sim (\text{P}) \sim (\text{P})$$

It is the monophosphates of each of these (abbreviated UMP and AMP) that occur as constituents of ribonucleic acid, RNA. In the triphosphates above, the bond between ribose and phosphate is an ordinary chemical bond, but the bonds between the two additional phosphate groups are unusually reactive; these "high-energy" bonds are shown by the symbol ~.

The symbol (P) is used here and in other chapters for the combining form of the phosphoric acid radical. The acid itself is H_3PO_4 or

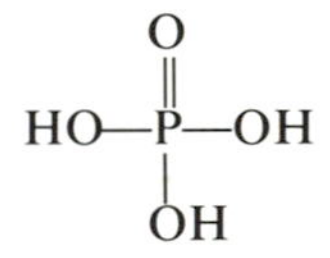

Thus each of the first two (P)s of UTP or ATP is

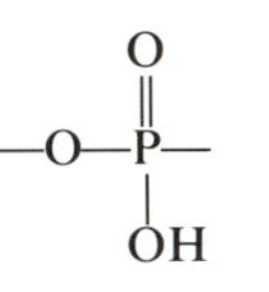

while the third is

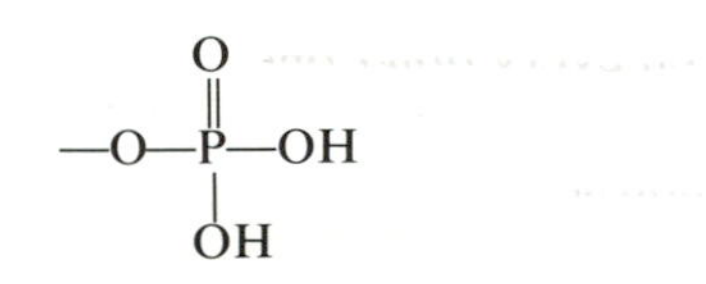

The other two bases (besides uracil and adenine) in RNA are cytosine and guanine. As to the names, note that uracil and cytosine ribosides are called *uridine* and *cytidine*, while adenine and guanine ribosides are *adenosine* and *guanosine*.

At the pH values of plant cells, phosphoric acid is not present as free acid but as ionized salts, organic phosphates. However, this dissociation is omitted here for simplicity, and the symbol (P) represents a more or less dissociated phosphate group. While DNA and RNA each contain four different bases as stated above, single types of nucleotides act as intermediaries in the synthesis of the di-, tri- and poly-saccharides. We will take up sucrose formation as the first example.

When UTP combines with glucose phosphate, each of the molecules loses one phosphate group, to form a diphosphate ion, pyrophosphate (derived from pyrophosphoric acid, $H_4P_2O_7$). In exchange, the glucose residue takes the place of the lost phosphate on the nucleotide (now uridine *di*phosphate) to form uridine diphosphate-glucose:

$$\text{UT(P)} + \text{G-1-(P)} \rightleftharpoons \text{UD(P)G} + \text{(P)—(P)}$$

In the resulting UDPG, the glucose molecule is in a very reactive state, and in presence of a suitable enzyme it leaves UDPG to form a new glucose compound. In this case the enzyme is *sucrose synthetase* and the new reaction is to combine with fructose to form sucrose:

$$\text{UD(P)G} + \text{Fr} \rightleftharpoons \text{UD(P)} + \text{G-Fr (sucrose)}$$

Although both of these reactions are reversible, their equilibrium position is in the direction of sucrose synthesis.

The fructose is formed from glucose by a

change in the bonding of the sugar ring, as shown (it takes place when these hexoses are in the form of phosphates):

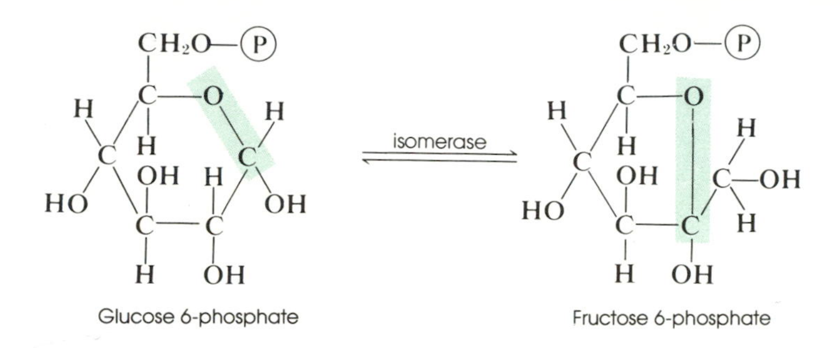

Glucose 6-phosphate   Fructose 6-phosphate

The fructose 6-phosphate then loses its phosphate group by hydrolysis (the hydrolytic enzyme is a *phosphatase*) and the free fructose then reacts with UDPG in the manner just shown.

The sugars in their phosphorylated forms are much more reactive than in the form of the ordinary sugars that we know. Corn sugar, which is a concentrated syrup of glucose, for instance, or honey, an equally concentrated mixture of glucose and fructose, can stand on the shelf for months, while in contrast, many sugar phosphates survive in the cells for only a few minutes.

Sucrose is not only stable in the plant, it is the principal form in which carbohydrates are carried from the photosynthetic area to other parts of the plant. In some plant families the transport sucrose is supplemented by larger molecules, containing three and four of the C_6 sugar molecules respectively, instead of the two that make up sucrose.

b. FORMATION OF STARCH AND CELLULOSE, POLYSACCHARIDES

Starch. Much of the photosynthetic product is deposited after a minute or two of light in the insoluble form as starch $(C_6H_{10}O_5)_n$. Starch, as noted in Chapter 5, is prominent in chloroplasts and is also deposited in colorless parts of plants such as seeds, roots, tubers, and fruits. Starch is formed by a pair of reactions comparable to those for sucrose, but mainly using ATP rather than UTP, which reacts more slowly. Unlike the sucrose synthesis, starch formation involves only one type of sugar, namely *glucose*, which again reacts as the phosphate, glucose-1-phosphate:

$$\text{ATP} + \text{G-1-}ⓟ \rightleftharpoons \text{ADP-G} + ⓟ - ⓟ$$

Then n molecules of ADPG react together, under the influence of an enzyme, to join up all the glucose residues:

$$n\text{ADPG} \rightleftharpoons n\text{ADP} + \text{G}n \text{ (starch)}$$

This second step takes place only in presence of a small amount of preformed starch called a "primer," in which the enzyme is tightly bound to the starch granules (see Chapter 21 supplement for more on the primer).

Starch makes an ideal storage material for plants because it is (a) very easily formed, since ATP is so available in photosynthetic cells, and (b) equally easily converted back to glucose. This reconversion is a double hydrolysis, that is, first the insertion of a molecule of water between every *other* one of the 6-carbon residues in starch to produce many molecules of the 12-carbon sugar maltose, followed by the insertion of a further water molecule between the two 6-

carbon halves of maltose to release glucose. Two enzymes participate, β-amylase in the first step and maltase in the second:

$$(C_6H_{10}O_5)_n + \tfrac{1}{2}nH_2O \xrightarrow{\beta\text{-amylase}} \tfrac{1}{2}\, nC_{12}H_{22}O_{11} \text{ (maltose)}$$

$$C_{12}H_{22}O_{11} + H_2O \xrightarrow{\text{maltase}} 2\ C_6H_{12}O_6 \text{ (glucose)}$$

In germinating seeds the first reaction tends to go faster than the second, so that maltose accumulates, hence the term "malt" for the germinating barley used in making beer (see Chapter 20).

Since starch occurs so widely as a deposit material in plants, it is the major source of carbohydrate food for human beings and most animals. While β-amylase produces maltose, another enzyme, α-amylase, converts starch to dextrin, a viscous material used in one form as the gum on postage stamps. Amylase enzymes are found in human saliva, so that starch foods can be quickly converted to sugar and thus, via oxidation or fermentation, yield supplies of ATP for synthetic chemical reactions. Among the many products thus made are the fats, and it is for this reason that "slimming" diets are always low in starch and sugar.

Cellulose and Its Relatives. We saw above that sucrose is formed by the reaction of glucose phosphate with UTP, and starch by the corresponding reaction with ATP (also more slowly with UTP). A third very similar reaction uses another nucleotide, guanosine triphosphate (GTP), to make different chains of glucose molecules called *cellulose*. These are long chains, characteristically deposited as microfibrils of indefinite length, making up the fibrous structures that are so characteristic of higher plants (Fig. 21-1). The formula for cellulose is the same as that of starch, $(C_6H_{10}O_5)_n$, where n is indefinite and in the hundreds or even thousands, while the value of n for starch is usually under 300. However, the manner in which the glucose units are linked in cellulose differs a little from that of starch, and its long polysaccharide fibrils are strong and tough (see supplement to Chapter 21).

Cellulose makes up around one half of the structural material of most higher plants, so that it is by far the commonest organic compound in the world. In many grasses, however, cellulose is partly replaced by a similar fibrous polymer, made up of chains of the 5-carbon sugar xylose, called xylan. The 5-carbon sugar arabinose furnishes similar chains called araban, and the two are often found combined together. Both occur in the cell walls. In woody plants there is a third cell wall constituent of quite different chemical nature, namely lignin; lignin impregnates the spaces between cellulose fibrils so that the whole wall of the xylem cells of trees is a complex of interpenetrating materials.

The algae have slightly different polysaccharides for their cell walls; some are based on mannose instead of glucose, and because they also contain acid groups they are less fibrous and more gelatinous, particularly in the brown and red algae.

2. THE BREAKDOWN PROCESSES

a. GLYCOLYSIS: THE CHEMICAL UNITY BEHIND DIFFERING BIOLOGICAL PROCESSES

The step in photosynthesis in which two 3-carbon molecules (phosphoglyceraldehyde) condense to a 6-carbon sugar phosphate is really a reversible reaction:

$$2\ C_3H_5O_3\text{-}Ⓟ \rightleftharpoons C_6H_{10}O_6\text{-}Ⓟ_2$$

where Ⓟ stands for the phosphate group as

before. In photosynthesis the reaction goes from left to right, but in other parts of the plant it may go from right to left, and because the C_6 compound is thus split into two halves the process is called *glycolysis* (literally, sugar-splitting). The enzyme responsible is *aldolase*.

Glycolysis is the first step in the oxidation of sugars to carbon dioxide and water, for glycolysis leads by a series of steps to pyruvic acid, $CH_3CO.COOH$, which is then oxidized to carbon dioxide and water, along with the production of some ATP. Furthermore, pyruvic acid combines with the very reactive coenzyme A (see section 6 below) to form acetyl-coenzyme A, which by different condensations yields a wide variety of plant products.

To form pyruvic acid, $C_3H_4O_3$, from a 3-carbon sugar, which is $C_3H_6O_3$, involves a loss of two hydrogen atoms (section 5 below); these are normally oxidized to water. But if no oxygen is present the hydrogen atoms must be accepted by another substance. In other words, for every molecule of pyruvic acid formed, some other molecule must be hydrogenated, i.e., reduced. This is the basis of *fermentation*, which will be taken up in Chapter 20. Thus the $C_3 \rightleftharpoons C_6$ interaction lies near the heart of photosynthesis, of respiration, and of fermentation, all three.

b. OXIDATION: THE BASIS OF RESPIRATION

We saw that in photosynthesis electrons are provided from chlorophyll and replaced from water, the oxygen is released, and the protons of water are thus available to combine with other compounds. The protons and electrons are first accepted by NADP and then released to react with phosphoglyceric acid to reduce it to its aldehyde:

$$RCOOH + 2H^+ + 2e \rightarrow RCHO + H_2O$$

Oxidation is in many ways the opposite of these reactions, for in oxidation electrons and protons are removed from the *organic compounds*, and they react with free oxygen to *form* water.

The reactions in oxidation, as in nearly all biological processes, go step by step, $2H^+ + 2e$ being removed each time, because each step requires a different hydrogen-removing enzyme, a *dehydrogenase*. The last step, combination with oxygen, however, almost always uses the *same* enzyme system, that of the *cytochromes*. We take up the two types of the reaction—dehydrogenation and oxidation—in order.

Dehydrogenation. In this process the organic compound becomes attached to the dehydrogenase enzyme that is specifically adapted to it, and two of its adjacent H atoms transfer themselves to the enzyme. The two electrons that had been bonding the H atoms to C atoms of the organic compound come together to form an additional pair, resulting in a *double bond*. The enzyme then, usually within a fraction of a second, loses the 2 hydrogen atoms to an acceptor, usually NAD or NADP. For example, an aldehyde, R-CHO, combined with a molecule of water, that is, as $R\text{-}CH(OH)_2$, can be dehydrogenated to form an acid:

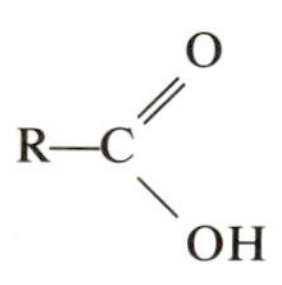

Acetaldehyde forms acetic acid:

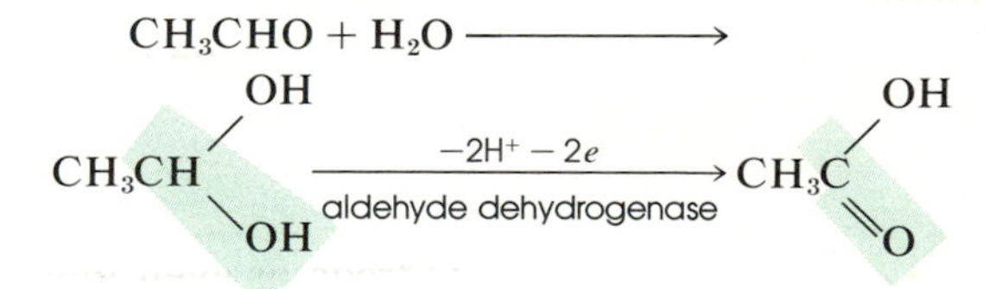

In the special case where the aldehyde happens to be phosphoglyceraldehyde, Ⓟ—$OCH_2 \cdot CHOH \cdot CHO$, we are seeing the exact reverse of the reduction process that occurs in photosynthesis (section 5 of Chapter 5), and as expected the product is phosphoglyceric acid, Ⓟ—$OCH_2 \cdot CHOH \cdot COOH$.

As soon as the two protons, with two electrons, are accepted by the NAD or NADP:

$$2H^+ + NAD^+ + 2e \rightarrow NADH + H^+$$

the dehydrogenase is freed from its hydrogen atoms and can pick up two more.

The hydrogen atoms, as protons, now pass from NADH or NADPH through a small series of acceptors, and thence to the four cytochromes, a group of iron-containing compounds related to the red pigment, hemoglobin, of blood. Through this group they ultimately meet and combine with oxygen, using several intermediaries, as shown here:

$$H^+ + e \rightarrow \text{NAD} \rightarrow \text{flavoprotein} \rightarrow \text{coenzyme Q} \rightarrow \underset{b, c, a, a_3}{\text{Cytochromes}} \rightarrow O_2$$

In the chain of carriers of protons and electrons from metabolizable organic compounds of the plant to the oxygen of the air, each stage becomes reduced and then reduces its next in line, passing on the ($H^+ + e$) until the electrons react with the cytochromes and the protons join them to convert O_2 to H_2O.

Combination with Oxygen: The Cytochromes. The name *cyto-chrome* means *cell-color*. They are red-brown compounds found in almost all cells, plant, animal and microbe. Their structures are basically comparable to that of chlorophyll (see Fig. 6-1). Their special function revolves around the iron atom each of them contains, for the two states of iron in combination, namely ferric and ferrous, differ by one electron:

$$Fe^{3+} + e \rightleftharpoons Fe^{2+}$$

Solutions of simple ferrous salts are easily oxidized to the ferric state, as shown by adding a drop of hydrogen peroxide to a green solution of ferrous sulfate; it turns brown because it has been oxidized to ferric sulfate. Similarly solutions of ferric salts are fairly easily reduced to ferrous, but the presence of the organic part of the cytochrome molecule makes these changes take place much more rapidly. Cytochromes can be oxidized and reduced many times in a second. Most organisms contain cytochromes of four types; three are designated (with not much originality!) *a, b,* and *c* (Fig. 6-2), whereas the fourth, a_3, is cytochrome *oxidase*.

When the dehydrogenase starts the protons and electrons on their passage through the train of acceptors, beginning with NAD or NADP, it is the electrons that finally pass to the cytochrome system. The essence of the role of cytochromes is for each of them to take up an electron by change of the valence of the iron atom from ferric to ferrous, then to pass this electron on to another cytochrome; thus the first goes back to the ferric state. The process is repeated at each stage. Cytochrome *b* is reduced first, then passes the electron to cytochrome *c*, which passes it on to cytochrome *a*. Cytochrome *a* is closely associated with *cytochrome oxidase* (cytochrome a_3), and the oxidase has a special property that the others do not possess; it can pass its electron to *oxygen*. When an oxygen atom has received two electrons in this way, it can combine with two protons and form water; the protons come from the dehydrogenase chain mentioned above:

$$2(\text{Cyt.Ox.})Fe^{2+} \rightarrow 2(\text{Cyt.Ox.})Fe^{3+} + 2e$$

$$2e + \tfrac{1}{2}O_2 + H^+ \rightarrow OH^-$$

$$OH^- + H^+ \rightarrow H_2O$$

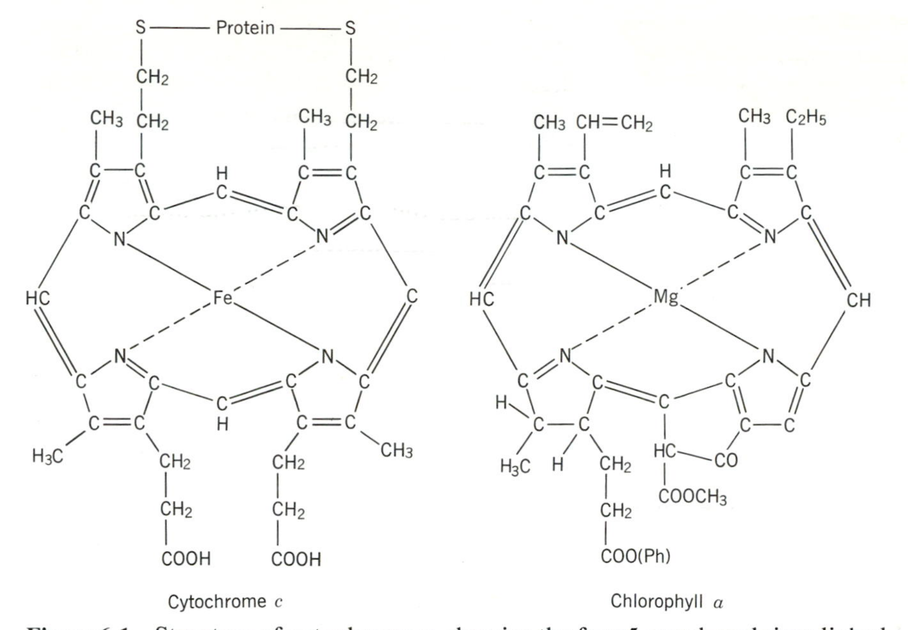

Figure 6-1 Structure of cytochrome c, showing the four 5-membered rings linked through single carbon atoms, much as in chlorophyll, but with different side chains and, most important, iron instead of magnesium at the center. Also in chlorophyll one of the acid side chains is esterified with the long-chain alcohol phytol (shown as Ph). Hemoglobin (see Chapter 9) is very like cytochrome c, but instead of being linked to protein by the two S atoms, hemoglobin is linked through the iron atom.

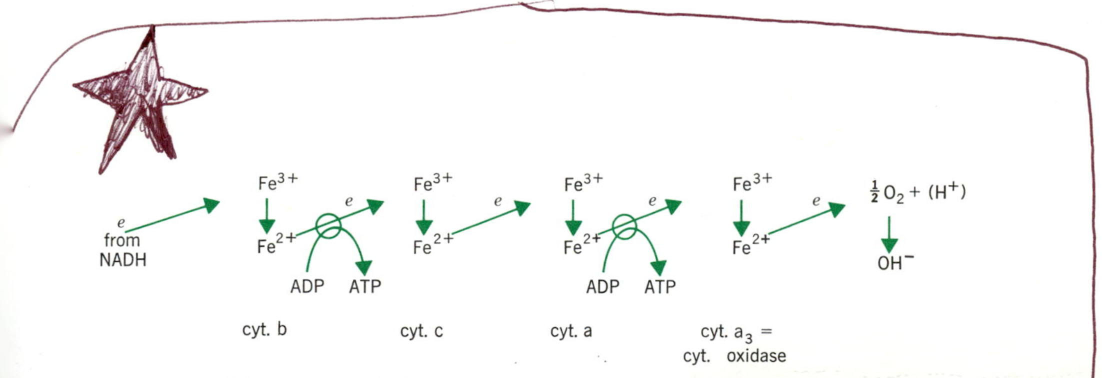

Figure 6-2 The passage of electrons through the cytochrome system to oxygen. At two of the transfers, at least, ATP is formed. (ATP is also formed at one of the electron transfers before the electron enters the cytochrome chain, i.e., to the left of the diagram, but this is used up in the working of the chain). After each ferric atom (in the cytochromes) has been reduced by the electron to the ferrous state, it gives off its electron to the next in line and so reverts to ferric. This happens hundreds of times a second. This chain of events produces nearly all the energy for animals, fungi, and the colorless parts of plants.

Thus the cytochrome system is a chain to accept electrons from the train of coenzyme acceptors and pass them to oxygen (Fig. 6-2).

The Significance of Respiration. Basically the above is the process of *respiration.* All aerobic organisms respire (i.e., they consume oxygen) and (with very few exceptions) all use the cytochrome system, though the actual cytochrome structures differ slightly from one kind of organism to another. The universality of the system raises the questions: Why do organisms respire, and why do they all use nearly the same system? The answer to the second question is that very few molecules have the oxidase property, that is, can pass electrons to oxygen. Evidently this is a property which Nature has been able to invent only rarely. As to the first question, in passing electrons along to one another, the cytochromes also form ATP from ADP and phosphate (Fig. 6-2). Here we see a striking parallel to photosynthesis, for in that system, too, ATP is formed where the electrons pass through cytochromes. Thus respiration in colorless cells carries out one of the functions of photosynthesis in green cells in that it is the route whereby ATP is formed, and since ATP is used in so many biological syntheses, respiration is therefore the source of chemical energy.

Organisms that do not possess cytochromes can make small amounts of ATP by fermentation (Chapter 20); these are *anaerobes* and they are relatively few in number or types. They do not include any green plants, and naturally their modes of life are somewhat limited.

It is still not clearly understood just how the change of iron valency in cytochromes activates ADP to make it combine with phosphate to form ATP.

There is another striking parallel to photosynthesis. Just as the chlorophylls are held in a special membrane-enclosed organelle, the chloroplast, so also the cytochromes are held in another membrane-enclosed organelle, the *mitochondrion.* Like the chloroplast it has an outer double membrane, and its interior contains a folded mass of membranes (Fig. 6-3).

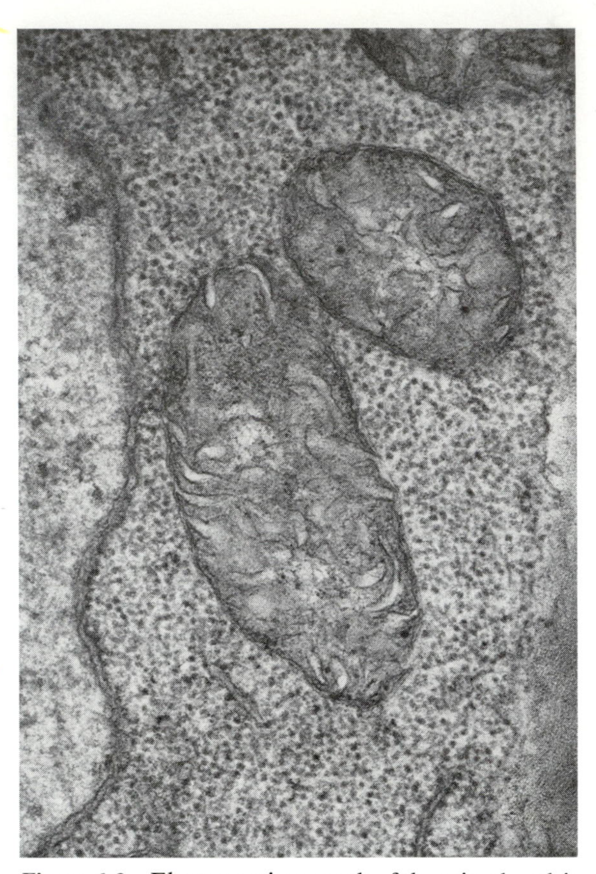

Figure 6-3 Electron micrograph of the mitochondria in a root cell of bean *(Phaseolus vulgaris).* The outer membrane of each is double; the folded membranes inside are single. The cytoplasm surrounding them is packed with ribosomes (dots). On the left is the nucleus, enclosed in its double membrane.

However, the appearance of these is quite unlike that of the thylakoids of the chloroplast. They are apparently formed by infolding of the inner membrane to produce a large internal surface, along which the cytochromes lie in close contact with one another, so as to be able to pass electrons along freely. At the highest magnifications of the electron microscope one can make out tiny outpocketings that may be where the oxidizing particles are held. Isolated mitochondria suspended in a suitable medium can oxidize many organic compounds about as fast as the tissue from which they were prepared.

Mitochondria are about the size of some bacteria, and for this reason it was suggested, as noted in Chapter 2, that mitochondria evolved from parasitic bacteria that became enclosed in the cell of a primitive organism, much as an amoeba ingests bacteria. Mitochondria indeed do contain DNA and some ribosomes. The idea is attractive as helping to explain the course of cellular evolution, but as yet it is hard to prove. For instance, no bacteria are known now that look like mitochondria.

c. THE LOSS OF CARBON DIOXIDE IN THE OXIDATION PROCESS

The characteristic sign of human respiration is the production of water vapor; Albert Szent-Gyorgyi said in 1938, "The cell knows but one fuel; hydrogen." But when hydrogen is successively abstracted from organic compounds to form water, the compounds become first converted to organic acids, and then a further dehydrogenation introduces a double bond adjacent to the COOH group, and that makes the molecule unstable, so that it loses CO_2. For example:

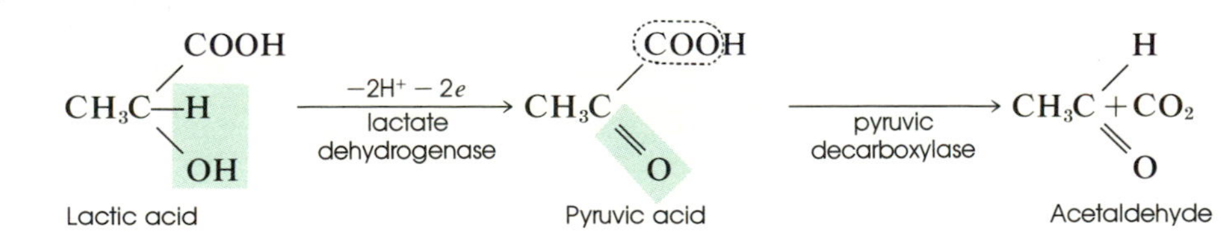

When CO_2 comes off in respiration it always arises as a result of such dehydrogenation, because the double bond adjacent to the COOH group tends to withdraw electrons from the COOH and thus sets the CO_2 free. An oxidizable organic compound in successive steps loses hydrogen ($2H^+ + 2e$), then its acid group becomes unstable and loses CO_2; then it loses more hydrogen, and so on until nothing is left but CO_2 and water. In this process, which is described in the Supplement at the end of this chapter, pyruvate reacts with coenzyme A and the resulting acetyl-coA undergoes a remarkable cycle of oxidations, yielding many molecules of ATP on its path to CO_2 and H_2O.

3. FEW MAIN ROUTES: MANY DESTINATIONS

Although plants contain a bewildering array of different substances, they are mostly formed via a few main pathways. We have just seen that one major group, the polysaccharides, are formed generally through intermediate compounds of sugars with nucleotides, which are purine or pyrimidine bases combined with ribose phosphate. We shall here follow five main pathways in plants which lead to other major products that are not carbohydrates.

a. THE ORGANIC ACIDS

Figure 6-7 shows that several organic acids are produced in the normal course of oxidation; of these, malic and citric acids are found in relatively large amounts in many plants and plant parts, especially in fruits. Malic acid derives its name from the classical generic name of the apple, *Malus,* and citric acid from the genus of the orange and lemon, *Citrus*. (Apples are now usually treated as a species in the genus *Pyrus*).

There are two additional routes to these acids, which account for their accumulation in larger amounts than would be expected of an intermediate in the oxidation cycle. The first of these is by the direct addition of CO_2 to phosphoenolpyruvic acid, as in the C_4 variant of photosynthesis. The second route is by a modified version of the Krebs cycle, whereby isocitric acid, instead of losing two molecules of CO_2 (reactions 4 through 7 of Fig. 6-7), is split directly into succinic acid, $C_4H_6O_4$, and the 2-carbon acid *glyoxylic acid,* $C_2H_2O_3$, or

$$\begin{array}{l} \text{CHO} \\ | \\ \text{COOH} \end{array}$$

This 2-carbon acid then condenses with a molecule of acetyl coenzyme A to form the 4-carbon molecule malic acid. In this way an extra molecule of malic acid (and of its metabolic successors) is produced, and the acids can accumulate. By this route oranges and lemons (and some cultivars of pears) become rich in citric acid, and apples, pears, and grapes in malic acid.

The formation and accumulation of acids is very marked in the leaves of succulent plants, such as the *Crassulaceae* (stonecrops), *Cactaceae,* and the pineapple. In these plants, the stomata (Chapter 7) are largely closed by day and open at night; CO_2 is thus taken in, in the dark, and malic acid is accumulated from the reaction between CO_2 and phosphoenolpyruvic acid. By morning the leaves have become very acid to the taste. At first light, however, the CO_2 begins to be given off and in a few hours most of it has entered into ordinary C_3 photosynthesis (Fig. 6-4). This system, called Crassulacean Acid Metabolism (CAM), enables such plants to survive arid conditions, minimizing water loss under the hot sun. The thickness of the leaves also helps to decrease water loss. The process can be likened to that in the C_4 plants (Chapter 5) but there the CO_2 fixation is separated from the C_3 system by a few cells, that is, *in space,* whereas in the Crassulacean acid metabolism the CO_2 fixation is separated from the C_3 system *in time*.

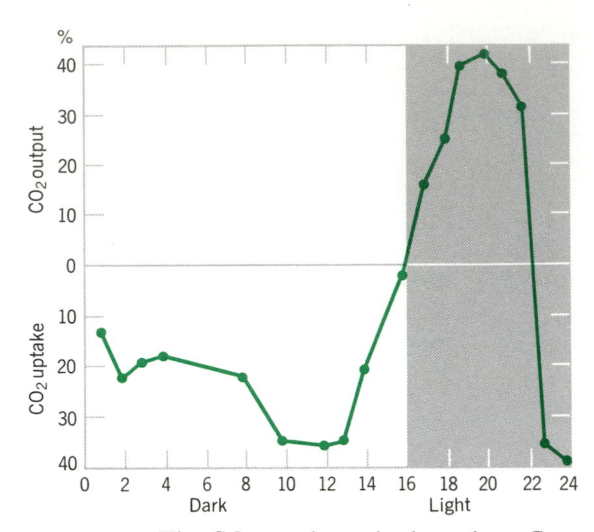

Figure 6-4 The CO_2 uptake and release by a Crassulacean plant *(Kalanchoe blossfeldiana)* growing on cycles of 8 hours light – 16 hours darkness. During the dark period, CO_2 is taken up but reaches saturation just before the end; when the lights come on some CO_2 is evolved, but then photosynthesis takes over and there is a net CO_2 uptake again. The values plotted are percent of the CO_2 that came into the vessel. The Kalanchoe is seen in Figure 10-14. (Data from Gregory, Spear, and Thimann, 1956.)

b. THE LIPIDS

The fats and oils, which are technically called lipids or lipides, appear as droplets in many cells, especially in chloroplasts, and are the main storage materials of fatty seeds like those of cotton, sesame, castorbean, and some palms (see Chapter 27). The related *phospholipids* are a prominent part of cell membranes. These compounds are formed from acetyl coenzyme A by successive condensations and hydrogenations; the coenzyme A is set free again and recycled continuously. The final products are usually acids with 16 or 18 carbon atoms, for example, $C_{17}H_{35}COOH$, stearic acid.

Besides occurring in small amounts as free acids, these acids combine with glycerol, $C_3H_8O_3$, to form fats (Chapter 27). Since glycerol has three OH groups the typical fat has three of the long C_{15} or C_{17} chains.

If one of the OH groups of glycerol bears a phosphate group, only two of the fatty acid molecules can combine, giving a *phospholipid*.

c. THE TERPENOIDS

Molecules of acetyl coenzyme A can condense in another pattern, which leads to branched carbon chains instead of straight ones. Three molecules of acetyl coenzyme A condense to form mevalonic acid (shown in Fig. 26-2). Because the branched structure is that of the hydrocarbon *isoprene*

$$CH_2{=}CH{-}C(CH_3){=}CH_2$$

and they are of composition C_5H_8, these derivatives are called *isoprenoids* or *terpenoids*. They include the essential oils (Chapter 26), the "juvenile hormones" of insects (which keep specific insects from molting into the adult form), the carotenoids and the related vitamin A ($C_{20}H_{27}OH$), the gibberellins (potent plant hormones—Chapter 10), and the steroids (from which the animal sex hormones are derived). Other polymers form the plant resins and rubber discussed in Chapter 26.

d. PHENOLIC COMPOUNDS AND THE FORMATION OF THE BENZENE RING

Quite different from any of the above are the derivatives of benzene, C_6H_6. This peculiar hydrocarbon owes its stability to the fact that six carbon atoms just fit together nicely to make a stable ring. The six hydrogen atoms lie outside. Of the four valences of each carbon in such a ring, two could join to other carbon atoms and the other two to hydrogens; this would make a "saturated" ring, C_6H_{12}. However, in the benzene ring there are only six hydrogen atoms and as a result the spare valence electrons are considered to be mobile double bonds, or perhaps to form a sort of cloud above and below the plane of the ring. The formula could be written in either of these two ways:

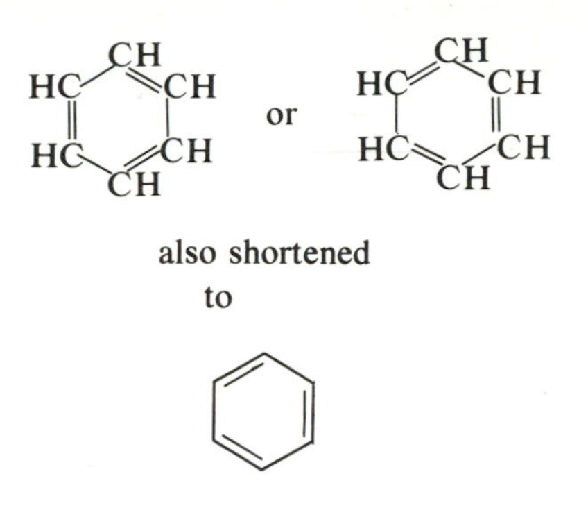

for the bonds are not really alternately double and single but rather are mobile and are symmetrically divided between the carbon atoms. If one of the H atoms is replaced by a hydroxyl group, the compound is C_6H_5OH, which is phenol, while the acid C_6H_5COOH is benzoic acid. The number of benzene derivatives (called *aromatic* compounds, though some of them smell anything but sweet) that has been found in plants is legion; it includes simple phenols and phenolic acids, (e.g., salicylic acid, whose acetyl derivative is the well-known aspirin), the pigments of flowers, three of the amino acids of proteins, many dyes and drugs, and the structural material of woody cells — lignin.

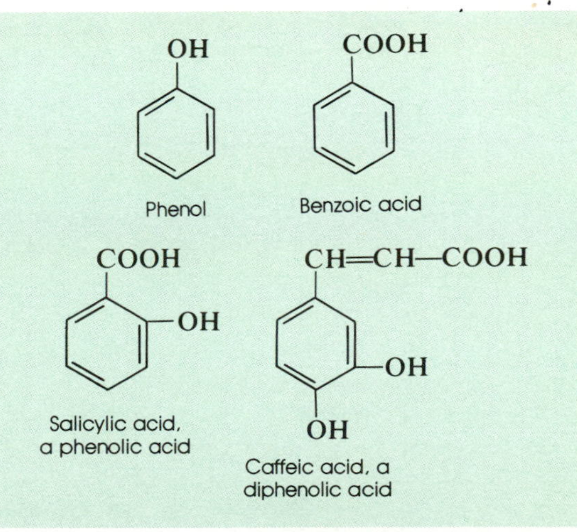

Many of these aromatic compounds contain a chain of three carbon atoms linked to the benzene ring, as in caffeic acid (above), which has two of the phenolic —OH groups. This C_6—C_3 pattern appears in many plant products. When the —COOH of the C_3 chain is adjacent to one of the —OH groups, they usually lose a molecule of water between them and form a second ring, using the oxygen atom as a link:

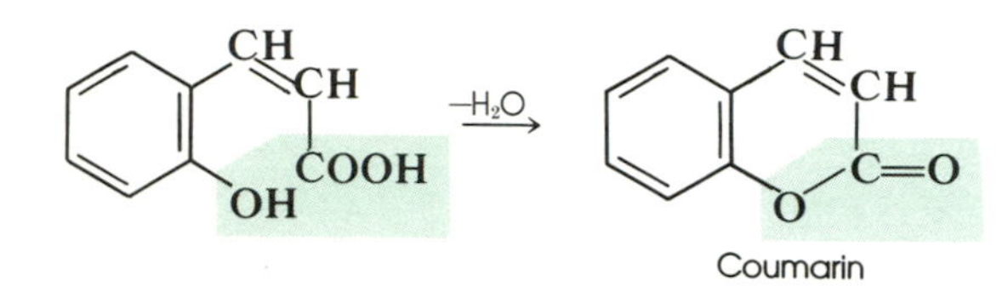

Coumarin

Now we have a double-ring compound and there are many variations on this theme too. The one shown, coumarin, gives the sweet smell to new-mown hay.

When an additional molecule of phenol is attached to the second ring, a family of three-ring compounds results, and this includes most of the red, yellow, blue, and purple pigments of flowers, and the reds of autumn leaves. The yellow ones are *flavonols* and the others are *anthocyanidins*. Typical structures are shown below:

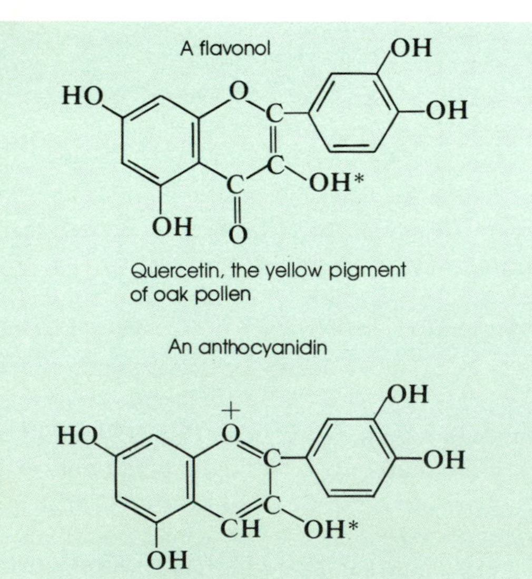

Quercetin, the yellow pigment of oak pollen

Cyanidin, the blue pigment of "Bachelor's Buttons"; also the red pigment of dark red roses

Most of these compounds (as they occur in the plant) have the —OH that is marked with a star (above) substituted with a molecule of sugar, sometimes glucose, sometimes another. This makes them soluble in water, and for this reason some of them have been extracted and used for dyes. But the majority are readily bleached by oxidation, especially in light. Fustic, from the wood of *Rhus cotinus,* the European smoke tree (of the family Anacardiaceae), a relative of poison oak and poison ivy or from the tropical tree *Chlorophora tinctora* (Moraceae, the fig family), is one of the few that are fairly stable; it is a dull yellow. Yellow-brown onions, yellow flowers, red berries, and the like have all been used since classical times, but their dyes all fade in sunlight. The stable plant dyes, like indigo (deep blue) and madder (rose-red), belong to quite other classes of substances.

Most of the simple plant phenols are readily converted in air to diphenols (i.e., containing a second —OH group) by the common plant enzyme *polyphenol oxidase*. The reaction goes in two stages:

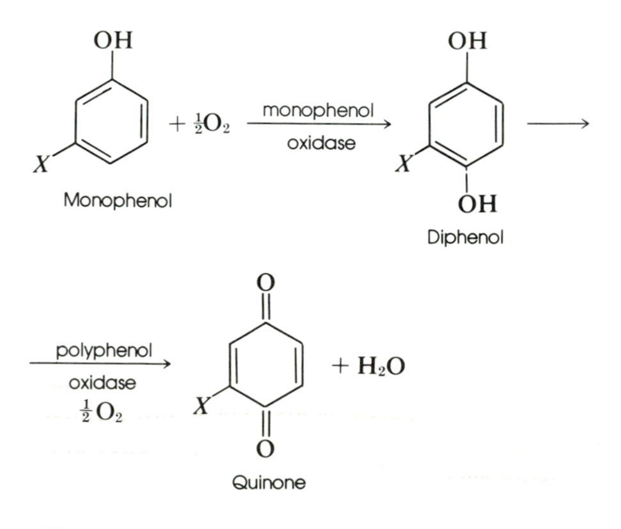

(*X* may be some additional substituent group.) The importance of this reaction is that the resulting quinones rapidly combine with each other to form brownish or blackish *polymers.* Freshly sliced apples soon turn brown with this reaction; potatoes even turn black. Keeping the

slices under water prevents the oxidation. Cooks add lemon juice, which contains the strong *reducing* agent ascorbic acid; this keeps the phenols reduced.

Some of the diphenols and triphenols occur as polymers combined with glucose to form *tannins*. These have played two important parts in civilization, furnishing ink and acting in the production of leather (Chapter 22).

Phenolic substances with a 3-carbon side chain, but this time with an alcohol (—CH_2OH) instead of an acid (—COOH), combine together into brown polymers called *lignin*. The composition of these differs in different plants. The complications in the making of paper pulp from wood, due to the presence of lignin, are described in section 4e of Chapter 22.

As noted in Chapter 4, lignin impregnates the cell walls of xylem, giving them rigidity both by its multiple molecular linkings and by the added mechanical strength. Because the cross-linkages make a sort of network instead of fibers, it complements the force of cellulose very well. It also causes the brown color of wood. But although lignin constitutes 40% of the weight of the world's forests, no one has found a major use for it other than the wood itself. Thus the six-membered ring constitutes an important chemical basis for plant life.

e. AMINO ACIDS

Lastly, in the Krebs cycle (Fig. 6-7), three of the intermediate keto-acids, namely pyruvic, ketoglutaric, and oxalacetic acids, are shown as giving rise to the bracketed compounds alanine, glutamic and aspartic acids. These are formed by combination with ammonia, in presence of a reducing agent and a specific enzyme:

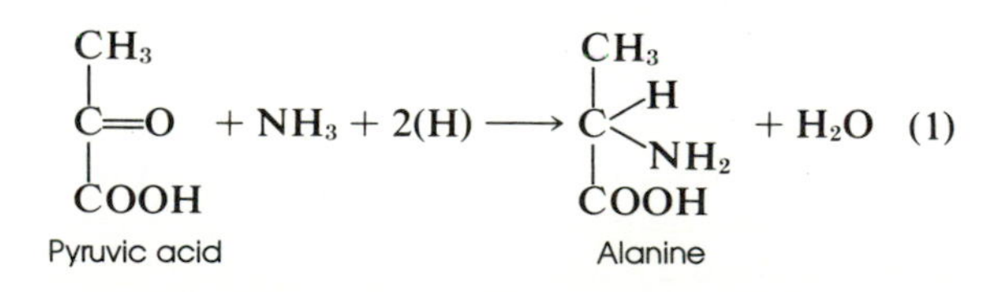

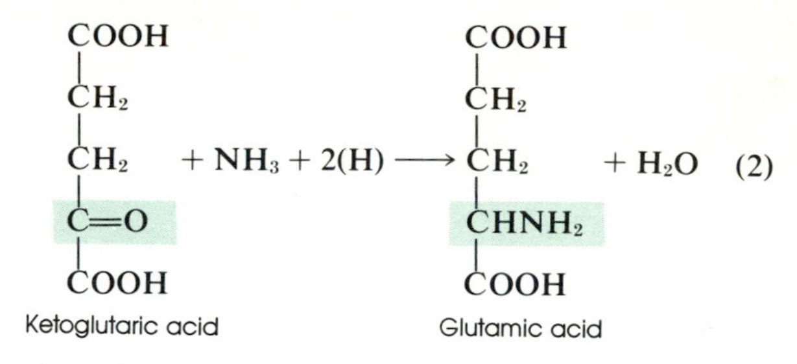

Reaction 2 has an intermediate step not presented here.

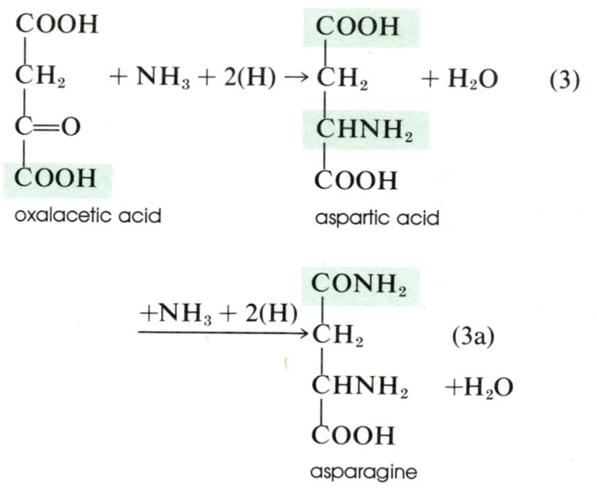

The product of reaction 3 is rapidly converted in many plants, especially in legume seedlings, into its amide, asparagine, as shown. This compound accumulates in seedlings grown in darkness, and even crystallizes out in extracts made from them.

Of the 30 amino acids found in plants, 20 are regularly combined into proteins (see Chapter 9). To form the other 17 amino acids from the initial three, their amino groups are transferred to the corresponding keto-acids, RCO.COOH. The dozen or so amino acids that do not become converted into proteins are irregularly distributed in plants, some being common whereas others are so far known from only one or two species.

Figure 6-5 puts together all these varied syntheses to give a general picture of the synthetic life of the plant. The left-hand wheel is driven by light energy and the right-hand wheel

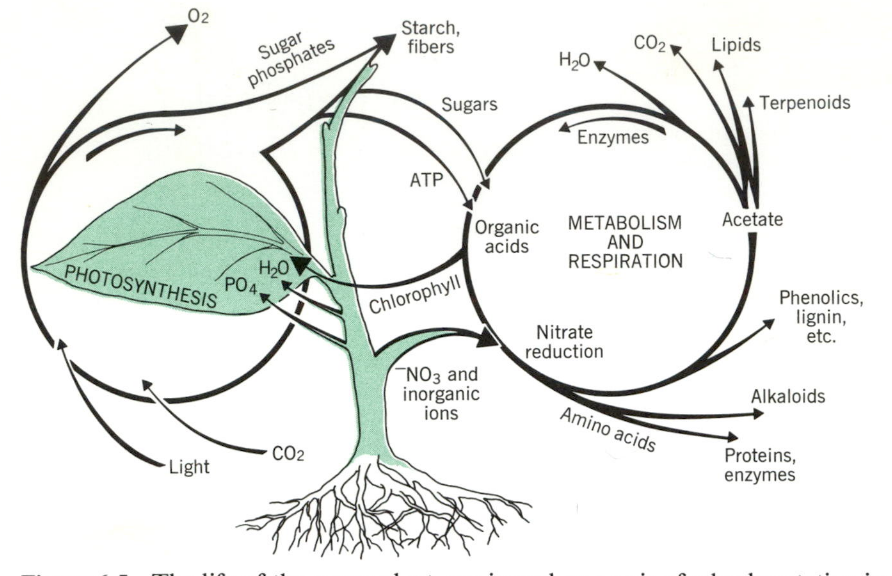

Figure 6-5 The life of the green plant, envisaged as a pair of wheels rotating in opposite directions. Photosynthesis, along with phosphate from the roots, produces sugar phosphates, of which some are deposited and some metabolized in the second wheel; nitrogen is contributed there from the roots.

is essentially run off it. Figure 6-6 shows in more detail the large number of plant materials and the routes of their formation. Many of these are treated further in the following chapters.

4. THE INTERACTION BETWEEN RESPIRATION AND PHOTOSYNTHESIS

As we saw in section 4 above, plants are respiring (that is, oxidizing) continuously, both in green and in colorless parts. This respiration naturally uses up some of the photosynthetic products; in the colorless cells the oxygen consumption must supply the needed ATP. Fortunately, at normal temperatures, the rate of respiration, as measured in the dark, is so much smaller than the rate of photosynthesis that there is a large net accumulation of organic matter. Since in respiration carbohydrates (sugars) are used up and carbon dioxide and water are the ultimate products, the overall equation is exactly the reverse of that for photosynthesis, namely:

$$C_6H_{12}O_6 + 6O_2 \rightarrow 6CO_2 + 6H_2O$$

The ratio CO_2/O_2 is called the *respiratory quotient*. If the above equation is precisely followed it would be 1.0, and values close to this are commonly found. Correspondingly the exchange of gases in photosynthesis, that is, the ratio O_2/CO_2, is also close to 1.0. However, in photosynthesis at low CO_2 levels, or at very high oxygen concentrations, the latter ratio, called the *photosynthetic quotient*, falls to around 0.5 or less. In other words, the oxygen produced is somehow being *withheld*. This observation suggests a close relationship between photosynthesis and respiration, which has been studied in recent years by means of isotopes.

Using the isotope $^{18}O_2$ in the air with ordinary oxygen, $^{16}O_2$, in the CO_2 supplied, it has been found that in some plants oxygen uptake in the light is higher than in the dark, and it even increases still further in high light intensities.

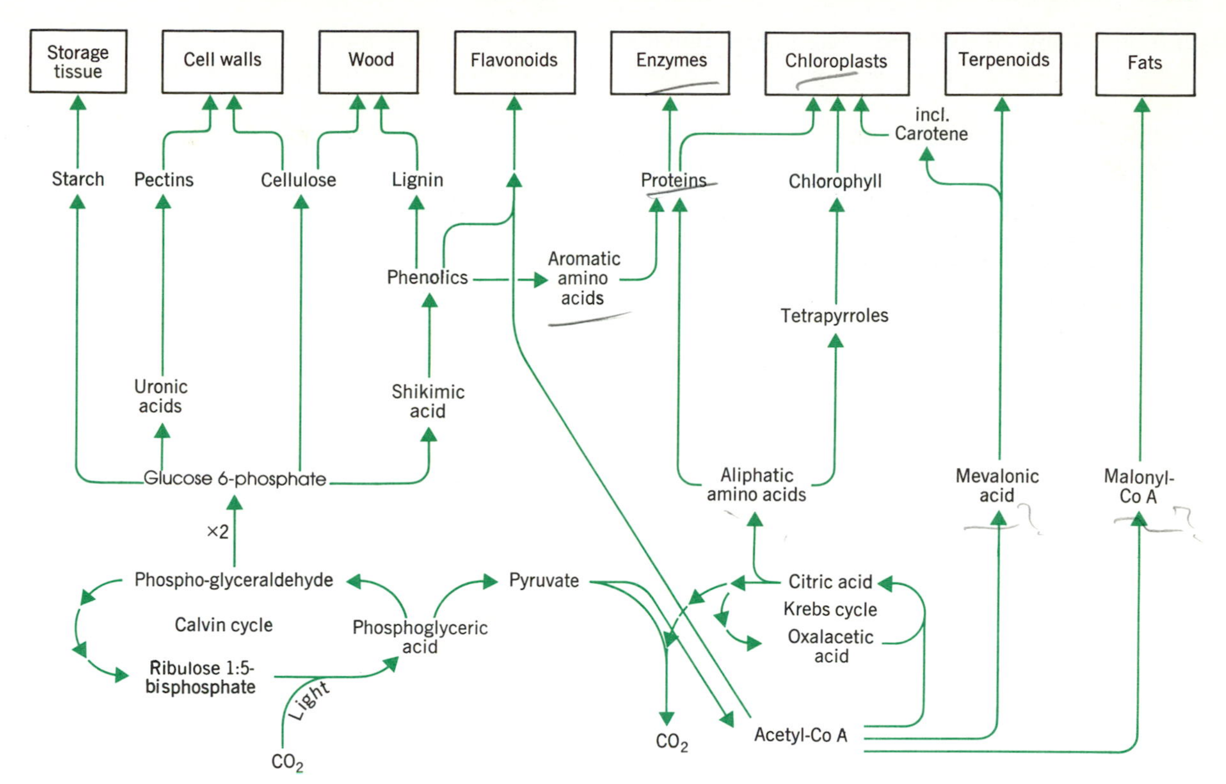

Figure 6-6 The main classes of plant materials and the routes of their formation. The two wheels of Figure 6-5 are discernible at the bottom of this figure.

Also, if the oxygen level is higher than that in ambient air, or the CO_2 concentration is lower, the oxygen uptake increases still more. The extra CO_2 that results from this oxidation above the level of that in the dark is called the *photorespiration.* The oxygen that is apparently being withheld in the light, and used in photorespiration, is producing organic acids instead of carbohydrates. The acids contain more oxygen than carbohydrates; for example, glycolic acid, which is one of the main products, has the composition $C_2H_4O_3$, whereas in carbohydrate the elements would have the ratio $C_2H_4O_2$. Glycolic acid can be further oxidized to glyoxylic acid, oxalic acid, and CO_2, while some becomes converted to the amino acids glycine and serine:

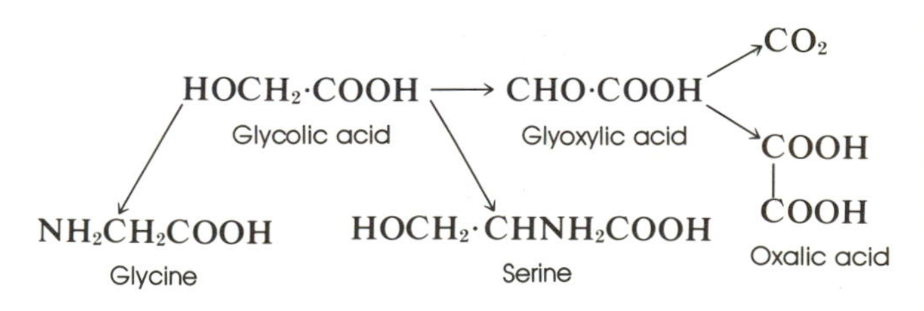

The CO_2 is partly recycled into photosynthesis. The rate of oxidation of these acids continues high after light is shut off, and for a few minutes after transfer to darkness an "outburst" of CO_2 therefore occurs.

All plants do not show photorespiration. One way to determine simply whether they do or not is to place the plants in light with their leaves in CO_2-free air, since if no CO_2 is supplied, any CO_2 that is detected must have been produced by respiration. Such experiments show that a large group of plants produce CO_2 in light, all at about the same rate as each other, while others do not produce any at all. Table 6-1 shows that maize and one out of two of the *Atriplex* (saltbush, a desert genus of the Chenopodiaceae) species studied give off no CO_2 under these conditions, while the two other plants produce more CO_2 in light than in darkness. The rates of respiration in the dark were all about the same. Because some of the CO_2 is recycled into photosynthesis, the rates of photorespiration are doubtless even higher for the last two plants in the table than those shown. In soybean leaves it has been found that the rate of photorespiration can be 30% of the maximum rate of photosynthesis.

These results bear on the calculated efficiencies of photosynthesis, for plants with low or zero photorespiration will obviously be more efficient accumulators of organic matter than others. Fortunately the simple balance sheet of Transeau, which was reproduced in Chapter 5, was based on corn, *Zea mays,* a C_4 plant, which has minimal photorespiration. Thus those data, though rough by modern standards, are not far off. Indeed, because plants with the C_4 system (like corn) have low rates of photorespiration, they are highly efficient at the warmer temperatures where they thrive.

One last point: the journey of carbon dioxide and water through the green plant begins with photosynthesis, proceeds via the formation of sugar, starch, and cellulose to bacterial decomposition of the dead tissue—or perhaps via use as animal food and conversion to glycogen or fat, or use in respiration in the animal—but in any case finally ends at carbon dioxide and water again (see section 7a of Chapter 14). It has been calculated that *on the average* a carbon dioxide molecule is thus cycled about every 300 years, and a water molecule every 3 million years. As exceptions, carbon atoms may be laid down as the wood of a redwood tree and go out of circulation for 2000 years, or if laid down as coal they may be out of circulation for 300 million years (see the chart shown in endpapers). The average, therefore, covers a very wide range.

TABLE 6-1

Photorespiration and Dark Respiration of Four Plants in CO_2-Free Air (The light intensity was 2000 to 3000 footcandles; temperature, 25° C. The measurements are expressed in milligrams of CO_2 set free per square decimeter of leaf surface per hour.)

Plant	*Dark respiration*	*Photorespiration*
Zea mays (corn)	3.0	0.0
Atriplex spongiosa	3.3	0.0
Atriplex hastata	3.0	4.2
Glycine max (soy bean)	3.0	6.0

Figures collected by Zelitch (1971).

SUPPLEMENT

Oxidation via the Krebs Cycle

We start with the 3-carbon compound $C_3H_5O_3$-P, a triose-phosphate, phosphoglyceraldehyde, the product of photosynthesis. First it undergoes dehydrogenation to phosphoglyceric acid, a step that is the reverse of what happens in photosynthesis. Phosphoglyceric acid then undergoes a migration of the phosphate group from the third carbon atom to the second, and then loses a molecule of water; this introduces a double bond and thus makes the phosphate group (like CO_2 in section 5) very unstable, so that it is quickly given off to a waiting molecule of ADP:

$$\underset{\text{3-Phosphoglyceric acid}}{\begin{array}{l}CH_2O\text{—(P)}\\ |\\ CHOH\\ |\\ COOH\end{array}} \rightleftarrows \underset{\text{2-Phosphoglyceric acid}}{\begin{array}{l}CH_2OH\\ |\\ CHO\text{—(P)}\\ |\\ COOH\end{array}} \xrightarrow{-H_2O} \underset{\text{Phosphopyruvic acid}}{\begin{array}{l}CH_2\\ \|\\ C\text{-}O \sim P\\ |\\ COOH\end{array}} + H_2O \xrightarrow{ADP} \underset{\text{Pyruvic acid}}{\begin{array}{l}CH_3\\ |\\ C{=}O\\ |\\ COOH\end{array}} + ATP$$

Hence in this step we see the formation of ATP without photosynthesis, the needed energy coming from the increasing energy of the phosphate group as the organic part of the molecule is successively modified. The passage of electrons to oxygen through cytochrome is one purely chemical way of making ATP, and the process described here adds another less productive way.

In the absence of oxygen, pyruvic acid becomes converted to alcohol in the process called *fermentation* (Chapter 20). But in the presence of oxygen, pyruvic acid is oxidized to CO_2 and water by the cycle of reactions shown in Fig. 6-5. Because this cycle was worked out largely by Hans Krebs and his colleagues at Oxford, it is known as the Krebs cycle. The 10 steps, numbered in Figure 6-7, are:

1. Displacement of the carboxyl group by coenzyme A, to form acetyl coenzyme A, or $CH_3CO{\cdot}CoA$. (The formula for coenzyme A is too complex to be presented here.)
2. Condensation of acetyl coenzyme A with the 4-carbon acid oxalacetic acid to form the 6-carbon acid citric acid, $C_6H_8O_7$. Coenzyme A is set free, and recycles at once.
3. Isomerization of citric to isocitric acid, with the same overall composition.
4. Dehydrogenation of this to oxalosuccinic acid, $C_6H_6O_7$.
5. Loss of one carboxyl group as CO_2, to form α-keto-glutaric acid, $C_5H_6O_5$.
6. Displacement of a carboxyl group by coenzyme A, releasing CO_2, in a reaction exactly similar to that of reaction 1 above.
7. Release of coenzyme A to form succinic acid.
8. Dehydrogenation to fumaric acid.
9. Addition of water to fumaric acid to make malic acid.
10. Dehydrogenation of malic to oxalacetic acid, thus completing the cycle.

In each of reactions 1, 5, and 6 a molecule of CO_2 is released, so that in all the whole molecule of pyruvic acid is used up. In each of reactions 1, 4, 6, 8, and 10 two hydrogen atoms are removed, and in each of reactions 2, 7, and 9 a molecule of water is added. Thus the total reaction in one "turn" of the Krebs cycle is:

$$C_3H_4O_3 + 3H_2O \rightarrow 3CO_2 + 10(H)$$

The hydrogen atoms are of course not liberated but are oxidized through the chain of cytochromes to water, $10(H) + 2\frac{1}{2}\ O_2$ making $5H_2O$. The summation of these reactions is thus

$$C_3H_4O_3 + 2\tfrac{1}{2}\ O_2 \rightarrow 3CO_2 + 2H_2O$$

In addition to the ATP that is formed in the cytochrome system, a molecule of ATP is formed in each "turn" of the cycle. All in all, the oxidation of 1 molecule of glucose, $C_6H_{12}O_6$, gives rise to over 30 molecules of ATP.

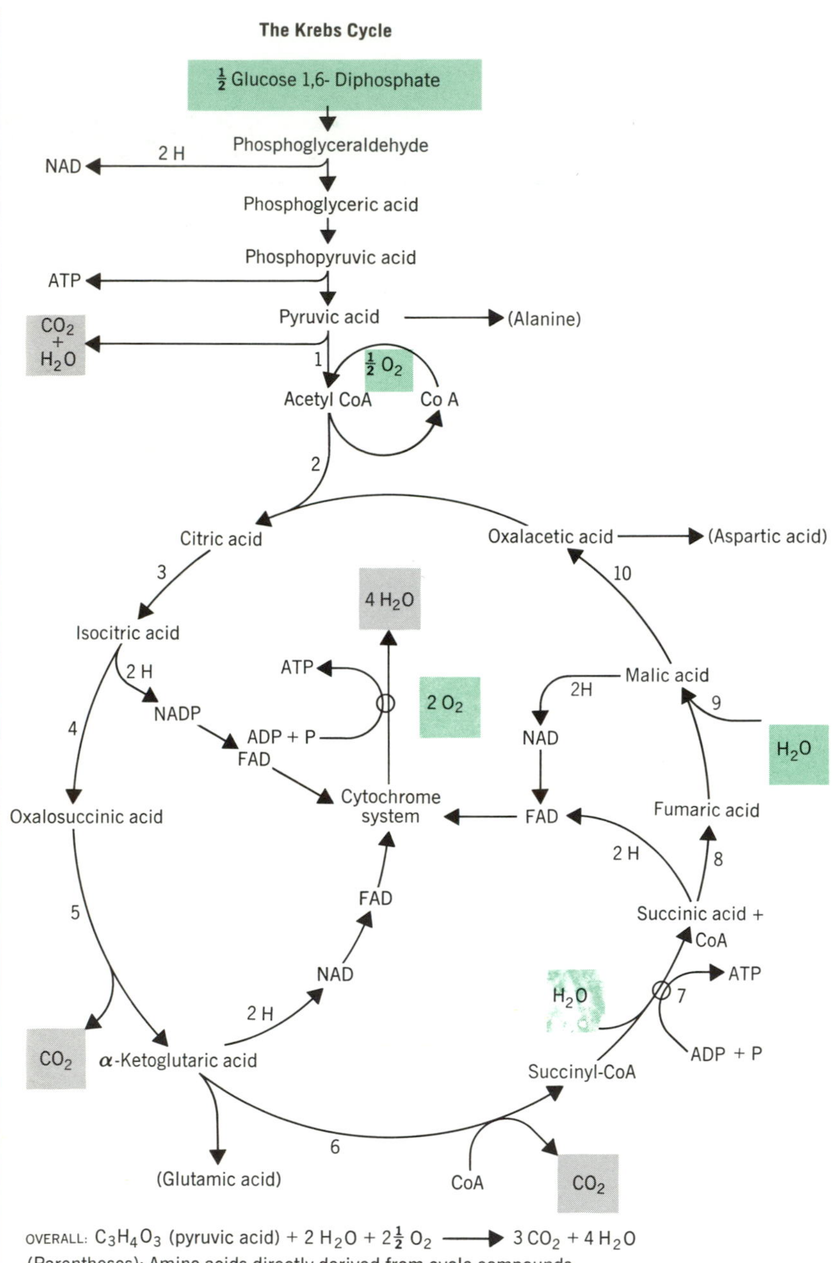

OVERALL: $C_3H_4O_3$ (pyruvic acid) + 2 H_2O + $2\frac{1}{2}$ O_2 ⟶ 3 CO_2 + 4 H_2O

(Parentheses): Amino acids directly derived from cycle compounds.

Coupling between oxidation and ATP formation shown as: ○

Input: Output:

Figure 6-7 The Krebs cycle, in which pyruvic acid is oxidized to CO_2 and water, yielding about 18 molecules of ATP (2 for each electron and proton – Fig. 6-2), plus 1 from phosphopyruvate and 1 from succinyl-CoA. The three amino acids in parentheses are those directly derived from the cycle; others are formed from them by exchange of the amino groups, as shown in section 6e.

CHAPTER 7

Water in the Life of the Plant

Since plants originally developed in water (Chapter 2), it is not remarkable that water continues to play a major part in their economy. Algae and other aquatic plants are, predictably, composed largely of water, and as long ago as 1804 de Saussure determined that terrestrial plants contained 80 to 90% water. Many seedlings, especially those grown in darkness, contain as much as 92% water; thus only 8% of their weight is dry matter. Animals generally have a much higher content of dry matter, although there are some remarkable exceptions among marine animals, jellyfish going as low as 0.2% dry matter or 99.8% water. Terrestrial animals, by being mobile, have cut themselves off from the source of water (frogs and toads return to it, of course, for part of their life cycle), while terrestrial plants, being rooted in the moist soil, live with a continuous supply of water flowing through them – in at the roots, out from the shoot. Some terrestrial plants – the mosses and liverworts – are generally restricted to moist places, for their "roots" (*rhizoids*) are not well developed, and the fusion of their gametes requires water. At the other extreme, those trees that can grow in semiarid conditions, like some of the oaks or acacias, have roots going down 8 to 10 meters and spreading out to beyond the area of the furthest branches.

1. FUNDAMENTALS

a. BASIC WATER RELATIONS OF CELLS

If we carefully flow a layer of water over a layer of heavy syrup, so that they do not mix, and put the vessel aside for a time, we shall find that, even though the layers are still clearly separate, the water layer contains a lot of the sugar and the syrup layer has been made thinner by dilution with water. The reason is that the sugar, like all substances in solution, has tended to diffuse from a higher to a lower concentration; all substances, given the opportunity, tend to equalize their distribution.* But not only does the sugar tend to equalize its distribution – so does the water. This may seem an odd aspect of water relations. For the pure water is, so to say, highly concentrated water, while the water in the syrup is diluted with sugar. So the water diffuses from a higher to a lower concentration of *water*. In other words, it flows into the

---

* In more precise terms, **entropy**, which essentially measures the disordiliness or randomness of a system, steadily tends toward a maximum. An equalized distribution of sugar is thus more **random** than a strong and a weak solution in contact. To lower entropy requires the expenditure of energy.

sugar solution and thus tends to equalize the distribution of water.

Inside the cell wall and appressed closely to it (see Fig. 7-1) is the cell membrane, which is composed in part of phospholipids, in which most water-soluble substances do not dissolve (Fig. 7-2). As a result, though it lets water through, along with some of the smallest molecules in solution — such as simple salts or urea, CON_2H_4, — it is not permeable to most larger molecules like sugars or peptides. The small molecules are able to pass through by way of the

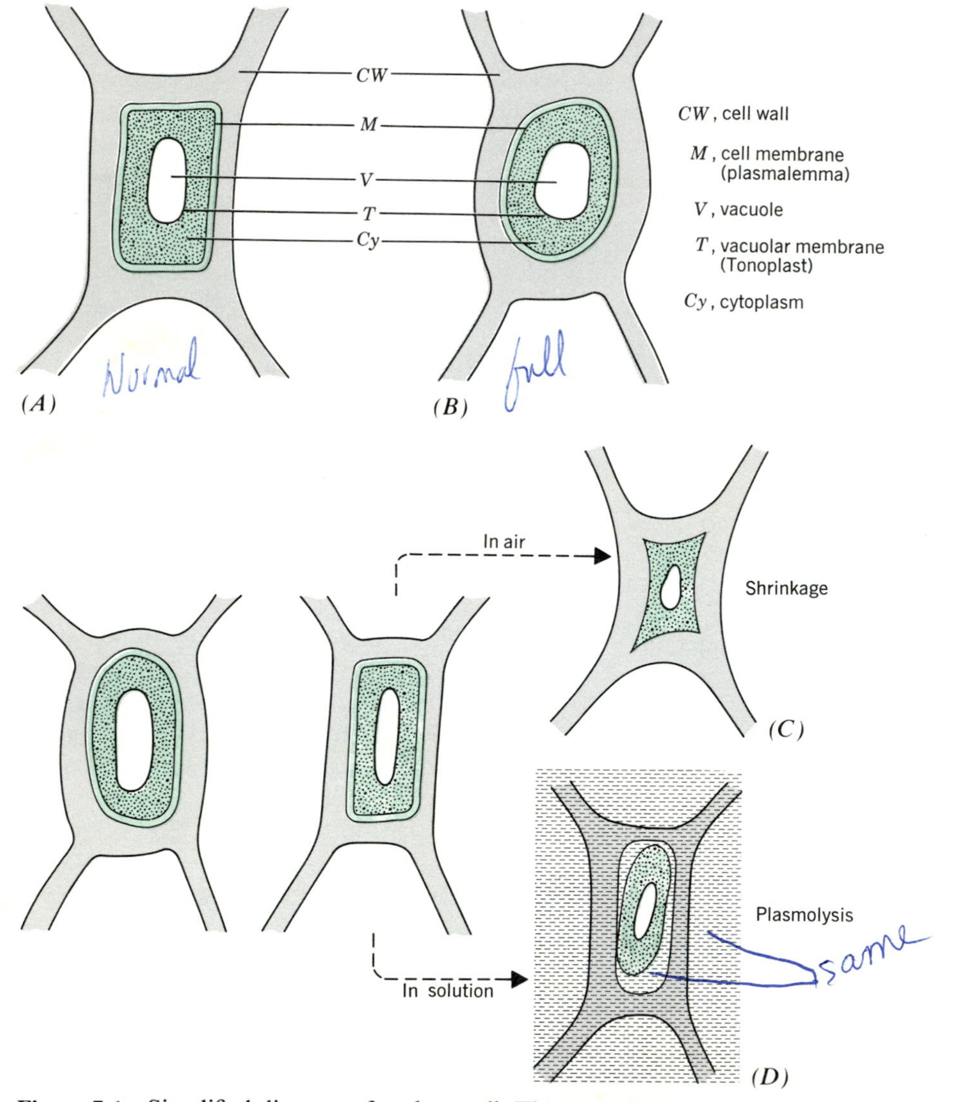

Figure 7-1 Simplified diagram of a plant cell. The space between wall and plasmalemma is only for clarity; actually they are in contact. Above, *A*. Normal state. *B*. At full turgor. Below, as the turgid cell at left loses water it first contracts to zero turgor, as in the center. In air it shrinks further as at *C*. (wilting); in a solution the solution enters and the cell contents shrink away from the wall as at *D*. (plasmolysis), described in section 2.

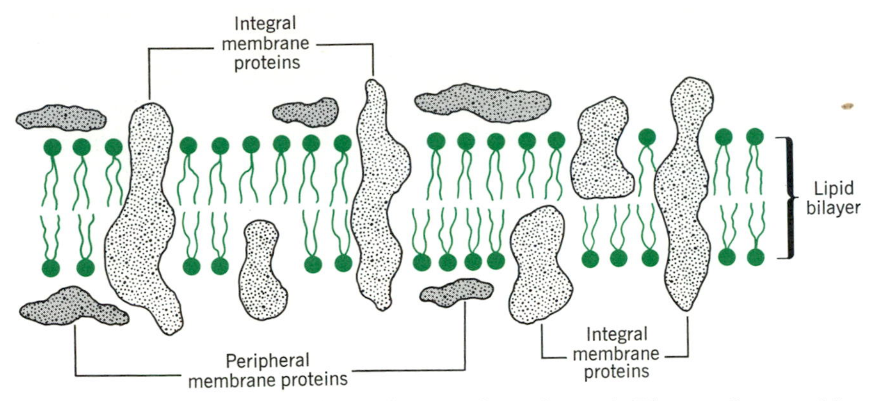

Figure 7-2 Model of the cell membrane (plasmalemma). The two layers of fatty acids have their long chains of -CH_2 groups stretched inward to make a "lipid bilayer," which is impervious to inorganic salts and to large molecules. Protein molecules in the membrane lie across it, making contact with the watery cytoplasm on both sides. (From Heslop-Harrison, 1978 after Singer and Nicholson, 1972.)

protein components. Hence the water can move from the soil solution, which is very dilute, into the root cells, that contain many solutes, but most of the solutes inside the root cells cannot move out. (How the solutes got there is described in the next chapter.) Such a membrane, permeable to water but not to most solutes, is termed *semi-permeable*.

It is clear that if there are appreciable amounts of solutes inside the cell and only traces of solutes outside, the concentration *of water* outside is greater than that inside, so water will enter. In so doing it will make the cell swell out (Fig. 7-1*A*). As this process continues, the cell wall will eventually not be able to stretch any further, so that its mechanical resistance will balance the tendency of water to enter. The cell is now said to be at *full turgor* (Fig. 7-1*B*).

The tendency of water to enter a solution is called *osmosis*. In general the amount that enters depends on the total concentration of solutes in the solution. Since we cannot very well add together the concentrations of different substances—some of which, indeed, like potassium chloride, are dissociated in solution to ions, each of which acts as a separate solute—it is usual to express the entering tendency of the water in terms of the pressure that would be needed to stop it. To visualize this, imagine a vertical tube closed at the bottom with a semi-permeable membrane, *M*; let there be sugar solution (*S*) inside the tube, and a vessel of water (*W*) outside, as in Figure 7-3. As water passes through the membrane into the tube the level of sugar solution will rise. Sugar cannot pass out, so water continues to come in. Eventually the level in the tube becomes so high that the pressure exerted by the weight of the water column on the membrane at the bottom is high enough to prevent any more water from entering. This limiting hydrostatic pressure is called the *osmotic potential*, π, of the solution. For a 1 molar solution* of a solute like sugar that is not dissociated into ions, π has the value of 22.4 atmospheres pressure or (approximately) 22.4 *bars*. Values in plant tissues are rarely as high as this, but in the phloem, where most of the sugar transport takes place, osmotic potentials

* The molecular weight of any substance expressed in grams is called a **mole**. When a mole is dissolved in a liter of solution, the solution is 1 molar.

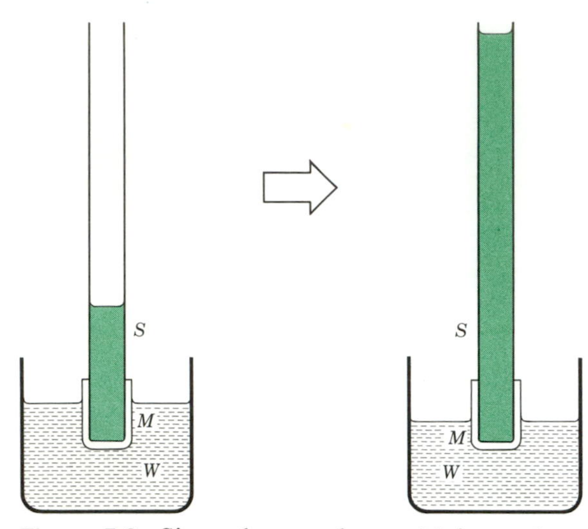

Figure 7-3 Since the membrane M is semipermeable, water W, enters the inner tube containing sugar solution, S, until the hydrostatic pressure stops it. The column height times the density of solution then measures the pressure exerted on the water passing through the membrane and hence the osmotic potential.

of 8 to 10 bars commonly occur. (This means that the concentration of phloem sap is about one-third molar).

Now let us apply these concentrations to a cell. At any time, the tendency of water to enter the tube in Figure 7-3 is determined by the difference between the osmotic potential of the solution at that time and the hydrostatic pressure exerted by the column of solution. Correspondingly, the tendency of water to enter a *cell* is, as we have seen above, equal to the osmotic pressure that the solution in the cell could exert (i.e., its osmotic potential) *minus* any pressure exerted against it by the tension in the cell wall (which we can call W). Hence the entering tendency is equal to $\pi - W$. The negative of this, $W - \pi$, is the tendency of water to leave the cell, and is called the *water potential.* This simple relationship helps very greatly to understand the water relations of plants.

There are two practical modifications of the water potential. The first is that, for acids, bases, and salts, π is increased by the degree of dissociation. For instance, K^+ and Cl^- act in this respect like two independent solutes. Calcium chloride, $CaCl_2$, in very dilute solution can give up to almost three times the osmotic potential calculated for its undissociated $CaCl_2$ form. (Note that dissociation is greatest in high dilution; for potassium salts at 0.001 molar concentration, as in plant nutrient solutions, dissociation is over 90%).

The second modification is that, for plants in soil, π is not exactly the osmotic potential of the cell contents, since the soil solution, though dilute, is not pure water, so the water potential of the roots is the difference between the water potential of the cells and that of the soil solution. As we shall see in Table 8-2, soil solutions are rarely stronger than 0.01 molar.

b. ENTRY OF WATER INTO ROOTS

The tendency of water to enter the root is complicated by the number of cell layers, from root hairs on the outside through a series of layers of parenchyma to xylem vessels or tracheids at the center (Fig. 7-4). The root hairs are extremely thin walled and thus quite elastic; when the root is in water or a typically dilute soil solution these cells absorb water rapidly, becoming turgid; their contents are now so dilute that the adjacent subepidermal cells, B-F, can absorb water from them and become more turgid, and *their* contents more diluted, in turn. Continuation of this process means that water moves across the root from outside to the center. In addition the cell walls, being continuous and readily wetted, conduct water through to the center like a cotton wick. In the center are vessel elements in the xylem, L, and the water now enters them for a special reason, discussed below.

The uptake of water can continue even when the soil is not particularly moist to the touch because the root system offers such an enormous surface through which water can enter (Fig. 7-5).

Table 7-1 on page 122 gives some idea of these very large surfaces, though the numbers are expressed in lengths only.

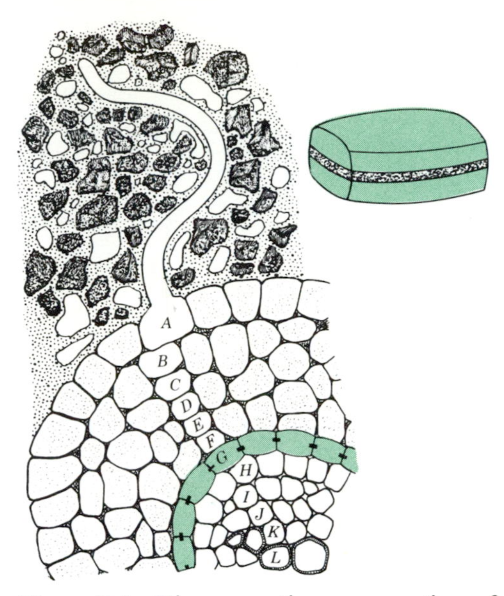

Figure 7-4 Diagrammatic cross section of a root, showing the series of cell rows *(B* to *J)* that intervene between the root hair *(A)*, in contact with the soil water, and the xylem parenchyma *(K)*, in contact with the xylem vessels *(L)*. The endodermis cells *(G)*, one of which is shown above, have a band of extra thickening that makes a tight seal separating the cortex from the vascular bundle.

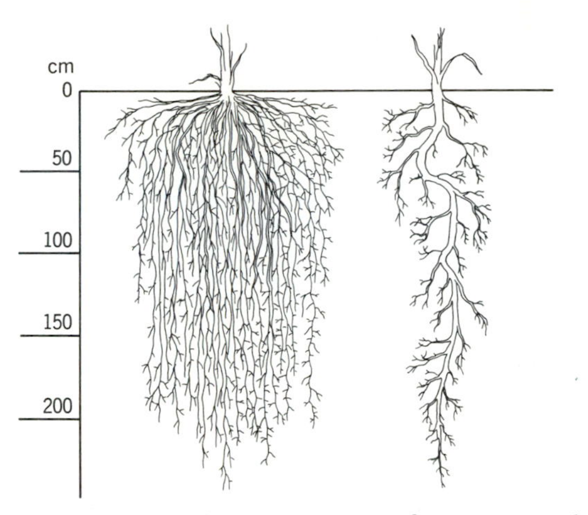

Figure 7-5 Left, root system of a mature wheat plant. The scale at left is in centimeters; the longest root is thus 7 feet long. In arid Australian soils it may go down to 20 feet (from Maximov, 1938). Right, tap root system typical of some plants of arid soils.

2. WILTING AND PLASMOLYSIS

The opposite situation to that just described occurs when a cell is placed in a solution of osmotic potential *higher* than its own contents. Water is now more diluted with solutes *outside* the cell than inside. Hence water will now leave the cell to enter the outer solution; the cell will lose any turgor it has and tend to become flaccid. A tissue in which the cells have become flaccid is itself limp and is called *wilted;* wilting also occurs when cells lose water to the air by evaporation. Plants or plant parts that have wilted often recover; the cells become turgid again if they are placed in water.

What happens when the flaccid cell continues to lose water to the outside solution? At first the cell continues to shrink, but soon the rigidity of the wall prevents it from shrinking further. It is now at *zero turgor.* As water continues to be lost, eventually there is no longer enough water to fill the cell. The result now depends on the circumstances; if the cell is in air any loss of water below the point of zero turgor means that the cell shrinks, as in Figure 7-1.

However, if the cell is floating in solution, it does not generally shrink; below the point of zero turgor the cytoplasmic membrane merely pulls in away from the cell wall and the resulting space becomes filled by the outside solution (Fig. 7-1*D*). For the cell wall itself presents little or no barrier to passage of the solute; it is only the membrane which is semipermeable. Such a cell is said to be *plasmolyzed.*

3. MOVEMENT OF WATER UPWARD

The molecules of water entering the root, whether coming along the cell walls or through the cell contents, are finally in contact with the xylem. These xylem elements are no longer living, their cytoplasm having been destroyed at the time they were differentiated (Chapter 10). They are simply tubes containing a very dilute solution of salts and nitrogenous compounds.

TABLE 7-1

Lengths of Root Systems

Plants	*No. of Roots*	*Total Length of Roots*
Tree Seedlings		
Loblolly pine		
6 months old	767	3.9 m
Dogwood		
6 months old	2,600	5.2 m
Locust		
4 months old	35,000	337 m
Cereal Plants		
Wheat at end of season		71 km
Rye at end of season		79 km
Rye, counting the fourth-order branchlets		624 km(!)

Since this is considerably more dilute than the contents of the living root parenchyma cells, it seems strange that water should move from such living cells into the more dilute contents of the xylem. The reason is that the water in the xylem is (generally) under *reduced pressure*—that is, under tension. The tension results from the fact that the xylem sap forms a continuous thread from the roots through the stem to the leaves; many end-walls of the xylem cells have been dissolved so that they offer no barrier to the sap and in each tube the sap column is unbroken, all the way to the leaves. From the surface of the leaves the water evaporates under the influence of sun or wind, and so more water is drawn up into the leaves to take its place.

The realization that water in the xylem can be under tension we owe to Harold Dixon and Eugen Askenasy, of Dublin, who in 1914 showed with models (a natural or artificial evaporating surface connected to a column of water, as in Fig. 7-6) that the water column could draw mercury up above atmospheric height and thus could stand a tension well

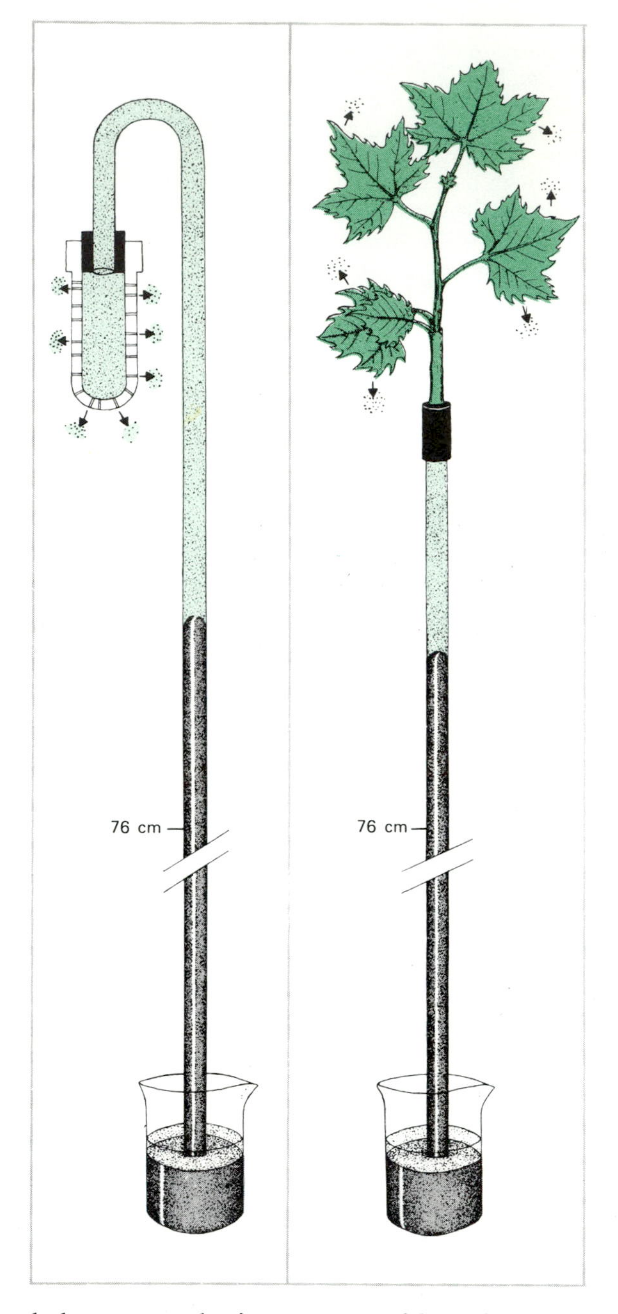

below atmospheric pressure without breaking. Later H. R. Bode of Germany, using the transparent stem of an *Impatiens* (jewelweed, an herbaceous dicotyledon) under the microscope, measured the width of the xylem vessels when the plant was losing water rapidly through its

Figure 7-6 Left, with this apparatus, Dixon and Askenasy in 1914 studied the cohesion of water under tension in the xylem. The porous pot of baked clay into which the tube fits tightly is a model for the transpiring shoot. After carefully filling all the interior with boiled water (to eliminate air bubbles), and dipping the lower end under mercury, the water evaporates through the pores of the pot and, as a result, the mercury is drawn up and finally exceeds the barometric height (76 cm) that the atmosphere supports. Thus the water column is now under tension, but does not break. On the right a leafy shoot is substituted for the clay pot, as in later experiments, and if air bubbles are excluded it shows the same phenomenon. The little arrows show the path of water vapor.

leaves, then cut off the stem at the base, and saw that the vessels expanded instantly by about 2 μ. Thus, their walls had been previously *drawn in* under the tension and consequently relaxed when the column was broken. A droplet of mercury at the cut surface was instantly drawn in. Even whole tree trunks are slightly drawn in during a warm or windy day, and relax or expand at night (Fig. 7-7).

One atmosphere of pressure supports a barometric height of 76 cm of mercury, and the density of mercury is 13.6, so that the corresponding column of water, whose density is 1.0, would be $76 \times 13.6 = 1033.6$ cm or about 34 ft. But the tallest trees are over 300 ft high, so that the water in their xylem must be under a tension of at least 10 atmospheres. Careful measurements of cells drying out have in fact shown that under some conditions water can stand tensions of around 200 atmospheres. Within the tree, therefore, the columns are unbroken and the sap is continuously present throughout.

One can see why the walls of conducting cells in the xylem are thickened; they have to stand these great tensions without tearing. The thick material, which is largely lignin, is laid down in characteristic patterns, as was shown in Figures 4-5 and 4-6. These thickenings are formed just before the conducting cell dies.

In the nonflowering plants, especially the coniferous trees such as pines, spruces and redwoods, the conducting cells of the xylem consist only of *tracheids*. However, the bordered pits between these cells are adequate to allow water to move through and to form a continuous column (although they do add some mechanical resistance). Perhaps it is on this account that trees of this group typically have

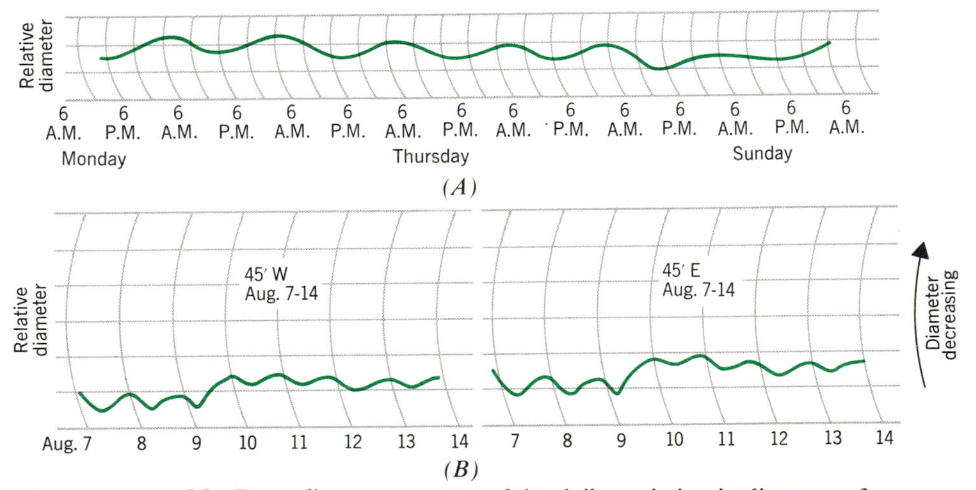

Figure 7-7 *A*. MacDougal's measurement of the daily variation in diameter of a pine tree trunk. The largest diameter is reached in the early morning (from MacDougal, 1936). *B*. Similar, but more recent, measurements on two sides (west and east) of a red pine *(Pinus resinosa)*. Note that the shrinkage *(rising* of the curve) comes a shade earlier on the east than on the west side (due to the morning sun). (From Kozlowski and Winget, 1964.)

small narrow leaves with a thick cuticle, allowing for only slow rates of evaporation.

It will be noted that the layman's idea of "the rise of sap in the spring" has little real meaning, for the sap is in the xylem all the way to the top of the tree throughout the winter. In very cold weather it may freeze solid and if so it will expel the dissolved air and the bubble will break the column. But new vessels in flowering plants are laid down just before the leaves expand in the spring. With the unfolding of leaves the loss by transpiration of course increases greatly, so that the sap does get replaced much faster than in the winter. Nevertheless there is a steady water loss from leafless twigs and even from cut or broken surfaces on warm or windy winter days.

4. LOSS OF WATER FROM THE LEAVES

In Figure 7-8 we see one of the many far ends of the conducting system, one of the small veinlets in a leaf. The bundle of xylem vessels has branched and rebranched and here only one is left. Its thickened walls are in contact with the thin walls of the leaf parenchyma cells, and thus with a number of spaces between these cells. Indeed, because there is so much intercellular space in leaves, this tissue is referred to as "spongy parenchyma" (Fig. 4-14). The water in the xylem passes through the wet walls into the cells and thence evaporates into the intercellular spaces, which are thus kept saturated with water vapor. Now a curious set of structures comes into the action.

As we saw in Chapter 4, the leaf epidermis is punctuated by a regular pattern of pairs of small bean-shaped cells, spaced at fairly regular distances from each other (Fig. 7-9).

Because the pair of curved cells with a space between is like a pair of lips it is called a stoma (Greek word for a mouth) and has the Greek plural *stómata*. The space between this pair of "guard cells" is variable and can be tightly closed or wide open. It communicates directly with an intercellular space in the leaf. In most dicotyledonous plants there are many more stomata on the lower side of the leaf than on the upper, but in monocotyledons the leaves tend to be held more upright and the stomata are distributed more equally.

The opening and closing of the stomatal space is the prime factor controlling *transpiration* or water loss. It is brought about by changes in the form of the guard cells, which are the result of changes in their turgor. In some mosses these cells are rectangular in section and when fully turgid they become more nearly spherical (Fig. 7-10*A*); this pulls them apart a little and opens the space. In most flowering plants the walls of the guard cells are thick on three sides so that when they are turgid it is the thin wall on the side away from the opening that is pushed out into the adjacent cell, and the cells thus curve away from each other (Fig. 7-10*B*). In the long narrow leaves of the grasses the guard cells are long and thin, but they are thick-walled except for their rounded ends; in turgidity these rounded thin parts expand and press against each other, forcing the straight thick-walled parts away from each other to make a long, very narrow slit between them (Fig. 7-10*C*). In all cases turgor of the guard cells causes opening; loss of turgor causes closing.

5. THE ENVIRONMENTAL CONTROL OF WATER LOSS

Let us see how this works in the field. There are two major environmental controls on transpiration.

a. LIGHT

The guard cells differ from the other cells of the epidermis in containing chloroplasts. There are several results of this. When light falls on the guard cells they start to photosynthesize, thus producing ATP. This reacts with the starch stored in these cells to cause

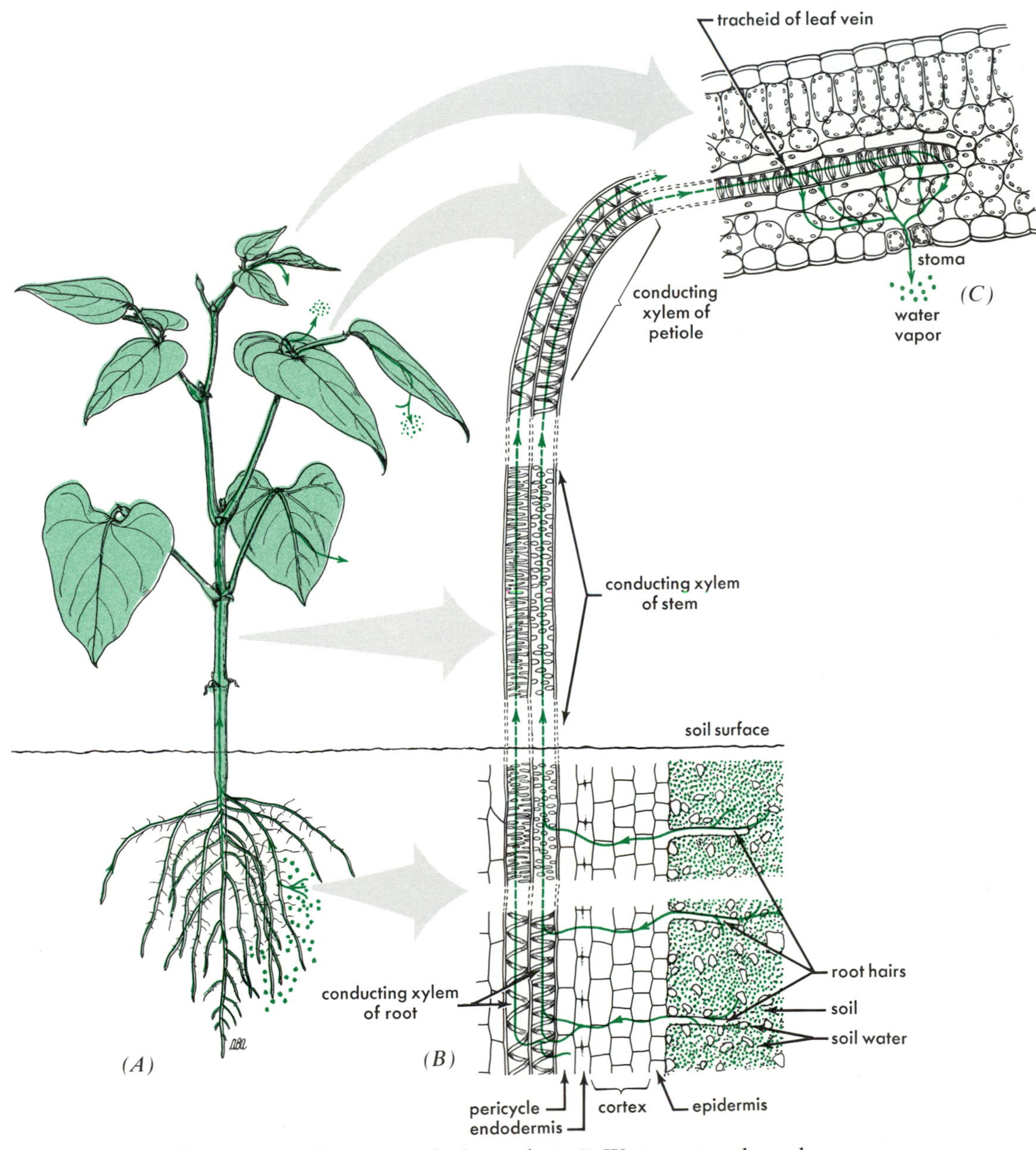

Figure 7-8 *A*. The water supply system of a bean plant. *B*. Water enters through root hairs across cortical cells, through endodermis and pericycle to the xylem of root, stem, and petiole. *C*. Greatly magnified, the far end of the system; the termination of vessels in a leaf surrounded by mesophyll cells in the spongy parenchyma. (From Rost, *et al.*, 1979)

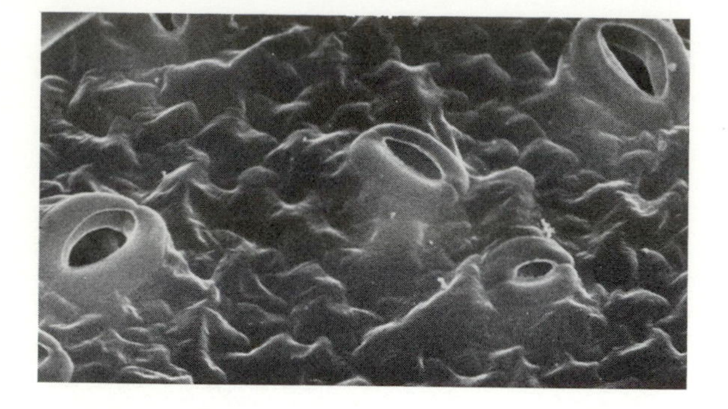

Figure 7-9 Scanning electron micrograph of open stomata on a cucumber leaf. In several stomata, the actual surfaces of the guard cells that meet and close the aperture can be seen below the surface.

partial breakdown to glucose phosphate, which increases the solute content of the cell. As a result water and potassium ions enter from the adjacent cells, which lose turgor somewhat while the guard cells gain turgor. This makes the guard cells swell and push back their neighbors. A curious consequence is that, while in most green cells the starch increases during the day, in the guard cells the starch actually *de*creases in the light.

The simple action of photosynthesis is not the whole story, however, because blue light is much more effective in causing opening of guard cells than red light. Yet for photosynthesis the effectiveness of these two zones of the spectrum would be about equal (Chapter 5, sections 2 and 3). Thus osmotic potential and other factors (below) participate importantly. But first let us note one important fact.

The changes in osmotic potential of the guard cells are large. Other epidermal cells show no corresponding change in the light. For example, in cyclamen leaves the following measurements have been made Table 7-2:

TABLE 7-2

Time	Stomatal Condition	Osmotic Potential in Bars*: Of Guard Cells	Osmotic Potential in Bars*: Of Epidermal Cells
7 A.M.	Just opening	14.6 bars	10.2 bars
11 A.M.	Fully open	31.0 bars	
5 P.M.	Just closed	18.5 bars	13 to 15 bars
12 midnight	Closed	13.5 bars	
12 noon	Fully open	23.0 bars	

*1 bar = 0.99 atmosphere

b. THE GENERAL WATER SUPPLY OF THE PLANT

During the night, as we have seen, water is steadily absorbed by the roots, so that by dawn all cells are well supplied. As a result the first light, often even before sunrise, is enough to cause immediate opening of the stomata. During the day the water loss normally exceeds the rate of intake (except on wet days) so that the guard cells slowly lose turgor and often close the stomata. Thus in dry weather we find stomata opening at first light and gradually closing in the afternoon; sometimes they partially reopen at night as turgor is built up again. Under extreme conditions, such as in Arizona, they can be closed most of the day and open most of the night (Fig. 7-11).

In addition to these environmental controls there is a hormonal mechanism. If the leaf loses enough water to make it wilt, the hormone abscisic acid is produced, and this fully inhibits the uptake of ions, especially K^+ ions (Fig. 7-12)

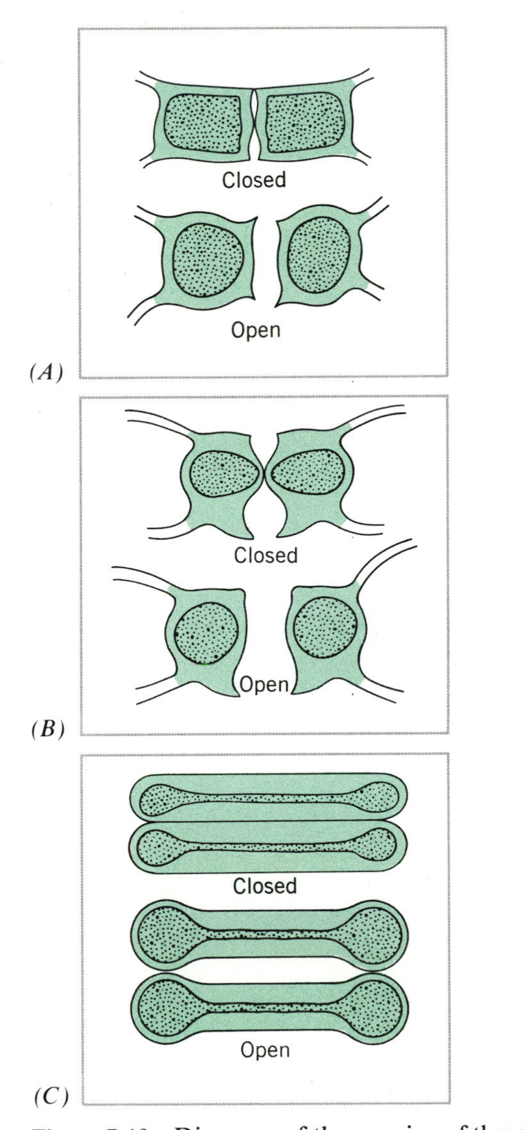

Figure 7-10 Diagrams of the opening of three types of stomata, *A* and *B* in section view, *C* in surface view. *A*. Thin-walled guard cells of a moss, becoming more nearly spherical (therefore shorter) when turgid. *B*. Typical dicot stoma with guard cells thickened on outer and inner sides, opening by pushing back on the adjoining cells, thus bending in the plane perpendicular to the paper. *C*. Surface view of stoma of a grass with a long, narrow opening between thick walls; it opens by turgid swelling of the flask-shaped end-parts.

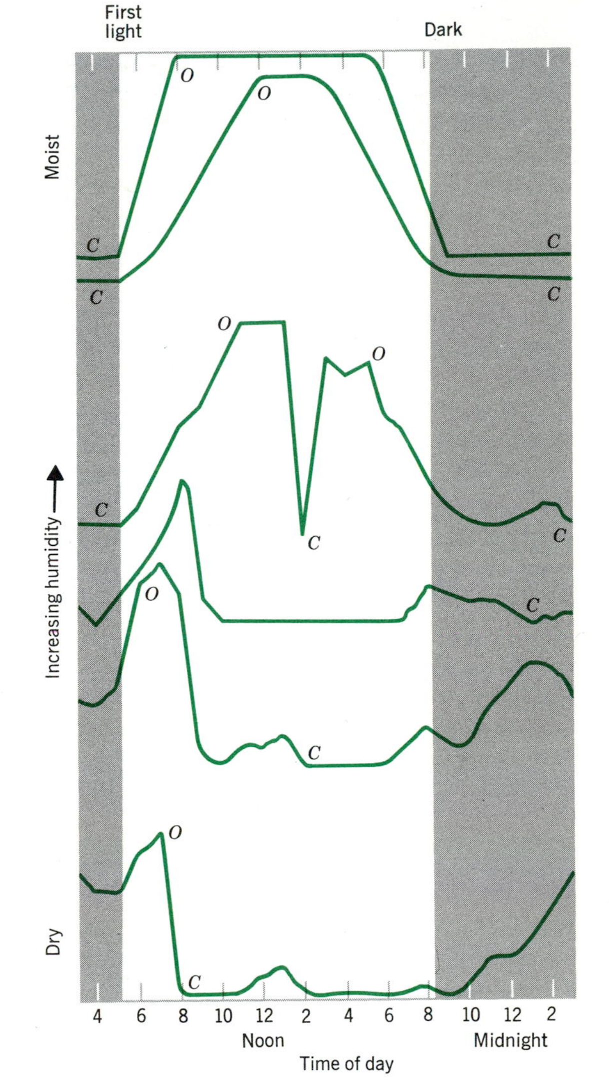

Figure 7-11 Stomatal opening on alfalfa leaves in Arizona. The aperture was recorded on a series of days throughout the season. From top to bottom the weather becomes hotter and drier. At first the stomata open at first light ($O =$ fully open) and close at night ($C =$ fully closed). Then there is temporary closure in the afternoon. Then gradually the daylight opening is curtailed and there is partial opening at night. Finally in the driest conditions (bottom) the stomata are mainly closed in the day and open at night. (From Loftfield, 1921)

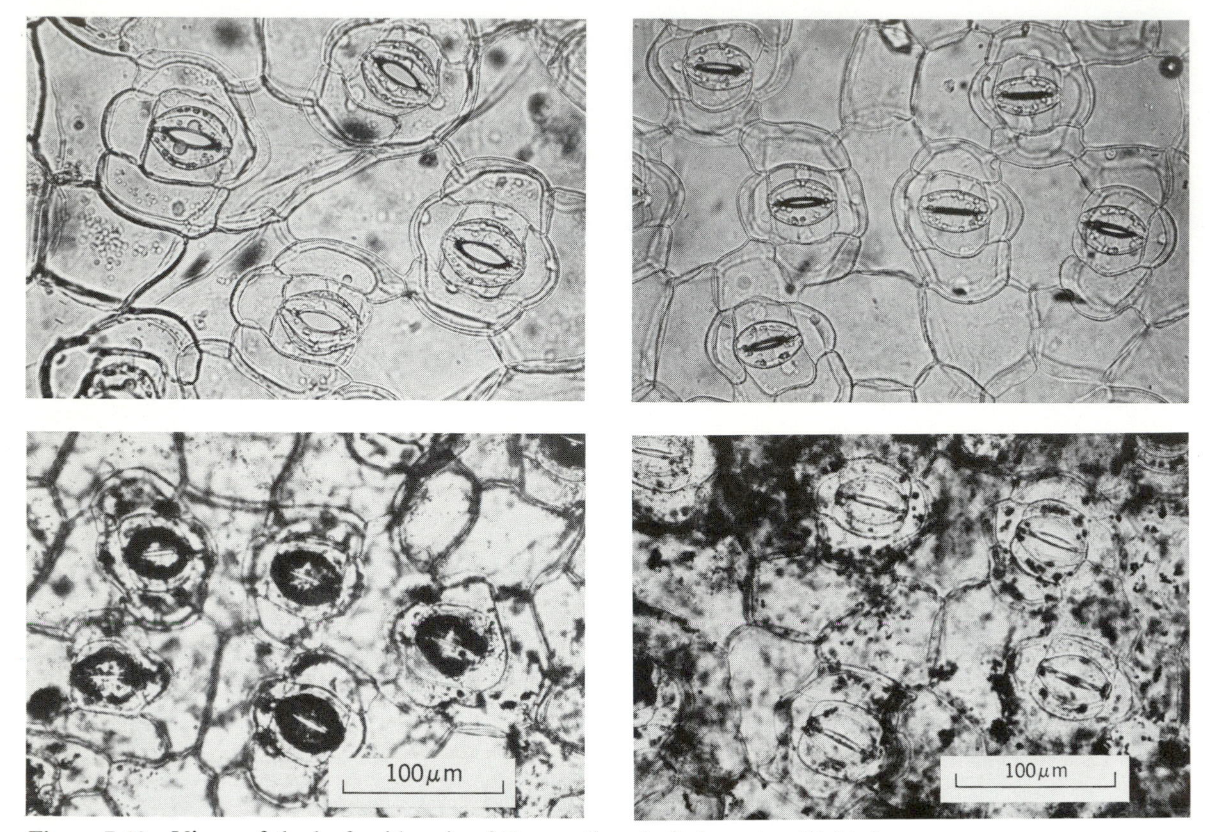

Figure 7-12 Views of the leaf epidermis of *Commelina;* Left, in water. Right, in abscisic acid solution. The two upper pictures are unstained, the two lower ones stained to show potassium localization. The open stomata on the left have potassium (black) concentrated in the guard cells; in the nearly closed ones on the right, the potassium has gone out into the epidermal cells.

into the guard cells and thus causes the stomata to close (see Chapter 10, section 3d). The hormone seems, however, to require the presence of carbon dioxide to act in this way, and this interrelates with photosynthesis because, in consuming carbon dioxide, photosynthesis prevents abscisic acid from acting and thus favors opening.

If in spite of stomatal closure the water supply continues to be deficient, the cells will finally lose their turgor and the leaf will *wilt*. During the night, with lowered temperature and higher humidity, the cells may regain their turgor and the wilting is reversed (Fig. 7-13). However, if the wilting becomes permanent, the leaf will soon die, and the whole plant may soon follow.

Some succulents, the so-called CAM plants, as we saw in Chapter 6, take in CO_2 at night to form malic acid and decompose this in the light to release CO_2 for photosynthesis, a modification of the C_4 system. The opening of the stomata at night fits this mode of life perfectly and consequently these plants grow well under arid conditions.

The whole system shows how tightly the

Figure 7-13 Wilting and recovery in leaves of squash *(Cucurbita maxima)*. Left, in the late afternoon of a hot day. Right, next morning.

different functions of a plant feed back into one another to give integrated control. Thus:

↑ Stomatal opening depends on guard cell turgor, and
Guard cell turgor depends (in part) on light, but
For the light to act depends on the general water (and ion) supply, and in turn,
The general water supply depends at least partly on the stomatal opening.

It also shows how photosynthesis, water relations, and the hormone system interact.

6. ROOT PRESSURE

One additional property of the whole plant completes this picture of the water relations. It is a common observation that, if the stem of a young plant is cut off, sap will ooze from the cut surface of the stump, sometimes for days. In 1727 the Rev. Stephen Hales studied this "bleeding" by attaching manometers to the cut stumps of grapevines and recorded *positive* pressures up to two atmospheres. He commented that "this force is near five times greater than the force of the blood in the great crural artery of a horse" (which he had previously measured). It appears that this root pressure is due to solutes in the xylem; the surrounding root cells act like the semipermeable membrane of an osmometer, so that the soil water enters and pushes up the column of sap. Root pressures are not observed in dry soil. In trees in the spring, the positive pressure can serve to refill broken water columns in the xylem, for a pressure of two atmospheres means over 20 m. of a water column. The volume of xylem sap required to refill the system is still quite small compared to the volume normally going out in transpiration.

Root pressure shows a daily cycle, reaching its maximum usually in the small hours of the morning (see Fig. 7-14). This cycle is largely set by the periods of light and dark that

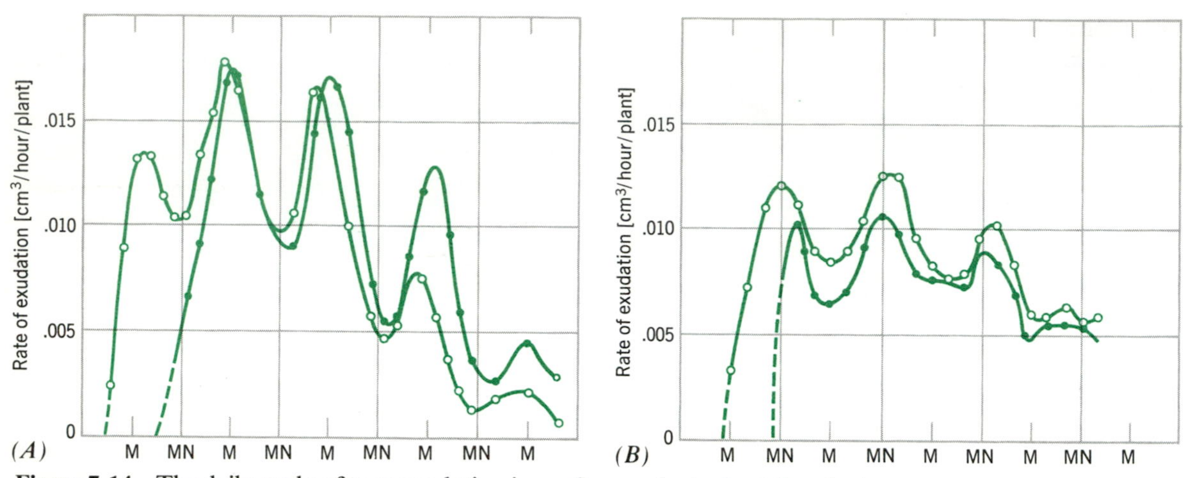

Figure 7-14 The daily cycle of root exudation in sunflower plants. In *A* the plants were grown on normal day–night cycles; in *B* with reversed cycles, that is, lighted at night and dark during the day. In both cases the peak falls at the middle of the light period. The roots "remember" the rhythm to which the shoot was exposed before it was cut off. In both *A* and *B* one shoot was cut off 12 hours later than the other but this treatment makes no difference to the timing of the cycle. M = mid-day, MN = midnight (From Grossenbacher, 1939)

the plant experienced before being cut off. After a few days the solutes become exhausted (since isolated roots do not photosynthesize) and the pressure gradually falls to zero. However, the stumps of large trees, with voluminous roots full of stored starch, will sometimes "bleed" for weeks when the trees are cut down in the spring. In intact plants, the pressure within the xylem in the daytime is normally negative.

7. PROBLEMS OF IRRIGATION

In a dry soil, water, which is a limiting factor for growth, must be supplied artificially. We water our gardens during a dry period without a thought of complications, but large-scale irrigation involves many interacting factors which sometimes make it difficult.

In the first place, the volume of water must be enough to penetrate as deeply as the furthermost roots. The roots of many plants tend to grow towards water, so that if the moist layer is only near the surface a shallow root system will develop and thus be sensitive to drying out later. On the other hand, the rate of supply must not be so great that the soil becomes waterlogged, for roots may die if the oxygen tension in the soil gets too low, and the associated accumulation of carbon dioxide also has a toxic effect on the roots of some plants. Thus soils that are to be irrigated must have good drainage.

The second consideration is one that proves of major importance when perpetually arid lands (deserts) are irrigated. Such soils commonly are warm and hence evaporation from them is rapid. As a result the solutes in the water—often chlorides—tend to accumulate in the soil, which may ultimately become too saline for plant growth. Also, if there are salts in the lower soil layers, evaporation may bring them to the surface. Thus for the irrigation of warm desert soils, the water available must either have a low salt content or else be large enough in volume so that the solutes are washed down into the level well below the roots. Yet the volume must not be large enough to make the soil waterlogged. The Nile, in Egypt, offers a prime example of a source of water both plentiful and of low solute content, so that the desert in the Nile Valley has been agriculturally productive under irrigation for about 4000 years (though waterlogging is now beginning to be a problem). But some regions of North America and of central India have been so plagued with salt accumulation that, after a few years of irrigation, agriculture has become virtually impossible (Chapters 15 and 31). Not only the total salt content but the alkalinity of the water (usually due to sodium and calcium bicarbonates) is also of critical importance. Three types of irrigation systems are shown in Figure 7-15.

(A)

(B)

(C)

Figure 7-15 Three types of irrigation. *A*, Furrow method. *B*, Trickle installation using plastic pipes, with a small hole by each plant. *C*, Sprinkler system using light-weight aluminum pipes.

CHAPTER 8

The Mineral Food of Plants

1. INORGANIC MATERIALS AS FOOD

Understanding of what plants need for growth (besides water and air) came from the study of plant ash. If plant tissue is dried and heated strongly till the organic matter is burnt off, the ash remains. In the eighteenth century several people, including the French botanist Duhamel du Monceau (1763), began to realize the value of plant ash as a fertilizer. Then, starting with Théodore de Saussure in 1804, botanists and chemists began to analyze the ash and to compare plants growing on different soils; de Saussure in particular discovered that plant ash contained the same salts as those in the soil and that these salts are indeed derived from the soil. Thus a knowledge of the types and amounts of minerals present in plants gradually developed and their importance came to be appreciated. When Justus Liebig, Professor at Munich, established in 1840 that some minerals, especially calcium phosphate, could be major limiting factors for the growth of plants in soil, the way was opened for seeing whether plants could grow in known mixtures of minerals. Franz Knop (1859) and Julius von Sachs (1860), both in Germany, finally succeeded in growing several plants to maturity in *nutrient solutions* whose composition was derived from the analyses of plant ash. Knop's own drawing of one of his experiments, the first successful ones in history, is shown in Figure 8-1.

Knop's success was most important in principle because it showed that inorganic salts can provide a complete medium for good, normal growth. Up through the eighteenth century it was believed that the organic matter of plants was directly derived from organic matter (humus) in the earth, or in manure used as fertilizer. But even in 1763 Duhamel, in his treatise on agriculture, had raised doubts about this and suggested that, instead, organic manures perhaps acted "by retaining the moisture which is necessary for vegetation, by lightening soils that are too compact, or by exciting through their fatty and oily contents a kind of fermentation in the soil. . . ." Then came the understanding of photosynthesis, with the implication that the organic matter of plants came primarily from CO_2 and water. This took the importance away from the organic matter of the soil, but left open the idea that the *inorganic* matter of plants came from the inorganic matter of the soil. The fact that normal plants could be grown with only air, water, and inorganic salts settled the question (Fig. 8-2).

A lingering regard for the value of organic fertilizers persists, however, to this day, and the virtues of "organic" vegetables and fruits are

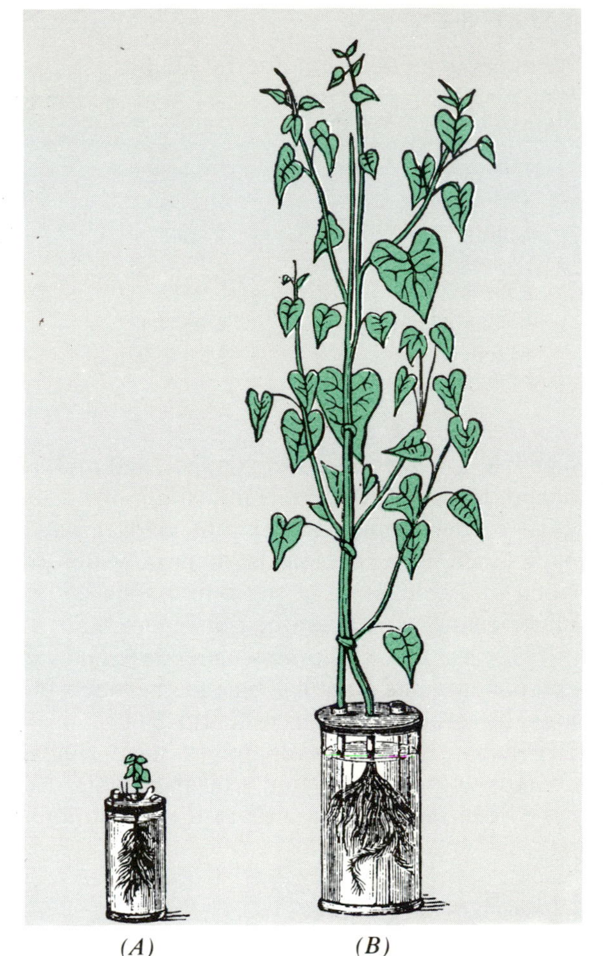

Figure 8-1 The first experiments on growth of plants in solution, by Knop, 1859. *A*. Without potassium. *B*. With potassium. (From Palladin, 1926)

Figure 8-2 The demonstration at the University of California that tomato plants grow, and yield fruit, equally well whether in fertilized soil, in nutrient solution, or in sand watered with nutrient solution. Fruit had been harvested for seven weeks before the photo was taken. Spacing, staking, and light were the same for all.

widely proclaimed. Yet over 30 years ago workers at Ithaca, New York, showed that tomatoes grown in a purely inorganic medium had at least as high a vitamin C content as those grown in several different soils. Recently a group in California showed that guinea pigs grew as well, and were as healthy, when fed bent grass (*Agrostis* spp.) grown in inorganic nutrient solution as on the same species and cultivar grown in organically manured soil. (Indeed, since so many animals derive their whole nutrition from pasture plants it is fortunate that leaves in general provide excellent animal nutrition—better in fact than many grains.) The organic matter (humus, or plant residues) in soil has quite other functions. When plant or animal material is added to the soil, microorganisms decompose most of it and thus liberate the inorganic salts, to be recycled into the roots of the growing plants. The parts that are resistant to decomposition remain as humus, and this gives a fibrous quality to the soil, helping to make it moist and workable, while its colloidal component increases the water-holding capacity of the soil (see Chapter 15). Duhamel's understanding was thus very close to the truth. In temperate soils humus may remain for many years, but in the moist tropics it disappears fast by microbial oxidation.

Now that we know that plant nutrients are inorganic, nutrient solutions have become a powerful tool for the study of exactly what salts are required, and how much of each.

The first point to note is the very small amounts needed. Indeed, one of the striking ways in which plants differ from animals is the very dilute form of their nutrition. Animals eat meat, fat, starch, and similar concentrated nutrients. Plants not only use as a source of carbon the carbon dioxide of the air, which is present at a concentration of only 0.03% or 300 parts per million, but their supply of minerals is similarly dilute. The total concentration of salts in the soil solution (i.e., the solution bathing the soil particles and the plant roots) is around 0.1%, and the concentrations of all the individual ions together in the *optimum* mineral solution of Table 8-2 is 0.015 molar. The levels of copper and molybdenum can be even 100 times lower still. Yet this very dilute solution satisfies all the plant's needs for inorganic nutrients, and in fact concentrations much higher than these tend to inhibit growth.

Over the years it has become clear that the mineral elements needed by plants fall into two groups—those required in relatively large amounts (designated *macronutrients*) and those needed at only a few per cent of those levels (*micronutrients*). They are listed in Table 8-1. Iron falls between the two groups but nearer the macro group. All these elements must be present if soils are to be fertile. In addition to the 13 elements in Table 8-1, micro-amounts of sodium are needed by marine algae and perhaps by a few shore-loving land plants.

TABLE 8-1

Mineral Elements Required by Higher Plants

Macronutrient Elements		*Micronutrient Elements*
Nitrogen		Manganese
Phosphorus		Boron
Sulfur		Zinc
	Iron	
Potassium		Copper
Calcium		Chlorine
Magnesium		Molybdenum

Knop's solution contained KNO_3, $Ca(NO_3)_2$, KH_2PO_4, and $MgSO_4$, with a trace of ferric phosphate. But as chemical methods improved it became clear that there were impurities in the salts used (or in the water) that were of critical importance to plant growth. First manganese and then boron (borate) were recognized in this way; later zinc, copper, and molybdenum were added. The composition of a very good modern nutrient solution (Hoagland's), made with pure chemicals in pure water, is given in Table 8-2 and the importance of the minor elements is illustrated in Figure 8-3.

Several other solutions using different salt components but a similar balance between the ions, have been used in different laboratories. Fertilizers, if they are complete, have similar composition. Since virtually all the plants that have been tested grow well in these solutions,

Figure 8-3 One of the demonstrations of Hoagland and Arnon of the great importance of the minor elements. Asparagus plants on the right are growing in a full nutrient solution but with the omission of zinc, copper, manganese, and molybdenum; those on the left have these elements added in concentrations of 1 part in several million parts of solution.

TABLE 8-2

Hoagland and Arnon's Nutrient Solution (see Fig. 8-2)

Salt			
Macronutrients	*Weight in Grams per Liter*	*Millimolar (mM)**	*Micromolar (μM)*
Potassium phosphate, KH_2PO_4	0.14	1	–
Potassium nitrate, KNO_3	0.51	5	–
Calcium nitrate, $Ca(NO_3)_2 \cdot 4H_2O$	1.16	5	–
Magnesium sulfate, $MgSO_4 \cdot 7H_2O$	0.50	2	–
Ferric tartrate, $Fe_2(C_4H_4O_6)_3$	0.55	1	–
Micronutrients	*Milligrams per Liter*		
Boric acid, H_3BO_3	2.5	–	40
Manganous chloride, $MnCl_2 \cdot 4H_2O$	1.5	–	8
Zinc chloride, $ZnCl_2$	0.10	–	0.8
Copper chloride, $CuCl_2 \cdot 2H_2O$	0.05	–	0.3
Sodium molybdate, Na_2MoO_4	0.05	–	0.2

* A 1 millimolar solution contains the mol. wt. in milligrams dissolved in 1 liter; e.g. KH_2PO_4 has mol. wt. 39 + 2 + 31 + 64 total 136, hence 0.136 grams per liter.

flowering and setting fruit, it is clear that all the needed ions are present and at about the right levels. There is some variation between species, and there are some special cases. For instance, sugar beets grow best with added chloride, and legumes need a trace of cobalt (for the reason, see Chapter 9).

With such known nutrients one can see the effects of deficiency or excess of any given ion, and then use these observations (with due reserve) to deduce some possible causes of failure of crop plants in the field.

2. THE ROLES OF INDIVIDUAL IONS

While our knowledge of the nutrition of plants is still imperfect, we do know why most of the mineral nutrients are needed. Carbon, hydrogen, oxygen, and nitrogen, of course, are the materials of all organisms, and nitrogen presents a specially complex system, which is deferred to Chapter 9. Here we describe the main roles of the inorganic nutrients.

a. THE MACRONUTRIENTS

Phosphorus is always present as phosphate. When phosphate combines with an organic molecule it is like an energetic youngster on a family walk taking his or her father by the hand – activation is the result. This is how ATP (also UTP and GTP) act; they transfer phosphate groups to sugars, acids, and other organic compounds and this activates them to take part in both synthetic and breakdown processes. Phosphate is also needed for *nucleic acids,* RNA and DNA, which, as we have seen, have the structure base–sugar–phosphate, and for the *phospholipids* in cell membranes. In these last, the phosphate group is at the end of a long molecule, where it can mix with the aqueous cell contents, while the long fatty acid chain lies in a double layer in the middle of the membrane, as shown in Figure 7-2.

Phosphate deficiency, as might be expected from its many functions, shows itself in a variety of forms—dwarfing, reddening due to the accumulation of anthocyanins, browning of leaves, and small imperfect fruits. In soil, most of the phosphate is either adsorbed on the surface of the soil particles or else is present as basic phosphates of iron or aluminum (Chapter 15). As a result, the aqueous soil solution is very low in free phosphate, but some is set free by the organic acids excreted by roots (cf. p. 140).

Sulfur is mainly present in soil as sulfate ($^{=}SO_4$) but usually is in plants as organic sulfur compounds ($^{=}S$). The sulfate in soils must therefore be reduced (i.e., hydrogenated) to the sulfide level in the plant before it can enter most organic compounds, and the reduction is a complex one requiring a reaction with ATP. Thus phosphorus and sulfur functions are interrelated. Hydrogen sulfide, H_2S, is formed by some bacteria in wet soils and is a common air pollutant; yet it has been shown recently that in low concentrations it can be absorbed by leaves and can furnish their sulfur requirement.

Sulfur in plants is needed for four *coenzymes*—coenzyme A, biotin, thiamine, and lipoic acid. It is present in two amino acids, *cysteine* (sometimes in the form of its dimer cystine) and *methionine,* and also in a few special compounds like the strong-flavored substances of onions, garlic, and mustard, and the bland *agar* from algae, which is a carbohydrate containing *sulfate* groups.

In connection with sulfur the case of selenium (Se) is interesting. Small amounts of selenium are essential for the formation of vitamin E in animals, and it is often taken up from soil by plants too. In some areas of the Dakotas and Wyoming, where the calcium sulfate occurs mixed with calcium selenate ($CaSeO_4$), it is present in the plants in fair amounts. Wheat grown there may have up to 0.1% of its dry weight as selenium, and such grain is toxic to cattle. A few legumes, especially *Astragalus,* (locoweed) flourish on this seleniferous soil and act as indicators that selenium is present.

Potassium (K^+) activates many plant enzyme systems. In particular it activates phosphate-transferring enzymes (kinases), which transfer phosphate groups to and from ATP. In this way potassium exerts control over carbohydrate metabolism and over the synthesis of polysaccharides, proteins, and nucleic acids, for virtually all these processes depend on phosphate transfer (Chapter 6). In some enzymes it can be replaced by ammonium ion (but not by sodium, which plays very little role in plants). Perhaps because of these actions, potassium greatly promotes cell enlargement, especially elongation, and plants in nutrient solution or in tissue cultures thrive best on a medium high in potassium (Fig. 8-1).

There is a balance between potassium ions, K^+, and the divalent ions Ca^{2+}, Mg^{2+} and Mn^{2+}. High K^+ lowers the uptake of these ions into both roots and leaves. This effect came into prominence recently when a fall-out of radioactive strontium, ^{90}Sr, took place in Britain. Strontium reacts like calcium, and its uptake into pasture grasses was found to be lowered by adding potassium.

Within the plant, potassium is highly mobile. Its concentration in a given leaf may vary even during the day; generally it tends to reach a maximum in the late afternoon. When the stomata open, potassium enters the guard cells and increases their turgor (Chapter 7). In potassium deficiency, on the other hand, K^+ migrates from the old to the new leaves, so that the potassium is limited to the youngest leaves, and the others turn brown.

Potassium deficiency shows in general by decreased elongation or smaller fruits but more specifically by brown areas on the leaves. In dicots these are separate brown spots that gradually merge; in monocots a brown "scald" develops on the edges and tips of the leaves.

Calcium. The insolubility of many calcium salts, especially the phosphate and sulfate, dominates the effects of calcium. As an ion, it balances the action of potassium, reducing the

K^+ uptake into plants. It decreases the elongation of shoots, but curiously enough it promotes the elongation of roots. It is immobile within the plant, and when the leaves age, losing amino acids and potassium, the calcium is left behind as insoluble carbonate, silicate, or oxalate. Also, the middle lamella of the cell wall contains gelatinous calcium pectate. This material is hydrolyzed by pectinases, both when leaves fall in autumn and when they are attacked by soft-rot bacteria.

Calcium carbonate in soil has a major role in controlling acidity. It neutralizes the organic acids produced by roots and by decaying organic matter; thus calcareous soils commonly remain at pH between 7 and 8, while noncalcareous soils, especially if they support a heavy growth of grasses or conifers, can become as acid as pH 4. The neutrality favors the bacterial oxidation of ammonia to nitrate, and, besides, calcium ions promote the growth of these "nitrifying" bacteria (Chapter 9). Farmers in the north, where soils tend to become acid, therefore often add lime. However, too much calcium lowers the solubility of the salts of other metals, so that deficiency of manganese, iron, or zinc may appear. These deficiencies occur also in naturally alkaline soils.

In northern Europe and the eastern United States, some industrial areas experience *acid rain*, of pH 4 to 4.5. This tends to leach calcium from the soil and acidifies lakes and rivers; however, it seems to promote the growth of pines. The acidity it contributes compares with that produced by agricultural fertilizers.

In calcium deficiency plants are dwarfed, the roots stunted and root hairs distorted; the leaves turn pale green, especially at the margins. The symptoms are not highly characteristic, and the deficiency is mainly limited to very sandy soils.

Magnesium ions (Mg^{2+}) activate many enzymes, including several that catalyze critical steps in carbohydrate breakdown and one that hydrolyzes ATP. In a few of these manganese acts about as well as magnesium. Magnesium ions also control the aggregation and disaggregation of ribosomes, and since it is the aggregated form ("polysomes") that acts in protein synthesis, this control is far-reaching.

The most obvious role of magnesium is as a constituent of chlorophyll, which contains 2.6% Mg. But since chlorophyll is only 3 to 5% of the dry weight of a leaf, the Mg of chlorophyll represents only 0.1% of the leaf by weight. Thus the Mg in ionic form constitutes up to 10 times as much as that in chlorophyll.

In sandy soils where magnesium may be deficient, dolomite—a double carbonate of calcium and magnesium—is often applied. Magnesium deficiency shows up in many plants as disappearance of chlorophyll *between* the leaf veins ("intervenous chlorosis"), the veins themselves remaining green. Sometimes, as in cotton, there is excessive accumulation of red anthocyanin.

Iron is especially needed for the cytochromes, which act by the reversible change of valency from Fe^{2+} to Fe^{3+}. It also acts in photosynthesis, both in ferredoxin, which passes electrons from chlorophyll to NADP (Chapter 5), and in the two cytochromes that act as electron carriers in other steps. Iron is also present in catalase and peroxidase, two enzymes that react with hydrogen peroxide in many cells.

Although plant ash typically contains only 2 to 3 mg of iron per gram, iron controls chlorophyll formation. Iron deficiency shows in leaves as an overall lack of chlorophyll (*general chlorosis*) (Fig. 8-4*A*), and in roots as characteristic browning.

Both in soil and in plants iron is rather immobile; it remains in autumn leaves, with the calcium, when the mobile materials have been exported. Its salts are so often insoluble that when the soil is even slightly alkaline they are precipitated and adsorbed on the colloids, causing a tendency to chlorosis, even though the soil may analyse for an adequate iron content. In

(A)

(B)

(C)

(D)

Figure 8-4 Responses of tomato leaves to single mineral deficiences. All plants were grown in nutrient solutions. Leaves of closely corresponding age and location on the plant are compared in each case. Read each group from left to right. *A*. Control and minus iron. *B*. Minus manganese and control. *C*. Minus copper, control, and minus zinc. *D*. Minus molybdenum and control. Controls had the full nutrient solution.

preparing nutrient solutions, one usually adds citrate or tartrate to hold the iron in solution; otherwise it precipitates as ferric phosphate.

b. THE MICRONUTRIENTS

Manganese activates numerous enzymes that respond about equally to magnesium. But its best understood function is in photosynthesis, for manganese is an essential constituent of Photosystem II, which takes electrons from water to replace those emitted by P_{700} (Chapter 5). Correspondingly, most of the manganese in plants is found in the chloroplasts. Manganese deficiency therefore causes rapid decline in the rate of photosynthesis, and this results in a severe decrease in growth rate. It also appears as a decrease in chlorophyll, the exact symptoms differing in different plants: stripe chlorosis in corn, grey specks on oat leaves, intervenous chlorosis in many dicots (Fig. 8-4*B*), and generalized chlorosis on spruce needles. Like iron, it forms insoluble salts at alkaline pH, so that manganese deficiency is common on neutral or alkaline soils. In such soils, manganese is often supplied, not as salts, but as simple organic compounds.

Curiously enough, chlorosis can also result from *too high* a level of manganese. This is due to its antagonism with iron, for high manganese levels interfere with uptake of iron. For tomato plants, the Fe : Mn ratio in the soil for best growth is around 2 : 1, and similar ratios hold for other plants.

Zinc is present in two enzymes—alcohol dehydrogenase, which converts ethyl alcohol to acetaldehyde and vice versa (Chapter 6), and carbonic anhydrase, which establishes equilibrium between carbon dioxide and water:

$$CO_2 + H_2O \rightleftharpoons H^+ + {}^-HCO_3$$

These functions, while evidently important, do not quite explain why zinc plays such a critical role in growth, especially the growth of leaves. Several conditions, including the disease called "little leaf" of apricots and peaches, first observed on a sandy California soil, as well as mottled leaves in citrus, distorted leaves in pineapple, and narrow, curled leaves in tomato (Fig. 8-4*C*), are all due to zinc deficiency. The symptoms are most marked in full sunlight, that is, on the sunny side of trees. Sometimes the soil contains plenty of zinc but it is too tightly bound on soil colloids to be released by the roots.

The zinc requirement of fungi is somewhat higher than that for green plants, and historically it was the fungus *Aspergillus niger* that was the first organism for which zinc was shown to be needed (in 1905). Zinc is now known to be needed by animals too.

Boron, in the form of boric acid, was the first micro-nutrient to be recognized for higher plants (1917). Its role is somehow related to carbohydrate transport, for tissues highly dependent on transport, like the buds and twigs on the branches of trees, are the first to die in boron-deficient soils. Tissues formed by transport and accumulation of sugars, such as apples or beets, develop dry, corky layers when boron is deficient. If the condition is recognized early, spraying with boric acid or borax will correct it.

As with iron and manganese, there is an interrelation between boron and calcium; thus an excess of lime in the soil can cause boron deficiency.

Copper is needed only in tiny amounts—about 3×10^{-7} molar in nutrient solutions for land plants (Figs. 8-4 and 8-5) and less than that for aquatic plants and algae. Indeed, algae are so sensitive to copper ions that one spoonful of copper sulfate crystals suffices to kill the algae in a whole average-sized swimming pool.

Copper is present in cytochrome oxidase but not in the other cytochromes; it is also in the noncytochrome oxidizing enzymes—ascorbic oxidase, which oxidizes ascorbic acid (vitamin C) when plant tissues are crushed, and polyphenol oxidase, which oxidizes phenolic compounds to brown polymeric substances.

(A)

(B)

Figure 8-5 *A*. Plum tree seedlings growing in nutrient solution. Right, complete solution as in Table 8-2. Left, molybdenum omitted. *B*. Similar plum tree. Left, with insufficient copper (10 parts per billion). Right, with 100 parts per billion copper (as sulfate).

The latter enzyme controls the formation of lignin in woody tissues (Chapter 6, section 6).

Chlorine. The heavy metals are so often supplied as chlorides that the necessity for traces of chlorine was discovered only recently. It participates, along with manganese, in photosystem II.

Molybdenum probably has only limited functions, though these are certainly critical (Fig. 8-5). It is a component of two enzymes active in nitrogen metabolism, one for reducing nitrate to amino acids and one which "fixes" nitrogen of the air into organic compounds (Chapter 9). The lack of productivity of some Australian soils has been traced to deficiency of molybdenum, and fertilizing with *two ounces* of ammonium molybdate per hectare has increased crop yields there ten- to twentyfold. Atom for atom, the molybdenum here is producing as much useful solar energy as fissionable uranium is producing useful nuclear energy!

Hydrogen is not normally thought of as a nutrient element, but hydrogen ions (H^+) are critically important in liberating other ions from soil particles. Malic, citric, and carbonic acids are excreted from roots, and the hydrogen ions of these acids exchange with metallic ions on the surface of colloidal particles (see Chapter 15), liberating potassium, calcium, magnesium, and sometimes other elements too. Humus and other organic matter in soil tends to have the opposite effect, binding up both metal and phosphate ions.

The changes in pH during the growth of plants are very revealing, especially in nutrient solutions. When nitrogen is supplied as ammonium salts—usually $(NH_4)_2SO_4$—the solution goes acid, because the salt behaves like a mixture of ammonia and H_2SO_4, and the ammonia is taken up faster than the acid. Conversely, with calcium nitrate the solution goes alkaline due to the more rapid uptake of nitrate than calcium; the water furnishes H^+ to go in with the $^-NO_3$, and the ^-OH remains in the solution with the Ca^{2+}. As a practical matter the pH of nutrient solutions is often kept in balance, especially in long-term experiments, by dividing the nitrogen supply between ammonium and nitrate. Farmers sometimes use ammonium nitrate, too, but it requires careful handling.

3. THE ABSORPTION OF IONS BY PLANTS

The reason that plants can develop so well on the dilute solutions described above is that they have the ability to concentrate the solution within their tissues. A classic case is the enrichment of iodine in seaweeds; the iodine concentration in seawater is about 1 part per million, yet some algae contain 200 parts per million, and kelps even serve as a commercial source of iodine. Aquatic plants in pond water can have internal concentrations of K^+ ions 100 to 1000 times that present in the water. The ions so accumulated are present in the sap as free ions, so that there is no question of precipitation or adsorption as a physical explanation for the great difference in concentrations. Instead we have to deal with a truly biological process.

The ability to accumulate a higher concentration of ions internally than externally has three properties (Fig. 8-6): (1) it occurs only in oxygen, (2) it is very sensitive to temperature, and therefore not a physical process like diffusion, and (3) it requires using up the sugar or other substrate in the roots. When the sugar is used up, or if the oxygen supply fails, ions that have been absorbed may go out again into the solution. Taken together, these findings mean that the accumulation process requires the continuing expenditure of energy, doubtless in the form of ATP. For this reason the uptake and accumulation of ions by roots or tubers is prevented by cyanide, parallel to its inhibition of respiration (Chapter 6), since cyanide poisons the iron-containing cytochrome system, which furnishes most of the ATP to the non-photosynthetic tissue of roots. (In agriculture one can see why hoeing of soil is valuable, because it brings oxygen to the roots, and so enables them to accumulate mineral ions.)

The different ions are not accumulated to the same extent. Potassium and phosphate usually show the greatest ratio between internal and external concentrations. But some plants accumulate aluminum; beets accumulate chloride; in selenium-rich soils, as noted on p. 136, many native dicots and some grasses (including wheat) accumulate selenium; near the zinc mines in Belgium a species of *Viola* accumulates zinc. Thus the choice of ions is often specific. In seaweeds the sodium ion is often *excluded* by the same energy-requiring process as accumulation, using the ATP provided by photosynthetic phosphorylation. With the green alga *Ulva* growing in tanks, it was shown that in the dark the K^+ content steadily falls, but as soon as the light was turned on the K^+ content greatly increased and the Na^+ content *decreased*. Doubtless many oceanic plants operate in the same way.

Most land plants tend to exclude sodium also, but some, especially those that grow well in saline soils, accept the sodium. The ratio between potassium and sodium, K/Na, in the plant ash is very revealing in this connection; for a group of ordinary crop plants grown in a standard nutrient solution with added sodium the ratio K/Na in the ash ranged from 12 to 40, but in plants of saline soil grown in the *same* solution the ratio K/Na was between 1 and 2.

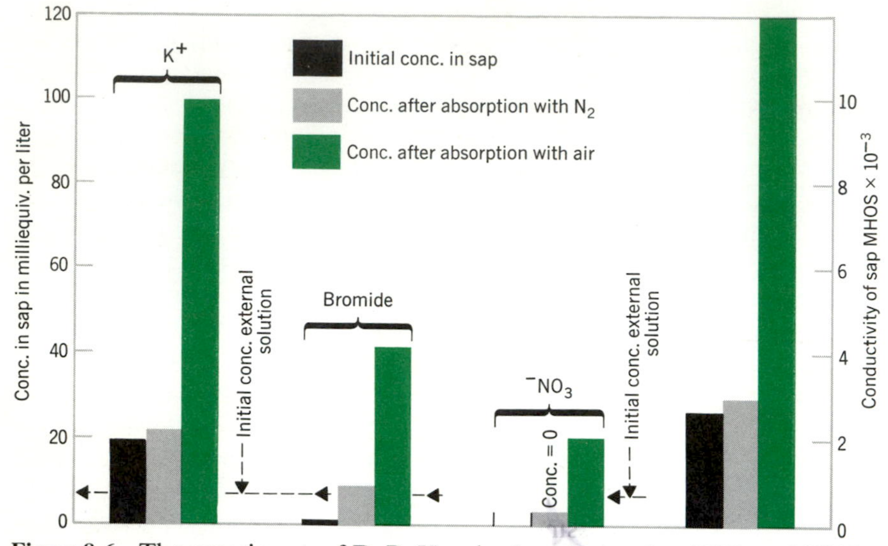

Figure 8-6 The experiments of D. R. Hoagland on the uptake of KBr and HNO_3 by barley roots. In the three groups of columns to the left, the K^+, Br^-, and $^-NO_3$ in the roots were measured chemically; the group on the right shows the conductivity of the root sap, which is essentially the sum of K^+ and Br^-. In all four groups the great effect of aeration is seen, since with aeration the internal concentrations of the salts are many times higher than they are in nitrogen, and still more than they are in the external solution. (From Hoagland & Broyer, 1936)

Plants of the New England roadside, on which salt is regularly spread every winter, show a similar tolerance.

This *selectivity* is usually considered to mean that for each ion there is a special carrier that brings it through the cell membrane. One carrier can sometimes serve for two ions; potassium and rubidium obey the same rules, and calcium and strontium make a similar pair. That there are such carriers has been demonstrated in bacteria; they are proteins called *permeases,* and are located in the cell membrane. As yet their existence has not been directly shown in plants. The carriers are evidently energized by ATP, and thus by oxidations (in colorless cells), and they use their energy in some way to bring the ion actively into the cell, against its concentration gradient. Inevitably, however, as the internal concentration builds up, the ions begin to leak out, so that a maximum internal level is reached. For many ions and many roots, the ratio between the internal and external concentrations at maximum varies from 1 to about 50, depending on the species, the temperature, and the absolute external concentration.

In addition to these selective, and energy-using processes, simple diffusion of ions also takes place across the cell membrane. It operates in both directions. Since soil solutions tend to be more dilute than cellular contents, such diffusion does not contribute much to normal mineral nutrition, and usually causes loss of ions (leakage) rather than uptake.

4. THE USE OF FERTILIZERS IN FARM AND GARDEN

Long before the mineral requirements of plants were known, or even before their need for minerals from the soil was realized, farmers knew that human and animal dung promoted the growth of crops. The use of dung is mentioned in the Bible. The North American Indians used to bury a fish in each hill of corn to increase the yield. In Europe and America farm manure, blood, and organic wastes were widely used in the eighteenth and nineteenth centuries; later Peruvian guano (bird manure) became popular. Much of the favorable effect of these materials will have been due to their nitrogen content, but the phosphate, potassium, and other minerals no doubt made important contributions. As we saw at the outset, organic matter as such was believed to be taken up by plants until the mid-nineteenth century, but from 1840 onwards the growing understanding of chemistry began to shed light on the inorganic components of fertilizers, due to almost simultaneous efforts in Germany, Britain, and France. When Liebig became convinced that plants obtained calcium, potassium, sulfate, and phosphate from the soil and needed them for growth (1840), he urged farmers to add these materials to their fields to replenish what was removed with the crop. Sir John Lawes, an English baronet who was farming his lands at Rothamsted (25 miles from London), found that bones—mainly $Ca_3(PO_4)_2$—were effective as fertilizer on some soils but not on others. He tried the acid phosphate $Ca(H_2PO_4)_2$, called "superphosphate," and found it excellent for turnips. In 1842 he started a factory for this material (with some misgivings as to whether a gentleman should enter the manure business) and thus initiated the chemical fertilizer industry. Shortly afterwards he converted his house and grounds into the world's first agricultural experiment station (Rothamsted). Here he and his colleague J. H. Gilbert carried out long-term experiments on fertilizers and their effects, and on the exhaustion of soil that occurs when the same crop is planted year after year without fertilizer. Here the role of boron was first discovered. Here, too, forty years earlier, the fixation of free nitrogen by leguminous plants was confirmed on an agricultural scale (Chapter 9). It was here that in 1856 the first "environmental impact" studies of fertilizer use were conducted, showing how selected fertilizers affect the botanical composition and yield of a mixed population of grasses, clovers, and weeds. The Park Grass Plots used for this work had been in grass for several centuries before Lawes began his work, and the soil type, aspect, management, and climate were the same for all of them—a perfect basis for critical experiments.

In France, J. B. Boussingault, like Lawes and Gilbert, used his country estate near Paris from 1838 onward for experimental plots, which were the first such plots in France. On a field scale he repeated and confirmed de Saussure's work on the uptake of salts by plants, but since most of his work concerned nitrogen metabolism it will be deferred to Chapter 9.

Farmers at first were slow to accept the idea that a hundred pounds or so per acre of dry salts, made in a factory, could substitute for the tons of farmyard manure they were accustomed to using. But the tests were so convincing that in 1854 English farmers collected money to build Lawes a larger and better laboratory. Nowadays chemical fertilizers are manufactured in millions of tons.

Both for inorganic and organic fertilizers on the market, the analyses for nitrogen, phosphorus, and potassium are normally stated on the container (obligatory in the United States). The data are the percent by weight of nitrogen (expressed as elemental N), of phosphorus (expressed as P_2O_5), and of potassium (as K_2O). The phosphorus and potassium are not expressed as the elements because they are not analyzed as such. Thus fertilizer labeled as 5-10-5 contains, per hundred pounds dry weight, 5 pounds of nitrogen, the amount of phosphorus in 10 pounds of P_2O_5, and the amount of potassium in 5 pounds of K_2O. The other materials may be organic or inorganic or both, depending on the source of supply; it is the nitrogen, phosphorus, and potassium that are most often found to be in limiting quantity. Analyses of the soil are sometimes made as a guide to what fertilizers should be applied, but their value is somewhat limited since crops have different requirements, and soils differ in their ion-binding capacities and pH.

A comparison of the qualities of inorganic and organic (animal) fertilizers is given in Table 8-3. Vegetable compost will be similar in most respects to the manure but is usually far lower in nitrogen content.

TABLE 8-3

Comparison of Inorganic and Organic Fertilizers

Chemical Fertilizers	*Animal Manure and Compost*
Require careful balance of constituents	More or less in biological balance
Variety of compositions available to suit different soils and crops	Composition essentially constant
Immediately available to roots (especially valuable for seedlings)	Require some time to complete decomposition in soil
Small bulk	Bulky and, if wet, heavy
Free from sodium (except traces of impurity)	Tend to have high sodium content (except in plant compost)
Make little contribution to soil quality	Provide humus (though if too low in nitrogen may sequester nitrogen from the soil)
Represent an investment of fuel energy	Less energy requirement, except for collection, drying, and transport

CHAPTER 9

Nitrogen Metabolism and the Nitrogen Cycle

1. THE GENERAL ROLE OF NITROGEN IN PLANTS

Plants contain a host of varied nitrogenous compounds; proteins, including enzymes, contain about 16% nitrogen, nucleic acids about 13.5%, and chlorophyll about 6%. Thus the element nitrogen enters into plant life in multiple ways, and its availability determines both growth and yield. On soils that yield adequate nitrogen (other conditions and nutrients being favorable) most plants appear dark green because they are high in chlorophyll, and in addition the younger tissues especially have a juicy or succulent quality. Conversely, on soils low in nitrogen plants tend to be paler green and more fibrous. In actual nitrogen deficiency the leaves become yellow.

As plant parts, especially leaves, get older their proteins break down and the resulting amino acids are largely transported back into the stems, trunks, or crowns. Along with this the chloroplasts are broken down and chlorophyll becomes oxidized, so that the symptoms of aging somewhat resemble those of nitrogen deficiency. The same thing happens when seeds are formed. Because seeds are usually rich in protein, much of the plant's nitrogen supply becomes accumulated in the seeds and fruits, while the leaves and stems correspondingly turn yellow. The golden yellow color of a field of ripe wheat is familiar to everyone. Less poetic, but essentially similar yellowing, is seen in most annual crops at harvest time.

Farmers know that seedlings, which are making rapid vegetative growth, require good quantities of readily available nitrogen early in the season. Usually ammonium salts, nitrates, or urea are used in developed countries, animal dung in others (and sometimes in developed countries, too, when available). With adequate nitrogen, plants can take in more potassium and phosphate; for example, the uptake of phosphate by wheat in its first 15 days is almost directly proportional to the amount of nitrogen supplied. Thus 200 to 300 lb of ammonium sulfate per acre,* or an equivalent amount of gaseous ammonia injected into the soil, or again an equivalent amount of potassium or calcium nitrate, is a typical requirement for acceptable yields of cereal grains. In pasture grasses where the total leaf area, rather than the grain, is the criterion, smaller dressings of nitrogen fertilizer are used, but unless some usable nitrogen is regularly applied, pasture quality deteriorates

* 100 lbs per acre equals 112 kg per hectare.

rapidly. Leaf and petiole crops, like cabbage, spinach, tobacco, and celery, need plenty of nitrogen.

The nitrogen taken up by plants goes back into the soil when the plant dies, or when it is eaten by an animal that excretes it (or dies also), so that the soil becomes a recipient of vast quantities of nitrogen compounds. Indeed at least two-thirds of all the total nitrogen in an ecosystem is in the soil at any time, even with a full canopy of herbaceous plants or trees. Some is in forms that are bound to soil particles, some in the organic residues, and some is present as free small molecules like ammonia or nitrate.

In the soil the proteins of the plants or animals, and the animal excreta, are broken down, pass through a series of changes, and the nitrogen is finally made available to the plant roots. A fraction of the combined nitrogen even exchanges with the free nitrogen of the air. In these reactions a variety of bacteria and fungi interact with the plants. All together these processes constitute the Nitrogen Cycle of nature. The cycle is presented in Figure 9-1, and the

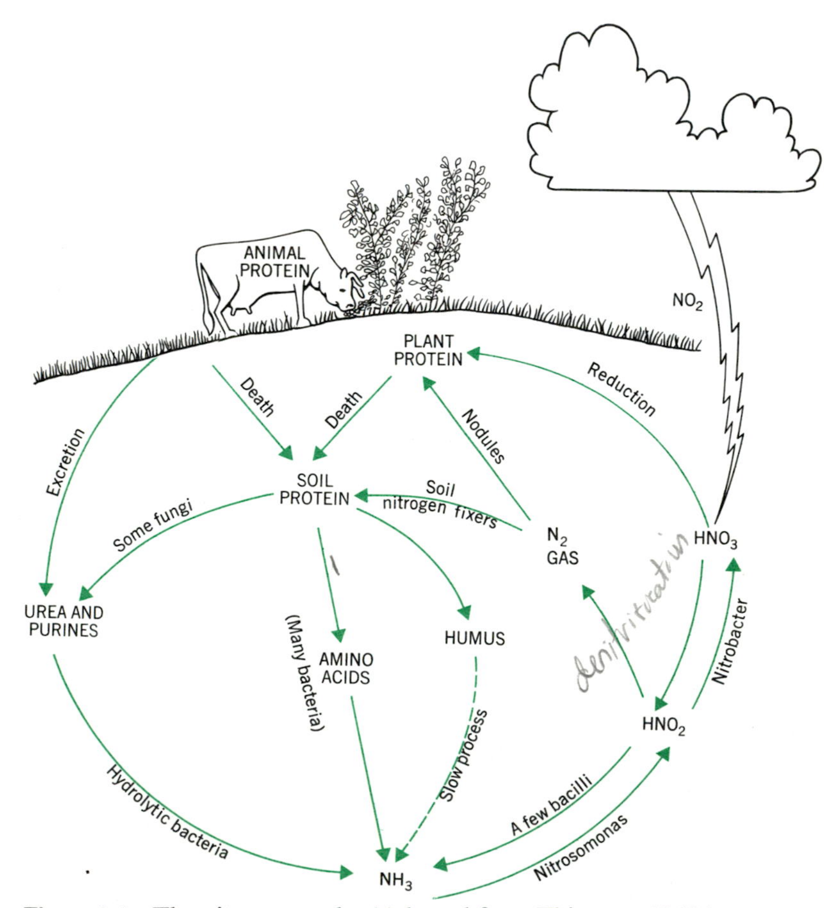

Figure 9-1 The nitrogen cycle. (Adapted from Thimann, 1963.)

remainder of this chapter is devoted to tracing out its various stages.

2. THE FORMATION OF PROTEIN IN PLANTS

We saw briefly just above that nitrate taken up from the soil by the roots gives rise to amino acids and thence to proteins. Here we consider the process in more detail. In nitrates the nitrogen is highly oxidized; potassium nitrate is

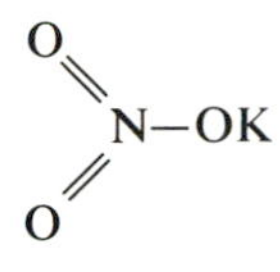

In amino acids the nitrogen is highly reduced; the amino acids can be thought of as ammonia, NH_3, with one of the H atoms replaced by a carbon-containing group, that is, $R\text{-}NH_2$; (the R also contains an acid group, -COOH). It follows that nitrate, after absorption into the roots, must be subject to rapid reduction. This is brought about by the hydrogen content of carbohydrates (e.g., glucose is $C_6H_{12}O_6$), which come down to the roots, via the phloem, from the green parts.

There are two phases in the use of nitrogen by plants—uptake and reduction. *Uptake,* first by the root hairs and then by the adjacent root parenchyma, is assisted by a special enzyme system that develops in the cells when nitrate is supplied. The entry of nitrate is slow for the first hour or two after nitrate is added, then—in barley roots at least—it becomes rapid for at least 16 hours. The first slow uptake is probably due to passive entry, the nitrate penetrating into the root cells along with the water, essentially a diffusion. The faster rate develops when the enzymatic process comes into play; this second system is *induced* in response to the presence of nitrate. This means, in essence, that of the many types of protein continually being synthesized by the cells, *one* is specifically made to fit the nitrate molecule.

What is interesting about this induction is that a second enzyme system, that for the reduction of nitrate to ammonia, is also induced, so that the accelerated uptake of nitrate is accompanied by an accelerated reduction of $-NO_3$ to $-NH_2$.

Reduction occurs in two steps. First the nitrate is reduced to nitrite:

$$HNO_3 + 2(H) \rightarrow HNO_2 + 2H_2O$$

(For simplicity nitrate and nitrite are written as nitric and nitrous acids.) The hydrogen for the reduction comes from the metabolic breakdown of the carbohydrates. In the second step the nitrite is further reduced to ammonia:

$$HNO_2 + 6(H) \rightarrow NH_3 + 2H_2O$$

We saw in Chapter 6 that the Krebs cycle produces, among other intermediates, a group of keto-acids, compounds of the general form

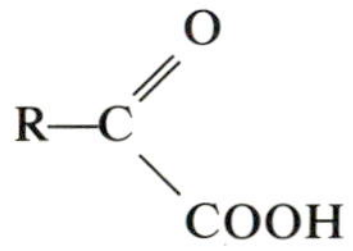

Some of these combine readily with ammonia, in the presence of specific enzymes. The combination produces amino acids.

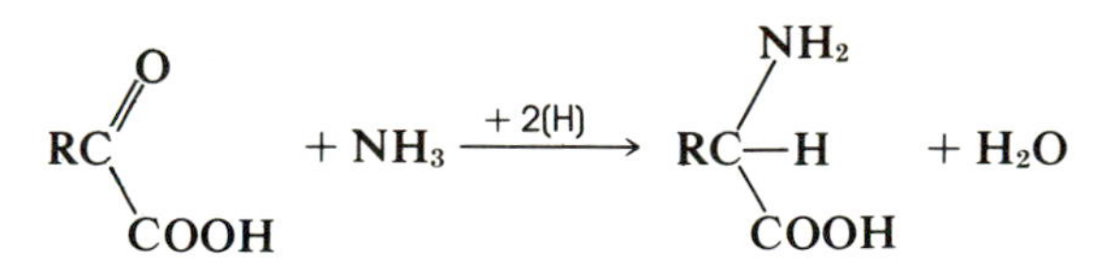

Only 3 amino acids are thus formed: alanine, aspartic acid, and glutamic acid. However, they can transfer their $-NH_2$ groups (amino groups) to other keto-acids, and by such transfers all the 20 amino acids needed to form proteins are produced. If we write the above 3 amino acids as containing R_1, and any of the other 17 as containing R_2, the transfer can be written as:

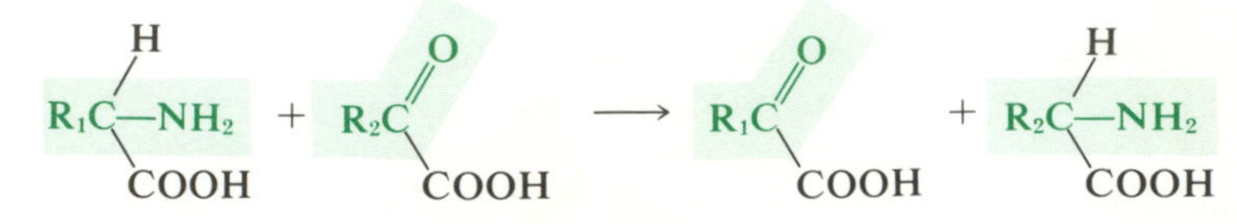

Of the 20, 3 are basic (arginine, histidine, and lysine), 3 contain benzene rings (phenylalanine, tyrosine, and tryptophan), 2 contain smaller rings (proline and oxyproline), 2 contain sulfur (cysteine and methionine), and the remainder have simple carbon chains with up to 6 carbon atoms.

Normally these amino acids are all formed in the roots, together with a few other amino acids that do not enter into proteins (and that vary with the plant species). If there is an extremely rapid supply of nitrate a little of it may be drawn up into the leaves with the movement of water (section 5 of Chapter 7); for this reason edible plants, like beets, spinach, radishes, and lettuce, that are richly fertilized, or pasture grass, rained on after a long dry spell, sometimes contain traces of nitrate or nitrite.

Photosynthesis in the leaves produces large quantities of ATP (Chapter 5), and this combines with the amino acids, which have come up from the roots, to produce phosphate derivatives, which are more reactive. They combine first with coenzyme A and then with specific types of nucleic acids (*transfer-RNA* or t-RNA). The resulting *aminoacyl-t-RNAs* are carried to the ribosomes of each cell (see Fig. 4-2). Here they are linked together into peptides, setting free the RNA again.

The formation of proteins from amino acids is described in more detail in the boxed supplemental material at the end of the chapter. Three different kinds of ribonucleic acids (RNA) direct the process, one to bring the amino acids to the site of synthesis, which is the ribosome, one to lay down the pattern dictated by the DNA, in which they are to be joined, and the one on the ribosome to hold the pieces together for the reaction to take place. For this purpose groups of ribosomes are joined into "polysomes."

Most proteins contain at least 100 amino acids joined together. Many of the proteins are enzymes, acting on specific substrates in a specific way, dehydrogenating, oxidizing, transferring groups, and so on. The latest international catalog lists about 3300 known enzymes, and more are being discovered all the time. When animals feed on plants they break down the plant proteins into amino acids and build them up again into new and different animal proteins, thus vastly increasing the number of different protein structures.

The syntheses of the nucleic acids, from the purine and pyrimidine bases plus ribose and phosphate, involve chains as complex as those for proteins, but they cannot be treated here.

3. THE RETURN OF NITROGEN TO THE SOIL

When plants and animals die their bodies, along with fallen leaves, are returned to the soil. The plant material forms *litter* on the surface of the soil. The excreta of animals are added to the soil as well. The main forms of nitrogen so returned are proteins, nucleic acids, and urea. These compounds are immediately attacked by fungi and bacteria, the attack being faster the warmer the soil. The simpler compounds are mainly oxidized, whereas much of the polymer material (cellulose, xylan, etc.) is hydrolyzed to smaller molecules and the rest is diverted to the form of stable, long-lasting brown complexes called *humus*.

There are some misunderstandings about the role of humus in soil. The fact that a dry, hard, or clayey soil is improved in its physical qualities by the presence of some organic matter, which decomposes only very slowly, often leads to the idea that the *organic* matter in the soil confers some chemical benefit or supplies some undefined but valuable adjunct to plant growth. As organic matter decays it liberates mineral ions, and the moisture-holding property of its polysaccharides and colloids make difficult soils tillable, increases pore space, and delays their drying out in the hot sun. But the nitrogenous compounds of the tissue, in their original organic form, are poorly absorbed by roots and hence do very little for the nourishment of plants; if it were not for the beneficial action of soil bacteria and fungi, which convert these materials from the organic to the inorganic form, most plants would soon starve from nitrogen deficiency. For it is only in the forms of ammonia or nitrate that roots can take up nitrogen at any useful rate. In thus rendering organic nitrogen available to green plants, the soil microbes carry out three major steps.

Step 1. The Breakdown of Polymers to Small Molecules

As described in the supplement, a protein consists essentially of a chain of *peptide* linkages, bearing side groups characteristic of the different amino acids. These peptide linkages thus bind the different amino acid structures together with loss of hydrogen and hydroxyl, achieved by way of the RNAs. Correspondingly the breakdown of these linkages is achieved by inserting molecules of water again to separate the nitrogen–carbon bonds. This is done by hydrolytic enzymes called *proteinases,* which are present both in plant tissues and in soil microbes. The products are the amino acids again.

The structures of the nucleic acids are based on groups of three compounds: a nitrogen-containing base, a 5-carbon sugar, and a phosphate group (Chapter 6, section 1). These *nucleotides* are linked into polymers through their phosphate groups by loss of hydrogen from one and hydroxyl from the other. As with the proteins, the nucleic acids are attacked by hydrolysis, which in this case acts at the phosphate bonds, liberating the nucleotides; then further hydrolysis sets free the bases and the 5-carbon sugars.

Step 2. The Release of Ammonia

The amino group, $—NH_2$, of the amino acids is set free when the amino acid is oxidized in the soil:

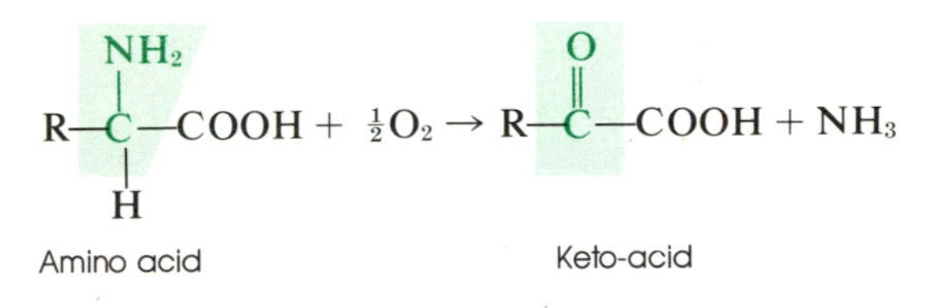

The products are *keto-acids,* which are further oxidized to carbon dioxide and water. Oxidations like this take place in most reasonably well-aerated soils. In very wet soils, like bogs, where the access of air is limited, the amino acids may be reduced instead, but again the amino group is released as ammonia:

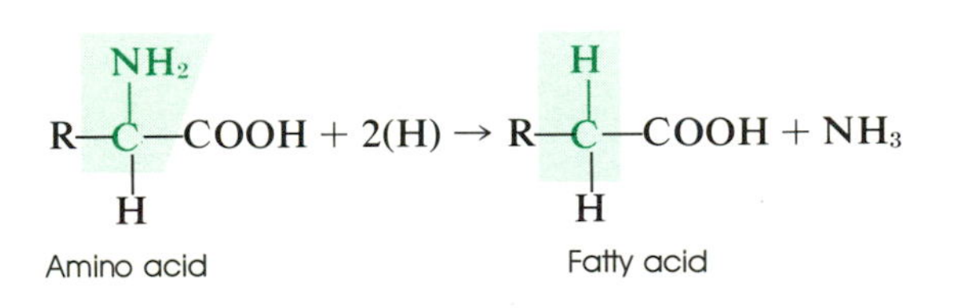

The resulting fatty acids are not broken down as rapidly; also some amino acids may be partly diverted to amines and other secondary products. As a result, bogs are much richer in organic matter than are aerobic soils, and they are also usually more acid. The ammonia set free does not come off as the gas NH_3 but is either neutralized by soil organic acids or else absorbed on soil particles; in both cases it forms the ion NH_4^+.

The reactions of the nucleotides set free from nucleic acids are more complex, but the net result of a series of oxidative reactions is again the release of ammonia.

The animal excreta are of simpler structure, and the simplest, as well as the one in largest amount, is *urea*, $NH_2—CO—NH_2$. It is easily hydrolyzed to CO_2 and ammonia by several soil microorganisms:

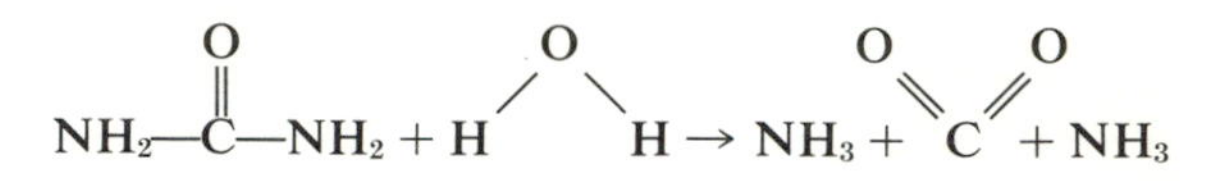

Step 3. The Oxidation of Ammonia

This step is the most characteristic, because it includes unusual biochemical reactions that can be carried out by only a few specialized bacteria. Solutions of ammonium salts do not oxidize to nitrate on simple exposure to air; yet, although ammonia would be the common product of the processes described above, analyses of agricultural soils in fact yield little or no ammonia. Instead, most of the soil nitrogen is in the form of nitrate. Only very acid soils, such as those under pine forests, contain much ammonia (as ammonium salts of organic acids). Soil must therefore bring about a special reaction.

In 1877 to 1879 two Alsatian workers, Schloesing and Müntz, set a solution of an ammonium salt to percolate slowly through a column of sandy soil mixed with chalk; after a few days the drops coming out at the bottom no longer contained ammonia but instead contained calcium nitrate. The NH_3 had been oxidized to HNO_3, which was then neutralized by the chalk. When boiling water was poured on the column the oxidation stopped. Since chloroform stopped it, too, they reasoned that the oxidation must be due to some kind of living organisms. Subsequent workers set up cultures with ammonium salt solutions together with other nutrients and inoculated them with soil; bacteria grew and nitrate was formed. More detailed study showed that some of the bacteria carried the reaction only part way, that is, to nitrous acid:

$$NH_3 + \tfrac{3}{2}O_2 \rightarrow HNO_2 + H_2O$$

Others oxidized the nitrous acid to nitric:

$$HNO_2 + \tfrac{1}{2}O_2 \rightarrow HNO_3$$

(In both cases calcium carbonate was present so that the acids were always neutralized.) However, these were mixed cultures and for a while no cultures containing only a *single* bacterial type able to carry out either one of these oxidations could be obtained.

In thinking about the problem, Serge Winogradsky, a Russian microbiologist, realized that such oxidations would yield a great deal of energy and surmised that with this energy bacteria might be able to *reduce CO_2 to organic matter*. In green plants the energy for CO_2 reduction comes from light, through photosynthesis; in these bacteria it would come from chemical reactions, i.e., *chemosynthesis*. The reason that pure cultures could not be obtained would then be that organic nutrients were always being added to the cultures, and these were supporting great numbers of other bacterial types that were not oxidizing the ammonia. The reasoning proved to be correct; by supplying only CO_2 (in the form of $CaCO_3$), but *no* organic compounds. Winogradsky was able to isolate pure cultures of two new organisms: *Nitrosomonas,* oxidizing ammonia to nitrite, and *Nitrobacter,* (see Fig. 9-1) oxidizing nitrite to nitrate. He thus not only solved the problem of the origin of nitrate in soil but also discovered a new way of life for microorganisms, namely to reduce CO_2 to the organic compounds of their cells with the energy (and the hydrogen) yielded by chemical oxidations; no other source of organic compounds is thus required. The series of reactions involved can be summarized in the form:

$$4(H) + CO_2 + \text{energy (ATP)}^* \rightarrow (CH_2O) + H_2O$$

* Of course, in 1891 Winogradsky did not know about ATP, but he thought in terms of transportable chemical energy.

The (CH_2O) represents the organic matter of the bacterial cells. Thus, these colorless organisms carry out a reaction very like photosynthesis. Indeed, in recent years it has been found that the intermediate steps from CO_2 to organic compounds are the same in these bacteria as they are in photosynthesis, phosphoglyceric acid occupying a key position.

Subsequently several other types of chemosynthetic bacteria have been discovered that oxidize sulfur or hydrogen sulfide to sulfate and utilize the resulting chemical energy to reduce CO_2 to the organic materials of their cells. Like the ammonia oxidizers, they need no other organic materials for growth.

The ammonia oxidizers not only do not need organic matter, their growth is actually inhibited by some organic compounds. Too much organic matter in the soil can thus delay ammonia oxidation. This might be useful in very well-watered soils, because nitrates are so soluble that they leach out easily while the ammonium ion (NH_4^+) tends to be adsorbed on to the negatively charged soil particles (Chapter 15). Experiments with specific inhibitors of ammonia-oxidizing bacteria have been claimed to show increases in crop yields. This may prove an interesting development.

4. THE FIXATION OF NITROGEN

a. FREE-LIVING BACTERIA

It is ironical that, though nitrogen and its endless chemical changes are the very stuff of life itself, nitrogen gas, N_2, is inert. Although it constitutes 78% of the atmosphere, it does not support a flame, does not combine directly with any but a very few active metals, and does not enter into the life of animals or any plants more advanced than the blue-green algae and bacteria. The French call it *Azote* (meaning, without life). It is only when it is converted from the "dinitrogen" form, N_2, to the mononitrogen forms like HNO_2 or RNH_2 that it becomes biologically active. The conversion from the inert gas form to the active combined form is called nitrogen *fixation*.

Toward the end of the last century it was established that pots of soil set aside in the laboratory for some weeks slowly gained in nitrogen content. Further study led to the isolation of two kinds of bacteria, one growing in presence of air and the other growing anaerobically (i.e., in the absence of oxygen); both of them converted nitrogen gas to the nitrogenous compounds of their cells. Both used sugars or related compounds as sources of carbon and energy, the anaerobe, *Clostridium pastorianum,* fermenting glucose to butyric acid (Chapter 20), while the aerobe, *Azotobacter chroöcoccum,* oxidized the sugars to CO_2 and water. Nitrogen gas, N_2, was the sole source of nitrogen for the bacterial proteins. Since that time numerous other bacteria have been found to fix nitrogen, and many (but not all) of the blue-green algae (Cyanobacteria) do it too.

In rice fields the moist soil is frequently covered with a mat of blue-green algae, which therefore make an appreciable contribution to the nitrogen nutrition of the rice plants. Cultures of active species of blue-greens are supplied to farmers in India to inoculate the soils of their rice fields. Since none of the higher groups of algae can fix nitrogen, and only traces of nitrogen are perhaps fixed by a very few special fungi, the ability to fix nitrogen is primarily a property of *procaryotic* cells.

Even among the procaryotes the power to fix nitrogen is by no means general, but it is of special interest that the *photosynthetic bacteria,* which contain chlorophyll of a slightly different kind from that in the algae and higher plants, all seem to fix nitrogen actively. Their photosynthesis (Chapter 5, section 4) is unlike that of the green plants since, instead of using water as a source of hydrogen (and thus releasing oxygen), they use either hydrogen sulfide,

H_2S (thus releasing sulfur) or the hydrogen of specific carbon compounds, such as malic acid.

b. NITROGEN FIXATION BY SYMBIOSIS: THE LEGUMES

One of the exceptions to the rule that only procaryotes appreciably fix nitrogen occurs in the legume family. This important group is described in Chapters 13 and 23, but here we are concerned only with their special nitrogen usage.

Long ago the Romans recognized that these plants have a special ability to "enrich the soil" or "renew its fertility." Early Chinese writings suggest that the Chinese knew about this property also. In the Middle Ages when *crop rotation* was practiced, one of the crops most often used was the common legume, clover. Alfalfa (or lucerne) is more often used now.

The legumes bear little outgrowths of tissue called *nodules* on their roots, and in these nodules there live masses of a special type of bacterium, the genus *Rhizobium* (*Rhizo-*, root; *bios*, life). The combination of plant and bacterium has the power of fixing nitrogen. The nodules thus represent a *symbiosis* between a procaryote, the *Rhizobium*, and a eucaryote, the leguminous plant.

The way in which the nodule is formed has been worked out in great detail. The *Rhizobia* enter through the tips of the root hairs, which are very thin-walled. They multiply within the root hair and spread in a sort of fine rope down into the cells of the root cortex (Figs. 9-2 and 9-3*A*). The auxin and cytokinin they secrete (see Chapter 10) stimulate some of these cortical cells, especially any that are tetraploid, to divide, and at each cell division some of the *Rhizobia*, which cluster around the nucleus, are carried into the newly formed daughter cells. One nodule cell is estimated to contain, when the process is complete, up to 50,000 *Rhizobia* (Figs. 9-3*B* and 9-4). As the number of bacteria-laden cells increases, the plant does its part by developing branches from a nearby vascular bundle, which supply the nodule with sugars and nutrient salts and remove the nitrogenous products.

There are several peculiarities about the composition of the nodule. Within each cell the *Rhizobia* are not free in the cytoplasm but are enclosed in little membranes, each of which holds from one to a dozen or so of the bacteria (Fig. 9-3*B*). Also the nodules contain a pink pigment that is a hemoglobin, like the oxygen-carrying hemoglobin of our red blood cells, but with an even stronger affinity for oxygen than ours. Since hemoglobin is not normally found in green plants, its synthesis is due to the symbiosis. The *hem* part, which resembles chlorophyll in structure, but with iron instead of magnesium, seems to be made by the *Rhizobia*, while the globin part, a large colorless protein, is contributed by the legume. Apparently the combined role of the hemoglobin and the membranes is to maintain a low level of oxygen within the nodule cells, making the local conditions around the *Rhizobia* reducing enough to hydrogenate the unreactive nitrogen.

Studies with isotopic nitrogen, $^{15}N_2$, show that the isotope enters first into ammonia, that is, the process is basically one of hydrogenation. Within a minute or two, though, the isotope is found in glutamic and aspartic acids, which are the main products of nitrogen fixation both in nodules and in free-living nitrogen fixers. As seen in section 2 above, these two amino acids have the special ability to transfer their amino groups to other keto-acids, so that soon all the 20 amino acids needed for the synthesis of the plant's proteins are formed. The legume plant can thus satisfy the whole of its nitrogen requirement through its root nodules.

This is not to say that all legumes are independent of external nitrogen supply under all circumstances. For one thing there are over 10,000 species in the legume family (Chapters 13 and 23). Of these less than 1000 have

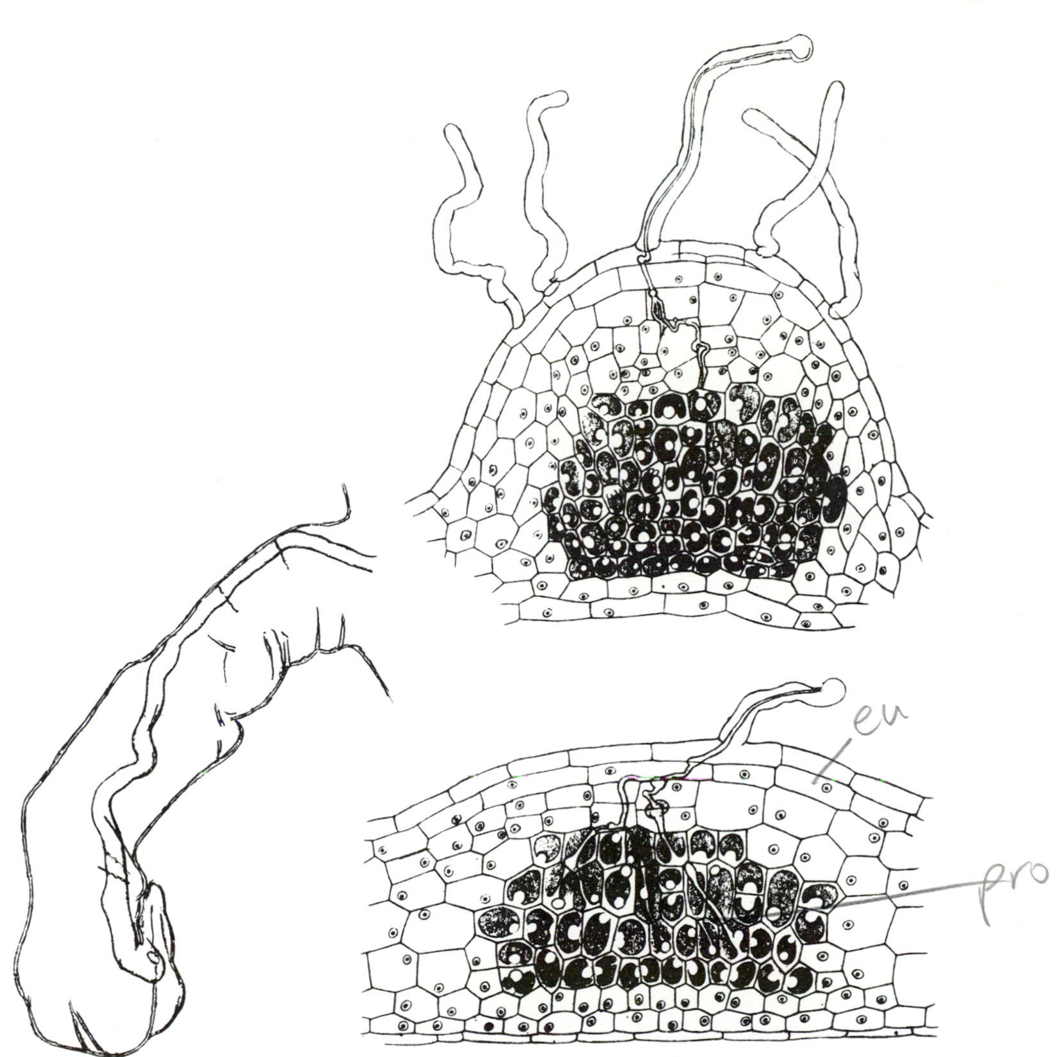

Figure 9-2 The first drawings of nodule formation by Frank in 1891. Left, a root hair of pea characteristically curled and containing a long *infection thread* full of the Rhizobia. This thread works its way into the root cortex and stimulates the cells there to divide and enlarge, forming a *nodule,* two growth stages of which are shown. The cells in the center of the nodule are packed with Rhizobia and are pink in color. (From Frank, 1891.)

been examined for nodules. Many of the large tropical trees (subfamily Caesalpinoideae) in the family do not seem to bear any nodules. Furthermore the nodules do not always supply enough nitrogen to give the maximal yields of agricultural crops like peas and beans. Many farmers get increased yields from soybeans, in particular, by supplying nitrogenous fertilizers to the young seedlings. Thirdly, in an area where a particular legume has not previously

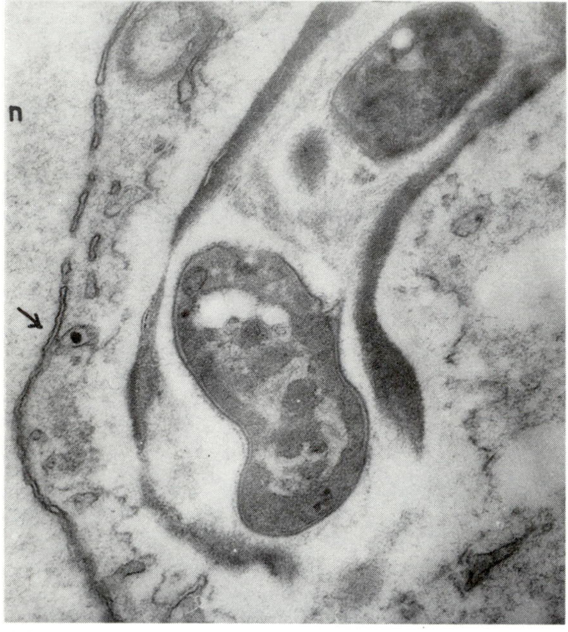

(A)

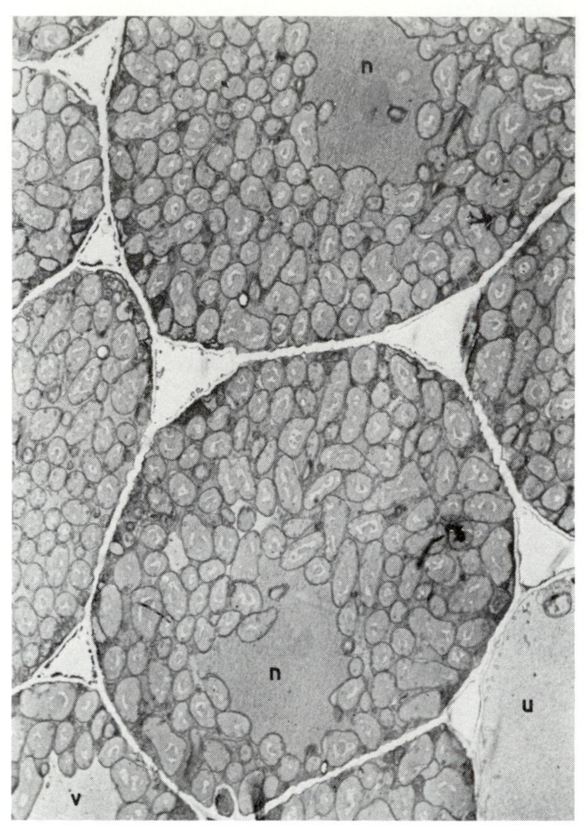

(B)

Figure 9-3 A. The tip of infection thread of a *Medicago* nodule with open end just about to release two Rhizobia. At far left is the nucleus *n*, with its porous membrane (arrow). × 38000. B. Bacteroidal cells of subterranean clover (*Trifolium subterraneum)* showing how the nucleus, *n*, is crowded by the Rhizobia. v, vacuole; *u*, uninfected cell. Magnification × 3260.

grown, the right strain of *Rhizobia* may not be present in the soil. Some species of *Rhizobium* are highly specific for one or a few leguminous hosts; *Rhizobium trifolii,* for example, is specific for clover (*Trifolium* spp.), while *R. leguminosarum* forms nodules mainly on pea and vetch. However, some wild legumes like *Astragalus* can nodulate with a number of different species. When soybeans were first introduced from China to Europe at the turn of the century they failed to nodulate, and only when a sample of soil from China was introduced into the bed did nodules subsequently form. For some years, both in Europe and America, farmers were supplied with cultures of *Rhizobia,* but as a rule they are widely distributed in agricultural soils.

C. OTHER CASES OF SYMBIOTIC NITROGEN FIXATION

Besides legumes, a number of other plants, mostly small trees or shrubs, bear extensive and irregular nodules on their roots, and several of these have been shown to fix nitrogen. They are distributed in a variety of plant families. Well-studied examples include the common alder, *Alnus* spp. (Betulaceae); the California lilac and New Jersey tea, both species of *Ceanothus,* (Rhamnaceae) a mainly Pacific coast group of shrubs; and *Casuarina,* an Australian genus known to grow well in poor soil. The small family Myricaceae, which is botanically close to the willows and poplars, contains *Myrica* and *Comptonia,* the so-called sweet fern (though it

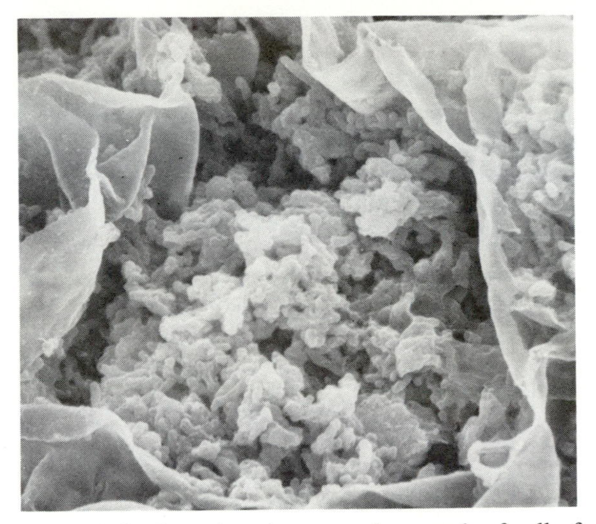

Figure 9-4 Scanning electron-micrograph of cell of a soybean nodule broken open to show the mass of Rhizobia inside. Magnification × 2900.

is not a fern). All of those mentioned have been proved to fix nitrogen. The symbiont often appears not to be a *Rhizobium,* but an organism belonging to the Actinomycetes, which are like bacteria but form long hyphae instead of short rods or spheres. The amounts of nitrogen fixed by these nonlegumes are often small, but they can be important in principle as showing that the symbiotic type of nitrogen fixation can occur in several different plant families.

There is an anatomical difference between the nodules of legumes and the nodules on some of these other families. In the legumes the bacteria enter and multiply in the cortex; the nodule develops there, and becomes connected with the plant's conducting system only later. In the nonlegumes that have been studied, the nodule appears to be initiated at or near the pericycle, probably due to a hormonal stimulus produced by the bacteria. Thus in its early stages it is more like a lateral root, which then later swells out instead of elongating. In *Comptonia* the nodule is indeed a mass of lateral roots (Fig. 9-5), tightly grown together and swollen into nodule tissue at the base. In *Casuarina* the lateral roots remain separate. In both types, legume and other, the nodule is rich in auxin, which evidently plays a major role both in its initiation and its development.

Figure 9-5 The nodule of sweet fern (*Comptonia peregrina*) comprises a mass of lateral roots grown together. Instead of Rhizobia it contains *Frankia,* (an Actinomycete) which also can form nodules on *Alnus* and *Myrica*. All fix some nitrogen. Magnification approximately × 8.

d. ENZYMES AND THEIR IMPLICATIONS FOR MINERAL NUTRITION

The enzymes bringing about the changes in nitrogen compounds are in several ways unusual. The enzyme active in reducing nitrates contains an atom of molybdenum in its large molecule. Another uses as intermediate the iron-containing compound *ferredoxin,* which acts as an intermediate in the steps of photosynthesis. For this reason chlorophasts in light can reduce nitrite to ammonia. removing excess nitrite from leaves rapidly. A similarity with nitrogen fixation appears because in all nitrogen-fixing organisms that have been studied ferredoxin (or another closely related compound) is again the critical intermediate. Ferredoxin seems to be able to accept or pass on electrons under the most strongly reducing conditions known.

The enzyme system for nitrogen fixation, as studied in the free-living nitrogen fixers, has several parts. One of these is ferredoxin, needed because it is the strongest reducing agent that plants have available, but another part contains *both* iron and molybdenum in its molecule. This and the nitrate reducers together explain why very small amounts of molybdenum can have disproportionately large effects in promoting plant growth. They also explain why legumes generally seem to have a high requirement for molybdenum. Peanuts in West Africa have shown a good increase in yield when only 28 g of molybdate per *hectare* were added to the soil. In some Australian soils, a similarly small amount of molybdate has raised legume yields by ten- or twentyfold.

Lastly, the *Rhizobia* require traces of cobalt for their growth. Green plants, as far as is known, do not require cobalt. Thus legumes growing on the nitrogen from their nodules have a need for cobalt, but when growing on nitrogen supplied as fertilizer they do not.

5. THE LOSS OF NITROGEN

It was seen above that nitrate, when taken up by roots, is reduced to ammonia and to the amino groups of amino acids. However, these reductions not only take place in plants, they can also be brought about by microbes in the soil. Put simply, the process is the use of the oxygen of nitrate or nitrite as hydrogen acceptor, instead of the oxygen of the air. The microbial reductions follow the same steps as the reduction of nitrate in the plant tissue (section 2 above). The similarity between the use of nitrate or nitrite and the use of oxygen of the air is borne out by the discovery that in these reactions one of the cytochromes, a form of cytochrome C, plays an essential part. The hydrogen comes ultimately from various organic compounds, being transferred by the intermediate participation of NAD or NADP, the universal hydrogen transfer agents, and by a second intermediary, which is a yellow protein referred to as FAD.

The reduction of nitrate to ammonia is carried out by only a few special bacteria. More commonly, soil microorganisms carry out an intermediate reaction that uses only 3(H) per molecule of HNO_2 and yields free nitrogen gas:

$$2HNO_2 + 6(H) \rightarrow N_2 + 4H_2O$$

This is *denitrification* (not quite the opposite of nitrification); it causes the loss of fixed nitrogen from the plant or ecosystem. The conditions under which denitrification can occur are worth some special consideration.

Nitrates are all very water soluble, so that they readily leach out into the lower layers of soil when not absorbed by roots. Also, because nitrate is a negatively charged ion, it does not adsorb readily on the soil colloids. In the lower layers of soil, conditions are less aerobic, especially if reactive organic compounds have been leached out too. In the presence of such lowered oxygen tension the nitrate begins to act as hydrogen acceptor and nitrogen gas comes off. Oxygen competes strongly with nitrate for the combination with hydrogen, so that denitrification only occurs when free oxygen is severely limited. Nitrous oxide, N_2O, is an intermediate in the reaction and traces of it are sometimes found in analyses of soil gases.

Whether denitrifiers can cause a serious loss of fixed nitrogen from the plant–soil ecosystem depends critically on the conditions. Oxidation to nitrate, which requires oxidizing conditions, has to be followed by the introduction of enough fermentable material to set up highly reducing conditions. Heavy fertilization followed by leaching, or the mixing of a clear stream with a deeply polluted one, offer obvious examples. Since the soil of bogs and marshes is largely anaerobic, if such an area receives nitrate leached out from a nearby farm, denitrification there would be expected.

6. THE NITROGEN CYCLE

The whole cycle of nitrogenous compounds in nature was represented in Figure 9-1. Only the most active organisms are shown in the diagram. No doubt many others participate but are not so prominent. The left-hand side of the cycle shows the breakdown processes, the right-hand side the stages between ammonia and the plant, and the right center the reactions involving nitrogen gas (and occasionally nitrous oxide). The reactions forming and decomposing humus are slow, so that humus nitrogen actually "locks up" a small amount of nitrogen in soil. Carbon dating of the age of some humus samples in England has indicated average ages of several centuries. The deposition of protein in tree trunks has the same effect, for trees can live for centuries, and, although the nitrogen content of wood is low, nevertheless the world's forests represent a considerable reservoir of locked-up nitrogen. The bacteria decomposing carbohydrates in the soil need nitrogen for their nutrition, so that the addition of organic material low in nitrogen can cause the bacteria to withdraw nitrogen from other sources close by, and this can cause additional "locking-up" of nitrogen, though usually not for very long periods. Too much compost in the soil can thus cause temporary nitrogen deficiency.

It is to be noted that the animal kingdom plays no essential part in the cycle. Most of the reactions shown were doubtless taking place on the earth before Cambrian times when the first primitive animals appeared.

Because nitrogen continually tends to be lost as a gas or to be impounded in forms unavailable to plants, the fixation of nitrogen by procaryotes and by symbiosis is critically important. Other sources of nitrogen to the biosphere are quantitatively much smaller. Lightning fixes some by forming nitrogen oxides, NO and NO_2, and these are then brought down to the earth by rain. Industry fixes an increasing amount by chemical synthesis, combining nitrogen with hydrogen to form ammonia (Haber process) and in a modified form of this to produce urea. This reaction provides the bulk of our nitrogenous fertilizer for today's intensive agriculture. Since the reaction only goes effectively at high temperatures and very high pressures, it is costly in fuel energy. (It is costly in plants too, for 6.5 mg of carbon are used per milligram of nitrogen fixed.) It requires metallic catalysts, which also offers a suggestive comparison with the participation of iron and molybdenum in the natural fixation process. It has been calculated, furthermore, that all the fixed nitrogen factories in the world together produce less than a quarter of the amount fixed in nature (Table 9-1). Probably most of the combined nitrogen that is now present in the biosphere came originally by biological fixation.

Researchers are occupied now with trying to develop symbiotic relations between important nonlegumes, like the cereals, and *Rhizobia* or related nitrogen fixers. As yet the intimate relations between the host plant and its symbiont are far from being understood. Specific proteins, called *lectins,* are excreted by the roots of legumes and these combine with an equally specific polysaccharide on the surface of the strain of *Rhizobium* that infects that legume species. It seems that such combinations hold the *Rhizobia* on the root hair and thus facilitate their entry. Hence to bring about rhizobial infection of nonlegumes both host and bacterium will need to be genetically modified in specific ways. The present approach is therefore to try to transfer appropriate genes to both partners. Since this cannot be done by ordinary crossing some more direct transfer of DNA will have to be tried. It may well be some years before we see dramatic developments in this field.

TABLE 9-1

Estimated Amounts of Nitrogen Fixed by Different Processes

All Processes (World Scale)	*Nitrogen in Millions of Tons (Metric) per Year*
All biological processes	175
Chemical processes producing fertilizers	45
Lightning, combustion, and ultraviolet light	40
Biological Processes	*Nitrogen in Kilograms per Hectare per Year*
Medicago (Alfalfa) in semitropical soils	Up to 400
Legumes with nodules in temperate climates	100 to 300
Alnus (Alder) with nodules	Up to 300
Kudzu in the tropics with nodules	60
Myrica (Wax myrtle) with nodules	35
Blue-green algae (Cyanobacteria) in tropical or semitropical rice paddies	About 20
Free-living bacteria in poor temperate soils	0.5 to 1.0

SUPPLEMENT

Protein Synthesis in Living Cells (Plant or Animal)

Protein synthesis requires four components:

1. **The necessary amino acids, 100 or more but only 20 different kinds.**
2. **A source of energy (ATP and GTP) for the reactions involved.**
3. **Enzymes to carry out the process.**
4. **Pattern-directing and holding mechanisms.**

It could be compared to the production of a piece of cloth, which similarly requires four components: the threads, the weaver's energy, the weft and shuttle, and the pattern.

In green tissues the energy comes from photosynthetic phosphorylation (Chapter 5), while in colorless tissues such as roots (or animal cells) the energy comes from the cytochrome system, located in the mitochondria. In either case the resulting ATP first combines with the amino acids to produce phosphate derivatives, which makes them more reactive (the amino acid shown is alanine):

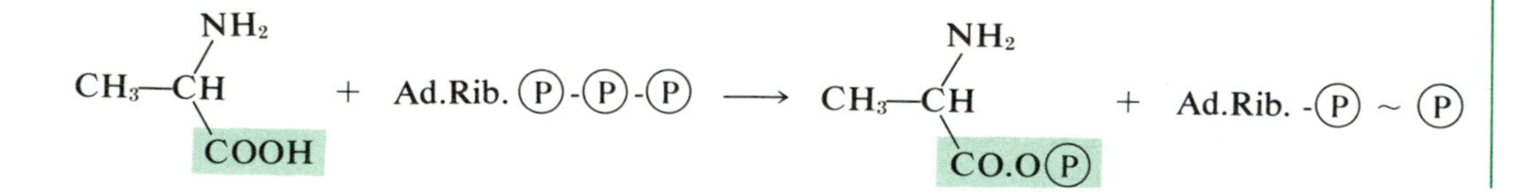

The aminoacyl phosphate then exchanges its phosphate group for that of coenzyme A (through its sulfur atom):

$$CH_3{-}\underset{\displaystyle CO.O\,\text{Ⓟ}}{\overset{\displaystyle NH_2}{CH}} + CoA.SH \longrightarrow CH_3{-}\underset{\displaystyle CO.SCoA}{\overset{\displaystyle NH_2}{CH}} + \text{Ⓟ}.OH$$

The Ⓟ stands for the phosphate group as in Chapters 5 and 6. Note the exchange of a phosphorus for a sulfur linkage; this occurs often in biological reactions.

Now the aminoacyl-coenzyme A combines with specific types of ribonucleic acids called transfer-RNA's (tRNA's); there is a different one for each of the 20 amino acids. The carbon containing groups are shown here as R_1 and R_2. The resulting *aminoacyl-tRNAs* are carried to the *ribosomes* of each cell (cf. Chapter 4, Fig. 4-1). Here they are linked up into chains in a specific order, losing the RNA again. The ribosomal RNA (*r*-RNA) is the site of linkage. This process continues until it has produced a molecule of molecular weight 30,000 or more; this is a protein. Some proteins have molecular weights up to 20 million. Each protein contains different proportions of the 20 amino acids, bound together by the peptide linkage, $-R_1-CH-CO-NH-R_2-CH-CO-NH-R_3-$etc., so that there are almost infinite numbers of different proteins. If a protein contains 200 amino acids and there are 20 different ones to choose from, the number of possible proteins would be 20^{200}, an unimaginably large number. The actual choice of amino acids, and the order in which they are attached, depends on the genes in the nucleus, which sends messengers (m-RNA's) to the ribosomes to dictate the specific pattern. Each segment of the nuclear DNA produces a corresponding m-RNA, which goes out from the nucleus to the ribosomes.

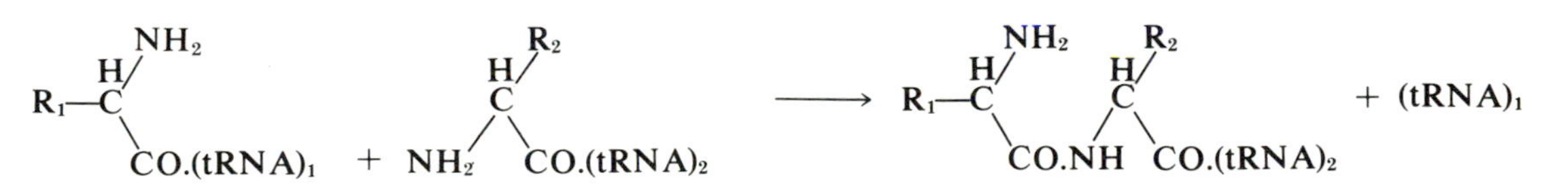

CHAPTER 10

Growth and Development and Their Control

The characteristic pattern of growth of plant cells is that they first divide, the nucleus undergoing mitosis; then each daughter cell increases in size. In massive tissues like fruits and tubers this enlargement takes place more or less equally in all directions—isodiametric enlargement—while in cylindrical structures like stems and roots the enlargement is mainly in one dimension—elongation. In either case it is the increase in *size* that is the basis of growth, and, because cells can also enlarge or shrink reversibly by osmosis (Chapter 7), growth can be defined as *irreversible increase in volume*.

1. GENERAL SCHEME OF THE GROWTH PROCESS

The organization of the apex of the shoot was described in section 7 of Chapter 4. Cell division takes place in the meristem (mainly) and the cell enlargement (especially elongation) that follows pushes the apex continually upwards. As it moves forward and the young leaves are successively left behind, each leaf-base carries with it a small fragment of meristem. This fragment becomes an *axillary bud,* which usually forms a few tiny leaf primordia and then remains more or less undeveloped, in the axil of each leaf (Fig. 4-1). In some plants such as bougainvillea, the axillary bud shows as no more than one or two small cells. In others it enlarges and may even open up somewhat, especially towards the base of a shoot (brussels sprouts offer an extreme example). In any event, it remains inhibited so long as the terminal apex is active. When the terminal bud stops growing, especially in trees, a few lateral buds often grow out into branches.

The meristem of a typical *root* (Fig. 10-1) differs in several respects, as seen also in section 8 of Chapter 4. It has a root cap, which the shoot apex lacks, and the lateral organs or "secondary roots" are produced some distance behind the meristem. Since the structure is so much simpler than that in the shoot, one can make out clearly the files of cells giving rise to the xylem, the phloem, and the root cortex, respectively. Just as in the shoot, the daughter cells away from the apex elongate and thus push the apex forward, in this case down into the soil.

The *cambium* is a secondary meristem forming a cylinder the whole length of a dicotyledonous stem and located just outside the xylem (see Fig. 4-15 and Fig. 22-4). The cambium cells of the shoot divide in the tangential

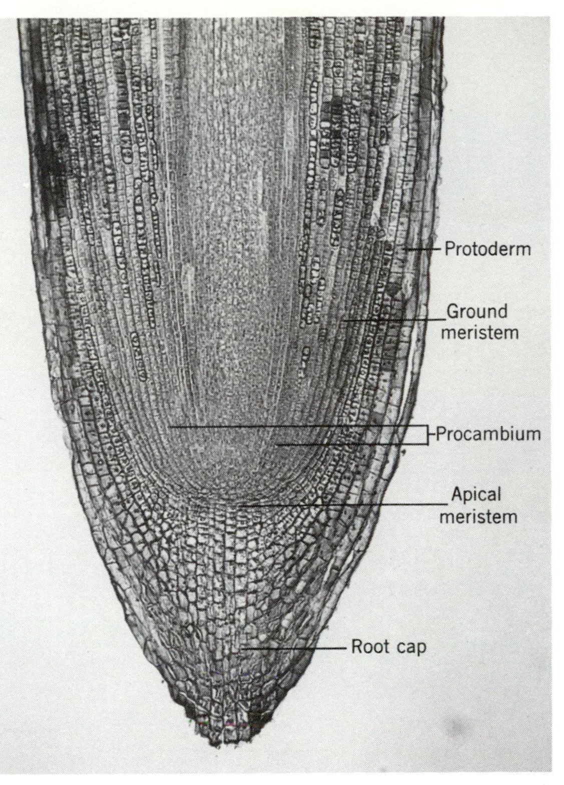

Figure 10-1 Root tip of a willow (*Salix* sp.) in longitudinal section, showing the cap, the meristem, and the main cell layers.

plane to form xylem internally and phloem externally. In the root the xylem and phloem are formed on alternate radii (Fig. 4-15).

The function of the xylem in transpiration (Chapter 7) is to conduct a very dilute solution upwards to the leaves and buds. The conducting function of the sieve tubes is almost the converse, for they carry a relatively concentrated sugar solution (usually about 0.3 *M* sucrose) out of the green parts down to the colorless tissues of the stem and roots, as well as outwards to the growing fruits.

It is tempting to draw a parallel between the xylem and phloem of the plant and the arteries and veins of a vertebrate, but the parallel is not at all close, for not only is the xylem sap dilute and phloem sap concentrated but—more importantly—the *bi-directional* movement of saps in plants is quite different from the continuous *circulation* of blood in vertebrates. Also plants have nothing analogous to the pumping system of the heart and its valves. About all that can be said is that both are supply systems, and that plants and animals have solved their supply problems in quite different ways.

2. THE GROWTH-CONTROLLING HORMONES

Growth is, of course, dependent on a supply of nutrients, especially *sugars,* formed in photosynthesis, *amino acids,* formed from sugars with the intake of nitrogen via the roots, and finally, since growth means increase in volume, *water*. However, such nutritive factors are seldom directly in *control* of growth. On the contrary, most phases of growth and development are under the control of *hormones*. These are defined as special regulatory substances that mainly exert their effects at some distance from the sites of their production. So far, five types of hormones are known, with structures shown in Fig. 10-2.

Auxin is primarily a single substance, indole-3-acetic acid (IAA), but there are a number of related compounds, esters, glycosides, aldehydes, and nitriles that can be converted to indoleacetic acid by enzymes in plant tissue and thus can in many cases act as auxin precursors. Also phenylacetic acid, of much less biological activity but somewhat higher concentration, functions as an auxin to a smaller extent. For experimental and practical purposes, synthetic auxins with structures similar to that of IAA are often used. They include naphthalene-acetic acid and 2,4-dichlorophenoxyacetic acid, "2,4-D." Auxins seem to be indispensable for all growth; tissue cultures cannot be grown without some auxin in the medium.

Gibberellins or gibberellic acids (GA), unlike auxin, are a group of over 50 very closely related compounds, all with the same basic structure of five linked rings, three in one plane

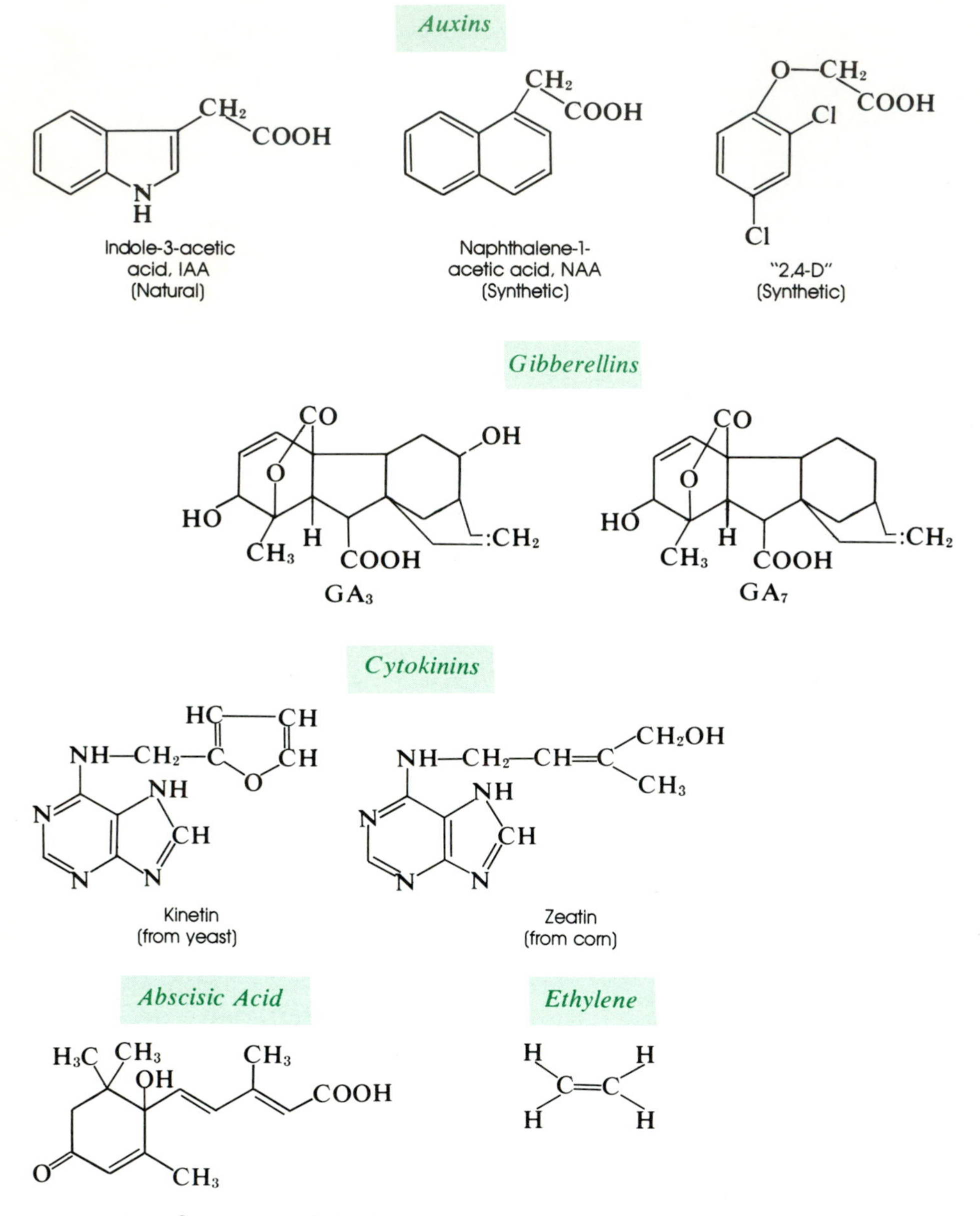

Figure 10-2 Structures of the known growth hormones of plants and some of their synthetic analogs.

and two perpendicular to this (see formulae, above). The name comes from that of the fungus from which gibberellin was first prepared, *Gibberella fujikuroi,* and the structure belongs in the group of terpenoid derivatives (Chapter 26).

Cytokinins are a small group of four or five naturally occurring compounds and some synthetic relatives; all are derivatives of adenine. They are named for their action on cell division – *cytokinesis*. *Abscisic acid* is a terpenoid acid related somewhat to carotenoids. Its growth inhibiting functions are shared to some extent by several phenols and glucosides.

Ethylene, C_2H_4, is strangely enough a gaseous hormone, capable of moving in the intercellular spaces as gas but from cell to cell in solution. Also, unlike other hormones, it can move via the air from one plant to another.

3. THE ACTION OF HORMONES IN GROWTH AND DEVELOPMENT

It is now clear that the phases of a plant's life are controlled by one, or often several, of the above hormones and perhaps by others still undiscovered. We can get a conspectus of their actions by following the plant's life stages in order.

a. SEED GERMINATION

The seeds of the cereals, of which barley is an example, consist of starch-rich endosperm tissue, largely dead cells, with a thin outer layer of living *aleurone* cells, and bearing at one end the *scutellum,* to which is attached the *embryo* (Fig. 10-3). When the seed is soaked and germination begins, the starch and protein in the endosperm are slowly hydrolyzed and the resulting sugars and amino acids travel to the embryo to provide nutrients for its growth. But if the embryo is removed, then even prolonged soaking will not cause the endosperm materials to hydrolyze. The embryo must therefore secrete something that causes the hydrolysis. Indeed, an extract from germinating barley embryos does cause starch hydrolysis in "de-embryonated" seeds. But the activity in this extract is heat stable and so can hardly be an enzyme; Japanese and Australian botanists found it to be a gibberellin, like that made from cultures of *Gibberella.* What the gibberellin does is to stimulate the *aleurone* cells, which then start to secrete amylase and other hydrolytic enzymes. Thus within the space of a few millimeters (the barley seed averages 8 × 2 mm) and within the time of a few hours following soaking, a four-

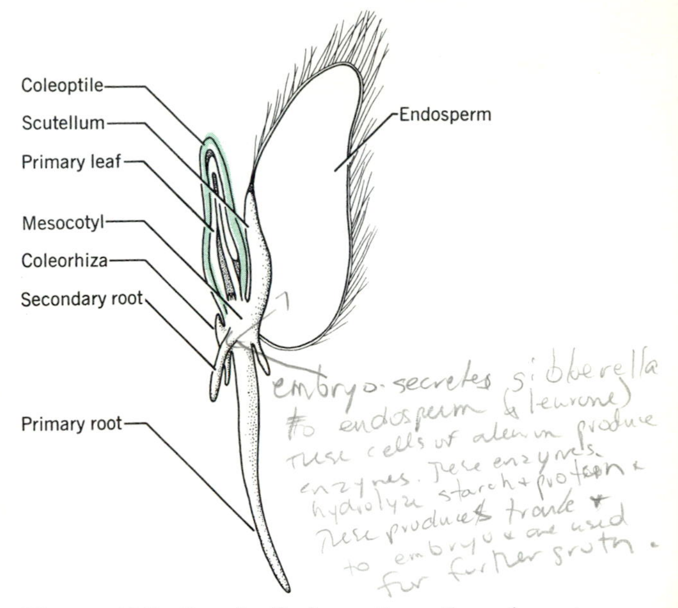

Figure 10-3 Longitudinal section through oat (*Avena sativa*) seedling after 30 hours' germination, at the beginning of the period of rapid elongation. Barley is very similar. (From Went & Thimann, 1937)

step chain of events is put in motion: (1) the swelling embryo secretes gibberellic acid, which diffuses to the aleurone, (2) the aleurone cells become activated to produce the enzymes, (3) the enzymes diffuse into the underlying endosperm, hydrolyzing the starch and proteins there, and (4) the hydrolytic products travel back to the embryo, which uses them at once for cell material and thus for growth.

Other cereal seeds have similar systems, but in most dicotyledonous seeds the situation is less clear. However, gibberellic acids do promote germination in two special types of dicotyledonous seeds, namely seeds whose germination is promoted or inhibited by light. The "Grand Rapids" cultivar of lettuce germinates only in light; red light is the most effective. Gibberellic acid, however, causes full germination in darkness (Fig. 10-4*A*). On the other hand, seeds of many desert plants, including *Phacelia* and *Nemophila,* germinate best in darkness and are inhibited by strong light. Gibberellic acid allows full germination of these seeds in light

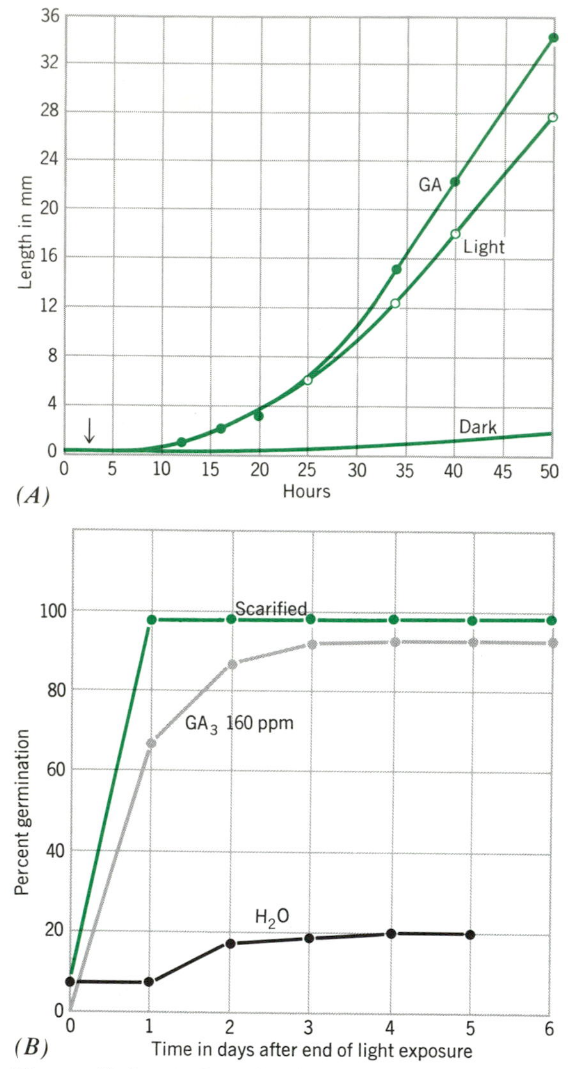

Figure 10-4 *A*. Germination of Grand Rapids lettuce seeds at 25°, given 2 minutes red light (at the arrow) or soaked in gibberellic acid (GA), or in continuous darkness. (Data from Ikuma & Thimann, 1960) *B*. Germination of seeds of *Phacelia* after exposure to bright light. The seeds in gibberellic acid germinate almost as well as those whose seed coat has been weakened by scarifying. (From Chen & Thimann, 1964)

(Fig. 10-4*B*). In both these types it appears that the mechanical restraining influence of the seed coat is the limiting factor, for germination without gibberellin can be brought about by cutting the seed coat at the point where the embryonic root would emerge, or by weakening it with enzymes that attack polysaccharides. Probably the action of GA is, as in the cereal seeds, to stimulate hydrolysis of stored polymers in the cotyledons, so that the resulting concentration of small molecules exerts a high enough osmotic potential to draw in water and so strain or force the tough seed coat open.

The way the gibberellic acids were discovered shows another of its hormonal actions. In Taiwan, some rice plants are afflicted with a fungus disease in which the leaves and stems grow abnormally tall and pale, and the grain yield is decreased. A Japanese agronomist, Eiichi Kurosawa, isolated the causal organism, *Gibberella fujikuroi,* and grew it in a nutrient medium. On reinoculating it into the roots of a healthy rice plant he reproduced the disease. What was remarkable, though, was that the (sterilized) nutrient medium, in which the fungus had grown, would also reproduce the disease. Thus the symptoms are due to an excretion product of the fungus. The problem was taken up by chemists, first in Tokyo and then in Britain, and from the *Gibberella* growth medium a group of compounds, GA_1 through GA_7, was isolated. In tiny amounts these compounds caused extreme elongation in a variety of test plants, both monocots and dicots (Fig. 10-5). But the gibberellic acids were not only fungal products, for soon several flowering plants were extracted and found to yield similar compounds. Of the over 52 GAs known (many from higher plants) generally the most active are those whose formulae were shown in Figure 10-2, GA_3 and GA_7.

b. THE ORIENTATION OF THE SEEDLING AND ROOT

The nutrient mobilized from stored polymers in the seed can generally only suffice for a short period of growth, so that it is essential that the young shoot reach the light to start photosynthesis, and correspondingly also that the young root find the deeper, moister layers in the soil from which to take up water. The shoot

(A)

(B)

Figure 10-5 Effect of gibberellic acid on the elongation of young plants of A wheat and B false dandelion (*Crepis leontodontoides*). In each case, the untreated control is on the left. The wheat plants were grown with their roots in the solution, the *Crepis* plant was treated once 7½ weeks before the photo was taken. The response of rice is similar to that of wheat. *A* is a historic photo, for it is the one that drew the attention of the western world to the importance of gibberellins.

grows toward a source of light and *away from* the influence of gravity; the root grows *towards* the source of gravity. These orienting growth curvatures are called *tropisms*—*phototropism* and *geotropism*, respectively—and are considered positive if toward the stimulus, negative if away from it. It was in studying them that the first plant hormone, auxin, was discovered.

This discovery came about through the experiments of Charles Darwin and his son Francis on the curvature of coleoptiles (seedling shoots, see Fig. 10-3) of certain grasses towards light. A few seconds of bright illumination from one side will cause an easily measurable phototropic curvature that develops over the next 90 minutes or so. Thus the coleoptile is about as sensitive to light as photographic bromide paper. The Darwins noted that if the extreme tip of the coleoptile is covered with a cap of foil this curvature does not take place, yet in the exposed plant the curved zone that results is located mostly below where the cover would have come (Fig. 10-6*A*). It follows that the tip *detects* the light but it is the more basal part that *responds*, or as the Darwins put it, some "influence" travels from the tip to the part below. Boysen-Jensen found that even if the tip were cut off and stuck on again, curvature toward light still occurs. Páal later found that a similar curvature could be caused, without light, if the tip were cut off and then replaced asymmetrically (Fig. 10-6*A*). Evidently the tip must be secreting a substance that travels down to accelerate the growth of the part directly below; if the tip rests only on one side, the substance goes into only one side of the coleoptile and thus a curvature results. (A curvature is due to a difference in growth on the two sides of an elongating organ). That this "growth substance" could diffuse outside of the plant was later shown by placing the tips on little blocks of agar or gelatin. After an hour or two the agar or gelatin, if applied asymmetrically to other coleoptiles (whose tips had been cut off), caused a curvature (Went in Fig. 10-6*A*). The angle of curvature was proportional, up to a limit, to the number of tips used and the time

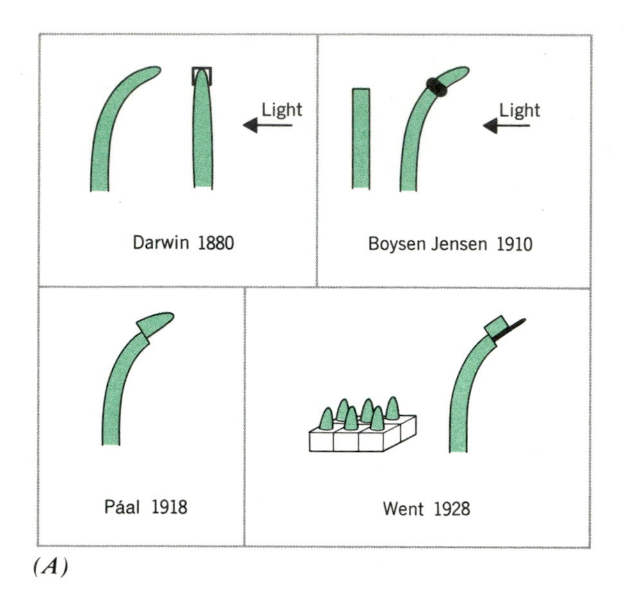

(*A*)

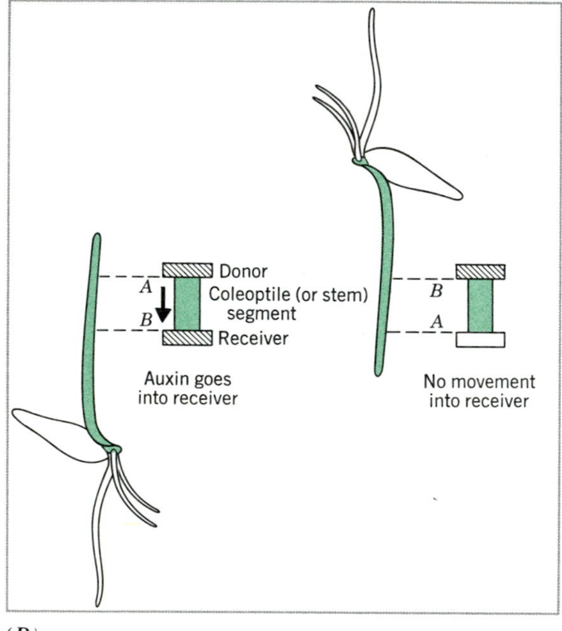

(*B*)

Figure 10-6 *A*. The critical historical experiments leading to the concept of the control of elongation by auxin. Note that the experiments of Páal and Went were carried out in darkness. (From Thimann, 1957) *B*. Diagram of the method of studying auxin transport through a coleoptile segment. It moves only in the apex-to-base direction. (*A*, apical end; *B*, basal end).

they had stood upon the agar. This reaction provided a bioassay by which, still later, the *auxin* was isolated chemically and found to be indole-3-acetic acid.

The explanation of the phototropism must be that the light from one side causes the auxin to be diverted in its downward transport, so that, instead of symmetrically promoting the growth of all sides of the coleoptile, the auxin goes in a greater amount to the shaded side and makes that side elongate faster than the lighted side. Direct measurements show that the shaded side can deliver two to three times as much auxin to receivers as the lighted side. If a sliver of mica is put through the tip perpendicular to the direction of the light, so that transport from the lighted to the shaded side cannot occur, the amounts of auxin coming out of the two sides remain the same, and such plants do not curve toward the light. Because of the downward movement of the auxin, illumination of the lower parts will not cause curvature higher up.

Geotropism, the upward curvature of shoots away from the earth, has the same basic mechanism as phototropism; an excess of auxin travels to one side—in this case to the lower side—producing an upward curvature (see Table 10-1). The proof was first given in 1930 by measuring the curvatures of test plants to which the auxin coming from the upper and lower sides was applied. Then 33 years later radioactive auxin was applied to horizontal coleoptiles, and the distribution of radioactivity so found agreed exactly with the old data. Gravity apparently brings about this lateral movement by the sinking of relatively heavy organelles in the sensitive cells.

The downward geotropism of roots seems to be more complex. The detecting organ in this case is the root *cap,* which is not the only source of auxin, since some comes down the root from the shoot. Nevertheless the principle of auxin asymmetry is the same here as in the shoot. The crucial difference is that a small increase in auxin content in the root causes a

TABLE 10-1

Percentage of Auxin on the Upper and Lower Sides of Coleoptiles Placed Horizontally

	Lower Side	*Upper Side*
Oats (*Avena sativa*)		
By bioassay in 1930	62	38
By radioactive IAA in 1963	60.2	39.8
Corn (*Zea mays*)		
By radioactive IAA in 1963	69.7	30.3

decrease in growth rate. The reason for this may be that auxin above a certain level in root cells causes the emission of ethylene, and ethylene is known to inhibit elongation of root cells. If seedling roots are allowed to dip into an auxin solution, their growth can be drastically inhibited, and the "threshold" concentration (that which causes the smallest measurable inhibition) is very close to that which begins to give rise to ethylene (Fig. 10-7).

Plants show a few other tropisms, but they are less well worked out. Tendrils of climbing plants curl around anything they touch. This results from a "contact stimulus" that somehow sets up an asymmetry of elongation in the tendril. Also, roots will sometimes curve toward water—*hydrotropism*—but this mechanism is unknown.

c. THE TRANSPORT OF HORMONES IN SHOOT TISSUE

As we saw in Chapter 4, all higher plants have structural polarity, an apex and a base. Both the main axis and every branch show apex/base polarity, and this polarity is not due to gravity but has its source in the *polar transport* of auxin. Figure 10-6*B* shows a simple demonstration of auxin transport. If we cut a segment from an elongating stem or coleoptile and apply auxin to the apical cut surface, auxin will be transported through to a receiver at the basal cut surface. But the reverse application, that is, to the basal cut surface, will not yield auxin to an apical receiver. Thus the polarity of transport is a property of shoot tissue, and it derives from the polarity of *each cell.* An auxin molecule within a cell will be preferentially transported out of its basal end. In stems, leaves, and petioles the polarity is often not complete, but it is always strong so long as elongation continues. In tubers and fruits it is less marked.

The importance of this polarity is seen in nearly all auxin functions, but not in gibberellin actions, for gibberellins are transported almost equally in both directions in growing shoots. The transport of cytokinins from cell to cell is rather slow and shows little polarity, but some cytokinin is formed in roots and travels upward in the xylem. In roots auxin polarity is dual, for it travels from the base of the shoot down *toward* the root apex, yet usually not quite *to* it, whereas auxin that is formed in the root cap travels back into the meristem and just a few millimeters into the elongating zone, that is, in the direction opposite to that from the shoot.

d. HORMONAL CONTROL IN THE MATURE PLANT

Although auxin controls elongation of cells in most growing parts, such an effect of auxin can rarely be shown in leaves. Leaves do form auxin, and do transport it polarly just as in stems, coleoptiles, and petioles, but leaves show little response to applied auxin. The geometry of growth in a dicot leaf, which is not

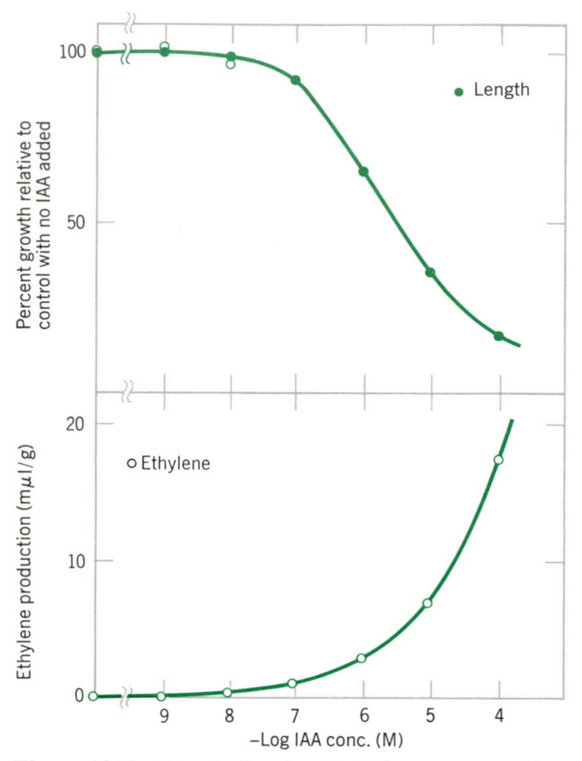

Figure 10-7 Graph showing that the concentrations of indoleacetic acid (logarithmic scale) that decrease the growth rate of the pea root are almost exactly those that induce the production of ethylene. (From Chadwick & Burg. 1967)

elongating but spreading in a plane, makes small increments of growth hard to detect. But since monocot leaves, which do elongate, also show this lack of response, we must deduce that auxin is, in general, *not the limiting factor* in growth of leaf tissue.

The situation is different for the gibberellins. They do not much affect the growth, or the tropisms, of coleoptiles. But they control, and greatly stimulate, the elongation of both the leaf-blade and the leaf-sheath of monocot plants (see Fig. 10-5). Indeed, one bioassay for gibberellins is to apply a droplet of test solution to the base of the leaf-blade of a cereal plant and measure the resulting elongation of the leaf-sheath. Note that here, in contrast to the tropisms, growth is accelerated above the point of application. The assay is much more effective on dwarf plants, simply because the untreated controls grow much less.

This leads to an important question. Do dwarfs owe their short stature to insufficient gibberellin? The answer is that many of them do, but not all. An example is shown in Fig. 10-8; here a corn mutant called dwarf-1 has been treated with 50 micrograms (0.00005 g) of GA_3 and as a result has become indistinguishable from its normal tall relative. Dwarf forms of dicotyledons, for example, peas, also respond to gibberellins.

A different aspect of form control is the development of lateral buds on the main shoot. In most dicots, these buds, which represent little nests of meristematic cells carried down from the flanks of the apical meristem, remain undeveloped so long as the main shoot is elongating. When elongation slows down or stops, some lateral buds may develop, especially if far below the apex. In a sunflower the repression of these buds is firmly maintained all the way down the stem; on the other hand, in such plants as the bush bean the lateral buds develop at many nodes. Control of this tendency to develop is partly genetic, partly environmental (e.g., a high level of nitrogen fertilizer and full sunlight both favor lateral bud development), but largely hormonal. Removal of the terminal bud usually causes immediate development of one or more lateral buds. This is the basic principle of pruning, that is, to release lateral buds in desired locations. Replacement of the terminal bud by auxin, if done at once (before bud growth has started), will reinstate the inhibition. The concentration of auxin needed for inhibition is comparable with that secreted by the apical bud, so the phenomenon represents the normal control exerted by the apex over lateral buds, termed *apical dominance*.

But apical dominance has two aspects – the inhibition and its release. For if we apply a cytokinin to the lateral bud, whether it is being inhibited naturally by the apex or artificially by applied auxin, the inhibition is countered and the bud grows out (Fig. 10-9). Along with this

Figure 10-8 Four corn plants of the same age. From left, normal control; normal treated with gibberellin; dwarf-1 control; dwarf-1 treated with gibberellin.

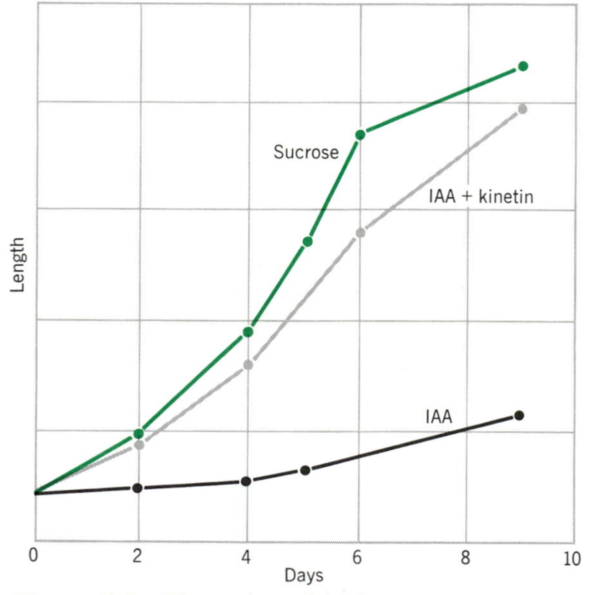

Figure 10-9 Elongation of buds on pea stem segments floating on sucrose solution is inhibited by indoleacetic acid (3 ppm) and the inhibition is almost completely reversed by kinetin (5 ppm). The plot shows the length of the lateral buds against time floating. (From Wickson & Thimann, 1958)

release comes the differentiation of vascular tissue to supply the bud with water and nutrient.

The antagonism between auxin and cytokinin in controlling the development of buds is clearly pictured in the development of tissue cultures. Figure 10-10 shows that, with a fixed level of auxin in all cultures of tobacco pith tissue, buds develop at intermediate levels of kinetin; at levels above these there is a more generalized inhibition of all growth. Thus both bud formation and bud outgrowth depend upon a balance between auxin and cytokinin.

Another aspect of growth and development is the formation of roots. A seedling usually forms a fixed pattern of initial roots—four in the oat or barley seedling, one main root with lateral branches in the pea or bean. Roots are added thereafter as the plant develops. The process of root formation is best studied with stem cuttings. The formation of roots on grape cuttings was shown in 1924 to depend on the presence of young leaves or growing buds (Fig. 10-11*A*) and to occur always below them. The influence

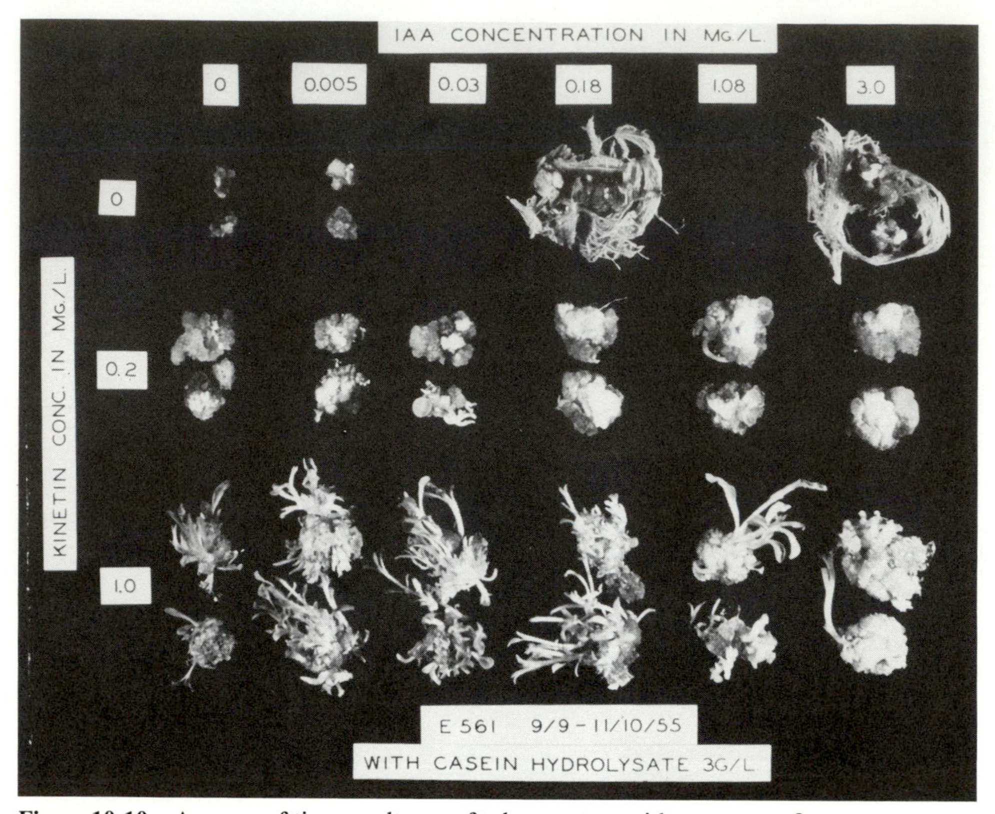

Figure 10-10 A group of tissue cultures of tobacco stem pith, grown on five concentrations of indoleacetic acid and two of kinetin (values shown at top and left). The lower kinetin concentration promotes callus growth but the higher one causes the formation and outgrowth of buds.

of these organs was later traced to their active formation of auxin. Applied auxin has therefore been used by nurserymen to promote root formation on cuttings ever since auxin was chemically identified in 1934 (Fig. 10-11*C*). On those species that root readily, auxin treatment increases the number of roots per cutting, whereas on many species that root only rarely or with difficulty auxin can often induce regular formation of a few roots. The induced root begins with a few cell divisions in the pericycle and these multiply into a mass of small cells that pushes its way out of the stem (Fig. 10-11*B*). Since the transport of auxin is polar from apex toward base, we can see one simple reason why roots are normally formed at the base of the stem. Cytokinins, which oppose the action of auxin in the outgrowth of lateral buds, also oppose it in the outgrowth of lateral roots on a main root.

4. FLOWERING AND ITS CONTROL

a. THE INFLUENCE OF AGE

The pattern of flowering is usually characteristic. Some small desert plants flower in a few weeks from seed germination; other herbaceous plants (*"annuals"*) flower at some time during their first season, whereas other (*"biennials"*) grow vegetatively for one season and flower only after the winter is over. Many of

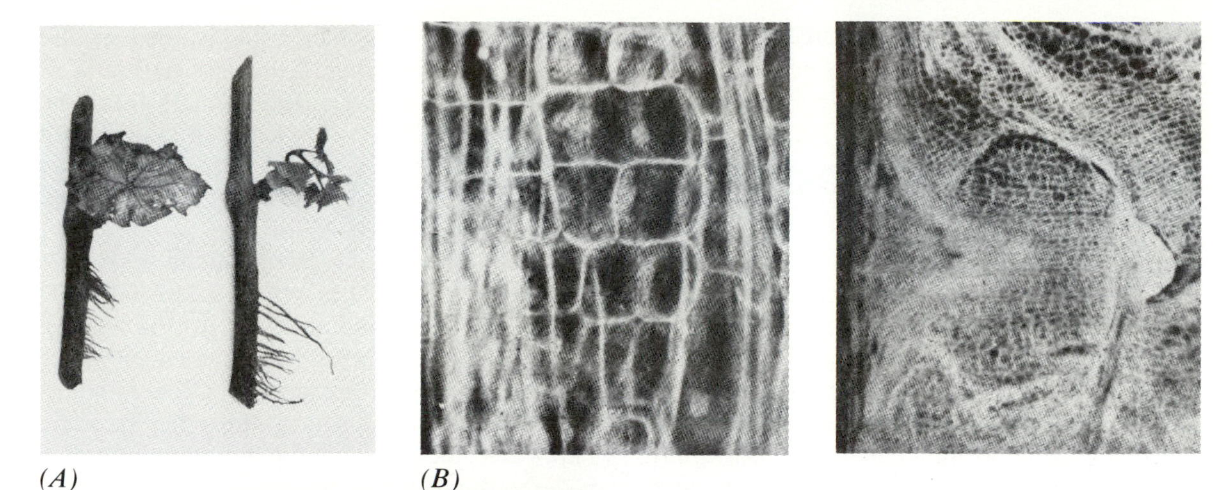

(A) (B)

(C)

Figure 10-11 *A*. Grape cuttings forming roots on the side where a young leaf or a growing bud is located, and directly below it. *B*. Cuttings of coffee sectioned to show (left) the first tangential divisions in the pericycle, (right) the formed root initial ready to elongate. *C*. Cuttings of privet treated with naphthalene-acetic acid (right) or water (left).

these will not flower unless the winter temperature has fallen to freezing, or close to it. Still other plants are *perennial* and having once flowered will continue to flower year after year. A few, like the agaves and palms, will not flower for many years—then produce massive blooms and die. Pine and fir trees, which are not flowering plants but bear their seeds in cones, take up to 10 years to form the first cones and thereafter behave as perennials. Oaks may even take 20 years to form their first flowers.

b. THE INFLUENCE OF COLD

The biennial plants that require a cold spell in order to form flower buds include an important group of cereals, "winter" wheat, oats, and barley, which are planted in the fall; they grow as rosettes at that time and then the following spring produce flowering shoots. These cultivars can be grown only where the winter temperatures are not severe enough to kill the young plants. "Spring" varieties are sown in the spring as soon as the soil is warm enough, and come into flower the *same* season. Early in this century it was discovered that, if the seeds of the winter types were first allowed to germinate and the small seedlings exposed artificially to cold for a few weeks, they could be planted in the spring and would flower; they had thus been converted to spring-flowering types. The term for this change is *vernalized*. This procedure was rediscovered in the 1920s in Russia by Lysenko and made the basis for a whole theory (since discarded) of required stages in plant development. Although this theory does not hold in general, it is true that many biennials can be exposed to cold when quite young and will then flower afterwards. Also the requirement for a cold spell is not restricted to biennials, for many varieties of apples and peaches flower only after a sufficiently low temperature has been reached.

C. THE INFLUENCE OF LIGHT AND DAY LENGTH

In some plants the time of onset of flowering depends on the length of the day. One type forms flowers only when the day length is *greater* than a certain minimum (often around 12 hours, depending on the species); another type flowers only when it is *less* than a certain maximum (often 10 to 11 hours), while a third type can flower on any day length. The first group is called long-day (LD) plants, the second short-day (SD) plants, and the third day-neutral (DN). In general the time needed to develop flowers is as shown in Figure 10-12; below the minimum day length for LD plants, or above the maximum day length for SD plants, the time to flowering is infinite, while for each there is a range of day length in which flowering comes on rapidly. The SD plants do not really require short days, but long nights; that is, they require *long dark periods,* and, if the dark period is interrupted by light, even a few minutes' duration, near the middle of the night, they will not flower. LD plants, on the other hand, can flower with such "night interruption," or even in continuous light (Fig. 10-12).

Roughly speaking, SD plants in temperate climates flower in the spring or late fall; LD plants in summer and early fall. In the tropics, where day length varies little, some plants have become sensitive to variation of a few minutes. There, however, other conditions may be more important; for instance, the flowering of coffee is determined largely by the time of rainfall. In subtropical areas in Australia or the Caribbean, sugarcane flowers only in the shortest days of winter, but there day length and temperature are interrelated, for the cane flowers only in years when the temperature falls low enough. A number of rice cultivars—"winter" varieties—require a succession of short days to flower and therefore have been unsuccessful in the southern United States, where the only growing

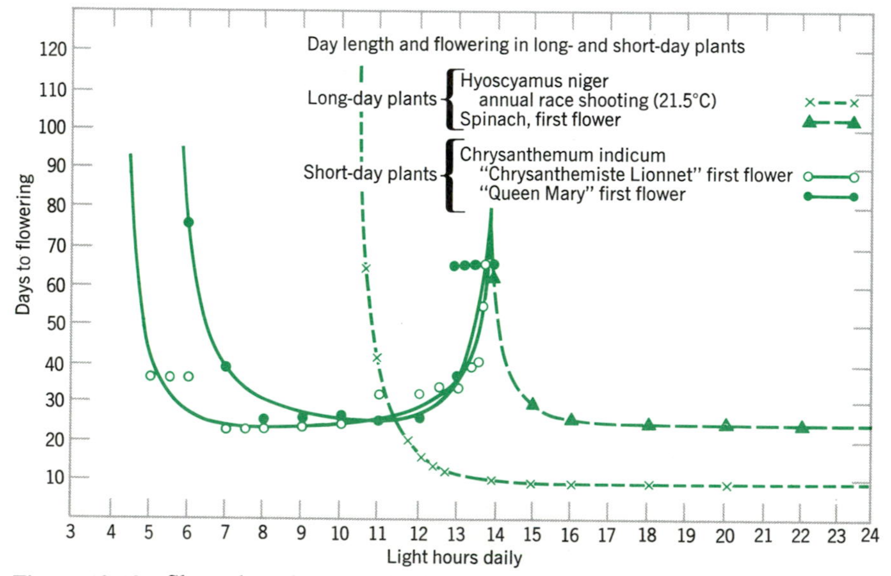

Figure 10-12 Short-day plants (solid lines) flower soonest on day lengths of 8 or 10 hours; long-day plants (dashed lines) on day lengths of 15 hours or more. (From Melchers, 1952)

season comprises long days. They yield well in India, where the winter temperature is high enough for good growth in short days.

What the day length does is to *induce* flowering, that is to change the vegetative meristem into a reproductive one. Once induction has taken place the formation of visible open flowers may go on in a day length that is no longer inductive. The cocklebur, *Xanthium pennsylvanicum,* can be grown on long days, exposed to a *single* short day of 8 hours, and then returned to long days; it will then form flower buds. The rye-grass *Lolium perenne* is a parallel case among LD plants; it flowers after a *single* 15-hour day, even if returned to short days after the exposure. Several other plants require a *number* of short days or long days in succession in order to induce flowering.

The influence of day length is not limited to flowering. Long days favor vegetative growth in many types of plants. When LD plants are kept on short days, they not only fail to flower, they also fail to elongate, forming a "rosette" in which the leaves develop normally but the internodes do not elongate. If the long night is interrupted with even a very short light exposure (red light will do), the internodes start to elongate and a tall plant results. Similarly many trees and bushes make most of their elongation in long days (Fig. 10-13). Tropical trees, however, tend to grow in spurts or flushes, often independent of day length and occurring two or three times a year. The formation of a bud, with many small leaves or bud scales close together and no internodal elongation between them, can be thought of as a kind of rosette, and indeed it is often the result of the short days that follow the fall equinox. Here a new hormone comes into play, *abscisic acid,* for this compound is produced in greater amounts in short days and it promotes the development of the bud scales and thus the formation of a winter bud.

In all the discussion above, the term "light" has been used with the implication of white light. Actually the most active light is red, of wavelength about 660 nm. Blue will often function but less effectively. If after exposure to red light the plants are exposed to a red of longer wavelength, around 730 nm ("far-red"), the effect of the red light is usually reversed. The same reversal can be observed with those seeds described in section 3a whose germination is promoted by red light; if the red is quickly followed by far-red, germination is prevented again. This behavior has been traced to a pigment related to the phycocyanin of the cyanobacteria, called *phytochrome*. This compound in its red-light-absorbing form is activated by the red light but can be reconverted to its initial state again by exposure to far-red:

$$P_r \underset{\text{far-red}}{\overset{\text{red}}{\rightleftharpoons}} P_{fr}$$

Figure 10-13 Young plants of white birch (*Betula*) after 8 weeks on day lengths of 24, 18, 16, and 10 hours, and (right) 10 hours with a half-hour night interruption, which evidently is interpreted by the plants as a lengthened day.

The form P_{fr} is unstable, decomposing in the dark; so the reversal is only complete if exposure to far-red comes quickly. It is believed that P_{fr} is the active form, but its activity is shown in different ways. If a SD plant is given a few minutes of red light near the middle of the long dark period, P_{fr} is formed and flowering is prevented, but if the red exposure is followed by far-red, flowering can be reinstated (Fig. 10-14). Similarly, if an LD plant held on short days is given red light in the dark period it can respond by flowering, as though it added the red exposure to the short day and thus apparently it had experienced a long day. Again a far-red exposure will nullify the effect of the red light.

What is remarkable about this is that it is not necessary to expose the whole plant to the inductive day length; it is not even effective to expose the particular apical meristem that is to produce the flower buds. What must be exposed are the *leaves*. With the cocklebur in long days, all leaves but one can be removed, and seven-eighths of that last one also removed; still one short day will cause flowering (see Fig. 10-15). Thus there is an analogy with the phototropism of the cereal coleoptile, where the tip must be exposed to light but the response (curvature) is located further down. As with the concept of auxin as a transportable growth hormone, so with flowering, many workers conceive of a flowering hormone, *"florigen,"* formed in leaves and transported to the responding bud. Unfortunately no such florigen has yet been clearly demonstrated. Grafting experiments seem to show that the amount of the "influence" increases within the recipient plant and have led some workers to conceive of the florigen as a sort of virus or a self-multiplying nucleic acid. But in the absence of proof the subject of *florigen* must be left open.

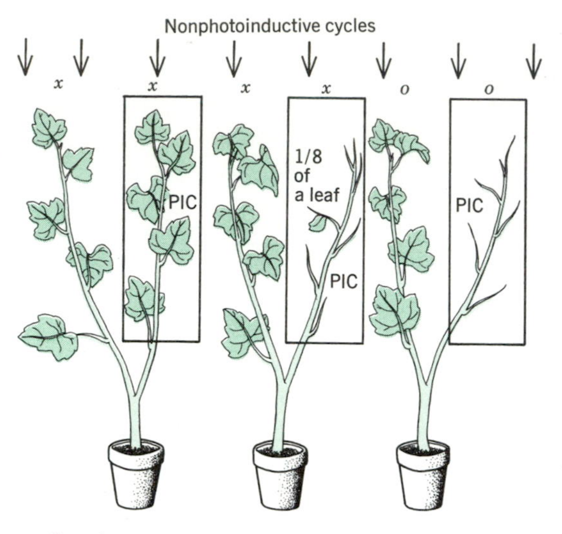

Figure 10-15 Cocklebur (*Xanthium*) plants pruned to two equal shoots, one being on long days, the other on short (i.e., photoinductive cycles, PIC). In the plant at left with a whole shoot on PIC the apex of the shoot in long day flowers (X = flower, O = vegetative). In the plant at center with only 1/8 of a leaf on PIC, the long-day apex also flowers. At right, however, with no leaves on the shoot on PIC, the long-day apex is vegetative. (Modified from Hamner, 1942.)

Figure 10-14 Plants of *Kalanchoë;* left, grown on short days, i.e., 16 hours darkness each night, flowering freely; center, nights interrupted in the middle with 1 minute red light, wholly vegetative; right, red interruption immediately followed by 1 minute far-red light.

d. THE INFLUENCE OF HORMONES

On the other hand, flowering is often modified, and sometimes induced, by the known hormones. When plants are in day-lengths close to the limit—for example, when the LD plant *Hyoscyamus niger* (henbane) is grown on short-day with extremely weak supplementary light during the night, application of *auxin* will cause flowering. In the pineapple and lichee nut, either *ethylene* or a rather high concentration of a synthetic auxin will induce flowering. This high level of auxin actually operates by causing the production of ethylene. Pineapple growers in Hawaii and Puerto Rico regularly use this method and thus can accurately control the time and the amount of flowering.

More notable is the fact that many SD plants are induced to flower in long days by applying *gibberellin*. LD plants that form rosettes in short days are caused by gibberellin to "bolt" or form tall scapes, and some of these flower. However, gibberellin is not *"the"* florigen because it does not cause all LD plants to flower. Analysis of flowers for their hormone content does not establish any causative role and, besides, the hormone content of flowers is very complex. For example, the young tassels of corn (Fig. 16-8) contain five different gibberellins, plus abscisic acid and several inhibitors.

It seems, therefore, that flowering is under a many-sided controlling system in which several hormones interact. Research is very active in this area, and a striking new development might come at any time.

5. THE SENESCENCE AND ABSCISSION OF LEAVES

Five stages can be distinguished in the life of a leaf: (1) initiation; (2) outgrowth and development of shape; (3) formation of chloroplast pigments (usually coupled to stage 2); (4) senescence and loss of chloroplast pigments, sometimes together with formation of red anthocyanin pigments; (5) abscission (in dicotyledons and in palms). Hormonal control becomes evident mainly in stages 4 and 5, for although the *rate* of stages 1, 2, and 3 varies with temperature, and stages 2 and 3 require light, yet the occurrence of these stages is genetically programmed in the plant.

When leaves have been mature for some time they begin to lose their protein content and their chlorophyll. The process can be artificially hastened by darkening the leaves, and indeed the *senescence* of the lowest leaves on an axis is almost certainly hastened by the shading that these leaves undergo from the leaves above them. Nevertheless senescence will occur, though more slowly, as a simple function of age, even in full sunlight. Leaves of "evergreens" require more than one season to senesce.

Basically the process is one in which synthetic processes are slowed down and processes of hydrolysis (see Chapter 6) are accelerated, so that proteins are hydrolyzed to amino acids, and polysaccharides to reducing sugars (Fig. 10-16). The resulting amino acids are transported downwards—in dicots back into the stem, in monocots down to the roots. In this way the plant recycles much of its nitrogen content. At the same time the hydrolytic and oxidative enzymes attack the chloroplasts and cause the chlorophyll to break down.

The principal hormonal control is exerted

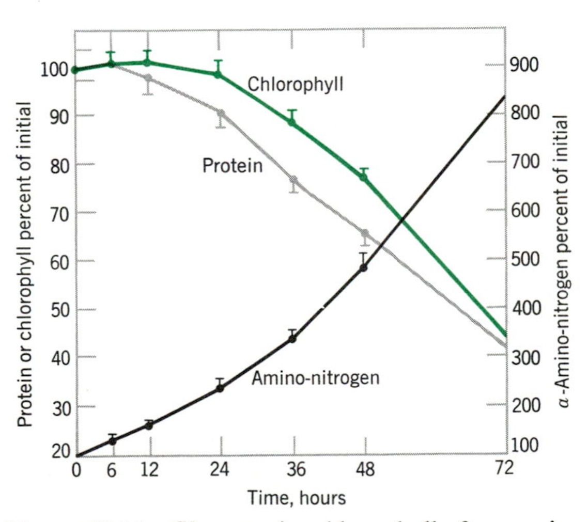

Figure 10-16 Changes in chlorophyll, free amino acids, and protein in oat seedling leaves, cut off and floated on water in the dark. The short vertical lines measure the probable error of each point. (From Martin & Thimann, 1972.)

by cytokinins, for these compounds oppose senescence, promoting synthetic processes. Thus they favor the syntheses both of proteins and of polysaccharides. Accordingly they maintain the polymers more or less intact, and keep the leaves green. Since some cytokinin is produced in roots and transported up to the leaves, the roots exert some control over leaf senescence. In a very few plants, however, gibberellic acid functions to maintain greenness like the cytokinins.

As the leaves age, the flow of auxin into the petiole from the leaf-blade decreases, and this has an important result. Cells in a line across the base of the petiole begin to divide, forming two or three rows of small cells at right angles to the axis of the petiole (Fig. 10-17). These new cell walls then hydrolyze so that the cells come apart, leaving the petioles attached only by the few strands of xylem; in the first slight breeze the petioles break and thus the leaves fall off. For this process of *abscission* auxin acts as an inhibitor, for if auxin is applied to the blade before aging has gone too far, abscission is largely prevented and the leaves can stay on for some weeks longer.

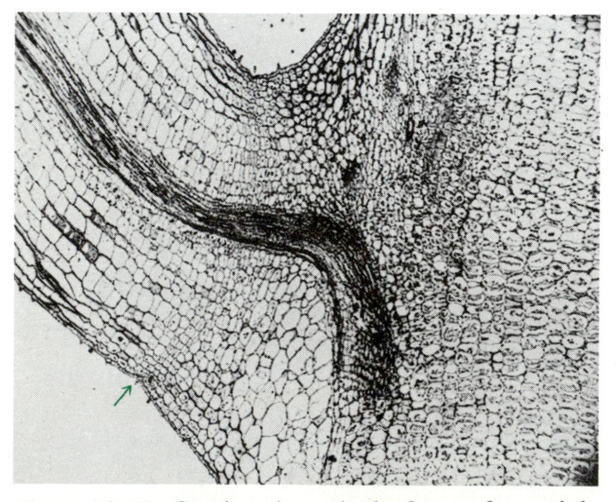

Figure 10-17 Section through the base of a petiole of *Coleus* just before abscission. The cells of the abscission layer (arrow) are rapidly dividing.

Ethylene is naturally formed in the nodes, and it acts as promoter in abscission; that is, it acts in opposition to auxin, for exposure of plants to ethylene causes very rapid and complete abscission. For this reason gas is not usually supplied to greenhouses (or, if used for heating, no gas pipes go through the greenhouse itself). Leaks of gas mains in city streets have often caused defoliation of the street trees. The effect of ethylene is triple, for (a) it decreases the auxin production in the leaf, (b) it gradually slows down the transport of auxin in the petiole, and (c) it stimulates the separation of cell walls in the petiole base described above. The separation is largely due to formation of the enzyme cellulase, which hydrolyses cellulose. The activity of cellulase increases just before abscission; ethylene promotes this increase and auxin inhibits it.

6. MATURATION AND ABSCISSION OF FRUIT

After fertilization of the ovule by the pollen nuclei (Chapter 4), the zygote (diploid) and the endosperm (triploid) begin to grow; the zygote grows with rapid cell division, but in the triploid tissue the nuclei alone divide at first and only later are separate cells formed. This first *acellular* stage of the endosperm is liquid in some plants (e.g., the "milk" stage of maize and chestnuts); the extreme example is that of coconut milk, which is the liquid endosperm of the biggest of all seeds. In the stage of liquid endosperm the cytokinin, the auxin, and the gibberellin content are all high; the cytokinin called *zeatin* was indeed isolated from the milk stage of *Zea mays*. The small amount of auxin contained in the pollen grain probably contributes to, or provokes, these initial growth stages, and the rich production of growth hormones in the endosperm certainly supports the growth of the embryo. The smallest embryos, isolated from the immature seed soon after fertilization, can only be grown in cultures to normal size and form if coconut milk (or other liquid endosperm) or a cytokinin is added.

The growing seeds secrete auxin, and perhaps gibberellin, to the ovary walls or the receptacle and thus promote its increase in size (see Chapter 18). In the strawberry, where the achenes (which are one-seeded ovaries) are on the outside of the receptacle, it has been easy to show that removal of these from the young "fruit" prevents its further enlargement, whereas application of auxin to the achene-less receptacle reinstates its growth (Fig. 10-18). With more surgical difficulty (because of the interior position of the seeds), similar facts have been established with the apple. The relation is not universal, though, for in the tomato the increase of auxin in the seeds comes mainly after fruit growth has slowed down.

In the last stages of fruit growth, the ovary tissue begins (usually rather sharply) to produce a fruit-ripening hormone, the gaseous hormone ethylene; this is formed from the amino acid methionine by a new enzyme system that comes into action at that last stage. The production of ethylene was discovered from the observation that lemons would ripen in a storage shed heated by a kerosene stove, but not in one with modern central heating. The active agent in the stove fumes was ethylene. When the vapors given off by ripe apples were liquified in 1934 ethylene was identified in them. Oranges and lemons, as it turned out, produce very little ethylene, but most other fruits form enough ethylene to ripen themselves or other fruits nearby. As shown in Table 10-2, the amount of ethylene needed for ripening is well below the concentrations actually formed in the tissues of most fruits; only in citrus fruits (oranges and lemons) is the concentration barely adequate. The many changes – softening, starch hydrolysis, loss of acidity, and others – brought about by ethylene are discussed in Chapter 18.

Figure 10-18 The control of growth in the strawberry by auxin coming from the achenes. Left, normal fruit; center, a fruit from which all the achenes have been removed; right, another fruit without achenes but with auxin applied.

Ethylene has another effect in many fruits; it inhibits the transport of auxin. Thus decreased amounts of auxin move down the peduncle and as a result an abscission layer forms, as with leaves. Hence, the fruit falls off. If the tree is sprayed with auxin at this stage the fruit (especially apples) can be made to stay on for two to six weeks more. There have been two earlier stages in the life of the fruit when it might have abscised. The first is soon after formation, if it fails to be fertilized; many tiny fruitlets fall at this time. The second, occurring especially in rosaceous fruits, is called the June Drop. This drop is not due to ethylene but probably to a hiatus in auxin formation after the endosperm has become cellular and before the embryo and seed have begun their auxin production. In some fruits abscisic acid plays a part here, too, promoting abscission. The hormonal influences throughout the life of a typical fruit in the northern hemisphere are summarized diagrammatically in Figure 10-19. It must be remembered that dry fruits like those of the cereals do not abscise readily, and indeed have been selected over the centuries for this "nonshattering" property. The hormonal basis for this property is not yet known.

With the seed on the ground and ready to germinate, the plant's life cycle is complete and we have returned to the point at which this chapter started.

TABLE 10-2

The Ethylene Content in Parts per Million of the Air in the Intercellular Spaces of Mature Fruits, in Relation to the Concentration Needed to Cause Ripening. Data of Burg, 1962

Fruit	*Concentration Found Internally Just Before Ripening*	*Concentration Needed To Cause Ripening When Applied Externally*
Apple	80	0.2
Banana	6 to 40	0.2
Cherimoya	100 to 370	1 to 4
Lemon	–	0.025 to 0.05
Orange	0.08 to 6	0.1
Pear	250 to 500	0.4
Tomato	4 to 6	0.2

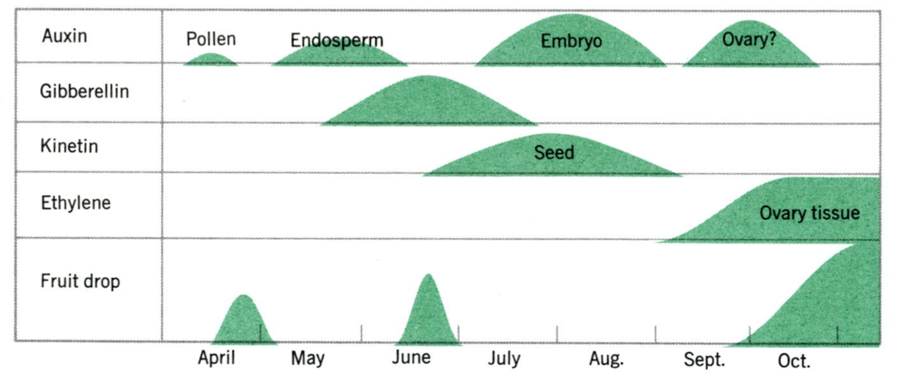

Figure 10-19 Schematic diagram of the production of the four main plant hormones during the life of a typical pomaceous fruit and its relationship to the three periods of abscission in the North temperate areas. (From Thimann, 1965.)

CHAPTER 11

Genetics of the Diploid Plant

Genetic mechanisms provide a constant source of variation that enables plants to cope with changing environmental conditions. If we think back to the chapter in which we discussed the discovery of agriculture (Chapter 3)—and later when we study the cereals (Chapter 16)—we will see that our ancestors selected plants that most suited their needs from a variety of those available. They then selected particular characters among those plants, such as grains in inflorescences (spikes) that do not shatter when they are harvested, and strains best adapted to different climatic conditions and soils. They also apparently observed better yields from some natural hybrids and selected them. Despite centuries of our ancestors' selection of useful variations among plants, an understanding of the role that variation played in evolution did not come until 1859 with the publication of Charles Darwin's epic, *The Origin of Species*.

Significant to our understanding in this book is that Darwin drew ideas for his evolutionary theory from both the natural scene and experimentation with plants and in breeding animals (Fig. 11-1). Darwin saw that organisms vary greatly in nature and that those individuals with traits more favorable for certain environmental conditions are likely to breed more successfully than those that do not possess these traits. Thereby over long periods of time slow but steady changes in the traits of individuals in populations occur. In contrast, breeders can select one or a few traits of interest (although

Figure 11-1 *A*. In Charles Darwin's greenhouse in England. *B*. HMS *Beagle*, on which Darwin sailed for five years, shown here in the Straits of Magellan on a trip in which his observations were very significant to his formulation of the theory of evolution.

(A)

Figure 11-2 *A*. Gregor Mendel in the garden at the monastery in Brno (in what is now Czechoslovakia). *B*. A diagrammatic summary of the characters Mendel selected for his experiments with the garden pea (*Pisum sativum*)—which fortunately was an excellent choice to demonstrate the particulate nature of the gene.

Flowers	Ripe pods	Seeds
Axial or terminal	Inflated or constricted	Round or wrinkled Yellow or green
Stems	**Unripe pods**	**Flowers**
Long or short	Green or yellow	Red or white

(B)

there are definite limits as we shall see later) and hence the selection process may take a relatively short time. Thus Darwin was led to compare *natural selection* and *artificial selection*. In natural selection, the entire organism must be fit to live in the total environment in which it lives, and this had led to natural selection's being referred to as the "survival of the fittest."

Despite Darwin's clarity in discerning the role of selection in evolution, he still knew nothing about the *cause* of hereditary variation. Along with most of the practical animal breeders of the day, he regarded hereditary substances as fluids, and thus the intermediate nature of hybrids as a mixture of parental fluids. This concept of *blending inheritance* led him, however, to wonder how a trait could skip a generation and reappear again. The answer to this question came with the work of the Austrian monk, Gregor Mendel (Fig. 11-2*A*), who showed that heredity is determined by *particulate units*.

1. MENDELIAN GENETICS

a. GENES IN DIPLOID ORGANISMS

In 1865 Gregor Mendel presented the idea that variations are transmitted as "unit determiners." Mendel selected the garden pea (*Pisum sativum*) for this study for several reasons. First, peas are annual and thus he could obtain several generations quickly. Secondly, because they had been under cultivation for centuries, he knew that a number of true-breeding strains were available. They were true breeding because they are usually self-pollinated, but Mendel also knew that they could be cross-pollinated. Third, the garden pea provided clearly contrasting characters with which to work. Mendel selected seven of these characters, shown in Figure 11-2*B*, and proceeded to cross plants with contrasting characters. For example, he took the pollen from a red-flowered plant and used it to pollinate the stigma of a white-flowered plant, having carefully removed the pollen-containing anthers of the white-flowered plant so it could not self-pollinate. He also made the reverse cross and later harvested the seed (Fig. 11-3).

b. DOMINANT AND RECESSIVE TRAITS

One of Mendel's first observations was that of *dominance*. He found in the two crosses just described that all of the plants in the first hybrid generation or *first filial generation* (written F_1) were red-flowered. What had happened to the character for white flowers? To try to understand this, he allowed F_1 plants to self-pollinate and planted the resulting seed the next year. Some of the plants of the second filial generation (F_2) had red flowers and some had white (Fig. 11-3). Then Mendel did something unusual for biologists of his time: he counted the number of each (Table 11-1). He thereby laid the basis for collection of quantitative data that he could analyze statistically. The results showed that he had about three red-flowered plants for every white-flowered one. You'll note from the table that the ratio is not *exactly* 3 to 1. However, he perhaps was somewhat lucky, because his ratios are amazingly close. As we shall see, these ratios are determined by two chance events: (1) separation of the chromosomes at meiosis (Chapter 4), and (2) recombination of eggs and sperm at fertilization. This and six other experiments with contrasting characters convinced Mendel that the F_1 plants contained "whatever it was" that determined red or white flower color, but that the determiner for red color masked the determiner for white color. In other words, the red determiner or gene was *dominant* over the white determiner or gene, which was *recessive*. He found that the gene for round seed was dominant over the gene for wrinkled seed; the gene for yellow seed color was dominant over the gene for green seed color, and so on.

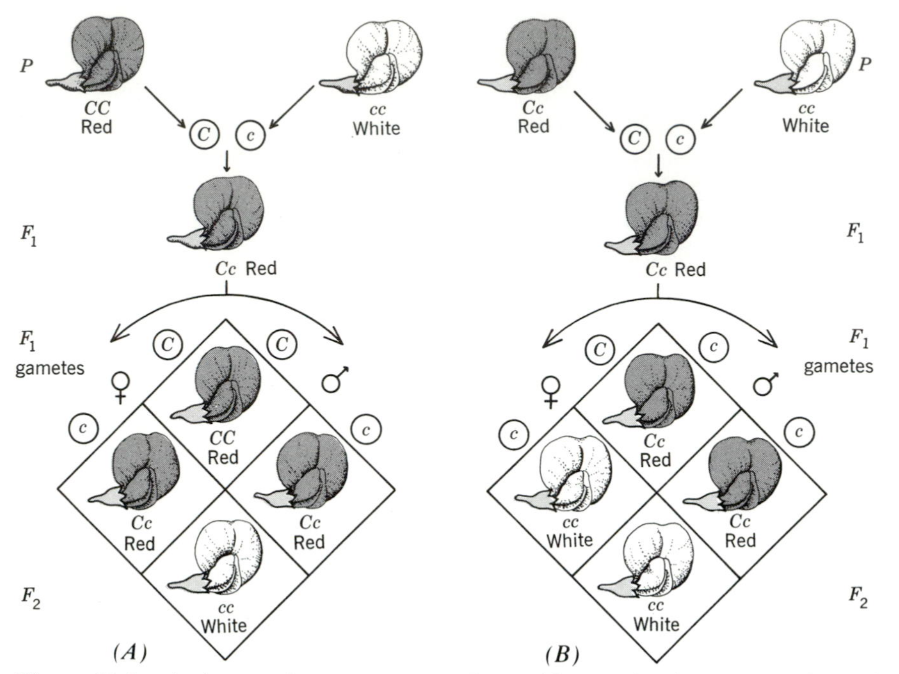

Figure 11-3 *A*. A cross between a pea plant with two dominant genes for red flower (*CC*) and one with two recessive genes for white flowers (*cc*). The genotype of the offspring in the F_1 generation is *Cc*, but the phenotype is red. Since the *C* allele is dominant, it determines the phenotype. The recessive trait (white) appears only when the recessive *c* allele is present from each parent. The ratio of dominant to recessive phenotypes is thus always expected to be 3 to 1. *B*. Testcross between a pea plant with red flowers (*Cc*) and one with white (*cc*). Although the red-flowering plant is phenotypically identical to the *CC* plant shown in (*A*), results of this testcross show that it is heterozygous for this gene.

TABLE 11-1
Mendel's Results Derived from Crosses Involving the Seven Single-Character Differences Shown in Figure 11-2

Parent Characters	F_1	F_2	F_2 *ratio*
Round × wrinkled seeds	All round	5474 round: 1850 wrinkled	2.96 : 1
Yellow × green seeds	All yellow	6022 yellow: 2001 green	3.01 : 1
Red × white flowers	All red	705 red: 224 white	3.15 : 1
Inflated × constricted pods	All inflated	882 inflated: 299 constricted	2.95 : 1
Green × yellow pods	All green	428 green: 152 yellow	2.82 : 1
Axial × terminal flowers	All axial	651 axial: 207 terminal	3.14 : 1
Long × short stems	All long	787 long: 277 short	2.84 : 1

From these types of crosses Mendel drew the conclusion that particulate units determine inherited characters and that these units exist in pairs. He knew that the original parents were true breeding, and, to explain his observations, he proposed that the original red-flowered parents had the genetic characters *CC* (*C* denotes redness), whereas the original white-flowered parents had the genes *cc* (*c* denotes whiteness). At the time of hybridization each

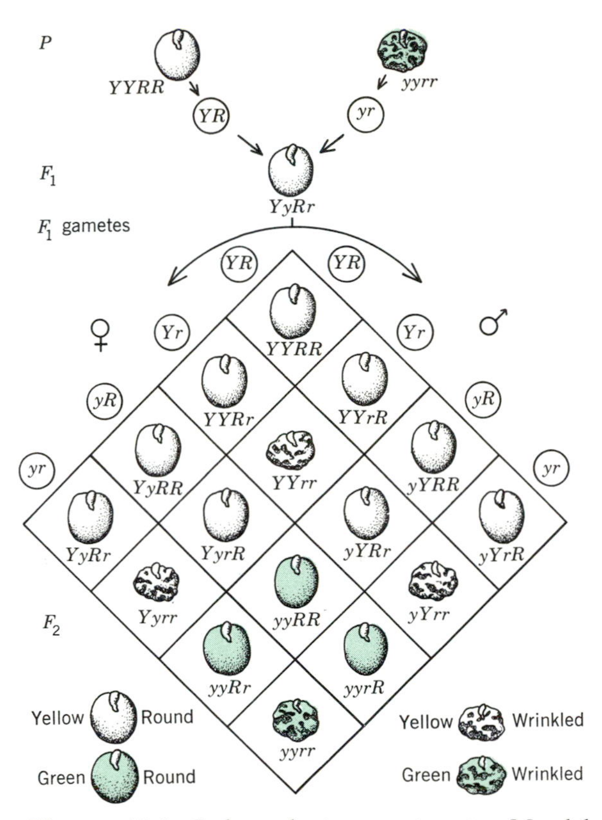

Figure 11-4 Independent assortment. Mendel crossed a plant having yellow (*YY*) and round (*RR*) peas with a plant having green (*yy*) and wrinkled (*rr*) peas in one of his experiments. The F_1 generation was all yellow and round, but in the F_2 generation, the recessive traits reappeared. They also appeared in new combinations. In a cross such as this involving two pairs of alleles of different chromosomes, the expected ratio in the F_2 generation is 9:3:3:1 with 9 plants having the two dominant traits 1 having the two recessive traits; and 3 and 3 representing the alternative combinations of dominant and recessive.

parent contributed one half of the pair (Chapter 4). As we would say now, the gametes (eggs or sperms) from the red-flowered parent contain only *C*, whereas the gametes from the white-flowered parent contain only *c*. Thus the F_1 generation contains *Cc* but the plants are all red-flowered because *C* is dominant over the recessive *c*. When members of the F_1 are crossed or are self-pollinated, either parent can produce two kinds of gametes (*C* or *c*). The *C* from one parent can combine with *C* or *c* from the other parent to produce *CC* or *Cc;* the *c* of one can combine with *C* or *c* to produce *cC* or *cc*. On the average there will be one *CC* for each two *Cc* for each one *cc*. Again, because of dominance there will be three reds to one white flower.

c. PHENOTYPE AND GENOTYPE

Mendel thus recognized that the genetic makeup of the plant and its physical appearance could be different. This again is due to dominance. A plant may have what we now call *genes CC* or *Cc* and still be red-flowered. Only by the appropriate cross will the presence of the *c* gene be known. Many years after Mendel's work, the terms *genotype* (the genetic composition) and *phenotype* (the way an organism appears) were introduced to clarify this situation. Because the genotype and phenotype may not necessarily be the same (because gene pairs may be different), two additional terms were coined. A gene pair consisting of a single kind of gene (e.g., *CC* or *cc*) is called *homozygous* and a pair of unlike genes (*Cc*) is called *heterozygous*.

How can you tell whether the genotype of a plant with red flowers is *CC* or *Cc*? As you can see in Figure 11-3*B*, you can tell by crossing it with a white-flowered plant and counting the progeny of the cross. This experiment, which was also performed by Mendel, is known as a *testcross*. It is also known as a *backcross*, because it is often, though not necessarily always, carried out by crossing a member of the F_1 generation with the recessive parent.

d. SEGREGATION (MENDEL'S FIRST LAW)

These experiments established that hereditary characters are determined by discrete "units" that appear in *pairs* (now called *alleles*), one of each inherited from each parent. Each time the adult offspring produces sex cells, which are haploid, the pair of factors (alleles) is separated or segregated, one member of each pair going to one sex cell or gamete and the other member going to another gamete. This principle of segregation is sometimes known as Mendel's first law. Also this concept of a unit determiner or particulate gene explains how a characteristic could persist from generation to generation without being blended with other characteristics and how it would seemingly disappear and then reappear in later generations (one of the problems that so perplexed Darwin).

Dominance was first viewed as the complete covering up of the action of one allele by another. However, dominance of this type has been shown not to be an essential aspect of Mendelian inheritance. Dominance may exist in varying degrees; that is, many characters show partial or *incomplete dominance*. Thus, in flowers such as snapdragons, the gene for red is not completely dominant over white, so that the color of the flowers is pink in the F_1 generation plants from a cross of a red-flowered plant and a white-flowered plant (Fig. 11-5). Actually, incomplete dominance is probably the rule rather than the exception. Incomplete dominance, however, still conforms to the principle of segregation, as is shown when the F_1 generation is crossbred. The traits sort themselves out again with one red-flowered (homozygous) plant to two pink-flowered (heterozygous) plants to one white-flowered plant.

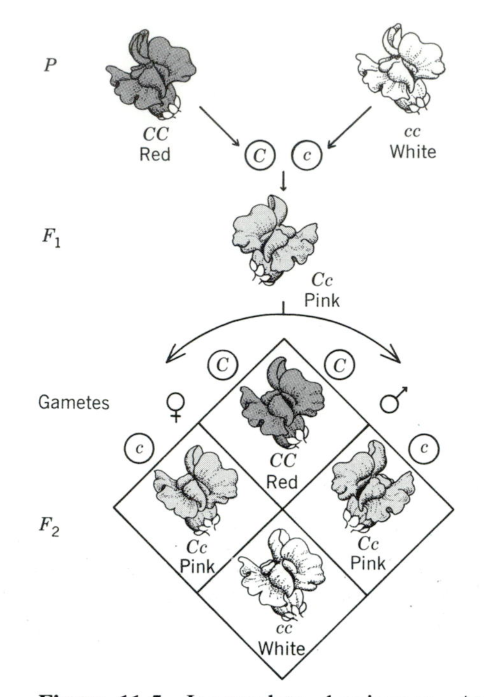

Figure 11-5 Incomplete dominance. At first sight the production of pink offspring from red and white parents looks like a case of blending inheritance. However, when the pink-flowering plants are crossed, the results show that, although the gene products blend in the heterozygote, the genes themselves remain discrete and segregate according to the Mendelian ratio.

e. INDEPENDENT ASSORTMENT (MENDEL'S SECOND LAW)

Mendel also crossed plants with two contrasting characters such as yellow round seed (*YYRR*) with a plant with green wrinkled seed (*yyrr*). All of the F_1 seeds were yellow and round (*YyRr*). When these were planted and allowed to self-pollinate, they produced 566 peas—315 yellow round, 101 yellow wrinkled, 108 green round, and 32 green wrinkled. This is almost precisely a 9 : 3 : 3 : 1 ratio (Fig. 11-4). The explanation that Mendel offered for these data is known as the principle of independent assortment of genes at reproduction or Mendel's second law. Assuming the genes *Y, y, R, r* become sorted out independently of one another some time during the production of the gametes, the gametes would have the following makeup: *YR, Yr, yR, yr*. Random combination of the eggs and sperms results in seeds in the F_2 having a ratio of approximately 9 yellow round, 3 yellow wrinkled, 3 green round, and 1 green wrinkled. Eventually Mendel worked with all seven sets of contrasting characters that he selected and showed that each was assorted independently of the others.

Mendel's work, published in 1866, was completely ignored. Mendel personally wrote to the famous German botanist Carl Nägeli, but he paid no attention to the data. It took 34 years before Mendel's papers were rediscovered by three scientists. The first was a Dutch botanist, Hugo de Vries, who in 1900 found Mendel's paper in writing up conclusions from his own studies. Karl Correns, a German scientist, and Erich Tschermak of Vienna also were doing experiments that led to the same conclusion as Mendel's; they discovered Mendel's papers within months after de Vries. Thus, not until 1900 was the value of Mendel's work appreciated and he was only given the credit he deserved long after he had died in 1884.

Why did Mendel's work lie neglected so long? There are many reasons, but perhaps one of the most important is that biologists were looking at heredity from a different viewpoint, a perspective that relied more on natural history rather than a quantitative approach. Darwin himself found a 3 : 1 ratio in crosses he made, but it meant nothing to him. To Mendel it meant a great deal; it was the key to understanding the unit character of inheritance.

After the rediscovery of Mendel's work the study of heredity sped up exponentially. The years from 1900 to 1920 saw not only the confirmation of Mendel's work but the extension of his basic ideas in several important directions.

f. LINKAGE AND CROSSING-OVER

Although the principle of independent assortment of genes in meiosis is one of the cornerstones on which the understanding of genetic systems has been built, there are exceptions to this rule. In fact, such an assortment occurs only when the genes are located on different chromosomes. Since the chromosomes of plants may contain thousands of genes and the chromosomes are limited in number, each chromosome must carry a large number of genes. If two genes are located on the same chromosome, independent assortment of them does not occur and they are said to be *linked*.

To illustrate the concept of linkage, consider the crossing of two plants, one having the genotype *AABB* and the other *aabb*. The F_1 phenotype will be *AaBb*. When the F_1 produces gametes, if they were independently assorted, the gametes would be *AB*, *Ab*, *aB*, and *ab* in equal numbers. However, if both dominant genes were located on one chromosome and the recessives on its homologue, only two kinds of gametes would be formed, *AB* and *ab*. Instead of the expected dihybrid 9 : 3 : 3 : 1 ratio in the F_2 generation, a 3 : 1 ratio would result, because the two genes are inherited as though they were a single gene. A number of genes that tend to be inherited together form a *linkage group*. In corn about 360 different sets of *alleles* (one of the two or more alternative, contrasting genes that may exist at a particular locus on a chromosome) are known, falling into 10 linkage groups representing the 10 pairs of chromosomes.

However, genes originally linked on the same chromosome do not always remain linked. In fact, complete linkage, in which gametes contain genes from only one chromosome of a particular pair, is not common. Genes located within one chromosome may be exchanged for corresponding genes of the homologue by a process of *crossing-over* (Fig. 11-6), which occurs at an early stage of meiosis. During the pairing process, the chromosomes tend to twist around one another. At this time two paired, nonsister chromatids of homologous chromosomes may break apart at corresponding points. Broken ends then become joined in new sequences, and two rearranged or crossover chromatids are formed. The genetic consequences of crossing-over, therefore, are the exchange of hereditary material between chromatids of homologous chromosomes. Genes formerly associated in one member of a pair of chromosomes may thus be exchanged to the other member of the pair. Figure 11-6 shows that genes located at different loci (or positions) on a chromosome

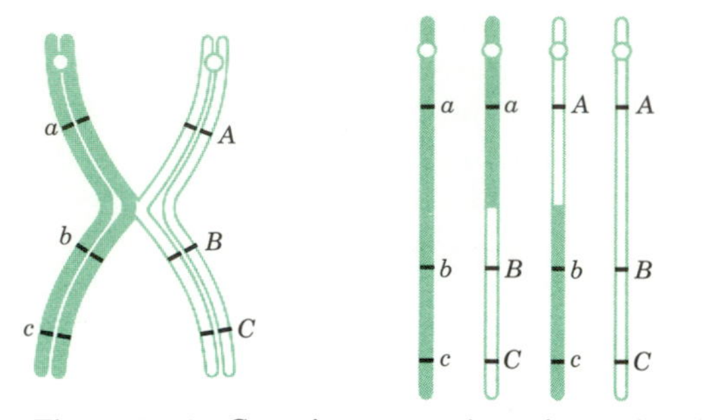

Figure 11-6 Crossing-over takes place when breaks occur in chromatids at the beginning of meiosis while the chromosomes are paired, and the broken end of each chromatid joins with the chromatid of an homologous chromosome. By this means alleles are exchanged between chromosomes.

are increasingly likely to be separated by crossing-over as the distances between them increase.

2. CHEMISTRY OF THE GENE AND HOW THE GENE WORKS

Because of the central importance of understanding the nature of the genetic material to our considerations of gene behavior at all levels of organization, we now discuss briefly the chemical attributes of DNA and the genetic code.

a. DNA—THE GENETIC MATERIAL

Chromosomes, you will recall from Chapter 4, are complexes of DNA (*D*eoxy ribo *N*ucleic *A*cid) and protein. There was accumulating evidence for the role of DNA as the genetic material based on a variety of kinds of information. Some of these are: (1) DNA is present and found almost exclusively in the chromosomes of all cells (demonstrated by specific stains); (2) generally the body cells of plants and animals (diploid) contain twice as much DNA as found in the sex cells (haploid); (3) DNA isolated from bacterial cells can endow other bacterial cells with new genetic characteristics; and (4) when viruses infect bacteria, the virus DNA is the only part that enters the cell and it acts to direct the formation of new viruses.

Despite all of this evidence, it was not until James Watson and Francis Crick in 1951 made their discovery of the structure of DNA that its genetic role came to be accepted. Watson and Crick knew that, only if the DNA molecule could be shown to have the size, configuration, and complexity required to code the incredible store of information needed by living organisms, as well as to make exact copies of this code, could it actually be the genetic material. By studying x-ray photographs of a DNA preparation, they deduced that the molecule was in the form of a spiral or helix, and that the helix was doubled.

A ladder twisted into the shape of a helix, keeping the rungs horizontal (Fig. 11-7*A*), provides a crude model of a double helix. The two

Figure 11-7 *A*. A representation of the double-stranded helical structure of the DNA molecule, drawn in side view as though it were contained in an imaginary cylinder with the center of the axis shown in dotted lines. The outward backbone of this molecule consists of two chains (shown as a ribbon) made of desoxyribose sugar (S), each linked to a phosphate group (P). These nucleotide chains are held together by hydrogen bonds. Adenine (A) and thymine (T) form two hydrogen bonds, shown as III; cytosine (C) and guanine (G) form three. The nitrogen base pairs are stacked flat one above the other at intervals of 3.4A* and rotated 36° with each step. Each chain makes a complete rotation every 34A and there are 10 base pairs in this rotation. These chains are separated by 120° in one direction and 240° in the other direction around the circumference of the cylinder. As a consequence two grooves occur along the side of the molecule with one being wider than the other. The chains also run in opposite directions. (From Ayala, 1976) *B*. Schematic diagram showing how one double stranded molecule of DNA could be replicated into two double-stranded molecules of DNA. (* 1A = $^1/_{10}$ nm or 10^{-7} mm)

railings are made up of alternating sugar and phosphate molecules. The rungs of the ladder are formed by the nitrogenous bases—adenine (A), thymine (T), guanine (G), and cytosine (C)—one base for each sugar-phosphate, with two bases forming each rung. The paired bases meet across the helix and are joined together by hydrogen bonds in such a way that A always pairs with T and G always pairs with C. It follows that DNA has as many A as T bases and as many G as C bases; that is, the ratios A/T and G/C are each equal to 1. In contrast (A + T)/(G + C) may vary from organism to organism.

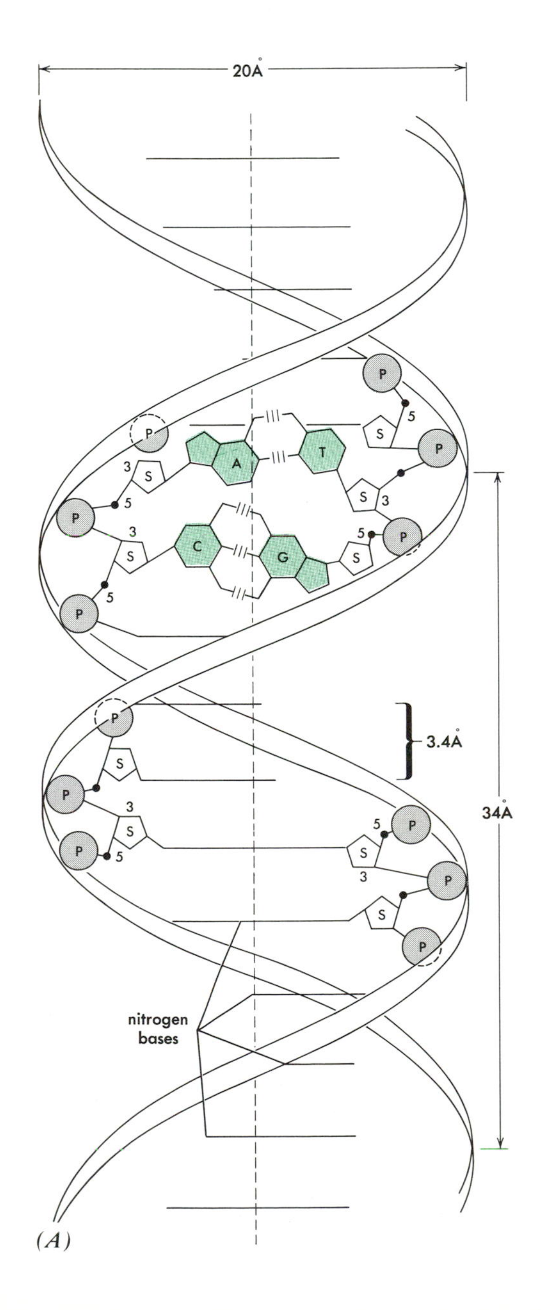

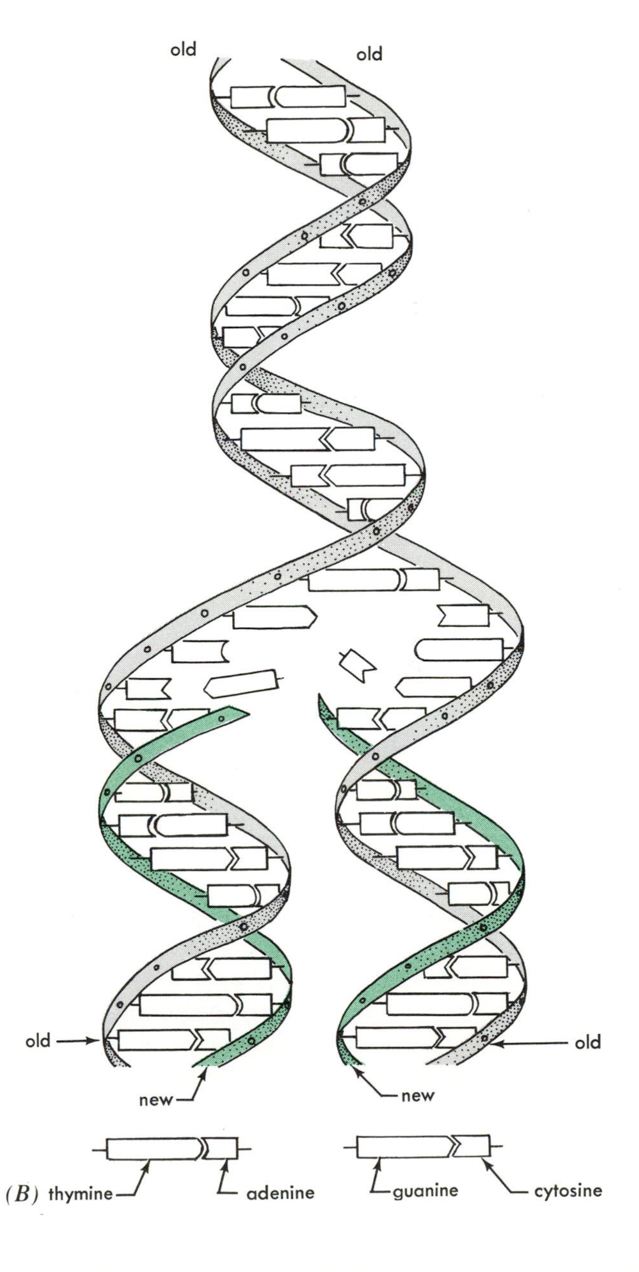

The fact that the two nucleotide chains in the double helix are complements of one another answered the age-old question of how hereditary information is duplicated and passed on for generation after generation. The sequence of bases along one of the strands dictates the sequence along the complementary strand because of the strict determination of base pairing (C only with G, A only with T) between the two DNA chains. Replication begins with an unwinding process; in fact, the molecule simply "unzips" down the middle, the bases breaking apart at the hydrogen bonds (Fig. 11-7*B*). Each chain then serves as a template for a complementary strand. Two new double helices are formed that are identical to each other and to the parental double helix.

b. THE GENETIC CODE

Although the discovery by Watson and Crick elucidated the chemical nature of the gene and indicated the way it duplicated itself, the question as to how the information in the gene influenced structure and function was unanswered. If the enzymes, which are proteins, with their 20 amino acids, were the "language of life" (using the metaphor commonly used by geneticists in the 1940s), the DNA molecule with its four nitrogenous bases could be thought of as a code for this language—the *genetic code*.

Actually the concept of a genetic code has been useful not only as a dramatic metaphor but also as a working analogy. The problem has been approached by biochemists using the methods of cryptographers. The four bases may be considered as letters in the genetic alphabet. Specific sequences of letters of an alphabet in any language make up words; a sequence of words makes up phrases or sentences that convey information. Analogously, one may think of nucleotide sequences as genetic "words." The genetic endowment of an individual organism then may be considered analogous to a "book" made up of DNA words. There is practically no limit to the different words or messages that can be encoded in the long DNA chains.

The basic units of information are not individual bases, but groups of three consecutive bases. Because there are four bases, the number of possible different groups of three bases is $4^3 = 64$. A DNA chain with 600 nucleotides (a probable length for the information segment of a structural gene) has 200 groups of three bases. The number of potentially different messages contained in chains of that length is $21^{200} = 10^{264}$, a number considerably greater than the number of atoms in the known universe.

3. GENE INTERACTION IN DIPLOID ORGANISMS

Many developmental steps usually intervene between primary gene action and ultimate phenotypic expression. Since the expression of a gene only takes place through its production of a protein enzyme, and since proteins are extremely diverse, they mediate large numbers of reactions in organisms. Enzymes and other proteins do not usually, perhaps ever, work in only one highly specific way in the organism. A long, stepwide process produces various biochemical effects that may influence the phenotype in many ways. The capacity of a single gene to control enzymes central to many characteristics of an organism is called *pleiotropy* (Fig. 11-8). Thus a change in a single gene, for example, producing an enzyme involved in lignin synthesis, may cause a great number of different and unpredictable effects throughout the plant.

Furthermore, research with a large number of genes has shown that a group of genes may be responsible for producing a single character. These genes work "in teams": each gene is a distinct unit that may have one or more alleles, but all members of the series must be present in the dominant condition to produce a given character. Flower color is frequently controlled by such combinations of genes; 20 to 30 or more steps may be involved in the synthesis of a single pigment (Chapter 6). Each of these steps requires a specific enzyme and each enzyme is determined by a gene.

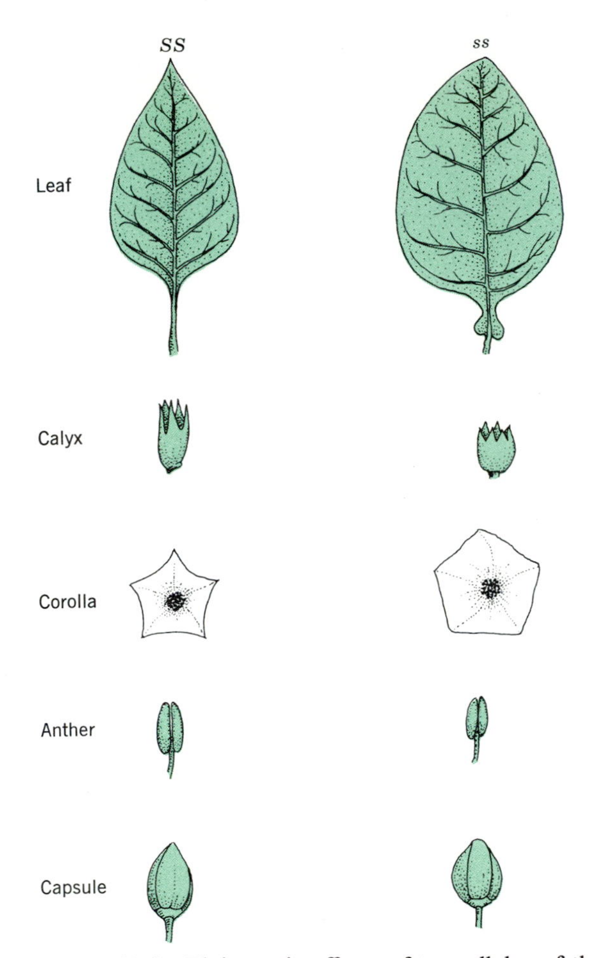

Figure 11-8 Pleiotropic effects of two alleles of the gene *S* on different organs in the tobacco plant (*Nicotiana tabacum*). (From Stebbins, 1959)

The team of genes may also act on some quantitative feature of an organism such as fat content of various seeds, corn-ear length, or skin color in man. In these cases many genes interact to produce a continuous pattern of variation. Such quantitative characteristics are determined by the combined effects of many pairs of genes. Some of the effects of these genes are additive and some may work in opposite directions. The first experiment demonstrating such *polygenic* inheritance was done on color intensity in wheat grains. Figure 11-9 shows the phenotypic effects of various combinations of genes controlling the color in the wheat kernels.

Finally, the effects of gene action can be modified by environmental conditions. Enzymes are usually active only within a certain range of temperature, pH, and other factors. Consequently the gene determining a given developmental sequence may have one phenotypic expression under one set of conditions and a different expression under another.

4. MUTATION

As we indicated earlier, the process of evolution depends on the occurrence of hereditary variation. If DNA replications were always perfect, life could not have evolved and diversified; the same kinds of organisms (no other being possible) would be living today that existed 3 billion years ago. Changes in hereditary material are known as *mutations*. This term has a complicated history of use. It was first used by de Vries, one of Mendel's rediscoverers. In working with evening primroses (*Oenothera*) de Vries found a number of traits that appeared abruptly and were not present in either parental line. He thought that these new characters represented changes in genes because he found they were and would be heritable; he concluded therefore that new species must arise by single steps, and not from slow environmental selection working on genetically produced variation. de Vries spoke of this *sudden hereditary change* as a *mutation* and the organism carrying it as a *mutant*. It is now known that a mutation can involve as little as one change in a single nucleotide pair (*point* or *gene mutations*) or it may involve larger changes such as *chromosomal mutations*. Unfortunately, only 2 of about 2000 changes in the evening primrose seen by de Vries were *gene mutations*. The others were due to new genetic combinations or extra chromosomes rather than abrupt changes in any particular gene. Although most of the examples of de Vries were thus invalid, his recog-

Parents: $G_1G_1G_2G_2 \times g_1g_1g_2g_2$
(dark green) (white)

F_1: $G_1g_1G_2g_2$ (medium green)

F_2:	*Genotype*	*Phenotype*	
1	$G_1G_1G_2G_2$	Dark green	
2} 4	$G_1G_1G_2g_2$	Medium-dark green	
2	$G_1g_1G_2G_2$	Medium-dark green	
4} 6	$G_1g_1G_2g_2$	Medium green	15 green
1	$G_1G_1g_2g_2$	Medium green	to
1	$g_1g_1G_2G_2$	Medium green	1 white
2} 4	$G_1g_1g_2g_2$	Light green	
2	$g_1g_1G_2g_2$	Light green	
1	$g_1g_1g_2g_2$	White	

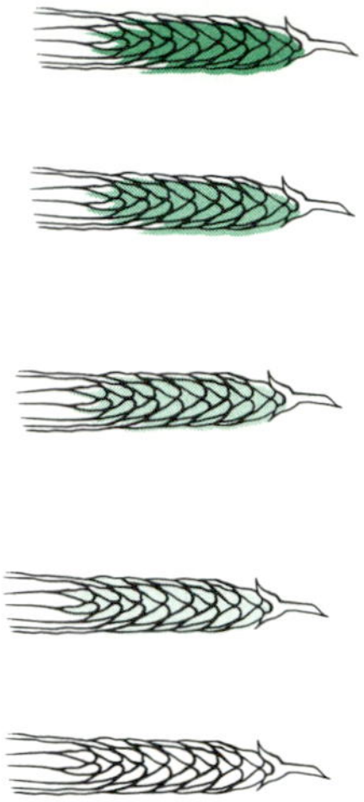

Figure 11-9 The genetic control of color in wheat kernels. Nilson-Ehle in Sweden designed the first experiment that showed how many genes interact to produce a continuous pattern of variation in plants. Here one sees the phenotypic effects of various combinations of four genes controlling the color intensity in wheat kernels. (Adapted from Raven, Evert, Curtis, 1976)

nition of the role that "mutation" plays in producing variation was extremely important in the history of genetics.

a. POINT OR GENE MUTATIONS

A point or gene mutation occurs when one nucleotide is altered and the new nucleotide sequence is passed on to the progeny. The change may be due to the substitution of one or more nucleotides for others or due to the addition or deletion of one or more nucleotides. Substitution in the DNA nucleotide sequence of a structural gene generally results in changes in the amino acid sequence of the protein encoded by the gene. This will have greater or lesser effect on the organism depending on whether the biological function of the protein is severely affected.

Point or gene mutations result from naturally occurring causes. However, rate of occurrence of gene mutations may be increased by high-frequency radiation and a variety of chemicals (known as *mutagens*). The rates of spontaneous mutation vary from organism to organism and from one gene locus to another within the same organism. In viruses and bacteria, and other unicellular organisms, recorded mutation rates occur at about one in a million per gene per cell division.

Depending upon how we view them, point or gene mutations are rare or almost ubiquitous events. The mutation rates of individual genes are low, but each organism has many genes and species are made up of many individuals. If we assume that typical multicellular organisms may have 50,000 pairs of genes and an average mutation per gene occurs once in 10^6 divisions, then on the average each individual would have a $2 \times 50{,}000 \times 10^{-6}$ chance, or 1 in 10, of *one* new mutation per life cycle. Of course many of the mutants would not yield viable seeds. Nevertheless, when all individuals comprising a species are considered, the frequency of occurrence of new mutations is high, even in a given single gene locus. Thus the potential of mutations to generate new genetic variation is rather large.

Furthermore, although gene mutations do provide a large pool of variability, it is important to remember that most modern *natural* species and races have lived as successful populations, well-adjusted to their environment for thousands or millions of generations. Thus nearly all of the mutations occurring in a successful population would lower their adaptation to the accustomed environment and therefore would be rejected by natural selection, unless the environment were changing. Gene mutations may be an important source of variation if slight or barely perceptible modifications get built into the system (Fig. 11-10). Also evidence from agricultural plants shows that some mutations increase adaptation to particular environments. Hundreds of mutations have been produced and tested in cultivated barley. Most of them, as would be expected, are deleterious. Nonetheless, a small number (less than 1%) have increased productivity. Other mutations have resulted in other valuable traits of the plant—such as stiff stems that support the weight of the ripening grain. These mutations, however, do not improve the plant under all conditions. For example, mutations leading to early heading and stiff stems are valuable in the moist climate of northern Europe, since moisture accumulating on the heads bends the stem over to the ground. On the contrary, in the United States the ripening grain is exposed to hot, dry winds. If the stem is too stiff and brittle, the winds break up the heads and scatter the grain over the field. In the language of the agronomist, the mutations valuable in Sweden because of resistance to *lodging* are disadvantageous in the United States because of susceptibility to *shattering*. This is only one example illustrating that any single mutation generally does not improve the overall fitness of an organism but only its adaptation to particular environments.

Actually most geneticists studying evolutionary processes think that the primary limiting factor on the supply of variability for the action of natural selection is not the availability or rate of occurrence of gene mutations but restrictions on gene exchange and recombination imposed by the mating structure of populations and structural patterns of chromosomes. Natural selection directs evolution not by accepting or rejecting gene mutations as they occur but by sorting out new adaptive combinations from the gene pool of variability that has been built up through the combined action of gene mutation, chromosomal mutations, genetic recombination, and selection over many generations.

b. CHROMOSOMAL MUTATIONS

Since most genetic characteristics of organisms are determined by groups of genes, changes within the chromosome produce a significant source of new groups of linked genes that result in phenotypic variation. Chromosomal mutations (Fig. 11-11) include: (1) those that decrease (*deletion*) or increase (*duplication*) the number of genes in the chromosomes; (2) those due to rearrangement of genes in the chromosomes: *inversions* (when a block of genes rotates 180° within a chromosome) and *translocations* (when blocks of genes exchange their location in the chromosomes); and (3) those that change the number of chromosomes: *haploidy* (when there is only one set of chromosomes) and *polyploidy* (when there are more than two sets of chromosomes).

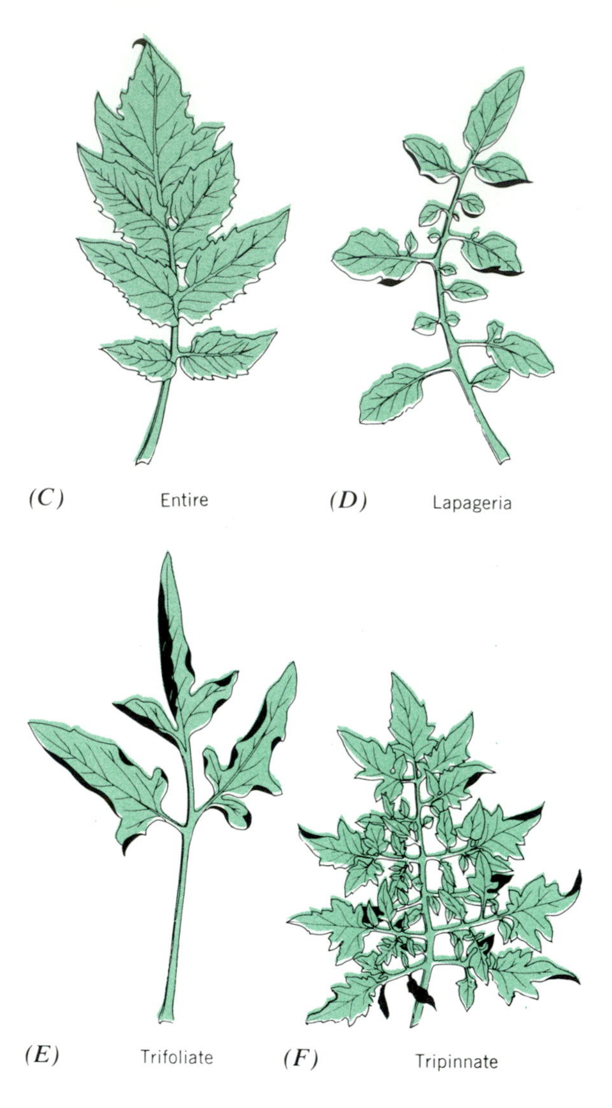

Figure 11-10 Examples of mutations. *A*. Common wild sunflower. *B*. Mutant sunflower known as sun gold. *C, D, E*, and *F*. Different leaf shapes induced in leaves of tomato by single gene mutations. (Adapted from Weir et al.)

Deletion, or removal of a gene locus or a group of loci from a chromosome, is most often lethal in a homozygous condition, although occasionally the functions performed by the normal alleles at the missing loci can be taken over by the genes in some other part of the chromosomal complement. The *duplication* of a gene locus may cause an imbalance of gene activity that will reduce the viability of the organism. However, some organisms can tolerate duplica-

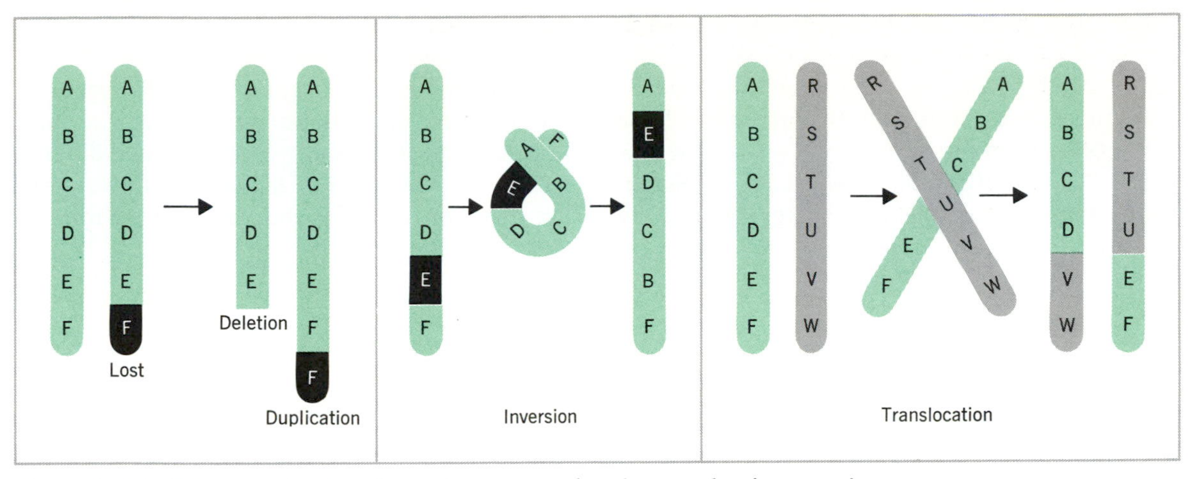

Figure 11-11 Diagrams to show how chromosome breakage and union can give rise to the four principal structural changes that chromosomes undergo; deletion, duplication, inversion, and translocation.

tions of chromosomal material. In fact, duplications may play a role in the evolution of organisms having an increased variety of gene-controlled activities.

Because normal homologous chromosomes pair exactly gene by gene, if an individual is heterozygous for an *inversion,* the chromosome containing the inverted segment often pairs with its normal counterpart in such a way that the two are always transmitted as a single unit. *Translocations* occur when two non-homologous chromosomes may break simultaneously and exchange segments. If an organism becomes homozygous for such a rearrangement, some of its genes have been transferred to a completely different chromosome and the linkage relationship of genes is radically altered. In addition, if it remains heterozygous for translocations, particular combinations of genes located on non-homologous chromosomes may be kept together and this causes them to be inherited as though they were genetically linked. Both inversion heterozygotes and translocation heterozygotes exist in natural populations and are probably significant mechanisms in retaining adaptive gene combinations in populations. Inversion heterozygosity, however, is more widespread than translocation heterozygosity.

Most races of the evening primrose (*Oenothera,* the genus studied by de Vries) in eastern North America are heterozygous for a series of segmental interchanges involving all 14 of their chromosomes, so that at meiosis they form a ring of 14. New races can be formed by occasional crossing between preexisting races or by abnormal segregation of chromosomes in the rings. The evening primroses are a striking example of a highly evolved and complex chromosomal behavior that achieves the same objective that human plant breeders have with hybrid vigor (see Chapter 12).

C. SOMATIC MUTATIONS

Mutations, either *genic* or *chromosomal,* may occur in somatic (body) cells as well as in the sexually reproductive cells. For example, *bud mutations* originate in this manner. A single branch may bear leaves, flowers, or fruits different from those on other parts of the plant. Some of our most valuable fruits, such as apples, seedless grapes, and seedless oranges have originated in this way. Bud mutations have often been found in ornamental plants such as roses, chrysanthemums, and dahlias. Sometimes these mutations can be reproduced by seed but generally it is necessary to propagate them vegetatively. The African violet (*Saintpaulia*), a popular houseplant, is well known for the number of somatic mutations that have arisen from leaf cuttings. Such mutants differ in size, form, and color of leaves and flowers and are maintained by vegetative reproduction.

PART 3
PLANT POPULATIONS AND ECOSYSTEMS

CHAPTER 12

Genetics of Diversification in Plant Populations

1. EVOLUTION OF PLANT DIVERSITY—BEHAVIOR OF GENES IN POPULATIONS

In Darwin's day, when the concept of "blending inheritance was widely accepted, the question arose as to why rare traits could continue to appear. The question persisted after Mendel's discovery and recognition of "unit inheritance", but the question was rephrased as to why dominant traits did not ultimately "drive out" the recessive ones.

a. HARDY–WEINBERG LAW

An answer to this question came in 1908 when a German physician, W. Weinberg, and an English mathematician, G. H. Hardy, simultaneously discovered that, in a large population in which there is random mating and in which the forces that change the proportions of genes are not operating, the original ratio of dominant to recessive alleles will be retained from generation to generation. This generalization is known as the Hardy–Weinberg law.

Another way of stating the Hardy–Weinberg law is that, when a population is in genetic equilibrium (i.e., when gene frequencies do not change), genes continue to be reshuffled by sexual recombination, and individual variations originate from this source. Understanding this equilibrium has important genetic applications. However, since gene frequencies do not change, from an evolutionary perspective it also is necessary to understand factors which produce deviations from this equilibrium.

b. APPLICATIONS OF THE HARDY–WEINBERG LAW

Natural selection operates on the phenotype. Therefore a recessive gene is exposed to natural selection only in an individual homozygous for that gene. Because a smaller proportion of recessives is exposed in homozygotes, its removal by natural selection slows down. This is an important fact for plant breeders in search of certain genes in wild populations to be introduced into cultivated areas. For example, a recessive gene determining disease resistance may be rare but will tend to be maintained in the natural population.

Also, if the allele is a rare one, it is present mainly in the heterozygous state and this results in an important consequence: if the heterozygote (*Aa*) is for some reason superior to either homozygote (*AA* or *aa*), then both alleles will be maintained in the population. This phenomenon is called *heterozygote superiority* (Fig. 12-1).

Hybrid vigor or *heterosis*, which apparently results from the plants' being heterozygous at many more loci than most natural varieties, may be the result either of heterozygote superiority or of the fact that hybrids are less likely to be homozygous for deleterious alleles. Hybrid corn, a dramatic example of this superiority of a hybrid over either of its parents in several characteristics, is discussed in detail in Chapter 16.

C. EXCEPTIONS TO THE HARDY–WEINBERG LAW

Since evolution could not occur if all populations were in the Hardy–Weinberg equilibrium, what conditions could result in deviation from the equilibrium? First, if a particular gene is mutating, the population cannot remain in equilibrium for the gene in question. Second, in a small population, the random loss of individual genotypes may eliminate one of the alleles from the population. Third, if individuals with particular genetic characteristics leave a population or enter it in a proportion different from that in which they are already represented in the population, the gene and genotype frequencies will obviously change. *Natural selection,* however, is the major factor causing exceptions to the Hardy–Weinberg law. Selection, as indicated earlier, provides the basis for most of our current understanding of evolutionary change. Mutations and recombination provide the variation, but changes in populations take place as a result of selection of the phenotypes. In any variable population, some individuals will leave more progeny than others. Therefore, certain genes become more frequent in populations and others become more infrequent. By continually eliminating certain genotypes from a population under certain sets of conditions, selection channels the variations in traits produced by mutations and recombination. In this way, selection results in individuals that can survive better under given conditions than those that were rejected. Extreme variants will be selected against; they will not survive because of their extremeness and thus natural selection can bring stability into a population (Fig. 12-2). This *stabilizing selection* minimizes the extremes in a population, although extreme types continuously reappear because of normal variation. Then, if the environment changes so that certain extreme forms are better adapted than the previous norm, gene frequencies will shift toward these extremes. This *directional selection* will cause a shift in the normal frequency. The population as a whole will have evolved toward the favorable extreme types. In some cases environmental changes may give a greater advantage to two or more extremes within a population, causing a shift in at least two directions. The norm will be selected against; this *disruptive selection* is so called because it disrupts a stable pattern and leads to evolution in several new directions.

The graph depicting these changes is misleading because it shows such shifts as taking place at one specific time. But shifts in gene frequency are dynamic and the amount of variability allowed depends both on the particular characters being selected and the nature of the environment. However, some species are evolving rapidly, such as some of the oak trees we discuss later in this chapter. Redwood trees, which are evolving slowly because they are an ancient group, are now only relics of a much larger population restricted to a narrow range of environmental conditions (Chapter 14). The chance of surviving something as drastic as a major climatic shift is low. Considering the extensive variation in many populations, the existence of new and unoccupied habitats can result in rapid evolution. This process of organisms' becoming adapted to new niches is called *adaptive radiation.* Populations exploit and increasingly adapt to new habitats by genetic

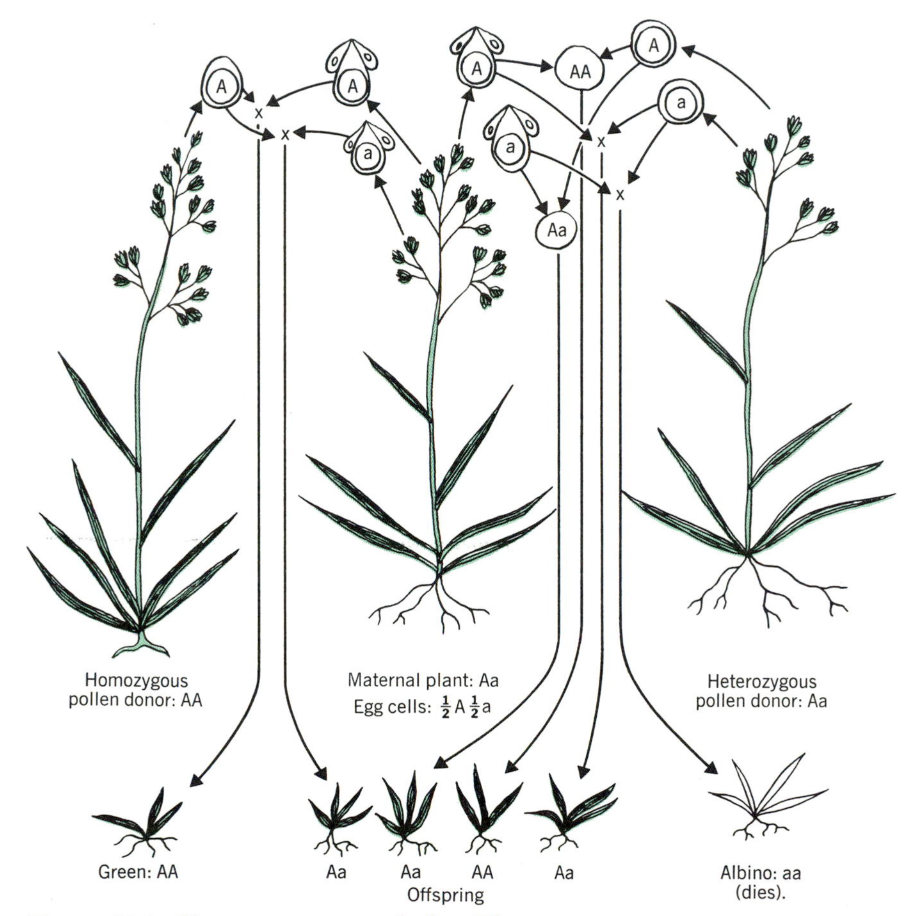

Figure 12-1 Heterozygote superiority. The superiority of heterozygotes for either genes or chromosomal segments is important because it maintains an extra supply of genetic diversity in populations. If a population exists in a relatively homogeneous habitat, so that all individuals are competing with each other, two opposite alleles that control very different phenotypes are not likely to be retained in the population indefinitely. One phenotype will usually have selective advantages over the other, and the allele controlling it will therefore increase continously in frequency. There are three possible cases: (1) If the heterozygote has a lower adaptive value than the homozygote for the superior allele, the inferior allele will gradually be driven out. (2) If the heterozygote is equal in adaptive value to the superior homozygote (a situation that usually exists for dominant and recessive alleles at a locus), the recessive allele will remain in the population at a low frequency and will be a part of its "mutational load." (3) If the heterozygote is superior to either homozygote, the inferior allele will be retained at a fairly high frequency in the population, even if genotypes homozygous for it are lethal. This latter situation seems to be common in natural populations of both plants and animals. The orchard grass in Israel (*Dactylis glomerata judaica*) shows how cross-pollination between individuals of a population that have normal green color but are heterozygous for an albino gene (*a*) can yield occasional lethal homozygotes (*aa*) among the seedlings raised from seed gathered from wild plants cross-pollinated in nature. (From Stebbins, 1966)

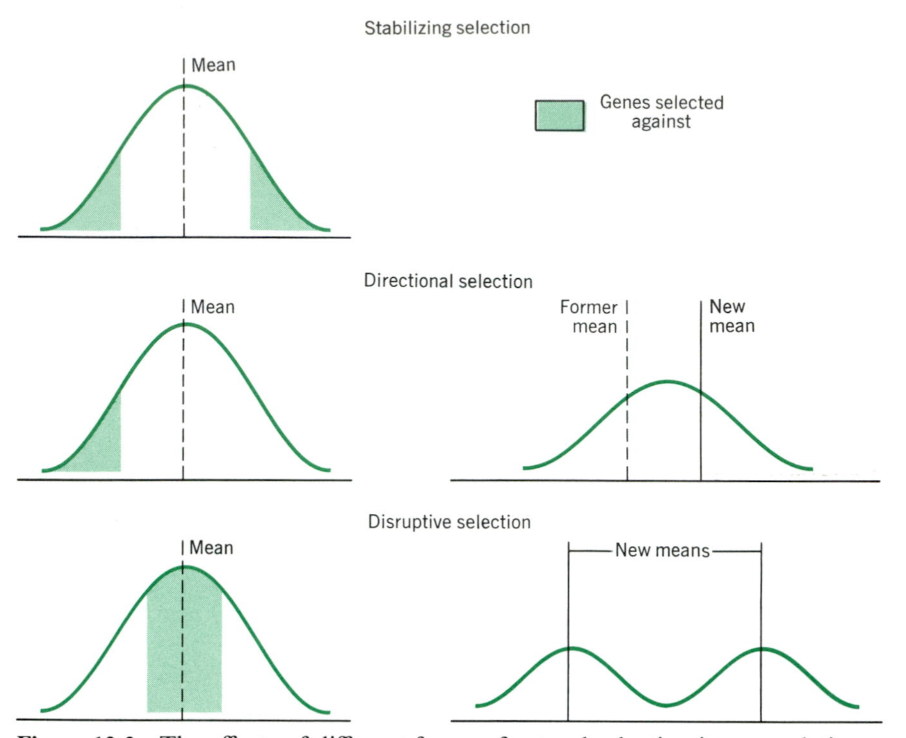

Figure 12-2 The effects of different forms of natural selection in a population.

variation. We discussed the adaptive radiation of the primates in the evolution of our predecessors (Chapter 2). Darwin also noted this phenomenon in the genus *Scalesia,* a member of the generally herbaceous sunflower family, which was treelike on the Galapagos Islands (Fig. 12-3).

The response of the individuals in a population to selection is affected by many of the principles of genetics just discussed. Since it is generally thought that the phenotype is being selected, the relationship between the phenotype and the environment determines the reproductive success of an organism. Because most traits in natural populations result from the interaction of many genes, individuals of a similar phenotype can have very different genetic properties. When some trait is strongly selected for, generally there is an accumulation of genes that contribute to this trait as well as elimination of those that work against it. Not merely a simple accumulation of one set of genes, with an elimination of their alleles, but complicated interactions between them help to determine the course of selection in a population. Because of gene interactions, we can measure the phenotypic effects of a particular gene only in a specific genetic context. As selection for particular genes changes the representation of the other genes in the population, the effects of selected genes become gradually changed. Thus the value of the genes in determining a particular trait, or the fitness of the organism as a whole, also changes. It should be remembered, however, that selection must result in a successful organism. Strong selection for one trait may be prevented because it would lead to the accumulation of so many undesirable side effects that the organism would no longer be able to survive or might lose its fertility.

Also of great importance are the limits to selection imposed by the need of the individual organism to meet vastly different environmental demands throughout the year. For example,

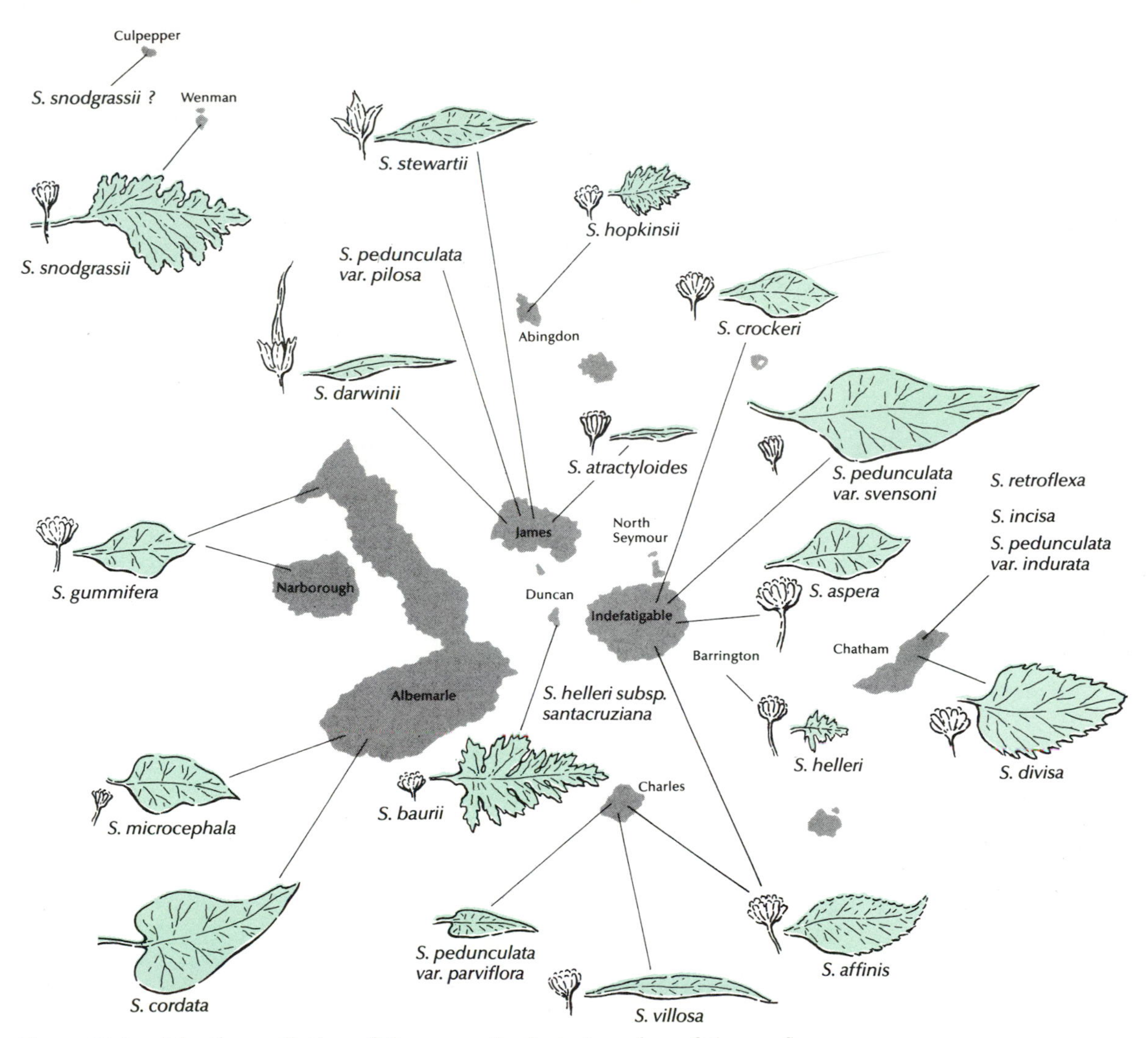

Figure 12-3 Adaptive radiation of the genus *Scalesia* (member of the sunflower family) was described by Darwin on his trip to the Galapagos Islands. *Scalesia* looks like a large, woody sunflower; most of the species are restricted to specific islands. (From Carlquist, 1965)

plants living in the arctic must grow and photosynthesize rapidly in order to store sufficient carbohydrate during the short summer to survive the long, severe winters and likewise be able to respond to the first signals of favorable growth conditions. A miscue in responding to a midwinter thaw might lead to their demise. In another harsh environment, the desert, seeds must germinate whenever sufficient water is available, but not germinate when too little water is available. This requires physiological fine tuning for survival and, evolutionarily, desert plants have met these conditions in a variety of ways.

Changes in environment have occurred since the beginning of Earth, but today many of the relatively rapid changes in natural populations have resulted from the activities of humans. The ability of higher plants to evolve relatively rapidly in conditions normally considered unfavorable may be illustrated by adaptation of grass plants to certain mine tailings in Wales that contain up to 1% lead and 0.03% zinc—conditions usually toxic to plants (Fig. 12-4). These tailings are from mines no more than 100 years old and the dumping areas are generally bare of plants except for one member of the bent grasses (species of *Agrostis*). This grass also grows in pastures near the mines, and experiments were designed to see just how the grass had adapted to the unfavorable soil conditions. Samples of plants were taken from the mine tailings area and a neighboring pasture and grown together in both normal pasture and lead-mine soil. In the pasture soil the lead-mine plants were definitely smaller and more slow-growing than the pasture plants. On the lead-mine soil, the mine plants were normal; the pasture plants on the other hand did not grow at all; all had misshapen roots and half of them died within three months. However, a few (3 out of 60) of the pasture plants showed some resistance to the lead-rich soils. They probably were similar to the plant originally selected in the development of the lead-resistant race.

We humans have not only created environments that have influenced the direction of natural selection of plants, but, as previously pointed out several times, we have intervened by deliberate and intentional processes of selection. These may go in various directions, sometimes depending upon our whims. For example, we have selected specific kinds of corn for a variety of purposes such as for boiling, roasting in the ear, or popping; for flour quality; for making dyes, hominy, or beer as well as unfermented beverages—and even for religious and ceremo-

Figure 12-4 Mining areas in Wales (Trelogan), contaminated with heavy metals such as lead, zinc, and copper, also are typically low in essential nutrients (nitrogen, phosphorus, and potassium). Because few plants can adapt to these soil conditions, the landscape often appears dune-like as shown above. Bent grasses (*Agrostis*) resistant to these conditions are in the foreground.

nial purposes. We select barley for food, beer, and livestock feed, for processing in grinding equipment, and for ease of harvesting. We apparently enjoy "the unusual" and thereby select and preserve plants, on occasion, that probably would have little chance of survival under natural conditions.

Of course, we have modified the part of the plant that interests us most. If the crop is a root or grain, it is this part that deviates the most from the wild plants. *Brassica oleracea* (mustard family) is a dramatic example of our influence in modifying a single species in six ways (Fig. 12-5). Selection for an enlarged terminal bud produced the cabbage; for inflorescence, the cauliflower; for the stem, kohlrabi; for lat-

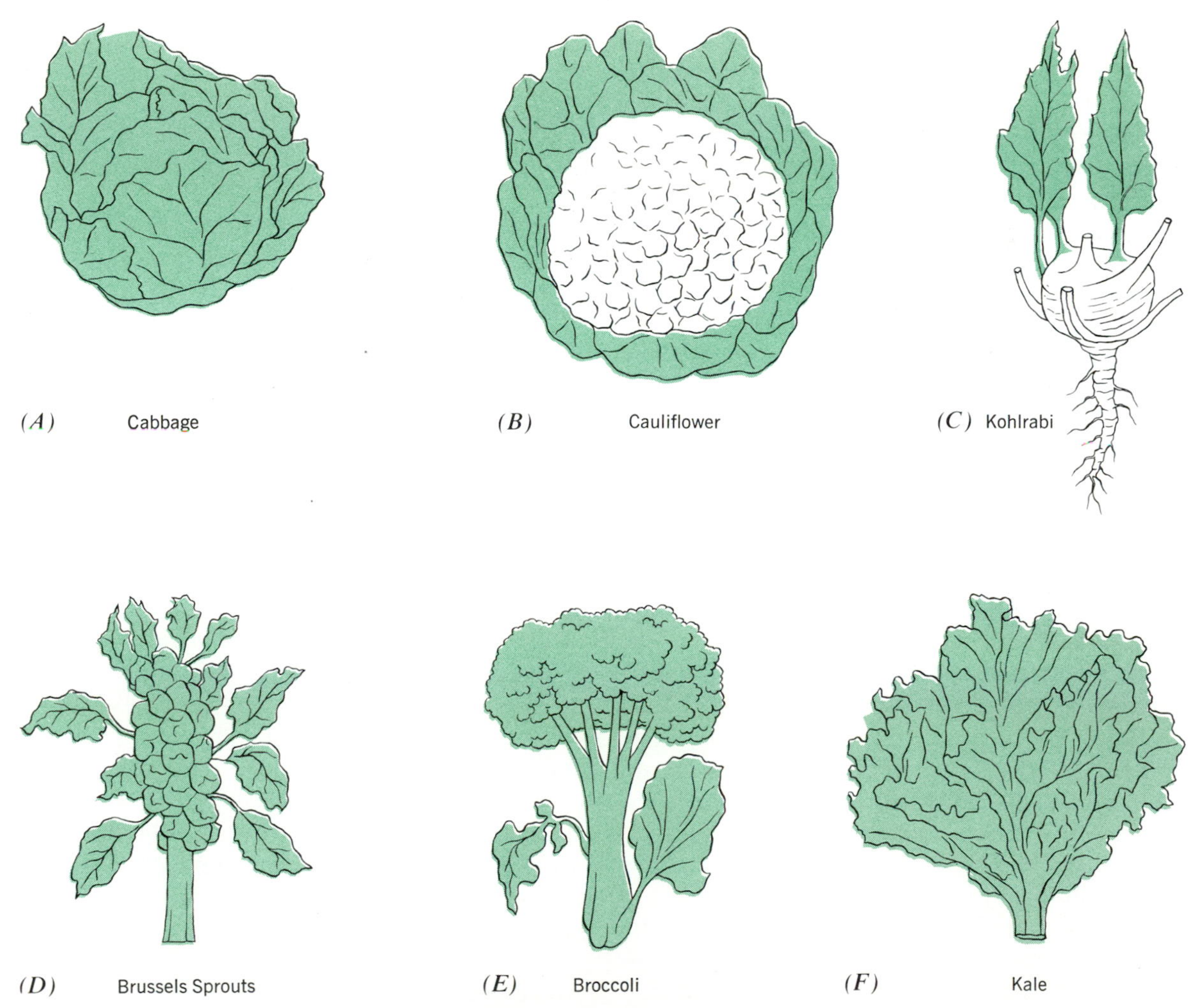

Figure 12-5 Selection has produced six separate vegetables from a single mustard species *Brassica oleracea*. *A*. Selection for enlarged terminal buds produced the cabbage. *B*. For inflorescences, the cauliflower. *C*. For the stem, kohlrabi. *D*. For lateral buds, brussels sprouts. *E*. For the stem and flowers, broccoli. *F*. For leaves, kale.

eral buds, brussels sprouts; for the stem and flowers, broccoli; and for the leaves, kale. Kale is closer to the wild type than the others.

2. DIFFERENTIATION OF POPULATIONS

Environmental conditions occur as a complex mosaic and individual plants may be selected from each other within the mosaic. The movement of genes between the isolated populations therefore is naturally limited and the populations respond to selective pressures from local habitats. Despite variation provided from gene flow and from the immigration of individuals from different populations, it cannot counter the effects of selection by local conditions. Therefore, the distances that may effectively isolate populations, such as with pollen dispersal, depend, of course, upon the various possible agents bringing about the cross pollination. For example, if the plant's pollination depends upon hummingbirds, the pollen may be carried perhaps 100 m or more. However, if it depends upon some insects, plants may be isolated at distances as small as 15 m. If wind is the pollinating agent, as in many grasses and coniferous trees such as pines, little pollen may be effectively dispersed more than 500 m from the parent. Dispersal beyond 1000 m in crop plants is rare. Also most of the pollen that is transported for hundreds of kilometers and is absent for days is not genetically viable. Pollen of corn, white barley, and many grasses typically do not survive more than 24 hours. This does not mean that two oak trees separated by 50 m are out of genetic contact but that each will be influenced mainly by conditions in its local environment in producing successful offspring.

Under differing environmental demands, plasticity in the phenotypic response shows up among genotypically identical organisms. This plasticity is much greater in plants than in animals, because the open system of growth of plants can be more easily modified in various ways that produce striking differences between individuals. In fact, most gardeners are aware of impressive changes that environmental conditions can make in the appearance of various species of plants. Leaves that develop in the shade are quite different from those that develop in the sun on the same plant (see Chapter 5). Also the leaf form can be affected by day length. Thus it is not unexpected that many botanists suggested, until about 1930, that much of the natural variation in plants was due to environmental effects rather than being genetically determined.

a. ECOTYPIC AND ECOCLINICAL DIFFERENTIATION

The question as to whether the differences between populations of plants living in different habitats are genetically or environmentally controlled was first answered unequivocally by the Swedish botanist Göte Turesson. Turesson transplanted "races" of 31 plant species in southern Sweden into his experimental garden and found that in the majority the differences in height, flowering time, production of seed, and so on were genetically controlled; environmental modifications predominated in just a few. Turesson called the genetic races, differentiated with respect to particular habitats, *ecotypes*.

The work of Turesson on ecotypes was expanded by workers in California. They established three transplant stations along a transect from the coast across the Sierra Nevada in which they used plants that could be propagated asexually so that genetically identical individuals could be grown at all three locations. This California transect was ideal for displaying the effects of environmental contrasts in producing ecotypic differentiation. They studied numerous plant species but those of *Potentilla* (rose family) and *Achillea* (sunflower family) were given some of the most intensive treatments (Fig. 12-6). Because there are striking contrasts in environmental condi-

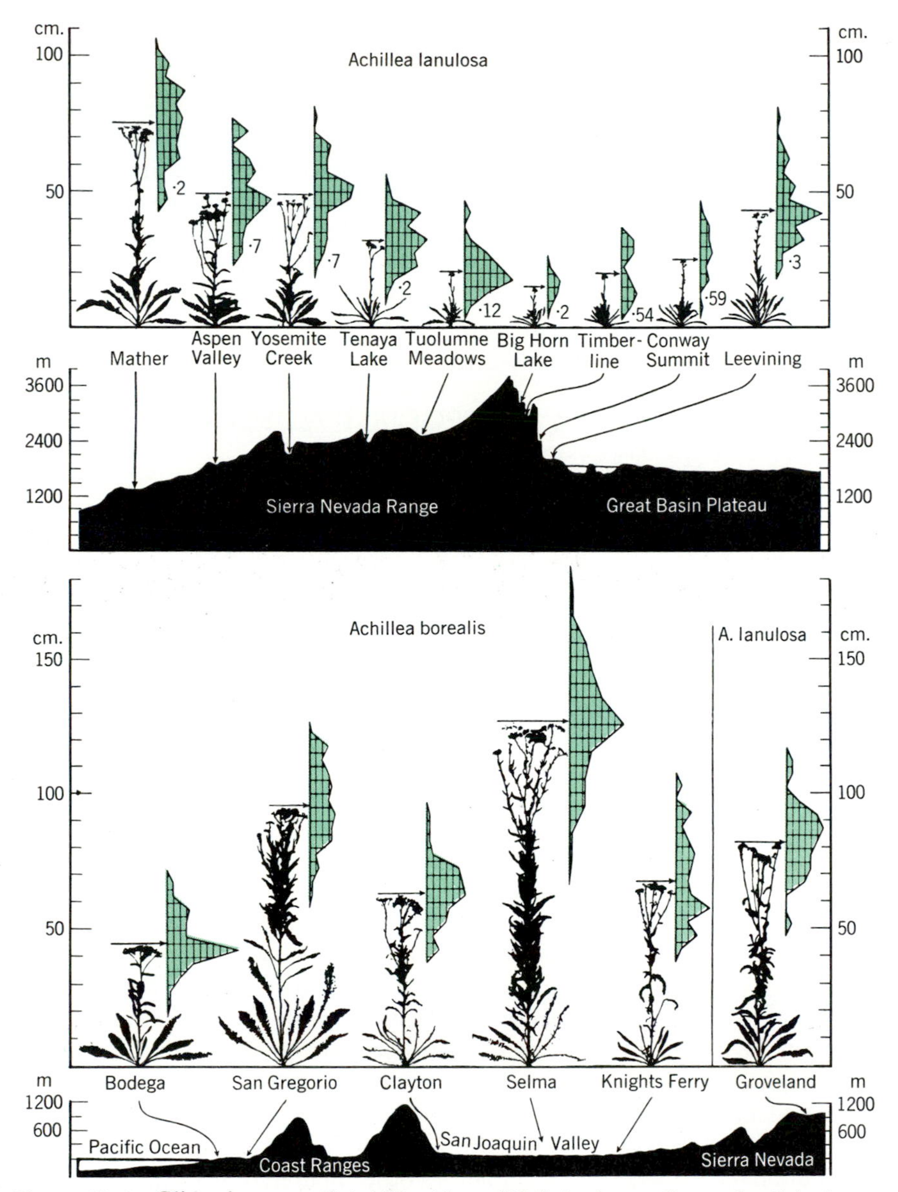

Figure 12-6 Climatic races of *Achillea* (the millfoil, in the sunflower family) along an altitudinal transect in California. Arrows indicate the origin of the population on an east-west profile across central California. The frequency diagrams indicate variation in height within the populations in the gardens of Carnegie Institution at Stanford, and the plant specimens with arrow above represent the means. The figures to the right of some frequency diagrams indicate the number of nonflowering plants. (From Clausen, 1962)

tions across this Sierran transect, it is not surprising that plant species such as *Achillea lanulosa* and *A. borealis* developed clearly defined ecotypes. When these species of *Achillea* (belonging to the widespread *millefolium* complex with circumboreal distribution) from different sites of approximately 330 m intervals were grown side by side in the coastal transplant garden at Stanford, different racial differences became apparent (Fig. 12-6). Also when cloned individuals of these various ecotypes were grown at the Stanford, Mather (midaltitude) and Timberline (high altitude) transplant stations, differences in physiological tolerance were revealed. For example, coast range ecotypes grew actively year round at Mather (1400 m). At Timberline (3050 m) these ecotypes usually die during the first winter because of the short growing season there. From studies of other species along the transects, it appears that different species of plants growing together are more similar to each other physiologically than to other populations of their respective species. These physiological characteristics of ecotypes are determined by many (sometimes even hundreds) of gene loci. The sharply defined ecotypes just described for *Achillea* only occur where there are sharp environmental breaks. If the environment changes gradually, populations of plants may reflect this likewise and then are called *ecoclines*. They are particularly characteristic of areas such as parts of Europe or the eastern United States where rainfall and temperature gradients may extend over great distances.

b. PHYSIOLOGICAL DIFFERENTIATION

The physiological basis for ecotypic differentiation is important to understand with regard to both native plants and crop plants. *Oxyria digyna* provides an example of physiological differentiation that occurs over its wide range in arctic and alpine habitats. Experiments using plants from high latitudes in Greenland and Alaska, as well as from the high altitudes in Colorado and California, have shown great differences in metabolic potential between the populations. For example, the plants of northern populations had higher respiration at all temperatures than southern populations, and high-elevation plants near the southern limits of their range attained photosynthetic light saturation at higher light intensity than plants from low-elevations in their northern range.

Knowledge of these physiological bases of ecotypic differentiation is important in our introduction of seeds from another area. For example, many important timber trees, such as *Pinus sylvestris,* are ecotypically differentiated as to photoperiod. Therefore, extreme caution must be used in selecting seed for establishing plantations of forest trees in different parts of the world.

c. REPRODUCTIVE ISOLATION

When two populations are unable to form fertile hybrids with one another (i.e, they are reproductively isolated), they no longer can influence each other's evolution (Table 12-1). Such a step, therefore, is very important not only in differentiation but in divergence of populations.

Chapter 13 discusses how reproductive isolation has been used by many biologists as the basis for defining "species." However, many morphologically and ecologically distinct groups of plants ("species") – particularly long-lived groups such as trees and shrubs – can hybridize. In contrast, many (certainly not all) individuals that belong to "species," and varieties within them, in herbaceous and short-lived groups are sterile when crossed. It would be expected that annual plants would change more rapidly than longer-lived ones, since each year they become established by seed. This naturally heightens the selection pressures and causes populations to diverge rapidly from each other.

Factors other than hybrid sterility can keep populations of plants distinct (Table 12-1). One is ecological isolation, which will prevent hybridization because the two groups of plants occupy distinct and isolated habitats. For example, in California throughout their ranges, *Quercus agrifolia,* the coastal live oak, and *Q. wislizenii,* the interior live oak, generally are separated by their habitats. *Quercus agrifolia* occurs primarily in coastal woodland or forest border habitats in a more southerly distribution

TABLE 12-1

Summary of the Most Important Isolating Mechanisms That Separate Species of Plants

A. Prezygotic mechanisms
 Prevent fertilization and zygote formation
 1. Habitat. The populations live in the same regions but occupy different habitats.
 2. Seasonal or temporal. The populations exist in the same regions but are sexually mature at different times.
 3. Mechanical. Cross-pollination is prevented or restricted by differences in reproductive structures.

B. Postzygotic mechanisms
 Fertilization takes place and hybrid zygotes are formed; these are inviable, giving rise to weak or sterile hybrids.
 1. Hybrid inviability or weakness.
 2. Developmental hybrid sterility. Hybrids are sterile because reproductive organs develop abnormally or meiosis breaks down before it is completed.
 3. Segregational hybrid sterility. Hybrids are sterile because of abnormal segregation of the gametes of whole chromosomes, chromosome segments, or combination of genes.
 4. F_2 Breakdown. F_1 hybrids are normal, vigorous, and fertile, but F_2 contains many weak or sterile individuals.

pattern than *Q. wislizenii,* which occurs in conifer forests and associated chaparral in a more northerly and easterly distribution pattern. However, the two ranges overlap in central California. Not only do the two habitats tend to keep the two groups isolated, but the time when pollen is produced also separates them. In *Q. agrifolia* pollination occurs between February and April; in *Q. wislizenii* between April and May. Here again there is some overlap in pollination time. In consequence, populations of the two species tend to remain distinct, but, where the ranges overlap, hybridization can occur. We discuss this example further with regard to introgressive hybridization.

Geographical isolation, like habitat barriers within a region, prevents crossing of two populations that are morphologically similar. The sycamore or plane tree is a commonly cited example. *Platanus occidentalis* occurs in the eastern United States and *P. orientalis* in the Mediterranean region (Fig. 12-7). Hybridization of these two is occurring as a result of introduction of *P. orientalis* as an ornamental in the United States.

3. HYBRIDIZATION

Hybridization constitutes one of the best means for genetic recombination at the population level. The term "hybrid" is used in several ways. Often geneticists define a hybrid as a cross between two individuals differing in one or more gene pairs—cross-pollinated plants *within* a species. From the standpoint of the origin of species, however, hybrids are of greatest importance when they are crosses between individuals belonging to *different* species (Fig. 12-8). Despite isolating mechanisms (Table 12-1) hybrids do occur in nature and the way in which recombination of parental characteristics occurs is of considerable evolutionary importance. Especially where environmental changes occur, the individuals of hybrid origin may often present genetic combinations better suited to the new environment than either parent. Also, they may be able to colonize some habitat where neither parent could grow. When the habitats of the parental species are adjacent and sharply distinct, establishment of the hybrids is not likely, but where the habitats intergrade or are disturbed there may be opportunities for establishment. Here the recombination of genetic material shared by the two species has a greater potential for producing progeny well-adapted to the different habitats than does change within a single population. This phenomenon is well illustrated in California in the woody shrubs *Ceanothus* (California lilac) and *Arctostaphylos* (manzanita).

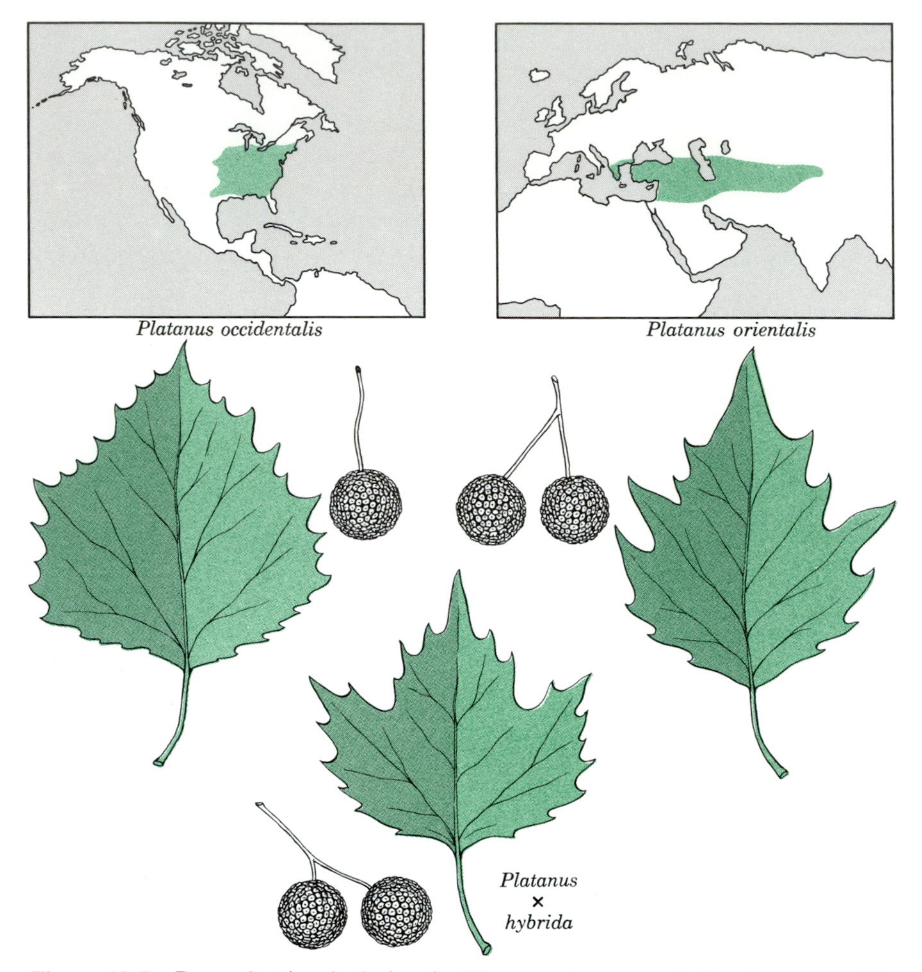

Figure 12-7 Reproductive isolation in *Platanus* (sycamore) due to geographic barriers. The western sycamore in eastern North America is morphologically similar to the eastern sycamore in Europe. As a result of the introduction of *P. orientalis* into the United States as an ornamental, introgressive hybridization is occuring with *P. occidentalis*. Formation of fertile hybrids indicates that the populations are only reproductively isolated due to geographic isolation.

a. INTROGRESSIVE HYBRIDIZATION

It is frequently difficult to determine whether populations with intermediate morphological characteristics originated from hybridization between two species or simply are expressions of variation within a single species. When two species are hybridizing, but the hybrids are rare, these hybrids will not have much chance to cross with other hybrids but will cross with one of the parent species. As a result of repeated backcrossing with rare hybrid individuals, one or both of the subsequent generations of the parental species may become modified. This phenomenon is called *introgressive hybridization* and is an excellent mechanism for introducing new variability into a species. If the

Figure 12-8 Oak species often hybridize where ranges overlap. Leaf forms of some hybrid oaks in California, with parent species shown in black. A. Cross between *Quercus wizlizenii* and *Q. kelloggii*. B. Hybrid swarm, produced by *Q. garryana* (left) and *Q. dumosa* (right), C. Random assortment of possible leaf and growth forms when tree and shrub species hybridize; *Q. douglasii* (left) and *Q. dumosa* (right) with two intermediate forms in the middle. Photos show *Q. chrysolepsis* (left) and *Q. palmeri* (right) which may cross when ranges overlap (Cavagnaro, 1974).

changed individuals are more suited to survive than the unchanged ones in part of the range of species, introgression may be favored selectively. However, the barrier to hybridization between the two original species may not be affected by the introgression that is changing one or both of them, and therefore they may continue to evolve along separate routes. Each may still be adapted to its own habitat, but new sources of variation may enable it to occupy new habitats. To return to the example of *Quercus agrifolia* and *Q. wislizenii,* in central

California where the ranges overlap, there is evidence for introgression of genes from *Q. wislizenii* into *Q. agrifolia* at the woodland–redwood forest border, and *Q. agrifolia* traits also appear in the redwood forest species of *Q. wislizenii* that are close to that border. However, the two species apparently remain better adapted to their respective habitats, although they obviously have incorporated genes from each other. Crop evolutionists consider that there has been "full-scale operation of introgression" in *all cases* of initially domesticated plants whose distribution was expanded with human help—when they came into contact with new wild races or species within the genus. This introgression has been documented for corn, rye, oats, sorghum, tomato, chili peppers, and others. Introgressive hybridization greatly complicated the concept of "place of origin for crop plants," because instead of a delimited area of a single progenitor there are "blurred boundaries."

Both stabilization of hybrid populations and introgressive hybridization are dependent upon fertility of the hybrids. Even if hybrids are sterile, their propagation may be possible either by asexual reproduction or following polyploidy (section 5, below). Hybridization followed by careful selection by plant breeders, of course, is one of the primary ways to obtain the desired traits in cultivated crops. We shall be discussing this process in many of the cultivated plants.

b. HYBRIDIZATION OF SOMATIC CELLS

Work on plant tissue cultures, especially by F. C. Steward at Cornell University, led to the finding that very small groups of cells, and, in liquid cultures even single cells, would grow into whole plants if provided with the right mineral, organic and hormonal nutrients (cf. Fig. 10-10). If treated with a cellulose-dissolving enzyme these single cells lose their cell walls and become naked *protoplasts*. Then it was discovered by Japanese workers that these protoplasts could be cultured, would form new cell walls and would eventually also yield whole plants. If such protoplasts are gently shaken before they have formed new walls they will often fuse together into a single "cell." In this way cells of two different species have been made to fuse—"somatic hybridization."

When this phenomenon was first discovered, protoplasts of two tobacco species, *Nicotiana tabacum* and *N. langsdorffi,* were fused, and resulted in a hybrid plant that was indistinguishable from the hybrid produced by ordinary cross-pollination. Naturally hopes were aroused that all sorts of crossings, even between different genera or families, could be hybridized somatically, though of course such crosses could never be made by cross-pollination. Such hopes have not been fulfilled as yet. Georg Melchers in Germany has achieved a somatic hybrid between two members of the Solanaceae family, a tomato and a potato, and workers in Russia have recently reported a hybrid between two members of the Cruciferae, a *Brassica* and an *Arabidopsis*. In general such distant crossings do not grow into live hybrid plants, but active attempts continue in many laboratories.

Meanwhile, tissue cultures are proving useful in other ways. If immature pollen is used as inoculum, the resulting cultures are haploid and produce haploid plants. Then, if mutations occur, even if they are of recessive type, the haploid plants will show the mutant character; numerous agricultural groups in mainland China are now raising haploid crop plants and screening for mutants. In any case it is obviously easier to screen tissue cultures, of which thousands can be grown in one petri dish, than to screen whole plants occupying hectares of field space.

Isolated protoplasts are being used to screen rapidly for resistance to pathogens. Recently clonal populations regenerated from single leaf-cell protoplasts of a variety of potato have shown enhanced disease resistance and other economically important characteristics (Fig. 12-9). Haploid protoplasts can also be

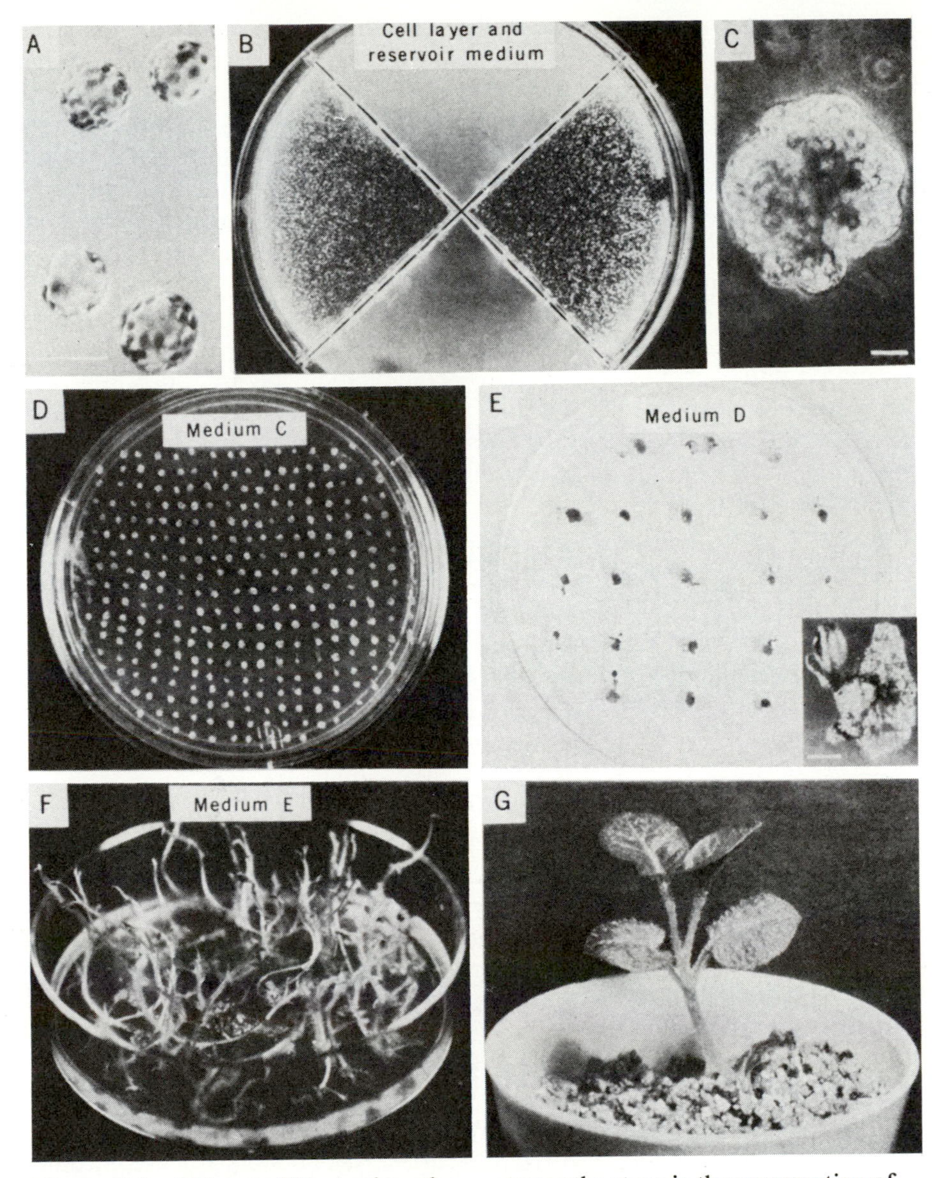

Figure 12-9 This set of illustrations demonstrates the steps in the regeneration of a shoot from a single protoplast from mesophyll cells of potato. *A*. Freshly isolated protoplasts. Scale bar, 50μm. *B*. Fourteen days after protoplasts cultured on medium. Top and bottom compartments contain reservoir medium and are connected with cell-layer medium by multiple holes. Developing protoplast-derived calli (undifferentiated tissues) are present in the left and right segments of the petri dish. *C*. A single protoplast-derived callus from the plate in part *B* at high magnification. Scale bar. 100 μm. *D*. "Conditioned" calli 14 days after transfer to medium *C*. *E*. Emergence of shoots from protoplast-derived calli after eight weeks on medium D. A shoot primordium at higher magnification is shown in the inset. Scale bar, 1 mm. *F*. Shoot elongation and rooting in petri dish (100 by 25 mm) containing medium E. *G*. Transplanted regenerated plant in 40-mm pot. (From Shepard *et al.*, 1980)

grown in the same way as diploid ones and are particularly useful in the search for valuable mutations, because the mutations cannot be masked by the dominant allele as they are in the diploid protoplasts.

Also of tremendous potential is the possibility of introducing selected DNA molecules into the protoplasts. Such DNA might be of plant, bacterial, or viral origin. For example, a shoot from a cloned tumor of tobacco (Havana 25), which had been induced by infection with a bacterium, was recently grafted into a healthy tobacco host plant. The flowering shoot and tumor that developed at the graft union both contained genetic information derived from the bacterium. Possible models have now been developed for the direct transfer of tumor-inducing DNA (rather than the tissue) from bacteria to plant DNA. Thus, work in recombinant DNA, or "genetic engineering," as it is often called, may one day be as important for agricultural crops as it already has become in medicine.

4. ASEXUAL REPRODUCTION

Asexual reproduction and polyploidy provide mechanisms to circumvent the problem produced by sterility of the progeny of many hybrids. Most simply, asexual reproduction consists of the production of rhizomes (such as in grasses) or runners (as in strawberries) (Chapter 19). Another more complicated means is *parthenogenesis,* forming seeds from embryos that are entirely derived from the mother plant (i.e., the megaspore does not go through meiosis). The most flexible breeding systems are those in which asexual reproduction is prevalent but occasional hybridization occurs, to produce variable and new combinations of genes. Thus a plant may proliferate in a local habitat to which it is well adapted and yet provide potential variation to meet environmental changes. Hundreds of "species" of hawthorn (*Crataegus*) and blackberries (*Rubus*) which occur in the eastern United States are asexually reproducing descendents of complexes that occasionally hybridize. Another striking example is *Poa pratensis,* the Kentucky blue grass that occurs throughout the northern hemisphere. Occasional crossing with a variety of related species has produced hundreds of asexual races, each adapted to the particular environmental conditions of the area in which it grows. By this means new genotypes are constantly produced and the best are preserved by vegetative reproduction.

This combination of hybridization and asexual reproduction is widely used in such crops as potatoes, sugarcane, bananas, and grapes.

5. POLYPLOIDY

Hybrids are often sterile because the chromosomes of the two species involved are unable to pair properly with one another and thereby undergo normal meiosis. *Polyploidy* provides a means of circumventing this particular problem of sterility. Polyploids are cells or individuals with more than two sets of chromosomes. In polyploid plants there may be three, four, or more sets of chromosomes and such plants are called triploids ($3n$), tetraploids ($4n$), pentaploids ($5n$), hexaploids ($6n$), etc., with all of the chromosome numbers usually constituting multiples of some basic haploid number.

a. NATURAL POLYPLOIDS

Tetraploidy, the most common type of polyploidy, arises from doubling of the chromosome number of a diploid plant, usually as a result of a "mistake" in mitosis so that the chromosomes divide but the cell does not. If these cells then divide by mitosis to produce a new individual, that individual will have twice the number of chromosomes of its parent. The chance fusion of two diploid gametes then produces a *tetraploid* plant. Doubling may occur in either the zygote (proembryo) or somatic cells (vegetative bud). In the proembryo the chromosomes become duplicated at mitosis, but a new cell does not form and the new nucleus contains $4n$ chromosome number. Doubling of the chromosome number in the cells of a vegetative bud may result in a tetraploid branch, the rest of the plant remaining diploid. A triploid usually arises by the fusion of a diploid and a haploid gamete. Apples are a good example. Whereas most commercial apple varieties are diploid (with 34 chromosomes), some varieties such as Baldwin and Gravenstein have 51 chromosomes in their somatic cells. These are triploids, having probably arisen by fusion of an unreduced or diploid egg (with 34 chromosomes) and sperm (with the usual set of 17 chromosomes). If chromosomes in a triploid plant double, a hexaploid plant is produced; if a tetraploid doubles, an octaploid is produced.

Plants of polyploid origin under natural conditions are relatively common. They occur in some algae and mosses and throughout the vascular plants. Among the flowering plants estimates from various botanists range from 30% to as high as 50%. Polyploidy appears to be most common in herbaceous perennials and least common in woody angiosperms.

Polyploids are usually classified in two major groups: (1) *autopolyploids,* which originate by increase in the number of chromosomes within a single species (*auto* =self), either from self-pollination of individuals or cross-pollination between individuals; (2) *allopolyploids,* which originate following hybridization between individuals of different species (*allo* = other; Fig. 12-10). If polyploidy occurs in a sterile hybrid, each chromosome from both parents will be present in duplicate and may be able to pair with its duplicate chromosome. In this way meiosis will be normal and fertility re-

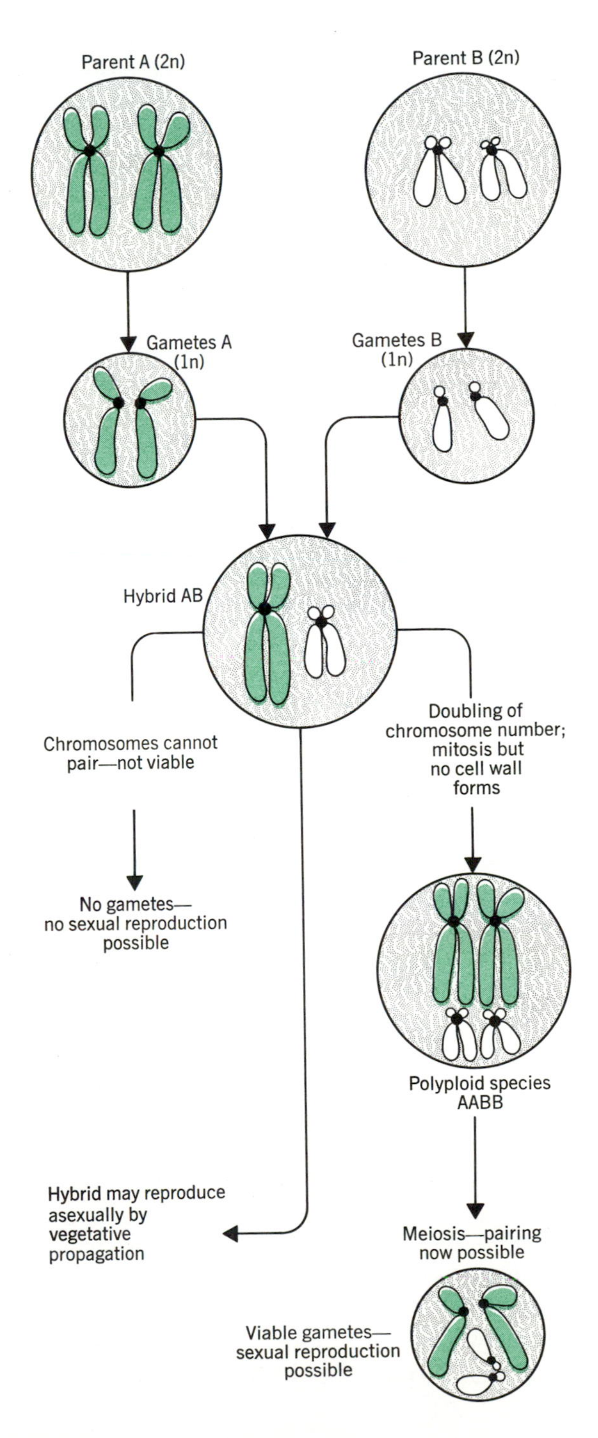

Figure 12-10 Chromosome behavior following the formation of a hybrid and allopolyploidy, which results in the doubling of the chromosome number in the hybrid. Because the chromosomes lack homologous pairs, the hybrid cannot produce viable gametes. The polyploids, however, are successful, because they now have homologous pairs of chromosomes. (Adapted from Jensen and Salisbury, 1972)

stored. Allopolyploids cannot be crossed with either of the original diploid parents, or if rarely they do cross, they form hybrids that are largely sterile. Allotetraploids may be produced by crossing autotetraploids of two different species. Further intercrosses and backcrosses lead to mixtures of allo- and autopolyploidy that have been called *segmental allopolyploidy*. The most successful natural polyploids are allopolyploids, and some geneticists think that most natural polyploids are segmental allopolyploids.

One of the best known examples of a natural allopolyploid that has arisen in historic times is the salt marsh grass (*Spartina townsendii*), a tall perennial grass with creeping rhizomes. The species, first noticed in southern England about 1870, spread rapidly and by 1906 had covered thousands of acres along the coast there. Then it crossed the English Channel and began to expand along the coast of France. Now there is evidence that this species arose suddenly by doubling of the chromosome number (122) following hybridization between *Spartina maritima* ($2n = 60$), a European species of salt marsh grass, and *Spartina alternifolia* ($2n = 62$), accidentally introduced into England from North America.

Many of our important cultivated plants are of polyploid origin. One of the most important ones is wheat and the complex history of polyploids in *Triticum aestivum* is discussed in Chapter 16. Noteworthy among others are sweet potato, sugarcane, coffee, banana, certain varieties of cotton, tobacco, pears, strawberries, many varieties of lilacs, tulips, hyacinths, and roses.

As at all levels of morphology, plants exhibit much genetic variability at the molecular level. Typically if one were to examine an "average" gene locus within a collection of diploid plants, 10 to 15% of individuals examined would probably be heterozygous. Thus a genetic variant at a particular enzyme locus is now directly detectable through use of specialized techniques (such as electrophoresis). One reason why plants maintain such high levels of variability may be that heterozygous individuals possessing alternative forms of a given enzyme with different biochemical characteristics are better able to cope with varied environmental stresses. Because plants are "tied" to their habitat by being rooted, such physiological flexibility may be of immense adaptive value. These advantages become even more apparent when one considers the implication of polyploidy in plants. For example, in a tetraploid plant, an alteration in any one of the four alternative genes leads in each such individual to a wide array of possible biochemical subunit groups. Tetraploid plants typically exhibit a far greater habitat tolerance than do their diploid parents. The generalized biochemical flexibility conferred by increases in ploidy, when they occur in a genetically variable species, appears to be one of the best explanations for the widespread occurrence and evolutionary success of polyploidy in plants.

b. INDUCED POLYPLOIDS

Induced or artificial polyploidy has become a significant process for both plant breeders and evolutionists. Although use of various chemical agents and extreme changes in temperature had been used to induce polyploidy prior to 1937, at that time an alkaloid (*colchicine*) was discovered to be more uniformly effective than any other means. Colchicine is obtained from the corm and seeds of the crocus (*Colchicum autumnale*) in the lily family. At low concentrations and with proper length of exposure, colchicine causes chromosomal doubling in many plants. The doubling occurs during *mitosis*. Prophase appears normal but the microtubules and the spindle fibers of metaphase are imperfect or missing so that the cell plate does not form. The chromosomes therefore do not separate to opposite poles and a new cell wall does not form. Autopolyploids may be produced by direct chromosomal doubling and allopolyploids by the same process following hybridization. Seeds, seedlings, or a shoot apex of older tissue may be treated with colchicine, or it may be placed on the bud or axil of a leaf to produce a tetraploid branch.

Figure 12-11 Polyploidy induced by colchicine in snapdragon. Inflorescence of diploid (left) and greatly enlarged tetraploid (right).

Autotetraploids have been induced in nearly all agriculturally significant genera. The doubled cells are regularly larger, resulting in larger size of certain tissues and organs, particularly in pollen grains, seeds, and flowers. Partial fertility and slow rates of genetic segregation reduce the commercial value of some induced autotetraploids. Nonetheless, increased flower size has made it important to the ornamental breeder and increased fruit size has made it equally interesting to fruit breeders (Fig. 12-11).

Several natural autopolyploids are agriculturally important. Among the autotetraploids are the potato, coffee, peanut, alfalfa, and orchard grass. The autohexaploids include sweet potato, plum, and timothy grass. Autotriploids rarely produce functional sex cells and the resulting sterility accounts for the economically valuable seedlessness of triploid bananas, limes, watermelons, and so on. In other triploid species, such as azaleas and lilies, sterility insures long flower life. Triploid aspen trees are giants that grow better and more rapidly than either diploids or tetraploids.

Allopolyploids are often larger and more vigorous than the diploid parents from which they arose; however, this is not always true. Allopolyploids have several advantages over autopolyploids. First, they capture whatever unusual hybrid features the two diverse genomes contribute. Diploid segregations and meiotic regularity both arise from the tendency of chromosomes in the allopolyploid to pair preferentially in bivalents rather than in multivalents as they do in autotetraploids.

Even without the use of chemicals, breeders have occasionally made artificial hybrids between genera which probably would not occur under natural conditions. Hybridization between the cultivated radish (*Raphanus sativus*) and the cultivated cabbage (*Brassica oleracea*) provides a dramatic case. Both species have 18 chromosomes in meiosis but the chromosomes were so different that they remained unpaired and were distributed irregularly to the gametes. In the polyploid there were 36 chromosomes in the vegetative cells and 18 similar pairs were formed regularly at meiosis. Thus the hybrid cell had all the chromosomes of the radish and the cabbage in its cell and these functioned normally to give the polyploid a relatively high fertility. The only disappointment for the researchers was that the hybrid had leaves characterizing the radish and the root of the cabbage—the inverse of what they had sought! However, an example of a man-made hybrid between genera that appears promising is *Triticosecale,* between wheat (*Triticum*) and rye (*Secale*). It is commonly called Triticale (Fig. 12-12). A number of polyploids have originated as weeds in habitats created by humans, and oftentimes, as we know only too well, they have been extremely successful.

(A)

Figure 12-12 An example of intergeneric hybrids; the wheat-rye (*Triticum-Secale*) called triticale. *A*. Spike of variety Siskiyou triticale. *B*. Grain of Siskiyou triticale. C. Grain of variety Anza wheat for comparison. Siskiyou is the first public variety of triticale to be released in the United States, and not only has good-sized seed but has higher protein content than wheat (see Chapter 16 for more details).

(B)

(C)

6. ORIGIN OF DOMESTICATED PLANTS

Determination of the possible origin of most of our basic food plants is difficult because they originated in prehistoric times. Nonetheless, much evidence recently has been revealed by plant explorers, cytologists, systematists, and archeologists. Information regarding the origin of domesticated plants is of more than purely academic interest; it frequently is of importance in plant breeding. For example, the genes and gene complexes existing in the ancestral forms and nearest relatives of a cultivated plant may be used in hybridization for plant improvement.

Cultivated plants, like wild plants, arose through evolutionary mechanisms such as mutation, recombination, and hybridization, followed by polyploidy and selection. As Darwin pointed out, evolution has proceeded much more rapidly under domestication than it did under natural conditions. Variations that would have little chance to survive under natural selection are preserved under artificial selection because they are valuable to human beings. They are protected by cultivation, under which competition with other species is largely eliminated. Under cultivation, also, favorable mutations have been observed and propagated.

Improved taxonomic understanding of crops has shown that chili peppers, cotton, tobacco, beans, squash, and amaranth, among others, each comprise two or more cultivated species, grown for some product and used in some way, but usually with their own characteristic geographic distribution. This poses the problem of whether speciation preceded domestication (implying that the individual cultivated species were domesticated independently from wild ancestors that were already specifically distinct) or whether speciation followed domestication (in which case each crop may have been domesticated once only). The answers to this question may well be different for each crop. For example, in cotton the data strongly suggest independent domestication of the two cultivated species of New World cotton from distinct wild ancestors. On the other hand, domestication of potatoes has been followed by increasing differentiation to the verge of speciation within the cultivated populations. Chili peppers are more complex with speciation in some instances preceding and in other instances accentuated by domestication.

Many New World cultivated plants, as just discussed in this chapter, are polyploids and here the direction of evolution from diploid to polyploid is unequivocal. However, the question of multiple origin arises again in polyploid crops. If their ancestral diploids were widespread, then hybridization, chromosome doubling, and successful establishment of resultant polyploids could have occurred more than once and in more than one place.

7. GENETICS OF WEEDS AND THEIR ROLE IN PLANT DOMESTICATION

Because of the significance of weeds to agriculture and their probable roles in plant domestication, it is important to understand the genetic background for their evolution. Much controversy surrounds the definition of weeds, but for our purposes a useful one is "a generally unwanted organism that thrives in habitats disturbed by humans." After the development and spread of agriculture, native vegetation was destroyed locally and replaced by the crop plants and weeds that were adapted to the new artificial habitats. We want the crops and try to encourage them; usually we do not want the weeds and we try to discourage them. However, because both are adapted to the same habitats, practices that favor crops also tend to favor weeds. In fact, the evolution of weeds

Figure 12-13 *Echinochloa crus-galli* var. *oryzicola* is a crop mimic restricted to cultivated rice (*Oryza sativa*). The mimic is native to Southeast Asia but now has been introduced to many rice-growing regions of the world, such as California (as shown above). The usual method of introduction is as a seed contaminant of cultivated rice. It resembles ("mimics") rice so closely in plant color, stature, and leaf morphology that it escapes hand weeding. (Compare Fig. 16-7).

often parallels that of crops. Both weeds and crops often begin with a common progenitor and many cultivated plants have one or more companion weed races, for example, einkorn, barley, sorghum, rice, oats, potato, tomato, pepper, sunflower, carrots, radish, and lettuce. In some cases, the weed races are clearly distinguishable from the cultivated ones, whereas in others their appearance is so similar that they are called *crop mimics* (Fig. 12-13). Even de Candolle in 1884 noted that distinguishing generally wild ancestors from weedy descendants of a crop is often difficult.

Wherever we go we are accompanied by various plants, whether we want them or whether we ignore them or even thoroughly dislike them. A good example is crabgrass (*Digitaria sanguinalis*), which is such a prolific seed producer that it was once cultivated for its seeds in central Europe. Was it cultivated because it was so aggressive or did it become aggressive because it was once cultivated? Actually, one person's weed is another's crop and vice versa. Some crops originated from weed ancestors whereas others have developed with various intermediate states into weed races. Plants thus move in and out of cultivation; in this complicated process they are domesticated and abandoned; they may escape, migrate, hybridize, form polyploid complexes, and evolve new races.

At the other extreme are truly accomplished weeds that are not closely related to cultivated plants but that follow humans despite our continual warfare against them. For example, the dandelion (*Taraxacum*) has been subjected to efforts at eradication costing millions of dollars and an untold amount of labor. Heavy applications of salt to streets and highways in wintertime have resulted in strong selection for salt-tolerant genotypes of roadside plants in some regions.

Wherever they occur weeds have a variety of characteristics that enable them to persist in disturbed habitats. Weeds generally produce large numbers of seeds and have special adaptations that prevent these seeds from germinating simultaneously. The seeds may remain viable for long periods, and their dormancy results from various special adaptations that may be broken in several ways. For example, light-sensitive seed buried in the soil may remain viable for many years, but will germinate when brought to the surface by disturbance. Many weeds are very persistent, having rhizomes or deep tap roots, and are protected by thorns, stinging hairs, or chemicals. Whatever the adaptations may be, single or elaborate, they tend to fit the weed into a particular human habitat, frequently with such success that they continually frustrate our attempts at their eradication.

On the other hand, sometimes aggressiveness has its virtues. After the conquest of California, the Spaniards introduced European livestock, which caused severe problems of overgrazing. The Californian grasses had not evolved under heavy grazing pressures and were easily obliterated. Thus the native grasslands of the coast and foothills have been virtually replaced by an aggressive annual weed flora from the Mediterranean adapted to livestock. The wild oats (*Avena fatua* and *A. barbata*), filarees (*Erodium*), mustards (*Brassica campestris* and *B. nigra*) and weed radish (*Raphanus sativus*) are among the most common and conspicuous. The California climate is well suited to these annuals and productivity of the coastal and foothills grasslands on a sustained basis is probably higher than before the weed flora was introduced.

CHAPTER 13

Organizing Diversity: How Plants Are Classified

1. GENERAL PROBLEMS OF CLASSIFICATION

The diversity among living things is almost infinite. Except for those that are asexually reproduced, no two individuals are alike in all details and often the variation seems bewilderingly continuous. But fortunately there are gaps and discontinuities. Recently anthropologists have suggested that evolution of the human mind seems to have been closely related to the perception of discontinuities in nature. Their study of "folk taxonomies" has been significant in interpreting the logical processes in the human mind, because it appears that by our very nature we are classifying creatures, that is, we categorize our observations based on patterns of similarities. In fact, some scientists have suggested that, even before the advent of hominids, the ability to classify must have been a component of fitness in biological evolution. Regardless of whether behavior is learned or instinctive, organisms must be able to perceive similarities in stimuli for survival. Thus the recognition of similarities in observed patterns may be as old as the earliest forms of sense perception in living organisms.

The origin of the *science* of classification goes back to the writings of the ancient Greeks, and the general problems of classifying are the same for all branches of knowledge. How similar do things have to be in order to put them together in the same category or how different in order to be separated into different categories? In other words, what are the defining properties of the class? It must be emphasized that classification is a human process; *we* are the ones who decide upon the defining properties of our class, be it a spoon or a fungus. After defining the set or class we then give it a name. Further, we organize the named classes into an ordered classification system. It follows that with the same evidence and experience two scientists can still rightfully disagree as to the defining characters of a taxonomic class and, correlatively, as to the assumptions upon which the system of classes is based.

By thus providing a basis for communication the classification process has been one of the keys to social evolution, cultural progress, and scientific advancement. We find that the various classification systems that have developed over the last 2000 years are distinguished first of all by their *purpose,* and this, of course, is related to the *philosophical concepts* of the time. Furthermore, it is useful to have several classification systems, since they may serve different roles. Actually, the classification systems we use today represent quite a struggle in dealing with our increasing knowledge of natural diversity. In addition to genetic information, we now have electron microscopes (which help us observe ultrastructural features), sophisticated chemical means of isolating and identifying plant compounds, and computers to aid in our analyses of complex patterns of variability. This new knowledge also broadens the role of the system by providing explanations for taxonomic relationships and implications from them that can be of importance to all areas of biology and other fields as well.

It is probably fortunate, as our ancestors began to classify and name organisms, that they did not realize that there are probably 10 million kinds of living things on earth. By now we may have described only about 15% of them, due to the exceedingly large number of such groups as tropical insects, nematodes, and fungi. As indicated recently by several experts, we probably will have to use electronic retrieval systems even to characterize some of these organisms (let alone get them fitted into existing taxonomic systems) before they are extinguished due to the growing human population.

In our early history, as well as among many primitive peoples today, the number of named classes of plants ranged between 250 and 800. This number persists regardless of the richness of the environment—with Eskimos in a small Arctic tundra flora or with Peruvian Indians living in the Amazon rain forest. We can ask the question: Are these the greatest numbers that the human memory can easily handle? It is interesting too that our ancestors concentrated on relatively large, clear differences and were prepared to overlook the differences that seemed minor or that distinguished members within the major groupings.

2. EARLY CLASSIFICATION SYSTEMS

The first classification systems are called "artificial" because they were based on similarities that might put unrelated plants into the same category for handy use in identification. For example, certain plants might be grouped by their edible parts, or their use as drugs or for perfumes and so on. As mentioned in Chapter 3, the first deliberate plant classification has been ascribed to the Greek Theophrastus, who in 300 B.C. (2300 B.P.) wrote down a so-called "folk taxonomy." He did not develop a true ordered system, but he detected major differences among plants such as those distinguishing angiosperms and gymnosperms or monocotyledons and dicotyledons. He also intuitively recognized many groups that today we consider families, such as the legumes, grasses, composites, and crucifers.

The Romans were excellent practical agriculturalists but contributed little to the development of botanical science. The only Roman botanist of note (other than Pliny) was Dioscorides, the army surgeon of Nero, who wrote one of the earliest herbals (an illustrated list of plants and their properties emphasizing medicinal aspects). In Dioscorides' herbal, *De Materia Medica,* 500 medicinal plants were described and illustrated. This manuscript was laboriously copied for the next 1500 to 1600 years and had so much influence that it was difficult to use a medicinal plant if it were not listed; it has been continuously updated and still exists today.

After the introduction of printing into Europe in the fifteenth century, herbals began to appear in profusion. In fact the period from approximately 1470 to 1670 is known as the Age of Herbals, and their appearance led to an increased interest in botanical knowledge. They actually represented a record of European folk taxonomies of this period and are a charming mixture of fact and fable replete with illustration. Several influential herbals are discussed in Chapter 28.

3. THE CONCEPT OF GENUS AND SPECIES: THE BINOMIAL SYSTEM

The period of the herbals was also the great era of exploration of the world and large numbers of plants quite different from those known in Europe were being brought back for description and even for cultivation. By the end of the sixteenth-century European botanists were facing the fact that there were indeed a great many interesting plants in the world.

Most plants of interest, of course, had been given common names, but it soon began to be recognized that even for the simplest purposes common names are inadequate. For example, a pine in the United States is not the same thing as a native pine in New Zealand or Australia—in fact, in the three places, the name "pine" is used to refer to plants that belong to quite different families (i.e., Pinaceae, Araucariaceae, and Casuarinaceae, respectively). A violet and a pansy belong to the same genus (*Viola*) in the Violaceae, but an African violet (*Saintpaulia*) even belongs to a different family (Gesneriaceae). Also when different languages are involved, the problems of communication become even more hopelessly complex. For this reason botanists began in medieval times to refer to organisms by Latin names, because Latin was the scholarly language of the time. As classification systems developed, organisms were first grouped into *genera* (singular: *genus*) and then identified by descriptive Latin phrase names, known as *polynomials* (literally meaning "many names"). By the end of the seventeenth century the first word of the polynomial designated the name of the genus to which plants belonged. Thus all oaks were identified by polynomials beginning with the word *Quercus*, the phrases describing kinds of daisies began with *Aster*, and those referring to elm trees as *Ulmus;* all of these are generic names. In other words, the classes used were the same as those in the folk taxonomies but the *names* were being formalized.

A major simplification in the system of naming organisms was developed by the Swedish physician and professor, Carl von Linné (Fig. 13-1*A*) usually known by his latinized name of Carolus Linnaeus (1707-1778), who hoped to classify *all* of the plants and animals in the world according to their genera! This seemed a possibility in his day. In 1737 he pulled together many scattered records in his *Genera Plantarum*. Then in 1753 Linnaeus published a two-volume work, *Species Plantarum*, which contained descriptions of all known plant species. He used the polynomial designation for each species, but altered them if necessary so that they would be comparable to other species in the same genus. Despite the fact that he regarded the polynomials as the proper name of the species, he also put opposite the description a single word in the margin of the book. This word, together with the generic name, made a convenient shorthand for referring to the species. Thus in *Species Plantarum*, one of the yuccas, which had been designated as *Yucca foliis serrato filamentosis* (filamentous serrate-leaved yucca) was described under *Yucca* while "filamentosa" was put in the margin, making *Yucca filamentosa* its binomial name, which we use today. (The change from "filamentosis" in the polynomial to "filamentosa" in the binomial was necessary for agreement of the endings of the Latin words).

The convenience of this innovation became so clear that Linnaeus and all subsequent authors replaced the polynomial names by *binomial* ("two-name") ones. This binomial system is the one used today. Thus *Species Plantarum* was the beginning of scientific botanical nomenclature. The earliest binomial name applied to a species is accepted as correct and

(*A*)

HEXANDRIA MONOGYNIA. 457

Yucca foliis aloës. *Bauh. pin.* 91.
Yucca indica, foliis aloës. *Barr. rar.* 70. *t.* 1194.
Cordyline foliis pungentibus integerrimis. *Roy. lugdb.* 22.
Habitat in Canada, Peru. ♄.

2. YUCCA foliis crenulatis ſtrictis. aloifolia.
Yucca foliis crenulatis. *Vir. cliff.* 29.
Yucca foliorum margine crenulato. *Hort. cliff.* 130. *α.*
Yucca arboreſcens, foliis rigidioribus rectis ſerratis. *Dill. elt.* 435. *t.* 323. *f.* 416.
Aloë, yuccæ foliis, cauleſcens. *Pluk. alm.* 19. *t.* 256. *f.* 4.
Aloë americana, yuccæ folio, arboreſcens. *Comm. præl.* 64. *t.* 14.
Habitat in Jamaica, Vera Cruce. ♄.

3. YUCCA foliis crenatis nutantibus. draconis.
Yucca foliorum margine crenulato. *Hort. cliff.* 130. *β.* *Hort. upſ.* 88.
Yucca draconis folio ſerrato. *Dill. elth.* 437. *t.* 324. *f.* 417. *Comm. præl.* 42. 67. *t.* 16.
Draconi arbori affinis americana. *Bauh. pin.* 506.
Cordyline foliis pungentibus crenatis. *Roy. lugdb.* 22.
Habitat in America. ♄.

4. YUCCA foliis ſerrato-filamentoſis. filamentoſa.
Yucca foliis lanceolatis acuminatis integerrimis margine filamentoſis. *Gron. virg.* 152. *Trew. ehret. t.* 37.
Yucca foliis filamentoſis. *Moriſ. hiſt.* 2. *p.* 419.
Yucca virginiana, foliis per marginem apprime filatis. *Pluk. alm.* 396.
Habitat in Virginia. ♄.

ALOE.

1. ALOE floribus pedunculatis cernuis corymboſis ſubcylindricis. perfoliata.
Aloë foliis caulinis dentatis amplexicaulibus vaginantibus. *Hort. cliff.* 132. *Hort. upſ.* 86. *Roy. lugdb.* 23.
α. Aloë africana cauleſcens, foliis magis glaucis caulem amplectentibus & in mucronem obtuſiorem deſinentibus. *Comm. præl.* 68. *t.* 17. *rar.* 44. *t.* 44.
β. Aloë africana cauleſcens, foliis minus glaucis, dorſi parte ſuprema ſpinoſa. *Comm. præl.* 69. *t.* 18.
γ. Aloë africana cauleſcens, foliis glaucis latioribus & undique ſpinoſis. *Comm. præl.* 70. *t.* 19.

Ff 5 δ. A-

(*B*)

Figure 13-1 *A*. Carl von Linné (Linnaeus), who developed the system of nomenclature that facilitated the recording and classifying of species. His first major expedition was in Lapland in 1732; five years later he posed for a portrait in Lapp dress, holding *Linnaea borealis* renamed in his honor. *B*. Sample page from Linnaeus' *Species Plantarum,* showing the polynomial description of *Yucca foliis serrato-filamentosis*. Since *filamentosis* was in the margin, this species became *Y. filamentosa* in the binomial system.

later names are rejected. The rules such as those governing the application of botanical names to plants are embodied in the International Code of Botanical Nomenclature, which is revised at successive International Botanical Congresses held every five or six years.

A species name thus consists of two parts – the generic name and the specific epithet. When reference is made to the entire group of species comprising that genus, the generic name may be used alone. The genus *Pinus* (pine) includes some 90 species of trees occurring primarily in the Northern Hemisphere. Some species like *Pinus ponderosa,* important throughout western North America for lumber, have an extensive native distribution, while *Pinus radiata,* which has been widely planted for timber throughout areas with a Mediterranean climate, is native only in three small areas in central California. A specific epithet is meaningless when written alone; for example, *ponderosa* or *radiata* could refer to other species in different genera. For this reason, the specific epithet is always preceded by the name of the genus (or its initial), that is, *Pinus ponderosa* or *P. radiata*. Species are sometimes further divided into subspecies or varieties, making the names of some plants consist of three parts.

Linnaeus not only was the originator of the binomial system, and thus of our present scientific nomenclature, but he also was a great focus for scientific activity of his time. He continuously published thousands of descriptions of organisms, and specimens from all corners of the globe were sent to him to name. In his herculean efforts to collect and classify, he laid the foundation for the work of the evolutionists of the next century. Without the order that Linnaeus introduced into the confusion of our emerging knowledge of the biological world, little headway would have been possible.

Over the years, quite different views have been held regarding the nature of species and the way species should be defined. Actually, the Latin word *species* means "kind" and, of course, there are many ways in which to define "the kind" of organisms. The classical definition relies on morphological similarity, but it is now clear that organisms are constantly evolving and, therefore, relationships among their "kind" are constantly changing. Groups of populations may be very similar, whereas others are less so. Thus there are different degrees of similarity among populations referred to as species, depending upon the group of organisms and the rate at which they are evolving.

"Species" among higher animals are commonly defined as a group of individuals that can interbreed with one another but not with individuals of another species. This criterion is the basis of the "biological concept" of a species but it has not proved as useful in most groups of higher plants. When attempts are made to cross plants from different morphologically defined species, in some cases individuals from these may not cross or the hybrids may be more or less infertile. Among herbs, such as *Clarkia* (in the evening primrose family), even different populations within a single species often produce sterile hybrids when crossed. On the other hand, among the oaks, birches, and willows, for example, most species may be crossed with any other to produce fertile hybrids (Fig. 12-8). Since all species of birch may freely interbreed (although many are now ecologically isolated), if we were to apply the biological species concept, there would be only one species of birch and probably only two or three species of oaks. This seems too radical a view for most plant taxonomists. Generally speaking, the greater the morphological differences between the individuals being crossed, the less fertile their hybrids are likely to be; nevertheless, in plants it is difficult to correlate the ability to interbreed with a concept of species.

To some it thus appears that the criteria for defining "species" may be different for different kinds of organisms. Since genetic recombination is not frequent among certain algae, one would not necessarily expect that "species" in these groups would be defined in the same way as in birches or mammals.

There are further problems with the species concept among cultivated plants. Both Darwin and de Candolle in the nineteenth century called attention to the striking and discontinuous variation within species of cultivated plants. Taxonomists of their day had generally elevated these variants to the rank of species. With the ideas on Mendelian genetics, these species were correctly evaluated as simply lines differing in only one or a few major genes. Then with attention focused on the process of speciation and the role of reproductive isolation, even fewer, but more variable, species were recognized. The effect of these changing species concepts is well illustrated by chili peppers (*Capsicum*). In the nineteenth century over 60 species were described, differing mostly in color, size, shape, and pungency of fruit (Fig. 26-7). In 1923 they were reduced to a single species, but now five different species of domesticated chili peppers are recognized, mainly by floral characters.

In sum, leading plant taxonomists and evolutionists today emphasize that, whereas *processes* of evolution are universal (as explained in Chapters 11 and 12), the products are highly individualistic as a result of inherent differences in genetic systems, sociology, and selection pressures. Thus the concept of species in plants can only serve as a *tool for characterizing diversity* in the most mentally satisfying way.

4. MODERN CLASSIFICATION SYSTEMS

a. THE HIERARCHICAL SYSTEM OF CLASSIFICATION

Linnaeus grouped the genera into 24 orders in a totally "artificial" system. He published a fragment of a "natural system" (*Classes Plantarum*) in 1783, but he recognized that the data were insufficient for a complete system. When Linnaeus spoke of a "natural" system of classification, he was referring to one that would outline the plan presumed to have been followed by the Creator. In 1789, Antoine de Jussieu made the first real progress in grouping genera into "natural orders"—the *families* we recognize today—to reduce the number of classes that had begun to emerge. The family has now become a focal point in systems of angiosperm classification, with 300 to 400 plant families depending upon viewpoint, and this number puts us back in the range of the folk taxonomies!

After the recognition of *families,* these were grouped into *orders,* orders into *classes,* and classes into *phyla* or *divisions*. These classes may be subdivided or aggregated into a number of less important ones, such as *tribes* (*subfamilies*) or *superfamilies, subclasses,* etc. An example of the manner in which a species is ordered into the taxonomic hierarchy is given in Table 13-1.

TABLE 13-1

The Place of Rice *(Oryza sativa)* in the Taxonomic Hierarchy of the Classification System Adopted in This Text

Kingdom—Plantae
 Division—Tracheophyta
  Subdivision—Spermatophytina
   Class—Angiospermae
    Subclass—Monocotyledonae
     Order—Graminales
      Family—Gramineae (Poaceae)
       Genus—Oryza
        Species—sativa

NOTE: The names of orders usually end in "ales" and families with "aceae" (although not always, as is indicated here). There is a proposal to change all family names to "aceae" endings but names such as Gramineae and Umbelliferae are so well established that now both names are often indicated (as in this book). When a population of species is sufficiently different morphologically, it may be called a *subspecies* or *variety.* The subspecies is generally considered to deviate more from the species norm than the variety, but some taxonomists consider "variety" as equivalent to "subspecies" and would eliminate the rank of "variety." In other cases, it is designated by a trinomial (three names) such as *Oryza sativa var. japonica.* A population deviating not quite enough to be called a variety is called a *form,*—*Taxus cuspidata* f. *densa* (an erect form) and *T. cuspidata* f. *nana* (a low spreading form). It is important to point out that the botanical or taxonomic "variety" just defined should not be confused with the horticultural "variety." The plant breeder or horticulturalist keeps an eye out for any interesting variation and where it is maintained in cultivation it has been referred to as a "variety." The name "cultivar" is presently gaining international acceptance for this "horticultural variety." A group of generally similar cultivars is often considered a *race.* By convention generic, specific, subspecific, varietal, racial and form names are written in italics whereas the names of families, orders, classes, and other taxa are not. The 'cultivar' is set off in single quotes.

Three Kingdoms—plants, animals and minerals—were recognized in Linnaeus' time. Organisms were considered to be either plants or animals. Animals moved, ate plants or other animals, whereas plants did not move and did not eat things. When microorganisms became recognized (with means such as microscopes enabling them to be seen and described), bacteria and fungi were grouped with plants and protozoans with animals. However, difficulties in classification began to arise with organisms that had characteristics of both plants and animals. For instance *Euglena* is a form that moves (swims) and yet has chlorophyll and manufactures its own food. Thus by 1930 it was recognized that a two-kingdom system was insufficient for grouping all living organisms. Recently a five-kingdom concept has been presented that recognizes that the most fundamental distinctions in the living world are those between procaryotes and eucaryotes, and unicellular and multicellular lines. The procaryotes, which have either solitary or colonial unicellular organization (with small ribosomes and low levels of cellular organization) are classified as Kingdom Monera and include blue-green algae and bacteria. The eucaryotes, with large ribosomes and high levels of cellular organization (mitochondria, plastids, organized nucleus, etc.) include both unicellular and multicellular forms and comprise the other four kingdoms. Kingdom Protista includes all organisms traditionally regarded as protozoans (one-celled animals) and all eucaryotic algae as well as the slime molds (often regarded as relatives of the fungi). Thus Protista constitute a heterogeneous assemblage of unicellular and multicellular eucaryotes with diverse forms of nutrition—photosynthesis, absorption, ingestion, or combination of these. The three kingdoms made up primarily of multicellular organisms (Plantae, Fungi, and Animalia) differ fundamentally in their mode of nutrition. Generally speaking, green plants (except parasites and saprophytes) manufacture organic food, fungi absorb it in solution, and animals ingest particulate food and convert it to solution for absorption. The fungi have traditionally been classified as plants, but there now is considerable evidence for their independence. They have little in common with algae other than their low level of differentiation. Thus the Kingdom Plantae in this system includes all green organisms more specialized than, and derived from, the green algae, i.e. the bryophytes and vascular plants. Despite these attempts in establishing a new concept of kingdom, no classification scheme as yet has been considered completely satisfactory. In all schemes, there are important exceptions to the "rule"; indeed, our concept of "plants" is so well established that, even in botany texts in which the five-kingdom approach has been considered an advance, bacteria, blue-green algae, fungi, and some of the protists are still discussed as though they were "plants" (Table 13-2).

Additionally, if different botany textbooks are perused, numerous suggestions for the organization of plants into the different classes within the taxonomic hierarchy will be found. It is important to realize that there is no "right one" as such—even though some systems appear more logical, or at least seem to display relationships between plants more effectively than others. In this book we have adopted commonly held groupings of algae, fungi, and nonvascular and vascular plants and then related these to the five-kingdom concept.

b. EARLY NATURAL SYSTEMS

In the early part of the nineteenth century there were numerous attempts to set up classification systems that would display "natural affinities." One of the most difficult problems is to decide what characters are the most important in displaying this relationship. Every character is potentially important taxonomically; that is, no character has any inherent, fixed importance. This is a point emphasized by the numerical taxonomists today who would not give weight to the importance of characters but consider them all of equal consequence. How-

TABLE 13-2

Classification of Major Groups of Living Organisms Traditionally Regarded as Plants and Their Relation to the Five-Kingdom Concept

	KINGDOM MONERA
PROCARYOTES	Bacteria, including Cyanophyta (blue-green algae)
EUCARYOTES	***KINGDOM PROTISTA***
ALGAE	Division Chlorophyta (green algae)
	Division Phaeophyta (brown algae)
	Division Rhodophyta (red algae)
	Division Chrysophyta (diatoms and golden-brown algae)
	Division Xanthophyta (yellow-green algae)
	Division Pyrrophyta (dinoflagellates)
	Division Euglenophyta (euglenoids)
FLAGELLATE HETEROTROPHIC FORMS	Division Gymnomycota (Slime molds)
	Class: Myxomycetes (true slime molds)
	Class: Acrasiomycetes (cellular slime molds)
	Division Mastigomycota
	Class: Chytridiomycetes (chytrids)
	Class: Oömycetes (water molds)
	KINGDOM FUNGI
FUNGI	Division Eumycota (true fungi)
	Class: Zygomycetes (bread molds)
	Class: Ascomycetes (sac fungi)
	Class: Basidiomycetes (club fungi)
	KINGDOM PLANTAE
BRYOPHYTES	Division Bryophyta (nonvascular plants)
	Class: Musci (mosses)
	Class: Anthocerotae (hornworts)
	Class: Hepaticae (liverworts)
SEEDLESS VASCULAR PLANTS	Division Tracheophyta (vascular plants)
	Subdivision Psilophytina (whisk ferns)
	Subdivision Lycophytina (club mosses)
	Subdivision Sphenophytina (horsetails)
	Subdivision Filicophytina (ferns)
SEED PLANTS	Subdivision Spermatophytina (seed plants)
GYMNOSPERMS	Class: Cycadinae (cycads)
	Class: Ginkgoinae (ginkgo)
	Class: Coniferinae (conifers)
ANGIOSPERMS	Class: Gnetinae (vessel-containing gymnosperms)
	Class: Angiospermae (flowering plants)
	Subclass: Dicotyledoneae (dicots)
	Subclass: Monocotyledoneae (monocots)

ever, most botanists have felt that some characters may be more stable or "conservative" than others. For example, the size, color, and often the shape of leaves can be altered by changes in the environment, but flower and fruit characters generally are not as altered by environmental changes. Also when *taxa* (any class or category) are arranged into groups on the basis of all evidence, frequently the most consistent differences seem to be in the flowers. Furthermore, the flowers (and fruit) are so diverse that there are many characters to employ without prejudice. Therefore, most systems tend to be based primarily on flower and fruit characters with, of course, the addition of evidence from all other characters possible, from chemistry to cellular ultrastructure. This wealth of evidence is one reason that the construction of a taxonomic system is so difficult, for it is the attempt to synthesize all known knowledge about organisms and thereby organize biological diversity into an ordered, understandable whole.

There was literally a "parade of natural systems" between 1824 and 1850. Actually these systems, in displaying natural relationships ("affinities"), had the data laid out for later phylogenetic interpretation through the concept of genetic lineage. However, they did not become philosophically a part of plant classification until the theory of evolution was presented, because phylogeny actually consists of the evolutionary history of a group of related individuals.

C. EVOLUTIONARY SYSTEMS

With the publication of Darwin's *Origin of Species* in 1859 an interest in the role of evolution in taxonomy was begun. Then came the problem of rearranging *taxa* to reflect the presumed relationships. In many cases there were not many changes from some of the previous natural systems at the generic level and below; most of the rearrangement came with regard to families, orders, and classes.

In making over the previous natural systems of classification of angiosperms to fit with the concept of evolution, it was now necessary to analyze two *additional* important issues: (1) to discover the ancestors of the flowering plants ("that abominable mystery," as Darwin so aptly put it), and (2) to determine which characters were primitive and which were advanced. To answer the first question unequivocally requires fossil evidence, and since flowers are delicate and seldom fossilize, the information has still not been forthcoming. Fortunately, however, recent morphological study of the abundant early angiosperm fossil pollen and leaves has provided evolutionary trends that support evidence from the reproductive structures of presently existing plants that the seed ferns are the ancestors of the flowering plants. These are the oldest and least specialized gymnosperms, more like cycads (Fig. 24-9) than conifers. But the determination of primitive and advanced characters is not as easy as it might seem. Simple structures may be primitive or the simplicity may have evolved by reduction and, therefore, be advanced. Correlatively, complex structures are not necessarily the most advanced or evolved. These points will become more clear as we discuss some of the most prominent evolutionary systems.

One of the two most widely accepted evolutionary systems was developed by Adolf Engler and his associate Karl Prantl at the University of Berlin (1897-1936). They concluded that the most primitive flowering plants among both dicots and monocots were small, simple, and wind-pollinated, in fact, they had no petals and very few stamens and pistils. They also thought that the monocots and dicots had arisen independently from an ancestral stock of extinct gymnosperms. The living plants most nearly like this ancestor in the dicots were wind pollinated (trees and shrubs including such plants as willow, poplar, oak, alder, and birch). The simple "catkins" (see Chapter 4) of these plants resembled the cones of gymnosperms. In the monocots the primitive group included the simple flowered, wind-pollinated cattails, burweeds, and screw pines. Engler and Prantl also thought that modern angiosperms were composed of many fragmentary lines of evolution. The evolutionary course was believed to have been from flowers having no petals to those with separate petals and finally to those with coalescent petals (Chapter 4). The primary focus in their evolutionary sequence was upon petals, the relative position of the ovary being considered secondary.

The Engler and Prantl system published in 20 volumes (*Die Natürlichen Pflanzenfamilien*), was by far the most well-known and widely circulated organization of the plant kingdom up to its time. Plant specimens were arranged in nearly all the world's *herbaria* (collections of dried and pressed plant specimens) according to this system, and it has been used in the vast majority of *floras* (descriptions of plants of a particular region) that have since been published. The phylogenetic conclusions of this system, however, have now generally been discounted in favor of viewing the position of the ovary as primary, with the development of the corolla considered secondary. This view was taken up by George Bentham and Joseph Dalton Hooker at Kew Gardens in England and presented between 1862 and 1896 in their *Genera Plantarum* (Genera of Plants). Then between 1897 and 1915 Charles E. Bessey, at the University of Nebraska, presented a system growing logically from that of Bentham and Hooker (Fig. 13-2). In addition to emphasizing the relative position of the ovary, rather than corolla characters, he thought that the most primitive angiosperm flower had numerous sepals, petals, stamens, and pistils and was insect pollinated. The living group with flowers most closely related to this ancient angiosperm "ancestor" occurs in the order Ranales, which includes plants such as the buttercups, magnolias and tulip trees. The magnolia flower is close to the "cone" of their presumed ancestor (Fig. 13-3). Evolution from this point was thought to have proceeded basically in three directions: (1) a monocot line, (2) a dicot line characterized by inferior ovaries (perigyny or epigyny), and (3) a line characterized by superior ovaries (hypogyny). In each of the three lines, evolution proceeded in a *parallel* fashion by (1) *reduction* in the number of flower parts, (2) *fusion* of the petals in the corolla, (3) development of an *irregular corolla,* and (4) *separation* of *sexes* into separate flowers on a single plant (monoecious) or on separate plants (dioecious) (see Chapter 4 for description of terms). Although these were the basic characters used in establishing his system, Bessey's "dicta" contrasted many additional characters as either primitive or advanced. Bessey's system was summarized in a figure displaying the *orders* that seemed to resemble a cactus and thus has been referred to as "Bessey's cactus" (Fig. 13-2*B*). This visual presentation of Bessey's ideas has aided several generations of students to grasp evolutionary relationships among the flowering plants and apply them in identifying plants they encountered.

Figure 13.2 *A*. Charles E. Bessey (1845-1915), American botanist noted for his dicta regarding evolutionary trends among the angiosperms.

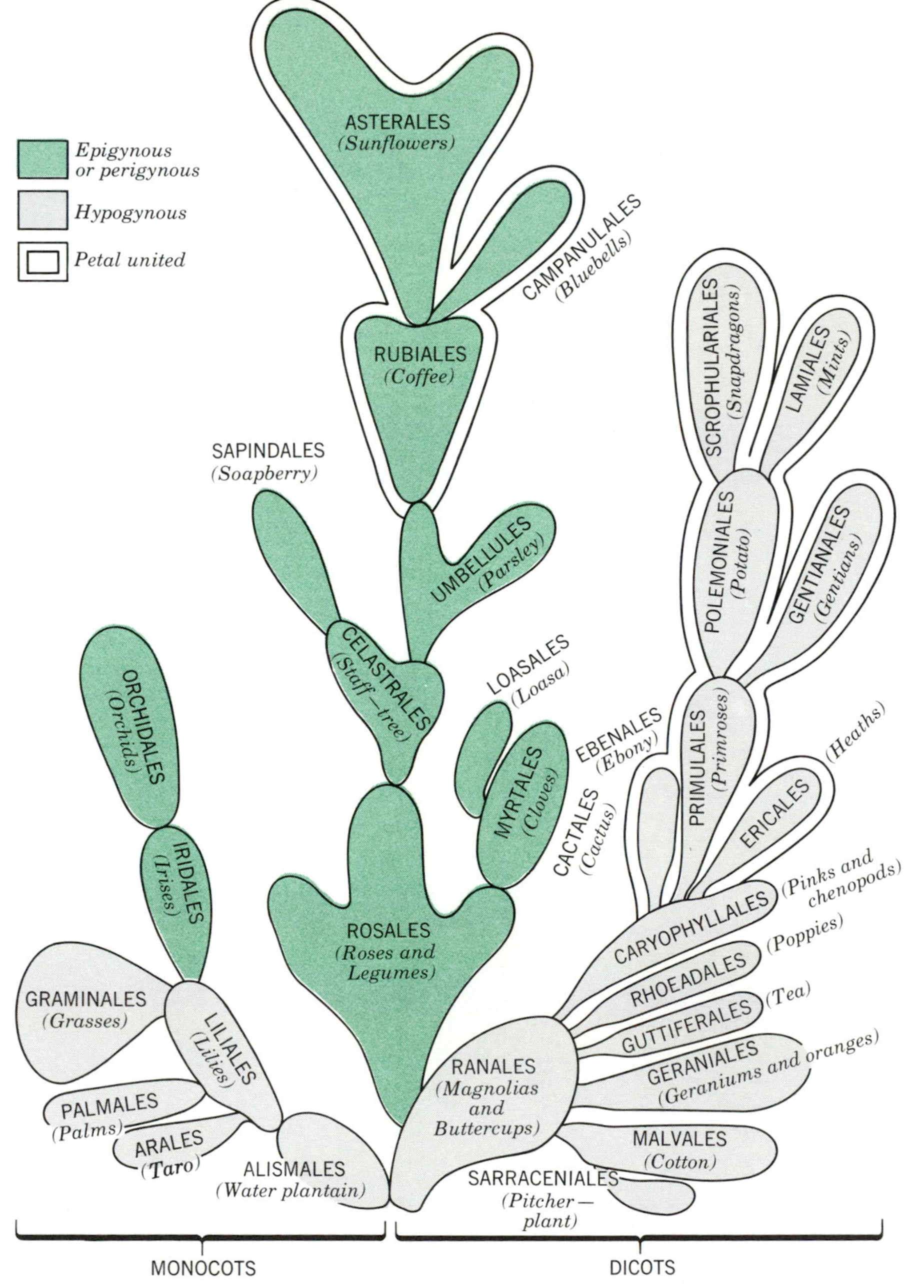

B. A modification of Bessey's chart ("cactus," as it is often called) by Lyman Benson showing relationship of orders of flowering plants. The monocotyledons are shown on the left branch of the cactus, epigynous and perigynous dicotyledons on the center branch and the hypogynous dicotyledons on the right branch; sympetaly (petals united) is also indicated. (Modified by Lyman Benson from Bessey's original chart, 1915.)

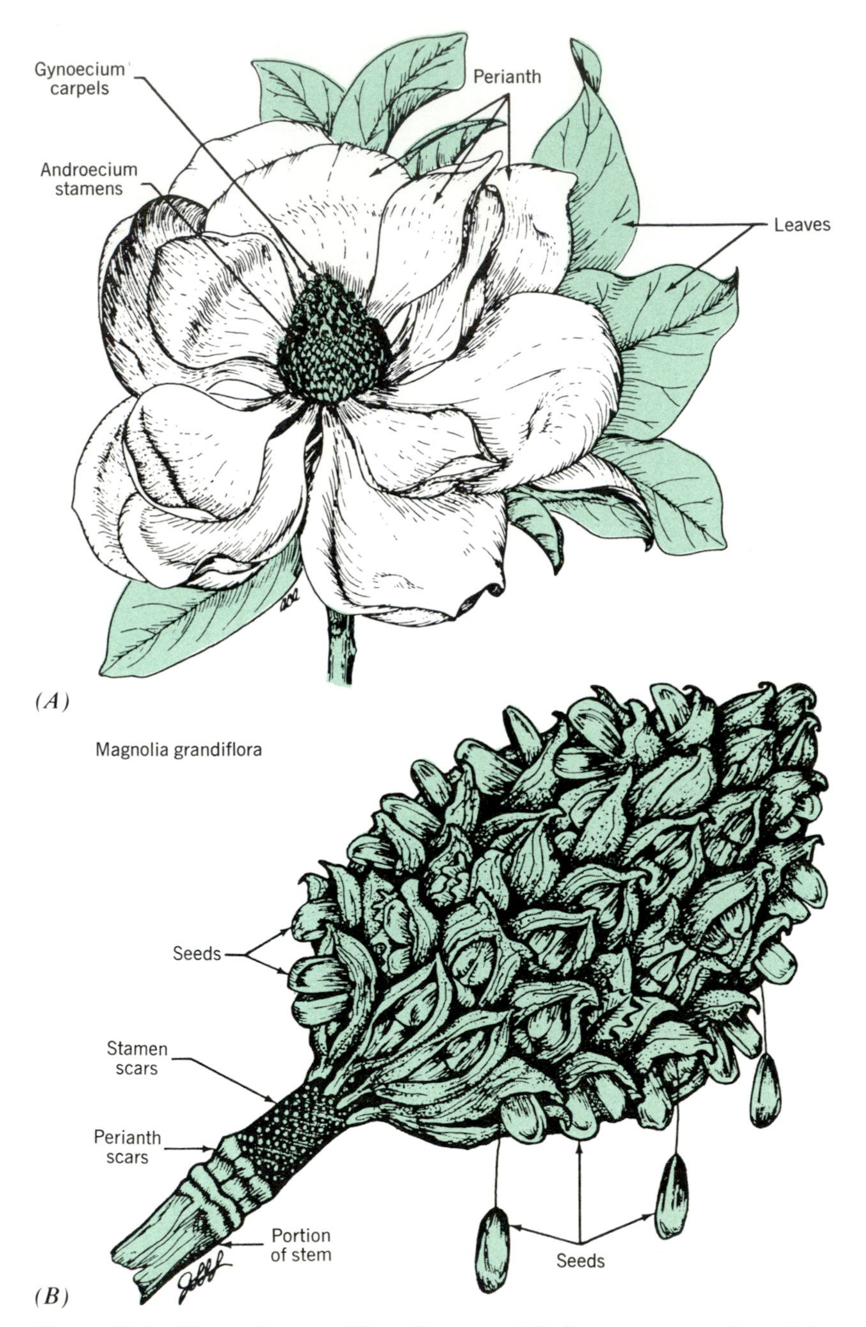

Figure 13-3 *Magnolia grandiflora* flower and fruit, a representative of the order Ranales, the living group with flowers considered by Charles Bessey as most closely related to the ancient ancestor of angiosperms. The fruit is similar to the cone of the gymnosperm, putative predecessor.

Obviously, developing a satisfactory evolutionary system of classification for the flowering plants is a difficult task and far from complete. A number of systems have been presented in recent years, and although the origin of the angiosperms remains an "abominable mystery," several lines of evidence continue to support the primitive nature of the order Ranales. Most current workers are also convinced that primary arrangement according to ovary position, as in Bessey's system, rather than corolla characters, as in Engler and Prantl's, results in a clearer reflection of natural affinities. Yet, it must be admitted that ovary position is not always consistent – inferior ovaries tend to appear "here and there" in every evolutionary scheme for the angiosperms.

Both Arthur Cronquist (New York Botanical Garden) and Lyman Benson (Pomona College, California) have devised recent original systems based, however, on many of Bessey's tenets. Cronquist's system was conceived independently but was so similar to those of Armen Takhtajan and Walter Zimmermann (Russian and German evolutionists, respectively) that they published jointly. Cronquist considers the angiosperms to form a division called Magnolophyta. The monocots and dicots then comprise classes within the division; the dicots are in turn divided into six subclasses. Whereas Bessey has two principal lines of evolution in the dicots from the Ranales, Cronquist has four. These systematists realize that no modern group is actually descended from another living group, as it appears to be implied from the visual presentation of phylogenetic trees. The only real hope of establishing true relationships between orders lies in fossils, but their record is woefully incomplete. The alternative is to deduce relationships from living groups and the meager fossil evidence we have.

Benson was strongly influenced by Bessey's "cactus" as a student, but has attempted to organize the angiosperms to produce another visual presentation that perhaps is not as misleading as a "phylogenetic tree." Rather than connecting the various groups as in the cactus, he outlines each group on a surface divided into six major areas, implying that the distance between groups is proportional to the degree of relationship we know from presently available evidence. Although Benson's chart is not visually like Bessey's cactus, the two systems are based upon many similar relationships.

Generally speaking, classification into orders and families is less controversial than arrangement of these taxa into an evolutionary system. Nevertheless Cronquist, for example, recognizes 74 orders with 354 families, whereas Benson recognizes 69 orders with 295 families (with 12 "uncertain" ones). Again, as we have emphasized throughout this chapter, all taxonomic classes and systems are mental constructs. As Cronquist has aptly put it, "Ultimately the taxonomist must produce a treatment that appeals to the mind as the best conceptual organization of the diversity which exists in nature." All these treatments obviously differ in varying degree with the taxonomists' conceptions and perspectives.

5. SOME ANGIOSPERM FAMILIES OF SPECIAL IMPORTANCE TO US

A brief description of the families that include our most useful plants, to be discussed later in the book, displays some of the variations found in flowering plants and demonstrates the multiple characteristics used in delimiting taxonomic groups. This survey covers only 22 families, less than 10% of the total – yet these include almost one-half of the species of flowering plants. The families will be grouped first as either monocotyledons or dicotyledons and then according to the main trends generally recognized in floral evolution: (1) reduction in number of flower parts, (2) fusion of floral parts, and (3) development of an irregular (bilaterally symmetrical) flower. Although the general trend is from hypogyny (superior ovary) to epigyny or perigyny (inferior ovary), it is necessary to divide the dicots into those that are hypogynous and those that are epigynous/perigynous, and to discuss the other floral evolutionary trends within these groups. By so ordering these selected families, we can obtain a general idea of their evolutionary status.

a. MONOCOTYLEDONOUS FAMILIES

Liliaceae (Lily Family). The lily family belongs to the order Liliales, which is often thought of as representative of all monocots. Its 200 genera and 3500 species occur mainly in warm temperate and tropical zones. The plants are characterized by simple leaves with parallel-veins, hypogynous flowers, three sepals, three petals, six stamens in two cycles of three each, and three carpels coalescent and forming a compound pistil, usually with a single style (Fig. 13-4). The sepals and petals are similar and often frequently colorful. They form bulbs or corms (Chapter 19). Some of the most commonly known genera are *Allium* (including onion, chives, and garlic), *Asparagus, Colchicum, Lilium, Tulipa,* and *Yucca*.

Araceae (Arum Family). The relatively large arum family (about 150 genera and 1400 to 1500 species) is tropical to subtropical but also occurs in temperate regions. Plants are mostly herbaceous, bearing rhizomes or tubers, with milky or watery, pungent sap. The inflorescence is a simple spadix, subtended by a spathe (or leaflike bract) that is generally large and brightly colored; flowers are small, unisexual or bisexual, often with fetid odor (Fig. 13-5). The genera most used are *Colocasia* (taro) and *Alocasia,* for their starchy tubers in Asia and Africa (Fig. 19-12). Also the fruits of *Monstera* are eaten in many tropical regions. Representatives of about 35 genera are cultivated as ornamentals, some of the most commonly known being *Calla, Dieffenbachia, Philodendron, Anthurium,* and *Caladium*.

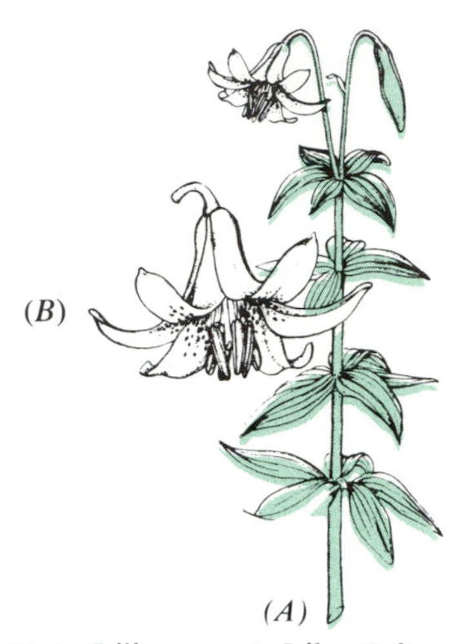

Figure 13-4 Liliaceae. *A*. Lily *(Lilium canadense)* plant, with *(B)* individual flower showing floral parts in multiples of three. (Adapted from Lawrence, 1951)

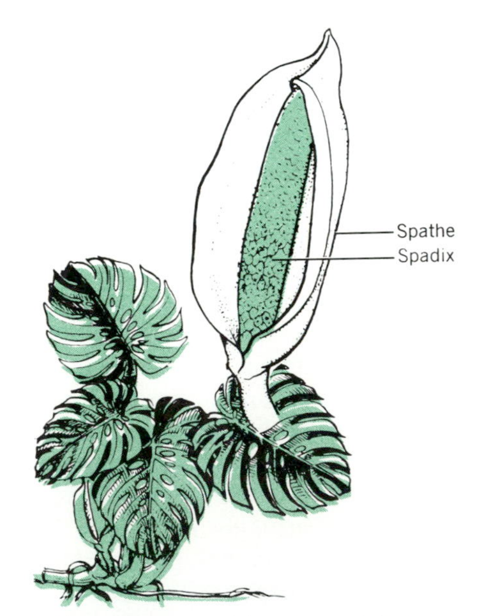

Figure 13-5 Araceae. *Monstera deliciosa*. From Bailey, 1976)

Palmaceae or Palmae (Palm Family). Next to the grass family, the palm family contains some of the plants of greatest practical importance. They are sources of such products as coconut and other oils, copra, dates, rattan cane, raffia, ivory nuts, and carnauba wax. There are 210 genera and 4000 species, distributed throughout the tropics and subtropics. The palm flower does not differ much from the lily flower; it has generally six stamens (in two whorls) and three carpels. One significant difference is that palm flowers are often unisexual, being either monoecious or dioecious. Often the base of the inflorescence is subtended by a spathe (like flowers in the Arum family). The endosperm is abundant and either oily or bony. Palms may be divided into fan palms, with a simple palmate leaf, and feather palms, with pinnate leaves. Some of the most commonly known genera are *Cocos* (coconut) (Fig. 18-9), *Phoenix* (date palm) (Fig. 18-12), *Elaeis* (oil palm) (Fig. 27-6), *Raphia* (raffia palm), and *Copernicia* (carnauba palm) (Fig. 27-8).

Poaceae or Gramineae (Grass Family). There are 700 genera and 8000 species of grasses, which are worldwide in distribution. They are mainly herbs but do include canelike plants such as the bamboos. The characteristics of the grass flower are described in Figure 16-1. This family probably includes more *individual plants* (not species) than any other flowering plant family. Grasses are the dominant plants in the prairies but they also occur virtually wherever plants of any kind can grow in both wet and dry habitat—in deciduous and coniferous forests, chaparral, arctic and alpine tundras, and deserts. Furthermore the grasses are the most important plant family to humans because they include such staple food plants as wheat (*Triticum*), rice (*Oryza*), oats (*Avena*), barley (*Hordeum*), rye (*Secale*), maize (*Zea*), sorghum (*Sorghum*), the various millets, sugarcane, and bamboo. The pasture and range grasses are even more numerous than the cereals and are also of tremendous value to our domestic animals.

Musaceae (Banana Family). The banana family is a small one of wide distribution in the tropics (5 genera and about 150 species). They are mainly large herbs, often treelike in appearance (Fig. 18-8). The flowers are usually unisexual, and the plants monoecious. The most important genus economically is *Musa,* which contains a number of tropical species grown both for their edible fruits (banana) and fibers (Abacá—Chapter 21). *Heliconia* and *Strelitzia* (bird of paradise) are also known as ornamentals.

Zingiberaceae (Ginger Family). The ginger family, known as the source of "ginger root" from plants of the genus *Zingiber,* and of cardamon seed from *Elettaria,* contains 47 genera and 1400 tropical and subtropical species. They are primarily perennial herbs with horizontal or tuberous rhizomes. The inflorescence can be a compact spike or an open raceme, or the plant can bear solitary flowers, each flower or cluster of flowers subtended by a conspicuous bract.

Orchidaceae (Orchid Family). The orchid family, a predominantly tropical group, is the third or possibly second largest plant family—the estimated number of species varying from 10,000 to 17,000. The orchid flower is unique; the parts are epigynous, with three sepals, and three petals quite different from the sepals (Fig. 13-6). At least one petal is greatly different from the others, producing a characteristic insect landing platform that protrudes forward. The style is attached to the only stamen (rarely are there two stamens). They also have a waxy mass of pollen (pollinium), which is transported as a unit by insects to the next flower. The fused stamen–style and the pollinium are almost—but not quite—unique to the orchids. Both characteristics also occur in the milkweeds, a family of dicots with no other distinguishing features in

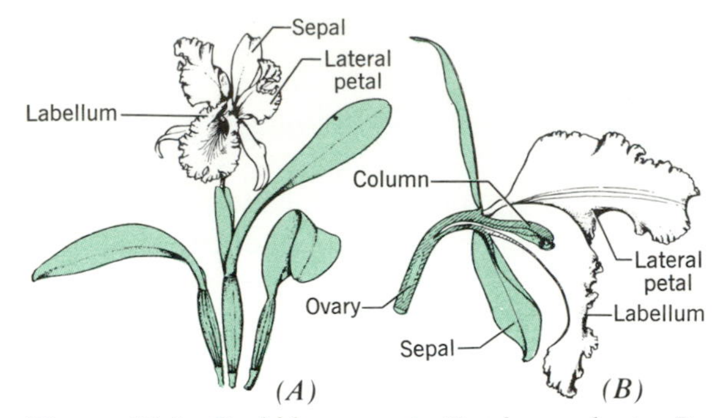

Figure 13-6 Orchidaceae. *A Cattleya* plant. *B.* Flower in vertical section.

common with the orchids. This is considered a striking example of parallel evolution or the recurrence of the same characteristics in an unconnected hereditary lineage. The perianth tube above the inferior ovary is twisted about 180° so that the flower is "upside down." The seeds are extremely minute, the embryo consisting of only a few cells, but each flower produces a large number of seeds (sometimes millions), which may make the pollinium with its large number of pollen grains a necessity. Most orchids in the tropical rain forest grow as epiphytes on the trunks or branches of the trees, often having aerial roots that absorb water. Our main use of orchids is as ornamentals, such genera as *Cattleya, Cymbidium,* and *Epidendrum* being well known. Vanilla flavoring is derived from the tropical genus *Vanilla* (Chapter 26). In temperate areas orchids are terrestrial and include such plants as lady slippers (*Cypripedium*) and coral root (*Corallorhiza*).

b. HYPOGYNOUS DICOTYLEDONOUS FAMILIES

Malvaceae (Mallow Family). The mallow family is composed of about 82 genera and 1500 species of herbs, shrubs, or trees distributed over most of the earth but particularly abundant in the American tropics. The flowers are actinomorphic and hypogynous, and are distinguished by the many stamens' being united to form a tube around the pistil. The sepals are united and the corolla is also tubular. The most important plants economically are cotton (*Gossypium hirsutum*) (Chapter 21) and okra (*Hibiscus esculenta*). Other *Hibiscus* species are grown for their fibers. Many genera are grown as ornamentals, such as *Hibiscus, Althea* (hollyhock), *Malva* (mallow), and *Sidalcea.*

Moraceae (Fig Family). The Moraceae are largely pantropical, with about 73 genera and over 1000 species of monoecious or dioecious trees or shrubs with milky latex. The family is important for edible fruits such as figs (*Ficus*) (Fig. 18-11), mulberries (*Morus*), breadfruit, and jackfruit (*Artocarpus*) (Fig. 18-10*A*). The fibers of *Cannabis* are used for hemp and the leaves and staminate flowers of the same plant for the drug marijuana. Hops (*Humulus*) are grown for their fruits used in flavoring beer (Chapter 20).

Euphorbiaceae (Spurge Family). The spurge family (283 genera and 7300 species of herbs, shrubs, and trees) is essentially cosmopolitan in distribution, but particularly abundant in the tropics. It is distinguished by a milky latex and hypogynous small unisexual flowers aggregated into small inflorescences that superficially resemble individual flowers. Products include rubber (*Hevea*) (Fig. 26-20), castor oil (*Ricinus*) (Fig. 27-7), tung oil (*Aleurites*), and cassava (*Manihot*) (Fig. 19-9). Also, *Poinsettia,* crown of thorns (*Euphorbia* spp.), and croton (*Codiacum*) are examples of those grown as ornamentals.

Rutaceae (Rue or Citrus Family). The citrus family (about 140 genera and 1300 species, widely distributed in temperate and tropical regions) is distinguished by having clear dots in the foliage, by the lobed ovary's being elevated on a disc, and by glands containing aromatic essential oils. It is noted for the kumquat (*Fortunella*) and the various fruits in the genus *Citrus* (orange, grapefruit, tangerine, lime, citron, and lemon). Ornamentals include the prickly ash (*Zanthoxylum*), rue (*Ruta*), orange jessamine (*Murraya*), and cape chestnut (*Calodendrum*).

Papaveraceae (Poppy Family). Herbaceous annuals or perennials (23 genera and about 250 species), mostly distributed in subtropical or temperate regions, members of the poppy family are readily distinguished by the combination of actinomorphic bisexual flowers, a usually crumpled corolla, numerous stamens in several whorls, capsular fruit dehiscing by pores or valves, and often a colored or milky latex (Fig. 24-6). *Papaver somniferum* is used for opium, obtained from the latex of unripe capsules. Of ornamental value are scores of species such as the Oriental poppy (*P. orientale*), Iceland poppy (*P. nudicaule*), blue poppy (*Meconopsis*), and California poppy (*Eschscholzia*).

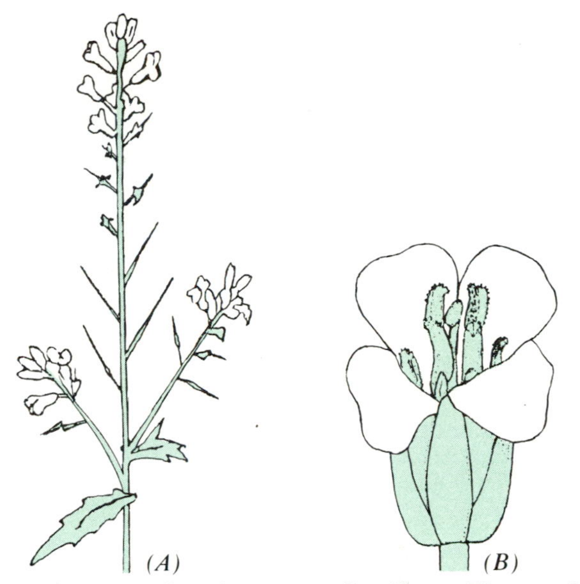

Figure 13-7 Brassicaceae or Cruciferae. Flower of winter cress *(Barbarea vulgaris)*.

Brassicaceae or Cruciferae (Mustard Family). The mustard family (350 genera and 2500 species) of annual, biennial, or perennial herbs occurs primarily in cool temperate regions. Most of the flowers are regular with four sepals and four petals spread out to form a cross (hence the name *Cruciferae*), six stamens (the two outer ones shorter than the four inner), and a characteristic dry, dehiscent fruit (Fig. 13-7). Many members of the mustard family have a mildly peppery odor or taste. Cabbage, broccoli, cauliflower, brussels sprouts, rutabagas, and turnips are familiar cultivated varieties of *Brassica oleracea* (Fig. 12-4). Other cultivated members include radish (*Raphanus*), watercress (*Rorippa*), horseradish (*Armoracia*), sweet alyssum (*Lobularia*), stocks (*Matthiola*), and candytuft (*Iberis*).

Figure 13-8 Solanaceae. *A*. Tomato plant *(Lycopersicum esculentum)*. *B*. Flower of potato *(Solanum tuberosum)*. (Adapted from Weier *et al.*, 1974)

Solanaceae (Nightshade Family). The Solanaceae, with about 85 genera and perhaps 2500 species of herbs, shrubs, and trees, are abundant in tropical America but also are well represented in temperate regions. They have sympetalous, usually regular, flowers with a superior ovary (Fig. 13-8). Over half the species belong to the very large genus *Solanum*. Many useful members include the food plants potato and eggplant (*Solanum*) and tomato (*Lycopersicon*); others of note are red pepper (*Capsicum*), tobacco (*Nicotiana*), and the drug plants belladonna and atropine (*Atropa*), stramonium (*Datura*), henbane (*Hyoscyamus*) (Chapter 24) and ornamentals from many genera such as *Petunia, Lycium, Solanum,* and *Schizanthus*.

Lamiaceae or Labiatae (Mint Family). The mint family, with about 200 genera and more than 3000 species, occurs throughout most of the world but is particularly common in the Mediterranean region. These plants are mostly herbs with square stem and simple, opposite leaves, and an irregular, commonly two-lipped, partly united corolla (Fig. 13-9). The mints are familiar because they are the source of aromatic oils. Many of the well-known herbs used in cooking such as sage (*Salvia*), thyme (*Thymus*), mint (*Mentha*), rosemary (*Rosmarinus*), marjoram (*Origanum*), basil (*Ocimum*), and savory (*Satureja*) belong to the Lamiaceae (Fig. 26-4). Also of importance are lavender (*Lavendula*) and horehound (*Marrubium*), whereas the ornamentals include genera such as *Nepeta, Stachys, Coleus,* and *Salvia*.

Figure 13-9 Lamiaceae or Labiatae. Peppermint (*Mentha piperita*).

c. EPIGYNOUS OR PERIGYNOUS DICOTYLEDONOUS FAMILIES

Rosaceae (Rose Family). The rose family is large with about 115 genera and 3200 species of trees, shrubs, or herbs, distributed worldwide but particularly abundant in eastern Asia, North America, and Europe. Its members have the parts of the flower in fives, being either epigynous or perigynous; the fruit is an achene, pome, or drupe (see Chapter 18). It includes many temperate-zone fruits such as cherry, prune, peach, nectarine, apricot, almond (all spp. of *Prunus*), apple and pear (*Pyrus*), quince (*Cydonia*), blackberry, raspberry, loganberry (spp. of *Rubus*), and strawberry (*Fragaria*). Among many ornamental trees and shrubs are cotoneaster (*Cotoneaster*), hawthorn (*Crataegus*), rose (*Rosa*), flowering quince (*Chaenomeles*), and mountain ash (*Sorbus*).

Fabaceae or Leguminosae (Pea Family). The pea family is one of the three largest families – represented by 550 genera and perhaps 13,000 species of herbs, shrubs, and trees of cosmopolitan distribution. The three subfamilies (Mimosoideae, Caesalpinioideae, and Papilionoideae) are treated by many botanists as families called Mimosaceae, Caesalpiniaceae, and Papilionaceae, and those who accept them as separate families treat the three as a single order; taxonomic relations remain unchanged whether they are considered subfamilies or families. The Mimosoideae contains about 40 genera of primarily trees and shrubs that are distributed almost exclusively in the tropics or subtropics. The flowers are radially symmetrical, borne in closely packed heads, the petals small or absent, and the stamens 10 or numerous (Fig. 13-10*A*). *Mimosa* and *Acacia* are two of the very well known large genera

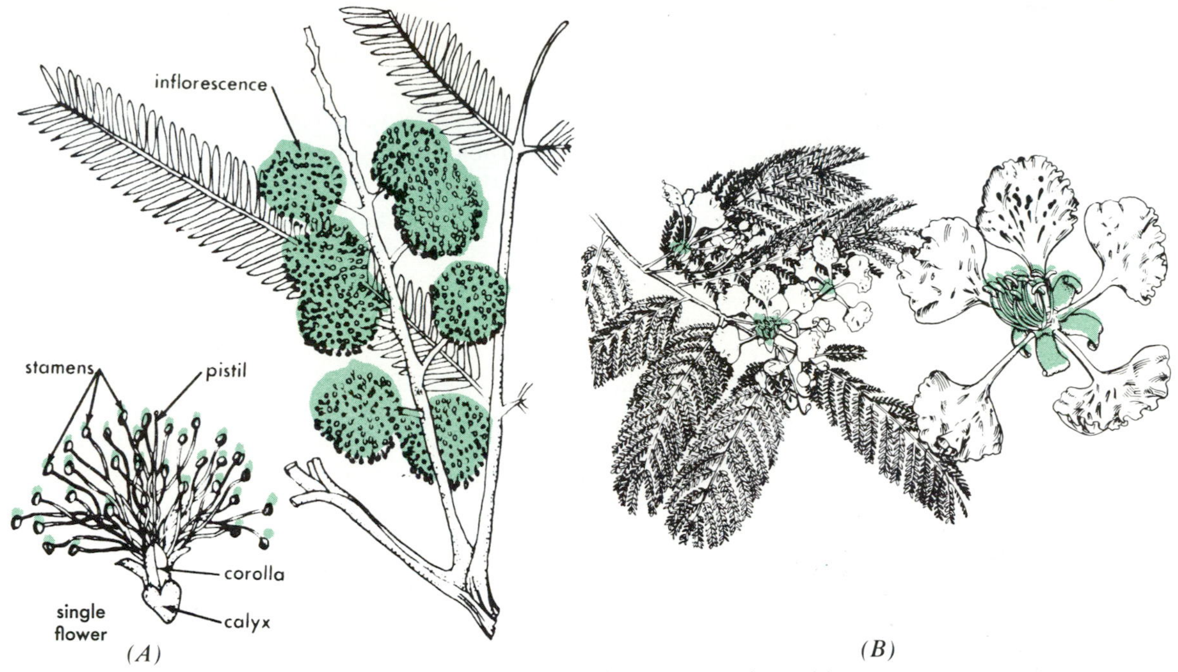

Figure 13-10 Subfamilies of the Fabaceae or Leguminosae. *A*. Mimosoideae. Inflorescence of *Acacia* (with single flower). *B*. Caesalpinioideae. Flowers of the "Flamboyant" *(Delonix regia)*. (From Bailey, 1976)

(with about 350 species each) native to the Old World, but planted throughout the tropics and subtropics for a variety of uses. The acacias particularly produce valuable products including gum arabic (*A. senegal*) and phenols for tannins and dyes (e.g., *A. catechu*). Other mimosoids, such as *Samanea,* (*Pithecolobium saman*) are prized as ornamentals. The Caesalpinioideae have flowers that are not quite symmetrical, usually with large petals, and borne in spikes or racemes (Fig. 13-10*B*). They include 135 genera and constitute many of the dominant rain forest trees of the New World and African tropics. Genera such as *Hymenaea* (Fig. 14-9) and *Copaifera* are used for both their wood and resins (Chapters 22 and 26). *Delonix* (e.g., *Delonix* or *Poinciana regia,* the "Royal Poinciana" or "Flamboyant") and many species of *Cassia* are grown for their beauty along streets. The majority of legumes of temperate regions, including peas (*Pisum*), beans (*Phaseolus*), soybeans (*Glycine*), and others, the forages such as clover (*Trifolium*) and alfalfa (*Medicago*) as well as herbaceous ornamentals like *Lathyrus* (sweet pea) and lupines (*Lupinus*) belong to the Papilionoideae (Chapter 23). Some papilionoids occur as trees in the tropics, along with abundant caesalpinioid and mimosoid genera. *Erythrina* is one of the best known because of its use as a shade tree in coffee and cocoa plantations (Chapter 25). Others such as *Butea frondosa* ("Flame of the Forest") and *Amherstia nobilis* are handsome ornamentals in the tropics.

Myrtaceae (Myrtle Family). The myrtle family includes about 80 genera and 3000 species of tropical shrubs or trees. The plants are distinguished by their essential oil-containing glands in the leaves, the great number of stamens, and the inferior ovary with coalescent carpels. Of economic importance are the guava

(*Psidium*), clove (*Eugenia Caryophyllata*) (Fig. 26-11) allspice (*Pimenta dioica*), bay rum (*Piracemosum*), *Eucalyptus,* and the bottle brushes (*Callistemon* and *Melaleuca*).

Cucurbitaceae (Gourd Family). The Cucurbitaceae include about 100 genera and 850 species of climbing or prostrate annual herbs, primarily tropical and subtropical in distribution but also extending into temperate zones. Stems bear tendrils and have unisexual flowers with an inferior ovary (Fig. 13-11). Important sources of food and ornamentals are pumpkin and squash (*Cucurbita*), cucumber, gherkin, and muskmelon (*Cucumis*), watermelon (*Citrullus*), dishcloth gourd (*Luffa*), and chayote (*Sechium*).

Apiaceae or Umbelliferae (Parsley Family). The parsley family has epigynous flowers, and coalescent carpels, but other flower parts are usually in fives. The characteristic inflorescence is an umbel (Fig. 13-12). The family (about 125 genera and 2900 species) occurs mostly in the northern hemisphere. Many plants in the Apiaceae produce essential oils that smell and taste good, some contain resins and others have alkaloids in lethal quantities (especially in the roots and fruits). Among the most useful plants are carrot (*Daucus*), parsley (*Petroselinum*), parsnip (*Pastinaca*), caraway (*Carum*), dill (*Anethum*), and anise (*Pimpinella*).

Rubiaceae (Madder Family). The Rubiaceae are largely tropical or subtropical with nearly 400 genera and 4800 to 5000 species. They are usually trees or shrubs, sometimes vines, and infrequently herbs. The flowers are bisexual, usually epigynous, actinomorphic, sympetalous, and funnelform. The Rubiaceae are most important as tropical crops such as coffee (*Coffea*) (Fig. 25-4) and quinine (*Cinchona*). Examples of ornamentals are gardenia (*Gardenia*) and madder (*Rubia*).

Asteraceae or Compositae (Sunflower Family). The composites constitute the largest family of flowering plants, with the genera estimated to be 950 and the species probably 20,000. They are distributed over most of the earth and in almost all habitats. Most are herbaceous, although a few tropical ones are trees or shrubs. In most parts of the temperate zones, from 10 to 15% of angiosperm species are composites. The family belongs in the Asterales, which is considered to be the most advanced of the epigynous dicots; their characters are different from those of the Ranales in every significant way. All of the flowers have inferior ovaries and the limited number of petals, anthers, and carpels are all coalescent (Fig. 13-13). The stamens arise from the corolla tube. The family gets its name from the arrangement of flowers into compact *heads,* each of which appears to be an individual flower (see Chapter 4). Thus what appears to be a single flower is really a "composite flower." Numerous disc flowers form the center of the head. Each of these is a radially symmetrical small flower; sepals are absent, often replaced by a hairlike pappus. Around the center of the disc flowers are ray flowers, each being bilaterally symmetrical with

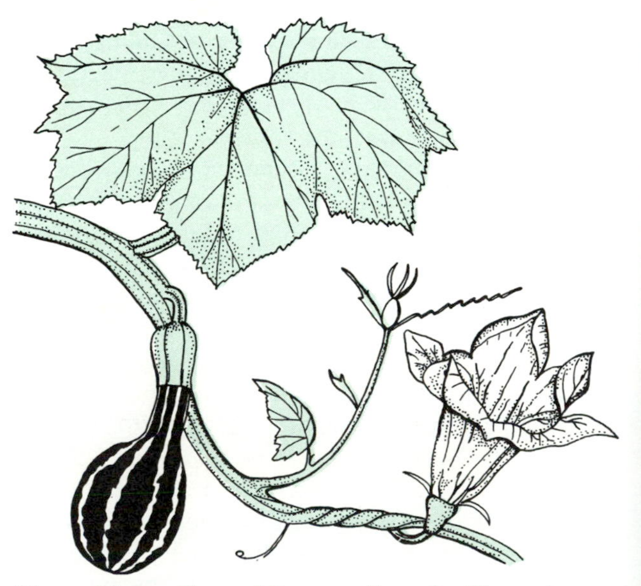

Figure 13-11 Curcurbitaceae. Squash *(Curcurita pepo)* flower and fruit.

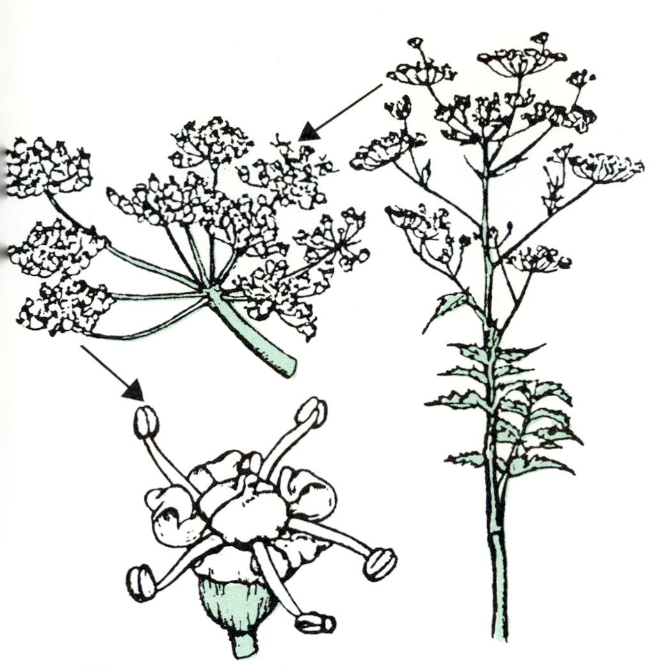

Figure 13-12 Apiaceae or Umbelliferae. Parsnip *(Pastinaca sativa)*.

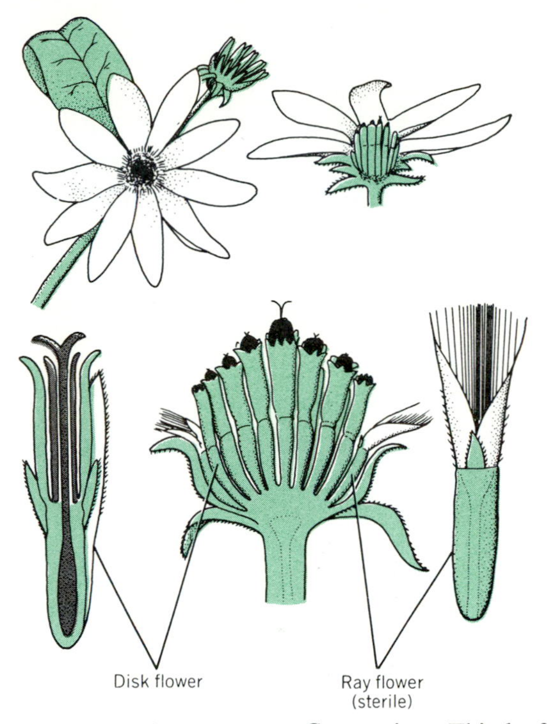

Figure 13-13 Asteraceae or Compositae. Thin-leaf sunflower *(Helianthus decapitalus)*. (From Wilson, Loomis, Steeves, 1971)

its corolla extended into a single long, straplike petal. These are arranged around the disc, similar to separate petals in a single buttercup flower. The entire head has many bracts (special floral leaves) around the base (the involucre) that look like the sepals on a single flower. Either disc or ray flowers may be absent from the head. Sexes may be in separate flowers on the same plant or in separate flowers on different plants; then the head is unisexual. We utilize fewer members of the family than might be expected from its size. Some genera of consequence for food are lettuce (*Lactuca*), globe artichoke (*Cynara*), and sunflower (*Helianthus*) for its seeds. The insecticide pyrethrum is obtained from *Chrysanthemum coccineum* and the red dye safflower from *Carthamus tinctorius*. Genera commonly used as ornamentals include perennial asters (*Aster*), chrysanthemum (*Chrysanthemum*), cinerarea (*Senecio*), marigolds (*Tagetes*), and zinnia (*Echinops*).

CHAPTER 14

Plants in Communities and Ecosystems

Alexander von Humboldt, one of the greatest scientific travelers at the beginning of the nineteenth century, was impressed with the tendency of plants to occur in repeatable groupings (i.e., *communities*) wherever there were similar environments. He noted that the growth forms of the associated plants determined the appearance of the vegetation. He also found that climbing a mountain in the tropics was analogous to passing either further north or south of the equator.

Then in 1863 Kerner von Marilaun, in a classic paper on the plant life of the Danube basin, gave a dynamic concept to the community: "In every zone the plants are gathered into definite groups which appear as developing or as finished communities. . . ." Another dimension was added to this concept in 1935; A. G. Tansley at Oxford University thought more emphasis should be placed upon the interrelation between the community and its environment. He thus defined an *ecosystem* as a community and its environment considered together as a functional system through which energy and matter flow. Thus the complex, ever-changing relationships that exist among the multitudes of organisms and their environment are seen as a whole in the ecosystem. For example, a redwood forest is an ecosystem that differs from other forests not only in its structure and composition but in its adaptation to a foggy coastal environment and its effects on its own microclimate, its rate of productivity, and its manner of circulating nutrients between the soil and organisms in the community.

1. COMMUNITY STRUCTURE AND COMPOSITION

The form and structure of a plant community, that is, its *community physiognomy,* can range from a floating population of aquatic plants to a giant redwood forest or tropical rain forest. Thus, as Humboldt recognized, plant communities are characterized by the growth forms of the individual plant species that compose them. This spectrum of growth forms determines both horizontal and vertical structure. Plants exist in the community in horizontal patterns, which may be expressed as a mosaiclike occurrence of species. Vertical differentiation, or *stratification,* results because different species reach different heights above ground or depths below the water surface. In a forest the vertical structure involves a gradient of growth forms such as trees, shrubs, herbs, mosses, and so on, in adap-

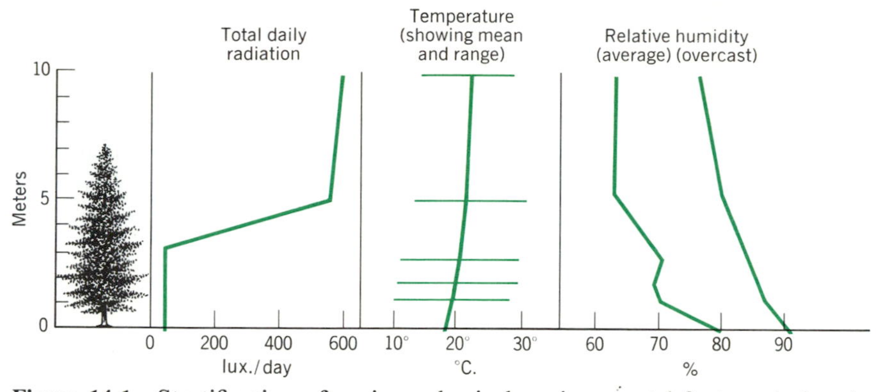

Figure 14-1 Stratification of various physical environmental factors during the summer in Europe within a dense fir *(Abies)* plantation whose physiognomy is relatively simple. For a more complex forest (e.g., the tropical rain forest shown in Fig. 14-2) the effects of the different strata would be roughly additive, although each stratum might not have as great an individual effect as that of the very dense fir foliage. (Adapted from Clapham, 1973; data from Geiger, 1966)

tation to differing amounts of light, temperature, and humidity (Fig. 14-1). The most complex vertical structure occurs in the tropical rain forest (Fig. 14-2); thus, different animals, such as birds and butterflies, also occupy different levels of the vertical gradient. Shrublands and grasslands also have mixtures of plants at different heights that produce a less obvious vertical structure. Each species has its own place in vertical and horizontal space, time, and relationship to other species of a given community. The species *place* in a community in relation to other species is its *niche*. Species evolve toward difference in niche, thereby reducing the competition among them. Generally speaking, no two species will occupy quite the same niche, using the same resources at the same time and place and subject to the same limiting factors in a stable community.

2. DEFINING THE COMMUNITY

The redwood forest of coastal California and southern Oregon illustrates the problems in defining a plant community. Here redwood (*Sequoia sempervirens*) is the dominant tree and, through its influence, the redwood forest appears similar throughout its range. However, when one plots the distribution of the associated species, no one has a range completely coincidental with it. Most species range beyond the distribution of the redwood and a few species occur only within a portion of the redwood's range. Thus the unity of the redwood

Figure 14-2 Profile diagram showing the four strata of a complex rain forest in Guyana. Three tree strata are shown; herbs and shrubs are uncommon. (From Weier, et al., 1974)

forest *community* is based upon the dominant species, *Sequoia sempervirens*. Other species also characterize the community in different portions of the forest in which redwood is still present. Since there is an excellent fossil record of *Sequoia* and many of its common associates, we can add a historical perspective to the understanding of the community. The story then becomes even more complex. Several different evolutionary lines have evolved, diverged, and come together, producing the aggregation of species that occur together in the redwood forest today. When considering the plants associated with the redwoods during the early Tertiary, they are different from those associated in the middle Tertiary and still different from those occurring in the forest today.

A major factor determining the distribution of the redwood forest today is the coastal fog, which cuts transpiration rates (and thereby helps to ensure adequate water supply) during the dry summers in the Mediterranean climate of California and Oregon (Fig. 14-3*A*). There are many kinds of environmental gradients, such as moisture, temperature, and light, that determine which species coexist with redwoods. Some of these conditions are produced by the redwoods themselves; they modify the habitat of other species growing with them by creating gradients of light and moisture and depositing litter (Fig. 14-3*B* and *C*). The degree of overlap of the distribution of species along these gradients determines how we define communities. Distribution of species along a hypothetical gradient is shown in Figure 14-4*A*, indicating how an ecologist might decide to define a community type. In Figure 14-4*B* is shown the distribution of the major tree species associated with redwood and their distribution along a moisture gradient in the northern range of the redwood forest (Humboldt Redwoods State Park). Some ecologists call the forest a redwood community if redwood is present but not necessarily the dominant tree, whereas others would subdivide the forest into several community types. In the drier sites where madrone (*Arbutus menziesii*), Douglas fir (*Pseudotsuga menziesii*), and tan oak (*Lithocarpus densiflora*) dominate the redwood, the forest would be called a "mixed evergreen community" rather than redwood. It becomes evident that, when a community is not dominated by one to several plants, the difficulty in defining it increases. For example, in the Great Smoky Mountains (part of the Appalachian mountain chain in Tennessee) the eastern deciduous forest reaches its greatest diversity with no clear-cut dominants evident among the numerous tree species. Species form a shifting series of combinations along various environmental gradients and thus present a complex, continuous pattern (Fig. 14-4*A*). However, even at various points along the gradient, species with overlaps in ecological amplitudes can be usefully referred to as separate communities. But how the community is defined is dependent upon the particular aggregations the ecologist chooses to recognize.

As was evident in the discussion of dominance, communities also differ greatly in the numbers of species, that is, their *species diversity*. There is a broad trend of decrease in species diversity from the lowland tropical forest through temperate to arctic and alpine ecosystems. Additionally diversity may occur as special products of evolution. For example, two interacting organisms—such as plants and herbivores—can make possible increases in each other's diversity, and their diversities can increase in parallel through evolutionary time (*co-evolution*). Although the influence of herbivores on plants may not always be obvious, it is profound. Herbivores may control the reproductive ability of the plant by destroying its leaves (photosynthetic surfaces) or by directly eating its reproductive organs. Pollination relationships are a very specialized plant–herbivore interaction where a particular part of the plant is eaten and pollination takes place. Here the herbivore is attracted and is thought to have produced a great diversity in angiosperm flowers. Herbivores are also thought to be an impor-

(A)

(B)

(C)

Figure 14-3 The redwood community. *A*. One of the principal environmental factors that determines the distribution of redwood *(Sequoia sempervirens)* is fog, which helps decrease transpiration, thus assuring an adequate moisture supply during the dry summers along the Pacific coast. Several other tree species often grow with the redwood (e.g., Douglas fir, tan oak, alder, California bay); each species is distributed along a moisture gradient (Fig. 14-5*B*). Where redwood is dominant, sword fern (*Polystichum munitum)* is a characteristic understory species *(B)*. In *(C)*, note white-barked alder along the river and mats of the herbaceous sorrel *(Oxalix oregana)*. The sword fern and sorrel are considered members of the redwood community even though their total range of distribution is not coincident with the redwood tree, *Sequoia sempervirens*. They occur as understory in a redwood forest only where their environmental requirements are met; otherwise they are members of other communities.

tant "organizing force" within the community.

3. CLASSIFICATION OF COMMUNITIES

Communities may be classified by a number of characteristics, such as growth-form dominance, species dominance, and species composition. As noted above, the boundaries between community types will be somewhat arbitrary depending upon the characteristics chosen by the ecologist to define the community. As with classification of species (Chapter 13) there is no single correct way to classify, and several different systems of classification of communities have developed that are useful for different purposes. For example, plant communities may be classified by their structure or physiognomy—the dominant growth form of the uppermost stratum or the stratum of highest coverage in the community. A major kind of community characterized by plant physiog-

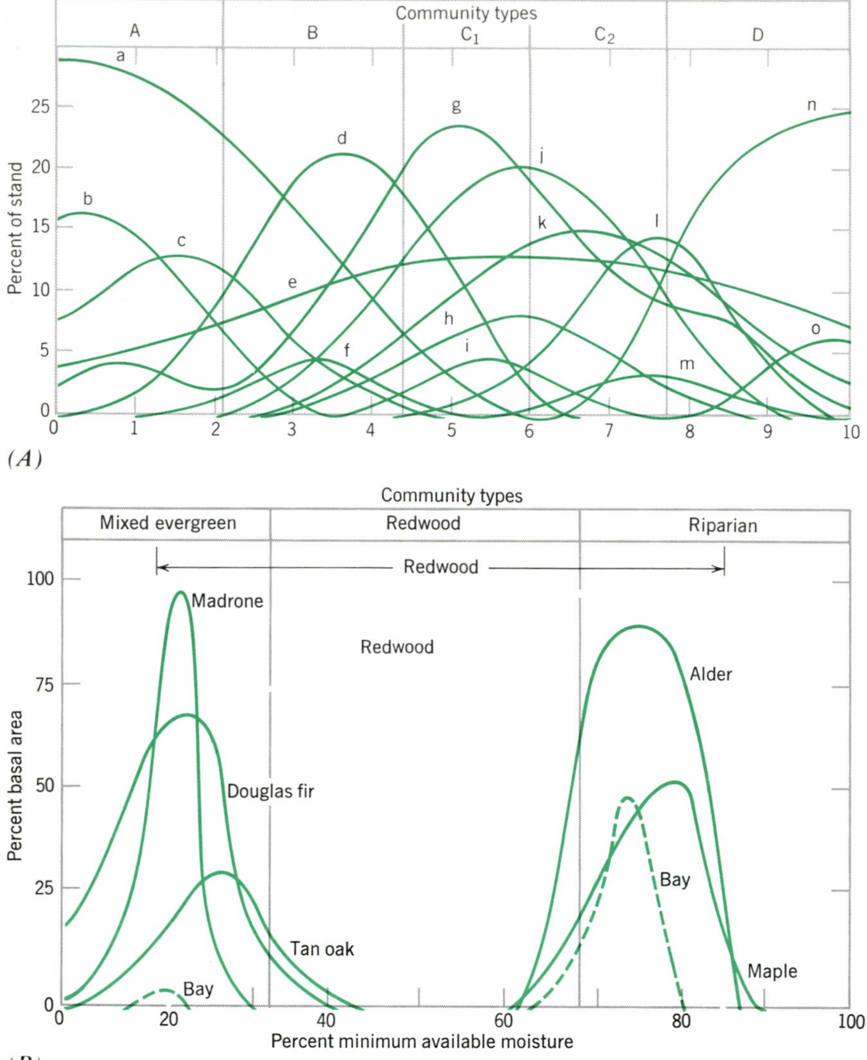

Figure 14-4 *A*. Distribution of species (percent of stand) along a hypothetical environmental gradient such as moisture or light (0 to10). These are curves drawn for illustration but may be compared with curves from work done in the Appalachian Mountains. Despite each species's (*a* to *o*) having a characteristic distribution curve for this specific environmental gradient, it is clear that it is possible to segregate groups of species to be recognized as communities. (Adapted from Whittaker, 1954) *B*. The redwood forest as an example of the problem of defining communities. When major tree species are plotted along a moisture gradient, Douglas fir, tan oak, and madrone comprise greater basal area (i.e., cover) than redwood, whereas in the moderately moist (mesic) sites redwood is the sole tree dominant; in the wet sites such as along streams and in ravines, alder, maple, bay have greater basal area than redwoods. Note that ecotypes of bay occupy both the extremes of wet and dry sites. Some ecologists would refer to all of these groupings of species as a "redwood community"; others separate the drier type with high cover of Douglas fir, tan oak, and madrone as a "mixed evergreen" community. Others also would distinguish a "riparian community" within the redwood forest. (Adapted from Waring & Major, 1964)

nomy is called a *formation,* or a *biome* if it also includes animals. Physiognomic classifications are generally used in describing communities or ecosystems of a continent or the world; thus they usually appear on maps. Also they are widely used by climatologists, soil scientists, and geographers—as well as by ecologists.

Communities may also be defined by their dominant plants. Although dominance types are one of the easiest ways of classifying communities, they are often not the most satisfactory way—as indicated by our contrasting example of the redwood forest and the mixed deciduous forest in the Smoky Mountains. Furthermore the incredible diversity of species in tropical rain forests makes the dominance concept inapplicable there. Communities may also be classified by their floristic composition and then are called *associations*. Classification by this characteristic is most useful where small areas can be studied in detail, as for example, in central Europe where only small patches of natural vegetation remain amid agricultural ecosystems. Likewise, forests can be classified according to the kind of site or habitat, a concept valuable for foresters.

The community patterns used in classification systems are usually those that have relatively stable conditions or have arrived at what the ecologists refer to as *climax*. In certain areas, however, the climax community may have been destroyed or may not have developed as yet. Here the communities and environmental conditions go through a progressive development of parallel and interacting changes. This change is called *succession*.

Ideally, the complexity of communities in a landscape, therefore, involves a pattern of intergrading climax communities. Local discontinuities of topography and soil, as well as disturbance and successional development, interrupt the continuity of climax communities. However, it is generally possible to recognize a climax community that is most widespread among the stable communities of the area; this principal community type for the area is the "prevailing climax." The prevailing climax is adapted to the climate of the area, and by means of prevailing climaxes we can relate communities to environment in terms of broad geographic or altitudinal gradients in relation to climate. Examples of three gradients are shown in Figure 14-5.

4. MAJOR KINDS OF ECOSYSTEMS IN THE WORLD

Of the many types of ecosystems that may be recognized (Fig. 14-6), seven major ones are presented here. Of course, they were present before humans began cultivation; Figures 14-7 and 14-8 present the amount of land area that has been cultivated.

a. TROPICAL RAIN FORESTS

More species of plants and animals live in the tropical rain forest than in all of the other world ecosystems put together. Not only is there a large number of organisms in the tropical rain forest, but their interrelationships are more complex than those found between plants and animals in other ecosystems.

Rainfall is abundant (generally from 200 to 400 cm per year). Although precipitation varies from month to month, there is generally no pronounced dry season. Most of the plants are woody and evergreen. The trees are usually taller than those in temperate forests, often reaching 40 to 60 m in height. Further, the trees are diverse, with seldom *fewer* than 80 species per hectare (10,000 m^2 = about 1 hectare), in contrast with temperate forests where there are rarely more than a dozen in a similar-sized area. Trees in the tropical rain forest are remarkably homogenous in appearance, although they occur in three to four stories or layers in the forest (Figs. 14-2 and 14-9). They usually branch only near the crown. Their leaves are large, leathery, and dark green; their bark is thin and smooth. Since their roots are generally

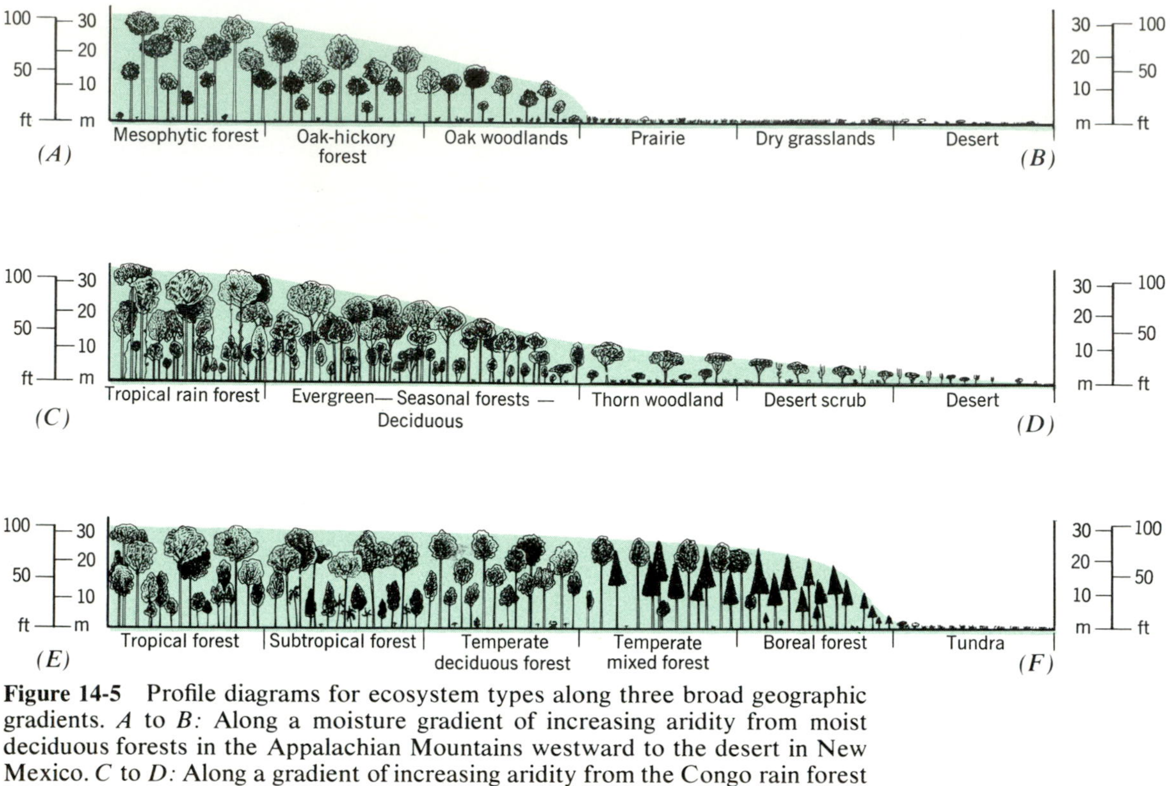

Figure 14-5 Profile diagrams for ecosystem types along three broad geographic gradients. *A* to *B:* Along a moisture gradient of increasing aridity from moist deciduous forests in the Appalachian Mountains westward to the desert in New Mexico. *C* to *D:* Along a gradient of increasing aridity from the Congo rain forest to the Sahara desert in Africa. *E* to *F:* Along a temperature gradient from the tropical seasonal forest in Mexico along the Pacific coast northward in forest climates to arctic tundra in Alaska. (*A, C, D, E, & F,* adapted from Whittaker 1974; *B* from Beard. 1944)

shallow, they often have buttresses or stilt roots at the base of the trunk to provide a firm anchorage (particularly in swampy soils). Because little light penetrates to the floor of the tropical rain forest, there are relatively few herbs there. On the other hand, many plants grow on the branches of other plants in the better lighted zone far above the forest floor. These *epiphytes* (plants growing on other plants), which have no direct contact with the soil, obtain water and minerals from the rain and humid air in the canopy of trees. Along with the epiphytes and climbing vines, animal life is most abundant in the treetops.

There are three major areas of the world characterized by tropical rain forest (Fig. 14-6). The largest is that in the Amazon basin of South America with outliers in coastal Brazil, Central America, and eastern Mexico. In Africa, a large area occurs in the Congo basin with extensions in coastal West Africa. The third major area of rain forest extends from Sri Lanka to eastern India and Thailand, the Philippines, the large islands of Malaysia, and a narrow strip along the northeast coast of Queensland, Australia.

The tropical rain forest now forms about half of the forest area of the earth but increasingly large areas of this forest are being destroyed. The rapidly expanding human population in the tropics has made the traditional tropical agricultural practices of clearing and short-term cultivation very destructive, be-

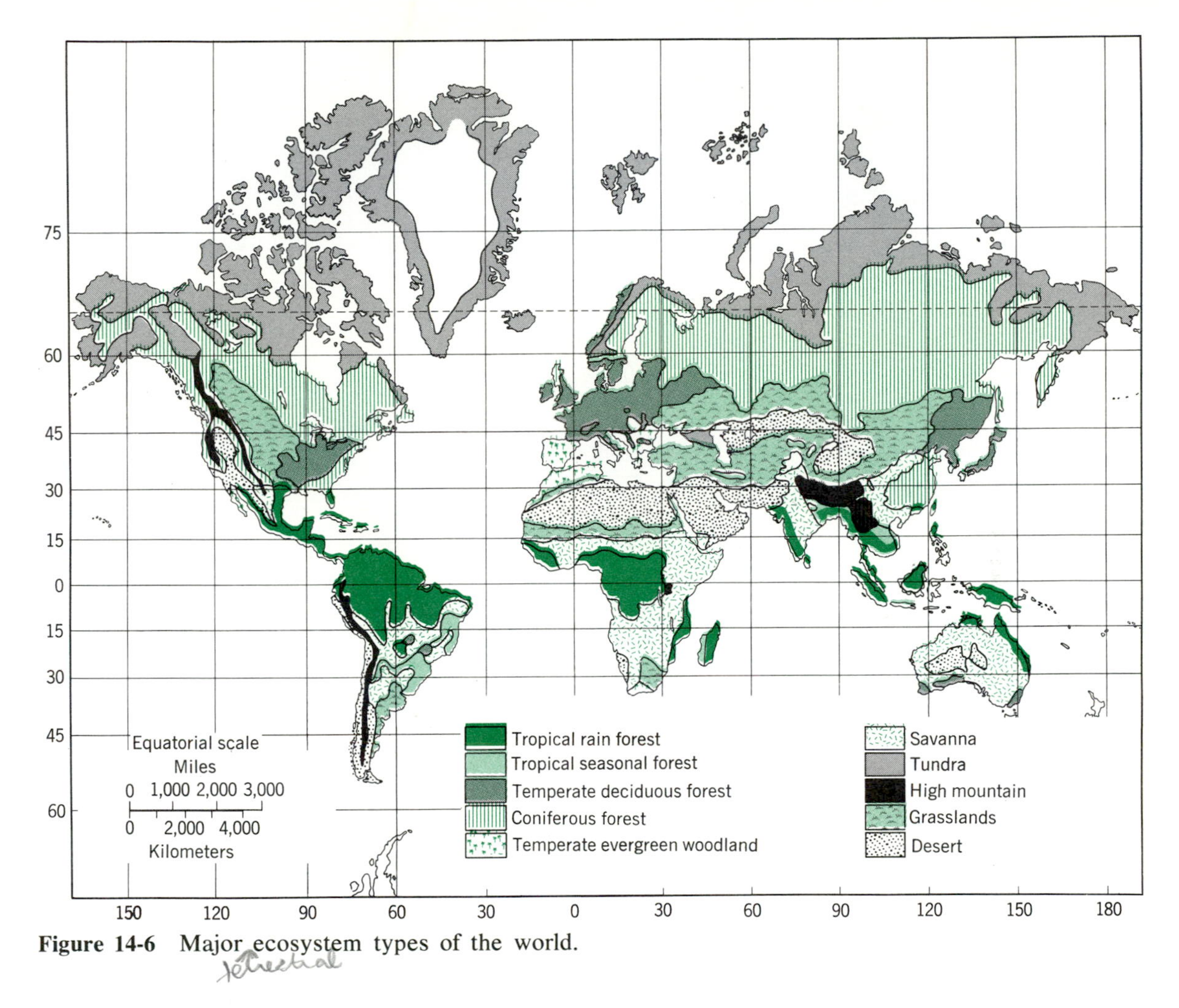

Figure 14-6 Major ecosystem types of the world.

cause they are now being conducted on such a wide scale. It is estimated (1980) that 40% of the rain forest has now been cut (38% in the New World, 52% in Africa, and 63% in India, Sri Lanka, and Burma). At the current rate of destruction (20 million ha/yr) it is estimated that little rain forest will have survived in the Philippines and India by the end of 15 years, and much of that throughout the tropics will be gone by the turn of the century. Also (see Chapter 15), soils supporting the rain forest tend to be nutrient poor, most of the essential minerals apparently being cycled more-or-less directly within the biomass. Therefore, more understanding of this very fragile ecosystem is needed to aid with development plans for remaining natural areas such as the Amazon basin. Many tropical scientists think that emphasis should also be put upon development of forest products and the simulation of natural forest conditions in cultivation patterns.

b. SAVANNAS

Savanna vegetation is transitional between that of the equatorial tropical rain forest and the

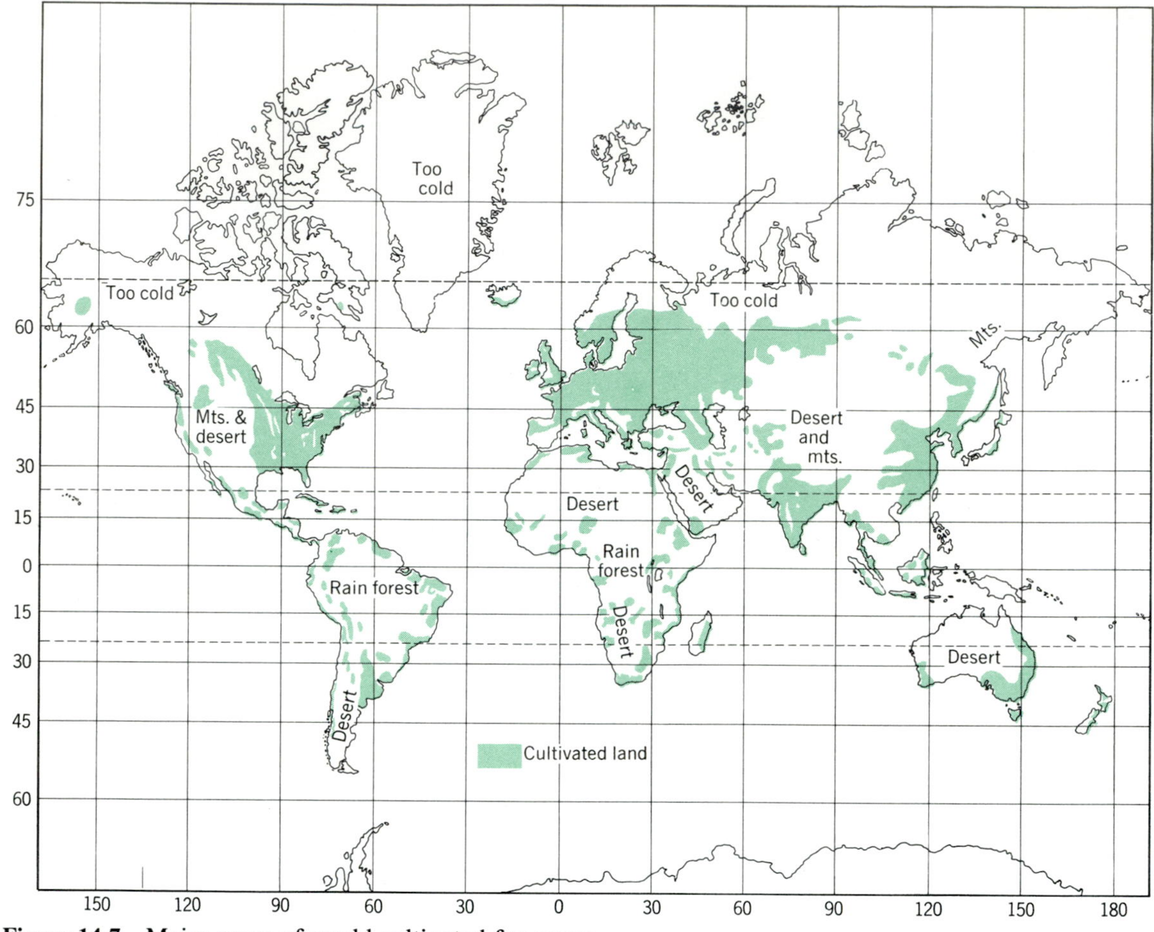

Figure 14-7 Major areas of world cultivated for crops.

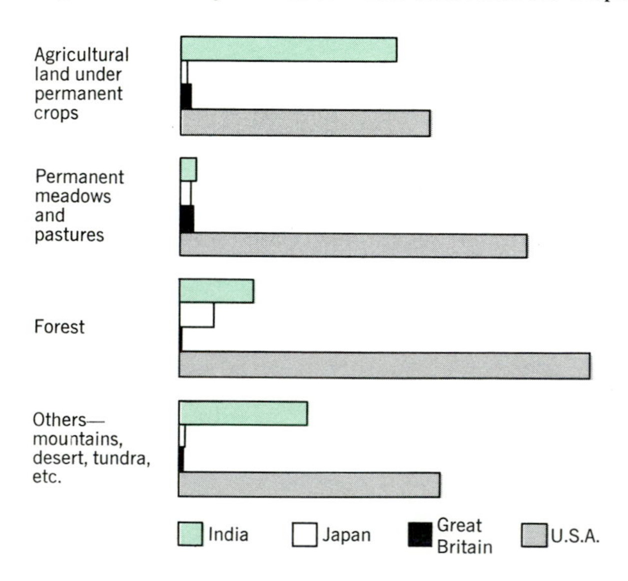

Figure 14-8 Differences in land use in the United States, Great Britain, India and Japan. The scale of acreages is the same for all countries. Note the differences from one country to another, between proportions of the available land devoted to permanent crops, grazing, and forest. (Adapted from Edlin, 1967)

Figure 14-9 Central Amazonian tropical rain forest, with four strata of trees.

desert. In contrast to the rain forest, savannas have less annual rainfall (85 to 145 cm), usually marked by periods of drought. There is also a wide fluctuation of average monthly temperatures, partially due to the seasonal rainfall.

Savanna communities may range from highly scattered trees and open thorn forests to more densely grouped trees. Because of the generally dispersed distribution of the trees, the understory is well illuminated and perennial herbs (mostly grasses) are common (Fig. 14-10). Annual herbs and epiphytes are rare. The trees found in savanna communities generally have a thick bark and are adapted to drought and effects of constant fires. Almost all of them are deciduous, losing their leaves at the beginning of the dry season and many of them flowering in a leafless condition.

As conditions become drier along a geographic gradient, the trees are replaced by grassland, or are more widely scattered. Savannas are most extensive in Africa (Fig. 14-6), where they support the richest fauna of grazing animals in the world. As pointed out in Chapter 3 this is the ecosystem type in which the hominids evolved. Similar but less extensive and less rich communities occur in Australia, South America, and southern Asia. Some of the African and Australian savannas occur in climates too dry for forest. However, soil conditions or fire, or even both, rather than climate, probably result in the occurrence of savannas in less arid climates such as in South America. Both sa-

Figure 14-10 Savanna in East Africa with flat-topped leguminous trees and grass understory. The giraffe has a special niche in this vegetation in browsing on trees that other animals cannot reach.

vannas and temperate grasslands are subject to fires, which affect the structure of the community and its extension into climates that might otherwise support forest.

c. DESERT SHRUBLANDS

The vast deserts of the world occur in the zones of atmospheric high pressure bordering the tropics at about 30 to 40° north and south latitude, and they extend further in the interiors of the large continents (Fig. 14-6). Most deserts are characterized by less than 10 cm of annual rain, and in the deserts along the coast of Peru and Chile, rain may not fall at all for several years. The Sahara Desert, which extends from the Atlantic coast of Africa to Arabia, is the largest one in the world, and Australia's desert, which covers about half of the continent, is next in area.

The annual distribution of rainfall in deserts usually reflects that of adjacent areas. On their equatorial side, rains occur in the summer; on their poleward side, in the winter. Between the two, as in the Arizona desert, two annual peaks of rainfall result in usually two periods of active plant growth—one in the winter and one in the summer—with different plants active in each.

Annual plants called *ephemerals* are most important in number and kind in desert and semiarid areas. They germinate quickly, grow to sufficient maturity to produce one or more seeds per plant, and then die. These seeds can survive in the soil for long periods of drought—sometimes over many years. Most deserts are also characterized by shrubby growth, which is particularly true in temperate areas (Fig. 14-11).

Figure 14-11 Desert landscape in southwestern United States with giant saguaro cactus, palo verde trees, cholla cactus, and other desert shrubs.

Many of the taller plants in the warm deserts are either succulent (such as cacti and euphorbias) or have small leaves that either are long lasting or are shed during unfavorable seasons (Fig. 14-11). Photosynthesis usually takes place in the stems as well as in the leaves during dry periods. Woody plants either have widespread, surficial roots that effectively absorb the periodic rainfall or have deep root systems that tap groundwater resources.

Desert vegetation may become highly productive when water is made available. Temperatures are ideal for plant growth and sunlight is abundant due to sparse cloud cover. The farms of the Imperial Valley in California, for example, are capable of producing sufficient food for large populations and many cities have grown up in the deserts of the southwestern United States. The reclamation projects of Israel are also exemplary; however, irrigation may bring serious problems (Chapters 5, 7, and 15).

d. GRASSLANDS

Grassland ecosystems are varied, some being related to savannas, others to deserts, and still others to temperate deciduous forests. Grasslands generally occur over large areas in the interior portion of continents, often between deserts and temperate woodlands where the amount of rainfall is intermediate between the two ecosystems (Figs. 14-6 and 14-12). Examples are the steppes of Eurasia, the veldt of Africa, and the pampas of Argentina. In North America, there is a transition from the western, more desertlike short-grass prairies (Great Plains) through the moister, richer, tall-grass prairie (Corn Belt) to the eastern temperate deciduous forest. Perennial bunch and sod-forming grasses are dominant, but when the other herbs are flowering these grasslands become a colorful sight.

Grasslands have traditionally been heavily used for agriculture, both for pasture and for growing grain crops. It is interesting to remember that agriculture arose in the Near East in areas where wild wheat and barley were naturally distributed and in Central and South America where the ancestors of maize grew. Also some of the most productive cultivated grasslands occur in areas formerly occupied by temperate forests. Large areas of rain forest and seasonally dry forests are now being cut throughout the New World tropics to establish African grasses in an effort to increase grazing land. For many reasons these plans are highly controversial.

The great grasslands of the world (as the savannas) have been inhabited by herds of grazing animals associated with a variety of large predators. Natural grasslands are now a rarity, having given way to cultivated fields. The large mammals also have been hunted almost to extinction and survive primarily in refuges. The floral composition of the large areas of range-

Figure 14-12 Prairie grassland on the plains in Oklahoma.

land has been greatly changed by introduced species, and these areas are now occupied by herds of domesticated animals.

e. TEMPERATE DECIDUOUS FORESTS

The temperate deciduous forest ecosystem attains its best development in areas of warm summers and mild, rainy winters—hence the most extensive forests occur in eastern North America, Europe, and eastern Asia (Fig. 14-13). Annual rainfall ranges from about 70 to 260 cm, and the deciduous habit is probably related to the unavailability of water during most winter months. In the early spring, herbaceous plants grow abundantly on the temporarily well-illuminated forest floor. Later in the spring when leaves appear on the trees, reducing light on the forest floor, most of the early spring herbs have set seed.

An outstanding characteristic of the temperate deciduous forest is its similarity in composition throughout the Northern Hemisphere. In contrast, deserts of the world are composed of different groups of plants that have converged in their ecological characteristics. The plants (including oak, maple, beech, ash, basswood, and others) of the deciduous forests of Japan are similar to those in Europe and the eastern United States (Fig. 14-13). Additionally, the forests of western North America once contained many of these deciduous trees, but they were eliminated in that area during the late Tertiary period and during the Pleistocene as summer rainfall was reduced. In their place now are the magnificent conifer forests including the redwoods and Douglas firs, arbor vitae, spruce, and others (Chapter 22).

The deciduous forests of North America were largely destroyed with the coming of people from Western Europe. However, when agriculture began to develop in the prairie areas, where the soils were deeper and more productive, much of the cleared eastern forest land was allowed to revert to forest. After about a century, the vegetation in the northeastern United States seems partially to have returned to climax forest conditions.

Figure 14-13 Virgin mixed deciduous forest in Wisconsin with profuse understory vegetation. The large trees shown are left to right: maple, elm, black ash, and elm.

f. CONIFEROUS FORESTS

Temperate evergreen forests dominated by coniferous species occur in varied climates. Extensive needle-leaved evergreen forests occur in the continental climates of the western United States, including various species of pine in the drier habitats, Douglas fir in the moister habitats, and the mixed forest of firs, incense cedars, and Sierran sequoia in the Sierra Nevada. Many of these coniferous forests have a

Figure 14-14 Taiga forest broken by openings such as this bog with spruces and larches growing in a carpet of raindeer moss.

more open structure than deciduous forests (Fig. 14-14). In the humid coastal climates, however, the redwood forest and Olympic coniferous rain forest are closed and dense.

Subarctic and subalpine coniferous forests occur at the cold edge of the climatic range of forest distribution. These forests are characterized by severe temperatures and a persistent cover of snow in the winter. The Russian word *taiga* is frequently used for this vegetation because it extends over vast areas of the Soviet Union. These coniferous forests also are often called *boreal* forests because they occur around the world in the far Northern Hemisphere, extending southward at high elevations in the mountains (Fig. 14-6). The most common genera of trees include spruce, fir, pine, hemlock, and larch. Some deciduous species such as birch and poplar are also common. This forest, although lacking in diversity of species, represents one of the great world forest resources in terms of tree biomass. Because of the short summers, and therefore the too-short growing period for most crops, this forest has not been destroyed as rapidly as the tropical rain forests. The Soviet Union, Canada, and the Scandi[illegible]an countries are trying to meet the challenge of wise utilization.

g. TUNDRA

The taiga forest opens northward into scattered trees amid tundra. Tundras are arctic (or alpine) plains that are dominated by grasses, sedges, and dwarf shrubs (Fig. 14-15). Mosses and lichens often form a ground cover, being better developed here than in any other ecosystem. In many tundras the deeper layers of the soil are permanently frozen (*permafrost;* see Chapter 15), and only the surface thaws in the summer. Trees are absent except in the vicinity of rivers, where stunted forests may develop, or on sheltered slopes where permafrost may melt to fair depths. Also, bogs and muskegs are common because drainage through the soil is generally poor.

Tundra occupies about one-tenth of the earth's surface and is best developed in the Northern Hemisphere (Fig. 14-6). Large mammals such as caribou, reindeer, musk ox, bear,

Figure 14-15 Arctic tundra meadow dominated by cotton grass *(Eriophorum)* in interior of Alaska.

and wolf are common; however, diversity of plants and animals is low. Despite the adaptations of tundra organisms, the biome as a structural unit appears highly vulnerable to many human activities—and thus like the tropical rain forest, it is considered a particularly "fragile ecosystem." Perhaps its best utilization will continue to be the herding of such animals as the caribou and reindeer.

5. ECOSYSTEM FUNCTION

An ecosystem is a functional system of the community and its environment in which energy is transferred and materials (e.g., minerals and water) are circulated. Carbon, phosphorus, nitrogen, and water may circulate many times between the living and nonliving compartments of the ecosystem; this means that any given atom may be used over and over again. However, energy is used once by an organism and then converted either to heat, which is lost from the ecosystem, or to chemical bonding, which may persist but will in turn eventually be lost. For example, the chemical energy of lignin (Chapter 6) may last for centuries.

Associated with the flow of energy and nutrients is the concept of balance. Sometimes ecosystems are in balance; at other times they are not, because there may be net inflow or outflow of energy or particularly some material of critical importance to the ecosystem as a whole. Thus ecologists speak of *budgets,* which show the gross input, gross outflow, and net surplus or deficit for a factor in question. Most ecosystems have surpluses in some environmental parameters and deficits in others.

6. ENERGY FLOW IN THE ECOSYSTEM

The one-way flow of energy is a universal phenomenon resulting from the operation of the first law of thermodynamics, which states that energy may be transformed from one type (such as light) into another (such as potential chemical energy of food) but is never created or destroyed. The second law of thermodynamics states that in all processes involving transfer of energy there is degradation from a concentrated form (such as high-temperature heat) into a dispersed or unavailable form (such as low-temperature heat). Because some energy is always dispersed into unavailable heat energy, no transfer of energy will be 100% efficient.

All organisms at or near the surface of the earth receive solar energy; however, only a small fraction of the direct solar radiation is converted by photosynthesis to food energy for the biotic components of the ecosystem (Chapter 5).

a. PRIMARY PRODUCTIVITY

Because the light energy used by the plant is converted into chemical energy, it should theoretically be possible to determine the entire energy uptake of plants in ecosystems by measuring the total amount of carbohydrate produced. The *quantity* of organic matter present at a given time, per unit of the earth's surface, is the *biomass* or *standing crop.* This is usually expressed as dry grams per square meter or kilograms per square meter or metric tons per hectare. A metric ton (t) is 10^6 g; a hectare (ha) is 10^4 m^2; kilograms per square meter $\times$ 10 = gross metric tons per hectare. *Gross primary productivity* is the *rate* at which energy is bound or organic matter produced by photosynthesis, per unit of the earth's surface *per unit time;* it is most often expressed as dry organic matter in grams per square meter per year or energy in kilocalories per square meter per year. Unfortunately, it is not easy to measure gross primary production in most ecosystems because about 20 to 25% of the carbohydrate produced by photosynthesis is usually lost immediately through respiration. Thus, if we measure the total organic material actually present in the plant after a given time, we are measuring gross primary production less that used in respiration, or the *net primary production.* Although gas-exchange techniques enable measurements of gross primary production of ecosystems, these data are probably less reliable than those available for net primary production.

Moreover, it is easier to measure production of individual plants in small experiments in the laboratory than in large-scale field studies of entire ecosystems. Also, it is almost impossible in the field to distinguish plant respiration from that of other organisms such as microorganisms associated with green plants. Production is dependent on many parameters, including climatic conditions such as temperature, rainfall, and total solar radiation, as well as the availability of essential nutrients and the successional stage of the ecosystem.

We can make "ball-park" estimates of ecosystem production from different environments, and some interesting and consistent patterns emerge (Table 14-1). Although light and carbon dioxide are fundamental inputs to productivity of terrestrial ecosystems, they are thought less likely to be limiting for *whole* ecosystems than moisture and the effects of temperature, nutrients, and successional states of the ecosystem. These results are, of course, different from analyses of photosynthetic capacity of *individual plants,* as components of ecosystems, in which light and carbon dioxide are often the prime limiting factors (Chapter 5).

As well as using water for metabolic reactions, plants on land transpire large amounts of it. In fact, land plants appear to transpire water extravagantly. Many plants transpire 700 to 1000 g or more of water for every gram of dry matter net production. Some plants living in

TABLE 14-1

Net Primary Production and Plant Biomass for Land Areas

Ecosystem Type	*Area (10^6 tons)*	*Net Primary Productivity, per Unit Area ($g/m^2/yr$) Mean*	*World Net Primary Production (10^9 t/yr)*	*Biomass per Unit Area (kg/m^2) Mean*	*World Biomass (10^9 t)*
Tropical rain forest	17.0	2200	37.4	45	765
Tropical seasonal forest	7.5	1600	12.0	35	260
Temperate evergreen forest	5.0	1300	6.5	35	175
Temperate deciduous forest	7.0	1200	8.4	30	210
Boreal forest	12.0	800	9.6	20	240
Woodland and shrubland	8.5	700	6.0	6	50
Savanna	15.0	900	13.5	4	60
Temperate grassland	9.0	600	5.4	1.6	14
Tundra and alpine	8.0	140	1.1	0.6	5
Desert and semidesert scrub	18.0	90	1.6	0.7	13
Extreme desert, rock, sand, and ice	24.0	3	0.07	0.02	0.5
Cultivated land	14.0	650	9.1	1	14
Swamp and marsh	2.0	2000	4.0	15	30
Lake and stream	2.0	250	0.5	0.02	0.05
Total land	149.	888	115.27	13,858	1837

Modified from R. H. Whittaker *Communities and Ecosystems,* Second Edition, MacMillan Publishing Co., New York, 1975.

arid environments have special adaptations for more efficient water use (using 50 to 300 g/g of net production). However, this is only comparatively efficient! The availability of water for such loss becomes a major determinant of productivity on land. In arid climates, there is a linear increase in net primary productivity with increase in annual precipitation, but in humid forest climates the linear relation no longer applies. Then the productivity supported by a given amount of precipitation is affected by its seasonal distribution, by mean temperature, and by the temperature cycle (thus relating to evapotranspiration) as well as by availability of nutrients and successional status.

Looking at Table 14-1, we see that one can generalize about terrestrial net primary productivity in four ranges. A "normal" range in favorable climates on land appears to be 1000 to 2000 $g/m^2/yr$ and includes many forests, some grasslands, and many of our highly productive temperate crops. The middle range [250 to 1000 $g/m^2/$ yr] includes woodlands, shrublands, and grasslands that may be limited by water, low temperatures, and perhaps essential nutrients. The low range [0 to 250 g/m^2 yr] includes deserts, semideserts, and arctic tundra growing under even less favorable conditions. At the opposite extreme is the very high range [2000 to 3000 g/m^2 yr], which includes tropical rain forests, marshes, and successional communities in particularly favorable environments—as well as some intensively cultivated crops (such as sugarcane and rice). All of these latter communities tend to combine conditions favorable for high levels of productivity.

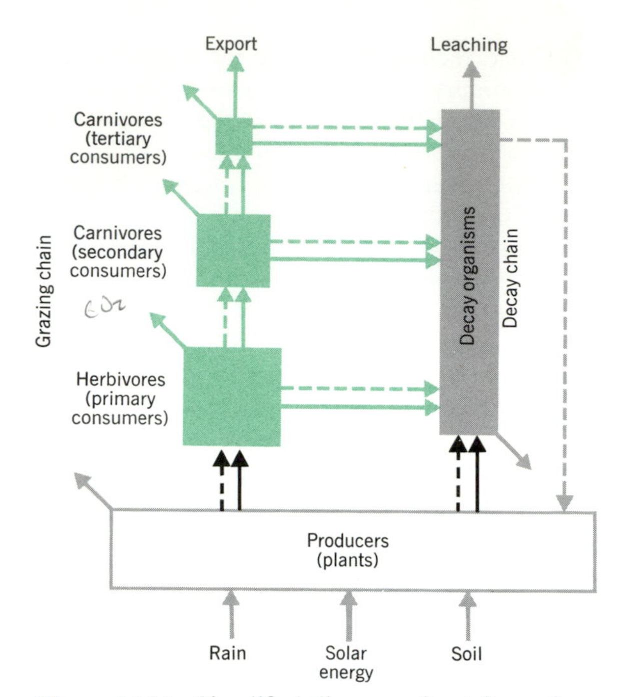

Figure 14-16 Simplified diagram of net flow of energy (solid arrows) and nutrients (dashed arrows) through a natural ecosystem. In a mature ecosystem almost all of the energy fixed by the primary producers, the plants, is dissipated as heat in the respiration of the plants, the consumers (herbivores and carnivores), and decay organisms (Arrows to side indicate heat loss from organisms). Almost all nutrients are eventually recycled.

b. THE GRAZING FOOD CHAIN

A much-used model for following the flow of energy and material through an ecosystem is called the *food chain* (Fig. 14-16). The gross production of a plant appears to be disposed of principally in three ways. First, as just mentioned, is respiration. Secondly, the plant may die and decompose. And thirdly, the plant may be eaten by animals (*herbivores*), which are called *primary consumers*. The total uptake of herbivores is very significant, because it represents the total amount of energy available not only to herbivores but also to animals in turn feeding upon them. This is termed *assimilation* or *gross secondary production* and is equivalent to the total plant material ingested by the herbivore less the material lost as feces.

As in plants, the food assimilated by animals can be stored as carbohydrate, protein, or fat, transformed into relatively simple substances, or rebuilt by the animal into much more complex organic molecules. The energy to perform these transformations is supplied by respiration. Also, as with autotrophs, the ultimate disposition of energy is by three routes: respiration, decay of organic matter by decomposer organisms, or consumption by carnivores.

Carnivores are animals that eat other animals. Those whose primary food source is herbivorous animals are termed *primary carnivores;* those whose main source of food is primary carnivores are termed *secondary carnivores,* and so on. Thus, much of the energy flowing through ecosystems can be described in terms of *trophic* (food) levels as just discussed:

plant → herbivore → primary carnivore → secondary carnivore

These levels and energy links between them comprise the *grazing food chain.* As a general model for the direction and extent of energy flow in ecosystems, the food chain is very useful. It demonstrates the amount of energy found at any trophic level, the transfer of energy from one trophic level to the next, and the amount of energy lost from the chain through either respiration or decay. Different ecosystems can be compared in their efficiency of energy transfer using this simple model. However, it should be pointed out that some species of organisms fit readily into specific trophic levels but others do not, and thus the food chain may not always be good for predicting the patterns of energy flow in some ecosystems.

As shown in Figures 14-16 and 14-17, *gross production* can be expected to drop rapidly in going up the trophic scale; that is, the amount of energy available to higher trophic levels is significantly less than that available to lower levels. This forms what has been called the *pyramid of productivity* (Fig. 14-17). Two other pyramids occur in Figure 14-17 as corollaries. The numbers of organisms generally decrease up the sequence of levels to form the *pyramid of numbers.* Pyramids of numbers, however, may be reversed when many small organisms feed on one large organism of a lower level (e.g., hundreds of insects may feed on a tree). *Biomass* for trophic levels also decreases up the sequence but biomass pyramids are less frequently reversed (although animals at times exceed the biomass of plants in plankton communities). The pyramid of productivity is the one of fundamental significance; the pyramids of numbers and biomass are less significant and less reliable consequences of the pyramid of productivity.

The percentage of released energy that is not incorporated into complex organic substances is lost to the ecosystem. In any living organism, the efficiency of energy conversion is fairly low, and the amount of energy lost from the ecosystem via respiration at any trophic level is a significant proportion of the gross production. As might be anticipated, the respiratory loss in plants of different ecosystems is highly variable. Estimates range from less than 15% to more than 60% depending upon the ecosystem. It is difficult to measure respiratory losses of animals in the field, because they move from place to place. Some measurements, however, have been made and, like those for plants, they are highly variable. There is a trend toward increase in respiratory loss at the higher trophic levels, which seems to be related to the energy expended by consumers in obtaining food.

There are several possible methods of comparing the percentage of energy transformed from one trophic level to the next. One of the most meaningful, and the one most commonly used, is *progressive efficiency* or *ecological efficiency.* This is defined as the ratio of production of any trophic level to the production of the trophic level preceding it. Roughly speaking, efficiencies in terms of percentage of energy transfer are about 1% at the autotrophic level (note Chapter 5) and about 10% at the herbivore level (comparing herbivore assimilation or respiration plus yield—including excretion—with gross primary production). Efficiencies at the third or carnivore level (comparing respiration and yield for primary carnivores and herbivores) can, but will not necessarily, be somewhat higher, that is, 15% or below. Because so much energy is lost in each transfer, the amount of food remaining after two or three successive

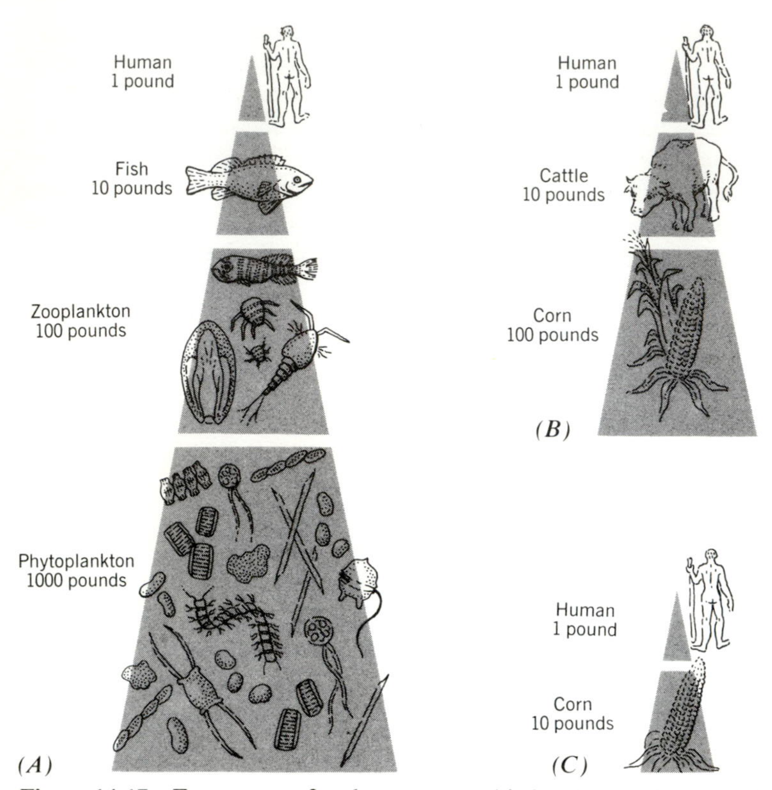

Figure 14-17 Energy transfers between trophic levels are often described as food pyramids. Gross production, that is, the amount of energy available to higher trophic levels, is less than that at lower levels. The taller the pyramid and the broader the base, the more primary production is required. This is known as the pyramid of productivity. The percentage of solar energy available at the four successive trophic levels in *A* is approximately 2%, 0.1%, 0.01%, and 0.001%, respectively. In comparing *B* and *C* note that the human–corn pyramid represents a relatively efficient energy transfer and shows why poor countries often favor vegetarianism. (From Laetsch, 1979)

transfers is so small that few organisms could be supported on food available at the end of a long food chain. Essentially, then, the food chain is limited to three or four trophic levels (Fig. 14-17).

The efficiency of human harvest is unlike the preceding ones, because we harvest, for example, only about 30% of net primary productivity when plant material is taken either as grain of cereal crops or wood from a forest. Higher efficiencies are possible in some favorable circumstances but lower ones also occur. Also with animals, only a fraction of a population can be harvested on a sustained basis—that fraction representing a surplus above the individuals necessary to maintain the reproduction and growth of a population itself. This limitation on harvest is one of the factors determining trophic level efficiencies; when human harvest exceeds the limitation, decline in productivity and yield may result—as indicated by overgrazing and overfishing.

c. DETRITUS FOOD CHAIN

The organic wastes and dead material derived from the grazing food chain (*detritus*) serves as the source of energy for organisms that are separate from the grazing food chain; these constitute the *detritus food chain* (Fig. 14-16).

In some ecosystems, considerably more energy flows through the detritus chain than through the grazing chain. The detritus food chain is a mixture of many different types of organisms (algae, bacteria, fungi, protozoans, insects, etc.) with overlapping, although different roles. In the detritus food chain the flow may be visualized as a continuous passage in contrast to the stepwise movement between the organisms that eat each other in the grazing chain. The typical decomposer lives in an environment in which it is surrounded by food particles which constitute the detritus. It ingests them in part to obtain energy for its own metabolism, then excretes the remainder as simpler organic molecules. Some soil organisms will ingest only certain types of material, whereas others consume most of the kinds of detritus available around them. The wastes from one also may be utilized by another. Gradually the complex organic molecules present in the detritus are degraded to simpler compounds, sometimes all the way to carbon dioxide, nitrate and water. In some cases, however, the detritus is only partially degraded to refractory substances called *humic acids* or *humus*, which form a stable part of the soil.

The soil is full of activity from decomposer organisms and much of the dynamics of the food chain is due to physical movement of the organisms. Thus, in terrestrial habitats, the physical movement of litter by mites and the plowing activities of earthworms aerate the soil. The efficiency of the detritus food chain depends on oxygen content, temperature, and available moisture etc. If oxygen is available, energy is obtained through respiration, as it is by the organisms of the grazing food chain.

Considerable energy is released by respiration in the detritus food chain. Decomposer organisms process great amounts of material, resulting in considerable energy lost as heat. This heat loss can well be demonstrated by comparing the temperature of a compost heap with the air temperature around it. The compost heap, which is a cultured, detritus-based ecosystem, will be much warmer.

If oxygen is not readily available, as in water-logged soils, release of energy must proceed by anaerobic means, that is, fermentation (Chapter 20). Although energy is released in sufficient quantity for metabolism of such microorganisms as bacteria and yeasts, much less energy is released by fermentation than by respiration (see Chapter 6). The differences between available energy derived from respiration and fermentation for the detritus food chain have important implications for both natural ecosystems and for those that have been altered by humans. First, because respiration is a much more efficient means of releasing energy contained in organic molecules, the breakdown of materials is much more complete under aerobic conditions. In fact, in the humid tropical forests, most of the detritus may be oxidized completely to carbon dioxide and water. In ecosystems where growth of microorganisms is less optimal, such as in northern conifer forests, the humus collects in relatively large amounts. Second, in anaerobic environments, the breakdown of detritus is still less complete, resulting in the accumulation of undegraded detritus in the form of peat, lingnin, and other products as well as soluble substances such as alcohols, acids, amines, and methane (CH_4). The characteristic odor of a marsh or polluted river is produced from partially decayed fermentation products released into the atmosphere where they ultimately become oxidized. Soluble compounds that enter aquatic habitats are generally oxidized by organisms there, but in some cases (such as bogs) the concentration of these waste products is so high that little diversity of life can be sustained.

7. NUTRIENT AND WATER CYCLING

Within an ecosystem materials circulate from the physical environment through producers, consumers, and reducers back to the environment. These processes of transfer and concentration have considerable consequences for both plants and humans. The term "nutrient" can be used for any substance taken into an organism that is metabolized or becomes a part of its ionic balances. The nutrients essential to plant life are discussed in Chapter 8. The cycling of nutrients and flow of energy are closely interrelated processes because energy is required to circulate nutrients. These materials occur in compartments or pools that have varying rates of exchange between them. Ecologically it is useful to distinguish between a large nonbiological pool in which exchange is occurring slowly and a smaller, more active biotic pool in which exchange is occurring rapidly. Thus, as with energy, the *rates* of movement or cycling of nutrient elements may be more important in determining the biological productivity than the actual amount present in any one place at any one time. Tracers (from dyes to radioactive isotopes) have helped greatly in determining rates of movement of nutrients from organism to environment.

Elements that have no known biological function also circulate between organisms and the environment. These may enter nutrient cycles linked with essential elements through chemical affinity or may be carried along in the same "energy-driven stream." Also toxins produced by human activity, such as insecticides and radioactive strontium, all too often circulate and may become incorporated into the tissue of animals. Nonessential elements, therefore, may be of considerable ecological importance if they occur in quantities or forms that are toxic, or if they react to make essential elements unavailable.

a. CARBON CYCLE

The carbon cycle shows a number of similarities to the flow of energy through the ecosystem, because energy in organisms is stored as fixed carbon. Carbon enters the food chain as CO_2 and is released by respiration as CO_2 into the atmosphere where it can be reused by plants. Thus the basic carbon cycle is simple (Fig. 14-18). However, not all carbon is respired; some is stored and some is fermented. The carbon compounds lost to the food chain after fermentation, such as methane, are oxidized to CO_2 by bacteria. Although carbon is deposited in sediments, erosion may expose these sediments and weathering of rock can oxidize this carbon. Some carbon remains stored in sediments but it may be replaced by CO_2 released from volcanic activity. Recently human activity has greatly increased the rate at which carbon is transferred from sedimentary form to CO_2. The carbonate system of the sea and earth's vegetation are efficient in removing CO_2 from the atmosphere, but the great increase in consumption of fossil fuels coupled with the decrease in "removal capacity" of the vegetation is beginning to have an effect on the atmospheric compartment. Apparently even modern agriculture and forestry can add to CO_2 in the atmosphere if the CO_2 fixed by crops does not compensate for that released from tree trunks and the soil. Although the rate of release of CO_2 by industry and agriculture is yet fairly small compared to the exchange with the sea, the CO_2 content of the air has slowly risen. As indicated in Chapter 30, it may be an advantage for growth of individual plants. However, since CO_2 acts as a greenhouse in letting light in but retarding the flow of heat out of the biosphere, considerable changes in climates could possibly result. These changes could be likely if the large reserves of fossil fuel were burned in a short time, releasing an estimated 4×10^{12} tons of CO_2. Thus the greenhouse effect presents another reason for careful use of fossil fuels.

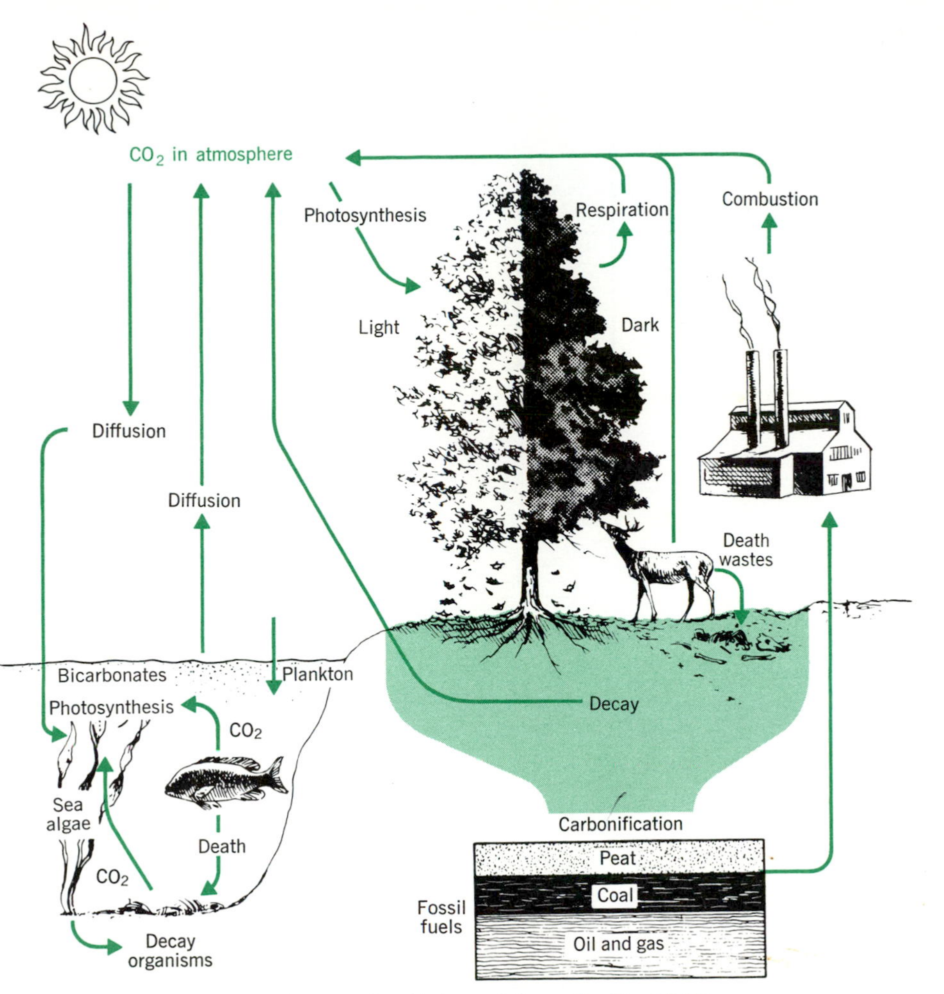

Figure 14-18 The carbon cycle. (Adapted from Smith, 1966)

b. NITROGEN CYCLE

Like the carbon cycle, the nitrogen cycle has an atmospheric reservoir but nitrogen is continually feeding into and out of this great reservoir from the rapidly cycling pool associated with organisms. This cycle is treated in some detail in Chapter 9.

c. PHOSPHORUS CYCLE

The phosphorus cycle is a critically important sediment-reservoir cycle (Fig. 14-19). The form in which phosphorus is found naturally in the environment is phosphate, which may occur either as soluble inorganic phosphate ions, as soluble organic phosphate, or as mineral phos-

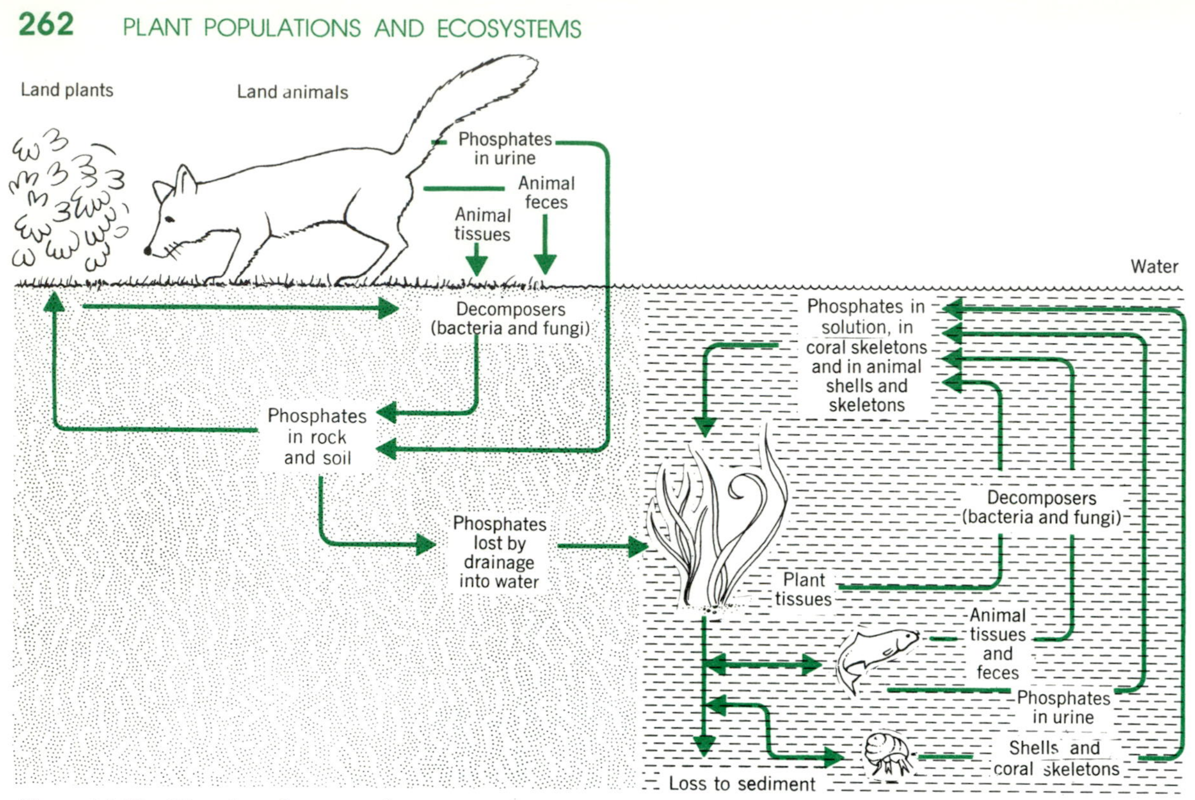

Figure 14-19 The phosphorus cycle.

phate. The ultimate origin of phosphate (as with all nutrients from the sedimentary reservoir) is crystalline rock. As the rock is weathered, phosphate becomes available to the plant generally as soluble ionic phosphate. This is taken up in the roots of plants (Chapter 7) and then incorporated into living tissue. It is passed along the grazing food chain in a manner similar to nitrogen and sulphur, excess phosphate being excreted in the feces. A dramatic example of the large amounts of phosphate excreted in feces is the guano deposit accumulated by birds on the west coast of Peru. These deposits once constituted a major world supply of phosphate. Phosphate can also be released from plants in forest and grassland fires.

In the detritus food chain, phosphate is liberated in its ionic form as large organic molecules as the organisms containing it are degraded. In this form, it can be taken up by organisms or it can be incorporated into a soil or sediments. Therefore, the organic phase of the phosphorus cycle is simple; on the other hand, the sedimentary phase is very complex, centering around the way in which inorganic phosphate, not immediately reused by living plants, is bound into and released from sediments. The amount of phosphate in the dissolved nutrients in water percolating through a typical soil is very low compared to that bound into the sediment. This is particularly obvious when compared with the amount of dissolved nitrate, which is high compared with the quantity of sediment-bound nitrogen. This binding results because some phosphates are not very soluble and can react chemically with particles in the soil or sediment. Some of these reactions bind phosphate so tightly that it cannot be used by plants. For example, in an aluminum-rich acid soil the following reactions take place:

$$Al^{3+} + H_2PO_4^- + 2H_2O \rightarrow 2H^+ + Al(OH)_2H_2PO_4$$

Soluble Insoluble

Similar reactions take place with calcium, iron, and manganese. Also phosphate can become incorporated into the crystal structure of the abundant clay minerals.

It has been shown by experiments utilizing radioactive ^{32}P that the flow rate through the organic phase is highly dependent on organic demand for phosphate. In lakes that have been intensively studied, the turnover time (the length of time it took for phosphate to move once through the organic cycle) was less than 10 minutes during the summer when demand was high, and substantially more in the winter when the demand was lower. Rate of flow is more difficult to measure through the sedimentary phase. Sedimentary turnover time in freshwater ecosystems can be measured in weeks, and in terrestrial ecosystems it may range up to 200 years!

The availability of phosphate for organisms depends upon the rate of cycling through both the organic and sedimentary phases. Thus, one of the primary reasons phosphate is frequently a limiting factor in ecosystems is the slowness of turnover in the sedimentary phase. The demands of the biota then by necessity must be met by cycling phosphate through the organic phase rather than by releasing phosphate from the sedimentary phase.

d. OTHER NUTRIENT CYCLES

As mentioned earlier, there are cycles for all of the other nutrients used by organisms, as well as some not used. Most of them are simple, such as the phosphorus cycle of Figure 14-19, with an essentially complete cycle in the sedimentary phase. The rate at which most nutrients cycle through an ecosystem and their availability to organisms depends upon their abundance, and a wide range of environmental variables for the area in question. We consider some additional specific cases in Chapter 22 in discussing the utilization of forest ecosystems.

e. WATER CYCLE

The flow of water through an ecosystem differs fundamentally from the cycles previously discussed. Eighty percent of the total solar energy absorbed goes to evaporate water, which condenses around particles of dust in the atmosphere and then falls as rain or snow (Fig. 14-20). Water is evaporated from the oceans, bodies of fresh water, or moist soil surfaces. It is also transpired from plants; in fact, the volume of water that passes daily through the roots, xylem, and stomata is far greater than the volume of the plant itself. A fair-sized maple tree has been observed to lose 400 liters of water *in a day*. If an annual rainfall of 40 in. (100 cm) falls on a cornfield, about a third or the equivalent of 40 cm is transpired back into the atmosphere. Thus plants have an important role in humidifying the atmosphere. Also the rate of water cycling between the earth's surface and the atmosphere is very rapid. The average turnover time for atmospheric water is about 11.4 days, or, in other words, the equivalent of all the water vapor in the entire atmosphere falls as precipitation and is reevaporated more than 32 times per year.

On the earth's surface about 62% of the annual precipitation returns to the atmosphere by both transpiration and evaporation, whereas some 30% is diverted into stream flow either by runoff from the saturated surface soil or by gradually percolating downward to the *water table* (the top layer of the groundwater). This groundwater is the vast natural reservoir that maintains the dry-weather flow of streams, helps maintain the level of lakes, and is the source of water for wells. Groundwater seeps or slowly flows to rivers and the ocean. Often, groundwater is tapped directly by deep wells and is an important source of water for irrigation, industry, and domestic purposes. In many parts of the world water requirements already exceed natural stream flow, and groundwater is being tapped faster than it can be replenished from precipitation. With increase in population more water will be required for all purposes and

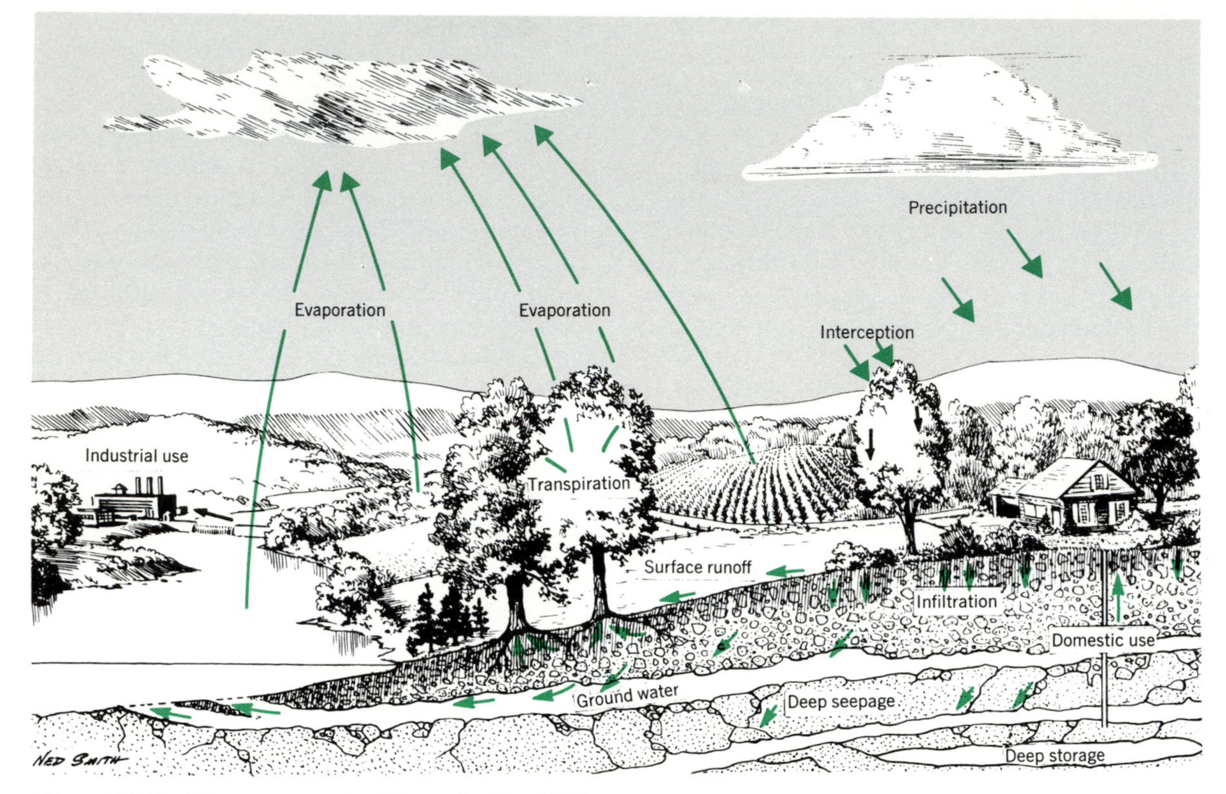

Figure 14-20 The water cycle. (From Smith, 1966)

many authorities think that water is one of several factors that will limit the number of human beings the earth can support.

8. ECOSYSTEM DYNAMICS: SUCCESSION

As natural communities functionally related to environmental conditions, ecosystems are dynamic. As previously noted, the reasonably directional and orderly development of parallel and interacting changes between communities and environmental conditions is called *succession*. Through the course of succession ecosystem characteristics such as community structure and composition, community energetics (productivity and biomass), and the nature of nutrient cycling change. This change results from modification of the physical environment by the community and culminates in a maturity or a relative steady-state condition called *climax*. Where there has been no previous soil development, that is, the vegetation develops on bare rock surfaces such as cliffs or those bared by glaciation or created by volcanic activity, flooding, and so on, the process of change is called *primary succession*. The process is relatively slow because it takes time to develop a soil profile (see Chapter 15). Where the vegetation has been destroyed due to clearing of the land for agriculture or from fire or other events, the process is called *secondary succession.* Development of a mature ecosystem takes less time from a secondary successional sequence than from a primary one because the soil is already developed.

Although the natural replacement of one community by another can be easily observed in many areas, and Anton Kerner von Marilaun had described it well in Austria in 1863, the greatest interest in this phenomenon developed in North America only after the beginning of the present century. Some of the first studies documenting primary succession were those occurring in bogs, lakes, or on sand dunes where the changes were clearly evident. Changes that took place following clearing of land for agriculture or after fire were some of the first noted secondary successional sequences. These are the kinds of changes we must understand in order to make wise decisions with regard to land utilization today.

In order to understand the successional process better, let us look at an example of primary and then secondary succession. First, let us examine a classic example—that of a bog lake. Bog lakes generally have developed in small depressions in the glaciated areas of northern Europe and the north-central and northeastern United States. They therefore tend to occur in the northern coniferous or mixed hardwood forests (Fig. 14-21). Although bog lakes differ from one another, the following description is a more-or-less typical successional sequence for the north-central United States. The primary sequence of events is shown in Figure 14-21. The open water of a bog lake is tea-colored from the humic acids derived from the surrounding floating vegetation mat characterized by peat mosses (*Sphagnum* spp.) and grasslike sedges. Other distinctive plants such as cotton grasses, orchids, and carnivorous plants (such as sundews) also occur on this mat. The mat grows by increasing both its thickness and its leading edge into the water. The thickening of the mat allows its occupation by shrubby members of the heath family. Many of the light-requiring plants of the open bog disappear in the shade of the shrubs and the mat of soil and roots becomes more massive until the bog no longer "quakes" when it is walked upon. The shrub stage then is invaded by several species of trees, of which the larch (*Larix laricina*) is most typical. Under the shade of the trees most of the shrubs die and are replaced by wet-forest undergrowth. The larch then is replaced by other trees such as spruces, beeches, and maples that dominate the prevailing forest of the area. Thus the first floating mat is "grounded" with a continuous deposit of peat supporting a mature and stable forest.

Our example of secondary succession is that which follows farm abandonment. It is usually called "old-field succession." The general pattern may be described by the following stages (Fig. 14-22). In the first and second years annual weeds dominate the newly forming communities (*Digitaria,* crab grass, and *Rumex,* sorrel, are good examples). Then perennial herbs such as goldenrod (*Solidago*) and broomsedge (*Andropogon*) and other perennial grasses are dominant for 10 to 15 years. This meadow stage may be followed by a shrub stage (from 15 to 20 until 30 to 40 years), or it may be skipped and pines may seed into the meadow. The grasses then succumb under the shade of the pines, as they close their canopy. The pine forest, which may then persist up to 100 years, is invaded by young oaks (*Quercus*) and later by hickories (*Carya*). With freedom from disturbance this forest returns to a mature oak-hickory forest in from 150 to 200 years.

Although each successional sequence may vary in its details, some general patterns of changes from successional stages to mature stages may be noted. Table 14-2 reduces the many ecosystem characteristics to the most essential ones. In the early stages of succession the rate of gross primary production (P) ex-

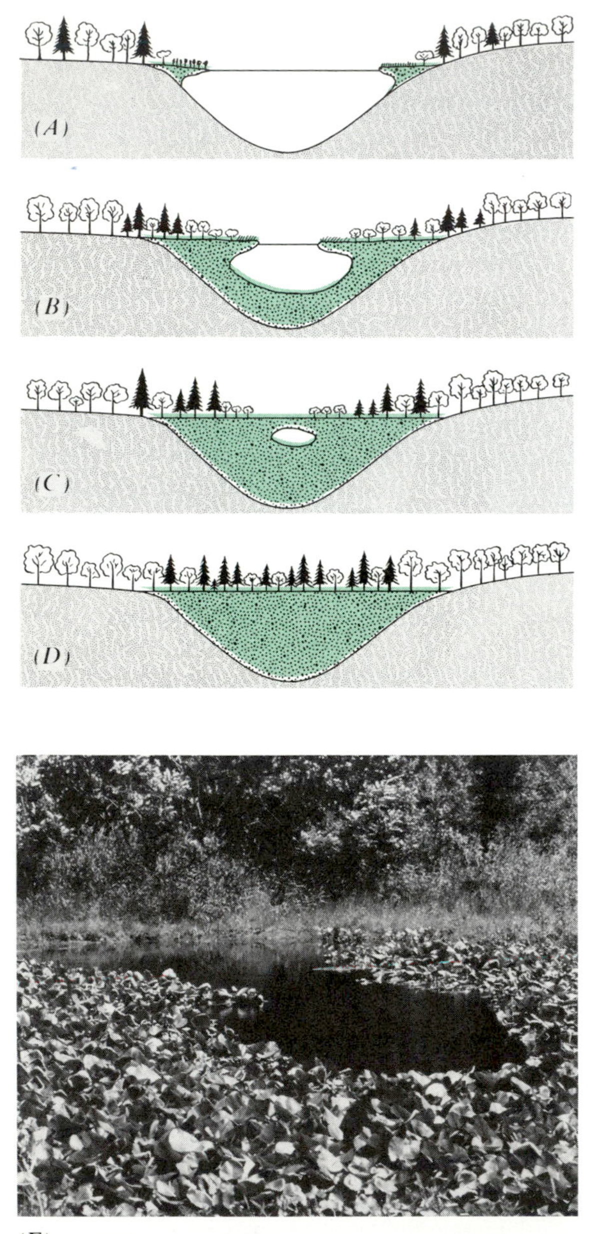

(E)

(F)

(G)

Figure 14-21 Bog lake succession. *A*. Floating mat of vegetation advances over the water surface in a small lake in a cool, humid climate. *B* and *C*. As the mat advances further and the lake ages, barely decomposed organic matter (peat) accumulates in the lake basin until (*D*) after some thousands of years the lake will be converted to forest. *E*. Photo of bog lake with various zones shown in *(A)* through *(D)*. *F*. Plants with surface-floating leaves that grow across surface of pond and choke out bottom-dwelling plants. *G*. Grasses, sedges, and cattails *(A, B, C)*, which are the peat formers preceding the invasion of shrubs and forest.

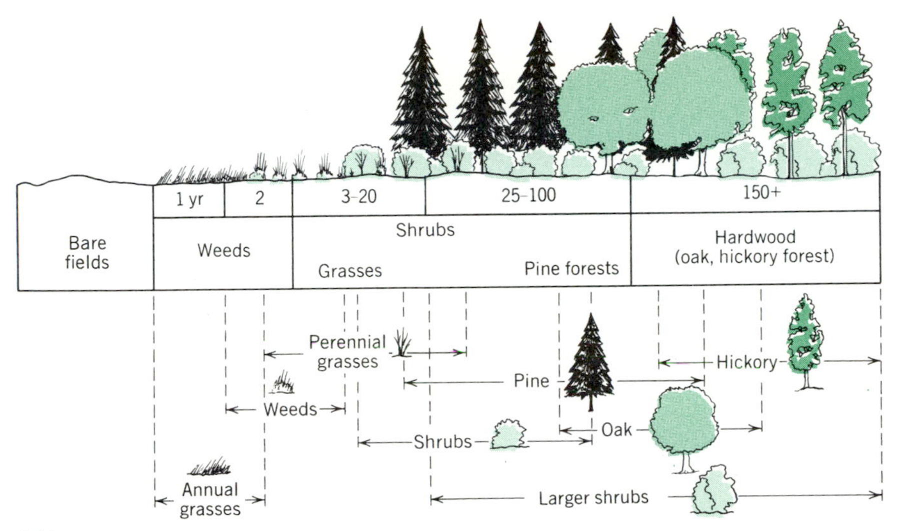

(A)

Figure 14-22 *A*. Diagram of typical stages in secondary succession following abandoment of cropland in North Carolina. B.-E. Photos illustrating stages of this old-field succession B. Tall weed stage, C. Perennial grass (broomsedge) has eliminated weeds, but loblolly pine is invading, D. Young stand of loblolly pine with plow furrows of abandoned field still visible, and E. Pines now succeeded by oak-hickory climax forest.

(B)

(C)

(D)

(E)

TABLE 14-2

Summary of Changes of Some of the Essential Ecosystem Characteristics from Early to Mature Stages of Succession

Ecosystem Characteristics	Successional Stages	Mature Stages
	Community Energetics	
1. Gross production/community respiration (P/R)	Greater or less than 1	Approaches 1
2. Gross production/biomass (P/B)	High	Low
3. Biomass supported/unit energy flow (B/E)	Low	High
4. Net community production	High	Low
5. Food chain	Linear, predominantly grazing	Weblike, predominantly detritus
	Community Structure	
6. Total organic matter	Small	Large
7. Inorganic nutrients	Extrabiotic	Intrabiotic
8. Species diversity (variety and equitability)	Low	High
9. Patterning of diversity (stratal and spatial)	Poorly organized	Well organized
	Nutrient Cycling	
10. Nutrient cycles	Open	Closed
11. Exchange rates between organisms and environment	Rapid	Slow
12. Role of decomposers in nutrient regeneration	Unimportant	Important

Modified from E. P. Odum, *Science* 164, 1969.

ceeds the rate of community respiration (R) so that the P/R ratio is greater than 1. Later it decreases to 1; that is, the energy fixed tends to be balanced by the energy cost of maintenance (total community respiration) in the mature ecosystem. As long as P exceeds R, organic matter and biomass (B) will accumulate with the result that P/B will tend to decrease. Thus in theory, the amount of biomass supported by the available energy flow (E) increases to a maximum in the mature stages. Therefore, the net annual community production is generally high in young stages and becomes lower in mature stages of succession. With the development of the ecosystem, changes also occur in the food chains. These networks are relatively simple and linear in the early stages as a consequence of low diversity. Primary production is used predominantly by way of grazing chains. In contrast, food chains become complex webs in mature stages with most of the biological energy flow following detritus pathways. For example, in a mature forest less than 10% of annual net production is consumed in the living state; most of it is utilized as dead matter (detritus).

In terms of community structure, changes in species diversity are among the most controversial trends. Distinction between different kinds of diversity is important since they may not follow parallel trends in certain successional lines. One kind of diversity is the variety of species (as either species/number ratio or species/area ratio); another kind is equitability or evenness in the proportion of individuals among species. Although an increase in the variety of species occurring together, along with increased evenness of distribution within the species, can generally be assumed during succession, other community changes may work against these trends. For example, increases in the size of organisms, or in the length and complexity of life histories, or in interspecific competition are trends that may reduce the number of species that can live in a given area. Therefore, whether or not species diversity continues to increase during the successional sequence appears to depend on whether the increase in potential niches resulting from increased biomass, stratification, and other results of organization exceeds the countereffects of increasing size and competition. There have been no catalogs as yet for all of the species in any sizable area and certainly not for a successional sequence. Data presently are available only for trees, insects, birds, and others. The cause-and-effect relationship between diversity and stability is not yet clear and needs to be studied in considerable detail. If indeed it can be shown that biotic diversity increases physical stability within the ecosystem, then it will become an important guide in a variety of conservation practices.

Another important trend during succession is the tightening of the cycling of the major nutrients such as nitrogen and phosphorus. Mature systems, in contrast to successional ones, have a great capacity to take up and hold nutrients within the system.

CHAPTER 15

The Soil As a Plant-Related System

We have seen how de Saussure (1804) demonstrated that the mineral contents of plants were derived from the soil and how the subsequent understanding of what elements are essential contributed to the development of fertilization practices used in modern agriculture (Chapter 8). At the turn of the twentieth century, however, V. V. Dokuchaev, K. D. Glinka, and others in Russia recognized that soil is more than a source of mineral nutrients and water for the plant, and founded soil science upon the precept that soil is a complex, dynamic system. In fostering this systemic approach to the study of soils, these Russian pioneers, along with E. W. Hilgard and Hans Jenny at the University of California, developed an understanding of the interacting factors involved in formation of a soil and its maintenance. They pointed out that soil is formed through the weathering of rocks at the earth's surface. This decomposed mineral material provides a substrate for the growth of organisms that in turn modify that substrate. Thus the primary succession of an ecosystem (Chapter 14) proceeds from pioneer plants on a bare rock surface to a highly developed distinctive community on a biologically molded soil. Both the vegetational and soil development, of course, occur within a time framework. Thus, there is a close general relationship between the major types of climate, vegetation, and soil. This understanding led to the recognition of basic soil-forming processes and classification of major soil types or "great soil groups" based upon characteristics of the soil profile.

1. THE SOIL PROFILE

As a result of the leaching action of the water on the rocks of the earth's surface and the developing organic material, a *soil profile* develops in which three basic zones or *horizons* may generally be recognized (Fig. 15-1). These zones may be seen if you dig a trench in any well-vegetated area. At the surface there is the litter of dead or decaying plant parts, and underneath are zones, some of which are sharply defined whereas others merge gradually one into another. These zones have been formed by weathering processes working down from the surface to unweathered rock below. The thickness of earth affected by these processes constitutes the soil. Underneath is the rock from which the soil was made, that is, the *parent material.*

Underneath the litter of plant parts, the top horizon of mineral soil is colored and structured by the organic particles mixed into it by soil animals, such as earthworms, or by roots and

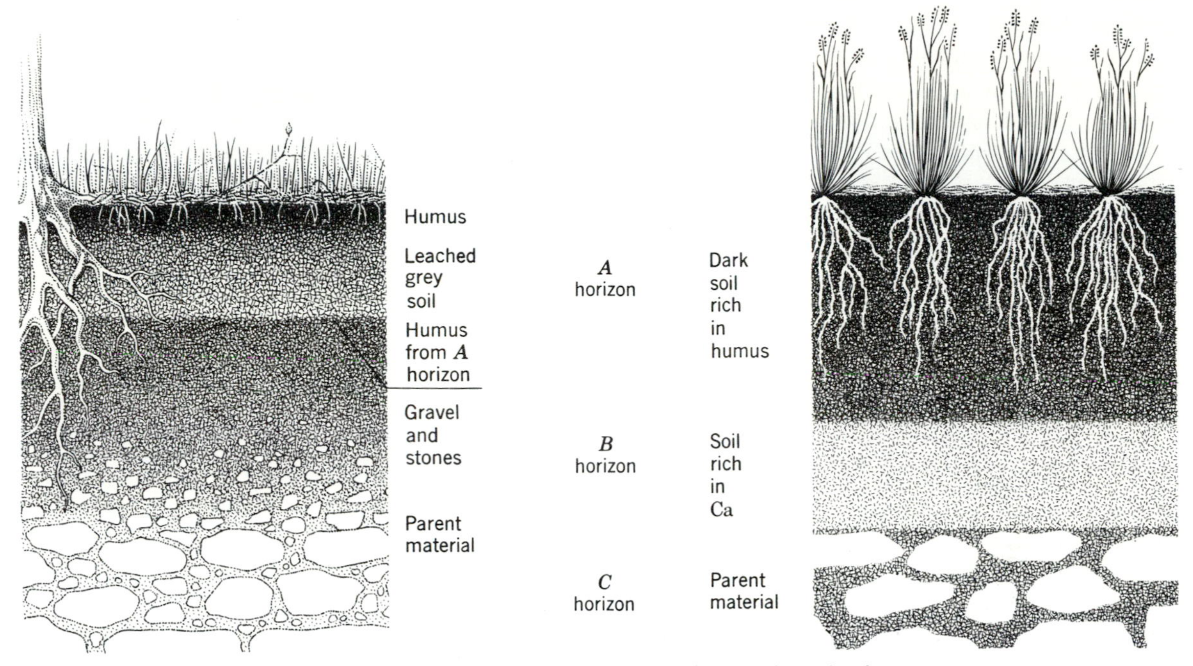

Figure 15-1 Schematic soil profiles showing the three basic horizons: the *A* horizon or topsoil, the *B* horizon or subsoil, *C* horizon, the substratum of relatively unweathered rock or parent material. The profile (left) is typical of a podzol soil and (right) of a chernozem soil. (From Gruelach and Adams, 1976)

organic materials produced through decomposition. The percolating water dissolves both inorganic and organic compounds from the litter and inorganic materials from the decomposing rock (composed of minerals) and carries them to deeper horizons. Percolating water also removes fine particles, such as clay and colloids, and carries them downward. This zone at the top of the profile that is being constantly leached is called the *A horizon*. It also is what we refer to as "topsoil." Underneath it is a horizon or group of zones in which clay and colloidal particles have been deposited in the spaces of the deeper soil, as in a filter, and solutes are redeposited over the surface of the same filter bed. The details of this process of redeposition are not always known but its occurrence is evident. This region of deposition is called the *B horizon*. Underneath it is the essentially unaltered parent material, which is referred to as the *C horizon*. Beneath this is the bedrock itself. There may be several subdivisions of these three basic horizons by which individual soil profiles are defined and described. Characterization of the zones depends upon changes in physical characters, such as texture and structure, and chemical characters, such as accumulation of the colloidal materials and salts and the pH. Soils vary in thickness from less than 1 ft (30 cm) in arctic regions to more than 50 ft (almost 20 m) in the tropics. In temperate regions the average soil ranges in depth from 2 to 5 ft (60 to 150 cm).

2. TEXTURE AND STRUCTURE

Texture of a soil is determined primarily by the size of the mineral particles (and to some extent by the organic particles). The interaction of water and temperature determines the degree of chemical weathering, that is, the decomposition of rocks to different-sized particles. Most soils contain a mixture of particles of different sizes and are named according to those that predominate. The following classification, based on diameter of the fragments in microns, is used for the smaller-size particles:

Coarse sand	200–2000 μ (0.2 to 2 mm)
Fine sand	20–200 μ
Silt	2–20 μ
Clay	Less than 2 μ
Colloids	Less than 1 μ

Soils are then classified according to the proportions of clay, sand, and silt. A *loam* is a mixture of sand, silt, and clay particles that exhibits light and heavy properties in about equal proportions. In most cases, however, the quantities of sand, silt, and clay present require a modified name for the loam. Thus a loam in which the sand is dominant is classified as a *sandy loam;* in the same way there are *silt loams, clay loams,* and *sandy clay loams.*

The *structure* of the soil is the way in which sand, silt, and clay are lumped together in granules or blocks. These structural units result from the binding together of particles by either clay or organic colloids. The texture of a soil does not change much with usage, but the structure may be modified by poor management. For example, when the soil is wet, compaction of the fine-textured components easily occurs with use of heavy farm machinery.

Texture and structure are both of great importance in determining the amount of air and water in the soil. A good garden soil is a sandy loam, which typically has about 45% mineral and 5% organic particles, 25% air, and 25% water.

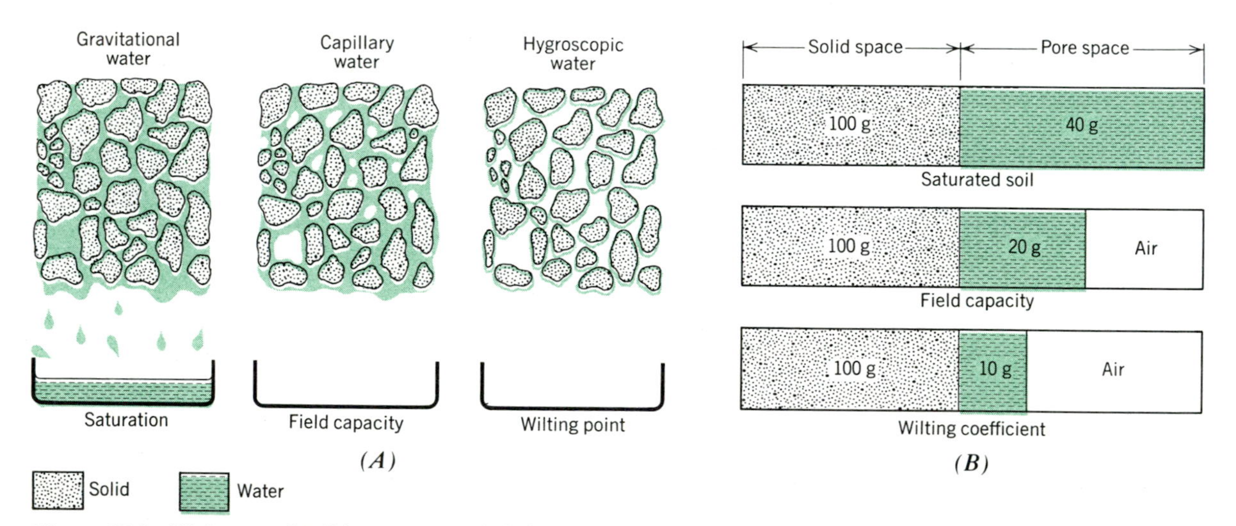

Figure 15-2 Volumes of solids, water, and air in a silt loam with good structure at different moisture levels. *A*. The situation when a representative soil is completely saturated, which usually occurs for short periods during a rain or during irrigation. *Gravitational water* will soon drain out of the larger pores; the soil is then said to be at *field capacity*. Plants will remove the *capillary water* quite rapidly until the *wilting coefficient* is approached. Permanent wilting occurs at this point even though there is still considerable moisture in the soil. *B*. Approximate proportions of water and air in each stage. (Adapted from Brady, 1974)

The importance of air in the soil is often underestimated. Roots respire and thus oxygen is needed, but simultaneously both roots and microorganisms are increasing the amount of CO_2 as they are depleting the O_2. The interchange of these gases may be impeded by inadequate pore space in the soil. If the particles are large, there will be adequate space for gas exchange, but if they are too small, this may cause difficulty for growth of some species. Gases may particularly be impeded in a heavy clay soil that becomes compacted when wet. Thus a flooded or inadequately drained soil has poor aeration. This may result in "root rot," in leaves' becoming chlorotic, in stopping growth, and ultimately in death of the plant. However, we all know that certain plants exist in bogs or swamps where roots are perpetually under water. These plants are both structurally and physiologically adapted to these unusual soil conditions. For example, they may develop large intercellular spaces in the stem and roots, allowing oxygen to diffuse downward from the aerial parts, as in the case of lowland rice.

Additionally, it is obvious to anyone who has germinated an avocado seed or sweet potato tuber, or rooted cuttings in a glass of water, that complete submersion of roots in water may not be injurious. In fact as we saw in Chapter 8 healthy plants can be cultivated in nutrient solutions, although it is often necessary to provide aeration of such culture solutions.

Water can be present around particles of soil as *hygroscopic* water, *capillary* water, or *gravitational* (free) water (Fig. 15-2). Hygroscopic water is bound so tightly to soil particles as a film that it cannot be used by plants, whereas gravitational water drains rapidly through the soil and out of the zone where roots occur. Therefore, capillary water is that most effectively used by plants. It is also the medium in which occur many of the microorganisms that are the decomposers (Chapter 14).

The amount of water that a soil holds against the gravitational force over several days is known as its *field capacity* (Fig. 15-2). Because of their larger surface, clay soils are able to hold a much greater amount of water against the force of gravity than coarser textured soils—often three to six times as much water as a comparable volume of sand (Fig. 15-3).

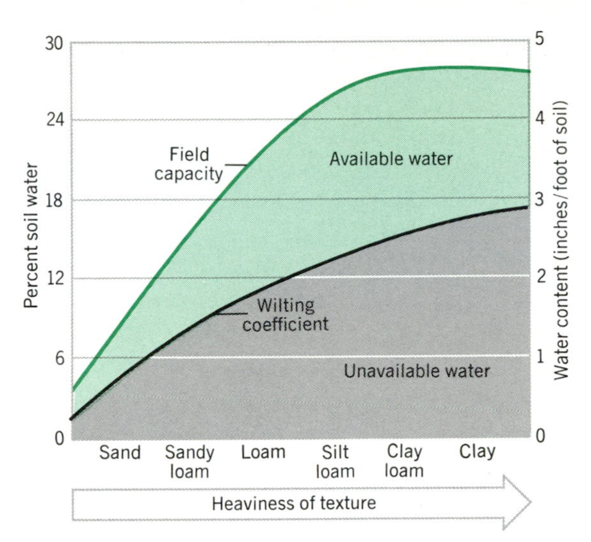

Figure 15-3 General relationship between soil moisture characteristics and soil texture. Note that the wilting coefficient increases as the texture becomes heavier. The field capacity increases up to the silt loams, then levels off. These are representative values; individual soils would probably have values different from these shown. (From Brady, 1974)

When a plant has used all of the available capillary water it exhibits symptoms that collectively are called *water stress* (Chapter 7). The moisture content of a soil at which a plant can no longer remove water from it is defined as the permanent wilting point or percentage (PWP) (Fig. 15-2). The permanent wilting point varies with the species; many native plants, for example, those adapted to arid conditions, can extract water from a soil that has reached the permanent wilting point for most other species. The bioassay test for PWP was worked out relative to the behavior of a cultivated plant, the sunflower, and thus is not always applicable for many native species, or even for other cultivated ones.

In a plant under water stress, photosynthesis usually decreases quite dramatically as the stomata close to decrease transpiration. As a result carbon dioxide, as well as water, may become a limiting factor for photosynthesis. This fact influences the planning of irrigation schedules for crops, so that active photosynthesis will continue and thereby produce high yields. Since rates of transpiration and photosynthesis vary with different kinds of crop plants, and under different climatic conditions, irrigation must be carefully regulated accordingly. However, as with all other limiting factors (Chapter 5), net photosynthesis and subsequent growth can be increased with available water only up to a point, beyond which more frequent application will produce no further increase in photosynthesis. In fact, prolonged flooding can induce root rot, a difficulty due to poor aeration.

3. THE CHEMICAL AND PHYSICAL ROLE OF COLLOIDS

We mentioned that structure of soils is in large measure dependent upon colloidal binding together of the mineral particles. Colloids are the seat of chemical activity in soils. Colloids may originate from either a rock (mineral) or an organic source. Soil colloids are characterized by (1) small size (less than 1 μ in diameter), (2) very large surface area per unit weight and, (3) their negative surface charge, to which cations (ions with a positive charge) and water are attracted. Both clay and humus colloids are important and the best agricultural soils contain a good balance between the two. On a weight basis, humus colloids have a greater (2 to 50 times) nutrient-, and water-holding capacity than clay colloids. However, clay colloids are generally present in larger amounts and are more stable than humus colloids; therefore, their total contribution is essentially the same. In form, the clay system is crystalline and made up principally of silica, aluminum, and oxygen, whereas the humus system is noncrystalline and consists of carbon, hydrogen, and oxygen.

Clay colloids are laminated but flimsy and fragile, like wet sheets of paper and cardboard, with torn edges and holes punched in them (Fig. 15-4). Their individual sizes and shapes depend upon their mineralogical organization and conditions under which they have developed. They are crystals made up of silicon and aluminum lattices with unsatisfied oxygen (O^{2-}) and hydroxyl (OH^-) valences exposed, thus attracting cations that can be bound weakly to the surfaces (Fig. 15-4). All clay particles, merely because of their small size, expose a large amount of external surface, but in some (Fig. 15-4) there are extensive internal surfaces as well. The internal interface occurs between the plate-like crystal units making up each particle. The external surface area of 1 gram of colloidal clay is at least 1000 times that of 1 gram of coarse sand. Although humic colloids are made up of organic debris (C,H,O) they too have unsatisfied O^{2-} and OH^- valences that attract cations. This surface bonding is known as *adsorption.* Because the bonding is somewhat loose, cations dissolved in the soil water can be exchanged for ions adsorbed onto the clay particles. This process is called *cation-exchange.*

Because most rainfall is lost either through transpiration, evaporation, or percolation, it is not possible for nutrients to persist dissolved in water, as in aquatic ecosystems. If there were no mechanism for nutrient storage in the solid fraction of the soil, soluble nutrients would be quickly leached out. But adsorption of these nutrients on to soil particles provides a means whereby they can be stored in a form available to living organisms. While all cations may be adsorbed by colloidal micelles, certain ones are especially prominent under natural conditions. For humid regions these in the order of their numbers are: H^+, Ca^{2+}, Mg^{2+}, K^+, NH_4^+ and Na^+. In fact, humid region soils are considered as having a calcium–hydrogen complex. Organic acids from the decomposing litter (see Chapter 14 for the role of microorganisms as decomposers) and exudates from roots release hydrogen ions. In the same way, soil water rich

(A)

Figure 15-4 *A*. Crystals of a silicate clay mineral (kaolinite) from Illinois; magnified about 1000 times. *B*. Diagram of a silicate clay crystal (micelle) similar to the above showing its sheetlike structure, its many negative charges, and swarm of adsorbed cations. To the right is the edge of the crystal which is enlarged to display the negatively charged internal surface to which cations and water are attracted. (From Brady, 1974)

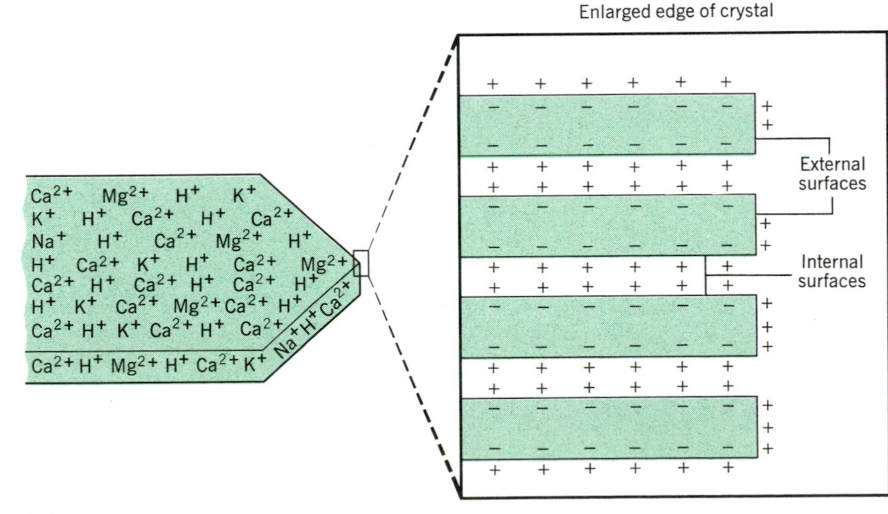

(B)

in dissolved carbon dioxide (from respiration of roots and organisms in the litter) and, hence, carbonic acid.

$$CO_2 + H_2O \rightleftharpoons H_2CO_3$$

may also supply hydrogen ions:

$$H_2CO_3 \rightleftharpoons H^+ + HCO_3^-$$

For example, Ca^{2+} adsorbed on colloidal micelles would tend to be replaced by H^+ because of mass action effects and also because under comparable conditions ionic hydrogen is adsorbed more strongly than ionic calcium. Calcium is then available in the soil water to be absorbed by the plant root, or it may be leached into the groundwater as calcium bicarbonate $Ca(HCO_3)_2$. Thus areas of high rainfall tend to have acidic soil because of greater availability of hydrogen ions to replace many of the other ions originally available on the colloidal particles.

If rainfall is low and evaporation is high, few acids are produced and ions are not readily leached to the groundwater. For well-drained arid-region soils the order of exchangeable ions is usually Ca^{2+}, Mg^{2+}, Na^+, and H^+ last. Thus arid-region colloids commonly are dominated by Ca^{2+} and Mg^{2+}, but when drainage is impeded and alkaline salts (e.g., sodium chloride and calcium carbonate) accumulate, the concentration of Na^+ is likely to exceed even that of adsorbed Ca^{2+}.

4. SOIL CLASSIFICATION

There are several different systems of classifying soils. The two most widely known and used are (1) zonal system and (2) comprehensive system (U.S. Soil Taxonomy). The zonal system emphasizes the genesis of soils and thus relates them to the characteristics of the major ecosystems in which they have developed. Some detail thereby has been sacrificed in description of soils from specific sites. On the other hand, the comprehensive system (developed by the U.S. Department of Agriculture) stresses precision in describing local field conditions but not the soil-forming processes. Its greatest utility lies in specialized, detailed research. Both of these systems are now commonly referred to in literature concerned with land-use planning, and it is advantageous for any citizen to know about them. Because in this book we are interested in a global analysis of ecosystem usage, we utilize the zonal system of classification. However, we often indicate the Soil Order from the comprehensive system so that you may become familiar with some names of the correlative taxonomic groups.

First, it is helpful to distinguish between a young or *immature* and a *mature* soil. Like an ecosystem, soils go through a "succession" in which the types of colloidal materials change from those derived directly from their parent material to those that are maintained in dynamic equilibrium with soil development. Also other soil characteristics, such as available nutrients, are correlated with the type of vegetation occupying the soil. If the vegetation is in an early stage of succession, then soil conditions will change with time as the vegetation matures. The maturation of both takes time, but, as with most natural processes, there is no exact timetable. It may take different lengths of time for a given soil type to develop on two different kinds of parent rock, or it may take different lengths of time for a given rock type to yield mature soils of different types in varying climatic regions and with different vegetative covers.

In areas with good drainage, almost all of the characteristics of mature soils are a function of the climatic regime, which also determines the major ecosystem type (and the soils may generally be distributed throughout the ecosystem, except where extreme differences in parent material occur). These are the *zonal* soils. The modes of formation of some mature soils, however, do not follow the regional pattern and these may be either *azonal* or *intrazonal*. Soils on very steep hillsides, generally little more than broken-up rock, are typically azonal. In this situation the rate of erosion is so fast that the normal soil-forming processes never have sufficient time to proceed. The most important azonal soil for human activities is the *alluvial* soil, which originates from sediments carried by rivers and deposited during periods of flooding. Unlike most other azonal soils, an alluvial soil is rich in colloids, and thus probably rich in nutrient materials. Most widespread alluvial soils are found in the floodplains of great rivers and have been among the most fertile soils throughout history. Unfortunately, many areas of alluvial soils, such as the delta regions of the Tigris and Euphrates and Nile, are warm areas in which the soils are easily leached. Flood control projects in these areas may then have the undesirable result of not allowing replacement of nutrients by flooding, when they have been leached out under the climatic regime.

Intrazonal soils develop from local geologic conditions that preclude normal soil-forming processes. For example, waterlogged soils are not aerated and thus the normal aerobic decomposition of detritus cannot take place. At the other extreme are soils formed in arid regimes characterized chiefly by their very high salt content. When there is improper drainage, the soil may become waterlogged following heavy rains. Because of the arid conditions, the excess water evaporates rather than draining off, and the salts contained in the water build up in the soil over years of evaporation to form salt lakes or salt flats (sometimes called by their Spanish name *playas*).

5. MAJOR SOIL-FORMING PROCESSES

The processes of soil formation modify weathered rock (parent material), eventually giving it the acquired characteristics that distinguish soil from parent material. The most important soil-forming processes that control the development of the major soil types of the world (Fig. 15-5) are the following.

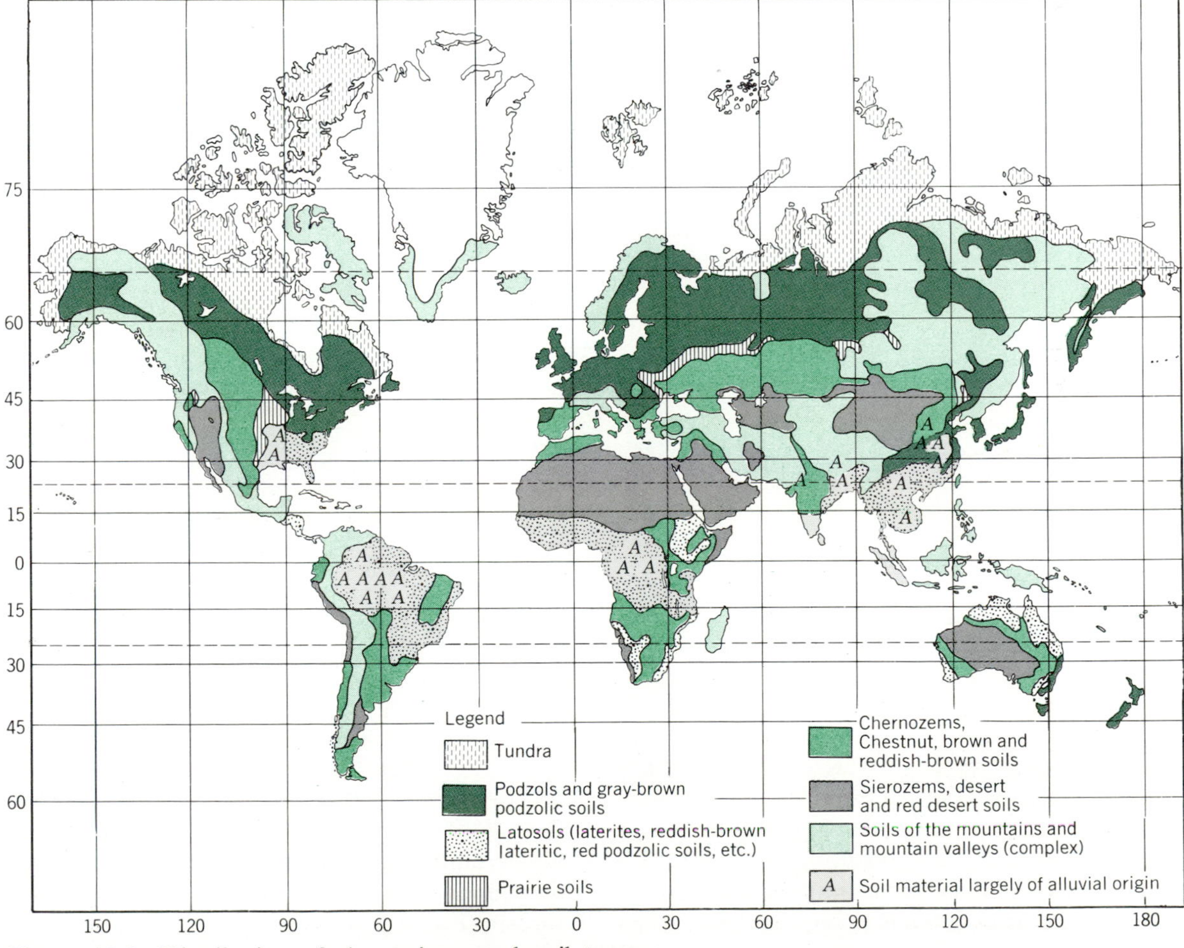

Figure 15-5 Distribution of the major zonal soil types.

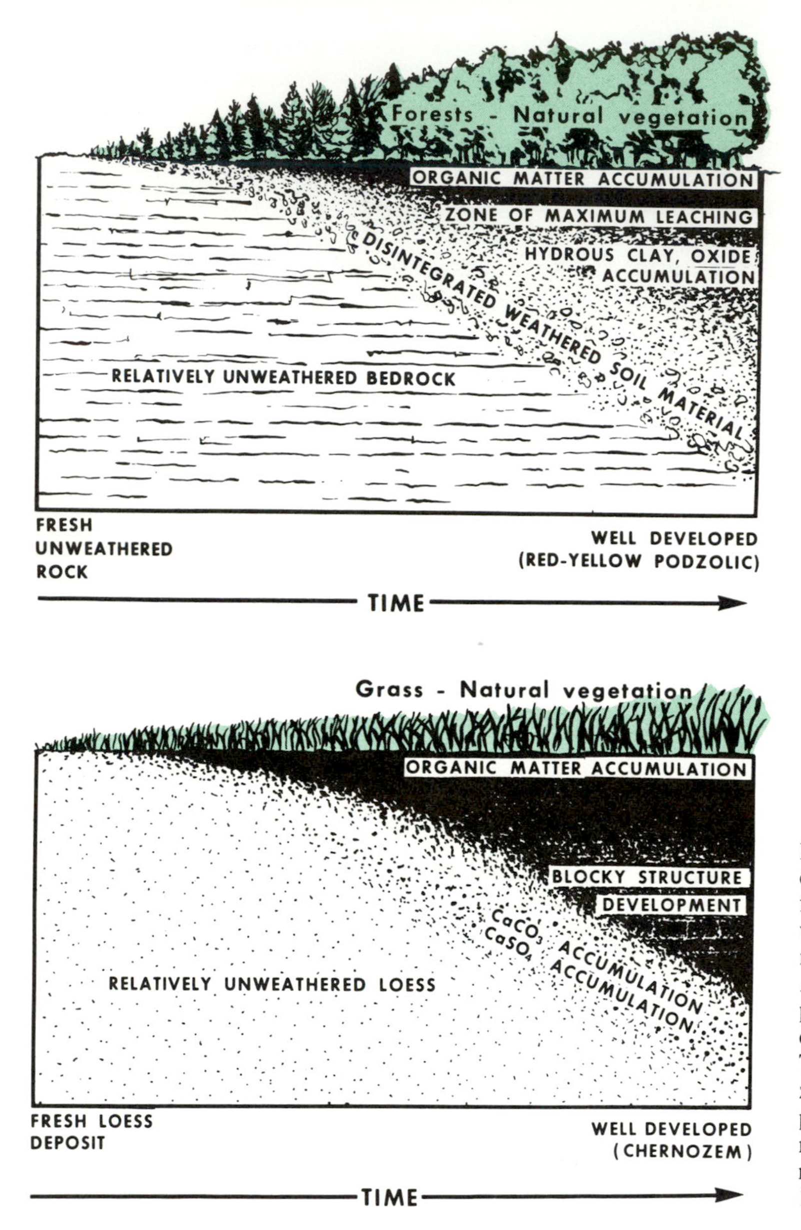

Figure 15-6 Diagrams showing the development of two typical soil profiles. *A*. A podzol developed from weathering in place of solid rock in a moist climate supporting a forest. *B*. A chernozem from wind-blown silt particles (called loess) in a relatively dry climate supporting grassland. Thus, the profile characteristics of a zonal soil type are dependent on the parent material, the vegetation, climate, and the time course of development. (From Buchman and Brady, 1969)

a. PODZOLIZATION

Podzolization is prevalent in cool, humid areas. A "true" podzol is differentiated from other members of the podzolic group by the severity of the process, resulting in the profile's being more distinct both in appearance and in its physical and chemical properties (Fig. 15-6). A *true* podzol characteristically develops under coniferous forest cover. Podzolic-type soils, however, occur under a range of vegetation covers including deciduous forest, pasture, and arable land where podzolization is not strongly developed, and even under some tropical forests (Fig. 15-5).

Podzolization involves the development of an extremely acid humus layer, resulting from slow accumulation of litter. Rainwater percolating through these surface horizons acquires carbonic acids, humic acids, and other organic acids as breakdown products from the plant debris and becomes strongly acid. Under these strongly acid conditions these breakdown products form complexes with iron that are then removed as the solutions percolate downward. In this state, iron is said to be "mobilized." In the same manner aluminum is also mobilized, leading to the breakdown of the aluminum silicate clay minerals. Therefore, there is a tendency for quartz (silica) to collect in the immediate subsurface, forming the bleached gray horizon from which the name "podzol" (meaning "ash" in Russian) is derived. The iron oxides, mobilized from the surface layers, together with the aluminum and organic matter are moved downward in the soil solution and then deposited in the *B* horizon.

In the production of the *gray-brown* podzolic soils, conditions are not as acid as in the true podzol. A richer soil flora and fauna aids in more rapid decomposition of the plant debris, and the unaltered clay is washed into the lower part of the *B* horizon.

b. FERRALLITIZATION

Ferrallitization (previously called "laterization") is the process in which the parent materials are changed into a soil consisting primarily of iron and aluminum sesquioxides. In the humid tropics, conditions of abundant rain, high temperatures, and long periods of weathering and soil formation have combined to produce a deep and strongly weathered parent material (Fig. 15-5). The rapid decomposition of litter and apparently the activity of mycorrhizal fungi maintain almost direct cycling of nutrients between soil and vegetation. In the moderately acid to neutral conditions in the soil, silica is removed in preference to iron and aluminum, which gradually accumulate and oxidize to the sesquioxides. Somewhat different conditions occur in tropical regions with a pronounced dry season. The soil is heavily leached in the rainy season but strongly dried during the dry season. These alternations tend to increase the rate of chemical activity and movement of soil constituents and can produce iron oxide crusts known as *laterites*.

c. CALCIFICATION

The process of calcification is characteristic of low rainfall areas, particularly in interior continental regions (Fig. 15-5). Leaching is slight, and although downward movement does take place the soluble compounds are not removed from the soil profile. The soils are wetted only from 1 to 2 meters down when the moisture begins to reevaporate. Calcium carbonate accumulates in a horizon at the general level where water tends to stop downward percolation. As these soils are relatively unleached, the colloidal particles are dominated by calcium ions and to a lesser extent by magnesium ions. The presence of these ions has a stabilizing effect upon the colloids and their movement in the soil is inhibited. Also the natural vegetation is generally dominated by grasses, which tend to have extensive root systems. When these die they provide large amounts of organic matter and consequently colloidal material to the soil. The aerial parts of the plants also return nutrient cations to the soil surface. Winter frost and summer drought combine to limit the rate of decomposition, resulting in the development of a deep and nutrient-rich *A* horizon.

d. SALINIZATION

In arid climates the rainfall is low, irregular, and generally insufficient to remove soluble salts to the groundwater. The resulting *salinization* is a process of enrichment of soils with salts, due to evaporation of moisture from the surface of the soil. Salts in solution are drawn upward by capillary action and are then redeposited as the water is evaporated. This results in a surface encrustation of salt called a *white alkali* soil. When sodium ions dominate the exchange sites of the colloidal complex, the soil structure becomes dispersed. The dispersed humus (even though present in small amounts) gives a dark color and these soils are called *black alkali* soils.

6. MAJOR WORLD SOIL TYPES: USE IN AGRICULTURAL ECOSYSTEMS

As discussed earlier, the environmental controls of zonal soils are primarily climatic and therefore also are related to major "formational" ecosystem types. We discuss here six major classes of soils, each of which has many subclasses and widespread distribution (Fig. 15-5).

a. SOILS OF ARCTIC REGIONS: TUNDRA SOILS

Soils in the arctic regions are very thin and tend to be wet because of lack of drainage through the lower permanently frozen horizons. The organic content of the topsoil, however, is high. So far little agricultural usage has been made of these tundra soils beyond uncertain gardening and extensive grazing of such animals as reindeer and caribou. Relatively few crop plants with the ability to grow at low temperatures, such as the native plants can tolerate, have as yet been developed.

b. SOILS OF TEMPERATE FORESTS: PODZOLS

Podzols (spodosols in the comprehensive system) are our best-known soils. Western civilization was built on the podzolic group of soils, which supported agriculture during the development of industrialization in such areas as the northeastern United States and most of Europe and of Asiatic USSR. *True* podzols primarily occur in geologically young landscapes; therefore they are "rocky" and are often mixed with swampy soils, because the northern areas were recently glaciated. The natural vegetation is the extensive coniferous forest. Generally, relatively small areas of arable soils are intricately mixed with barren areas. Nevertheless, true podzols are responsive to management and can be made agriculturally productive. They can grow vegetables, especially potatoes, cabbage, and rutabagas, small grains (particularly rye), and good pasture feed and hay for dairy cattle. After clearing the forest, the first step in altering the soil for agriculture is liming (addition of calcium and magnesium) because the nutrient cations have been leached. It is also necessary to add nitrogen, phosphorus, and potassium. Thus by proper handling, true podzol soils can be farmed and thereby a good balance can be achieved between forestry, recreation, mining (all possible while maintaining the forest cover), and agriculture.

Until 150 years ago the *gray-brown* podzolic soils supported the heartland of Western Civilization. They have developed generally under temperate deciduous forests. Perhaps more is known scientifically about this type of soil than any other. Its outstanding feature is the wide range of crops (cereals, grasses, legumes, root crops, fruits, and vegetables) that grow well in it. If well managed, it is dependable, but with poor farming (neglecting to fertilize or to rotate legumes), the soil deteriorates to low levels of productivity. Thus, on what was originally one kind of soil, such as the gray-brown podzol, it is possible to find either very poor soils or excellent ones that have been made even more productive than they were under natural conditions.

c. SOILS OF WET TROPICAL FORESTS: LATOSOLS

Latosols (oxisols) are similar to podzols in being strongly leached; in fact, the red-yellow "podzols" are transitional between the latosols and podzols. Because of severe weathering in the tropics and a great diversity of seasonal combinations of rainfall, temperature, and humidity, probably more soil types occur there than in all of the regions elsewhere. The most common type of soil there is a red latosol, which also has many variations. Although the annual fall of leaves in tropical forests is high, they decompose so rapidly that only a very thin litter remains on the surface at any time (0.1 to 0.6 kg/m^2). The nutrient content of the litter is high, and, combined with large litter fall and rapid decomposition, rapid turnover of the nutrients is implied. About 14% of the nutrients in the annual litter fall is released to the soil or vegetation, compared to 6% in temperate deciduous forest and 4% in arctic forests. The tropical rain forest thus has a relatively rich nutrient economy perched on a nutrient-poor substrate. The *A* horizon is composed of a top few cm of highly granular, dark reddish-brown clay, below which occurs a zone well permeated with roots and insect holes. The main *B* horizon is characterized by a blocky red clay. At varying depths there is often a heavy band of quartz pebbles and stones, marking an old stone line once uncovered by soil erosion and later covered by wash from higher land.

Because of heavy precipitation and high temperatures throughout the year, at least in the equatorial tropics, these soils have been heavily weathered and leached of mineral nutrients. Yet they sustain one of the most luxuriant types

(*A*)

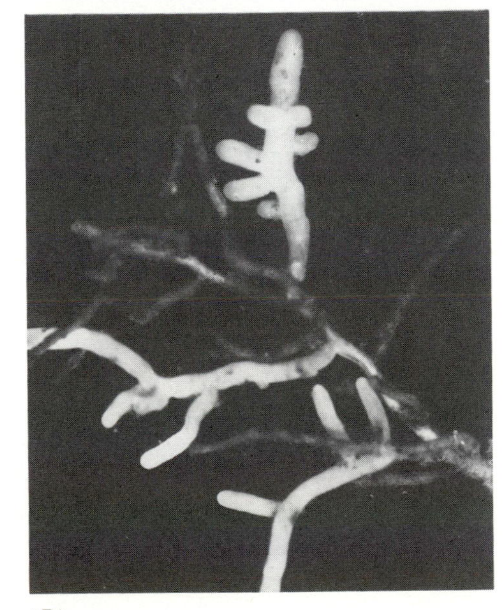

(*B*)

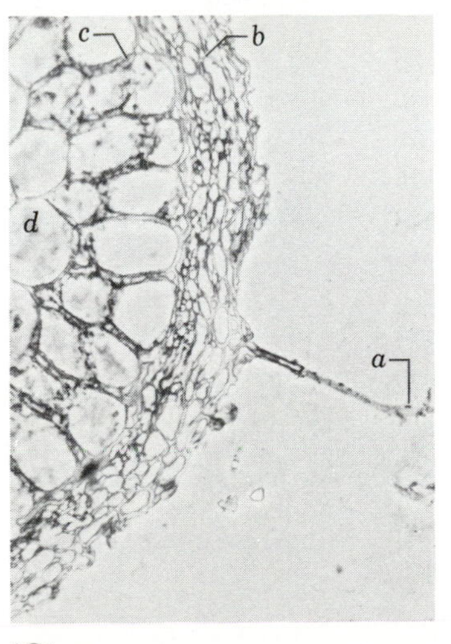

(*C*)

Figure 15-7 *A*. Ectomycorrhizal laterals on "old" roots of *Pinus strobus*. *B*. On *Betula alleghanensis*. *C*. Ectomycorrhiza of *Betula pendula* (×230): Hyphae of fungal partner radiating into soil (a); fungal sheath (b); Hartig net that facilitates exchange of substance between tree and fungus (c); normal cortex cells (d).

of vegetation known. How can this be, if mineral nutrients in the soil are deficient? It is probable that mycorrhizal fungi rapidly invade decomposing litter and recapture nutrients for the plant–fungus partnership so that nutrients leached from leaf surfaces are effectively captured by the network of roots and fungal filaments (Fig. 15-7). In this manner there is relatively direct or tight nutrient cycling, and little opportunity occurs for the loss of nutrients through the leaching process.

Under continuous cultivation this type of soil may require large and continuous application of chemical fertilizers—in fact, so much that it becomes economically infeasible. However, if properly managed, it can be cropped with little use of fertilizers. In this case the forest is cleared in part, making space for such small tree crops as rubber, cocoa, coffee, bananas or Brazil nuts (refer to Chapters 25 and 27). Then, as the crop-bearing trees grow larger, more forest can be removed. For food crops, the forest can be cut in strips or corridors for planting mixtures of corn, bananas, cassava, rice, and so on. After 4 or 5 years the exhausted soil is allowed to return to forest but in another 10 or 15 years the soil may be put into crops again. This corridor system is a scientific substitute for the method commonly practiced by natives of the tropics who clear a patch, farm it for several years until production drops drastically, and move on to clear a fresh patch (i.e., slash/burn or swidden agriculture).

The best soils in the tropics are the very young ones or those continually renewed by volcanic activity (ash) or by silty stream flooding (alluvial flats). In these situations the soil often can be used for crops without fertilizers and even without lying fallow under forest conditions.

Latosols can also harden and form a crust called *laterite* (Latin *later,* brick). Laterites often form on plateaus in areas with alternation of wet and dry seasons. During the rainy season the water table may be near the surface but in the dry season it may drop 10 m or so. A thick mass of heavy clay begins relatively close to the surface and gradually merges into the unweathered rock far below (as much as 10 m). When thoroughly dried from exposure, as happens in agriculture, the clay hardens irreversibly. This hard, slaglike laterite is very resistant to weathering and reconversion to soil. In fact, laterite clays have been used to make bricks and even statues by many civilizations (Fig. 15-8).

d. SOILS TRANSITIONAL BETWEEN WET AND DRY REGIONS: PRAIRIE SOILS

In the soils of the transition zone between wet and dry climates the leaching is not exten-

Figure 15-8 On exposure to air, laterite clay turns into a bricklike form of rock. This material is then cut into bricks and used for contruction as shown here in the gate to Angkor Thom, a major city of the ancient Khmer civilization. Laterite bricks are still used today for construction in tropical areas.

sive. Even some very soluble cations, such as sodium and potassium, may not be heavily leached into the lower *B* horizon. Such soils are best represented in the prairies of the United States and Canada, dominating in the "Corn Belt" there (Fig. 15-5). Prairie soils (mollisols) have high fertility and nearly as wide a range of crops as the gray-brown podzols. The grass cover has resulted in a rather high humus content, and a relatively dry climate has concentrated the minerals in the *A* horizon. Any tillage and exposure of the soil to the sun reduces the humus, but a rotation of corn, small grains, and mixed grasses and legumes can keep the humus at about 70% of the natural level.

e. SOILS OF SEMIARID FORESTS AND GRASSLANDS: CHERNOZEMS AND CHESTNUT SOILS

If rainfall fairly closely balances losses from evaporation and transpiration, a diverse flora of mixed grasses and other herbs may occur. This results in production of an abundance of organic material with a minimum of leaching, so that the *A* horizon (Fig. 15-6) is unusually thick and rich in organic material (up to 15% by weight). The high organic content produces a black color from which the soil derives its name "chernozem," meaning "black earth" in Russian. Because of the abundance of organic colloids with their high cation exchange system and low level of leaching, the soils are rich in all mineral nutrients, including nitrogen and sulphur. Calcification is typical of grassland soils, and a layer of calcium carbonate usually occurs at the base of the *B* horizon. Because of the resulting fertility and ease in cultivation, grassland soils are often the most productive in the world. They are found in the Great Plains of the United States and Canada, the steppes of the Ukraine and southern Asiatic USSR, and southern South America (Fig. 15-5). In all of these areas they enable agriculture to be so highly productive that they are called the "bread baskets of the continent." Because rainfall is both minimal and unreliable, however, these soils often need to be irrigated for maximum production.

The next driest areas beyond the chernozems contain the chestnut soils, which are somewhat like chernozems, but lower in organic matter and even less leached of calcium carbonate. Both of these soil types fall within the mollisols in the comprehensive system. The uncertainty of the climate and limited range of crops that can be grown make farming more hazardous here than on chernozem soils.

f. SOILS OF ARID REGIONS: BROWN AND RED DESERT SOILS

Desert soils (aridisols) develop in areas where evaporation and transpiration exceed moisture obtained from rainfall. A layer of cemented calcium carbonate often lies close to the surface of the thin soil. In the hot desert, the soils tend to be red, as a result of the presence of oxidized iron. Although there is little leaching in arid soils and, therefore, nutrients are available, the natural productivity of the vegetation is extremely low because of lack of water—which often also is not free of salts. Thus these areas are not useful for agriculture without extensive irrigation; yet it is not easy to make "deserts bloom" through irrigation. Even if there is not too much total quantity of salts, there may be an imbalance of those needed for plant nutrition. For example, as pointed out in Chapter 8, sodium chloride is toxic to most terrestrial plants. For successful irrigation, soils must not be too sandy or gravelly, in order to hold water reasonably well. However, they must be sufficiently well-drained so that excess water can leach away. A soil that is well-drained in the natural desert under 12 to 15 cm of rainfall may be swamped under irrigation with 90 to 120 cm. If drainage is poor, the soil is very likely to accumulate sufficient salt to kill cultivated plants, as is happening now in parts of Egypt.

PART 4

OUR USES OF PLANTS

A. PLANTS GROWN PRIMARILY FOR THEIR CARBOHYDRATES

CHAPTER 16

The Cereals: Historical Basis of Civilization

The beginnings of agriculture have been discussed in detail in Chapter 3. It was shown there that the small grains played a major part in the earliest days in the Near East. Corn evidently played a comparable part in Meso-America. So far back as we know, in oriental agriculture, rice is and has been the key plant. These three groups, the wheat and barley of the Near East, the rice of the Far East, and the corn (maize) of the Americas together show the importance of the cereals to human development.

Wheat flour is not easy to eat, so it was natural to make it into a paste with water. Such a paste, if rolled out thinly, could be baked into a very pleasant food, and our neolithic ancestors had discovered fire—and cooked by it—many thousands of years before agriculture began. Such "unleavened bread" is what the Israelites sometimes had in the Old Testament—flour paste rolled out thin and baked. (When thick it becomes bricklike and almost inedible.) However, if the moist paste is put aside for some days it quickly becomes infected with microorganisms, particularly with yeast, which will grow in it to produce gas bubbles (see Chapter 20); these *lift,* or *leaven,* the bread so that much thicker masses of dough can remain edible after baking. The Israelites had this leavened bread too, and so did the Egyptians as early as 6000 B.P. In Pompeii (A.D. 79) the Romans were baking small loaves like ours today.

Other cereals, for the reasons discussed below, do not lend themselves to making bread easily and have to be used in different ways. Some can be cooked and eaten directly. Popcorn can be eaten after popping and for this reason was no doubt the prominent type in the Americas in early times. All have the great advantage that they can be stored dry.

1. THE GRASS PLANT

Cereals are grasses and are therefore classified in the family Gramineae (or Poaceae). The older name Gramineae comes from the Latin *gramen* "grass", in turn akin to the old Indo-Aryan *ghra* ("grow"); the name Poaceae comes from one of the genera, *Poa*. The grasses belong to the monocotyledons and thus have the typical long, narrow, bladelike leaves with parallel veins (Fig. 16-1 and 16-8). The spreading, fibrous root system (Fig. 7-5) takes up water very effectively and doubtless helps to account for the success of the grasses under such a variety of habitats. The stems are very unlike the stems of the dicotyledons, for while the blade of each leaf is flat, the basal part forms a closed cylinder (leaf-sheath) that grows out of the stem at a clearly defined, slightly swollen node (Fig. 16-1). The stem produces a series of these nodes with relatively long internodes, until finally it elongates further and bears the flowers, which protrude above the top of the youngest leaves. Thus the stem bears no vegetative branches in this structural form, though short branchlets, *spikelets,* bearing flowers, form part of the infloresence. Under some conditions elongation of the stem is postponed (see below) and the leaves just form a rosette close to the ground. True branches called *tillers* arise only from the basal crown, usually after the first flower-stalk has elongated to its maximum. As seen in Figure 16-1, the basal crown in some grasses also produces horizontally growing *rhizomes,* that is, underground stems (Chapter 19), from which new shoots arise at some distance from the parent plant.

The flowers of the grasses are small and incomplete, lacking the sepals and petals. Most have just three stamens (rice and the bamboos have six), which hang down when mature (Fig. 16-1*A*, 16-8). In the small cereals each flower is borne between two bracts, the *lemma* and *palea,* which age rapidly, becoming brownish and semitransparent by the time the pollen is mature (Fig. 16-1*B*). The style has two stigmas, often brushlike; there is one ovule in each ovary (Fig. 16-1*C*). When ripe the single ovule has the dry ovary wall adherent to it so that the "seed" or grain is really a whole fruit—termed a *caryopsis* (Chapter 18). It often bears a long bristle or awn. The flowers are borne on the spikelets, from 1 to 12 on each, with one or two bracts or *glumes* at the base of each spikelet (Fig. 16-1*A*). In some species the bracts bear long awns also. The succession of spikelets constitutes the inflorescence, a *spike* or *panicle,* whose shape, with or without the many awns sticking out and with the spikelets of varying lengths in different species, is highly characteristic. The spike in corn differs from the above in several respects (section 6).

The structure of a grain of oats is shown in Figure 10-3 and that of a grain of corn below in Figure 16-8*E*. The embryo is borne at one end, supported on the shield-shaped *scutellum* and comprising a first shoot or *coleoptile,* with a node at its base from which the true leaves arise, and one or more primary roots. Most of the space is occupied by starchy *endosperm,* and near the outer tough seed coat are a few layers of *aleurone cells,* whose special function in germination was described in Chapter 10.

Many of the small-fruited grasses are grown for pasture (see section 8 below); among the large-fruited members are the all-important *cereals*—named for Ceres, the Greek goddess of grain, whose portrait (Fig. 16-2) is usually shown with heads of wheat.

2. THE CEREALS IN GENERAL

The importance of the cereals for humans rests partly in their sturdiness and good roots, partly in the large grains, with their compact content of starch and protein, and partly in the extreme dryness of these grains, so that they are easy to store. Also, because the grains are nearly all free from bitter-tasting glucosides, they can be eaten without elaborate processing.

The cereal grains generally contain 10 to 16% protein and the rest is fiber and starch; this material is the endosperm, which serves as the first food for the developing embryo. A peculiarity of most cereals is that the phosphate is present as *phytin,* inositol hexaphosphate, from which the phosphate is not readily liberated in

Figure 16-1 Left, the typical grass plant. In wheat and barley, the spikelets are much shorter, and the stolon and rhizomes are absent in wheat. (From Edlin, 1967) Right, the flower of oats (*Avena sativa*). *A*. Each flower is subtended by two glumes; *B* and *C*. The lemma and palea are two inner enclosing bracts that open up to allow the three stamens and the two brushy stigmas to protrude. There is a single ovule. Opening is caused by the swelling of the two scales (lodicules) at the base.

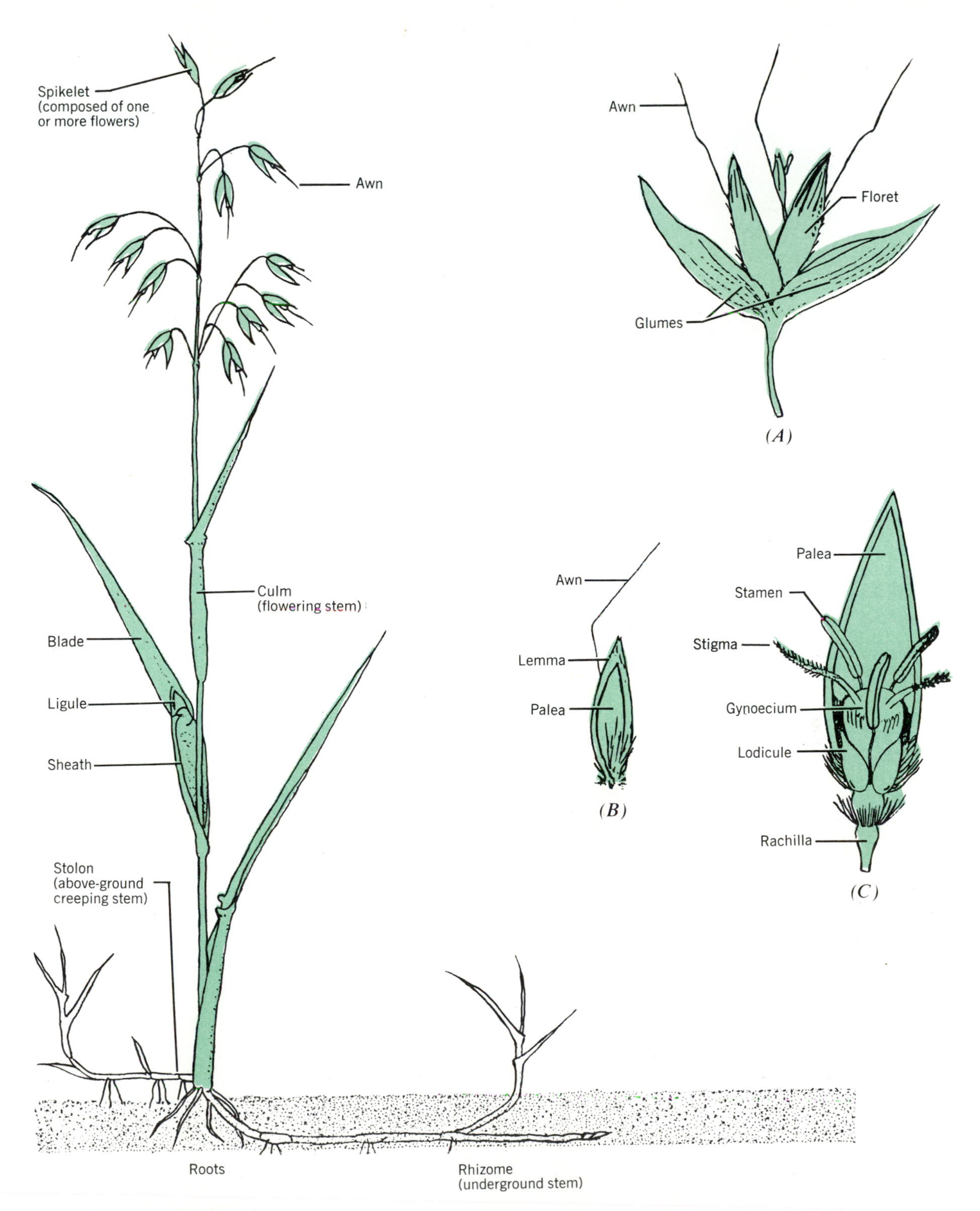

Spikelet (composed of one or more flowers)
Awn
Culm (flowering stem)
Blade
Ligule
Sheath
Stolon (above-ground creeping stem)
Roots
Rhizome (underground stem)
Awn
Floret
Glumes
(A)
Awn
Lemma
Palea
(B)
Palea
Stamen
Stigma
Gynoecium
Lodicule
Rachilla
(C)

Figure 16-2 A Greek terra-cotta portrait of Ceres (called Demeter by the Greeks) with wheat ears. She was the goddess of agriculture and mother of Persephone, whose excursion into Hades and return gave us the seasons winter and spring.

human digestion. Since the phytin is largely broken down on fermenting, raised breads are more nourishing than raw cereals.

World production of the cereals is listed in Table 16-1. Together they constitute 70% of the world's food crops. Here we will consider the major cereals in their individual detail.

3. WHEAT, "STAFF OF LIFE" FOR THE WEST

The origin and history of wheat, like that of many of the major plants that undergird our civilization, is complex, for its wild ancestors include three different genera; *Triticum, Agropyron,* and *Aegilops*. These all occur in Asia Minor and northeast Africa, which is where wheat originated (see Chapter 3). Two species of wild wheat, *Triticum aegilopoides* and *T. dicoccoides* (wild emmer) can still be found in Ethiopia, Syria, and Iraq. The various natural hybrids that gave rise to our present wheats have been selected over the centuries for larger grains, for holding on to the ripe grain ("non-shatter"), and for disease resistance. We know the Romans especially selected for size of grain, though unfortunately this is often due more to other factors than to heredity. In general, large seeds are characteristic of shade-loving or shade-tolerant plants. Where two species of the same genus have different light requirements, the seeds of the shade-loving species average three times the weight of those of the sun-loving form. Early wheats may therefore have grown

TABLE 16-1

Approximate World Production of the Major Cereals (in Millions of Tons Annually)

Cereal	*Production*	*Main World Locations*
Rice	280	90% in the Orient
Wheat	265	N. America, USSR, Argentina, Central Europe, Australia, France
Corn	250	N. and S. America, S.E. Europe, India
Barley	110	Whole Northern Hemisphere
Oats	50	Midwest N. America, N. Europe, N. USSR
Sorghum	45	China, India, Midwest U.S.A., E. Africa
Rye	35	USSR, N.E. Europe
Millets	30	India, Pakistan, Central Africa

in partial shade, probably in the open oak woods.

Wheat must have always been rather rewarding to grow. Herodotus, the Greek historian (484 to 425 B.C.), who traveled widely in the Near East, records that wheat returned "200 to 300-fold" to the grower, that is, each grain sown produced 200 to 300 grains. The Egyptians named their seasons after the wheat crop; they were called "Flood," "Sprouting of the Seed," and "Harvesting of the Grain."

A major part in the evolution of wheat has been played by polyploidy (Chapter 12). It has taken the work of a number of botanists from different countries to elaborate the fascinating story of the origins of domesticated wheat. Taxonomists early recognized that there were three groups of species of wheat on the basis of appearance of the plant. Then it was shown that these three groups were characterized by different chromosome numbers, the diploids having 14 chromosomes, the tetraploids having 28, and the hexaploids 42 (Fig. 16-3). Thus the wheat species formed a polyploid series with a base chromosome number of 7.

The diploid wheats, with two sets of chromosomes, are assumed to be the most ancient of the series and the base from which this crop developed. Two diploid species are recognized, *Triticum boeoticum* (wild einkorn) and *T. monococcum* (cultivated einkorn). Wild einkorn, native to southern Europe and the Near East, is one of the presumed parents of cultivated wheats. Einkorn (= one grain; the Latin name *monococcum* also refers to the occurrence of one seed per spikelet) is thought to be a cultivated form of the wild einkorn with slightly larger seeds and less fragile heads. Einkorn is a low-yielding species but still is cultivated to a limited extent in the Near East and parts of Europe.

The seven wheat species are considered to have arisen by hybridization of wild einkorn with another wheatlike Mideastern wild grass, *Aegilops speltoides,* followed, as frequently happens with such crosses, by doubling of the hybrid chromosome number to give a tetraploid form. Among these tetraploids is a wild "emmer," *T. dicoccoides,* and emmer, *T. dicoccum* (the Latin name here refers to two seeds in a spikelet). Both of these have been identified in the charred wheat grains found at Jarmo (Fig. 3-5). The presence of tetraploid rather than diploid wheats there suggests that Jarmo was a fairly advanced agricultural settlement. Emmer was extensively grown in the early civilizations of the Mediterranean region to as far north as Britain. Since it retains the disadvantages of einkorn (brittle heads and attached glumes), it is used primarily (where it is still grown) for animal feed. The one tetraploid that is still widely grown is *T. durum,* which has hard grains containing the sticky protein gluten; it is used in making macaroni and other forms of pasta.

Hexaploid wheats are known only in cultivation and the principal forms are sometimes grouped under one species, *T. aestivum,* sometimes given separate specific names. All are thought to have developed by hybridization between a diploid wild grass *Aegilops squarrosa* (a common weed in the wheat fields in the Near East) and tetraploid wheat, followed by doubling of the chromosome number. These are the bread wheats and have firm stems that, like those of *T. durum,* tend not to shatter when reaped, and glumes that open readily to release the grain when threshed.

Its importance for bread-making lies in the nature of the grain protein, which is of two kinds, glutenin and gliadin; the latter is only moderately sticky but the mixture is maximally effective in holding the dough together when the yeast generates gas bubbles in it. The only other cereal with these qualities besides club wheat, which is also used for bread, is rye. Among the bread-wheat cultivars are a group with soft grains—doing best in moist climates and yielding 8 to 11% protein and a soft fine flour for

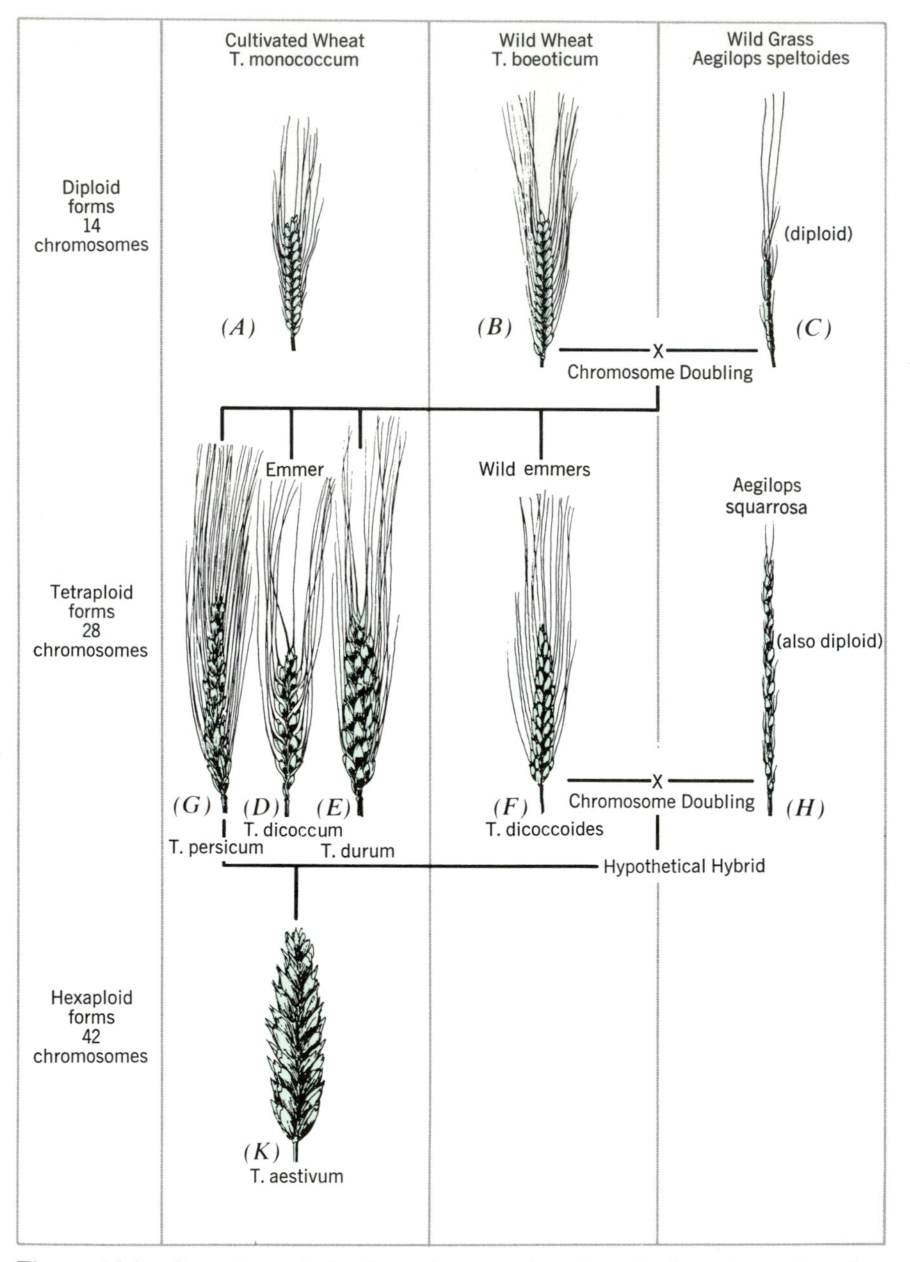

Figure 16-3 Genetic analysis shows how modern bread wheat has arisen by crossing and polyploidy. *Triticum boeoticum* crossed with *Aegilops speltoides,* both diploids, and the hybrid doubled its chromosome number to produce a series of tetraploids. One of these then hybridized with another diploid, *Aegilops squarrosa,* with chromosome doubling, to produce *T. aestivum,* bread wheat, a hexaploid, and *T. compactum*, club wheat (not all shown here). (From Boughey. 1971.)

pastry or French bread—and a group with hard, flinty grains, suitable for drier climates like Australia or Montana (U.S.A.). It has 13 to 16% protein and is the best type for ordinary bread. It is its use in bread that has given wheat its dominance in the Indo-European civilization.

The cultivars of *T. aestivum* also have two different types of flowering behavior: (1) "spring wheat" which, when planted in the spring, flowers in the same season, yielding grain in around 90 days, and (2) "winter wheat," which will flower only after it has experienced a cold period and is therefore planted in the fall. The latter is used where the winter is moderate, as in southern Europe, England, and the southern United States. Canada uses spring wheat. The Soviet Union went through a difficult time agriculturally, when Trofim Lysenko rediscovered the earlier-known fact that if winter wheat seeds are germinated they can be exposed to cold at once and may then flower in the same season like spring wheat. The treatment involved germinating the seeds and then holding them for several weeks in snow. It was called *vernalization* and it led Lysenko to a general theory that all plants, in order to flower, had to undergo a succession of phases—cold-requiring, darkness-requiring, and so on—that they could go through at virtually any stage of development. Lysenko's theory was officially accepted by Stalin and the Communist party in 1936, and since it incidentally involved a rejection of Mendelian genetics, it stultified Russian plant and animal breeding for a quarter of a century or more. The vernalization treatment was in any event difficult for farmers, since the moist germinating seed was awkward to plant and apt to go mouldy.

More recently developed is a group of *dwarf cultivars,* adapted to dry climates in which a full-sized plant would require too much water. These produce stiff spikes not over 18 in. high but with good grain yield, and have formed the basis for the very great increase in wheat production in Mexico during the last two decades. Norman Borlaug, in charge of the Rockefeller program there, received the Nobel Prize for this work in 1967.

One major criterion for wheat breeders is resistance of the plants to disease. The wheats are particularly susceptible to "stem rust," which drastically reduces the yield (described in Chapter 29). Each new selection has to be grown on the experiment station in a closed space where it can be dusted with rust spores and its sensitivity determined. The rust fungus, *Puccinia graminis,* unfortunately produces new recombinants and mutants from time to time, so that a wheat cultivar that was once resistant to the known rust varieties may after some years succumb to a new rust and have to be abandoned. Generally the breeders manage to keep ahead of the variation but the race never seems to end. *Aegilops* ($2n = 14$) is very resistant, and recently its gene for resistance was transferred to wheat, by hybridizing *Aegilops* with *T. dicoccum* to give a triploid ($2n = 21$); the chromosome number was then doubled by treating with colchicine. The resulting hybrid with $2n = 42$ was then x-rayed to produce fragmented chromosomes. Out of 6000 plants so treated, one was found in which the fragment from *Aegilops* that carried rust-resistance (but not many other genes) was incorporated in the nucleus of a hexaploid *Triticum.*

In a comparable way, several hybrids between durum wheat and rye (*Secale*) have been made by treating the rye ($2n = 14$) with colchicine and mating the resulting tetraploid with *T. durum* ($2n = 28$). The plants, called Triticale (Fig. 12-11) have good quality and the grains contain 17% protein.

4. BARLEY, OATS, AND RYE

The three cereals, barley, oats, and rye are plants of cooler, moister climates than wheat; thus, they are grown in northern and western Europe and northern North America. Together their production is about three-quarters that of wheat (Table 16-1). Most barley and rye cultivars are not polyploids. Barley can tolerate saline soils better than other cereals.

Barley (*Hordeum* spp.) has a history about as ancient as wheat. Two wild species, *H. spontaneum* and *H. maritimum,* can still be found in the "Fertile Crescent" and as far east as Afghanistan, and the various cultivars of *H. vulgare* cross readily with these. All have $2n = 14$. There are several tetraploid species, but they are seldom grown. *Hordeum spontaneum* and a cultivated form known as *H. distichum* have single spikelets arranged above one another in two "rows," whereas the cultivars most usually grown bear spikelets in groups of three on alternate sides, making six "rows" of grains. In general the barley spike looks much like that of wheat, though since both the glume and the lemma end in very long awns most barleys look more heavily "bearded" than most wheats. Unlike wheat, however, barley naturalizes very readily and is found as a widespread weed in both Europe and America.

A little barley is used for human food, the polished grains (known as pearl barley) being usually cooked like a vegetable. However, more than half the world's product is fed to horses and cattle, and about a third is fermented to beer. This process (Chapter 20) depends on brief germination, so that unbroken grains are essential. Like wheat, there are spring and winter forms of barley, the winter kinds being planted where the winter is not too extreme.

Oats (Avena). Like barley, oats are also grown in the north (e.g., Sweden, Scotland, Canada, the northern United States, and northern Poland). The starch in the oat grain, after the grains have been crushed ("rolled oats"), swells upon boiling in water to make a confluent jelly-like mass that can be eaten as porridge. In general the porridge is its main use in cooking. Like wheat, most oat cultivars have around 14% protein, but since the protein is not a gliadin or a glutenin, oat flour cannot be used by itself to make bread.

Historically oats are believed to have come into cultivation relatively late (e.g., around 4500 B.P.). The cultivated form, *Avena sativa,* is not known in the wild, but *A. fatua* and *A. sterilis,* two widespread wild forms, closely resemble it, and all three are hexaploid ($2n = 42$). There are diploid and tetraploid species that are of little importance as grains, though the tetraploid *A. barbata* is a useful range grass in the western United States. *Avena fatua* is a major weed in many areas of North America.

The spike of an oat plant differs in appearance from that of any of the wheat or barley cultivars in that the spikelets are very long and have up to seven grains well spaced along them, producing a very open, delicate-looking head (Fig. 16-1). The coleoptile, or first shoot of the oat, especially of the cultivar 'Victory,' is celebrated for its wide use in auxin studies (Chapter 10). Another cultivar with a similar name, 'Victoria,' when introduced in the United States, turned out to be subject to a special disease, caused by the fungus *Helminthosporium victoriae.* This fungus produces a toxin, victorin (a proteinlike compound), when it grows in the plant, and kills the host very quickly. Since the fungus is endemic in North America, surviving inconspicuously on some wild grasses, it prevents the use of hybrids based on 'Victoria' as cultivars here.

Rye (*Secale* spp) is more resistant than other cereals to winter cold and is therefore cultivated mainly in northern Europe, especially Russia. It was known to the Greeks and Romans but has not been found in excavations from earlier times, so it was probably a fairly recent adoption for cultivation. Again the major cultivar, *Secale cereale,* is not known in the wild, but is thought to be descended from a wild perennial species, *S. montanum.* Its ancestry is probably:

S. montanum (perennial)
↓ mutation
S. sylvestris (annual)
↓ three chromosome translocations
S. vavilovi (annual)
↓
S. cereale (annual)

All of these are diploids, $2n = 14$. Tetraploids have been produced by treatment with colchicine, but the seeds have such low fertility that they are not used. The spikes of rye usually have long awns, especially growing out of the lemmas, so that the heads look rather like those of barley, though rye usually has a slight purplish color due to an anthocyanin pigment.

Most rye grain is used for cattle feed, but because the grain protein is high in gliadin, rye flour can be used in making bread. In the Western world it is usually mixed with wheat flour, since rye alone makes a blackish and soggy bread. This "black bread," slightly bitter tasting, was common food in medieval Europe and is still made in a few places. Rye grain is also fermented and distilled to make rye whiskey, especially in Canada. Rye straw, being long and fine, is used in stables and for thatched roofs.

Rye is subject to the interesting and historically important disease ergot, in which infection by the fungus *Claviceps purpurea* produces a group of four powerful muscle and blood vessel contractants (Chapter 24). It is relatively uncommon now but still occurs occasionally in Russia and the Balkans.

5. RICE, STAPLE FOOD OF ASIA

Rice is a plant of warm, moist climates and is grown virtually wherever the climate will support it. Asia, particularly China, Japan, the Philippines, and India, is above all the home of rice (Fig. 16-4), and 90% of the world's rice is grown there. Its origin is probably in northeastern India or Burma but its early history is little known (Chapter 3). In China its importance in ancient times was such that at the annual ceremonies of sowing the crops, described in a manuscript of 4800 B.P., the Emperor Chen-Ning himself is said to have sown the rice; the other major plants were sown by the princes.

Oryza sativa, cultivated rice, is not found wild, though there are some 20 other species that do occur wild. *Oryza perennis,* found wild in Asia, Africa, and America, may be a parent

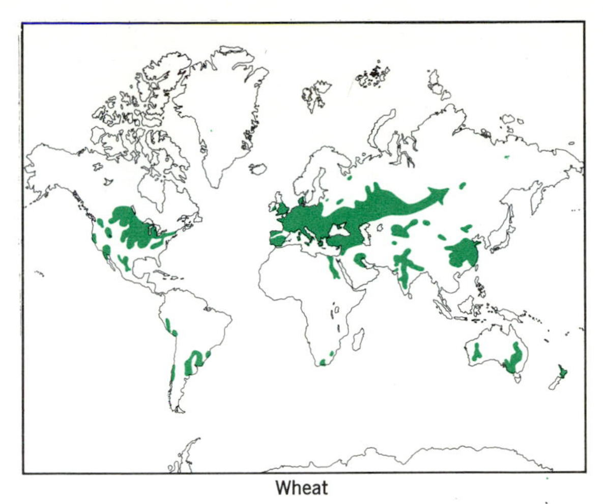

Wheat

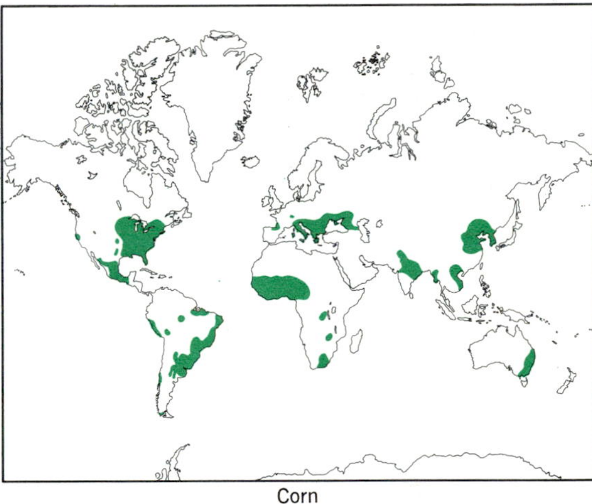

Corn

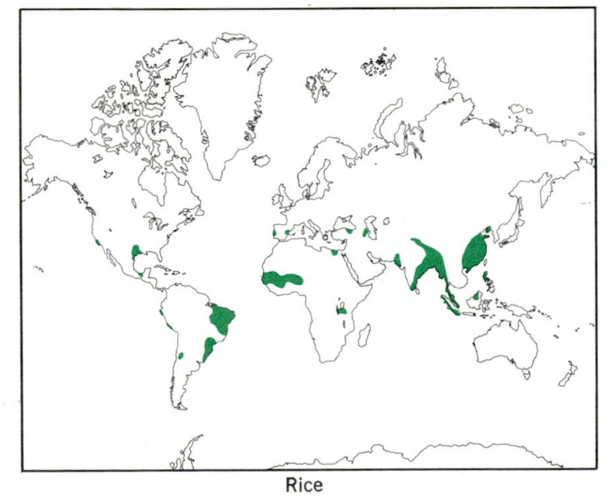

Rice

Figure 16-10 Maps of world cereal production. Note that on the whole there is little overlap between these 3 major cereals. (From Edlin, 1967)

of modern rice. Only one other species, *Oryza glutinosa,* is cultivated, and that in limited areas of Malaya and Java. Its grains contain less starch and more sugar than rice, and when cooked yield a sticky mass, as the name glutinosa suggests.

Besides Asian countries other important rice-growing areas are Egypt, Central Africa, and North and South America. Rice is even grown in Europe, in the Po River valley in Italy. Its introduction into the United States is said to be due to a sea-captain sailing an English boat home from Madagascar; storms blew him off course and he put in at Charleston, South Carolina, Here he gave the governor a small bag of rice he had brought from Madagascar, and this led to the establishment of rice in the Carolinas. Later rice growing spread to Louisiana and California. The United States, although its output is only 2% of the world's production, is now the greatest rice exporter; in other words, most of the oriental countries except Thailand, which exports a little, consume all that they grow.

Two or three thousand cultivars of rice, *O. sativa,* are known from different parts of the world. All have $2n = 24$ and thus are not closely related to wheat, oats, rye, and barley, all of which have $2n = 14$. The many rice cultivars fall into two main groups. *Japonica* types have short, plump grains that contain enough gluten to make them sticky on cooking. *Indica* varieties have longer, narrower, harder grains; after cooking the grains separate readily. Rice is the only major cereal whose grains can be directly cooked and eaten; oats, pearl barley, sweet corn, and some millet types are among the minor ones with the same advantage. Other cereals require grinding.

What distinguishes rice most sharply from all other cereals is the way it is grown. For most rice cultivars actually grow best with their bases in water. The rice field or "paddy," therefore, must be absolutely level and surrounded by low banks to keep the water in. In Asia, ploughing in the mud is done with the help of water buffalo (Fig. 16-5*A*) or even elephants. Usually the seedlings are raised in a separate small bed, where the seeds are planted in mud and the water allowed to flow in (to a depth of six cm or so) only as the plants elongate. The crop is usually started when the rains begin. The seedlings are transplanted into the mud below ten to twelve cm of water (Fig. 16-5*B*); here the rains maintain the water level, and the plants grow like other cereals. When the grains are fully formed the bank is knocked away at one point so that the water drains out. A few weeks later the stalks are cut off a little above the ground and dried in the sun. The fields are necessarily small, which militates against the use of elaborate harvesting machinery. For this reason much of the world's rice is not only planted and weeded by hand, but it is also harvested by hand (with a sickle). The rice crop is thus "labor-intensive," whereas other cereals, sown, weeded, fertilized, and harvested by expensive machinery, are "energy-intensive." In countries with a large rural population, each owning only a small tract of land, a labor-intensive crop is not out of place. In any case it is effective, for grain yields from rice average up to twice as much per acre as those from wheat, especially if the rice is thoroughly weeded.

In Japan the method is less labor-intensive, for the fields are drained just before harvest time and the harvesting is mainly done by small portable machines. But in the United States, where wages are high, the fields are leveled, water pipes laid in, the rice seeded by airplane, flooded when the plants are up several centimeters, drained for harvest, and cut by large harvesting machines (Fig. 30-2). Yields of around 5.7 tons per hectare are usually reached.

In India and elsewhere, a small amount of "upland rice" is grown on dry land, but the yields are lower than in the paddy. About 10% of the world's rice is grown in this way. Upland rice has been introduced experimentally into some areas of the Amazon basin—so far without much success, because of weed problems and infertile soils.

(A)

(B)

Figure 16-5 *A*, ploughing for rice with water buffalo in Laos. *B*, transplanting rice in the Philippines. The batches of young plants have been brought from a nursery. Rice culture in the orient uses large amounts of hand labor.

Rice seeds are freed from the chaff (glumes and ovary walls) by flailing or threshing and then winnowing in the air (Fig. 16-6), or by regular cereal mills; some of these have lately been introduced into Vietnam and the Philippines. The protein content of the average rice cultivar, when polished in the usual way, is about 7.5%, but the unpolished, brown rice varies up to 12%. (These figures are for the "dry" grain with 12% moisture.) However, the protein is of good quality, being high in lysine. The polishing rubs off the thin layer of cells that contains most of

Figure 16-6 Winnowing of rice in Indonesia by shaking to remove the chaff (bracts).

the thiamine (see Chapter 30), but the brown rice, unfortunately, looks unattractive and does not keep well. Thus polishing continues, and more varied diets have made the thiamine deficiency no longer common.

One of the main factors in determining yields is the number of tillers, which is much greater than in wheat or barley. The grain yield also depends heavily on the amount of fertilizer used. There is some natural input, because generally the water supports a good growth of blue-green algae, which fix appreciable amounts of nitrogen (Chapter 9). The stubble, plowed in or left to decay in the ground, returns some nutrient elements also. But addition of fertilizer before sowing has a major effect. In Japan a heavy dose of fertilizer is applied before sowing and a second, smaller application is made just as the panicles are appearing.

With many cultivars, especially the long-grained *Indica* types, high nitrogen produces tall thin stalks that droop over ("lodge") so that some dip into the water and the yields are therefore lowered. For this reason the Ford and Rockefeller foundations, bearing in mind that 60% of the world's population derives about half its energy from rice, planned an international group to study rice in its relation to fertilizers, culture methods, and varieties. A group of 20 Ph.D. biologists representing genetics, cytology, plant physiology, soil chemistry, and agronomy, with technical assistants, was set up in 1962 at Los Baños in the Philippines. The international staff collected the rice varieties from all over the world. The work with wheat in Mexico and West Texas had shown the great advantages of dwarf plants; there the principal advantage was in the reduced use of water, but at Los Baños it was realized that the dwarf plants, with large heads but stiff stems, would not "lodge" when fertilized, because the dwarfing gene prevents elongation. Dwarf forms also have the advantage that, although the number of spikelets per head is the same as in taller forms, the number of tillers is usually greater. Thus the Los Baños group set to work to create new high-yielding dwarfs. Three years later they produced IR-8, a dwarf form whose yield when grown by native farmers was 30 to 50% greater than the tall native strains. More importantly, the yield increased with fertilizer, giving up to three times that of the native strains (Fig. 16-7). In 1966 seeds were sent to India and Pakistan, where they produced record yields. Then genes for resistance to disease were bred into the plants: resistance first to a fungus disease, rice blast, then to leaf blight (a bacterial disease), and then to gall midge (an insect). This last strain (IR-22), with resistance to the three major diseases and high response to fertilizer, has now been widely planted and new varieties are being further developed from it. It is clear that world yields of rice will increase very substantially.

It was stated at the outset that wild rice is unknown. There is another cereal called wild rice, however, which is not *Oryza* at all. It is a different genus, *Zizania,* with a single species, *Z. aquatica*. It grows wild in swamps in eastern Canada and the northern United States, especially in Michigan. Like true rice it has six

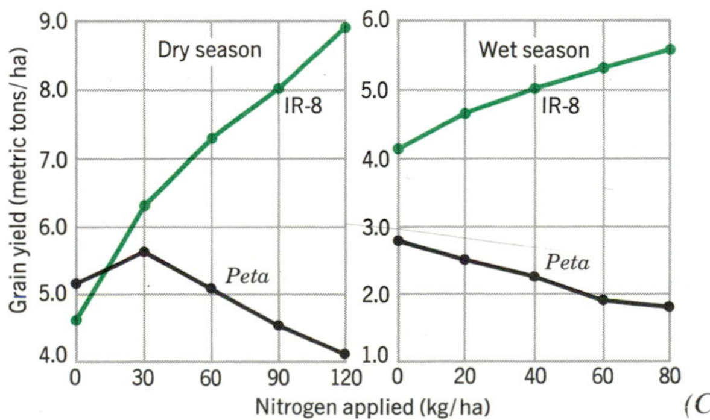

Figure 16-7 The first success of the International Rice Research Institute, the cultivar IR 8. *A*. The plants ready for harvest, with upright leaves, short stems, and multiple tillers. This field actually yielded 8500 Kg per hectare or 3.6 tons per acre. *B*. Border of a field with IR 8 on the left and a typical tall variety, Intan, on the right. The yield of IR 8 was double that of the Intan. *C*. The responses to nitrogen fertilizer of IR 8 and Peta, another tall variety, widely grown in the area. India's average yield is only 2 tons (*ca* 4000 Kg) per hectare. (From Chandler, 1969).

stamens and bushy forked stigmas. The long narrow seeds are usually collected by Indians in small boats and are cooked like rice—an expensive delicacy. A similar plant occurs in China.

6. CORN, AMERICA'S GRAIN

a. THE CORN (MAIZE) PLANT AND ITS HISTORY

In some respects corn is the New World analogue of rice. Figure 16-8 shows a corn plant and mature grain. Its origins go back as far or further, its true wild form is not known, and it has played as dominant a part in the development of civilization in America as rice has in Asia. Pollen claimed to be recognizable as maize has been found in old lake beds dated about 80,000 B.P. However, the earliest actual corn cob found was in the caves of Tehuacán in southern Mexico, dated 7000 B.P.; it is a tiny cob 2 cm long, which bore about 40 small kernels. Each kernel had a husk surrounding it, somewhat as the lemma and palea surround the wheat grain, but unlike present-day corn in which the husks are wrapped around outside the whole cob. Also it bore a staminate (i.e. male) spikelet at the *top of the ear*. Cobs from 5500–4300 B.P. were similar, but had evidently been selected for larger size. By about 3500 B.P. the husks surrounded the whole ear, as now, and the ears were up to 10 cm long (Fig. 16-9).

Although there has been controversy in the past, there is now growing agreement over the plants that gave rise to maize. It had been known since the last century that teosinte, a wild grass in Mexico, Guatemala, and Honduras, was its closest relative (Fig. 16-10). Teosinte and maize have the same chromosome number ($2n = 20$) and the two plants cross naturally to produce fertile hybrids. Teosinte belongs to the genus *Zea* and some taxonomists consider it to be the same species as maize (*Z. mays*). The idea that teosinte was the ancestor of maize is not a new one, but it was ignored for a number of years in favor of the idea that teosinte was a hybrid that had maize as one of its parents. According to this hypothesis maize was introduced by humans from South America to Central America and Mexico where it encountered the wild gama grass *Tripsacum* and hybridized with it. However, no hybrids between maize and *Tripsacum* have been found in nature and *Tripsacum* has a different chromosome number ($2n = 36$). Thus, although it is possible for humans to make the cross, special techniques are required to do it and no teosinte-like plants have been produced from the cross.

According to the most commonly accepted current view of the origin of maize, the 80,000-year-old pollen from the lake beds in Mexico City would have to be regarded as from an early, primitive form of maize. Although maize presumably was highly mutable, it must have required considerable conscious human selection to have produced the number of races that existed when it was discovered by the Europeans. The development of a wild grass into the unusual form that once fed most of the Americas was a remarkable achievement of prehistoric plant breeders.

The special qualities of corn may help to account for the exceptional prominence it always had in the New World culture. In Aztec Mexico, at the time of the Spanish conquest, the first ears of the corn harvest were offered to the corn goddess Cinteutl, in striking parallel to the way in which the first ears of the wheat harvest were offered by the Greeks to Ceres (Fig. 16-11).

b. CORN GENETICS

It is important to realize that a corn cob, whether it bears 40, or, like modern cultivars, 1000 grains, represents a *population,* that is, the individual kernels can be genetically quite different. If we cross a corn variety that bears all yellow grain with another that bears all purple grain, the first generation will bear all purple grain — the purple aleurone gene being dominant—and if then inbred the F_2 generation will give a typical Mendelian ratio of three

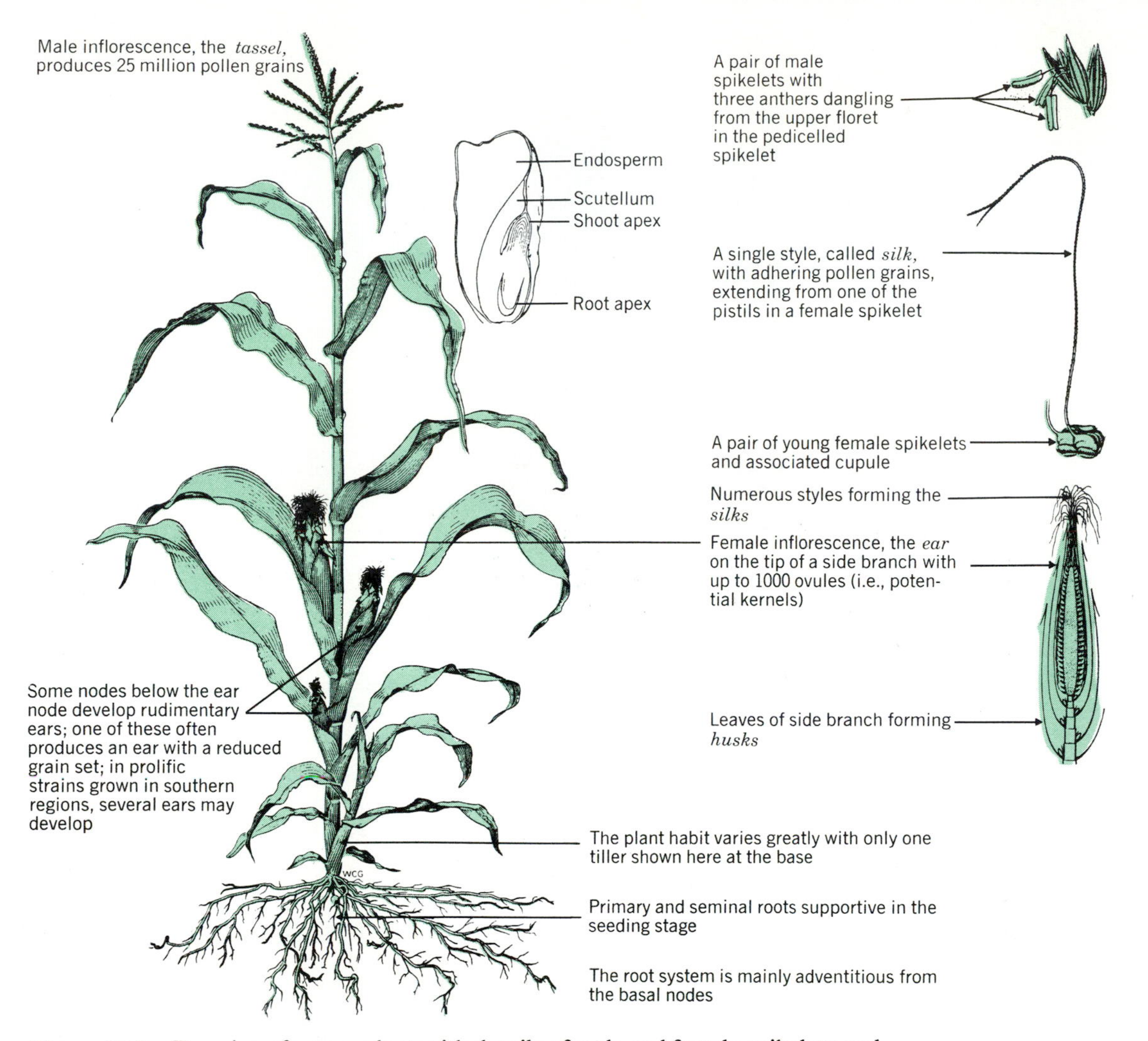

Figure 16-8 Drawing of a corn plant with details of male and female spikelets and of the ear; inset, section through a mature grain. (Drawing by W. E. Galinat, 1979)

purple grain to one yellow, on the same cob. Because of this heterogeneity of the corn ear, in the 1920s pure lines for genetic study were developed by inbreeding for five generations. Unexpectedly, the plants were successively smaller in each generation, but when the pure lines were interbred, the resulting hybrids were much larger plants with large ears (hybrid vigor, or *heterosis*). This was the basis for the introduction of "hybrid corn" for increased yields. Subsequently four pure lines were used and hybridized in pairs; the hybrids between these hybrids ("double hybrid") give even greater yields (Fig. 16-12, p. 305). With proper fertilizer the corn yields per acre in the United States are now three times what they were in 1920. The hybrid seed of course is heterozygous so that the farmer must obtain new seed each year from the growers.

The problem of preventing the selfing of the parental lines, which means insuring that all the grain was from crosses, was first dealt with

Figure 16-9 Ancestral types of corn, excavated from caves in Mexico and dated by archeologists. Left to right, wild corn, 7000 B.P., early cultured, about 5000 B.P.; cultivated forms, 300 B.P.; 1000 B.P.; 1500 B.P.

by detasseling (removing or bagging the male spike before it released any pollen), but this is highly labor-intensive and thus costly. A recessive gene was therefore introduced that makes the male flowers sterile; in one of the crosses it was arranged that the dominant allele reinstates the pollen fertility, so that the double hybrid can be produced. Unfortunately this gene, known as the Texas gene *T*, also confers sensitivity to the fungus *Helminthosporium maydii*, Southern leaf blight. In 1970, a wet season, this fungus caused 18% loss of the corn crop, especially in the southern states and in Illinois, which had the most rainfall. Further west, fortunately, the drier season minimized the attack. The two following seasons were fairly dry, and by 1973, through an accelerated program of growing two crops per year in Central America, other male-sterile genes had been introduced into the seed stock in place of the risky *T*. This incident points up the risks of having plants of the same genetic background spread over large areas, and has led to considerable restudy of the heredity of some major crops. The Irish potato famine (Chapter 29) is a classical example of this problem.

c. CORN AS A CROP AND A FOOD

Corn requires more rainfall than wheat, and most cultivars need a long growing season. Early-fruiting varieties suitable for New England or western Europe usually give only low yields. Some dwarf cultivars suitable for low-rainfall areas yield quite well, however, Mechanization, early planting, proper fertilization, and the use of insecticides against the corn borer and other pests have tremendously increased the yields of corn and decreased the amount of labor needed for it. (However, the mechanization naturally involves a heavy use of petroleum.) In 1800 it took 3 to 5 man-hours to raise a bushel of corn; in 1950, 0.34; in 1960 with the wide use of hybrids the figure was down to 0.03, and in 1974 it was estimated to be below 0.01.

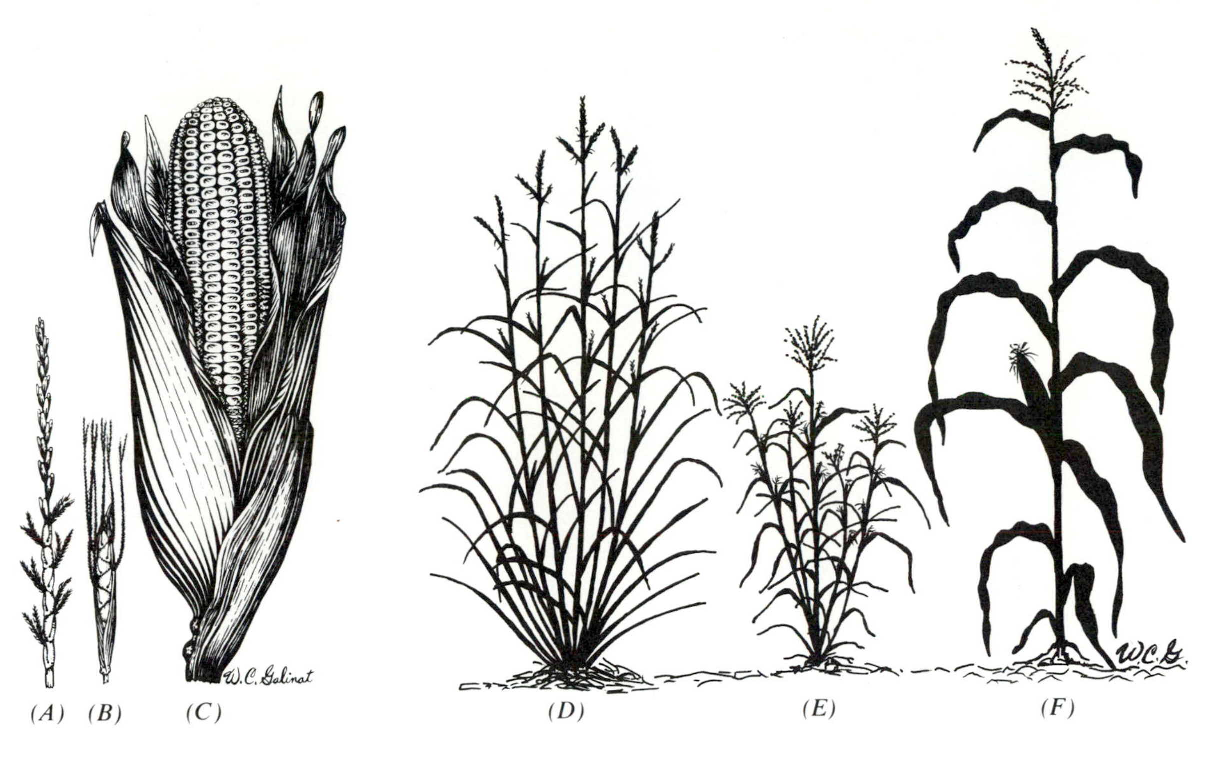

Figure 16-10 Left. Diagrams of fruiting spikes of *(A) Tripsacum,* with male flowers above, females below, on the same axis; *(B)* Teosinte, female only; and *(C)* Dent corn, female only, husks surrounding the spike. Right. Growth habits of the three. *D. Tripsacum, E.* Teosinte. *F.* Corn. (Drawings by W. E. Galinat) *G.* Close up of Teosinte plant in the field.

Figure 16-11 Peruvian pottery vessel with a "maize deity," probably Cinteutl (compare Fig. 16-2). Note the painted maize grains. Dated 500 to 1000 A.D.

The many corn cultivars fall into five groups: (1) flint corn (which was being grown by the Indians when the Pilgrims arrived), with hard starch; (2) dent corn – the external side of each grain is indented – which has hard starch in part of the grain and a soft cap, and which gives the highest yields; (3) flour corn, with soft starch; (4) sweet corn, in which much of the starch is converted to sugar (the sweetness can be further increased by breeding into it two or even three doses of the sugary gene); and (5) popcorn, whose starch expands on heating until the grain explodes. This last was probably the primitive type, since it eliminates the difficult task of grinding. The first three types constitute *field corn,* whereas the other two are the edible types.

Corn does not contain gliadin and therefore does not make good bread. In the United States 90% of the yield is fed to livestock, but in Central and South America it is ground, generally using flour corn, to make a thick paste that is baked into *tortillas*. Before grinding, the grains are softened by soaking in strong lime water; since corn is somewhat deficient in calcium this practice is nutritionally as well as mechanically useful. In southern Europe the rather granular paste is cooked as *polenta,* but corn is so low in niacin (one member of the vitamin B group) that a niacin-deficiency disease known as pellagra became quite widespread there about 50 years ago. With a more varied diet this has now largely disappeared again.

Figure 16-12 Diagram showing how two stages of hybrid vigor are used in modern corn breeding. Four homozygous strains are crossed in two pairs and the resulting hybrids are crossed again. (From Greulach & Adams, 1976)

Corn is deficient not only in both calcium and niacin, but also in the amino acids lysine and tryptophan. Farmers are always encouraged to bring in any new or unusual looking variants or mutants to university experiment stations. When one brought in some unusually opaque grains, analysis revealed, surprisingly, that the grains were high in lysine and in tryptophan (see Chapter 30). The gene involved, *opaque-2,* is now being bred into standard corn cultivars, and the resulting high-lysine corn will go at least part way to making corn a complete food. Other genes control the types of starch – the straight-chain amylose and the branched-chain amylopectin (compare Fig. 19-5). These starches have different industrial uses but are hard to separate. The *waxy* gene produces only amylopectin when it is present, and by contrast an amylose-extender gene raises the amylose content from the usual 27% to about 60%. Thus the selection and cultivation of particular genetic lines has great utility for industry.

7. SORGHUM, THE MILLETS, AND AMARANTH

Here we encounter a rather mixed group of cereals – some important, some minor on a

plants selfed for several generations to produce homozygous strains

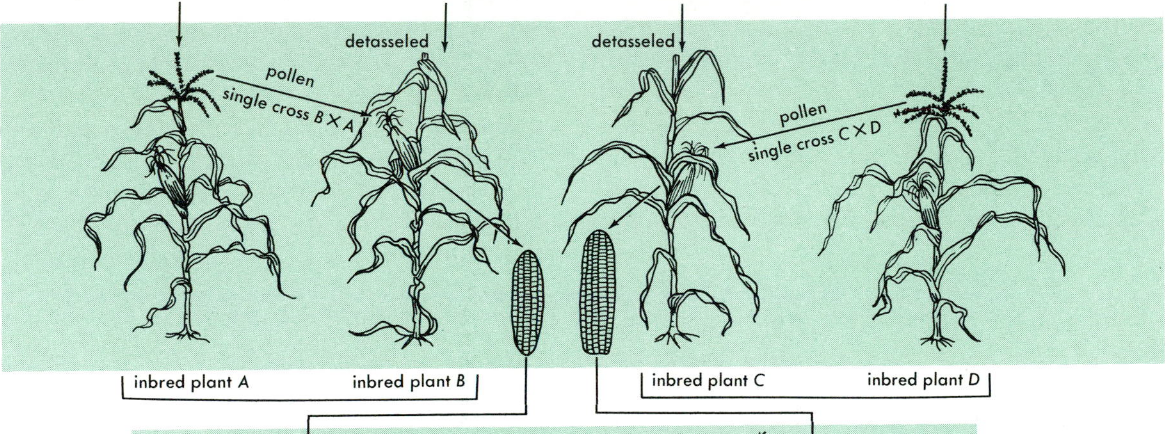

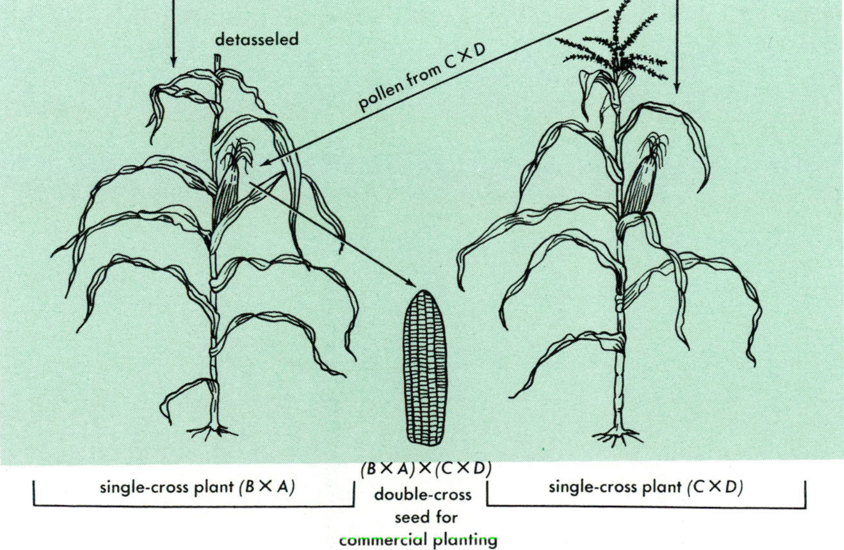

world scale, but each one of interest to some geographical group of mankind.

The *sorghums* together make up the world's fourth largest cereal in quantity. All the four types mentioned here are considered as variants of *S. bicolor* (*S. vulgare*), which probably originated in East Africa. It was grown in Egypt long before the birth of Christ, and later spread to India and China. The plant looks something like corn, but differs in that its flowers are perfect and are not borne on a cob but on short spikelets, like the small grains. It survives dry weather much better than corn, perhaps because its stomatal closing is more delicately adjusted to drought (Chapter 7) than is that of corn, and its root system is also very extensive. Thus it can be grown where the rainfall is not adequate for corn. The perfect flowers make hybridization difficult, but now male-sterile genes have been discovered and put to use.

Grain sorghum is a plant of stiff stalks and rather crowded "heads" of small round grains, many in each spikelet (Fig. 16-13*A*). The grains, which contain 12% protein, are ground up and eaten as mush or as dried cakes in India and Africa. Elsewhere, and especially in the United States where 20 million tons a year are grown, it is fed to livestock and to birds. In Mexico some sorghum flour is mixed with corn flour for tortillas. The grain is also sometimes extracted for starch or for oil.

The many cultivars of sorghum adapt to a range of climates; the group called "Milo" are grown in the United States as far north as New England. There are short-stemmed varieties for the hot, dry regions of west Texas, Nebraska, and the like, and in 1973 a new variety high in lysine was discovered, like the high-lysine corn. This will be interbred with standard varieties in the next few years and should prove nutritionally valuable in India and Africa. The foliage of most sorghum cultivars is good cattle feed.

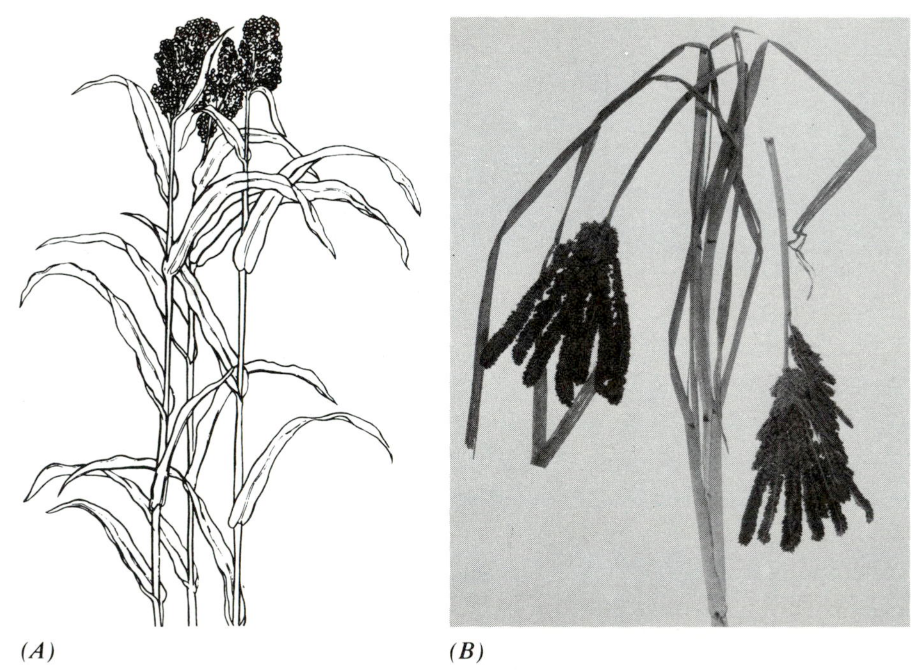

(A) *(B)*

Figure 16-13 *A.* Sketch of the grain sorghum plant. (From Edlin, 1967) *B.* Finger millet, *Eleusine coracana,* a lowland race.

Sudan grass, S. vulgare var. *sudanensis,* is grown wholly as a forage plant and is usually cut green for ensilage. Crosses with other forms of sorghum, called "Sordan," are proving very useful for this purpose in the southwestern United States. *Broom-corn* is a small-seeded type, the long, tough spikes of which are bound together to make brooms. *Sweet sorghum* is discussed in the following chapter.

Millet is a general term for a number of small-seeded cereal grasses belonging to different genera. As the "poor man's cereal" they are eaten mainly in India, Pakistan, Central Asia, and Africa south of the Sahara. Like sorghum the millets have been selected over the centuries for the ability to grow in regions of low rainfall. In India some 32 million hectares produce about 12 million tons of grain; compare this with rice, which can yield over 6 tons of grain per hectare. Millet grains can be roasted or boiled. In the West, however, millet is used only for forage and for birdseed.

The principal members of the group are: *Setaria italica,* foxtail millet, grown now mainly in the Near East and in China where it dates from 4800 B.P.; *Pennisetum glaucum,* pearl millet, grown widely in India and also in northeast Africa, where it is ground into flour. Discovery of male sterility in pearl millet has made possible the development of hybrid seed (as with corn) with resulting improvement in size of the head. *Panicum miliaceum,* proso millet, is the most ancient of all; its round grains firmly surrounded by the bracts have been found among remains of the Swiss lake-dwellers. The crop is now grown mainly in the Soviet Union and central Asia. All the *Panicum* millets are thought to have originated in China. *Eleusine coracana,* finger millet (ragi) (Fig. 16-13*B*) requires a cooler, moister climate than the other species and is therefore grown in southern India and the more favorable climates of East Africa. The boiled grain can be eaten like rice.

Amaranth does not really belong here, since it is a dicotyledon, but it is a grain plant and is believed to have been grown for grain, as well as for leaves, in Meso-America about 3000 B.P. The seeds are borne in close "heads" (Fig. 16-14), and the large leaves can be cooked for food. Some cultivars of amaranth are being developed now for use in several tropical countries, especially in India.

Figure 16-14 One of the grain amaranths, *Amaranthus hypochondriacus.* Amaranth is the only dicotyledonous cereal. It was grown as cereal in Meso-America in Aztec times and now is grown in parts of India.

8. FORAGE AND PASTURE GRASSES

The importance of the cereals—the grasses grown for edible seed—should not obscure the tremendous role of the grasses grown for edibility of the whole plant by animals. In North

America, more than two-thirds of the dairy and beef cattle feed is pasture and forage, mainly grasses. A similar ratio holds for the rest of the Western world. Indeed, of the whole land surface of the earth, more than a half is used for grazing; of this about one-third has been planted, and the rest is native vegetation, sometimes growing with minimal rainfall (Chapter 14).

Some of the minor cereals are valuable as forage; they include several sorghums, the pearl millets, *Pennisetum glaucum* and *P. typhoides,* and, in the tropics, guinea grass (*Panicum maximum*). In addition to these some 20 other grass genera are grown as forage. Eight of the major ones are shown in Figure 16-15. Selections of the grasses have been made to suit local conditions of (a) rainfall, (b) length of growing season, (c) tolerance to stressful conditions, especially heat, frost, and drought, and (d) digestibility by the animals being raised.

In selecting from introduced varieties and in breeding, two somewhat different characters are sought. A grass for hay should have a high dry matter content when it is harvested, not only so that it will be nutritious but also so that it will dry rapidly. Grass that is primarily for grazing has a quite different requirement. For horses and cattle will, by and large, eat about the same amount each day, so that the ideal pasture will grow at a constant rate throughout the season.

Digestibility is an important property, usually determined in a routine way in the laboratory, but obviously involving complex considerations such as concentration of the appropriate enzymes in the animal, and time spent in its digestive organs. In cattle most of the digestion takes place in the rumen, where cellulose and other polymers are fermented by bacteria, largely to propionic and succinic acids; methane is a major by-product (see Chapter 20). In horses somewhat similar changes take place in the enlarged cecum or small intestine, but less extensively, so that cattle digest cellulose better than horses. The leaf proteins are more readily absorbed. In the yellowing that occurs on aging or on making the grass into hay, leaf proteins are hydrolyzed to amino acids and thus readily released when the leaf is chewed (Chapter 30). If the cut grass is made into green ensilage some of the amino groups become liberated as ammonia by the ensilage bacteria, and thus part of the nitrogen is apt to be lost; this is particularly true when legumes are used for ensilage. Fortunately the pH becomes acid during the bacterial action and thus the loss of ammonia is held to a minimum.

A number of the forage grasses have been improved in recent years. Bermuda grass, *Cynodon dactylon,* has given rise to a very good variety "Coastal," which has high dry weight on harvesting. It grows rapidly and yields at least twice as much as ordinary Bermuda grass; in dry seasons, because of its drought-resistance, the ratio is even higher. It has been planted on 10 million acres in the southern United States. The use of male sterility in pearl millet (*Pennisetum glaucum*) has led to several hybrids that have both higher yield and better quality than the original strains. In several cases chance hybrids exhibiting typical heterosis have been brought into large-scale cultivation. In other instances the improved types are simply the result of careful selection of fast-growing or high-yielding individuals from a large planting. Lines of *Festuca* and *Bromus* are among the successful cultivars so obtained. Considering the immense areas that could be benefited, far more research on pasture grasses could be justified. Pasture plants for arid lands have hardly been touched yet.

Some natural meadow grasses spread by stolons and thus do not need to be raised from seed. They include pangola grass, *Digitaria decumbens,* and couch-grass or quack-grass, *Agropyron repens,* which is even a nuisance because its stolons invade tilled fields adjacent

Figure 16-15 Eight of the most important pasture grasses of temperate climates 1) Perennial ryegrass, *Lolium perenne;* 2) Italian ryegrass, *Lolium italicum;* 3) Timothy, *Phleum pratense;* 4) Meadow fescue, *Festuca pratensis;* 5) Cocksfoot, *Dactylis glomerata;* 6) Crested dogstail, *Cynosurus cristatus;* 7) Buffalo grass, *Bulbilis dactyloides;* 8) Kentucky bluegrass, *Poa pratensis*. (From Edlin, 1967)

to the meadows, sending up new shoots from every node as they spread.

It should be added that many natural meadows contain legumes and members of still other families as part of the ecosystem. Nutritious pastures are commonly made with mixtures of grasses and clover. The forage legumes are discussed in Chapter 23.

CHAPTER 17

Sugars and Sugar-Bearing Plants

The human tongue knows only four real tastes —sweet, sour, salt, and bitter; all the others are really odors. Our "sweet tooth" has therefore always played a major part in making foods attractive, and so the sugar-bearing plants are of great interest. Here we look at the formation of the sugars and their function in plants, the main plants of the world from which sugar is extracted, and finally at a new extra-sweet substance that is not even a sugar.

1. PRELUDE: THE INTERCONVERSIONS OF SUGARS FOLLOWING PHOTOSYNTHESIS

It was shown in Chapter 5 that the main product of photosynthesis is glucose phosphate, formed by the combination of two molecules of phosphoglyceraldehyde. This chapter, the following one on fruits and that on fermentation products (Chapter 20) will take up the sugars and fermentation products formed from glucose phosphate. As a necessary prelude, we must take a look at the chemical basis for the changes from primary photosynthetic products to the sugars and fibers.

The formation of sucrose from glucose 6-phosphate has been described in Chapter 6. Photosynthesis leads to the production of glucose phosphate, and, from this, four steps lead to sucrose. Each step is catalyzed by a specific enzyme (with those names we need not be concerned here). To recapitulate, the route to sucrose can be summarized as follows:

Step A. Conversion (isomerization) of glucose 6-phosphate to fructose 6-phosphate.

Step B. Hydolysis of the phosphate group from fructose 6-phosphate to liberate free fructose, and migration of the phosphate from glucose 6-phosphate to make glucose-1-phosphate.

Step C. Combination of glucose-1-phosphate with uridine triphosphate to form a uridine-diphosphate-glucose complex (UDPG).

Step D. Reaction of UDPG with fructose to form sucrose.

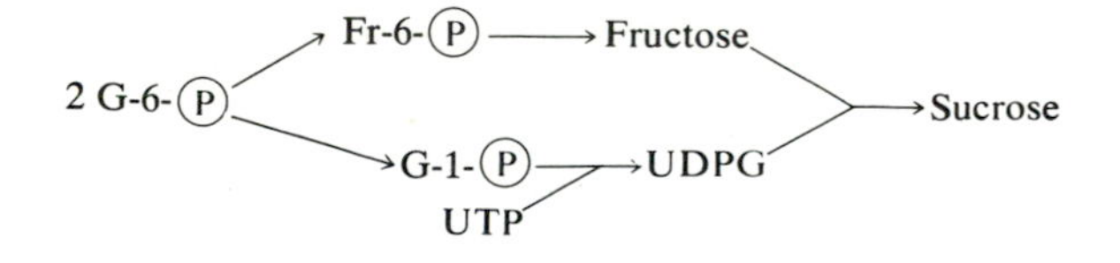

In this chapter we consider the function of the transport sugars and take up in more detail the plants that accumulate sucrose in large quantities.

2. THE TRANSPORT OF SUGARS IN THE PHLOEM

The sucrose arises in the parenchyma cells of the green leaf, adjacent to the vascular bundles. These, as we saw in Chapter 4, comprise groups of xylem cells and groups of phloem cells, separated by a layer of dividing cells that constitute the cambium, giving rise to phloem on the outer side and xylem on the inner. Phloem comprises three types of cells: sieve tubes, companion cells closely adjacent to them, and parenchyma around them (Fig. 4-7). Through the sieve tube there flows a rather concentrated sugar solution. In trees the rate of flow can be followed because the flow stops at night when photosynthesis stops; the resulting wave of sucrose moves at about 1 m/hr. Thus the sap takes all night to reach the base of a 10-m tree. The xylem sap, which moves upward and outward due to the transpiration pull, flows at much higher speeds (see Chapter 8). When the phloem is injured the sugar in the sieve tubes polymerizes very rapidly so that the sieve tubes become plugged with a polymer, which is a gelatinous material called *callose*. This system has been compared to the rapid clotting of blood at the surface of a wound, which similarly acts to stop the flow by blocking the blood vessels with a rapidly solidifying polymer.

There has been much controversy about the movement of the phloem sap. The clinching evidence that it is indeed true flow comes from an unexpected source, namely the feeding of aphids. For the aphid, resting on the stem, drives its fine-pointed stylet in through the epidermis, through the cortex—deeper and deeper until it strikes the sieve tube (Fig. 17-1). The reason its penetration stops at this point is probably that the aphid feels the sweet solution flow up through the stylet into its stomach. Now if the body of the unfortunate aphid is cut off, a tiny drop of phloem sap exudes from the cut surface at the base of the stylet. This shows that the sap is under some pressure. The high sucrose concentration—equal to about 0.3 M—shows that it is indeed phloem sap, for xylem sap is very dilute. There are small amounts of amino acids present also, and these supply the nitrogen metabolism of the aphid. The sucrose is generally more than it needs and it excretes it as "honeydew," which forms sticky spots on the stem; these give away the presence of infesting aphids and attract ants, who actively "farm" the aphids and consume the honeydew. Cars parked under an aphid-infested tree often develop similar sticky spots.

Thus the sieve-tubes, which through their sieve ends form a continuous supply system, contain a sucrose-rich sap that is under some pressure, probably due to the osmotic entry of water from neighboring parenchyma cells. As to the reactions leading to sucrose, the companion cells are the main seat of these, for they are rich in phosphate-removing enzymes, the *phosphatases* (Step B on p. 310). The continued flow of the phloem sap has two causes: (a) the supply from the leaves, and (b) the removal of the sucrose in the stem and in the roots. This removal is also due to enzymes, that transfer the molecule of glucose from the sucrose to sugar polymers which act as receivers in the adjacent cells. No large supply of energy is needed for this because the transfer of glucose from one glucoside (sucrose) to another (cell wall polymers, for instance) proceeds smoothly with minimal change in the bond energy. In this way the cells in the stem and roots, which contain no chlorophyll, receive a continuous supply of carbohydrate.

All in all the phloem provides a wonderfully effective supply system. Indeed, once it was evolved, presumably in the first land (vascular) plants, it has persisted almost unchanged throughout the plant kingdom. There is even a similar sieve tube system in the brown algae.

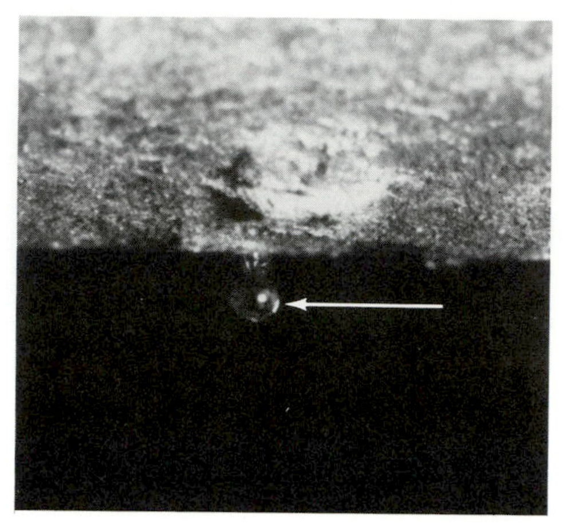

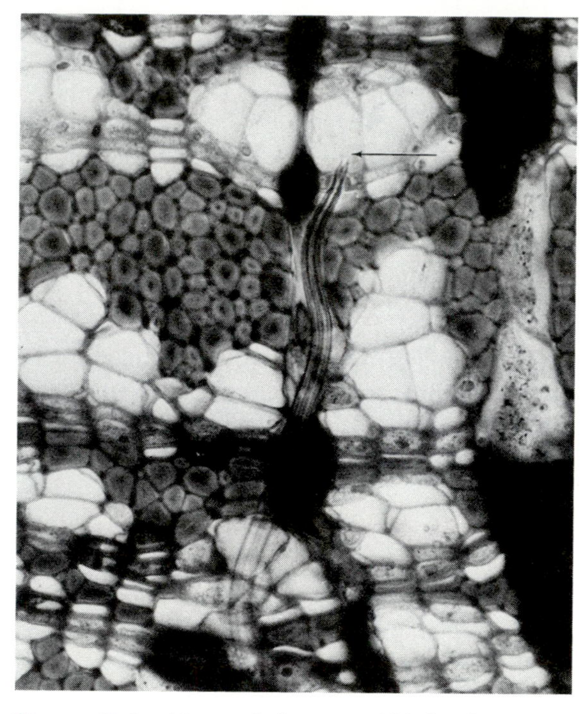

Figure 17-1 Above, left, an aphid feeding on a branch of basswood *(Tilia)*. It is about the size of a small house-fly. Right, droplet of honeydew exuding from the stylet after the aphid has been cut off. Below, section through the phloem of a *Tilia* twig showing two of the stylet tips actually inside a sieve-tube (arrow). Magnification 400×.

3. PLANTS THAT ACCUMULATE SUGARS

a. SUGARCANE

A few plants, besides using sucrose as a transport material, accumulate it in large quantities. Whereas most plants store their carbohydrate as starch, a few plants store it as sucrose. Among them the sugarcane (*Saccharum officinarum*) is the best studied, and the most important for mankind.

The cane is a very large grass, around 12 ft (about 4 m) and sometimes up to 18 ft tall, the leaves some 3 ft (about 1 m) long. The stalks are 5 to 8 cm thick and woody, often with whitish wax deposited on the outside (Fig. 17-2*A*). Sugarcane is native either to India or to New Guinea, and it is believed that the conquering army of Alexander the Great brought it back to the Mediterranean and to the Canary Islands. The cane was planted in Egypt by Arabs and the migration of the North African Arabs (called Moors) brought it to southern Spain in the eighth century. From here it made its journey to Central and South America with the invading Spaniards in the sixteenth century. Because the sweet syrup extracted from it was originally

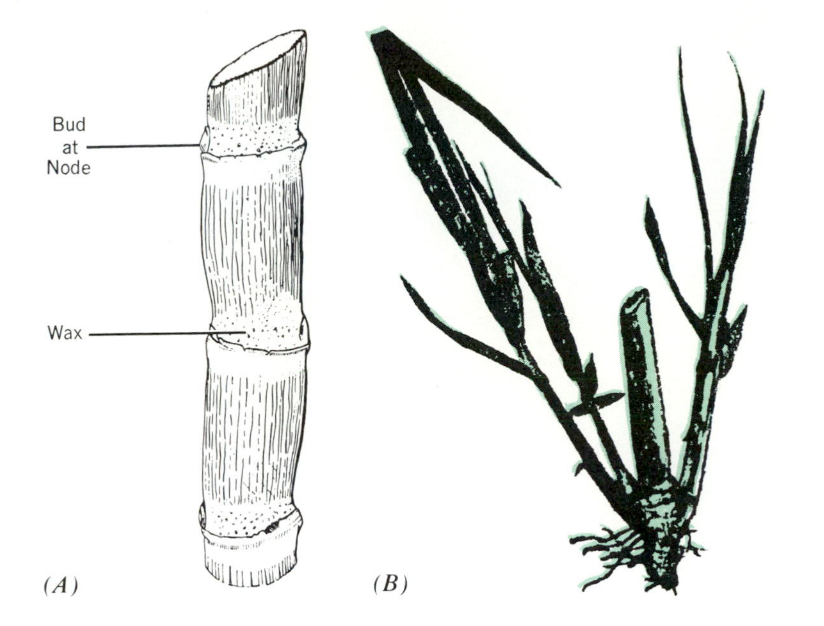

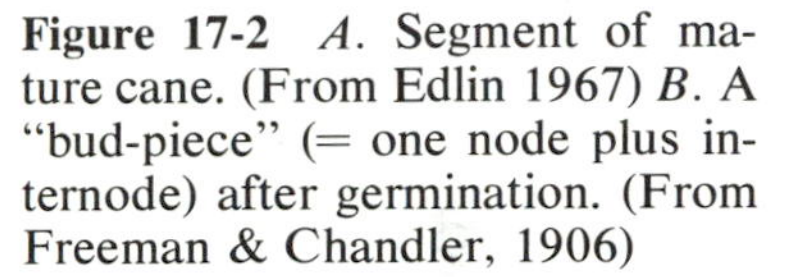
Figure 17-2 *A*. Segment of mature cane. (From Edlin 1967) *B*. A "bud-piece" (= one node plus internode) after germination. (From Freeman & Chandler, 1906)

used in medicine, it received the name *officinarum*. It is now a major cash crop in Cuba, Peru, and Brazil, which together account for almost half of the world production of cane sugar. The gift of sugarcane to the Americans has thus been of far greater value than all the gold and jewels that the Conquistadores took away with them to Europe. Since the cane is a plant of the moist tropics, it has flourished greatly not only in Central and South America, but also in the Old World tropics: southeast Asia, Oceania and northern Australia, India, the east coast of South Africa, and the islands of both the Caribbean and the Indian Ocean. In the United States it is limited by climate to Louisiana, Florida, and Hawaii.

Cane is planted by cutting the stem up into nodal segments as shown in Fig. 17-2*B*. Each segment bears one bud (or occasionally more). They are usually laid horizontally, bud uppermost, in shallow trenches, and they quickly form roots from the node. After about 18 months' growth the canes are ready to harvest. They are cut off, close to the ground, and new shoots, called ratoons, appear either from the crown or from buds on the spreading rhizomes. These buds have thus been released from inhibition by the removal of the main stalk (removal of apical dominance; see Chapter 10).

There is much variation in the detailed procedure of harvesting. In Hawaii and Australia the canes are cut by large machines (Fig. 17-3); in Cuba and India by hand with the machete. The leaves are generally cast on the ground to be browsed by cattle or to rot, thus returning most of their nitrogen content to the soil. Sometimes, as in Fiji, the workers set light to the field first, the green leaves burn fiercely (Fig. 17-3), and the cutting is made easier, though probably some of the nitrogen is lost in the process.

The canes are crushed between heavy steel rollers, rinsed, and recrushed, extracting some 96% of the sucrose in the form of a thick brown syrup of 12 to 15% sugar content (Fig. 17-4). The extract is concentrated, impurities are precipitated by adding lime water, and after filter-

Figure 17-3 Above, burning the cane before harvesting in Florida. Below, harvesting the cut cane.

Figure 17-4 A village sugar mill in Costa Rica. The crude cane juice, which has been boiled down to a thick syrup, is ladled out into a cooling trough as seen here. It will then be transferred into molds in which it forms blocks, 1 kg. each, of dark brown sugar ("tapa de dulce"). Notice the piles of crushed cane ("bagasse") at the back.

ing it is evaporated until crystals of "raw sugar" are formed and can be centrifuged off. These crystals, about 97% sucrose, contain a few inorganic salts, organic acids, gums, and traces of protein—also some sugars other than sucrose. The crystals are washed, redissolved in hot water, filtered through charcoal (which absorbs some impurities), and again evaporated under reduced pressure until they yield white sugar crystals. The whole process must be kept moving rapidly, for if the warm sugar solution has to be held for 24 hours or more in a tank, the polymer-forming bacterium *Leuconostoc dextranicus* grows very rapidly and can convert a large tank of sucrose solution into useless jelly, called *dextran,* almost overnight. For this reason, once the harvesting has begun the mill is kept running continuously night and day for the three to five months necessary to deal with all the cane in the area.

There are three by-products: (1) molasses, the uncrystallizable sweet mother liquor, which is either fed to cattle, used as a syrup, or fermented to rum (Chapter 20) whose characteristic taste derives from the small amount of molasses; (2) the cane fibers, called *bagasse,* which are compressed into building-board in some mills (in others they are simply burned to help drive the mill machinery); and (3) the wax on the surface of the cane, which is rubbed off during the grinding. Many mills do not find it economic to recover the wax, for the process requires extraction with a solvent, and other waxes can often be obtained more cheaply.

Because the leaves are returned to the soil, and because the stalks themselves contain little nitrogen or phosphorus, the harvesting is very conservative of the plant nutrients. In some cane fields in Cuba no fertilizer has been added within living memory, yet the sugar yields continue at about 5 metric tons per hectare. In Java and Hawaii, where fertilizer is added each year under scientific control, and where new canes are planted every three to five years, a regular yield of at least 25 metric tons per hectare is obtained. Compared with the yields of 2 to 5 tons per hectare of wheat or rice grains these figures are very high, but it must be remembered that cane is a C_4 plant (Chapter 5) and therefore more efficient in photosynthesis than the C_3 cereal species.

Sugarcane has been subjected to elaborate breeding work, and the hybrid canes produced have been selected for disease resistance, faster growth, lack of branching, and higher sucrose yield. The Dutch installed a sugar experiment station in Java as early as 1885 and one of the canes produced there is among the world's best. The British set up a similar station at Coimbatore in India, and another in Barbados, and the improved canes produced by the staff of these stations are also widely planted. The breeding has involved at least three related species of *Saccharum: S. spontaneum,* the common wild cane, *S. robustum,* and *S. sinense,* which originally came from south China.

For breeding, of course, the cane must flower and set seed. Flowering is favored by relatively low temperatures and short days; thus in the Northern Hemisphere it occurs commonly in January. However, since flowering is supposed to decrease the sugar yield it is not desired in a commercial plantation. Some cane fields are therefore illuminated by searchlights for a short period in the middle of winter nights to prevent flowering.

b. SUGAR BEET

The sugar beet (*Beta vulgaris*), totally unrelated to the cane, is of the same species as the red table beet, as also are Swiss chard and the mangold. All are considered to have been derived from the wild beet of western Europe, *Beta maritima,* which, as its name implies, is found near the shore, especially along the North Sea (see Chapter 19). The family is Chenopodiaceae, which means goosefoot (from the shape of the leaves of the common weed, *Chenopodium album*). The red beet has been eaten in western Europe since the fifteenth century and the mangold was grown for cattle feed by the Romans, its thick roots containing 2 to 4% sucrose. The possibility of industrial use of a beet, with 15 to 20% sugar, occurred in 1774 when Marggraf, a German chemist, demonstrated that its sugar was chemically identical with that from cane. A factory to extract the sugar was started in 1801. The subsequent rise of the beet sugar industry was attended by a curious interplay of circumstances. The British Navy's blockade of Napoleon cut off cane sugar imports to France and led the Emperor to encourage the development of beet sugar. After the exile of Napoleon canesugar could be imported into France again, now depressing the beet industry. Seed selection nevertheless continued and the steadily improved yields, gradually approaching 20% sucrose content, reinstated the industry, until by 1875, when the sugar beet was introduced into America, the sugar beet began to be competitive with cane. Now the beet, worldwide, provides a third of the world's total sugar production.

The beet plant is a biennial but is harvested at the end of the first year, when its sucrose content is maximal. Only for seed production is it grown for a second year. At the time of harvest it is in the rosette stage, with 8 or 10 large, crinkled, white-veined leaves and a stout, swollen root weighing up to five pounds, bearing two rings of small secondary roots (see Fig. 17-5). The sugar in the leaves is largely glucose and fructose and is converted to sucrose in phloem of the veins for transportation and storage. The plant grows well on a variety of soils, and its growth is promoted by moderate salinity. Since it is a plant of temperate regions it has been planted all over western Europe, as well as in many parts of the United States from Maine to California. Within the United States there are so many climates that a great number of strains or cultivars have been developed with different optima of day and night temperature. Beets yield up to 70 tons per hectare (fresh weight) with a sugar content of 15 to 16%.

Extraction of sugar from the beet is a little more complex than from the cane. The washed roots are sliced or shredded and steeped in hot water in a circulating system. The extracted tissue is then pressed to remove the remaining

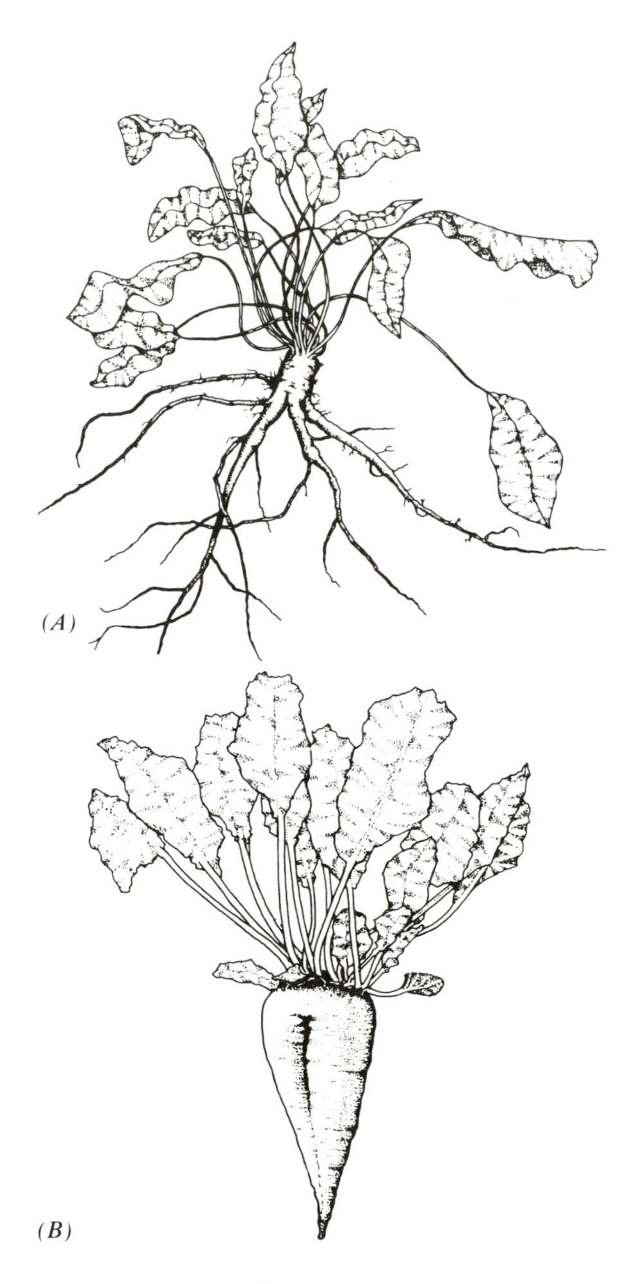

Figure 17-5 *A*. The sea beet (*Beta maritima*), ancestor of *(B)* the sugar beet. (From Edlin, 1967) *C*. Modern sugar beet with 21% sucrose content.

sugar solution. Extraction is virtually complete, but nitrogenous substances, pectin and gums are extracted too and have to be precipitated by adding lime, then bubbling carbon dioxide through to precipitate calcium carbonate, which carries down most of the impurities. The purification is done in much the same way as with cane sugar. Unlike cane molasses, the mother liquor from the beet is bitter tasting and hence difficult to use as a by-product. However, the extracted pulp is pressed into cakes for cattle feed.

5. THE DISEASES OF CANE AND BEET

Both cane and beet have been subject to diseases that have exerted major influences on their development as world crops. The most important for cane has been *cane mosaic,* which is shown by the formation of a mosaic of pale green or yellow areas within the leaf. It is carried by aphids. The first hybrids produced at the Java station grew well in spite of the mosaic disease, and so these "tolerant" plants, unfortunately, were widely distributed, carrying the disease with them. Later a natural hybrid between *S. officinarum* and *S. spontaneum* was found to be truly resistant to mosaic, and this laid the basis for the subsequent series of strains that led finally to POJ 2878, which now populates the cane-growing areas of about half the world. Infected canes of other cultivars may be partially freed from the virus by soaking the nodal "seed" segments in hot water, the time and temperature being carefully controlled, but this process seems to work better with other plant viruses than it does with cane mosaic.

A minor disease is a root rot caused by the fungus *Pythium.* Where it occurs this can be countered by introducing crop rotation, generally with two years of legumes between cane crops. In the Caribbean an insect called the froghopper, which bites the leaves, causing extensive local injury, can also be serious; this has called for considerable research on insecticides and on insect hormones.

The main problem with beets is another virus, *beet curly top.* This disease is most serious in the United States but occurs also in Turkey and some parts of South America. It affects other crops too, including tomatoes, beans, flax, melons, and spinach. The leaves roll up, the veins swell, growth of the root is inhibited, and its tissue becomes woody. Later the leaves turn yellow or brown and become tight balls; their form is wholly distorted. Since the virus is transmitted by the bite of leafhoppers, insecticide sprays have some effect if practiced on a wide enough scale, but the development of resistant varieties has been the most successful counter-action.

Another virus disease, *beet yellows,* in which the leaves turn yellow and thus decrease their photosynthesis (among other effects) is less widespread. However, one result is that the infected leaves tend not to transport their sugar to the root, so that, curiously enough, starch actually accumulates in the leaves. This virus is spread by the winged aphid *Myzus persicae,* which also feeds on peach leaves, transmitting the yellows disease of peaches, and on lettuce leaves, transmitting the lettuce mosaic. These three crops should therefore not be planted near each other.

6. SOME MINOR SUGAR PLANTS

a. SORGHUM SYRUP

The broad species *Sorghum bicolor,* which includes plants grown for grain as well as for several other uses, also has one variety *saccharatum,* which is grown for its sugar content. The sugar is principally in the stalk, as in the sugarcane. The cultivation is limited to the middle western and southern United States, usually to small farms for local use. At harvest the leaves are stripped off, the stalks crushed, and the syrup clarified and boiled down to concentrate it. The syrup is normally not taken to the point of crystallization.

b. MAPLE SUGAR

The sugar maple (*Acer saccharum*), native to Canada and the northeastern United States, is a large, handsome tree with a trunk approaching 130 cm. in diameter (Fig. 17-6*A*); the leaves make a great contribution to the celebrated "autumn colors" of New England by turning bright orange-red for a few weeks before falling. The sugar is obtained by drilling holes in the trunk, inserting a metal spout or "spile," and collecting the freely exuding sap in buckets or with a series of small pipes (Fig. 17-6*B*). The volume is surprisingly large, a good-sized tree yielding around a gallon a day, but the sap is usually only 3 to 5% sucrose (though 7% has been recorded). The phenomenon is peculiar in several respects: (a) it is most unusual for xylem sap, normally very dilute, to be so rich in sugar, (b) the flow is not, as one might guess, the result of root pressure but persists if the tree is completely cut off at the base and immersed in a tub of water, (c) the time of flow is sharply limited to the last part of the winter, when the days, in these northern latitudes, are warm but residual snow still lies on the ground and the night temperature falls below freezing. The process has been explained essentially as follows: Starch produced by the late summer's photosynthesis is stored over the winter in the xylem parenchyma or rays; when the temperature rises it is hydrolyzed and is released into the vessels by a

(*A*)

(*B*)

Figure 17-6 *A*. Sugar maple tree (*Acer saccharum*) with its characteristic five-pointed palmatifid leaves. (From Masefield *et al.*, 1969) *B*. Collecting maple sap in Vermont in late winter. The "spile" is driven about 2 inches into the xylem. Sometimes two or three spiles and buckets are placed on a large maple tree and a farmer will have an "orchard" of 100 or more trees, from which the buckets are emptied daily for four to six weeks.

group of special "contact cells." As soon as the leaves begin to open, negative pressure is established throughout the xylem by transpiration (see Chapter 7), and so the positive flow can no longer occur. Thus the maple is not really a sucrose-accumulating plant in the same sense as the cane, the beet, and the sorghum; it is really a starch-accumulating plant.

Acer rubrum, another maple species, and the paper birch, *Betula papyrifera,* also yield a sweet but dilute sap that is occasionally tapped in the same way. The whole idea of tapping large trees for sugar is an odd one, which the New England farmers might well never have discovered had the Indians resident in the area not shown it to them. Like the story of quinine and many other drugs, it furnishes a good example of how our ancestors learned to utilize plant resources when living under a restricted set of conditions.

C. PALM SUGAR

Palm sugar is in some respects the tropical equivalent of maple sugar, for it is obtained by cutting open the developing flower bud of the palm, when the sugary sap flows freely (Fig. 17-7). The flow rate is about as fast as that from the maple and can continue for several weeks. The sap, called *jaggery,* contains about 10% sucrose, and on being boiled down it crystallizes. Several genera of palms in India and South Asia are used, the two most popular being *Phoenix sylvestris,* the wild date (Fig. 18-10) and *Arenga pinnata,* called the sugar palm. Both of these, as well as at least one species of *Borassus,* are often planted specifically as sugar-yielding crops. The sap can also be left to ferment, forming palm *toddy*. The fact that so much sugar flows to the inflorescence recalls another fact, namely that some palms, like many agaves, die after flowering. Perhaps such death is due to total exhaustion of all carbohydrate.

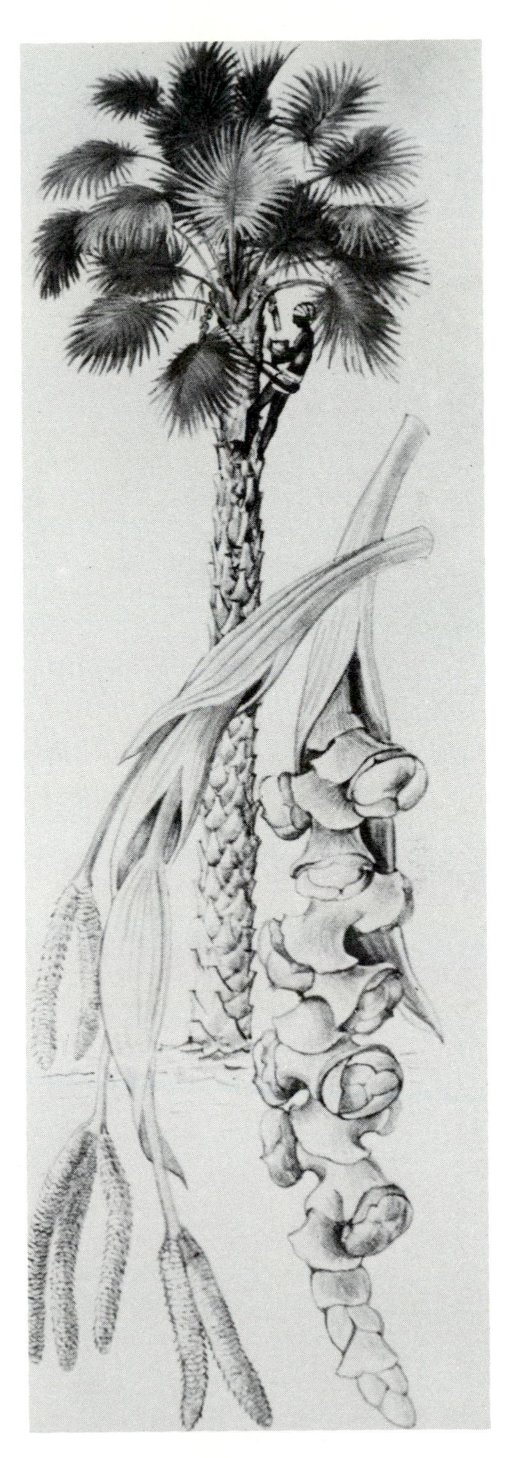

Figure 17-7 Palmyra palm *(Borassus)* showing the male and female spadix and the method of tapping. (From Masefield *et al.,* 1969)

d. TALIN

Because of the problems that have developed with synthetic sugar substitutes (saccharin and the cyclamates) it is interesting that an unusual *natural* sweetening agent has been discovered. Although it is chemically not a sugar, we include it here as a promising sugar substitute. It is many times sweeter to the taste than sucrose, but being a protein it is coagulated on boiling and loses its sweetness as a result; thus it is for use only in foods that are not to be cooked. Talin is produced by the West African plant *Thaumatococcus daniellii* (Fig. 17-8) and is already in limited use in Japan. Plantations are being developed for wider marketing, mainly in Europe.

Figure 17-8 *Thaumatococcus* plants, the source of Talin sweetener, being grown on a West African plantation.

CHAPTER 18

Fruits

Fruits have been esteemed by human beings since earliest days, when our ancestors lived from roots they dug and wild grains and fruits that they gathered. Most of the fruits, except for a few in the tropics, such as the banana, coconut, and avocado, have not provided either sufficient amino acids or fat to be a staple food. They have been the source primarily of sugar and vitamins and have added variety and flavor to the diet.

1. THE EVOLUTIONARY ROLE OF FRUITS

"True" fruits occur only among the flowering plants; the name *angiosperms* means "protected seed." This protection is furnished by the enclosure of the seed in the ovary. The function of the fruit in the angiosperm is not only to protect the seed, but also to provide its chief means of dispersal (Fig. 18-1). The size, color, edibility, and external form of fleshy fruits are related to their special dispersal mechanism of being eaten by animals. In some cases the animal eats the seed with the fruit, and, in fact, some seeds require the action of the digestive enzymes of an animal in order to germinate. In other cases the animal eats the fleshy part of the fruit and leaves the seed behind. The fruit may be eaten close to the plant that has produced it or may be carried away some distance, thereby effectively dispersing the plant. An important evolutionary mechanism seems to be involved in that conspicuous red, yellow, orange and purple colors (Chapter 6), serving to attract the animals, only develop after the fruit is mature. Often the green, immature fruits are unattractive, a situation that serves a useful purpose in discouraging animals from dispersing the enclosed seeds before they are fully developed.

The ripening of dry fruits involves some quite different processes from that of fleshy fruits. There is an extensive thickening of the secondary walls, followed by death and desiccation of these cells. Often then the fruits break open (*dehisce*) to shed the seeds since dry fruits are generally not as attractive to animals to eat, so that dehiscence helps to disperse the seed (Fig. 18-1).

Since the fruit is the normal reproductive structure of angiosperms, an infinite variety of fruits occurs in the wild and many have been chosen over the years for domestication. In fact, the number of species whose fruit is used for food is much greater than that of plants used for either roots, stems, leaves, or seeds.

2. THE TYPES OF EDIBLE FRUITS

Although the entire fruit may be formed from various parts of the flower, part of it always develops from the ovary. Thus the fruit is a result of changes in the ovary wall following fertilization, just as the seed is a result of modifications

(A)

(B)

(C)

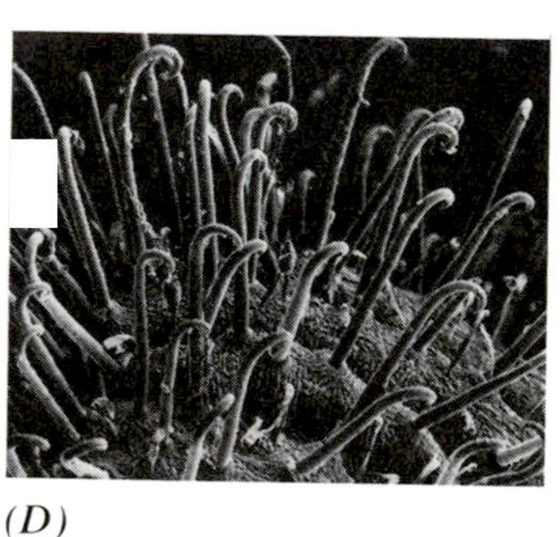

(D)

(E)

(F)

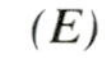

Figure 18-1 Fruit provides a variety of mechanisms whereby seeds are dispersed. *A*. Many seeds of fleshy fruits are eaten by vertebrates who deposit the seed in the feces. The fruit of the sausage tree *(Kigelia pinnata)* hangs on a long stipe, thus being more easily eaten by bats who disperse their seeds. *B*. Here the tree sparrow is eating the bright red fruits of the winterberry *(Ilex montana)*. It is perhaps not unusual that many ripe fruits are red, because this color is very conspicuous to birds and mammals. C. Many fruits have spines, barbs or hooks that catch on the hair or feathers of animals. Here are pitchfork seeds on a red fox's fur. *D*. A scanning electron micrograph of Enchanter's nightshade *(Circea)*. *E*. Milkweed *(Aselepius)* pods dehiscing to shed seeds with parachute-like wind dispersal mechanisms. *F*. Another familiar wind-dispersed seed, the dandelion *(Taraxacum)*.

in the ovule. Fruits are usually categorized according to the number of ovaries and flowers involved in their formation, the three main types being (1) *simple fruits* consisting of one ovary, (2) *aggregate fruits,* consisting of several ovaries from a single flower, and (3) *multiple fruits,* consisting of several ovaries from several flowers (Table 18-1). The last two types are clusters of simple fruits. In some fruits only a part comes from the ovary, the remainder being derived from the calyx or receptacle. These are known as *accessory fruits*. Many of our most familiar fruits and "vegetables" (which even though they are not called "fruits," actually have their origin from floral rather than vegetative parts of the plant) are simple or simple accessory types.

In addition, there are two basic types of simple fruits: dry and fleshy. In the dry fruits, the ovary forms a hard, usually thin, inedible coat for the seed; in the fleshy fruits the ovary and associated floral parts usually form a juicy, edible structure surrounding the seed.

a. SIMPLE FRUITS

Because of the variety in types of simple fruits, special names have been given to characterize groups of them (Table 18-1; Fig. 18-2). For example, a *berry* is considered a fleshy fruit

TABLE 18-1

Classification for Some Common Edible Fruits

Kinds of Fruit		*Examples*
Simple fruits	Fruits derived from pistil(s) in a single flower	
Fleshy fruits	Fruit wall or portion of it becoming fleshy	
Berry	Entire ovary wall ripens into fleshy pericarp	Grape, orange, cucumber
Drupe	Pericarp divided into fleshy exocarp and mesocarp, with stony or woody endocarp	Peach, plum, cherry
False berry	An accessory fruit, entire fruit wall fleshy, composed of pericarp and floral tube	Banana, blueberry
Pome	Like a false berry but pericarp chiefly fleshy; mesocarp and endocarp papery or bony	Apple, pear
Dry fruits		
Legume	Dehiscent fruit derived from simple pistil capable of splitting along two sutures	Pea, bean
Achene	Indehiscent fruit, seed coat of single seed not fused with fruit wall	Sunflower, buckwheat
Caryopsis	(Grain) Indehiscent with seed coat of single seed fused with the pericarp	All cereal grains
Nut	Large one-seeded fruit with woody or stony wall	Hazelnut
Aggregate fruits	Fruits derived from a number of simple pistils all from same flower. Some aggregate only, others accessory, for receptacle is part of mature fruit	Raspberry Strawberry, blackberry
Multiple fruits	Fruit derived from several or many flowers, consisting of whole inflorescence with many crowded flowers that coalesce as they mature	Pineapple, fig

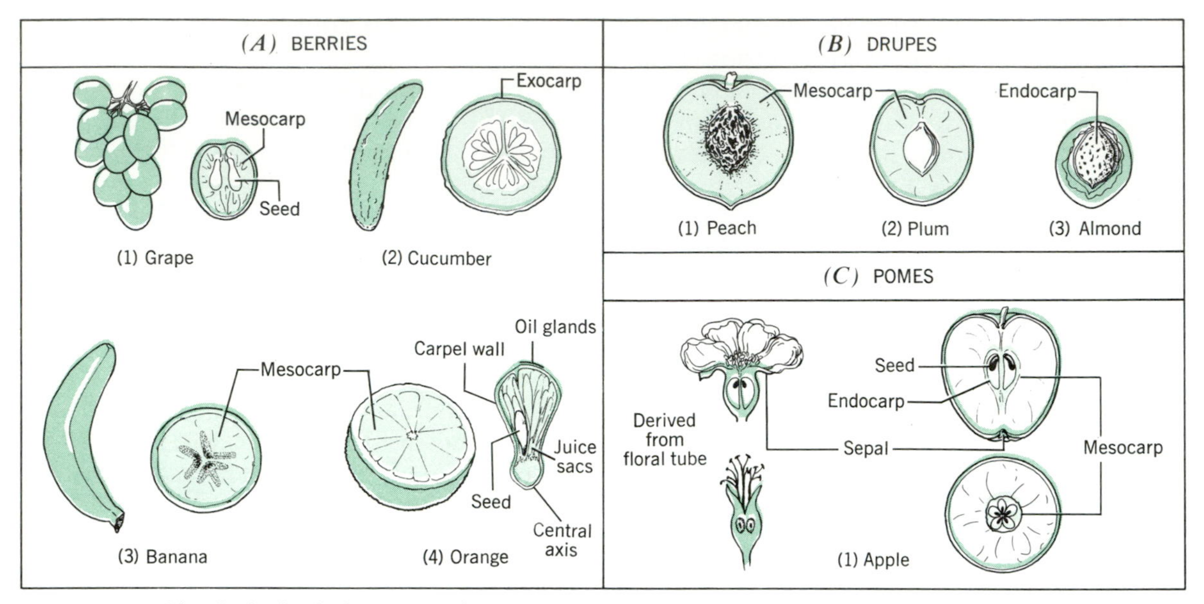

Figure 18-2 Simple fleshy fruits. *A*. Berries: 1, grape, 2, cucumber, 3, banana, 4, orange. *B*. Drupes: 1, peach, 2, plum, 3, almond. *C*. Pome: 1, apple.

in which some or all of the ovary wall becomes enlarged. One or more seeds are surrounded by a juicy flesh, each seed having its own hard coat. Berries develop from flowers with either simple or compound pistils. A date is a single-seeded berry; grapes and persimmons are several-seeded berries and a tomato is a many-seeded one. Pumpkins, watermelons, cucumbers, and other members of the gourd family (Cucurbitaceae) are berries with a hard outer rind (a *pepo*). Oranges, lemons, limes, and other citrus fruits are berries with a tough, leathery, separable rind and parchmentlike partitions (a *hesperidium*). In some cases, berries are accessory fruits, as in blueberries and cranberries. In the banana, which also is an accessory berry, the "peel" is formed from floral parts other than the ovary. The *drupe* (or stone fruit) is another simple, fleshy fruit with a "pit" surrounding the thin-walled seed. Peaches, apricots, cherries, plums, and almonds are drupes.

The *pome* is a simple, accessory fleshy fruit whose edible pulp is derived only in part from the ovary wall, a large portion of it arising from the receptacle and calyx. Apples and pears are pome fruits. The ovary in these fruits forms a section of the core, surrounding the seeds. The remains of the calyx and the stamens can often be seen at the end opposite the stem of the pome (Fig. 18-2).

There are also several basic kinds of simple dry fruits, which have their special names too (Table 18-1). The simplest is an *achene,* in which an ovary containing but a single ovule later forms a thin layer closely adhering to the single seed (Fig. 18-3). The fruits of the sunflower and buckwheat are achenes. In an achene the seed is attached to the wall of the fruit at only one point. If it is attached around its entire periphery, the fruit is known as a *grain,* or a *caryopsis.* Since externally a grain is very similar to an achene, both are often referred to as "seeds." Corn, wheat, rice, and in fact most all of the cereal crops (in the grass family) have this type of fruit (Fig. 18-3*A*).

A *nut* is a dry fruit in which all or part of the ovary wall forms a hard shell around the

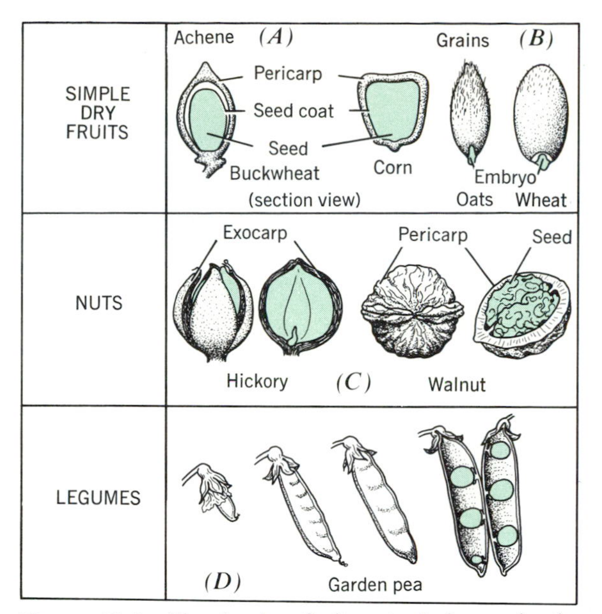

Figure 18-3 Simple dry fruits. *A*. Achene: buckwheat. *B*. Grain: corn, oats and wheat. *C*. Nut: hickory and walnut. *D*. Legumes: garden pea.

seed (Fig. 18-3*B*). In acorns, chestnuts, and hazelnuts the edible portion is the seed and the shell is the entire fruit coat. Such nuts are very much like enlarged achenes. In walnuts, pecans, and hickory nuts, the inner part of the ovary wall forms a hard shell, and the outer part becomes transformed into a leathery husk. A few of our familiar "nuts" are really seeds; this is true, for example, of Brazil nuts. An almond is not a true nut but the "stone" of a fleshy fruit closely related to peaches and apricots (Fig. 18-2*B*).

Achenes, grains, and nuts are all *indehiscent;* that is, they remain closed when ripe, retaining the usually few or solitary seeds. In contrast, *pods* or *legumes* are *dehiscent;* they split open when mature. A pod develops from a single pistil in which the ovules form a row of seeds attached to one surface of the ovary (Fig. 18-3*C*). At maturity the pod splits open along both margins. This type of fruit is characteristic of the pea family (Fabaceae or Leguminosae), members of which are often known as legumes for this reason. Except for string beans and Chinese snow peas most of these fruits are of little food value, the nutritious material being found in the enclosed seeds. The peanut is one of the few fruits that matures underground; it is also an exception among the legumes in that the pod does not split open when mature.

b. AGGREGATE AND MULTIPLE FRUITS

The strawberry, an aggregate accessory fruit, is a product of many carpels attached to a common receptacle (Table 18-1; Fig. 18-4*A*). The carpels individually develop into achenes while the receptacle becomes fleshy and sweet. In other words, the surface of the swollen red receptacle has many simple achenes attached to it. In blackberries and raspberries the receptacle does not become excessively enlarged, but there is an aggregate of separate carpels, each developing into a small, sweet, one-seeded drupe.

The mulberry, a multiple fruit, is the product of separate flowers, each of which produces a nutlet enclosed in the juicy calyx of the flower (Fig. 18-4*B*). These individuals become crowded together into the single mulberry. The fig is an enlarged, fleshy receptacle with a deep invagination in it. Many small flowers are attached "inside" this receptacle (Fig. 18-11). The pineapple originates in the fleshy axis of the inflorescence; fusion of the ovaries and sepals of adjacent flowers results in a large, compact, multiple accessory fruit.

During the formation of the fruit, the ovary wall becomes differentiated into three more or less distinct parts. A usually thin inner layer forms the *endocarp* and an outer one forms the skin or *exocarp*. Between the two layers a thicker *mesocarp* develops. In the peach (a drupe) the exocarp is the thin fuzzy skin and the sweet flesh is mesocarp. The peach pit is a stony endocarp enclosing the seed. An orange skin is exocarp and mesocarp, whereas the segments of the orange each derive from a carpel. Juice-filled vesicles in the segments are but

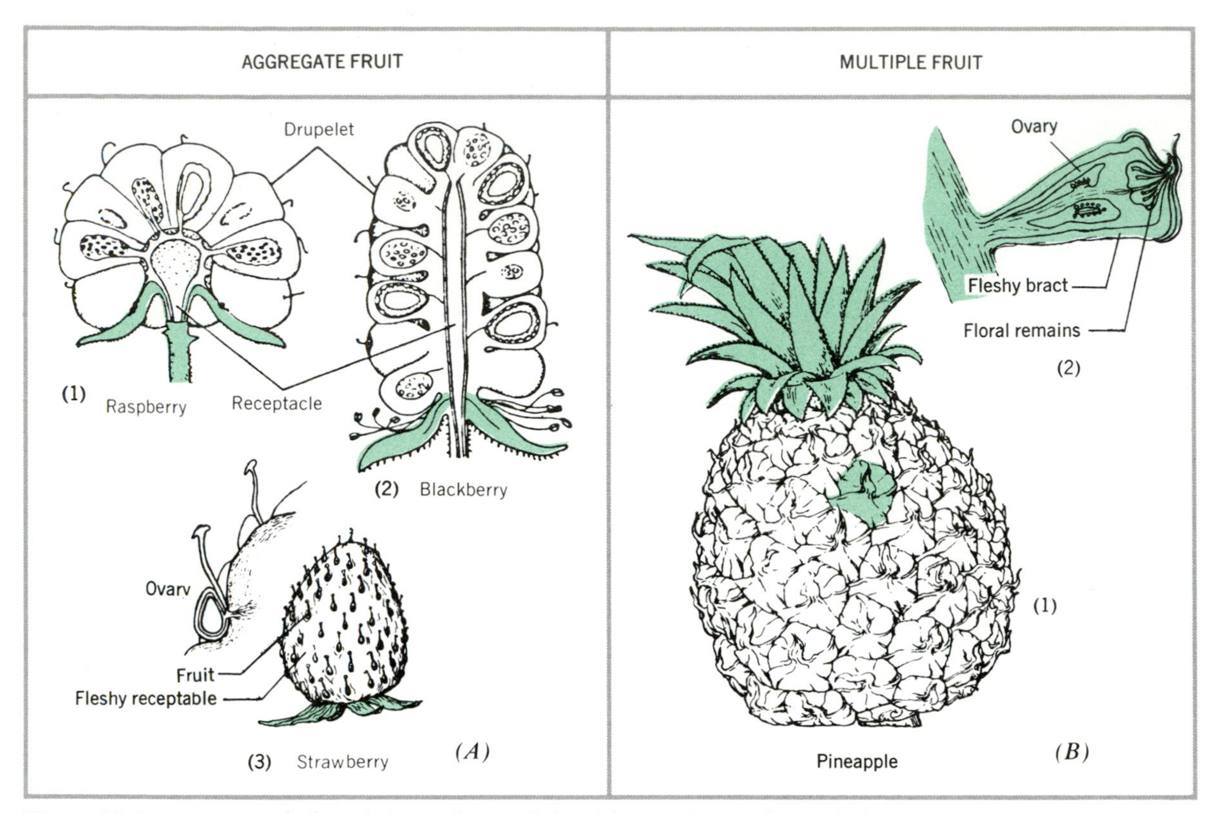

Figure 18-4 Aggregate fruits: *A*. 1, raspberry, 2, blackberry, 3, strawberry. Multiple fruit: *B*. 1, whole pineapple, 2, section of individual flower.

growths from the inner surface of the locule, a product of the endocarp. The cucumber rind is receptacle tissue that becomes fused with the exocarp whereas the flesh is mainly mesocarp and endocarp.

3. DEVELOPMENT OF THE FLESHY FRUITS

The ripening processes of fleshy and dry fruits are quite different and the development of some of the important dry fruits has been discussed in the chapters on cereals and legumes. Thus here we concentrate on the growth and ripening of fleshy fruits.

a. FUNCTION OF POLLINATION

The development of the fruit begins with pollination (Chapters 4 and 10). Although fertilization only affects the ovules directly, it indirectly also determines the growth of the ovary wall. Generally, if fertilization does not occur in at least some ovules (seeds), the flower dies and the ovary does not grow. Experimentally, by fertilizing ovules on just one side of an ovary in an apple workers have shown that the ovary wall and receptacle develop only on the side where the seeds also mature. Removal of the seed has the same effect (Chapter 10). The "setting" of fruit is determined by hormonal responses associated with pollination and fertilization. As previously pointed out, these processes involve multiple hormonal interactions. For example, there are small amounts of auxin in the pollen, and both auxin (IAA) and gibberellic acid (GA) are formed in the seed (the endosperm contains both IAA and GA and the embryo has IAA and possibly GA). The liquid endosperm of the coconut contains not only IAA and GA at high levels, but also cytokinin. The rich supply of hormones in the endosperm probably supports the growth of the embryo. Some of these hormonal effects can be initiated by applying hormones to the young fruit (Fig. 18-5).

If one applies auxin directly to the ovary or to the style it may grow into a fruit without needing fertilization by the pollen, even though it contains no seeds. Among the seedless or "parthenocarpic" fruit thus produced are crookneck squash, tomato, bell pepper, and watermelon (Fig. 18-6). Some fruits fail to respond to auxin but respond to gibberellin, for example the apple, pear, almond, and peach. Thus probably both auxin and GA are needed for fruit growth, and either one can be a limiting factor if some of the other is present. In tomatoes *either* auxin *or* GA will act, but if both are supplied the resulting fruit size is greater than that with either one alone.

b. FRUIT GROWTH

Once the fruit is "set" the ovary is converted into a fruit primarily by growth in size (cell enlargement), not change in shape. In many cases active cell division occurs before cell enlargement; in others there is little cell division first. The volume often increases more than the weight as the number and size of intercellular spaces increase. For example, about one-fourth of the volume of an apple is intercellular space. The hormonal control of this growth is described in Chapter 10.

Substantial amounts of carbohydrates, water, mineral elements, and organic acids accumulate with the increase in size of the fruit. Since most young fruits are green, they produce by photosynthesis some carbohydrate; however, most of the carbohydrate present is translocated from the leaves. Not only are these carbohydrates used in respiration and assimilation, but some accumulate as the fruit is growing. About 10 to 20% of the dry weight of most fruits is primarily carbohydrate. Starch accumulates in the pericarp of the ovary. Organic acids (e.g., citric, malic, tartaric) accumulate in vacuoles; they are probably derived partly from dark fixation of carbon dioxide by pyruvic acid or PEP acid and partly from drainage from the Krebs cycle.

c. FRUIT RIPENING

"Ripening" indicates the condition that makes the fruit useful or edible to animals, including ourselves. Actually this condition initiates the period of senescence in the development of the fruit. Ripening usually is completed before fruits fall off or are picked, but avocados and mangoes do not ripen on the tree and the ripening of bananas is promoted by picking.

Some characteristics associated with the ripening of fleshy fruits are obvious and others are less so. One of the most prominent changes is in the pectic compounds, which are gel-forming substances. These pectic compounds are derived from polysaccharides other than cellulose. There are three principal kinds: pectic acid, pectin, and protopectin—all polymers of (mainly) galacturonic acid, the acid formed when the CH_2OH group of carbon atom number 6 of galactose is oxidized:

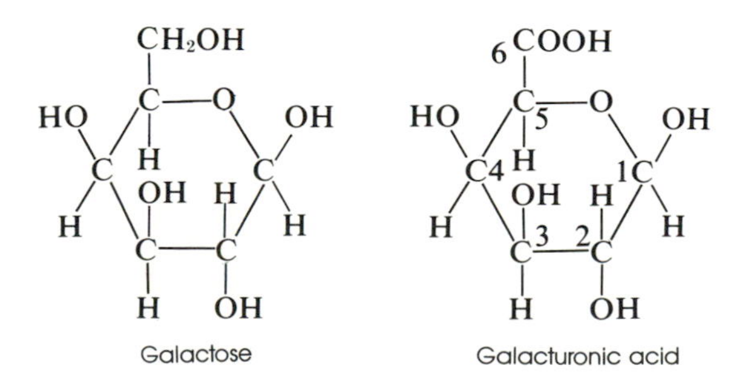

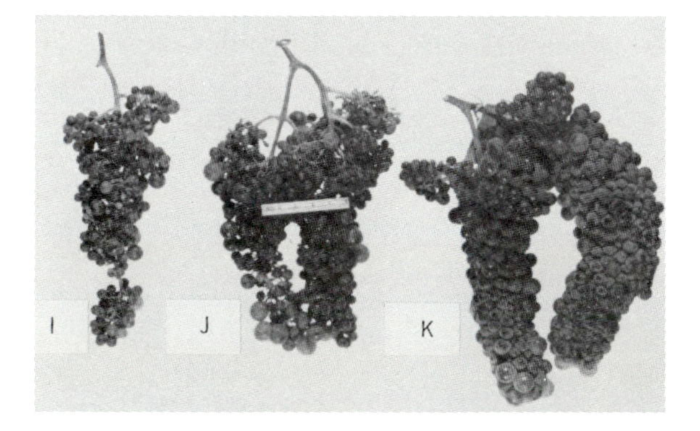

Figure 18-5 The effects of different plant hormones on grapes. A to *E*. Control (A) with effects of increasing concentrations of auxins (*B-E*) on Tokay grapes *F* to *H*. *F* after 59 days of treatment of Black Corinth grapes with gibberellin; *G* clusters girdled and *H*, control. *I* to *K*, Control *I* with effects of increasing concentrations of cytokinins on Black Corinth grapes *J* and *K*.

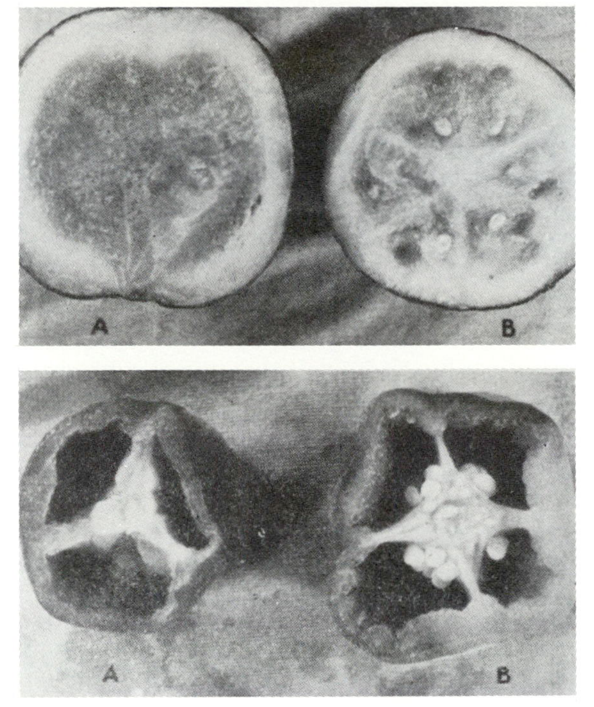

Figure 18-6 Parthenocarpic fruits induced by napthalene acetic acid. Top. Watermelon. A. Longitudinal section of NAA-treated watermelon without seeds, B. Cross section of natural-pollinated watermelon of the same variety with seeds. Bottom. Pepper. A. From NAA-treated blossom with no seed. B. From natural-pollinated pepper.

Protopectin is a large polymer in which these units, with their carboxyl groups methylated, are linked through their carbon atoms numbers 1 and 4 into chains of many hundreds of units:

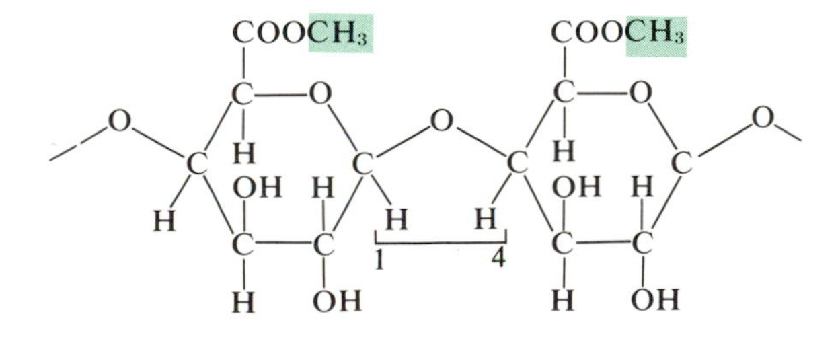

Pectin is a somewhat smaller molecule derived from it during development. Pectin is a major constituent of primary cell walls and may make up 15% of the total dry weight of the apple, and up to 35% of that of the lemon rind. It is a principal constituent of the middle lamella and acts as a cement that holds cells together (Chapter 4). As fruits ripen the methyl esters in pectin are hydrolyzed, forming pectic acid.

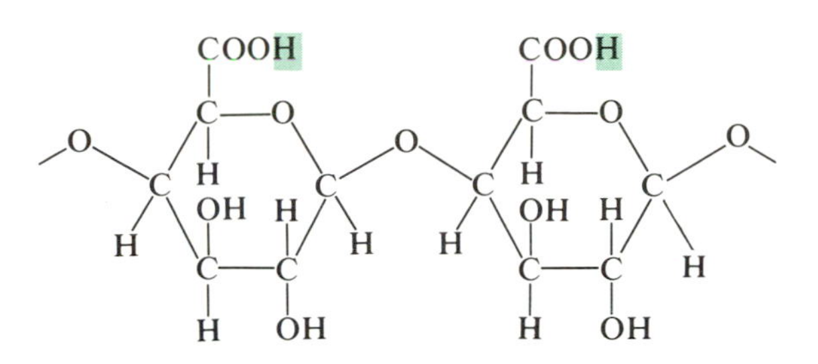

Much of this is present as calcium and magnesium salts, and since these are divalent ions they help to link the molecules together in pairs and thus contribute to the strength of the middle lamella. When the ripe fruits abscise (Chapter 10), it is the middle lamella that seems to dissolve first. In overripe apples, decomposition of the middle lamella may result in the fruits' becoming "mealy."

When small amounts of pectin are dissolved in concentrated sugar at an acid pH, the whole sets to a rigid gel. Thus to make good jellies and jams it used to be necessary to pick fruit at just the right stage of ripening (where the concentrations of sugar, pectin, and pH were appropriate). Now commercially prepared pectin (usually from apples or citrus fruits) may be used to insure against failure to have "just the right stage" of fruit ripening to set the jam or jelly.

Another characteristic change in ripening is the hydrolysis of starch to sugar, with consequent increases in the amounts of glucose, fructose, and sucrose. Likewise there generally is a decrease in the amount of organic acids; lemons are one of the few fruits in which organic acids continue to increase during ripening.

Also in some fruits (such as persimmons) astringent phenolics (tannins) that produce a puckering effect in the mouth polymerize, binding to the cell wall so that the unpleasant taste no longer occurs. In others volatile essential oils are synthesized which enhance the flavor and aroma. Change in pigments is also one of the most obvious differences occurring with maturation. The green chlorophyll usually breaks down (except in such cases as avocados and watermelons) unmasking the yellow pigments (xanthophylls, carotenes, and lycopenes); and at the same time red and purple anthocyanins are produced.

Changes in respiratory rates accompany the ripening process. As fleshy fruits complete their growth, there is first a gradual decrease in rate of respiration followed by a marked increase that is called the *respiratory climacteric* (Fig. 18-7). As pointed out in Chapter 10, the production of ethylene is closely associated with this climacteric. Ethylene also will induce ripening in fruits such as oranges, which normally do not have a climacteric. Ethylene content gets quite high in avocados, pears, and cherimoyas, whereas citrus fruits, which generally produce very little ethylene, ripen slowly. Ripening in these cases does not involve the considerable hydrolysis that comes with ripening of fruits with a climacteric. In a few cases, the application of auxin can also hasten ripening, presumably because it increases ethylene synthesis.

Temperature has a marked effect on the climacteric; generally, the higher the temperature, the sharper and higher the peak. Low temperatures tend to suppress or obliterate the climacteric. Also the ripening process can be

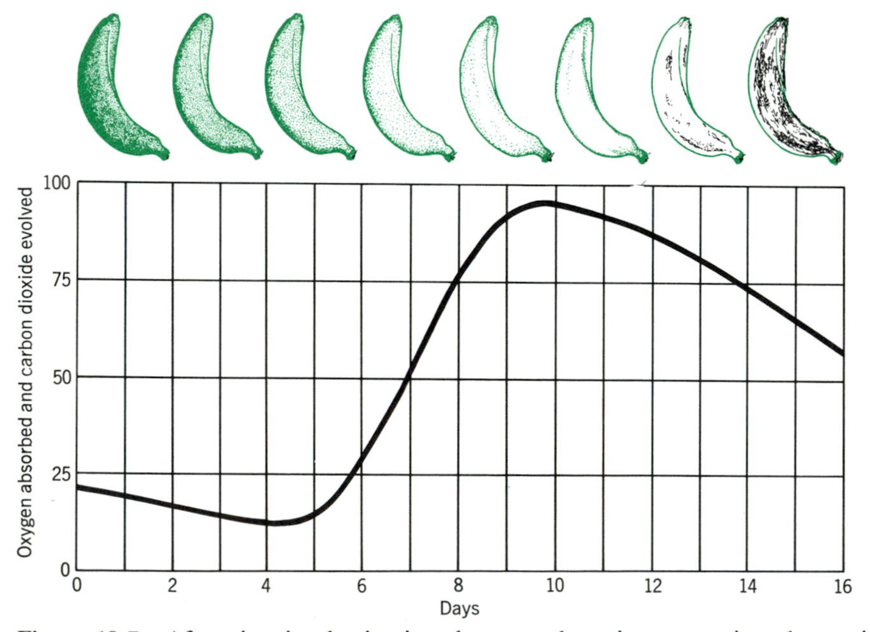

Figure 18-7 After ripening begins in a banana, there is progressive change in color from deep green to yellow. The respiration curve, in which either oxygen uptake or carbon dioxide evolution is plotted, shows that the fruit is ripe just after the high point or "climacteric." (From Biale, 1954)

slowed down by storage in atmospheres with less than the usual amount of oxygen (21% in dry air). If oxygen is too low, fermentation results, but 5 to 10% oxygen prolongs storage life and improves the quality of the fruit. Some fruit is stored in an atmosphere containing 5 to 10% carbon dioxide, which inhibits ethylene action and thus delays ripening.

4. SOME REPRESENTATIVE FRUITS

In the tropics, a wide assortment of wild species is still important to local populations but in temperate climates the once commonplace activity of collecting wild fruits and berries has now largely been replaced by mass marketing of especially grown cultivars. Even so, it is surprising how relatively little changed most fruits are from the ancestral types, compared to the tremendous changes resulting from domestication of the cereals. New types of fleshy fruits have arisen more from chance crossings and mutations than from genetic design. Breeding programs have been concerned primarily with improving characteristics already established in earlier selections, since large-scale crossing with long-lived perennial plants is extremely slow.

Although temperate fruits are more thoroughly developed as horticultural crops, specialized industries have developed to market such tropical fruits as bananas, pineapple, and papaya and subtropical ones such as citrus (Table 18-2). The volume of citrus and bananas entering world markets is on a par with that of temperate zone apples (about 20 million tons annually) but the grape leads all others in total production (about 50 million tons), largely because of its use in making wine. The coconut ranks second (28 million tons) after the grape —but much of its commercial usage is for copra (Chapter 27). Pears, peaches, apricots, plums, pineapples, dates, cherries, and figs rank next, essentially in this order of importance in world commerce.

a. TROPICAL FRUITS

A tremendous number of fruits are found locally in tropical markets and a few provide staple food for a large number of people—banana, coconut, and breadfruit.

Banana. The taxonomy of the banana is confusing but most cultivated bananas today are included within the species *Musa paradisiaca* var. *sapientum, M. sapientum* or *M. cavendishii* (Musaceae). Cultivars of the banana include the "plantains," which are cooked rather than eaten raw. A related species, *M. textilis,* (abacá), is grown for the leaf sheath fiber, not the fruit (Chapter 21).

The banana plant looks like a tree, but actually is a gigantic herb (probably the largest in the plant kingdom) which has a curious form of growth enabling it to attain a height of 10 m (Fig. 18-8). Below ground a rhizome bears roots that last many years. The ensheathing bases of the petioles form a pseudostem, the leaf blades being 3 or more meters long. Every 8 to 14 months a terminal, pendulous inflorescence comes up through the hollow "stem." The flowers are covered by large purple bracts, with female flowers toward the end of the inflorescence and the male ones below. The female flowers are generally sterile in the cultivated species and therefore the ovaries develop unfertilized. Thus, the fruit is botanically a parthenocarpic berry. Since it does not set seed, it is reproduced mainly by rhizomes or suckers.

Wild bananas were probably used wherever they occurred with human settlements, but one of the first steps in domestication was establishment of parthenocarpic, seedless diploids of *Musa acuminata,* which had to be propagated vegetatively. The next step was the production of triploids and tetraploids either from *M. acuminata* alone or in combination with *M. bulbisiana.* Such derivatives were produced on each side of the zone in which the diploid *M. acuminata* was grown from East India and Burma on the west to the Philippines

TABLE 18-2

Plants Commonly Cultivated for Their Fleshy Fruits

Common name	Scientific name	Family
Tropical regions		
Banana	*Musa paradisiaca* *Musa sapientum* *Musa cavendishii*	Musaceae
Mango	*Mangifera indica*	Anacardiaceae
Coconut	*Cocos nucifera*	Palmaceae
Papaya	*Carica papaya*	Caricaceae
Pineapple	*Ananas comosus*	Bromeliaceae
Breadfruit	*Artocarpus altilus*	Moraceae
Subtropical regions		
Avocado	*Persea americana*	Lauraceae
Lemon	*Citrus limon*	Rutaceae
Lime	*Citrus aurantifolia*	Rutaceae
Orange	*Citrus sinensis*	Rutaceae
Tangerine	*Citrus reticulata*	Rutaceae
Grapefruit	*Citrus paradisi*	Rutaceae
Fig	*Ficus carica*	Moraceae
Date	*Phoenix dactylifera*	Palmaceae
Temperate regions		
Apple	*Pyrus malus*	Rosaceae
Pear	*Pyrus communis*	Rosaceae
Peach	*Prunus persica*	Rosaceae
Apricot	*Prunus armeniaca*	Rosaceae
Plum	*Prunus domestica*	Rosaceae
Cherry	*Prunus* spp. (*avium* and *cerasus*)	Rosaceae
Strawberry	*Fragaria* spp.	Rosaceae
"Brambles" (raspberry, blackberry, loganberry)	*Rubus* spp.	Rosaceae
Grape	*Vitis vinifera* *Vitis labrusca*	Vitaceae
Watermelon	*Citrullus vulgaris*	Cucurbitaceae
Muskmelon	*Cucumis melo*	Cucurbitaceae
Cucumber	*Cucumis sativus*	Cucurbitaceae
Pumpkin and squashes	*Cucurbita* spp.	Cucurbitaceae
Tomato	*Lycopersicum esculentum*	Solanaceae

Figure 18-8 Banana plant *(Musa paradisiaca). A.* Young plant. *B.* Inflorescence. *C.* Leaf detail. *D.* Ripe fruit. (From Masefield et al., 1969)

and Borneo on the east. The names *M. paradisiaca* and *M. sapientum* refer to certain clones of the *M. acuminata* and *M. bulbisiana* hybrids. *Musa cavendishii* is a mutant of the *M. acuminata* triploid type.

Although bananas were first domesticated in Southeast Asia (being known in India more than 2000 years ago) they were introduced to Africa via Arabian traders before historic times. They were first encountered by Europeans in Africa during the fifteenth century and the name "banana" was first applied in Sierra Leone. Like sugarcane, the banana apparently was introduced to tropical America soon after the first voyage of Columbus. Today Latin America produces three-fourths of the world's bananas with Africa a distant second. Distribution of its cultivation today follows the same pattern as that of manioc and sweet potato.

Most dwellings in the humid tropics have a few banana trees around them to be used as a staple food. A ripe banana is quite wholesome, with a high carbohydrate content (23%), some fat, and a little protein (Table 18-3); it is also surprisingly high in potassium. It is eaten either raw, boiled, baked, fried, or pounded into a meal.

Bananas also provide a remunerative cash crop. On a good soil they can annually produce 4 tons of fruit per hectare. By picking them green and keeping them below 14°C during transportation, growers have made bananas one of the cheapest fruits available in temperate regions during the winter seasons. In the early twentieth century the banana fruit companies developed large corporate empires in Latin America (such as the United Fruit Company in Honduras), with extensive plantations, privately owned railways and steamships. Now these great estates are generally under local control, but bananas still constitute one of the basic items in the economies of many Latin American countries.

Coconut. The coconut palm is often referred to as "man's most useful plant." In fact, palms are considered to be the single most important group of plants in tropical economy—yielding not only edible fruit but abundant oils, waxes, fibers, and other products. The coconut palm (*Cocos nucifera*) provides the basic materials for survival of many South Sea Island inhabitants. Because the fruit floats and is easily transported by ocean currents, it is widely distributed throughout coastal areas in tropical and subtropical countries. The outer husk is waterproof and safeguards the inner nut or seed.

The coconut fruit is technically a drupe, with fibrous instead of fleshy mesocarp (Fig. 18-9). This fibrous mesocarp is called "coir" and is used in various ways as a fiber (Chapter 21). What we generally see in temperate zone markets is the shell (endocarp) covering the seed. The seed coat surrounds the endosperm, which when dried is called "copra"; when the endosperm is young, it is liquid ("coconut

TABLE 18-3

Constituents of Fruits Grown in Temperate, Subtropical, and Tropical Climates

				Minerals			*Vitamins*	
	Protein	*Fat*	*Carbohydrates*	*Calcium*	*Iron*	*Phosphorus*	*C*	B_1
Temperate zone								
Apple	0.3	0.4	12.0	7	0.4	12	6	0.03
Apricot	1.0	0.1	10.4	13	0.6	24		
Cherry	1.2	0.5	10.5	19	0.4	30	9	0.05
Grape	0.8	0.4	14.9	11	0.3	10	4	0.05
Peach	0.5	0.1	8.8	10	0.3	19	8	0.05
Pear	0.7	0.4	8.9	15	0.3	18	4	0.06
Plum	0.7	0.2	8.3	20	0.6	27	6	0.13
Strawberry	0.6	0.6	5.0	34	0.7	28	55	0.03
Subtropical								
Avocado	1.7	20.0	5	10	0.6	38	16	0.06
Date	2.2	0.6	75	72	2.1	60	0	0.09
Fig	1.4	0.4	20	54	0.6	32	2	0.06
Grapefruit	0.5	0.2	10	22	0.2	18	40	0.04
Lemon	0.9	0.6	8	40	0.6	22	50	0.04
Orange	0.9	0.2	11	33	0.4	23	50	0.08
Tropical								
Banana	1.2	0.2	23	8	0.6	23	10	0.04
Mango	0.7	0.2	17	9	0.2	13	41	0.06
Papaya	0.6	0.1	10	20	0.4	23	55	0.03
Pineapple	0.4	0.2	14	16	0.3	16	25	0.08

Adapted from J. A. Biale. "The Ripening of Fruit", Copyright May 1954 by *Scientific American*. All rights reserved.
NOTE: The numbers in the first three columns are in grams per 100 grams; those in remaining columns in milligrams per 100 grams.

milk"). Being a typical monocotyledonous plant, the fruit has three carpels. Although each one is represented by an "eye," only one of the fused carpels develops into a mature coconut. Thus one "eye" remains soft where the embryo occurs and pushes out when it germinates, the cotyledons remaining inside (Fig. 18-9).

The coconut is harvested throughout the year, since fruits reach maturity at different times. The inflorescence is produced in the axis of each leaf and bears many flowers—250 to 300 male and 20 to 40 female ones. The male and female flowers do not mature at the same time: they are cross-pollinated by insects and thus there is considerable variation in progeny.

Breadfruit. The breadfruit (*Artocarpus altilis*, in the Moraceae) is one of several useful species in this genus. It is native to the East Indies where it is a staple in the diet of the Pacific Island peoples. The handsome monoecious trees produce large, rough multiple fruits (Fig. 18-10*A*) resembling North American "hedge-apples" (*Maclura*). The fruit is starchy, nor-

Figure 18-9 Coconut palm *(Cocos nucifera)*. *A*. Immature fruits. *B*. Ripe fruit, above and seed, below. *C*. Opened seed or nut ($\times\frac{2}{3}$). *D*. Young plant growing from fruit. (From Masefield, et al., 1969)

mally 30 to 40% carbohydrate, and is usually eaten baked, boiled, or fried. It resembles potatoes in both taste and consistency.

This fruit became famous for the drama associated with its introduction from Tahiti in the Pacific to the West Indies. Breadfruit trees were thriving in Tahiti when James Cook, Joseph Banks, and botanist Daniel Solander arrived there exploring in 1769. Solander wrote that it was used as a substitute for bread by the ship's company, and even called it "one of the most useful vegetables in the world." In the eighteenth century Great Britain's sugar came primarily from its West Indies colonies and the labor employed in harvesting and processing the cane was supplied by African slaves. Therefore, the planters saw in the breadfruit a good, cheap food for their slaves and were moved to explore its use. These ideas greatly impressed Banks who urged King George III to outfit an expedition to introduce the breadfruit to the British West Indies. In 1773 William Bligh, Cook's sailing master, took command of the HMS Bounty, which was outfitted as a "floating garden" (Fig. 18-10*B*). After considerable trials in reaching Tahiti, Bligh and his crew remained six months there collecting plants and getting them well enough established to make the voyage to the West Indies. The idled crew became enamored with Pacific island life and

(A)

(B)

Figure 18-10 *A*. The breadfruit *(Artocarpus altilis)*. The starchy fruit that became famous with the drama associated with its introduction from Tahiti to the West Indies. (From Masefield, et al., 1969) *B*. Artist's conception of "The Mutineers turning Lt. Bligh and part of his officers and crew adrift from His Majesty's Ship the Bounty."

set sail with reluctance. Off the Friendly Islands, Bligh's ship was seized by members of the crew; Bligh and 18 crew members were cast adrift in the Bounty's open longboat with small rations (Fig. 18-10*B*). Almost miraculously Bligh pointed the boat westward and arrived on the Dutch island of Timor—5000 km away. He then returned to England via a Dutch vessel. The mutineers returned to Tahiti, and took their Tahitian women with them to Pitcairn Island in the hope of setting up a Utopian society.

Despite the "Mutiny on the Bounty," neither Bligh nor Banks was discouraged, and a new breadfruit expedition arrived in Tahiti in 1792. More than 2000 breadfruit were gathered and grown, and in 1795 Bligh's vessels had left 544 breadfruit for introduction in Kingston, St. Vincent—and over 1000 in Port Royal, Jamaica. An early nineteenth-century historian wrote that "the cultivation of these valuable Exoticks will, without doubt, lessen the dependence of the Sugar Islands on North America for food and necessities." Unfortunately these hopes never materialized because the problems of human taste had not been considered. The breadfruit trees thrived and produced large crops of fruit, but the slaves refused to eat the breadfruit, preferring the more familiar yams and plantains. For 50 years it was primarily fed to pigs and only in the middle of the nineteenth century did it come to be adopted as a human food in the British West Indies. This is a story that continually repeats itself with a variety of plant introductions and is frustrating in terms of feeding malnourished people in various parts of the world.

b. SUBTROPICAL FRUITS

Although the distinction between tropical and subtropical fruits is not clearcut, those we discuss here seem to flourish better in other than equatorial climates.

Citrus. Orange, tangerine, lemon, lime, and grapefruit are perhaps the most familiar members of the genus *Citrus* (family Rutaceae). All except the grapefruit, which apparently arose from hybridization between the orange and pumelo during the seventeenth century in the West Indies, are probably native to Southeast Asia, where they were selected before historical times. Many hybridizations probably have occurred, although all species of *Citrus* have the same diploid chromosome complement.

The citrus fruit (Fig. 18-2*A*), is an excellent source of vitamin C and various fruit acids (Table 18-3). The rind has many glands that are the source of essential oils (Chapter 26).

Most citrus today is grown in subtropical America, although southern Europe, Japan, North Africa, and the Near East are important producers. Oranges and tangerines are the most popular, constituting about four times the production of all other citrus fruit combined. Grapefruits, on the other hand, find their market almost solely in the United States.

Since citrus fruits ship well to distant markets, they are generally allowed to ripen almost fully on the tree, except for lemons and limes which are picked green. The demand for preserved citrus juice also is increasing. In the past most juice was preserved by canning, but today frozen concentrate (prepared by boiling the juice in vacuum pans and then freezing the concentrate) is more popular in the United States. More than half of Florida orange and grapefruit crop is now so processed. The expressed liquor from the processing residues (peel, pulp, and seed) is evaporated to yield oils and a sugary liquor. The press cake is dried separately for cattle feed. The essential oils are used for flavoring, perfumery, pharmaceuticals, and soap.

Avocado. The avocado, alligator pear or aquacate (*Persea americana*—Lauraceae) is native to the New World, but now has been introduced worldwide. It was an important staple food of the early Central American civilizations. Archeological studies at Tehuacán, Mexico, show it had been brought early from the tropical lowlands into this semiarid habitat. The fruit also was greatly increased in size by prehistoric selection. Today the avocado is used more as a delicacy than a staple food, largely because of its high percentage of digestible fat (as much as 30%, Table 18-3). Although some cultivars have been selected, there are fewer widely recognized varieties maintained by vegetative propagation than in citrus, apples, and other more commercially important fruit.

Fig. The genus *Ficus* (Moraceae) contains many wild, semidomesticated, and cultivated species of "figs." The fig fruit is an unusual structure (called a *syconium*) in which the unisexual flowers occur inside a pear-shaped, nearly enclosed receptacle (Fig. 18-11). The most commonly cultivated fig (*Ficus carica*) is native to the eastern Mediterranean area; it has been cultivated in the Holy Land for at least 5000 years and was recorded in Egyptian documents around 4700 B.P. During classical times figs accompanied olives and grapes as the main horticultural elements of rain-dependent agriculture in the Mediterranean region, providing fresh fruit in the summer and sweet dry figs through the year. Fig "trees" vary widely in size and shape, ranging from climbers to orchard trees about the size of an apple tree. Wild figs depend on birds and fruit bats to eat the ripe fruit and disseminate the seed, but the orchard trees are propagated by cuttings. Two cultivars, the Kadota and Mission figs, are self-pollinated or parthenocarpic, and can produce a nearly continuous sequence of fruits. The larger, more delectable Smyrna figs (var. *smyrniaca*), however, bear syconia with only female flowers and thus are dependent on pollen from other plants for fertilization. The wild or semicultivated caprifigs (var. *sylvestris*) have male flowers, and the larvae of a small wasp (*Blastophaga*) develop in the syconia of caprifigs and carry the

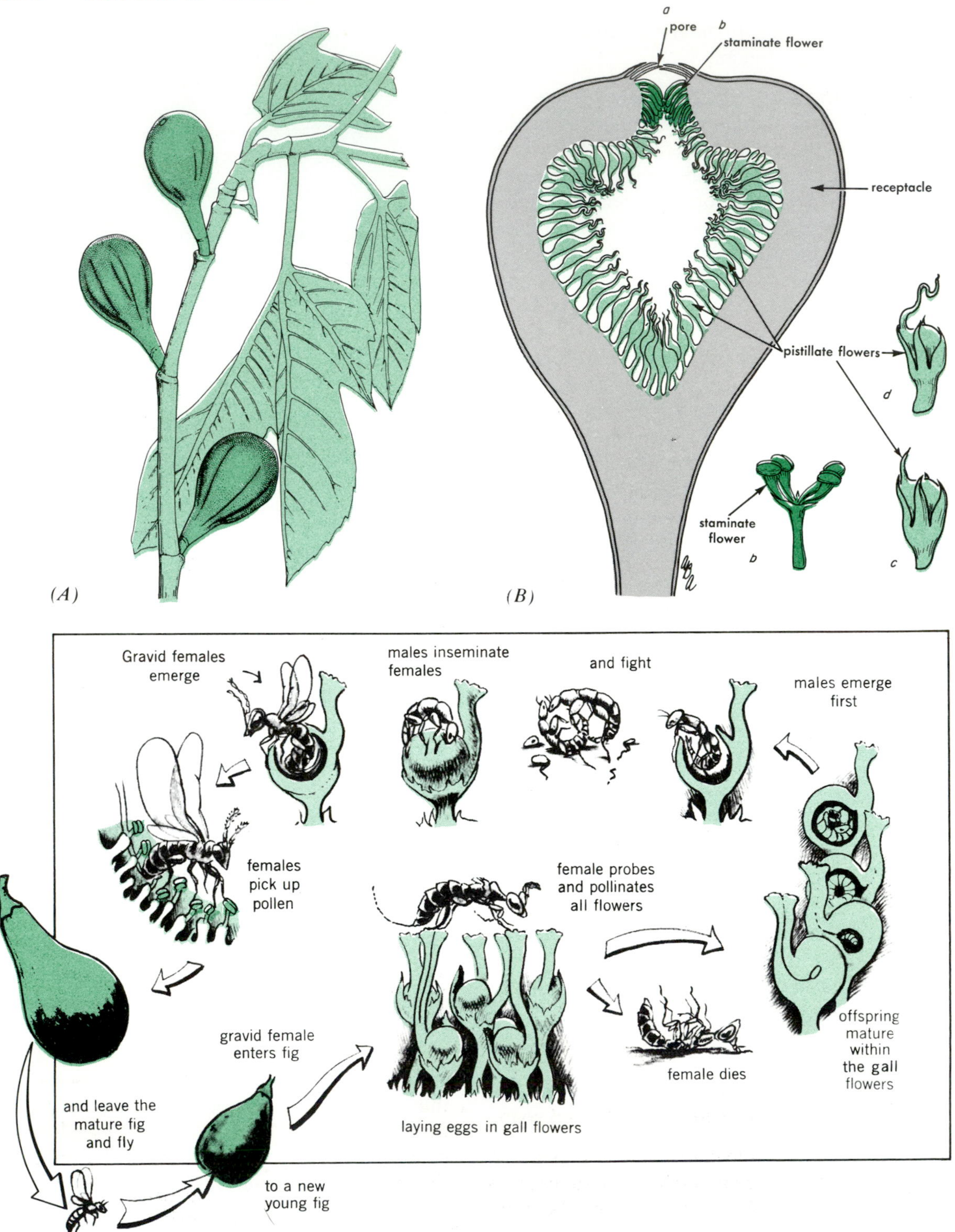

a
pore
b
staminate flower
receptacle
pistillate flowers
d
staminate
flower
b
c
(A)
(B)
Gravid females
emerge
males inseminate
females
and fight
males emerge
first
females
pick up
pollen
female probes
and pollinates
all flowers
gravid female
enters fig
female dies
offspring
mature
within
the gall
flowers
and leave the
mature fig
and fly
laying eggs in gall flowers
to a new
young fig

Figure 18-11 Fig *(Ficus carica) A*. Fruiting branch with syconia (From Scagel et al, 1969) *B*. Syconium in longitudinal section, showing the compound structure made up of the flattened axis of the inflorescence, which becomes a hollow inverted urn with a small apical pore opening to the outside. *(a)* Flowers line the inside of this urn. *Ficus carica* is monoecious and bears three generations of flowers and fruits in a year. The first contains staminate flowers *(b)*, formed chiefly around the pore, with abortive ovulate "gall" flowers *(c)* lower down. The gall flower has a rudimentary ovary, a short style with an open canal, and a single ovule incapable of forming a seed. As shown in the sequence below, female wasps enter the syconium and deposit a single egg in the ovule of each gall flower; here the larva hatch out, feed, grow, and undergo metamorphosis. The male wasps gnaw their way out, locate the gall flowers containing females, pierce the wall of the ovary, and fertilize the female within. The male wasps then die without leaving the syconium. By this time the fruit is ripe and the staminate flowers are shedding their pollen. The gravid female wasps leave the gall flowers and crawl out through the pore of the syconium, becoming dusted with pollen on the way. These wasps crawl about on the tree in search of young syconia in which to lay their eggs, which they find on the second generation of figs. These syconia, however, only contain normal ovulate flowers with long styles *(d)* in *(B)* above; the wasps try in vain to lay their eggs and in so doing deposit pollen on the stigmas. These syconia reopen to become fleshy and edible. In the meantime the third generation of fruits is developing. Females finally make their way into these and lay their eggs in the gall flowers, which are the only kind present. Here the larvae pass the winter, emerging in the spring to repeat the cycle. (Adapted from Hidy & Bennett, 1979)

pollen from the caprifig to the Smyrna. The emerging wasp, after pollinating the stigma, lays its eggs and dies in the fruit. Wasp and eggs are then absorbed by the fig. The process is known as *caprification* and apparently was discovered in Greece before the Christian era. The Greeks did not understand the biological facts of "caprification" but knew that caprifig stems had to be hung in Smyrna trees accompanied by suitable ritual and tribute to the Gods. Wild fig species are mostly pollinated by wasps too (Fig. 18-11). Portugal, Italy, Greece, and Turkey are the major fig-producing countries, with Israel also now becoming important. In the United States, most of the figs are of the Kadota type which originated, as a modification of an Italian cultivar, in California.

Date. Date (*Phoenix dactylifera*) is another example of the importance of the palm family. It has been cultivated in the Holy Land for at least 8000 years. According to the Muslims the date palm was made from dust left over after the creation of Adam and, therefore, is considered "the tree of life." The tree was widely distributed in the Mediterranean region by the Arabs, and its fruit has long been a staple in the diet of the nomadic tribes. It is very nutritious when dried, consisting of about 75% carbohydrate and 2% protein (Table 18-3). Dates may be eaten fresh, dried or pounded into pastes, mixed with milk (increasing the protein content), or fermented into arrak (which Pedro Texeria, a sixteenth-century traveler described as "the strongest and most dreadful drink ever invented").

The date palm is usually propagated from side shoots from the base of the trunk, and flowering begins in the fourth year. The trees are dioecious, the flowers of the female tree being dusted by hand with flowers from a relatively few male trees grown apart from the female trees. (Fig. 18-12) Date palms are generally grown in oases or under irrigation (Fig. 18-13). A single tree when mature may yield more than 100 lb of fruit yearly. The fruit is hand picked and then commonly pasteurized or sterilized. When commercially produced, dates are

Figure 18-12 Herodotus, a Greek scholar, pointed out in the 5th century B.C. that Arabs and Assyrians pollinated the date palm by hand. This relief, carved in limestone (884-859 B.C.) shows in stylized form two winged deities, holding the pollen-bearing flowers of the date palm in what has been interpreted as being the artificial pollination of the female date palm.

artificially ripened by "sweating" in incubators during which sugars develop and astringent tannins become insoluble.

The eastern Mediterranean region is the center of date cultivation, Egypt, Iraq, Iran, and Saudi Arabia producing two-thirds of the world's supply. Growing well with little care in saline habitats, but requiring relatively high temperatures, the date is being replanted in Israel and the Jordan valley where few other crops can be grown. The date is also cultivated quite extensively in southern California.

c. TEMPERATE REGION FRUITS

Many of the familiar fruits grown in temperate regions are members of the rose family (Table 18-2). The majority of the leading species arose in central Asia, where prehistoric people probably first selected for domestication the most appealing native apples, pears, plums, cherries, and apricots. Thus, most of these fruits are cultivars that were known by the time of the Greek and Roman civilizations.

Temperate zone fruits have never been basic to the human diet, since nutritionally they provide little food value (some sugar and negligible quantities of fat and protein). Yet their content of vitamins and minerals is an important adjunct to our diet (Table 18-3). Having high moisture content and delicate texture, most mature fleshy temperate fruits are highly perishable and thus generally seasonal. Many fruits can be held for some time by cold storage, but much of the fruit crop must be preserved by canning, drying, sugaring, or less commonly by salting or pickling.

Although many temperate zone fruits are seasonally important in world commerce, we

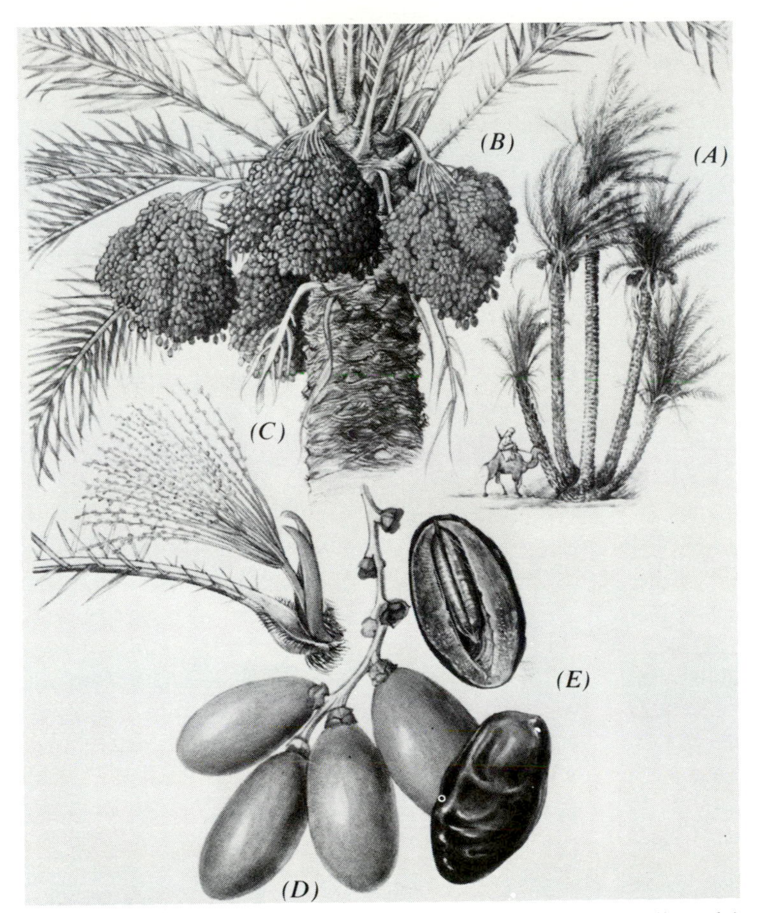

Figure 18-13 *A*. Date palm *(Phoenix dactylifera)* (small scale). *B*. Fruiting spadices. *C*. Female spadix. *D*. Ripe fruit. *E*. Dried fruit showing seed. (From Masefield, et al., 1969)

concentrate our discussion here on grapes and apples, which account for the largest share of the market.

Grapes. Many species of the grape genus (*Vitis*), in the family Vitaceae, are known in both the Old and New Worlds, and few plants over the ages have been esteemed as highly. The wine grape (*V. vinifera*) is closely associated with the beginnings of Western Civilization. This grape has had a long record of domestication, apparently being cultivated in southwestern Asia well before 7000 B.P. At that time it was brought into Palestine from the north, and inscriptions show that the wine grapes were grown in Egypt as early as 4000 B.P. Wine was a popular beverage in the Holy Land during Biblical times and is mentioned in practically every book of the Bible. The grape was also important in ancient Greece, and the Romans established many terraced vineyards.

In North America the Indians were using several native grape species before the arrival of the Europeans. Early explorers of the New World were impressed by the profuse occurrence of wild grapes; even the ancient Norse name for Viking discoveries in North America was "vinland." Size and quality of the New

World grapes, however, were not equal to those of *V. vinifera,* and as early as 1619 Lord Delaware had French stock of this species introduced into Virginia colony. For more than 200 years, however, attempts to grow *V. vinifera* failed in eastern North America, largely because of its sensitivity to cold. Finally, the successful "Concord" cultivar was developed in 1852, as a result of crossing of a native fox grape (*V. labrusca*) with the *V. vinifera* cultivar growing in a garden. The Concord grape today is still the leading "juice" grape in the northeastern United States (Fig. 18-14). Also wine grapes are grown around the Great Lakes where the winter low temperature is modified by this large body of water.

The greatest development of viticulture in North America, however, came with the settlement of California. The earliest grapes planted in California were brought by Jesuit Fathers about 1635 from Spain via Mexico, and soon the missions had thriving vineyards. Many varieties of *V. vinifera* were introduced from France and the Near East following statehood in 1850. The dry intermontane valleys in California are free from phylloxera aphids and mildew and, with irrigation, grapes of excellent quality can be produced. Today about 90% of the United States domestic crop is produced there from select cultivars grown on phylloxera-resistant root stock.

Although there are numerous cultivars today suitable for many regions of the world (over 5000 cultivars are presently known), grapes require a long summer and fairly mild winters (temperatures above −18° C or 0° F).

(A) *(B)*

Figure 18-14 *A.* White Thompson's seedless and *B.* purple Catawba table grapes ready for picking. Catawba grapes are also used for wine in the northeastern United States; Catawba was probably a parent of Concord.

The sugar content and various subtle flavors also develop best in sunny and dry climates. Grapes are propagated chiefly by cuttings, and the vines grow best on well-drained soils.

Two-thirds of the world grape production is European, especially from France, Italy, and Spain. In many localities as much as 90% of the crop is used for wine (see Chapter 20). In the Near East and California, however, production of raisin-grape cultivars is nearly three times that for wine cultivars and about four times that for table grapes.

Apples. Perhaps no other fruit tree in temperate climates has the importance of the apples. *Pyrus malus* (*Malus sylvestris, M. communis*) includes most of the cultivated fruit apples, although there are many wild species in the genus. The apple apparently was among the first fruits sought by Neolithic people in the wild. It appears to be native to the Caucasus Mountains of western Asia, where wild forms still grow, and it was then gradually adapted to orchards in northern Mesopotamia. In the New World the genus is represented by a confusing group of native crabapples which, in addition to introduced species, are used primarily as attractive ornamentals.

A ripe apple contains approximately 12% sugar and 1% fiber, with small amounts of protein, fat, and minerals (Table 18-3). It is generally harvested ripe or nearly so because immature apples ripen poorly after removal from the tree. On the other hand, apples left too long on the tree turn "mealy" because of the decomposition of the pectic portions of the cell wall. Ripe fruit stored in cool, well ventilated areas (to remove the ethylene) will keep for up to a year, permitting year-round marketing. However, only about one-half of the apple production is eaten as fresh fruit. Much of the remainder is processed into juice (cider). A

Figure 18-15 Apple trees in flower in orchard in northeastern United States.

controlled bacterial inoculation of cider results in acetic acid fermentation to yield vinegar, or it may undergo a natural alcoholic yeast fermentation to become "hard" cider (Chapter 20). The fruit residues at the cider mills are an important source of pectic gum. Other portions of the crop are converted into apple butter, jelly, and apple sauce.

The apple tree is a spreading, long-lived perennial able to bear fruit for as long as a century, (Fig. 18-15) although most orchards are renewed within a shorter interval. At present, however, small trees conveniently spaced for mechanized tending and harvesting are favored. Individual cultivars are propagated by budding or grafting on to selected rootstock, since apple tree cuttings root poorly. Apple trees are pollinated by bees but generally flower and set fruit abundantly only in alternate years. Sprays are routinely used in technologically advanced parts of the world, not only to protect the apple from numerous insect and fungus pests but to control set of fruit, its retention on the tree, or its fall where thinning is needed.

The apple is hardy in cold climates; in fact, it is ill adapted to tropical conditions because it requires low temperatures in order to flower. Apple orcharding, however, is favored by steady spring temperatures, which prevent early flowering and possible loss to frost. Recently new varieties have been developed for summer ripening. Climate throughout western Europe (particularly England, France, Italy and Germany) is excellent for apple growing and as a result it is the leading apple growing area with North America a distant second. In the United States the Pacific Slope of Washington, Oregon, and northern California is the primary growing area with Michigan, New York, and Virginia important in the East.

CHAPTER 19

Food from Underground

"Root crop" is a convenient catchall term for all edible fleshy underground organs; they may not actually be true roots but in many cases are modified stem tissue. Numerous underground storage organs from an assortment of wild species, representing several families, were eaten by our ancestors. In fact, in several parts of the world wild species of roots and tubers are still collected and planted. For example, a great variety of underground organs in the Oxalidaceae and Brassicaceae are used by the Peruvian Indians, and some from the Cyperaceae and Araceae are eaten by the Polynesians. As mentioned earlier in the discussion of the origins of agriculture (Chapter 3), it is possible that agriculture first developed through collecting and then simply replanting selected starchy underground organs, rather than by selection of grains. In the moist tropics this would seem likely, but of course the climate does not allow for preservation of this evidence.

The tonnage of staple root crops produced is only a little less than that of the major cereals; in fact potatoes alone provide more calories per acre than either rice, corn, or wheat (Fig. 19-1). Yet their worldwide role as a source of food is not as great as the cereals or legumes, because the food is less concentrated (i.e., there is more water per unit volume). As a result they do not transport or store as well and thus often primarily serve a local market. Also their main constituent is carbohydrate; they are low in protein and oil, although they may contain abundant vitamins and minerals.

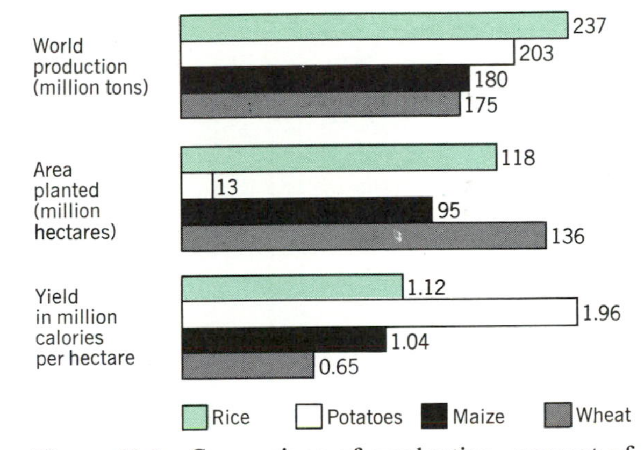

Figure 19-1 Comparison of production, amount of land planted and nutritional yield of rice, potatoes, maize and wheat. Although a less-concentrated food than the others, potatoes require much less land, and so their caloric yield per land area is the highest of the four. (Adapted from Edlin, 1967)

1. DISTINCTION BETWEEN ROOT AND STEM

Since both true roots and modified stem tissue may serve as underground storage organs, and the external form of the stem has the appearance of a root, we shall reiterate the differences between the root and stem. Details were given in Chapter 4. One of the obvious distinctions between the two is that the stem has a *terminal bud* enclosing the shoot apex (Fig. 19-2). Secondly, leaves and buds originate along the stem axis at regular intervals known as *nodes* (the segments between them are *internodes*).

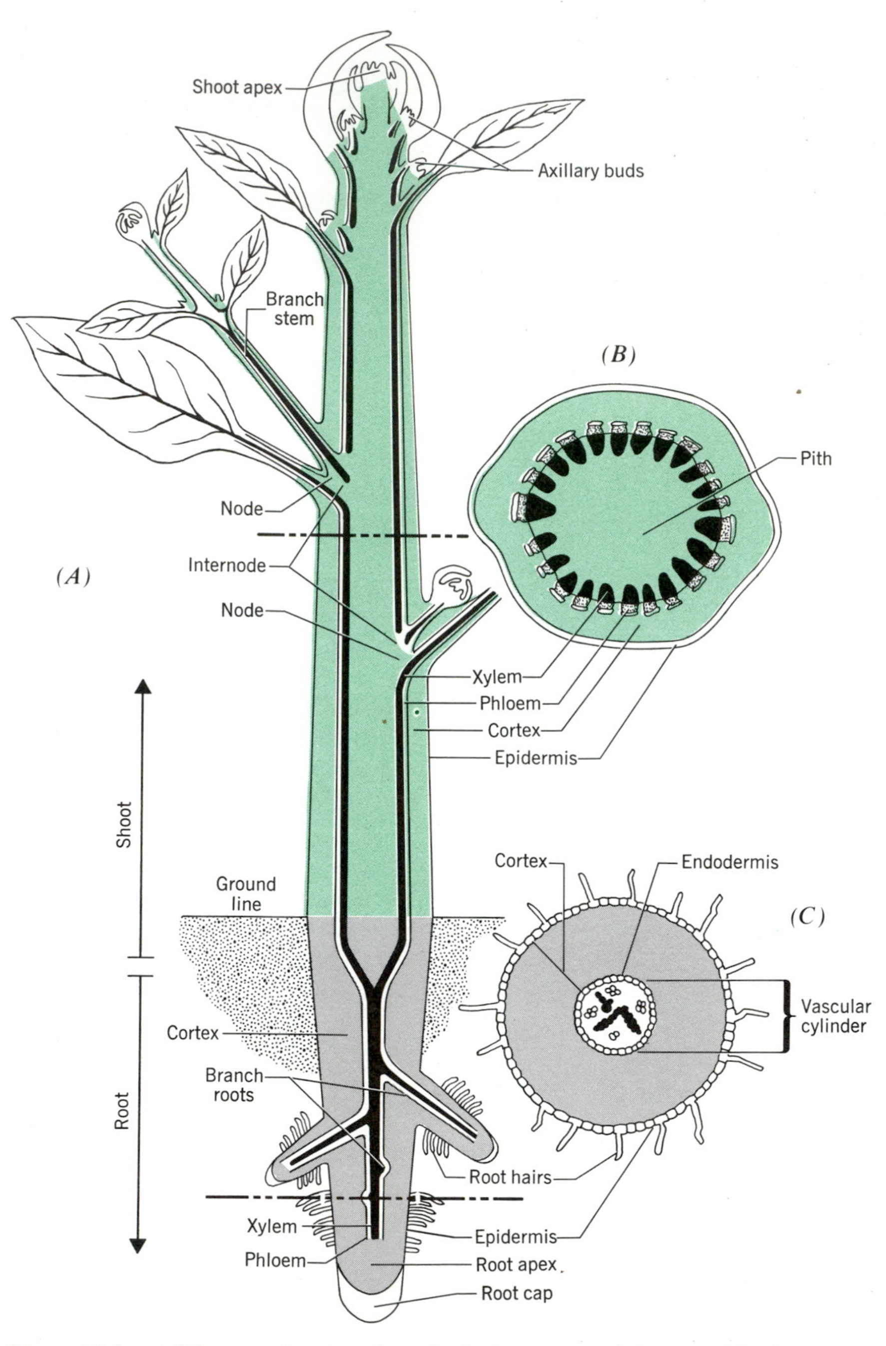

Figure 19-2 *A*. Diagram showing the principal organs and tissues of the body of a dicot flowering plant emphasizing the major distinctions between stem and root. *B*. Cross section of stem. *C*. Cross section of root.

Thirdly, branches arise in the axils of the leaves. Roots have none of these features. Moreover, the internal structures of stems and roots differ. For example, there is no pith tissue in the root but there is a very large area of cortex. Also, whereas stem branches normally originate near the surface of a main stem, branch roots arise from meristematic tissue situated some distance below the surface of the large root. Thus stems, no matter what their form or whether they are growing above or below ground, have external and internal characteristics that distinguish them from true roots. Although some plants may be propagated by means of pieces of roots, most plants may be propagated by stems, provided that these pieces include one or more nodes with their buds.

The first root of the plant originates in the embryo and is called the *primary root* or *taproot* (Chapter 4). This root grows directly downward, giving rise to secondary or *branch roots* along the way. This type of root system (i.e., one that develops from a taproot and its branches) is called a *taproot system* (Fig. 7-5). In monocotyledons, the primary root is usually short-lived and the root system develops from *adventitious* roots that arise from the stem. These adventitious roots and their branch or lateral roots give rise to a *fibrous root system* in which no one root is more prominent than the others. Taproot systems generally penetrate deeper into the soil than fibrous root systems.

All roots, even slender ones whose primary function is absorption, may store a certain amount of food temporarily. For example, when sucrose moves into root cells more rapidly than it can be utilized, it may be converted to starch and stored temporarily in the cortical cells. The sugar beet (Chapter 17) is an exception since it stores the sucrose unchanged. During a dormant season, relatively large quantities of starch are stored in tree roots, constituting a reserve that can be called upon when active growth is resumed. Arctic and alpine plants usually have large stored carbohydrate reserves in roots, which are quickly mobilized when the snow melts and the temperature is sufficiently warm for growth.

2. STEM MODIFICATIONS

Rhizomes, tubers, bulbs, and *corms* are important underground modifications of stems that serve as storage organs. *Rhizomes* are horizontal underground stems (Fig. 19-3). They may be slender with elongated internodes, as in Bermuda grass, or compressed and fleshy like iris, ginger, or asparagus. Usually roots and shoots develop from the nodal regions of rhizomes. *Tubers* are greatly enlarged, fleshy portions of rhizomes. The form is indicated by the name "tuber," which is derived from a Latin word meaning "lump." The Irish potato (Figs. 19-3 and 19-4), a typical tuber, shows the scar where the tuber was broken from the rhizome. The "eyes" of the tuber are buds; each group along the sides represents a lateral branch with undeveloped internodes. At the unattached "seed end" of the tuber the eye is actually a terminal branch on which only one bud is strictly terminal.

Bulbs appear as compressed modifications of the shoot and consist of short, flattened or disc-shaped stems surrounded by fleshy, leaf-like structures called scales (Fig. 19-3). They may enclose a shoot or flower buds. The common onion is a good example of a bulb, with the scales storing carbohydrates, principally sugars rather than starch. The bulbs of garlic comprise several small egg-shaped bulbils called "cloves," all of which are enclosed in a whitish skin. *Corms* are short, fleshy underground stems having a few nodes (Fig. 19-3). In contrast to bulbs, corms have more stem tissue (which contains stored starch) and relatively fewer scale leaves. The corm differs from the tuber in that it is the enlarged base rather than the swollen tip of a lateral stem; on its upper surface are sev-

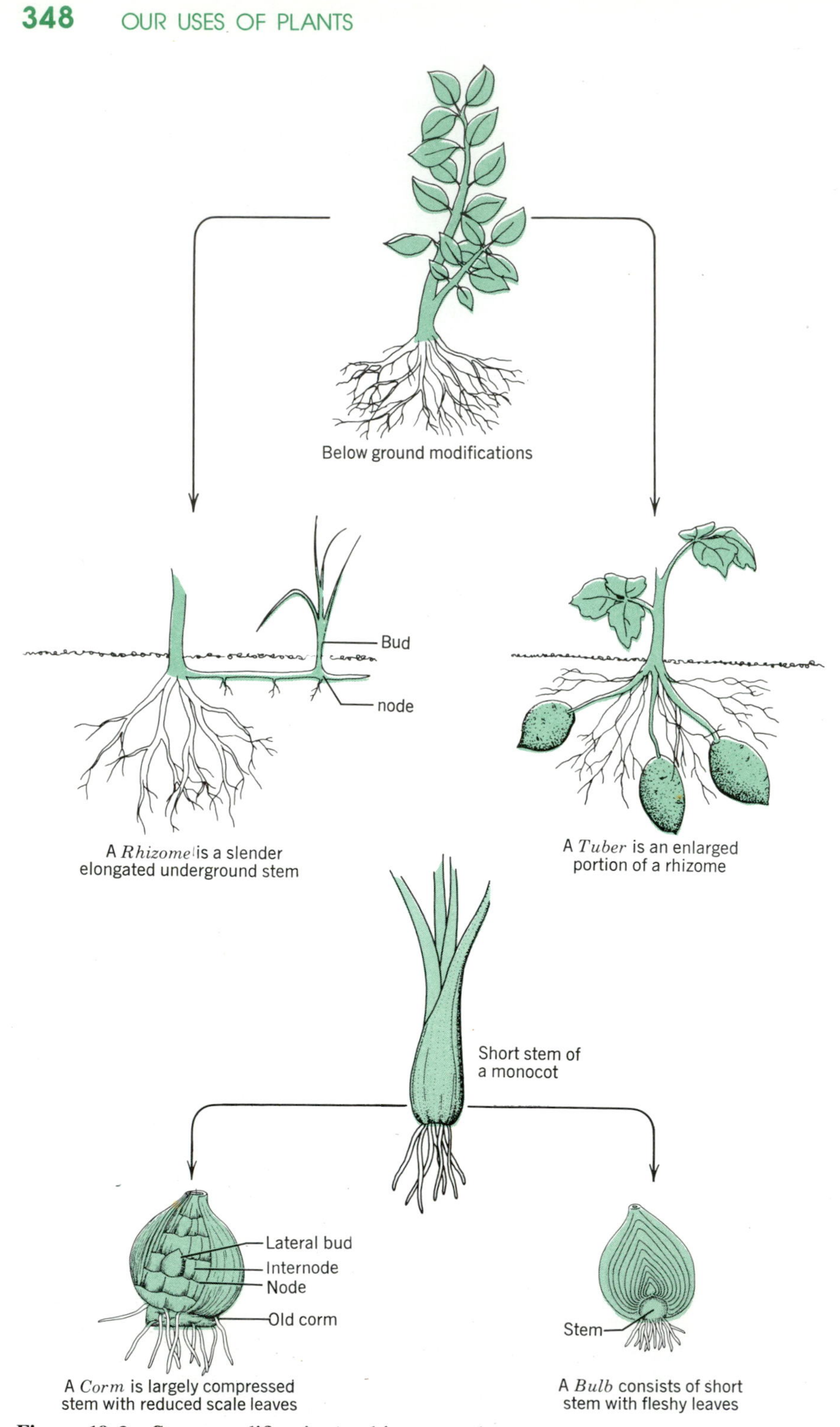

Figure 19-3 Stem modifications: rhizome, tuber, corn and bulb. Note that all have nodes and leaf-like structures.

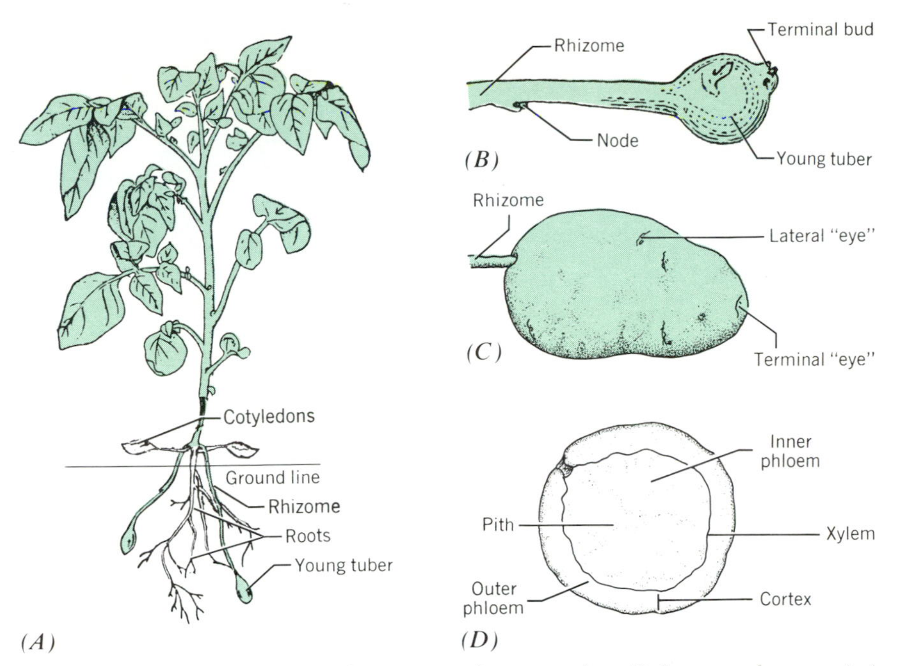

Figure 19-4 Development and anatomy of potato tuber *(Solanum tuberosum) A*. Plant showing origin of first tubers at tips of rhizomes, *B*. Stage in development of a tuber at tip of rhizome. *C*. External view of mature tuber. *D*. Cross section of tuber, showing both inner and outer phloem.

eral lateral buds and a terminal bud. Corms occur in ornamental plants such as gladiolus and crocus, and in taro, the tropical Asiatic food staple. Bulbs and corms are found only in monocotyledonous plants.

3. SYNTHESIS, STORAGE, AND BREAKDOWN OF STARCH

Carbohydrate is stored primarily as starch, which occurs as grains (up to 150 μ in size) inside plastids where the enzymes for synthesis occur. In the chloroplasts, the glucose from photosynthesis is synthesized into starch (composing relatively small grains) that is relatively transient, being converted to transportable sucrose usually within a few days. The interconversion between starch and sugars is treated in detail in Chapter 6. In the parenchyma cells of roots and other storage organs, as well as in many seedlings, the translocated sucrose is synthesized into starch in *leucoplasts* or *amyloplasts* (Fig. 19-5), which may remain for months before being hydrolyzed by amylase. Each leucoplast produces only one starch grain, so that the grains appear to be free in the cytoplasm, but electron microscopy shows them to have a double bounding membrane.

Although one might expect polysaccharides to be synthesized from any monosaccharide, most plants specialize in using pentosans and hexosans. Starch is composed of α-d-glucose residues of two different kinds—*amylose* and *amylopectin* (Fig. 19-5*A*, 19-5*B*, 19-5*C*).

Amylose is a long molecule (200 to 1000 glucose units; see boxed material at the end of Chapter 21) that forms an unbranched helix. It is soluble in hot water and therefore the kind of starch used commercially as "soluble starch" (e.g., for starching clothes). *Amylopectin* is a shorter molecule (containing 40 to 60 units) but forms a highly branched helix. It makes up 70 to 80% of the starch in most grains and legumes (almost 100% in corn and 30% in peas). Starch

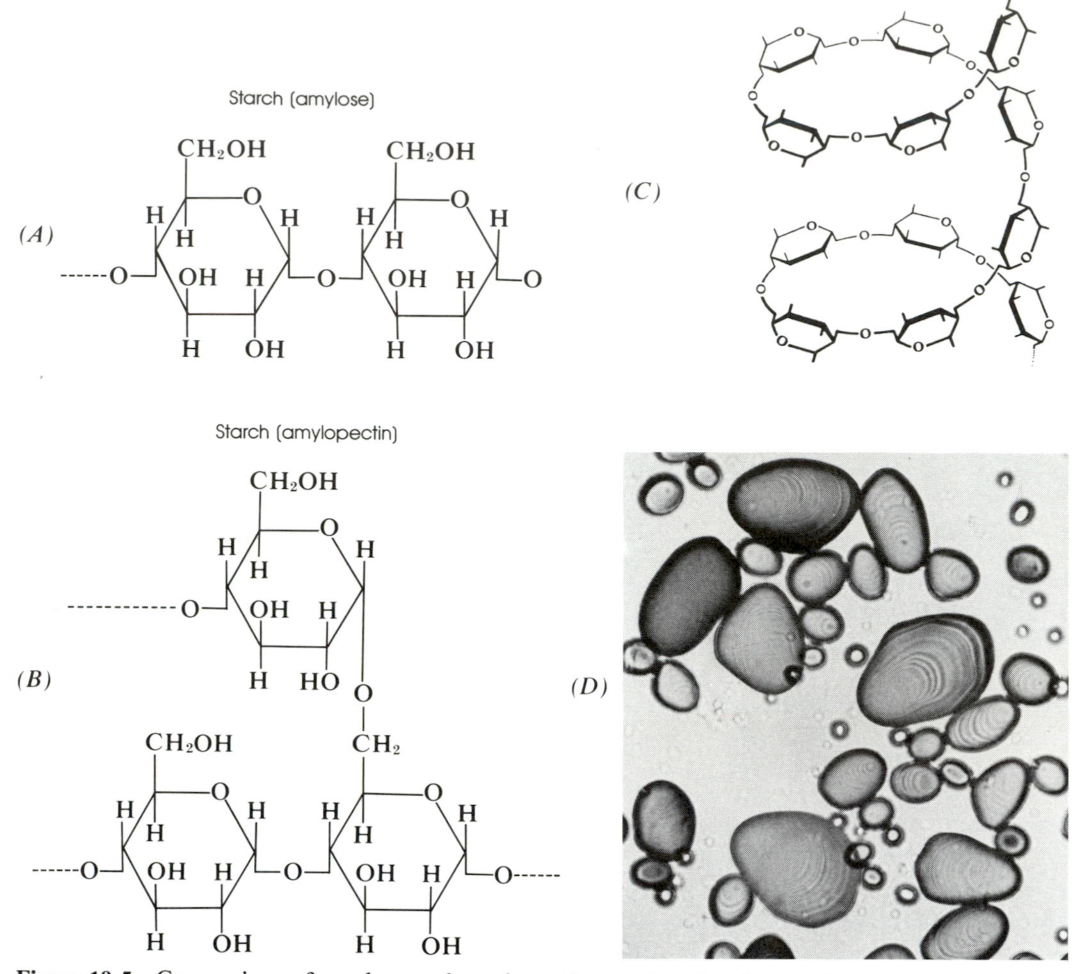

Figure 19-5 Comparison of amylose and amylopectin starch molecules. A single molecule of amylose *(A)* contains 1000 or more glucose units in a long unbranched chain. *(B)*, amylopectin is made up of 40 to 60 glucose units arranged in shorter, branched chains. Because starch molecules tend to form a helix *(C)*, they then aggregate into granules as shown by the photograph *(D)* of starch grains from the potato tuber (× 135) and in Fig. 19-6.

molecules, perhaps because of their helical nature, tend to cluster in granules. Starch is deposited in concentric layers in the grain, and there are characteristic patterns for different species (Fig. 19-6). It is thus possible to check such commercial products as corn and potato starch for adulterants.

The synthesis and hydrolysis of starch proceed by different pathways. As explained in Chapter 6, synthesis is brought about with the use of chemical energy (ATP) supplied in the chloroplast by photosynthesis, or in colorless tissues by respiration. The breakdown to sugar, however, is a relatively simple hydrolysis, brought about by the *amylase* complex of enzymes (β-amylase and maltase) that insert a molecule of water between each of the glucose residues, set out in detail in section b of Chapter 6). These conversions can be shown in a simplified diagram as follows:

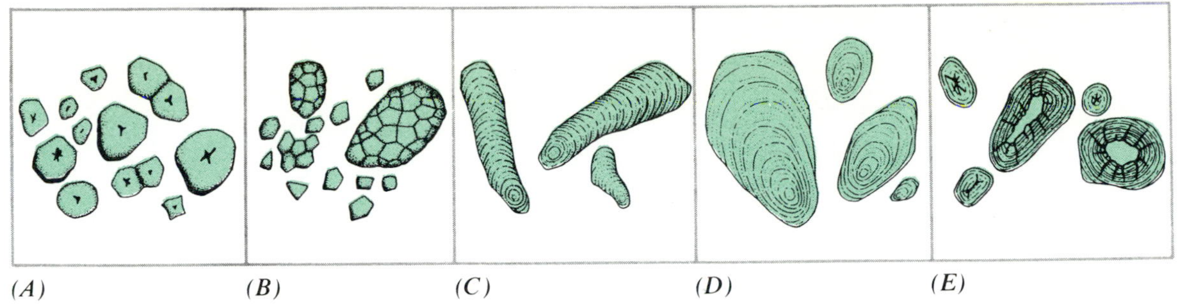

Figure 19-6 Starch grains from different plants showing characteristic patterns for each species: *(A)* Maize, *(B)* oats, *(C)* Banana, *(D)* Potato, *(E)* Wheat.

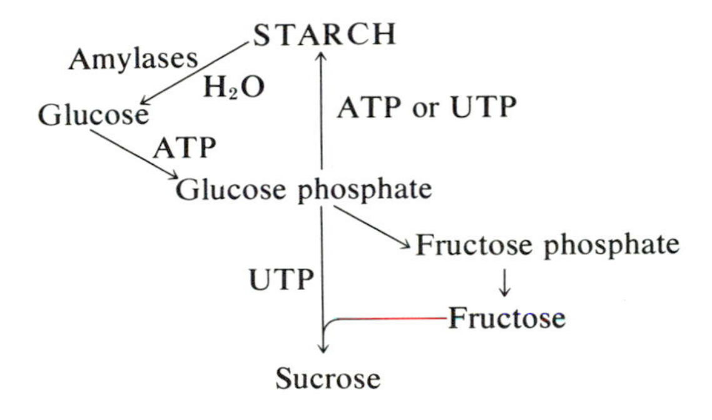

4. STORAGE IN TAPROOTS

a. ROOTS EATEN AS VEGETABLES

We eat as vegetables a good number of taproots such as beets, carrots, parsnips, radishes, and turnips (Fig. 19-7). The plants from which they come are generally *biennials;* that is, they live two years. They produce the fleshy taproot the first year and use the stored carbohydrate the second year to produce the aerial shoot and flowers.

The garden beet belongs to the same species (*Beta vulgaris*) as the sugar beet, mangelwurzels or "stock beets", and leaf beets (including swiss chard). These cultivated beets were derived probably from the wild species *Beta maritima,* which grows along the coasts of northern and southern Europe. It belongs to the family Chenopodiaceae, which has many members able to live under saline conditions in deserts. Small amounts of sodium chloride in fact promote its growth. Table beets were cultivated in the Mediterranean countries during "classical" times and probably long before that. These beets are not so rich in sugar as the sugar beet (Chapter 17) and also differ in color, shape, and taste.

Although *Beta vulgaris* is a biennial, as a garden vegetable it is grown from seed as an annual. In these fleshy roots the food is stored in parenchyma cells in alternate zones of xylem and phloem produced by successive cambiums, with resulting concentric rings of growth similar to the annual rings of a tree (Fig. 19-7E).

Turnips (*Brassica rapa,* in the mustard family, Brassicaceae) like beets are biennials forming swollen taproots in the first year and blooming the second. Originally native to Europe and western Asia, the turnip has become a cool-season crop throughout temperate regions. Unlike the beet, the food is stored almost entirely in the xylem portion of the root, outside of which is a small phloem and cortex region (Fig. 19-7C). The cells of the xylem are not lignified like the woody tissue of many plants and hence are soft and edible. Turnips were introduced into America by the first settlers, and their use was spread by both colonists and the Indians.

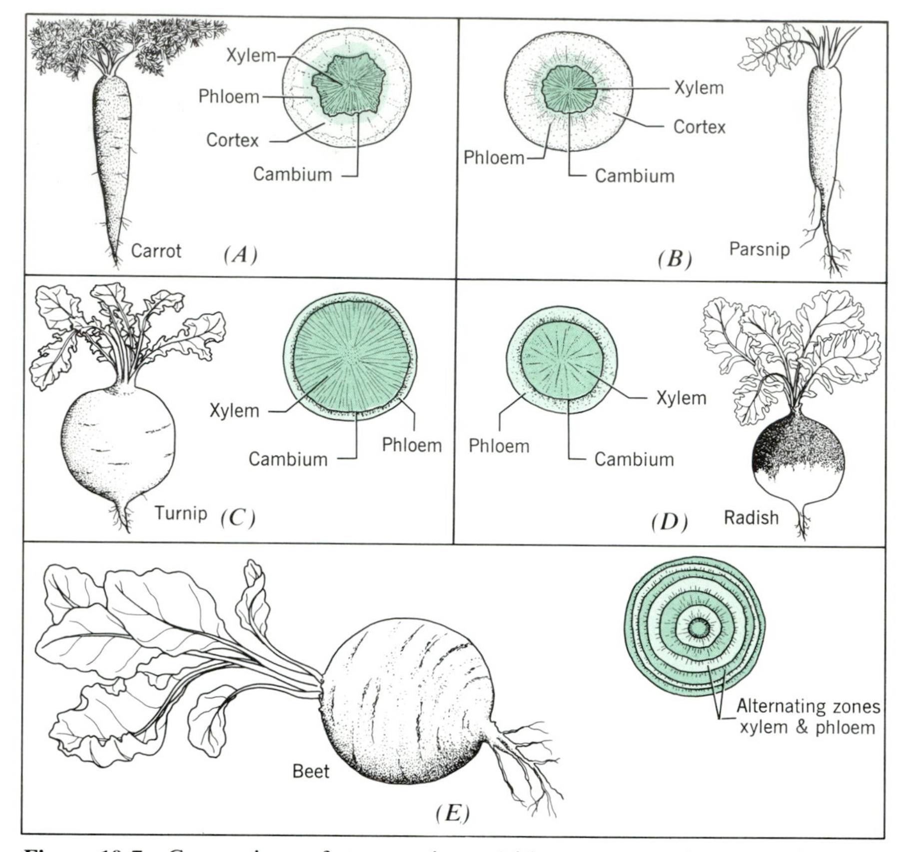

Figure 19-7 Comparison of storage tissue (either cortex, xylem, or phloem) in cross sections of *(A)* carrot, *(B)* parsnip, *(C)* turnip, *(D)* radish, and *(E)* beet.

Another member of the mustard family, the radish (*Raphanus sativus*), has a fleshy taproot at maturity similar to the turnip but smaller and with more food stored in a slightly larger phloem region (Fig. 19-7*D*). The common radish grows wild in the temperate regions of the Old World and was introduced into North America by the early settlers. It often escapes from gardens and becomes a weed wherever it is grown. Radish is one of the few economic plants long cultivated in both Oriental and Mediterranean regions. The species is presumably native to China and it is used in abundance there and in Japan. In Japan today it makes up about half of their vegetable tonnage; the Japanese radishes are as large as our turnips but very mild.

Carrots (*Daucus carota*), in the umbel family (*Apiaceae*), commonly grow wild over a large portion of the Old World. They have long been cultivated in Europe and Asia, and were early introduced into Virginia and New England. The cortex and phloem region of a carrot root is much larger than that of the radish or turnip and it is here that most of the food is accumulated (Fig. 19-7*E*). The central xylem portion is less nutritious and often becomes woody. There are numerous varieties of carrots varying in size, shape, and color. The yellow pigment, carotene, is sometimes extracted from the roots and used for coloring butter.

Another umbelliferous plant, the parsnip (*Pastinaca sativa*), still grows wild in the Mediterranean region, where it has been cultivated since Roman times. In the parsnip root the food storage region occurs in the cortex and phloem, which is proportionally larger even than that of the carrot (Fig. 19-7*B*). The sweet flavor of the roots develops only after exposure to cold.

b. ROOTS EATEN FOR CARBOHYDRATE: SWEET POTATO AND CASSAVA

Sweet Potato. The sweet potato (*Ipomoea batatas*), of the morning glory family (*Convolvulaceae*), is a twining, trailing perennial herb with a great thickened tuberous root (Fig. 19-8). *Ipomoea* is a large genus of about 400 species, known for many ornamental plants, but only *I. batatas* is used for food. It is a hexaploid ($2n = 90$) of uncertain ancestry but may have arisen by hybridization and polyploidy of such wild Central American species as *I. trifida* and *I. tiliacea*. The home of the sweet potato is probably Central America, and it was introduced from there into Spain by the early Spanish explorers. From Spain it was spread throughout the Mediterranean region and also carried by the navigators to the Pacific area. Here, however, it was found that the sweet potato had reached Polynesia in pre-Columbian times by unknown means. Interestingly enough, the word "potato" was first applied to *Ipomoea batatas* (from the Carib word for potato) but later was used more commonly for the stem tuber. If the "law of priority" in scientific nomenclature were followed in the case of common names, it would be necessary to call *Ipomoea batatas* by that name (potato) and find another name for the plant so much cultivated in Ireland and northern Europe. Now we do often call one "Irish potato" and the other "sweet potato." Numerous varieties of sweet potato are cultivated throughout the tropics. They may be divided into two groups based upon the amount of water and sugar present: (1) dry, mealy,

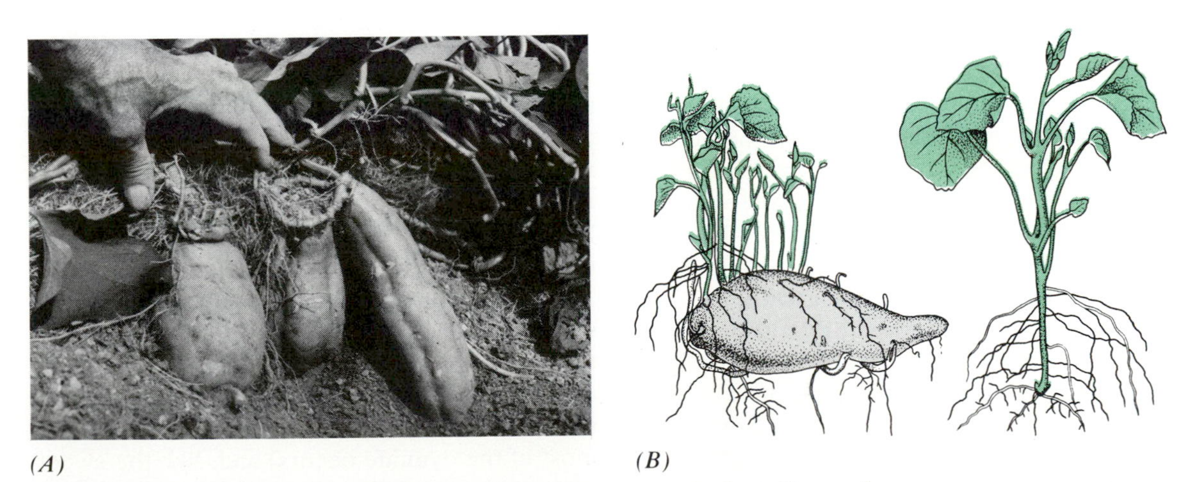

(A) *(B)*

Figure 19-8 *A*. Sweet potato *(Ipomoea batatas)* being harvested. *B*. Vegetative reproduction in sweet potato; (left) adventitious shoots growing from swollen storage root; (right) single shoot called "slip" removed, ready for planting.

yellow types, preferred in the northern United States, (2) watery "yams" (so-called), rich in sugar, and soft and gelatinous when cooked, which are more in demand in the southern United States and Mexico. Sweet potatoes contain both sugar and starch, which is stored primarily in the central xylem zone. They have considerably more calories than Irish potatoes; they are also high in iron and calcium.

The sweet potato has found more favor in the Far East than in its homeland or Europe. It is especially important in Japan, where its productivity is second only to rice on the intensively managed uplands. The sweet potato has proved temperamental in breeding programs and therefore is usually propagated from numerous sprouts (Fig. 19-8*B*) that originate at the head of a root when it is kept moist and warm. Because sweet potatoes do not keep well, they are sometimes sliced and dried. In Japan the dried sweet potatoes may be ground into meal, which can be cooked or fermented for alcohol.

Cassava. *Manihot esculenta* (Euphorbiaceae) is the staff of life for millions of tropical people where cereal and potatoes will not grow. Mandioca, manioc, yuca, tapioca plant, and sagu are other common names by which cassava is known. The tuberous roots are the mainstay of diet in numerous developing tropical areas for a number of reasons. First, they grow well in depleted soils, either in areas where the rain forest vegetation has been stripped or in savannah areas that have been persistently burned. Second, the roots are easy to plant, harvest, and store. Third, cassava can be planted at different times during the year, ensuring a year-round crop. Fourth, the calorie per acre yield is excellent; it yields more starch per acre (up to 20 tons or better on fresh roots) than any other crop and with a minimum of labor. Fifth, it is not subject to many diseases. Its drawback is that the roots consist primarily of starch (30 to 35% by weight) with only 1% protein and 1% fat. It does, however, contain calcium and vitamins B and C. Although the leaves contain 30% protein, they unfortunately have not been much eaten by local people.

Manihot esculenta, indigenous to eastern Brazil, represents a huge complex of cultivars that taxonomists have tried to separate into distinctive species with little success. The genus *Manihot* also includes many uncertain species that apparently hybridize with each other and are most simply handled taxonomically by being considered a part of the *M. esculenta* "complex." This is another example of the role of introgressive hybridization between wild species, races, and escaped cultivars. South American Indians must have domesticated *M. esculenta* for some time because the species is not known with certainty in the wild. Cassava remains have been found in archeological sites in Peru dating from 2800 B.P., and indirect evidence indicates that it was cultivated in Colombia and Venezuela at least 3000 years ago. The differences among the hundreds of cultivars are often slight but the group is generally divided into sweet and bitter types. The former are edible without preliminary treatment, but the latter contain cyanide (HCN) as a cyanogenic glycoside, which must be destroyed by heating or eliminated by expression prior to eating. Generally sweet types are grown in eastern South America and bitter types in the Amazon valley. Both types are grown in Africa and the Far East, having been introduced from South America during the seventeenth and eighteenth centuries.

The shrubby cassava plant has a ramifying root system, the main roots of which develop tuberous swellings a short distance from the erect stem (Fig. 19-9*A*). These swollen areas have a shallow periderm and cortex (where most of the cyanide occurs) and, like the sweet potato, an abundant starch-bearing xylem, which is not lignified.

Cassava is usually hand cultured and harvested. Typically, forest is cut and burned over and then stem cuttings of a few nodes' length are inserted into holes in the soil. Harvestable roots develop in as little as 8 months, but maximum yield of quality roots usually takes up to 16 months. If not all of the roots are dug up, the plant will continue as a "perennial" with new stems arising from the roots left in the ground.

Cassava is prepared in a variety of ways in different areas. When boiled, it has a "heavy" consistency, like incompletely cooked macaroni or spaghetti. Nonetheless, it is eaten in this manner wherever it is grown. In Africa the boiled root may be pounded into a thick paste called "fufu". In Brazil "farinha", the shredded and roasted cassava, is an important addition to most meals, being put on the table along with the salt, pepper and sugar. It also is incorporated into many dishes. The Amazonian Indians express the liquid by alternately constricting and extending a pulverized mass of cassava in large woven tubes called tipitis (Fig. 19-9*B*). These expressed juices are fermented to alcoholic beverages, and also may be used for meat sauces and in "West Indian pepper pot." The flour may be made into flat breads called bijú in Amazonia and "cassake" in the West Indies.

Cassava is also the source of *tapioca,* and this is the only form in which cassava is eaten by temperate-zone inhabitants. The tapioca "pearls" most Americans recognize from various desserts are cassava starch pellets forced through a mesh and heated at controlled temperatures while being shaken or stirred on a plate, a process causing swelling, gelatinization, and partial hydrolysis to sugar.

Recently cassava has been put to another use. Brazilian scientists now say that, by 1985, 20% of its motor fuel needs will be met by ethanol, produced by fermenting cassava (and sugarcane). Actually it is predicted that all cars in Brazil by the end of the century will be running with this gasoline supplement. The plan includes planting a greatly increased acreage of cassava as well as construction of industrial distilleries to mix sugarcane, which Brazil produces in great quantities, with cassava. The cassava–cane mixture is still in experimental stages, but at present the most feasible mixture combines six parts cassava with one part cane. Although cassava is easy to cultivate and harvest, costs are high for this ethanol production. However, they are much lower than Brazil's oil-import bill, since Brazil produces only 10% of its petroleum needs.

5. STORAGE IN MODIFIED STEMS

a. EDIBLE TUBERS: THE POTATO AND YAM

Potato. The common potato (often called white or Irish potato) is one of the world's most important foods. On a nutritional yield/acre it is perhaps *the* most important (Fig. 19-1). It is a member of a large family, Solanaceae, which includes such plants as tomato, eggplant, and tobacco (Chapter 13). Various wild *Solanum* species produce small potatolike tubers that have been dug by the Indians. One of these, probably the subspecies *andigena,* was domesticated in the Andes Mountains of Peru and Bolivia, prior to 1800 B.P., to become our domesticated *S. tuberosum*. The diploid chromosome complement in *Solanum* is 24, *S. tuberosum* ($2n = 48$) being an autotetraploid. Archeologists have found ancient pottery shards sculptured with figures of potatoes. *Solanum tuberosum* is still cultivated as a staple in its Andean homeland where, to better preserve its food value and build reserves against a poor harvest, it is made into "chuño" by trampling and drying during alternate freezing and thawing. This dried product is almost pure starch, and it is sufficiently important in the diets of the Andean Indians for there to be a

(A)

(B)

Figure 19-9 *A*. Displaying large tubers of cassava (*Manihot esculenta*) plant at an agricultural experiment station in eastern Amazonia. This tuber is an important item in the diet of many tropical peoples. *B*. Processing cassava to produce farinha (cassava flour) by Amazonian Indians. Here the liquid containing cyanide (in the bitter varieties) is expressed by use of a "tipiti." Although the tipiti was developed by the Indians, the same technique is used commonly by others throughout the Amazonian region.

common saying that "stew without chuño is like life without love." Tubers of some varieties of potato that are quite unpalatable when fresh can be used for chuño, and the tubers of *oca* (a species of *Tropaeolum,* a nasturtium, in the Oxalidaceae) are dehydrated in a similar manner. Dehydrated oca found in archeological remains may be distinguished from chuño by the different shape of the starch grains. An alcoholic beverage (chica) was also made from potatoes (as well as from *Quinoa* in the Chenopodiaceae). The use of potato for production of alcohol was rediscovered in Europe at a later date and has become very important in eastern countries for vodka (Chapter 20). All in all, the potato is the most important plant in the high Andes. Cultivated only in the highlands, it apparently was traded to coastal people in prehistoric times.

At the time of the discovery of America by Europeans, potatoes were cultivated from Chile to Colombia but not in either Central or North America (Fig. 19-10*A*). The first Spaniards to see potatoes in South America thought that they were a type of truffle, an underground fungus considered a great delicacy in Europe. Potatoes were sent to Europe before 1570 and were definitely unlike any food that Europeans had known previously. They were cultivated in gardens of Europe by 1600 and in Ireland by 1663. Reintroduction into the New World was made by 1612. Nowhere, however, was the potato widely planted until after 1700. One reason for its sudden prominence in Europe during the eighteenth century was that in Germany and Sweden the people were compelled by royal edict to plant it as a source of food. The potato soon became an important food staple throughout Europe and especially in Ireland, where it became responsible for population increases such that by 1841 the population of Ireland was about 8 million. Most of the people lived as peasants, some on large estates with English landlords, and their standard of living was so low that the potato was their only staple food. It has been stated that a working man ate from 12 to 14 pounds of potatoes daily. The Irish were actually more dependent upon the potato than the Andean Indians had ever been! Although crop failures had occurred earlier, disaster followed a series of events in 1845. After three weeks of wet weather, the plants became became blackened by the late blight *Phytophthora infestans* (see Chapter 29). Suffering from hunger was great that year, despite some efforts to import corn from America, but greater disaster was to follow. The crop looked promising the next year, but then completely failed, resulting in famine. The winter that year was severe and how many died is not absolutely known. It has been recorded that the living were so weak that they could not count the dead, let alone bury them. Then epidemics of typhus, scurvy, and dysentery struck due to the combination of cold, starvation, and accumulated filth. Great numbers of people began to leave the country, most going to America making Chicago a bigger Irish city than Dublin. Fortunately the potato crop was good in 1847, but in 1849 the blight appeared again on a large scale, and the suffering equaled that of earlier years. Between 1846 and 1851 an estimated 1½ million people died either from hunger or disease and another million emigrated (Fig. 19-10*B*).

It may appear strange that a plant from the New World tropics should have become so important in temperate Europe. However, it must be remembered that the potato is a highland crop from the tropical zone, grown at elevations in the Andes that were too cold for corn (maize). It will not yield well when temperatures average above 21° C. Today the potato is chiefly grown in Europe, with North America a poor second, and Latin America a distant third.

The potato tuber consists of an outer suberized skin (periderm) varying in color from red to light brown, a narrow cortex of small starch-rich cells, an adjacent narrow zone of conducting cells (phloem-xylem-internal phloem), and a large pith containing most of the starch-bearing cells (Fig. 19-4). Each cell con-

(*A*)

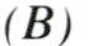

(*B*)

Figure 19-10 The potato is native to the Andes but after being taken to Europe greatly influenced the history of Ireland and the United States. *A*. Sixteenth century Spanish sketch of Inca potato harvest. The woman uses a hoe and the man at the left a hand plough. *B*. Funeral in Irish village in 1847 during the famine caused by late blight of the potato crop. Whole towns and villages had to be abandoned as people starved to death or, if lucky, managed to emigrate overseas.

tains starch grains that swell upon cooking and, according to some authorities, burst the thin cell wall.

The food value of the potato varies, depending on variety, growth conditions, storage, and handling. Analyses have shown variations from 70 to 81% water, 8 to 28% starch, and 1 to 4% protein with traces of minerals. In comparison to other familiar foods, the potato is fairly economical, the cost per calorie of nutrient being roughly the same as that of bread and margarine. In Europe much of the potato crop is fed to livestock, and a sizable portion is used for fermentation and other industrial purposes. Although corn starch has generally supplanted potato starch in the United States, potatoes mill more easily than corn. Industrial potato starch goes chiefly into sizing for paper and textiles and into confections and adhesives.

Since the potato is anatomically a stem, sprouts arise at the "eye" (Fig. 19-4). Therefore, sections of potato with a bud (eye) are generally used to propagate and maintain potato varieties. Potatoes saved for propagation (about 7% of the crop) are called "seed" potatoes, but of course they are not true seed (Fig. 19-11). To

Figure 19-11 "Seed" potatoes (sections of the tuber including the "eye" from which the sprout arises) being planted.

avoid virus infection, seed potatoes are mainly grown at the northern edge of their production zone, as in Scotland, Maine, or western Canada.

Because of the ease with which starch and sugar are interconverted, storage of tubers, especially those of the potato, has to be carefully controlled. If potatoes are held below about 6° C, the hydrolysis of starch to sugar still proceeds at a moderate rate, but the resynthesis of starch is almost arrested. The latter process requires a source of ATP or UTP, which in turn requires respiratory energy, and respiration is more temperature-dependent than hydrolysis. On the other hand, at high temperatures the carbohydrate can be respired away and the simultaneous loss of water causes the tuber to contract. Also the high temperature causes germination and elongation of the buds. Thus the best storage conditions require temperatures between 6° C and 15° C with not too low a humidity.

Only since about 1850 have there been serious attempts to improve the potato. Because there was a gradual decrease in yields, probably due primarily to tuber-transmitted virus diseases, new introductions have been secured throughout the Americas and a breeding program initiated.

The Yam. Although moist sweet potatoes are sometimes called "yams" in the United States, the true yams belong to the genus *Dioscorea,* and the cultivated ones are indigenous to the tropics. After cassava and sweet potatoes, they are the most used tropical lowland root crop, especially in West Africa, parts of Meso-America, the Pacific Islands, and Southeastern Asia. Yams, however, cannot ordinarily compete with either cassava or rice on a calories/man-day basis because of the energy expended in digging for the deep tubers.

Yams occur in a number of *Dioscorea* species and cultivars, with tubers varying in size from those of a small potato to others weighing as much as 100 pounds (Fig. 19-12). The yam is about 20% starch and as a food very similar to a potato. Coarse, mealy, crisp, mushy, and other types occur. Some are baked, boiled, or fried, whereas others are used primarily for soup.

Yam plants are climbing perennial vines with chordate leaves and inconspicuous monocotyledonous flowers (Fig. 19-12). Because they seldom fruit, they are usually propagated from tubers. The species most commonly cultivated is *D. alata,* which is probably native to China, but numerous other species are cultivated in various areas. An 8- to 10-month growing season is required for the crop to mature, so that it is not possible to cultivate it outside of the tropics.

Yams also contain such substances as saponins and alkaloids, and wild, inedible species from southern Mexico and China have been used recently as the source for a steroid precursor, diosgenin, used to make birth control pills.

b. EDIBLE CORMS

Corms and tubers of several genera in the family Araceae have long been used for food in the tropics, *Colocasia* and *Xanthosoma* being the most important ones. They are known by a long list of common names such as taro, dashien, cocyam, kolokasi, ocumo, and yautia. Also in the Araceae are many plants native to the tropics but known to people in the temperate zone as hothouse ornamentals—philodendrons, caladiums, and arum lilies. Most are either herbs or trailing vines, having characteristic large, heart-shaped leaves (called "elephant ears") and fleshy underground organs (Fig. 19-13).

The aroid most widely used for food, *Colocasia esculenta* (taro or dashien), may have originated in southeastern Asia (dashien means *de* Chine = from China). It has been

(A) (B)

Figure 19-12 *A*. Yams (*Dioscorea* spp) are climbing plants and in cultivation are trained to grow up stakes. *B*. The tubers seen in storage on an African farm, take about 10 months to reach maturity. The typical tuber is almost a foot long and as a food is only a little less nourishing than a potato.

suggested that its cultivation may have been carried in prehistoric times westward through India and eastward through the Pacific Islands as far as Hawaii. It was introduced into the Mediterranean region during the Greek era and spread southwestward across tropical Africa and finally within recent centuries to the West Indies and tropical America.

The food value of all aroid corms is similar; they are rich in amylose (20 to 25%) and have about 3% sugar (therefore, are more like sweet potato than Irish potato), about 2% protein, and a trace of fat. Usually the corms are baked like potatoes or crushed to make cakes; roasting and steaming are also relatively commonplace. The dashien, the *antiquorum* variety of taro, is the source of Hawaiian *poi,* which is made by steaming the corm, followed by crushing and natural fermentation (first by bacteria and then by yeast). Taro also contains large amounts of calcium oxalate. The starch granules are small, which contributes to their digestibility but makes their extraction for industrial purposes difficult.

(*A*)

(*B*)

Figure 19-13 *A*. Experimental plantations of taro *(Colocasia esculenta)* in Hawaii; it usually grows in marshy habitats. *B*. Corms of different varieties of this important starchy staple in the Pacific islands.

CHAPTER 20

Fermentation and Its Varied Products

1. STEPS IN THE UNDERSTANDING OF FERMENTATION

The idea that fermentation is anaerobic metabolism of carbohydrates was many years in developing. It came through the study of microorganisms. Until well into the eighteenth century it was believed that microorganisms could arise spontaneously when plant or animal materials were held for a while at a warm temperature. This happened even if they were boiled first. But in 1776 the abbot Spallanzani, at Pavia, showed that if fruit juice or bouillon were first completely sealed up in glass tubes and *then* boiled, no microbes appeared. The explanation must be that the microbes come in from the air, but the idea that microbes are constantly floating about, invisibly, in the air was a difficult one to believe and was not widely accepted. However, 60 years later Theodor Schwann, in Germany, made a careful examination of fermenting grape juice and observed that there were always microbes present, mainly of one type—oval cells that he thought probably belonged to the plant kingdom. As the grape juice slowly turned to wine, these "yeast" cells were seen to increase greatly in numbers. More important, he showed that the boiled juice would not ferment even if air was admitted, providing that the air had been passed through a very hot tube.

But the chemists Antoine Lavoisier, Baron Justus von Liebig, and Friedrich Wöhler (who in 1828 had synthesized for the first time a natural product, urea) had viewed fermentation as a purely chemical process, and the yeast as no more than a sort of catalyst. Lavoisier's student Joseph Gay-Lussac had shown that the process conforms to the Law of Simple Proportions and could be formulated:

$$\underset{\text{Glucose}}{C_6H_{12}O_6} \rightarrow 2\ CO_2 + 2\ \underset{\text{Alcohol}}{C_2H_6O}$$

They scoffed at the idea that fermentation could be due to the *life* of the yeast, and so again the matter remained unsettled.

Louis Pasteur (Fig. 20-1) cleared the matter up in his famous "Mémoire sur la Fermentation Alcoolique" in 1860. He confirmed Schwann's result that grape juice did not ferment if carefully freed from the yeast, but showed that ripe grapes normally have yeasts growing on the outside, because small wasps puncture the fruit and a little of the sweet juice exudes, to act as a growth medium for airborne yeast cells. Secondly, he showed that Gay-Lussac's equation was insufficient, for his careful analysis revealed glycerol, succinic acid, and fats in the liquid after fermentation was over. Thus fermentation could not be a simple chemical process. But most important, he became convinced that in general all fermentations are the

Figure 20-1 Louis Pasteur as a student at the École Normale in Paris, about 1844.

result of the organism's growing in the absence of air; "fermentation by yeast," he wrote, "is the direct consequence of . . . life . . . carried on without the agency of free oxygen." As to the conceptual difficulty that "spontaneous generation" must really be due to microbes floating in the air, he attacked this head-on by using filters for the air and showed that if they did keep the liquids from decomposing they actually trapped small cells or spores which could be recognized under the microscope. He was a dramatic speaker and writer, and his combination of biological and chemical methods carried the day.

For our present concern, it was the biochemical concept that was crucial. If metabolism is to go on without oxygen—that means without the normal acceptor for hydrogen—then the hydrogen must be accepted by some of the products of metabolism instead of by oxygen. In other words, some products must be hydrogenated and others *de*hydrogenated. We see this in the Gay-Lussac equation, for in sugar the ratio H : O is 2 : 1, in alcohol it is 6 : 1, and in carbon dioxide it is 0 : 1. One product contains relatively more hydrogen than before and the other contains less.

2. THE UNDERLYING BASIS OF FERMENTATION

The ways in which this rearrangement takes place in fermentations are readily grasped if we recall what happens in oxidation and photosynthesis. In oxidation, the protons and electrons from organic compounds, after passing through carriers and the cytochrome system, are combined with oxygen to form water (Chapter 6). In photosynthesis, water supplies protons and electrons (via photosystem II), which, after passing through several carriers, ultimately reduce phosphoglyceric acid to its related aldehyde (Chapter 5); the phosphoglyceraldehyde condenses to form hexose diphosphate. In the alcoholic fermentation one step of the process is exactly the opposite of that in photosynthesis, for hexose diphosphate is split into two molecules of phosphoglyceraldehyde and this then releases two protons and two electrons to form phosphoglyceric acid. The protons and electrons are then used to reduce *other organic compounds:*

$$\underset{\text{Hexose diphosphate}}{1/2\ C_6H_{10}O_6(H_2PO_3)_2} \rightleftharpoons \underset{\text{Phosphoglyceraldehyde}}{C_3H_5O_3(H_2PO_3)} \xrightarrow{H_2O}$$

$$\underset{\text{Phosphoglyceric acid}}{C_3H_5O_4(H_2PO_3)} + 2\ H^+ + 2e.$$

The hydrogenation occurs after the phosphoglyceric acid has been converted to pyruvic acid. This either accepts $2H^+ + 2e$ itself, to form lactic acid (p. 370), or else it first loses CO_2 to form acetaldehyde, which then accepts $2H^+ + 2e$ to form *ethyl alcohol:*

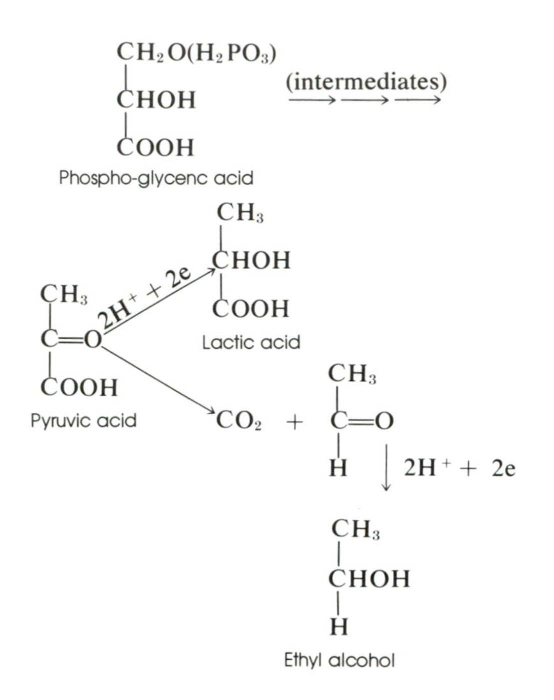

In the butyric fermentation (see section 6) the pyruvic acid is first converted to acetoacetic acid, and it is this that acts as hydrogen acceptor, becoming reduced to butyric acid. Thus the different fermentations are distinguished by the use of *different hydrogen acceptors*. The coenzyme NAD is the intermediate carrier in all these hydrogen transfers.

There are also fermentations in which the dominant process is not the transfer of hydrogen to acceptors but is simply hydrolysis, either of proteins to amino acids, of polysaccharides to simple sugars (which can be further fermented as above), or of aromatic complexes. These hydrolyses, of course, do not require oxygen but are often followed by hydrogen transfer reactions as well. The "fermentations" of many plant products, such as tobacco, tea, cacao, and vanilla are thus mainly mixtures of hydrolysis with some oxidation and often do not require the participation of microorganisms.

3. THE ALCOHOLIC FERMENTATION IN PARTICULAR

a. ITS ANTIQUITY

Alcoholic drinks have been made since ancient times, for alcohol is the simplest of stimulating drugs. Beer was made from barley in Babylon around 2000 B.C. (4000 B.P.) and was drunk by people of all classes there. From Asia it passed to Egypt and the ancient Egyptians were drinking it in 1500 B.C. (3500 B.P.). Other parts of Africa brewed beer from barley and also from millet. Beer was being made from millet in China by at least 300 B.C. (2300 B.P.) and perhaps much earlier.

The early history of wine is less certain, but its preparation was certainly well developed by the time of classical Greece (*ca.* 500 B.C.; 2500 B.P.). Dionysos or Bacchus, the god of wine, was said to have discovered the art of wine-making. Wine was made in other Mediterranean countries also, and where wine, which contains 10 to 14% of alcohol, was available, the use of beer, which contains only 4 to 8%, seems to have been looked down upon. To this day wine is the favored drink in countries where grapes will grow, while beer is usually preferred in more northerly climates. A kind of wine has been made from rice in China and Japan since ancient times.

A special case of the fermentation is presented by bread, for here the important product is not the alcohol but the carbon dioxide, which "raises" the dough. The old English word for yeast, *leaven,* and the French word *Levure* both refer to the *lifting* action of yeast, and so does the German word *Hefe* (*heben* is to lift). Without the gas bubbles the dough would harden to an inedible mass on baking. A few hours of fermentation is enough for this, and sugar is usually added to hasten the growth of the yeast. The alcohol is driven off by the high baking temperature. The "unleavened bread" of the Bible and the knackebröd of Scandinavia, being made without yeast, require rolling out the dough into very thin layers.

When the Egyptian Pharaohs were buried, models were sometimes placed in the tomb to aid in planning the deceased's household in the other world. At Thebes the excavations have brought to light some models showing brewing and baking in adjacent rooms of the palace (1200 B.C.). In one room grain is ground into flour and the flour is worked up into dough; the rising mixture stands in tall crocks to ferment, and from this some is poured off into jugs for drinking, while the rest goes into the next room, which is the bakery, with an oven. At Pompeii, the Roman city that was overwhelmed (and thus preserved) by the eruption of Vesuvius in 79 A.D., a Roman loaf is among the items preserved. Though externally charred, it is recognizable as a *raised* loaf.

Besides food and drink, the alcoholic fermentation is now producing fuel. Various types of agricultural waste (largely carbohydrate) can be fermented and then distilled to yield a crude alcohol to drive automobiles. In Brazil sugarcane residues and cane molasses are already used on a large scale in this way.

b. THE YEASTS: SACCHAROMYCES AND ITS RELATIVES

In spite of their great technical importance, the yeasts are rather poorly understood. They are single cells, oval to round, reproducing vegetatively by budding (Fig. 20-2), the bud growing out at an angle to the long axis of the oval. In *Saccharomyces cereviseae,* the yeast of beer and wine, the cells mostly remain joined in groups of two to about six, but in the related genus *Saccharomycodes,* chains of cells are formed that show obvious relationship to the separate hyphae of the Ascomycetes (see the discussion of fungus types in Chapter 29). Several species of the yeastlike organism *Hansenula* actually form true mycelium under some conditions. Also, one of the *Saccharomyces* relatives forms eight spores in the cell (Fig. 20-2*B*), like the eight-spored asci that typify the Ascomycetes (see Chapter 29). For these reasons the yeasts clearly belong to the *Ascomycetes.* However, *Saccharomyces* usually forms only two ascospores. Since there is no mycelium, the ascospores either fuse as they germinate, giving rise to a diploid generation, or they merely grow (without fusing) into haploid cells, and the haploid and diploid types appear indistinguishable. This behavior, combined with the fact that the nucleus and chromosomes are difficult to make out, has made the relationships of the yeasts obscure for many years. A further source of confusion is the formation in some fungi in a quite different class, namely certain Phycomycetes which form true mycelium when growing on a surface, of yeastlike cells when growing immersed in solution. There is also an organism called *Torula* that closely resembles the yeasts (a) in forming oval to roundish cells that reproduce by budding, and (b) by growing best in sugary solutions, but this organism has never been known to form ascospores. It is classed as an imperfect fungus, like the many mycelial forms which undergo no sexual phase. The *Torulae* are more aerobic than *Saccharomyces* and sometimes are responsible for spoilage of fermenting media that become exposed to air.

c. THE FORMATION OF BEER

As we saw in Chapter 16 the grains of the cereals are packed with starch-bearing cells—the endosperm—and on germination the gibberellic acid that is liberated catalyzes the protein-containing aleurone cells to secrete amylase. As a result, germinating grains of wheat, oats, and barley quickly become rich in sugars—first maltose, then glucose. Beer is therefore made from germinated grain, the sugar content of which suffices for a modest amount of fermentation. In western countries the grain used is barley; the barley proteins and minerals contribute much to the character of the brew.

The procedure is to wet the grain and to lay it out on the "malting floor" for three to four days at about 30° C. From the resulting "malt" the young rootlets are rubbed off; the sugary grains are soaked in water and mashed, and the

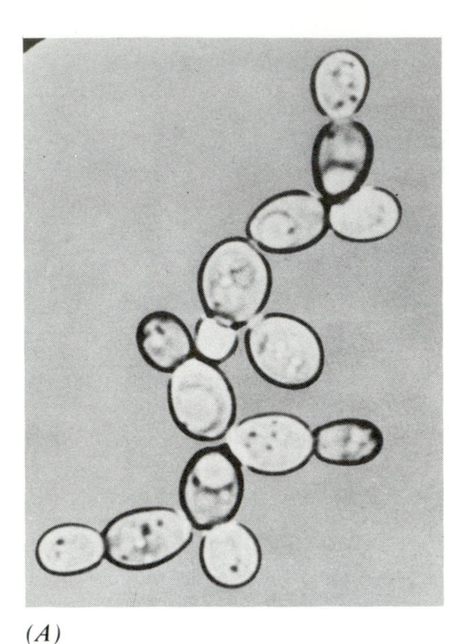

(A)

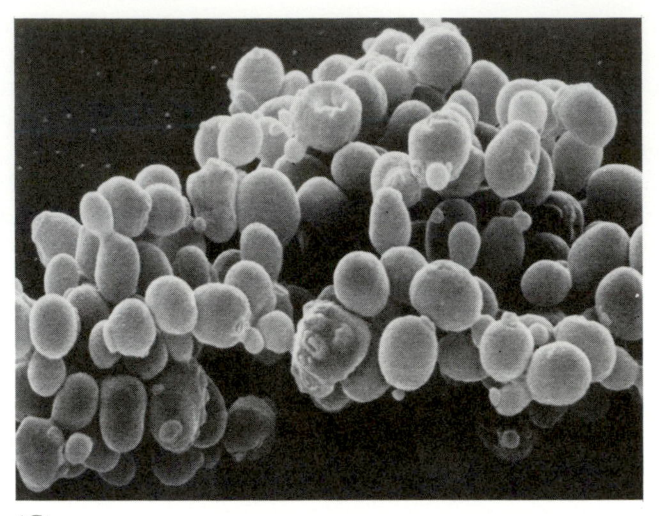

(C)

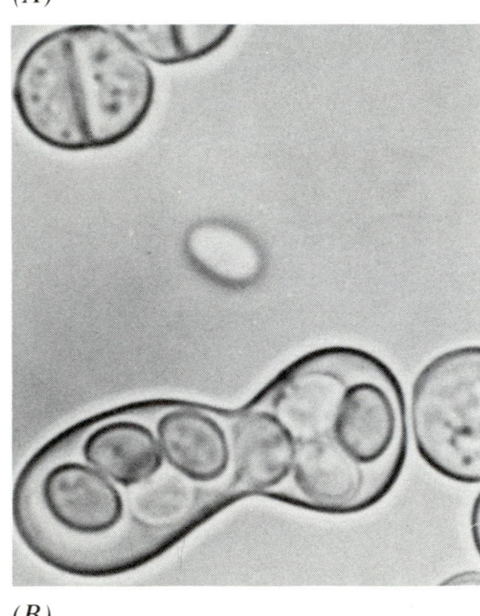

(B)

Figure 20-2 Yeasts under moderate magnification. *A. Saccharomyces cereviseae,* an ale yeast, multiplying by budding. (×1400). *B. Schizosaccharomyces octosporus,* showing cells multiplying by fission. and an ascus with 8 spores. *C.* Scanning electron-micrograph of *S. cereviseae,* with buds visible on several of the cells. (×1800).

solids strained off. The solution, now called *wort,* is boiled and filtered. Then the flowers of the hop plant are added. The hop, *Humulus lupulus* (or in Japan, *Humulus japonica*), is a tall, coarse climbing plant belonging to the Moraceae (fig family). Its flowers, borne in close-packed racemes (Fig. 20-3) contain two bitter-tasting resins, humulon and lupulon, which, being volatile, give an aroma as well as flavor to the beer. Since they are somewhat antibiotic, their presence also minimizes the growth of infecting *Torulae,* lactic bacteria, and of a bacterium, *Streptococcus damnosus,* whose name amply suggests the master brewer's reaction to the results of this growth in the beer! The quality and flavor of the beer generally depend more on the quality and flavor of the hops used than on the wort, which has less variation from one brewing to another. The best quality hops (i.e., with the highest aromatic con-

Figure 20-3 Female inflorescences and leaves of the hop, *Humulus lupulus*. The resin comes from glands at the base of the corolla.

tent) are grown in Czechoslovakia, in southeast England and in Oregon. After the hop tissues have been strained off, the yeast is "pitched" in, and the solution left to ferment in the cool for some days or weeks.

The different kinds of beer result from different procedural details. Ale is produced by a strain of yeast that floats up to the top of the fermenter, and lager by one that sinks ("top" and "bottom" yeasts respectively). Bock beer is darkened by adding molasses to the malt. Stout, which is still darker and slightly sweet, results from a secondary fermentation by a different genus, *Brettanomyces,* which is added after the first fermentation has subsided.

There are also several "beers" (sometimes called wines) not made from barley. Mead used to be made in western England from diluted honey. Chicha is made in the South American highlands from corn; to promote hydrolysis of the starch, human saliva is added to the wet grains. Pulque is made in Mexico, not from grains but from the sweet sap of an *Agave* species, and palm toddy is similarly made in several tropical countries from the exuded sap of the sugar palm trees (Chapter 17). In these last three cases the inoculum is carried over from each batch of fermentation to the next.

"Hard" liquors are made by distilling various kinds of beers (without hops). Although alcohol boils at 78° C and water at 100° C, aqueous alcohols like beer on boiling give a distillate that contains about 40% alcohol. Scotch whisky is distilled from fermented barley, bourbon from fermented corn (maize), Canadian whiskey from fermented rye, and vodka from fermented potatoes.

d. WINE

Wine is the fermented juice of the fruits of *Vitis vinifera* (Fig. 20-4). Some of the *Vitis* species native to America, especially *V. labrusca,* make good grape juice but not good wine (see Chapter 18). *V. vinifera* is native to the Mediterranean area; its grapes are rather small and either white or dark purple (Fig. 20-5 and 18-4). It has several thousand cultivars. The wine yeast is a form or cultivar of *S. cereviseae*. As noted above, the ripe wine grapes often have yeasts growing on the surface, in exuded droplets of juice. The same thing happens in nectaries, especially those on leaves, where the sugar-rich phloem sap exudes to the surface. Yeast cells can usually be found in these exuded droplets. Until the last century wine was usually fermented by the yeasts already present on the grapes, but now the juice is inoculated with pure yeast cultures. When the alcohol concentration reaches about 12% (or 2.6 M) the yeast becomes inhibited and the fermentation ends.

The flavors and types of wines depend in part on the cultivars of grapes used but also, as with beer, on the details of the procedure. Red wines are made from purple grapes, and after crushing the juice is left standing on the skins, which are thus decomposed (or may be secondarily crushed also) to set free the anthocyanins, (purple phenolic pigments; see Chapter 6, section 6) and tannins that are present in the skins.

Figure 20-4 Harvesting Cabernet Sauvignon grapes for red wine in the Médoc district of France.

The tannins produce the so-called "dry" flavor. If the dark red grapes are simply crushed and the juice strained off, the juice itself may be barely colored. Some champagnes are made in this way from red grapes and so develop only a very slight coloration ("pink champagne"). Most white wines are made directly from the crushed juice with small or variable contribution from the skins. The red types include famous names like Pinot Noir, Gamay, and Cabernet Sauvignon (Fig. 20-4); the best-known white cultivars are Chardonnay, Pinot Blanc, Muscat, and Chenin Blanc, Rosé wine is made from intermediate types, by removing the skins early or sometimes by blending whites and reds. The large scale on which some modern wineries operate is suggested by Figure 20-5.

Secondary fermentations and chemical changes play an important part in the flavor and aroma of wines. The grapes are harvested when their sugar content is at, or close to, its peak and the acid content is being oxidized slowly away. The main alcoholic fermentation begins at once on crushing and is over in a week or two at the usual warm autumn temperatures. In the secondary changes the malic acid decreases by a series of slow reactions, in which one of the carboxyl groups is removed and lactic acid is formed. Slower still are the reactions—occurring mainly after the bottling—in which the remaining acids are converted by the alcohol to esters, giving ethyl acetate, succinate, malate, lactate, and pyruvate. Also, since small amounts of other alcohols are produced (partly by metabolism of amino acids in the juice), esters of these higher alcohols appear. Most of these esters are volatile and have sweet "fruity" smells that, together with the compounds specific to the grape type being used, account for the bouquet and aroma of the wine.

Fortified wines such as port and sherry are made by distilling off the alcohol–water mixture ("grape brandy") and adding it to wine to bring

Figure 20-5 Chardonnay (white) grapes on the way to the press at a modern winery.

up the alcohol content to about 20%. Such strong liquors as brandy, rum, and tequila are single distillates. Champagne and other sparkling wines are bottled before the fermentation is quite complete; the bottles are stored with their necks downward at a slight angle, so that the accumulating yeast will collect on the cork. After a year or two the corks are allowed to blow out and fresh corks inserted and wired in place. On release of the pressure the dissolved carbon dioxide bubbles out.

Two major diseases of grapevines, *Phylloxera,* caused by aphid attack on the roots, and downy mildew, due to fungus attack on the leaves and berries, are described in Chapter 29. "Diseases" of the wine itself are due to bacteria and may cause bitter tastes by formation of acrolein, or excessive acidity by oxidation of the alcohol to acetic acid (vinegar). It was a request from the winemakers of France to Louis Pasteur to study these problems in bottled wine that led to the recognition of the role of bacteria and to his demonstration that they could be killed by heating the wine to 60° C for half an hour. This "pasteurization" was later extended to milk, making fresh, sweet milk available worldwide.

Saké, sometimes called rice wine, is a very different product. It is made in Japan from rice; the starch is hydrolyzed by a true Ascomycete, *Aspergillus oryzae* (rice is *Oryza sativa*). The ensuing alcoholic fermentation is brought about partly by this fungus itself and partly by a special yeast, *Saccharomyces saké.* Because of its origin from grain it contains neither anthocyanin nor tannin and is colorless.

4. THE FORMATION OF VINEGAR

When the alcohol of beer or wine is oxidized to acetic acid, the product is vinegar:

$$\underset{\text{alcohol}}{CH_3CH_2OH} + O_2 \rightarrow \underset{\text{acetic acid}}{CH_3COOH} + H_2O$$

The oxidation is due to rod-shaped bacteria, not forming spores, called *Acetobacter.* They are common in soil and in the air, and when alcoholic liquids are exposed to air they usually infect it. Commercially, vinegar is made in two ways.

In the Orleans process, wine is placed in shallow vats and small twigs are floated on the surface to make a stable support for the thin film of *Acetobacter* that develops. The clear vinegar is drawn off from a tap below without disturbing the oxidizing bacterial film. This vinegar is usually red, from the red wine used. In the "quick vinegar" process beer is trickled through barrels packed with beechwood shavings on whose surface the *Acetobacter* has formed permanent films. Air is blown through from the bottom and the rates of flow are adjusted so that the liquid coming through has completed the oxidation of its alcohol. Before Pasteur it was thought that the change was brought about by the shavings themselves, and "good quality" beech shavings were very much in demand. As would be expected, the *Acetobacter* species are highly aerobic and are rich in the cytochrome system.

5. THE LACTIC FERMENTATION

In the formation of pyruvate from glucose phosphate two protons are liberated and we have seen that the subsequent course of the fermentations depends on the acceptors of these protons. Biochemically, the simplest acceptor is the pyruvate itself; the product is lactic acid:

$$\underset{\text{Pyruvic acid}}{CH_3COCOOH} + 2H^+ + 2e \rightarrow \underset{\text{Lactic acid}}{CH_3CHOHCOOH}$$

The lactic fermentation is mainly of interest in connection with the animal product milk, which lactic acid cogulates to form cheese. The organism that makes pulque is a bacterium that produces both lactic acid and alcohol.

6. THE BUTYRIC, "BUTYLIC," AND METHANE FERMENTATIONS

Fermentations of quite different type from alcoholic and lactic are the butyric, "butylic," and methane fermentations. The first two are caused by a group of strictly anaerobic bacteria, the *Clostridia,* forming spores extremely resistant to heat. Although these vegetative cells can only survive and grow in the total absence of oxygen, their resting spores can survive prolonged exposure to air and therefore are widespread in soil. They possess powerful amylases and thus the butyric fermentations commonly begin from starch. The starch of roots and tubers is hydrolyzed to glucose and the glucose fermented to pyruvate as above. From here on, the reactions are quite different. The pyruvate is converted to acetyl coenzyme A as in the citric acid cycle (Chapter 6), and two molecules of this condense "head-to-tail" to form acetoacetyl coenzyme A. The hydrogen then goes to reduce the acetoacetyl coenzyme A to butyrate:

$$1/2 \text{ glucose} \rightarrow \underset{\text{Pyruvic acid}}{\underset{+2(H)}{CH_3COCOOH}} \rightarrow \underset{\text{Acetyl-CoA}}{\underset{+2(H)}{CH_3CO.CoA}} \rightarrow$$

$$\underset{\text{Acetoacetyl-CoA}}{CH_3CO.CH_2CO.CoA} \xrightarrow{4(H)} \underset{\text{Butyric acid}}{CH_3CH_2CH_2COOH} + CoA.H + H_2O$$

The product, butyric acid, has a smell like that of rancid butter, which can often be noticed when potatoes have rotted in the ground.

A variant of this fermentation became of industrial importance when a related organism, *Clostridium butylicum,* was found to produce butyl alcohol (butanol) instead of the butyric acid formed by *Clostridium butyricum.* In this case some of the hydrogen is used to reduce the second carbonyl group:

$$CH_3CO.CH_2CO.CoA + 8(H) \rightarrow$$

$$\underset{\text{Butanol}}{CH_3CH_2CH_2CH_2OH} + CoA + 2H_2O$$

Since this uses up the reducing power rapidly, some acetoacetyl coenzyme A remains unreduced; in consequence it becomes hydrolyzed to acetoacetic acid and then loses CO_2 to yield acetone:

$$\underset{\text{Acetoacetyl CoA}}{CH_3CO.CH_2CO.CoA} + H_2O \rightarrow$$

$$\underset{\text{Acetoacetic acid}}{CH_3CO.CH_2COOH + CoA.H} \rightarrow \underset{\text{Acetone}}{CH_3COCH_3} + CO_2$$

Both butanol and acetone are useful as commercial solvents, and hence the process became an important route for solvent production, especially in World War I. Chaim Weizmann, who worked up the strain of *Clostridia* most active in this butanol–acetone fermentation, later became the first President of Israel, a rare honor for a microbiologist.

One other fermentation deserves mention; its product is methane, CH_4. It represents the extreme case of the principle mentioned in section 2, namely the formation of a product enriched in hydrogen relative to oxygen. In principle, the organic substrates are dehydrogenated to CO_2 and some of the CO_2 is then hydrogenated to methane. The methane bacteria are strictly anaerobic, and the special nature of the fermentation rests on their unique ability to use CO_2 as hydrogen acceptor in the absence of oxygen. Methane is formed from plant material decaying in bogs; it is also formed in the anaerobic stage of the decomposition of sewage. It is formed in the rumen of cattle, where the ingested herbage is first fermented to organic acid that forms the substrate for special bacteria; a cow forms about 700 liters of mixed CO_2 and methane each day. After long neglect the methane process is now being considered as a possible source of fuel; however, in these fermentations the methane is of necessity mixed with CO_2. Thus with acetic acid we have:

$$C_2H_4O_2 \rightarrow CO_2 + CH_4$$

Hence the gas has rather low calorific value.

CHAPTER 21

Fibers, Textiles, and Fiber Plants

1. FIBER STRUCTURE AT THE CELLULAR LEVEL

The primary wall of plant cells is a deposit of fine cellulose fibrils that become matted together in a kind of felt (see Fig. 21-1). If the cell is longer than wide, as in most elongating stems and roots, then the predominant direction of the fibrils tends to be somewhat transverse to the long axis of the cell, averaging perhaps 60° or 70° to that axis, though varying widely. As the cell elongates, new fibrils are laid down at roughly the same average angle, but sometimes, if the rate of elongation is faster than the rate of deposition of new cellulose, the average angle may become steeper, that is, more nearly parallel to the axis. Mechanical stretching has a stronger effect in changing the alignment of the fibrils to the axis.

The carbohydrate needed to make these fibrils is apparently passed out through the outer membrane (the plasmalemma) in the form of vesicles, which are continually thrown off by the Golgi bodies (Chapter 4). Such cells, with only primary wall, remain alive whether or not they are enlarging. Between the cell walls lies

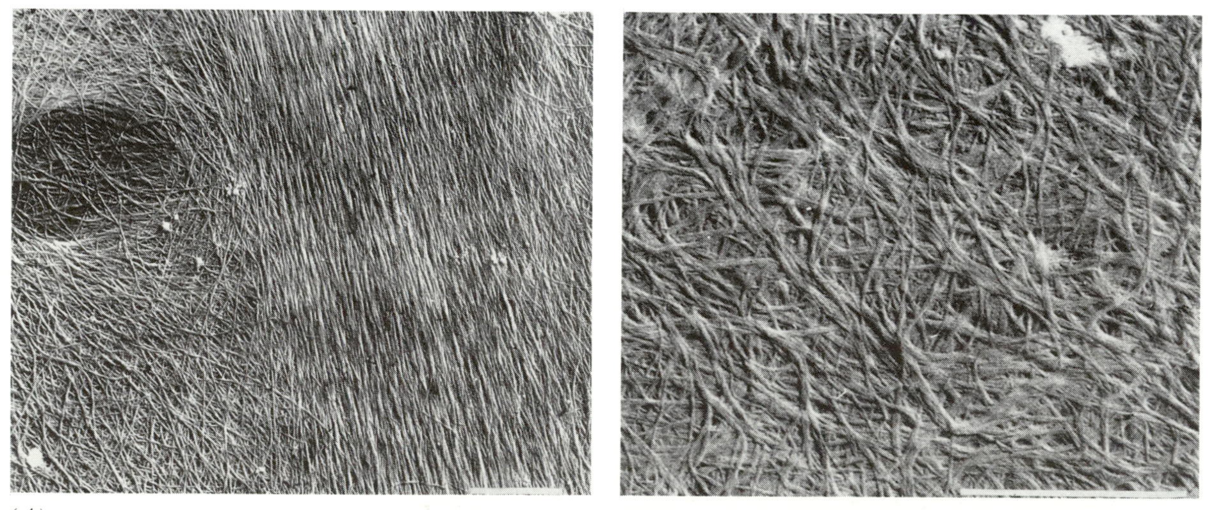

(*A*) (*B*)

Figure 21-1 The fine structure of plant cell walls. *A*. primary wall of a parenchyma cell of oat coleoptile with (at left center) a primary pit area, and (at right) secondary wall laid on it. *B*. primary wall of cell of a tulip bulb. The bar in each photo is one micron.

the middle lamella, which is composed primarily of pectins (Chapters 4 and 18).

Quite different is the behavior of cells that produce *secondary wall.* The material of secondary walls can be laid down only by cells that have stopped enlarging. An exception occurs in those cells that are to become the first xylem vessels in a young seedling stem, for these lay down secondary wall only in rings or spirals within the primary wall, and these rings or spirals can be pulled apart as the cell elongates. But in fiber cells, secondary wall is laid down over the whole area. What happens is that, first, a second layer of polysaccharide is deposited inside the primary wall. The average angle of the fibrils in this layer usually differs sharply from that in the primary wall. After this another layer is laid down, again at a different angle, and then one more. The resulting three-layered secondary wall is shown diagrammatically in Figure 21-2*A*. Sometimes there are up to seven such alternating layers. The orientation of the fibrils in the layers can be distinguished by viewing thin sections in polarized light, which is transmitted differently according to whether the plane of polarization is (more or less) parallel or perpendicular to the direction of the fibrils. As a result the cell walls in polarized light show alternate bright and dark layers, as in Figure 21-2*B*. This structure (seen directly in Fig. 21-3)

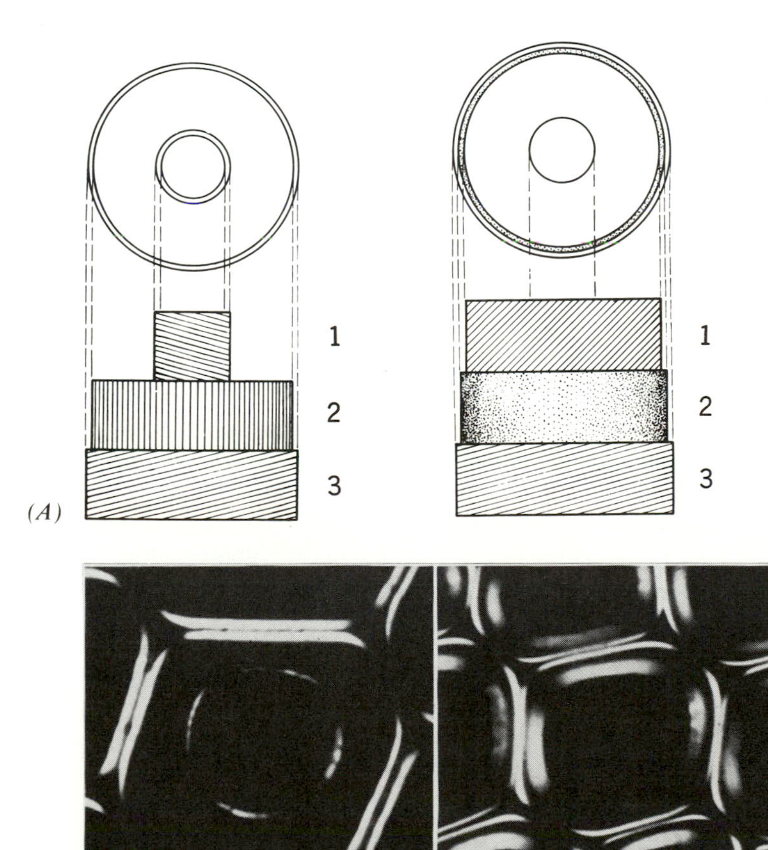

Figure 21-2 The essential structure of three-layered cellulose walls. *(A)* Simplified diagrams showing the orientation of the microfibrils. In the cell at the left, layer 2 is the thick one; on the right, layer 1 is thick and layer 2 is isotropic. *(B)* Corresponding cells in cross-section and viewed in polarized light, making the 3 layers alternately bright and dark. (From I. W. Bailey, 1940)

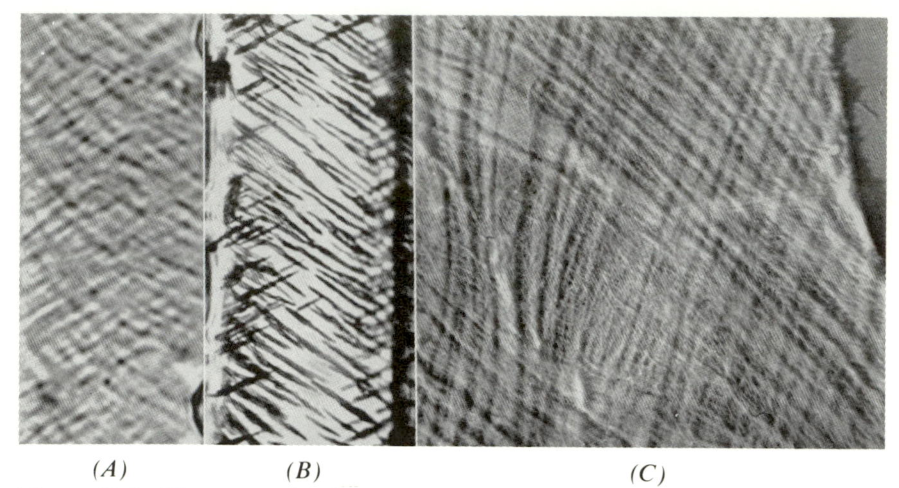

Figure 21-3 The crossed helix orientation of secondary walls in many woody plants. A and B, Two preparations made by depositing microcrystals of iodine, which follow the orientation of the cellulose fibrils. C. Direct electron micrograph (low power) of a cell wall of ash *(Fraxinus)*.

gives great strength to the fiber cells, much as the alternating orientation of the layers in plywood is responsible for its mechanical strength.

Because of this mechanical strength the strands of plant fiber cells have long been used for twisting into continuous threads. The choice fibers are those that can readily be separated from the adjacent tissues, whether in the form of whole vascular bundles, as in leaves, or in the form of groups of fiber cells usually occurring as a ring in the phloem parenchyma, the so-called "bast fibers" of stems. Even the more delicate fibers growing out from seeds have this quality of alternating layer structure. Fibers of commerce are made by first combing out the biological material to make the individual strands parallel and of comparable length and then spinning it to make a long continuous thread in which the individual fibers overlap so much that the whole thread gets great resistance to tension. Textiles are then made by weaving these threads in and out across one another in several different ways. Figure 21-4 shows three different methods, all used in ancient Peru and still in use today, especially by craft weavers.

The strength of fibers is measured by applying increasing amounts of tension until they break; the *tensile strength* is then given by the "breaking load" per unit of cross-section area.

The material of secondary walls is chemically more varied than that of the primary wall. In particular the polysaccharides are impregnated to different extents with *lignin*. As a general rule, the higher the lignin content, the browner the color and the poorer mechanical quality has the fiber. Because lignin is a netlike polymer based on phenol molecules (Chapter 6) its branching nonlinear arrangement does not contribute much to the fiber strength, and indeed for some uses, for example, for making white paper, the lignin, which is brown, is largely removed (Chapter 22).

Of the major fiber-forming plants, perhaps 30 or 40 have been put to human use. We take up the ones most widely used, from the three types—seed, stem, and leaf fibers (Table 21-1).

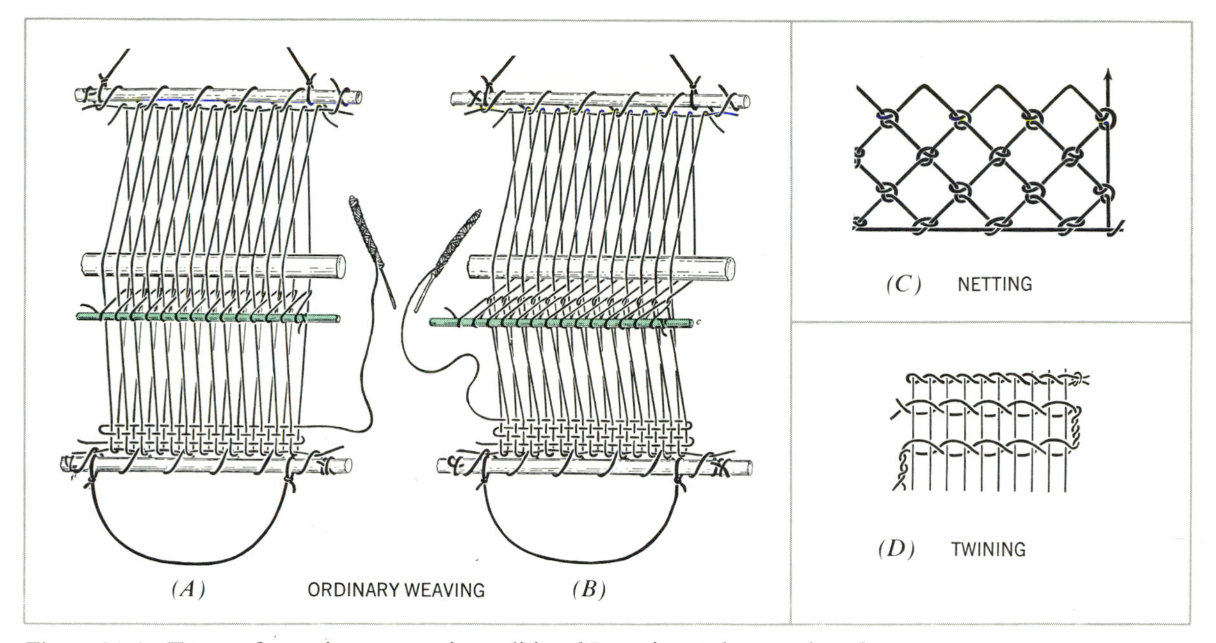

Figure 21-4 Types of weaving as seen in traditional Peruvian style. *A* and *B*. Ordinary "tabby" weaving; the vertical ("warp") threads that are down below in *(A)* are brought up in *(B)* so that each time the shuttle goes across it carries the "weft" thread over and under alternate wrap threads. *C*. Netting, with uniform meshes fastened with a simple knot. *D*. Twining; two weft threads twisted in pairs cross the warp either singly (above) or two at a time (below). (From Harcourt, 1974)

2. FIBERS FROM SEEDS AND FRUITS

A great many seeds bear hairs, which are in most cases unicellular. Often they are very short, but the longer hairs can be twisted together into usable fibers, which are tough and strong. Cotton hairs are said to be as strong as steel wire of the same diameter.

a. THE COTTON PLANT

The Malvaceae, the family that includes mallows and the hollyhock, also includes cotton, *Gossypium*. The palmate three-lobed leaves, the five spreading petals, the many stamens united into a tube but separating just below the anthers, and the characteristic style branching into multiple stigmas are typical of the mallow family (Fig. 21-5). What is not typical is the formation of the mass of very long hairs, that fill and eventually burst the ovary (called the "boll"). There are a great many species of cotton, some low bushes, some herbs, and some tall trees. The numerous Old World species all have the diploid chromosome number 26, whereas the two New World species, *G. hirsutum* and *G. barbadense*, have $2n = 52$ and are thus tetraploids. However, half of these 52 chromosomes are large, resembling those of the Old World species, and the other half are smaller, so that it is believed that somehow one of the 26-chromosome type got transported across the Atlantic in early times, crossed with another diploid mallow, which may no longer exist now, and then underwent spontaneous chromosome doubling (compare the wheat species described in Chapter 16).

TABLE 21-1

The Major Fibers and Their Production throughout the World

Fiber	*Genus and Species*	*Total Tons per Year (Approximate)*	*Producing Countries in Order of Importance*
Seed fibers (monocots and dicots)			
Cotton	*Gossypium hirsutum* *G. barbadense* *G. herbaceum*	12 million	U.S.S.R., China, U.S.A., India, Brazil, Pakistan, Turkey, Egypt, Sudan, Caribbean Islands
Kapok	*Ceiba pentandra*	25,000	Java, Thailand, India
Coir	*Cocos nucifera*	100,000	India, Java, Sri Lanka, South Pacific islands
Stem fibers (dicots)			
Flax	*Linum usitatissimum*	460,000	U.S.S.R. (more than half), Poland, Czechoslovakia, Germany, Hungary, Italy, France, U.S.A., Argentina, Belgium (the finest grade), Manchuria
Ramie	*Boehmeria nivea*	100,000	China, Japan, Taiwan
Jute	*Corchorus capsularis* *C. olitorius*	3 million	India, Bangladesh, China, Brazil
Jute-like fibers Kenaf and Roselle	*Hibiscus cannabinus* *H. sabdariffa*	300,000 30,000	India, southeast Asia, West Africa
Sunn hemp	*Crotalaria juncea*	100,000	India
Urena	*Urena lobata*	20,000	Brazil, Central Africa
Hemp	*Cannabis sativus*	300,000	U.S.S.R., Italy, Yugoslavia, Hungary, India, China
Leaf fibers (monocots)			
Sisal	*Agave sisalana*	550,000	Southern and eastern Africa
Hennequen	*A. fourcroydes*	200,000	Mexico, Cuba
Abacá (Manila hemp)	*Musa textilis*	180,000	Philippines
Bowstring hemp	*Sanseveria* spp.		Central Africa

Both diploid and tetraploid types of cotton have been used for fiber for over 4000 years. In the New World, the earliest fishnets found in excavations in Peru date from the days even before evidence of ceramic pots or of corn there, that is, from about 4400 B.P.; these have been identified as made from *G. barbadense*. In the Old World, numerous cotton textiles from about the same period have been discovered by archeologists in Ethiopia and Pakistan. Fabric from Mohenjo Daro, an ancient city site in the Indus valley, is from *G. arboreum*, which

Figure 21-5 The flower and open ("ripe") boll or ovary of cotton (*Gossypium hirsutum*).

is one of the two best fiber-yielding species among the 26-chromosome types. However, neither the ancient Egyptians nor the Greeks and Romans used cotton; they preferred linen, made from flax, which grows in temperate climates. Cotton is a semitropical plant, requiring a long warm summer when grown as an annual or a frost-free winter if grown as a perennial, as it is in Brazil. Cotton was not used in ancient China either, for the Chinese very early developed the making of silk, which was preeminent for the finer textiles, whereas for the coarser, stronger materials they used jute. Cotton growing in China, as in Egypt, dates only from the introduction of the New World species in relatively recent times.

The history of the Old World cottons, as elucidated from genetic and anatomical considerations, goes like this: From hot, dry southwest Africa or Angola a perennial form of *G. herbaceum* (race *Africanum*) spread across Africa to Mozambique, whence it was taken in the trade route to southern Arabia, as in the migrations of sorghum described in Chapter 16. Here it was probably domesticated for the first time, giving rise to a race named *acerifolium*. From there it was soon carried to northwest India where it gave rise to another new race, *indicum*. Theophrastus, in 350 B.C. (2350 B.P.) noted that cotton was cultivated both in Arabia and in India. The *indicum* race, by a single translocation, gave rise to *G. arboreum*. In the course of these travels the perennial forms were replaced by several annual cultivars; the importance of these is that they can be used in other climates, for they can die down in the winter and their seeds can germinate next spring. As a result annual forms can migrate northwards, and this gave rise to many central Asiatic forms. Marco Polo (1290 A.D.) reported cotton growing in Turkestan, where the winter is very cold, so that it must have been an annual form. He also noted perennial cotton plants 20 years old in Gujarat (west central India) where frosts would not generally occur. This complex history can be summarized thus:

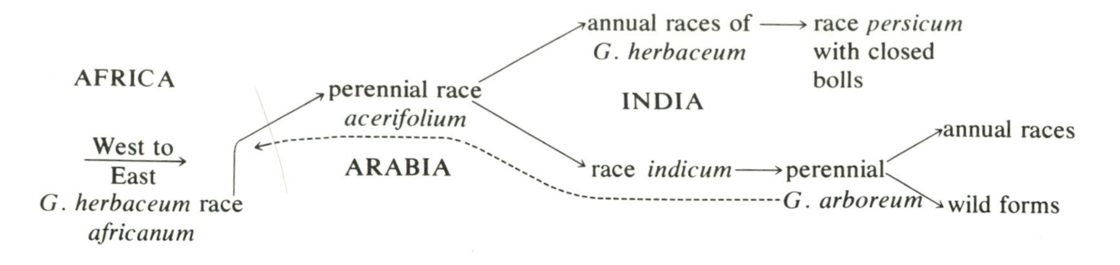

(The dashed line in the diagram shows the return pathway of the seeds.)

Later, both the cotton plant and cotton fabrics were exported along the same trade route as before, going from western India to Arabia and probably to East Africa. The seeds of wild cotton doubtless eventually made the same return trip.

The New World cottons cannot be traced in such detail, but their migrations and their formation of a number of distinguishable races are roughly similar to those of the Old World types. The *upland* cotton, *G. hirsutum,* originated in Guatemala or southern Mexico and spread into South America. Originally they may have been long-day plants, but the present upland cotton, which has given rise to at least four races, is cultivated in most of the world—South America, Africa, northern Australia, southern Russia, and parts of China, and thus may comprise both long- and short-day forms.

The second important New World species is *G. barbadense,* the one whose use dates back the longest. It probably originated in Colombia or Peru and has now spread over as wide an area as *G. hirsutum.* Its races include those with the longest and finest hairs, notably the one now called Egyptian, which is thought to be a hybrid, and Sea Island, which when grown in the smaller Caribbean islands or in Fiji Islands produces the finest and softest lint of all. Two varieties with coarser lint are *Lohan,* grown in Nigeria, and *Peruvian,* a perennial form. There are also several Polynesian "species," having deep yellow flowers and yellow to brown lint; these may be races of *G. hirsutum* or *G. barbadense.* A wild tetraploid, *G. mustelinum,* recently found in northeast Brazil, is probably a true species.

b. THE COTTON HAIR AND ITS USES

The seed of the cotton plant bears relatively long hairs, called *lint,* and much shorter hairs called *linters* or *fuzz.* In cultivated forms the lint hairs are easily pulled free while the fuzz remains attached; in wild forms the lint is firmly attached too. In the tetraploid (New World) species the lint is up to 50 mm long, whereas in the diploids it barely reaches 28 mm. The length of the lint largely determines the quality of the resulting thread. The processing has four steps: ginning, carding, combing, and spinning. In the gin, fine brushes pull the lint off the seed by drawing it through holes too fine for the seeds to pass. The lint has then to be straightened ("carded") and brought into parallel, uniform-sized groups ("combed"), which involves being drawn out by a series of rotating cylinders. Finally it is spun, or twisted into a long continuous thread, by feeding it under constant tension to a device that twists it as it rolls it up. In the old-style spinning wheel the tension is maintained by the hand of the spinner, and the twist is produced by a bobbin geared to the wheel, which is turned by pedals.

The hairs grow out from cells on the surface of the seed, immediately after fertilization. This truly remarkable process is pictured in Figure 21-6. After they have elongated, the walls of the hairs thicken continuously until only a very narrow lumen is left; this lumen may be 50 mm long but only 20 μ wide, a length-to-width ratio of some 2000 : 1. The

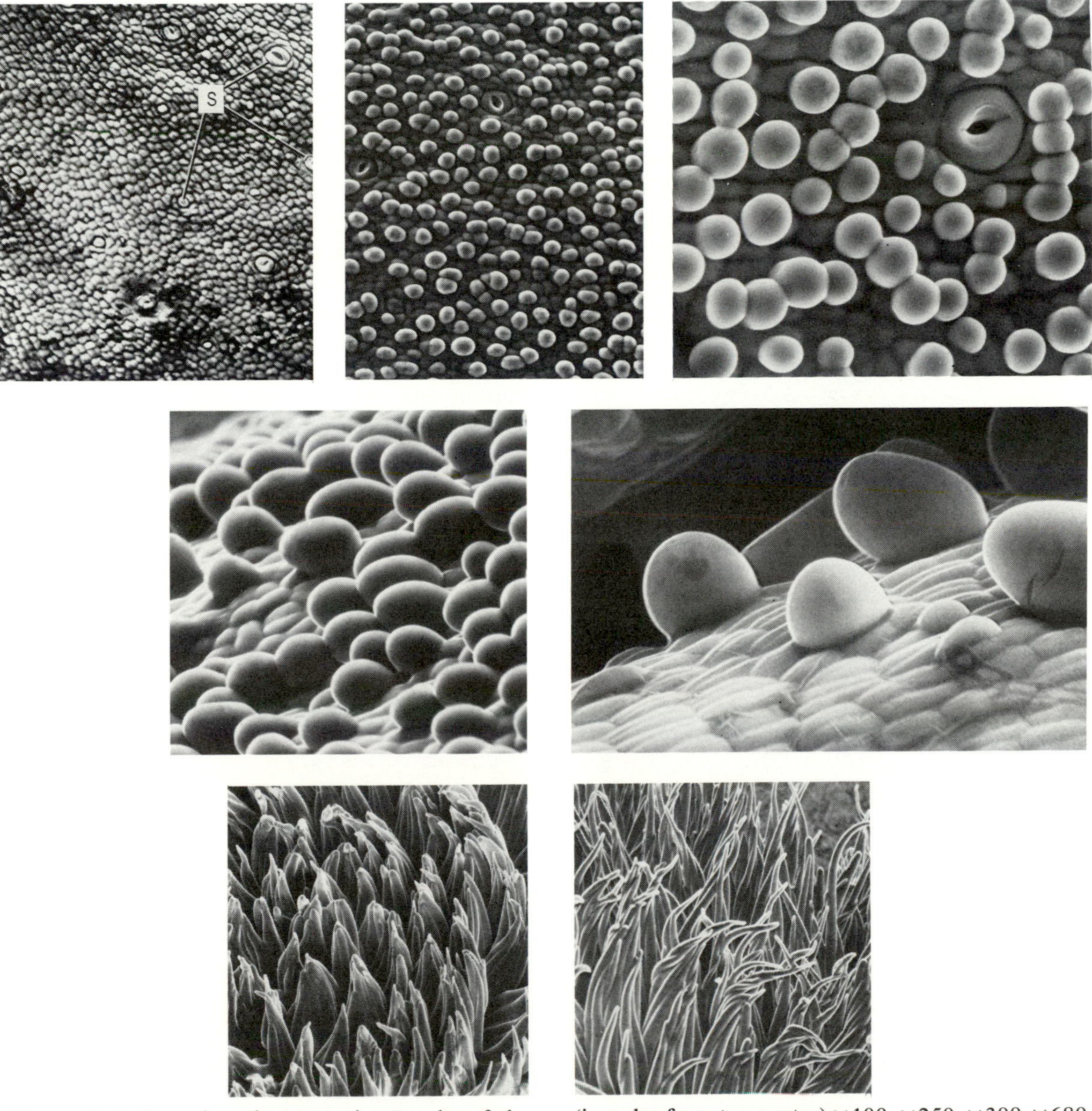

Figure 21-6 Scanning electron micrographs of the early growth of cotton hairs. At top, far left, the ovule surface (×70) shows stomata,S. The successive stages of hair growth are then seen at magnifications (in order from top center) ×100, ×250, ×300, ×680, ×50, ×40, the last being three full days after anthesis, when the tips are beginning to elongate and the cells are about 0.5 mm long.

deposition of the steadily thickening wall takes place in layers that are alternately denser and lighter, due to the effects of day and night; plants grown in continuous artificial light show no such layering in the hairs (Fig. 21-7). The many layers also alternate in the orientation of the constituent microfibrils, much like the alternating orientations of the secondary wall in stem fibers described above. This alternating structure has the same effect too, giving great tensile strength to the cotton hair (Fig. 21-7).

Almost all of the cottonseed is useful, for the linters are made into rayon (see Chapter 22) and the cottonseed is pressed to release much of the reserve material in the form of cottonseed oil. The remainder of the seed, which contains the protein, is pressed into a cake for cattle feed. A standard 500-lb bale of cotton corresponds to about 900 lb of seed and 80 lb of linters produced at the same time.

The commercial cotton types have been subjected to genetic selection and hybridization for over half a century, and work is continuing at experiment stations. It has been estimated that cotton yields in Africa have improved by an average of 4% per generation bred. The original Egyptian cotton was a perennial form of *G. barbadense,* brought to Egypt by Jumel in 1820, but soon an annual race was developed from it and, because the annual needed less irrigation, it became substituted for the original. From time to time heterosis (hybrid vigor) has been drawn into the picture (Chapters 12 and 16), and the hybrids yield about twice as much lint as the parent strains. Unfortunately the pollination is too uneven to allow general adoption. Nevertheless the Delhi experiment station in India has recently developed new crosses that show good hybrid vigor and are being tried on a large scale.

Cotton is the textile produced in by far the largest amounts worldwide (see Table 21-1). The advent of artificial fibers has only slightly lowered the world demand for cotton, and that only recently.

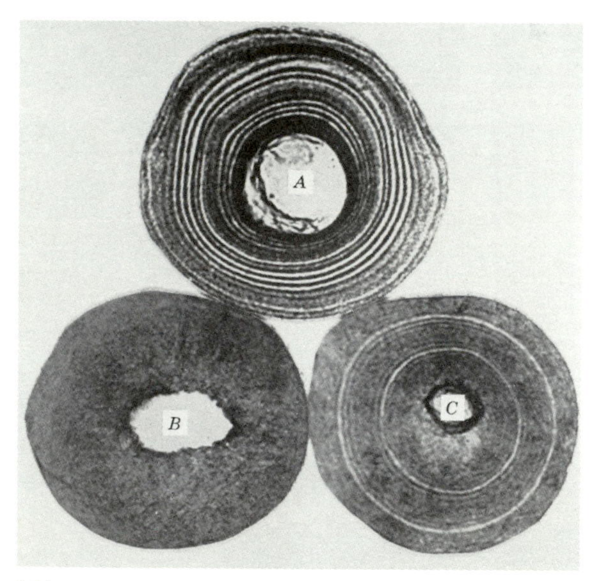

(*A*)

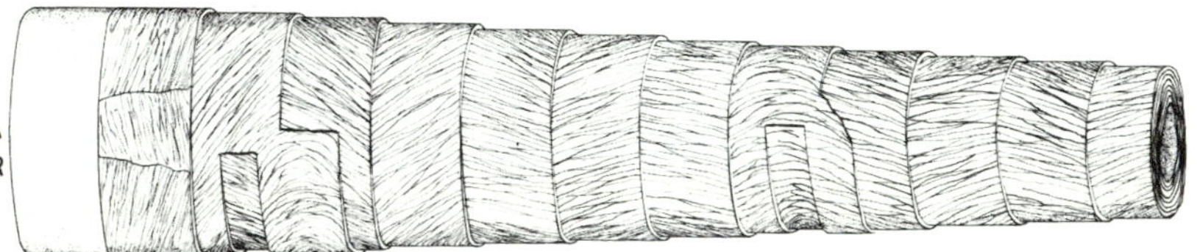

(*D*)

Figure 21-7 *A*. Cross sections of cotton hairs swollen in alkali to bring out detail of the structure. *A*, grown in normal photoperiods; *B*, grown in constant temperature and light; *C*, same but with two photoperiods added experimentally. The dark and light layers show differing densities. *D*. Diagram to correspond roughly with *A*. (*D*, USDA)

C. KAPOK

The kapok tree, *Ceiba pentandra,* is more tropical in habitat than cotton. Its family, the Bombacaceae, is largely limited to the New World tropics; it includes the genus *Ochroma,* which yields the lightest of all woods, the famous *balsa* (Chapter 22).

The kapok tree is grown almost wholly in Java, having been introduced from India by Hindu emigrants around 300 A.D. Dodonaeus' herbal (1618) has a picture of a kapok fruit, very much like those of today. The tree grows wild in India and Sri Lanka, but generally yields poor fiber. Attempts to introduce it into other areas with suitable climates have not been successful. Related species and varieties were started in some of the Caribbean islands and in Tanzania, but they were later cut down because they harbor the "cotton stainer" insect, and damage to the cotton crop was considered too dangerous. Kapok has been planted on a commercial scale in the Philippines and southeast Asia, but yields good quality fiber only when there is a long relatively dry season, as on the north coast of Java.

The trees flower at the onset of the dry season, and the long pods are picked when they are just beginning to split open, by men who climb the trees and take care to pick only the ripe fruits, leaving the unripe ones for a second harvest (Fig. 21-8). Skilled pickers can harvest

(A) *(B)* *(C)*

Figure 21-8 *A*. Armed with a long bamboo pole, the harvester gathers the ripe Kapok pods on a plantation in Java. *B*. When the pods have been opened and the floss pulled out, it is dried under netting (to prevent its blowing away). *C*. Fruit (pod).

the whole crop of a 30-ft tree in about 15 minutes.

The fibers grow out of the ovary wall, rather than out of the seeds as in cotton, but they become detached from the ovary when it is mature; this makes harvesting and separating from the seeds relatively easy. A hectare of seven-year-old trees usually produces about 500 kg of fiber, with 1000 kg of seeds, which are pressed to yield about 12% of their weight of a yellow oil similar to cottonseed oil (see Chapter 27). The pressed residue is sometimes used as cattle feed or as a low-grade fertilizer.

Not much scientific work has been done on kapok, other than the selection by the Dutch in the late nineteenth century, which led to development of the Javanese plantations. The fibers are about the same length as average diploid cotton fibers, but they contain some lignin. A major difference is that they are coated with a highly water-resistant cutin. This outer coat, together with the fact that the empty lumen is larger than in cotton (and hence the fiber is lighter), gives kapok its major applications. It is used as waterproof filler for mattresses, pillows, upholstery and softballs and especially for life jackets. A kapok-filled life jacket can support 30 times its own weight. Kapok has always been found difficult to spin and hence has not been made into textiles.

d. OTHER FLOSS-YIELDING PLANTS

Among the *Bombax* species, *Bombax malabaricum* of southern India has produced a kapoklike floss that is exported in modest amounts. Besides the kapok tree proper, other *Ceiba* species, about 20 in number, yield an inferior fiber, and one other genus, *Chorisia* (the "floss-silk tree"), yields a type of floss fiber used in Brazil. An unrelated group which yields a usable floss is that of the milkweeds of the family *Asclepiadaceae,* native to North America. The seeds of these species yield long silky hairs that are twisted into dental floss, and during World War II this material had a limited use as a substitute for kapok.

e. COIR

Coir is at the opposite extreme from cotton, for it is the very coarsest type of fiber and grows around the largest of all seeds, the coconut. The coconut palm, *Cocos nucifera,* is found on virtually all tropical seashores and is a major contributor to the tropical economies (Fig. 18-9). The fibers grow out from the ovary walls or "husks." They are separated from the nuts by splitting the husks open and soaking them in tanks of hot water until thoroughly softened. Beating with wooden mallets, or more often now in a simple beating machine, liberates the fibers; they are combed and dried. Most of the coir in use comes from India and Sri Lanka. In the Orient it is twisted into rope and twine, and in the West it is made into door mats ("coco-mats"), stuffing, and the coarser types of floor covering.

3. STEM FIBERS

As we saw in section 1, the phloem or "bast" of some stems contains groups of long, thick-walled fiber cells, and these, sometimes with some xylem tissue in addition, constitute four of the major fibers now in use. Flax and ramie are fine and white and mainly cellulose, with a little pectin, whereas jute and hemp are coarse and brownish and contain 10 to 15% lignin.

a. FLAX

Flax (*Linum usitatissimum*) belongs to the small family Linaceae, of worldwide distribution. Some *Linum* species, especially the blue-flowered ones, are garden ornamentals, but *L. usitatissimum* is the only one with useful fibers. It is a native of Asia but can be grown in cool climates almost everywhere (Fig. 21-9).

Linen from flax is perhaps our oldest fiber. The name "*usitatissimum*" means "the most useful." Linen was made, probably from a wild perennial plant, by the Swiss lake dwellers around 9000 B.P., and a piece of woven cloth dated 8000 B.P. has been excavated in Turkey.

By 5400 B.P. the Egyptians had learned to cultivate it in the irrigated Nile delta. Impregnated with resin, it formed the wrappings of mummies and thus has been preserved to the present day. The Egyptians were also able to dye linen and produced green, blue, yellow, and red materials. The fine drapery that we see in the ancient carvings of Egyptian royalty and goddesses (see Figure 21-9B) was all linen, for it was woven then with more threads to the inch than we use nowadays. Some linen from Egyptian tombs has 220 threads to each cm and is 150 cm wide! From Egypt, flax spread to Greece and Rome, and then all over Europe, as a cultivated crop. In modern times, Czechoslovakia, Poland, Russia, Germany, Belgium and, until recently, Ireland have been the greatest linen-producing centers, and Russia still grows over half the world's flax (though not the finer grades). The Russians have greatly increased their yields by developing several strains resistant to fungi. Poland, the northern United States, northern India, and an area around the Plate valley in Argentina also contribute.

The flax plant is grown as an annual. When planted close together to favor long, unbranching stems it grows to 1.5 m tall. Branching is promoted by high temperature—for example, the hot summers of Russia. It is harvested after flowering when stem elongation has ceased. The plants are pulled out of the ground so as to get the full length of the stem; they are dried and bundled and then subjected to *retting*. This is a process, common to virtually all stem fibers, in which the materials other than cellulose are decomposed by bacteria and thus partially removed. The simplest method, which is almost the same as was used by the Egyptians 5000 years ago, is to immerse the stems in tanks or small ponds for 5 to 10 days, the duration depending on the temperature. Aerobic bacteria first attack the sugars, proteins, and some of the pectins, thus using up the dissolved oxygen. Then anaerobes develop and rapidly ferment the starch, pectin, and some of the polysaccharide cell wall substances, so that the phloem parenchyma cells break up and the bast fibers become free. Retting can also be done, as formerly in Ireland, by leaving the stems in the field to become soaked by dew. In Italy retting has been done in tanks using "starter cultures" of known bacteria, with the temperature con-

(A)

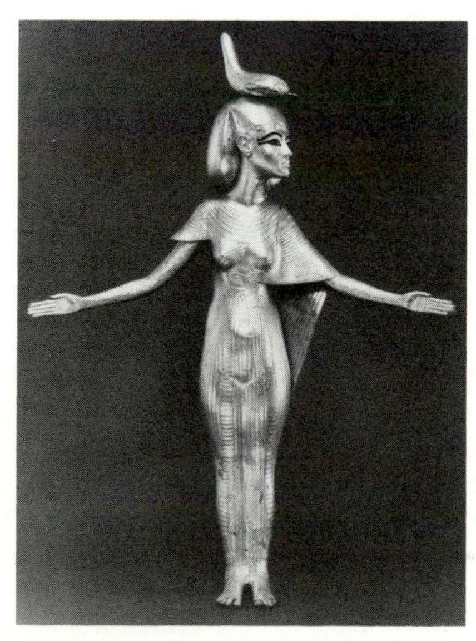

(B)

Figure 21-9 *A*. Flax plant. (From Edlin, 1967) *B*. The goddess Selkis or Serket, clad in the finest linen. (about 1344 B.C.).

trolled so as to give the maximum breakdown of the accompanying materials but minimum attack on the cellulose. With all methods, the excess water is squeezed out; then the retted stems are dried and crushed between rollers, when the decomposed material falls away and the fibers remain. Flax fibers can be up to a meter long, though length of the individual fiber cells is not over 5 to 6 cm. Each has a multilayered secondary wall as shown in Figure 21-2. After being beaten, washed, and combed, the retted fibers are dyed and spun into thread for weaving. Linen textiles are soft and lustrous and become still softer with repeated washing. Since it is water-absorbent, the coarser grades are made into hand towels. As is the case with cotton, the seeds yield a useful oil (Chapter 27).

b. RAMIE

Ramie is a beautiful, lustrous fiber from the stem phloem of *Boehmeria nivea,* a tall, tough perennial of the nettle family (Urticaceae). It has rather rough, two-lobed leaves; in the temperate climate variety these are characteristically white underneath (*nivea* = snowy), whereas in the tropical variety they are green. The tall stems, 2 to 2½ m high in carefully grown plantations, thicken to nearly an inch diameter at the base. The bark is stripped off by hand in long ribbons, and the fibers also have to be liberated by hand, because the retting bacteria do not attack the gum or resin that binds the fiber bundles together. It is these resinous materials that have prevented ramie from becoming a fiber of major importance, for they can be removed only by rather drastic treatment like soaking in hot alkali, followed by bleaching and washing with dilute acid. Discovery of a new method of "degumming" might at any time cause a revival of the use of ramie as a major textile.

The fiber cells can be up to 30 cm long and are pure cellulose. The spun fiber is stronger than cotton or flax and less affected by moisture. Much of it is used in China, where it is mainly grown, and the lustrous "China grass cloth" that reaches the West is only a small fraction of the ramie production.

c. JUTE

Jute is the stem fiber produced in the largest quantity worldwide. It comes from two species of *Corchorus,* belonging to the Tiliaceae, the same family as the European street tree, the linden (*Tilia europea*) which gave its name to the fashionable Berlin street "Unterdenlinden." More to the point, the jute plant is in the same family as *Tilia americana,* the American basswood, whose bark furnished fiber to several tribes of American Indians. They peeled the inner bark and boiled it with wood ashes to release the fiber, which they spun and wove. Since wood ashes are alkaline, the treatment has much in common with that accorded to ramie. Jute fibers are obtained more simply.

Jute is a tall annual, native to India and south China, where it is still mainly grown. It requires good rains and grows fast. In consequence, two or more crops can be harvested in a season. After flowering, the tall, straight stalks (see Fig. 21-10) are cut, dried, and retted in small ponds. Then the partly decayed cells of the bark are beaten off the fiber bundles into the water. These fiber bundles, located in the phloem and inner cortex, are strong and pliable at first but, since the individual cells are short and since also there are 10 to 12% lignin and some tannin present, they become darker and stiffer with time. Besides lignin and cellulose, they contain some xylan, the polymer typical of the stems of the grasses. (However, *Corchorus* is a dicot and hence unrelated to the grasses.)

Most of the fiber is woven into burlap, sackcloth, and tough twines. The process is simpler than for cotton, since the fiber bundles are used whole and not broken down into their individual cells. At the jute experiment station near Calcutta improved cultivars have been developed, but there has been little systematic hybridization.

d. THE HEMPS

The name *hemp* has been applied to at least three fibers, obtained from quite unrelated plants (Table 21-1). Most important is the stem fiber from *Cannabis sativa* of the fig family (Moraceae), whereas sunn hemp is from a legume and Manila hemp is a leaf fiber obtained from a banana-like plant. The leaf fiber sisal is also sometimes called sisal hemp.

The word *hemp* has a curious background. Its old German form *hanaf* suggests that another leaf fiber called *kenaf* has its name from the same source, and a still older form of the name is *hannafis,* which is a corruption of the Latin name *Cannabis*. Thus the English and Latin names, although apparently different, are forms of the same word.

(*A*)

(*B*)

(*C*)

Figure 21-10 *A*. Harvesting jute in Nepal. *B*. Extracting jute fibers from the retted stems after they have been beaten, Bangladesh. *C*. Carrying off the jute fibers to be worked into rope or cloth, Bangladesh.

Cannabis is thought to be native to China, where some is still grown. However, Russia and southern Europe produce most of the hemp today. It is dioecious, that is, has separate male and female plants; both produce fiber. *Cannabis* takes about four months to reach its full height of about 2½ m, and another month or so to form mature seed. The seed protein, *edestin,* is a good nutrient, and hemp seed was formerly ground for food; today it is mainly used for birdseed or pressed for oil. The seed is normally boiled before it is sold, to prevent germination, because the leaves and flowers secrete a resin that contains the narcotic tetrahydrocannabinol, known as marijuana, hashish, or bhang (see Chapter 24). Because of the illicit traffic in this material, hemp growers in the United States have to be specially licensed and to observe rigid regulations. Unlicensed hemp growers are subject to imprisonment.

Hemp is planted close to form tall, unbranched stems. The individual fiber cells may be up to 5 cm long but they develop in bundles that run the whole length of the stem, in the phloem. The male plants are said to yield the better fiber and are cut down first (Fig. 21-11). The stems are usually cut at the base by hand, tied in bundles, and retted in ponds like jute, then dried and crushed to release the fiber bundles. They are combed, spun, and woven like other textiles, or (more commonly) twisted into cord and rope. The best varieties, which are grown in Italy, are almost white in color and nearly as soft as linen.

The production of the different kinds of hemp is summarized in Table 21-1. *Sunn hemp,* the stem fiber from *Crotalaria juncea (Leguminosae)* is almost limited to northwest India, where it grows to about 2½ m high and bears racemes of bright yellow papilionaceous flowers (see Chapter 23). Like those of hemp and jute, the long, straight stems on retting give rise to long, firm bundles of fibers, most of which go into rope or, because it is resistant to water, into fishnets. A little is used in India for papermaking.

Kenaf and *Roselle* are stem fibers from the mallow family (Malvaceae), to which cotton also belongs. Kenaf, or Deccan hemp, is from *Hibiscus cannabinus* and roselle from *H. sabdariffa*. Both species are grown in India, southeast Asia, and China, though roselle is actually native to central Africa. Kenaf is a diploid ($2n = 36$); roselle is a tetraploid and like some other tetraploids it is of slower growth. Both are planted close, to yield unbranched stems, and are retted to produce rough, strong fiber similar to jute. Some cultivars of roselle are grown for food, the seeds being ground into flour, the leaves cooked as a vegetable, and the juice made into a fermented drink.

Another member of the Malvaceae, *Urena lobata,* is grown in central Africa and in Brazil to produce a coarse, jutelike fiber used for sacking and, in Brazil, for coffee bags.

4. LEAF FIBERS

Leaf fibers are derived from the long, narrow leaves typical of monocotyledons. No doubt many such plants have served from time to time

Figure 21-11 Gathering hemp in France.

as sources of fiber, for a great many monocot leaves have tough vascular bundles with fiber cells associated. However, just three species, two agaves and one banana, account for the largest part of the world's leaf fiber today (Table 21-1).

a. THE AGAVES

Agave sisalana yields sisal and *A. fourcroydes* yields hennequen. Botanically they belong to the Amaryllidaceae, the family of our garden daffodils and narcissi, but anything looking less like those familiar plants can hardly be imagined. They are very large plants with thick, spiny, semisucculent leaves forming enormous rosettes and bearing, after a number of years, huge flowering stalks (see Figs. 21-12 and 13) ending in a raceme of pendulous, somewhat lilylike flowers. Both species are natural polyploids. Several other *Agave* species, cultivated mainly in the high arid regions of Mexico, yield smaller amounts of similar fiber.

Agave sisalana is native to central America, and the Spanish explorers found it growing in Yucatan (southern Mexico) and being used as fiber. The pulp of the leaves seems to have been used as food around 7000 B.P. In the middle of the last century it was introduced into central and southern Africa, where it is now mostly grown. In Zululand there are tremendous areas covered with sisal plantations. Sisal is propagated vegetatively from plantlets that develop *in situ* in the inflorescence—an unusual habit, seen also in the mangrove. The plantlets drop to the ground and either establish themselves or are collected and planted.

Sisal does best in hot, fairly dry soil, where it grows slowly, so that it is 3 to 4 years before the lowest leaves are ready to be cut off. The leaves are then produced continuously for up to 40 years (about 300 leaves in all); then the immense inflorescence develops and the plant dies. By this time lateral outgrowths have developed from the crown, and these replace the dead parent.

Figure 21-12 *Agave sisalana* in Zululand, South Africa. Note the plantlets growing on the inflorescence.

Agave fourcroydes flourishes in still more arid conditions, being mainly grown on dry, stony, alkaline soils in the Yucatan peninsula and in Cuba. It grows more slowly than sisal and the leaves bear murderously sharp spines that are cut off at harvest (Fig. 21-13).

In both *Agave* species the leaves are run through a crusher, and the green, succulent tissues with the tough, cutinous epidermis are scraped off by hand or with simple machinery. The remaining fibers, which are the whole vascular bundles, are washed, dried in the sun, and brushed until free of residual tissues. The longer fibers are combed and then either twisted into rope or twine or woven into mailbags; the short fibers are usually sent to the paper mill. The waste leaf material is returned to the soil. Sisal and hennequen are both strong, rather harsh, brownish fibers, and they furnish most of the world's (non-naval) string and rope.

Another fiber plant closely related to the agaves is *Fourcrea gigantea,* which is grown on Mauritius and yields a fiber much like sisal, called Mauritius hemp. Since the main agricultural product of the island is sugar, the hemp is used almost wholly to make sugar bags.

(A)

(B)

Figure 21-13 *A*. *Agave fourcroydes* in Cuba. The basal leaves have been cut the previous season, and the spiny tips have now been cut off. *B*. Drying the scraped out fibers in Merida, Mexico.

b. MANILA HEMP

Manila hemp or abacá is from the leaves of a banana species, *Musa textilis,* a native of the Philippines, which is almost the only area where it is grown. Since it needs good rainfall but will not flourish in swampy ground, it is cultivated on hillsides. Propagation, like that of sisal and also that of the edible banana, is from lateral branches from the crown or the rhizome.

Like the edible banana, the plant grows up from a rhizome with a series of overlapping leaf sheaths, forming a "pseudo-stem," and through the center of this there eventually appears the flower stalk, growing to a height of 5 to 7 m (Fig. 18-8). When this flower stalk appears the plants are cut down and the leaf sheaths separated and cut into strips. These strips are then pulled under the edge of a sharp knife (usually made of bamboo), which scrapes off the leaf parenchyma, leaving the fiber intact. Simple scraping machinery has been introduced in some plantations. The individual strands, 2 to 4 m long, are light brown or buff in color, hence low in lignin and tannins and very strong. Manila hemp makes the finest qualities of rope, much used in seamanship. Manila ropes have held ships to the quay in most harbors of the world, but nylon is now replacing it. In Ethiopia a comparable species. *Musa ensete,* is similarly used.

C. BOWSTRING HEMP

Bowstring hemp or African bowstring is made from the leaves of a member of the Liliaceae, *Sanseviera metalaea,* also called *S. thyrsiflora* or. *S. guineensis.* The plant forms a rosette with stiff, erect, and rather thick leaves that yield a whitish fiber used in Central Africa for fishnets and coarse cloth.

SUPPLEMENT
Fiber Structure at the Molecular Level

The path of glucose from its production as glucose phosphate in photosynthesis to short chains of glucose residues, formed with the elimination of water between two glucose molecules, was traced in Chapter 6. In the formation of cellulose, which is the major polymer of fibers, glucose-1-phosphate (G-1-Ⓟ) combines with guanosine-triphosphate (GuTP), and the product then reacts with additional glucose to liberate the base, leaving the glucose molecules combined in chains as shown below: G-1-Ⓟ + GuTP → GuDPG + Ⓟ—Ⓟ ; *n* GuDPG + *n*G → *n*GuDP + G_n. In starch and cellulose the *n* glucose molecules are linked by loss of water between the 1- and 4-positions. In cellulose the linkage between the 1-position of one glucose molecule and the 4-position of the next has the beta-form, which makes the glucose molecules alternate in their positions in space:

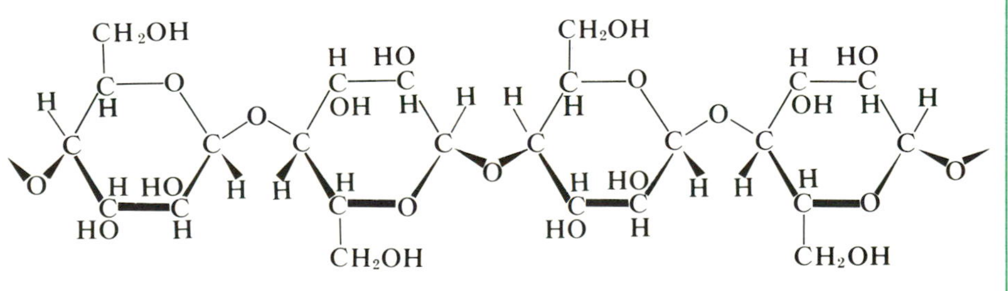

The opposite is seen in amylose, the unbranched chain form of starch, in which the glucose molecules are in alpha-linkage:

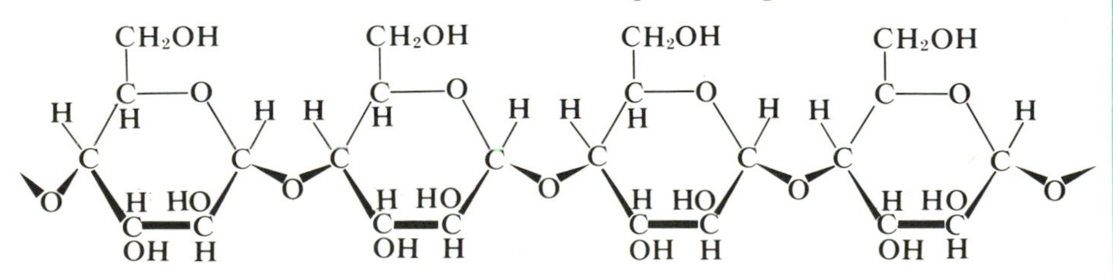

The balanced symmetry of the cellulose type of linkage seems to be one cause of its physical characteristics; the chains are of indefinite length, chemically inert, and of great mechanical strength. Furthermore, they are rarely, if ever, laid down as single macromolecules, but as overlapping microfibrils.

In the stems and leaf-fibers of the Gramineae there are comparable chains of the C_5 sugars xylose and arabinose; these are also linked by 1,4-beta-linkage, but the projecting sixth carbon atom is of course missing. Using the termination -*an* to mean *anhydride* (i.e., linked by loss of water), we call these xylan and araban. They can be combined in araboxylan. They occur in some of the fibers in families other than the grasses, notably in jute. Using the same terminology, cellulose is classed as a glu*can*.

As far as it is understood, the chains of sugar molecules seem to be built up in two steps. In the first step the enzyme plus the polymer (the "primer") reacts with the specific sugar–base complex, GuDPG, to form short chains of sugar molecules attached at specific points along its length. When these chains are long enough—four glucose units in the diagram—a less specific reaction takes over and several different sugar–base complexes start attaching glucose (or C_5 sugars in the Gramineae) to form relatively long chains, as we saw in Chapter 17. For starch formation (alpha-linkages) uridine diphosphoglucose, UDPG, is the initiating complex; for cellulose formation it is guanosine diphosphoglucose, GuDPG. With starch some of the chains become highly branched. In both cases adenosine diphosphoglucose, ADPG, is among the most active types in the second step. Analysis shows the presence of a glucose–protein compound in the cell membranes of some parenchyma cells, and this compound may be the actual primer molecule. It would constitute only a small fraction of the total material of the membrane, and hence would appear as only a very small fraction of the cell wall.

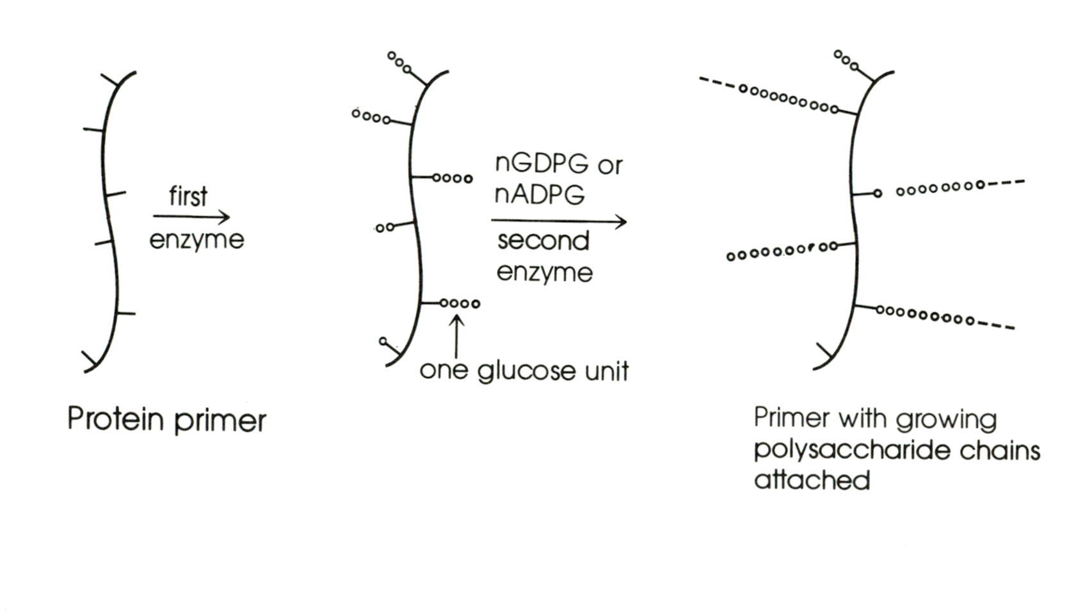

CHAPTER 22

Wood, Bark, and Forests

1. FORESTS AS A RESOURCE

Trees have been remarkably successful in plants' struggle for the occupancy of land. In fact, as pointed out in Chapter 2, herbaceous flowering plants did not arrive on the scene until relatively recently. Even today trees tend to dominate the vegetation wherever the rainfall is over about 60 cm and adequately distributed throughout the year (Chapter 14). Although much forest has been cleared by humans for agricultural or urban activities, forests still cover one-third of the land surfaces and 10% of the total surface of the earth. In terms of world net primary productivity, they account for 67% of all dry matter production on land.

There seems to be a universal trend in the attitudes of people toward forests as their nations develop from infancy to a high level of technology. Initially the forests appear to be an obstacle; land needs to be cleared to provide for agriculture and living space. During the next stage of national development, forests are storehouses to be exploited for construction materials and energy. Finally, they are recognized as a valuable, renewable resource that needs to be protected and maintained. Also today as the virgin forests become rapidly depleted, there is increasing interest in growing trees as a crop. Likewise, there is recognition that not all forests will, or should, become "cellulose factories." Conservation of soil and water resources, as well as recreation, is demanding some compromise in use of forest resources. These topics are discussed further in Chapter 31.

Knowledge about forest trees and their influence on the environment is meager as yet. It is still not clear how felling a forest may influence watershed stability and soil erosion or other changes in the soil. Of course, trees contribute considerable moisture to the atmosphere (Fig. 14-20). They are also atmospheric filters and major suppliers of oxygen. Furthermore, planting trees as shelters provides protection from drying winds, particularly in prairie or steppe regions. As a result, productivity of grain and of most crops so protected may be increased as much as 20% compared with unprotected crops.

There has been considerable speculation as to whether the severe decimation of forests east of the Mediterranean and along the north coasts of South America and Africa, where wood had been the prime fuel since the advent of human beings, may not have been instrumental in climatic changes there. Certainly once productive lands in these marginal climates have become progressively more arid and unremunerative, as have the Rajputana deserts of western India and those of South Africa. Is the change in climate because of loss of forest, or is loss of forest inevitable (even without human interference) because of climatic change? We cannot be sure but both influences are probably at work.

At one time Europe was almost totally

forested; today less than one-third is in forest, and this is for the most part regularly utilized and intensively managed. Except in Scandinavia and the central and southern mountains, little remains of Europe's indigenous forests. The British Isles, for example, were heavily forested as late as the seventeenth century; the English Oak (*Quercus robur*) and Scotch pine (*Pinus sylvestris*) were the trees that built the ships that ruled the world. Now only about 4% of the land is forested, primarily from plantings made through the years of ash, birch, larch, Norway spruce, and Douglas fir.

The boreal *taiga* forest (Chapter 14), however, still largely uninhabited, is one of the great native forests left today and includes nearly half of all the world's total conifers, which we have only partially exploited. The humid tropical forests are even a greater resource than the *taiga*. The great diversity of species has led to practical problems in managing and marketing the trees, but these forests are storehouses of natural wealth—not only for lumber but for resins, latex, oils, and also edible fruits.

2. SECONDARY DEVELOPMENT: THE TREE'S PRODUCTION OF WOOD AND BARK

Wood and bark are products of secondary development of stem and root tissue. Secondary growth occurs through the activity of additional meristems (other than apical meristems described in Chapter 4) that produce an increase in girth of the plant. The growth of these meristems, the *vascular cambium* and the *cork cambium,* causes an increase in thickness of the stem and roots and provides for increased conducting and supporting tissue for the plant.

The *vascular cambium* is a sheath of tissue one layer thick that is laid down late in the course of primary growth. It completely encircles the axis of the plant, passing between phloem and xylem. As discussed in Chapter 4 its products are secondary xylem inwardly and secondary phloem outwardly (Figs. 22-1 and 22-2). The vascular cambium is made up of two types of cells—*fusiform* and *ray initials*—which divide tangentially. The fusiform initials are much longer than wide and are thin-walled cells that are flattened tangentially. The ray initials are comparatively small cells that develop into radial bands of cells (Fig. 22-2) varying in height and width in different species of shrubs and trees. Rays more or less uniformly penetrate the xylem and phloem, the number often increasing as the xylem and phloem masses increase with age; in other species the same result is obtained by an increasing width of the rays by radial cell division within the ray cells in the cambial zone. The rays serve as a storage system for reserve food materials and for other substances.

Stages of differentiation of vascular cambium cells to form secondary phloem and secondary xylem are shown in Figure 22-3. Generally the inner daughter cell next to the phloem remains meristematic but, when new phloem elements are produced, an inner daughter cell retains the capacity to divide and the outer daughter cell gives rise to one or more phloem cells. Generally more xylem than phloem is produced in a season. Although cambial cells by repeated divisions are differentiating into xylem and phloem elements, some daughter cells remain meristematic just as some do in the apical meristem, and so there always is a cambium between xylem and phloem. The resulting cellular composition of secondary xylem and phloem is quite similar to that of primary vascular tissue.

The new layers of secondary xylem are laid down outside the old ones, which persist. In trees and shrubs they accumulate to form wood. Xylem vessels often function as a water-conducting system for several years, after which they remain only as durable supporting tissue. *Heartwood* is xylem that has ceased to function and has become filled with substances such as tannins and resins that help prevent decay. *Sapwood* is xylem that still conducts water, stores carbohydrates, and performs other vital functions.

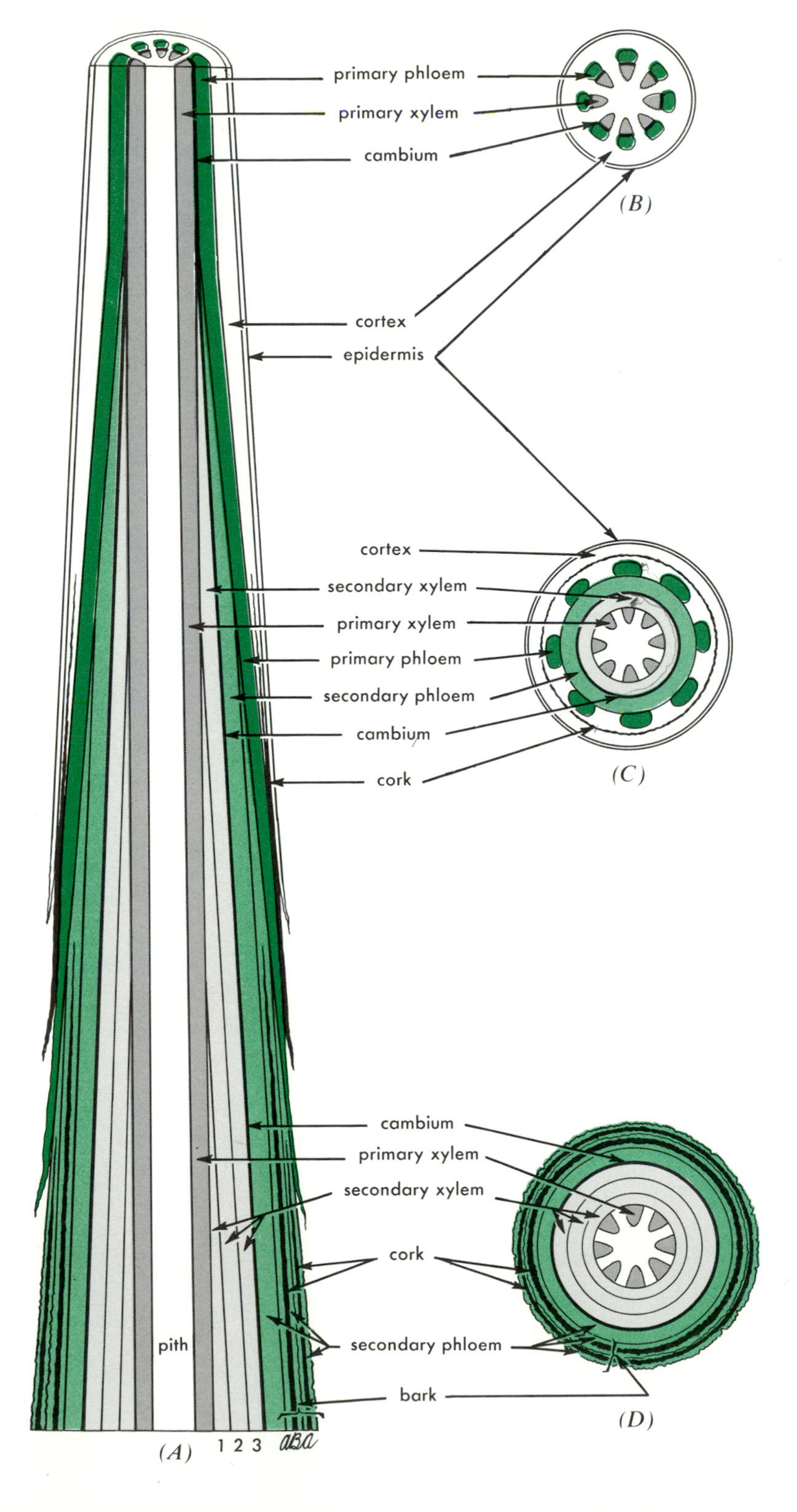

Figure 22-1 Changes in a stem as it increases in age and girth. *A*. Longitudinal section through a three-year old stem. *B*, *C*, and *D*. Cross sections at indicated levels.

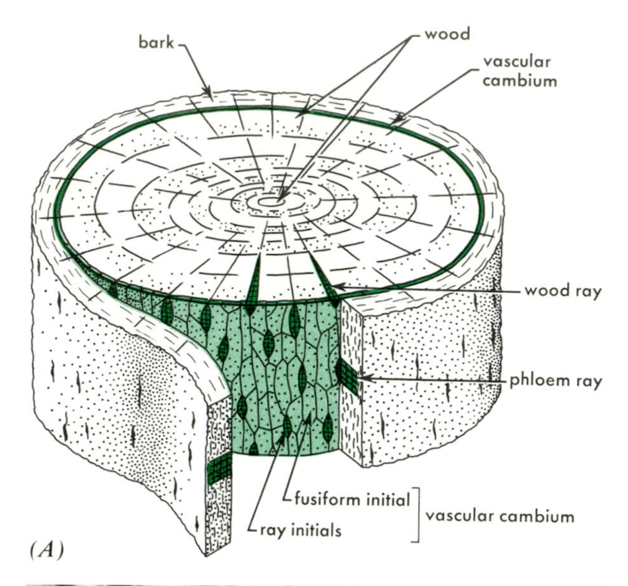

(A)

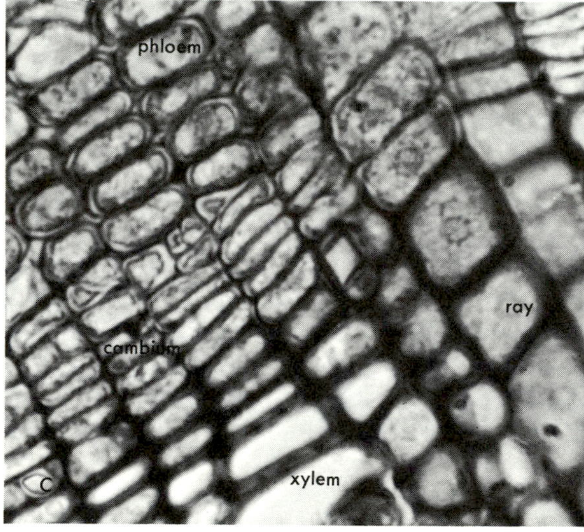

(B)

Figure 22-2 Vascular cambium. *A*. Diagram of cross-section of a woody stem showing cambium, its intitials, and the wood and phloem rays. *B*. Cross-section of quiescent cambium of locust *(Robinia)* (×3 to 3 ³/₄).

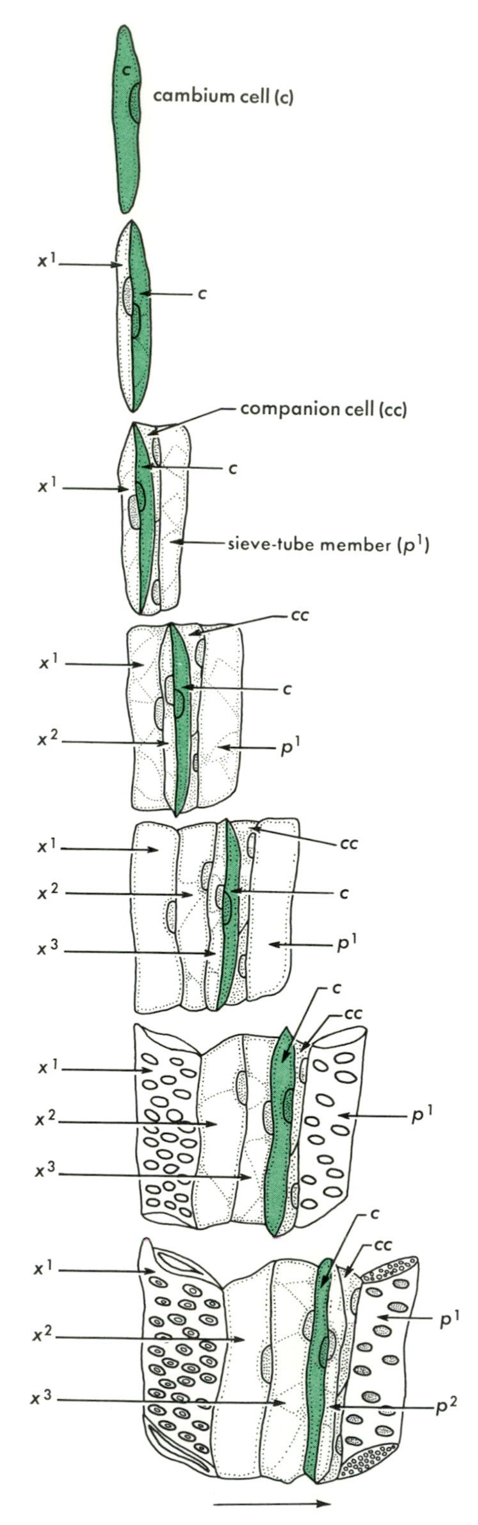

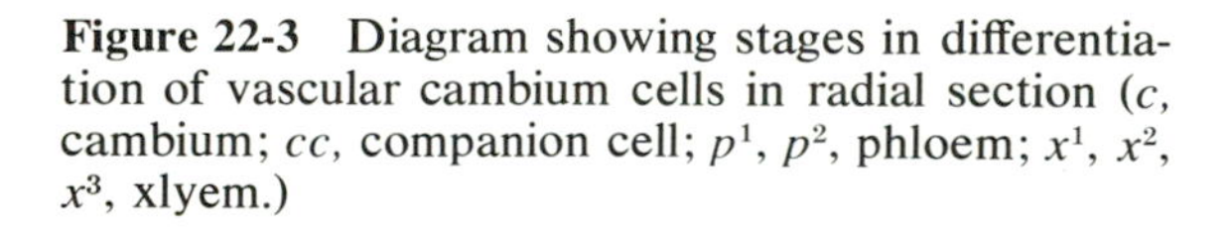
Figure 22-3 Diagram showing stages in differentiation of vascular cambium cells in radial section (*c*, cambium; *cc*, companion cell; p^1, p^2, phloem; x^1, x^2, x^3, xlyem.)

The vascular cambium divides only during the growing season of the plant (the spring and summer in most plants of temperate regions and the rainy season in plants of the desert and seasonally dry tropics). As a consequence, new layers of secondary xylem are laid down every growing season. At the beginning of the growing season, the cambial cells produce relatively large, thin-walled cells. Later in the season the new cells are smaller. In temperate climates, there is only one growth season per year, and the abrupt change in cell size between the end of one season and the beginning of the next clearly marks off an *annual ring* (Fig. 22-4). Using these annual rings, one can get some idea of climatic conditions and determine the ages of trees living in the temperate zone. In the wet tropics, where growth is continuous, it is impossible to date trees by this method because there are no differences in cellular growth to record.

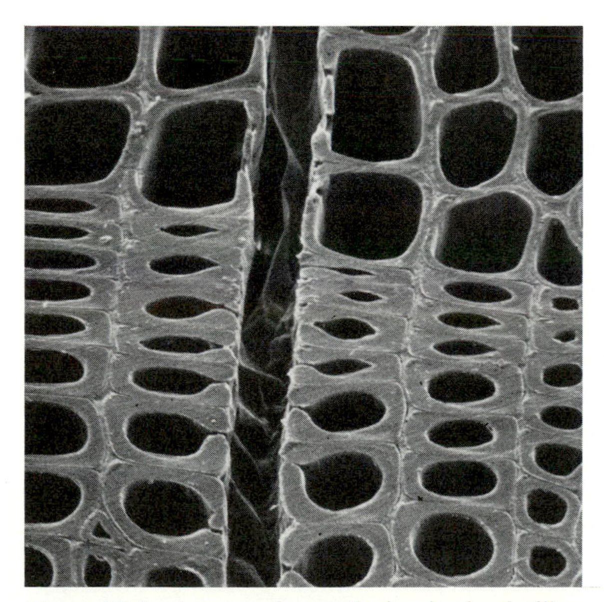

Figure 22-4 An annual growth ring is clearly illustrated in this cross section of *Pinus radiata* (Monterey pine). Small, thick-walled tracheids of the summer wood occur in the lower side; beyond the ring boundary the cells are larger and have thinner walls. A ray may be seen crossing the wood at right angles to the ring boundary through the left center (×550).

In plants in which the girth of stems and roots increases by secondary growth, the epidermis is eventually destroyed and replaced by tissue known as *periderm*. The periderm includes some of the tissue we usually think of as bark, but "bark" is often used to include all of the plant tissue from the outside of the cambium. Since the cambium is made up of the cells with the thinnest walls, it may be pulled away with all of the outer phloem tissue. The word "phloem" actually is derived from the Greek word for bark. The periderm generally includes three types of tissue, which are, from the outside in, the *phellum*, more commonly but less precisely called "cork," the *phellogen* or *cork cambium*, and the *phelloderm*, "cork-skin," which is a thin layer of parenchyma cells inside the phellogen.

Phellogen is a meristematic tissue that arises initially from parenchyma cells in the cortex of the stem or in the pericycle and cortex of the root (Chapter 4). Unlike the vascular cambium, it does not form a continuous cylinder but occurs as a series of disconnected plates, which are active for only a single season. In subsequent years, new plates of phellogen are formed at successively deeper levels. In a similar manner, phellogen forms in response to injury, producing bark to heal a wound in a plant's protective covering or periderm.

In the first year, phellogen in the stem generally is produced just beneath the epidermis (Fig. 22-5). Cells of the cortex divide tangentially to produce a phellogen layer, which gives rise to phelloderm on the inside and cork on the outside. Cork is dead tissue, and its cells are heavily impregnated with suberin, the same waxy substance that occurs in the casparian strip of the endodermal cells of the root (Chapter 4). They may be air filled or filled with tannins or resin.

After the first year, as periderm plates are formed at deeper and deeper levels, the periderm becomes differentiated from the parenchyma cells of the secondary phloem. Since it cannot continue to grow, new cork must be added as the plant increases in diameter. The

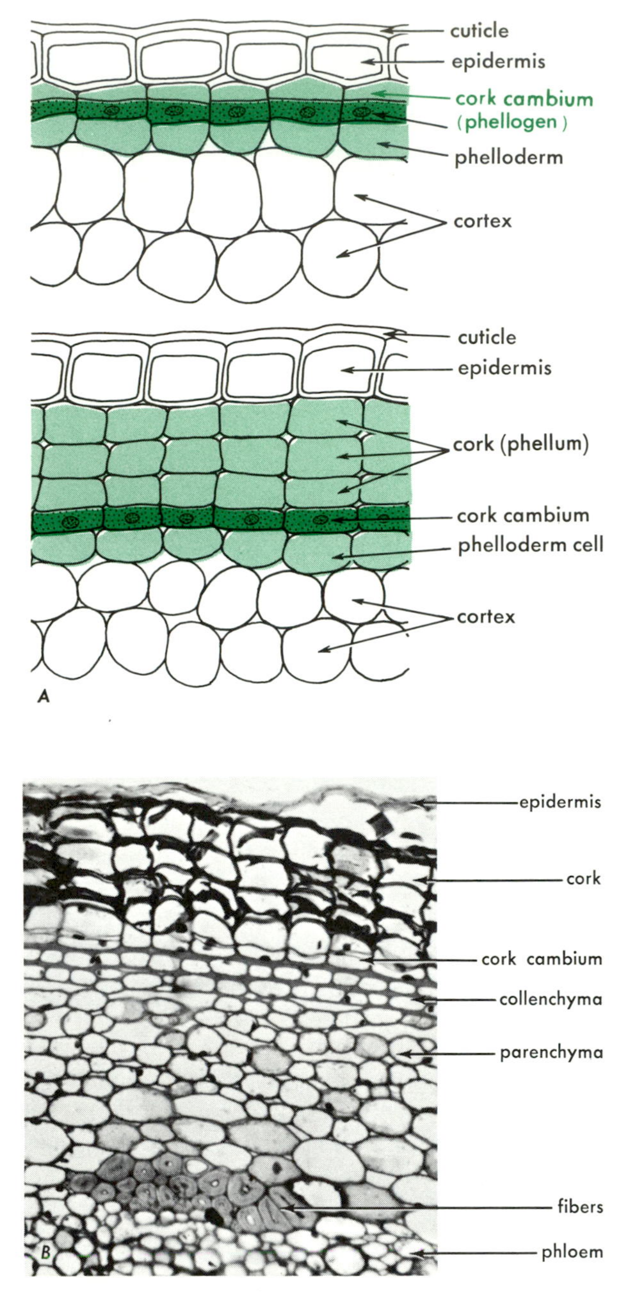

Figure 22-5 *A*. Origin of the first phellogen (cork cambium) and first layer of cork shown diagrammatically. Each growing season new cork cambia and their layers of cork form in a similar manner. *B*. Light micrograph of cork cambium and newly formed cork in *Sambucus* (elderberry) stem.

old periderm develops furrows and ridges, as in oaks and pines, or flakes off in strips as in birches or eucalypts. Masses of loosely packed cells in the periderm form areas known as *lenticels,* which are "breathing spaces" that allow passage of oxygen into the underlying layer of cells in the densely packed tissue. Lenticels are often formed beneath stomata when the periderm is first organized. In older stems, they characteristically occur in cracks in the bark, where the periderm is relatively thin.

3. PROPERTIES THAT MAKE WOOD USEFUL

Only a little more than a century ago wood was the primary material used for building houses, ships, vehicles and bridges, as well as a major fuel. Today metals, concrete, and plastics have taken the place of wood for many purposes. However qualities that insure continued use of wood are its high strength proportionate to its weight and the ease with which it can be cut, shaped, and fixed with simple tools. Wood is also a good insulator against heat loss. It is attractive and can be made more so by special cutting to expose particular grain or figure; it is easily stained, polished or painted. Furthermore it can be obtained in varied kinds for different purposes. Even where alternatives prove cheaper, people still prefer wood because it is homey, friendly, and a traditional material. It fulfills so many of our needs from cradle through an active life to the graveyard! Wood is able to play these roles because it is made up of varied cells. It is unique in being primarily composed of flexible, elongated cells that combine to make a rigid substance able to bear weights and resist strain. Thus it can hold nails and screws and withstand strains without breaking; wood also has fundamental characteristics that enable us to use it in reconstituted forms.

Many physical characteristics of wood are determined by the thickness of the cell walls and the relative size and proportions of the vessels, tracheids, and fibers. Gymnosperms are known as *softwoods* because their wood consists entirely of tracheids. Angiosperms, with both vessels and fibers, are known as *hardwoods*. This classification, however, does not mean that all angiosperms have "harder" wood than gymnosperms. Hardness of wood is a reflection of the sturdiness of the individual cell walls and other factors such as the amount of lignin they contain. The harder the wood, the better it will resist wear. Some species such as black locust and white oak are extremely hard. On the other hand, other woods are so soft that they can be dented with a thumbnail. The tropical balsam (*Ochroma*) is one of the best-known very soft woods (often being used to make rafts and lightweight model airplanes).

Tensile strength and *resistance to shear* are other important characteristics of wood. In some species the cells separate more readily than they do in others. Easy cleavage may be an advantage in splitting firewood, but a disadvantage in making handles, athletic equipment, and other items. Hickory, elm, maple, ash, and persimmon are strong woods with considerable bending strength and shock resistance. *Density* is controlled particularly by the number of fibers. Some woods, especially from tropical angiosperms, are so dense that they will not float.

Porosity is determined by the number and size of the vessels and tracheids. Porosity of wood is important in its ability to take paint or even resist decay. White oak (*Quercus alba*), valued for barrels used to age wines and whiskeys, has low porosity in spite of its being a hardwood with large vessel cells. This wood develops balloonlike *tyloses* that block the cell lumen especially in vessels, making the wood essentially impermeable. Tyloses also make black locust resistant to decay.

Durability is a feature of wood that varies greatly, depending upon the kinds of deposits in the cell wall of the heartwood. Where phenolic and terpenoid compounds accumulate, fungi and bacteria that cause decay grow poorly. The heavy impregnation of heartwood of members of the family Taxodiaceae, such as redwood (*Sequoia*) and bald cypress (*Taxodium*), make these species especially resistant to decay; to a greater or lesser degree so are various cedars *Cedrus* junipers (*Juniperus*), and black walnut (*Juglans*). At the other extreme, poplar (*Populus*) and basswood (*Tilia*) decay readily. Woods that absorb little water also may remain too dry for the flourishing of most decay organisms. Actually most woods preserve almost indefinitely if kept dry, as is demonstrated by wooden artifacts taken from Egyptian tombs. From the viewpoint of human usage, most woods can be made more durable by impregnating them with preservatives. These are of two main kinds: the creosotes or tar oils (derived by distilling coal tar) and poisonous mineral salts usually of copper, chromium, and arsenic. The timber may be steeped in one of the preservative fluids, or just heated and cooled or subjected to reduced pressure, which causes the preservatives to be drawn into the cells.

Certain types of wood have particular appeal because of the *grain* (orientation of cells), which determines the *figure* (design or pattern). This figure results not only from the inherent arrangement of the cells of the wood, but also from the direction of cutting. The most important structural features in the pattern result from the growth increments and the vascular rays, but the nature of the pattern varies according to whether the wood is cut on a tangential or a radial plane (Fig. 22-6). Cuts tangential to the circle of the log and at right angles to the longest dimensions of the vascular rays result in *plain-sawed* or *flat-sawed* lumber. The surface of such wood has a pattern or figure of stripes, concentric irregular parabolas, or ellipses caused by the difference in color or structure of the early and late wood of the growth increments (Fig. 22-7). Understanding the pattern on the surface of a plain-sawed board may be aided

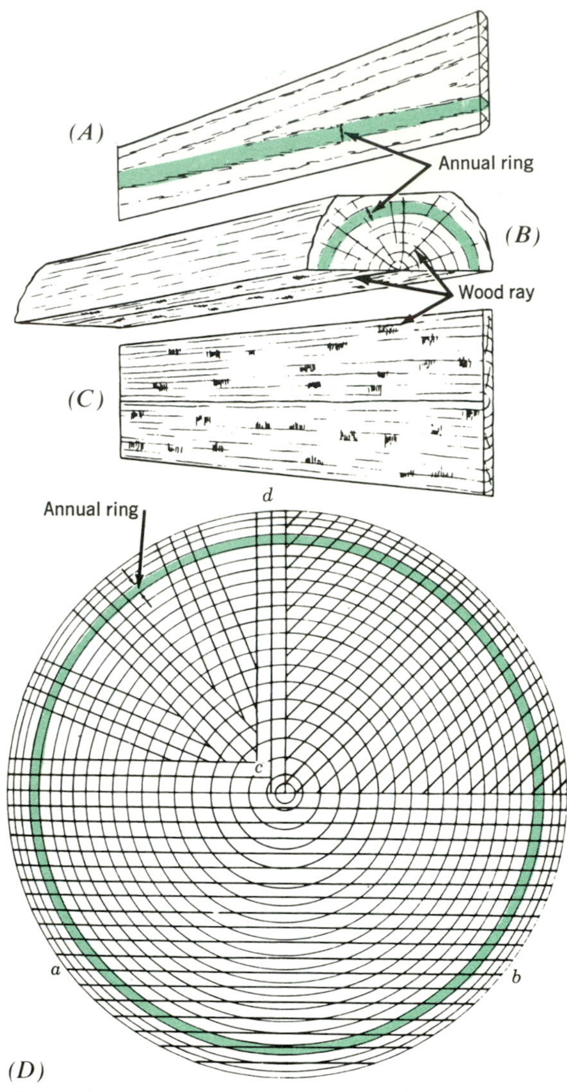

Figure 22-6 Methods of cutting a log into lumber. *A*. Plain-sawed or tangential section, also *a* to *b* in *D*. *C*. Quarter-sawed or radial section, also *c* to *d* in *D*. *B*. Cross section.

by recalling that each growth increment is an irregular cone that tapers gradually from the base of the trunk to the top of the tree (Fig. 22-1). A tangential cut through a log therefore exposes relatively broad surfaces of several growth layers. Since the cones are irregular in shape, there are many departures from a symmetrical geometric figure.

Quarter-sawed wood is cut along a radial plane approximately parallel to the vascular rays. If the rays are large, they appear as stripes or ribbons running across the surface. Oak is one of the most common hardwoods that is quarter-sawed because the rays are often large and conspicuous. Quarter-sawed softwoods, such as yellow pine and Douglas fir, are frequently used as flooring. The wearing quality of quarter-sawed wood is much superior to that of plain-sawed lumber because the harder areas of late wood are closely packed and exposed vertically to surface wear.

4. THE MANY WAYS WOOD IS USED

a. ENERGY

Wood has been our most important fuel until relatively recently when it gave way in importance in the developed world to fossil fuels. However, millions of people who live in or near forests still rely on wood for winter warmth and cooking their meals. If it is partially burned in a kiln that restricts the access of air, wood is converted into charcoal. This form of almost pure carbon burns at high temperatures when given ample air and is still used by many people for cooking. Until the invention of coke, charcoal was the only effective fuel for smelting iron and other metals.

Recently, consideration has been given in the Untied States to cultivating trees on a large scale for their energy. In the United States there are 110 million hectares of land suitable for these "energy plantations"—land not now used for farming or recreation. Furthermore, if only 5% of this land were used to grow plants for fuel, it could supply 4 to 7% of the total projected U.S. energy demand for 1985. Perpetually renewable fuel could compete successfully with costs of oil. Some power plants are now being built in Oregon that will run on wood waste.

In the proposed energy plantation, trees would be planted less than four feet apart and

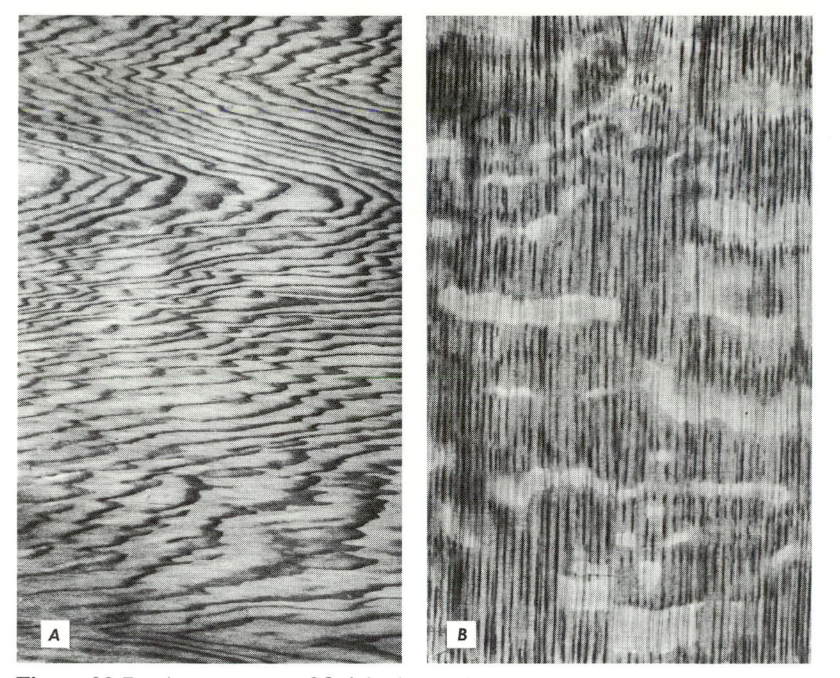

Figure 22-7 Appearance of finished woods. *A*. Plywood, a perfect tangential section. *B*. Quarter-sawed oak, a radial section.

would be harvested by machines similar to those that gather sorghum. It would take a 94,000 ha plantation to supply the fuel for a 500-megawatt power plant. The smallest area that would be practical for such an energy farm would be 12,000 ha. Electricity is only one energy product that such a plantation could provide, and in the long term it might not be the most important. Ethanol and methanol could be produced at near-market prices. Methanol and ethanol can be used to replace a fraction of gasoline (as discussed in Chapter 19).

b. LUMBER

The United States (25% forested) and Canada (26% forested) are the premier lumber-producing countries of the world, not only because they are richly endowed with forest but because of their mechanized techniques of logging and milling. There is relatively little sawlog production from the less developed parts of the world such as Latin America, Africa, and most of Asia. Here cutting is primarily still done primitively, large timbers mostly being hand trimmed and sawed.

In temperate zone forests pines constitute some of the most prized timber species. For example, *Pinus sylvestris* (Scotch pine) is the most important coniferous tree through northern Europe. *Pinus strobus* (eastern white pine) is perhaps the most famous pine in the history of the United States. Originally it occurred in extensive, nearly pure stands from the New England seaboard through southern Canada to the Great Lakes states (Fig. 22-8*A*). This white pine was highly valued for masts of the sailing ships of the British navy when northeastern America was still a colony (Fig. 22-8*B*). In fact, a penalty for cutting this magnificent tree on the royal reserves was one of the cumulative reasons for the rebellion of the colonies. After the Revolution exploitation of white pine was a significant factor in settlement as far west as the Great Lakes area. For more than a century thereafter, white pine remained the most useful,

(*A*)

Figure 22-8 *A*. Mature white pines (*Pinus strobus*) in New York, similar to the forest that greeted early colonists in New England. *B*. Reproduction of the cover and title page of "An Act for the Preservation of White and Other Pine Trees Growing in Her Majesty's Colony, 1710" showing the interest in this valuable tree even at that early time.

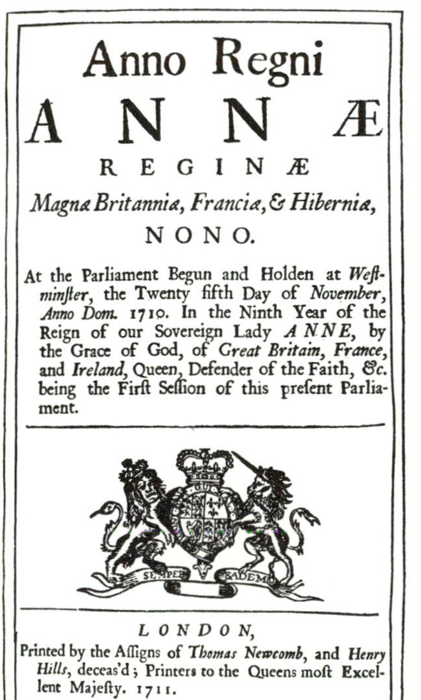

Anno Regni

A N N Æ

R E G I N Æ

Magnæ Britanniæ, Franciæ, & Hiberniæ,

N O N O.

At the Parliament Begun and Holden at *Weſtminſter*, the Twenty fifth Day of *November*, *Anno Dom.* 1710. In the Ninth Year of the Reign of our Sovereign Lady *ANNE*, by the Grace of God, of *Great Britain*, *France*, and *Ireland*, Queen, Defender of the Faith, *&c.* being the Firſt Seſſion of this preſent Parliament.

L O N D O N,

Printed by the Aſſigns of *Thomas Newcomb*, and *Henry Hills*, deceas'd; Printers to the Queens moſt Excellent Majeſty. 1711.

Anno Nono

Annæ Reginæ.

An Act for the Preſervation of White and other Pine-Trees growing in Her Majeſties Colonies of *New-Hampſhire*, the *Maſſachuſets-Bay*, and Province of *Main*, *Rhode-Iſland*, and *Providence-Plantation*, the *Narraganſet* Country, or *Kings-Province*, and *Connecticut* in *New-England*, and *New-York*, and *New-Jerſey*, in *America*, for the Maſting Her Majeſties Navy.

hereas there are great Numbers of White or other Sort of Pine-Trees, fit for Maſts, growing in Her Majeſties Colonies of New-Hampſhire the Maſſachuſets Bay, and Province of Main, Rhode-Iſland, and Providence-Plantation, the Narraganſet Country, or Kings-Province, and Connecticut in New-England, and New-York, and New-Jerſey, fit for the Maſting Her Majeſties Royal Navy: And whereas the ſame growing near the Sea, and on Navigable Rivers, may commodiouſly be brought into this Kingdom for the Service aforeſaid: Wherefore, for the better Preſervation thereof, Be it Enacted by the Queens moſt Excellent Majeſty, by and with the Advice and Conſent of the Lords Spiritual and Temporal, and Commons, in this preſent Parliament Aſſembled, and by the Authority of the ſame, That from and after the Twenty fourth Day of September, which ſhall be in the Year of our Lord, One thouſand ſeven hundred aud eleven, no Perſon or Perſons within the ſaid Colonies of New-Hampſhire, the Maſſachuſets-Bay, and Province of Main, Rhode Iſland, and Providence-Plantation, the Narraganſet Country, or Kings-Province, Connecticut in New-England, and New-York, and New-Jerſey, or any of them, do or ſhall preſume to Cut, Fell, or Deſtroy any White or other ſort of Pine-Tree

Eeeee 2 fit

(*B*)

and therefore the most cut timber in the New World. Only in the twentieth century has *P. strobus* become of secondary importance as a result both of the extensive cutting and the introduction of the white pine blister rust from Europe. In spite of disease and over-zealous lumbering activities, white pine has reseeded over much of its original range and the forest is now making a comeback. However, white pine still occupies only 2% of the area of the primeval forests. *Pinus strobus* demands a comparatively good soil. Therefore with repeated logging white pine is replaced by the less desirable jack pine (*P. banksiana*) and red pine (*P. resinosa*) until the soil is built up again. Reforestation has remained risky because of the white pine blister rust.

The longleaf pine (*P. palustris*) probably is the best known of southern pines in the United States and is the major source of lumber, pulp, and turpentine there. However, *Pinus ponderosa* (the ponderosa or western yellow pine) now provides more lumber than any other pine in North America. The trees are quite large, some attaining a diameter of 2.5 m, and it occurs in commercial quantities in every state west of the Great Plains. The wood is light but hard and strong; it is used for general construction, interior finish, and boxes.

Douglas fir (*Pseudotsuga menziesii*) is one of the finest woods of North America; it constitutes about one-half of the standing timber of the western forests and furnishes about one-fifth of the timber cut annually. Douglas fir occurs throughout the Rocky Mountains but attains its greatest size in Pacific coastal forests. Stands of trees there are rivaled only by the redwoods in amount of timber per acre. The immense size and height of Douglas firs permit production of large, knot-free boards in great quantity. Additionally, the wood is among the strongest relative to its weight, works well, and is noted for its durability. The species is in heavy demand by lumber and building trades and is used for all types of construction; it is the most used species also for making plywood. Another highly prized Pacific Coast tree is the coastal redwood (*Sequoia sempervirens*) (Fig. 14-4). The wood is soft, light, strong, easy to work, and particularly resistant to decay.

The oaks are the most economically important of all the temperate-zone hardwood genera. A number of species of *Quercus* occur in the northern deciduous forests and are put into two general groups: the white oaks (with rounded leaf lobes) and the red or black oaks (with pointed leaf lobes). *Quercus alba*, the white oak, is the most prized of the oak species (Fig. 22-9). The wood is about twice as heavy as white pine, durable and attractive. It has multiple uses such as cabinetwork, trim, flooring, and tight cooperage (as in wine and whiskey barrels).

For cabinets and furniture veneers, the most used individual species from temperate hardwood forests are black walnut (*Juglans nigra*) (Fig. 22-9) and black cherry (*Prunus serotina*). Other hardwood trees of commercial value are ashes, basswood, the hickories (particularly shagbark, *Carya ovata*), maples (especially hard or sugar maple, *Acer saccharum*, Fig. 22-9), sweet gum (*Liquidambar*), and yellow poplar or tulip tree (*Liriodendron*).

The tropical rain forests are a vast storehouse of tree products and, with active research, these trees should become increasingly utilized. In the past, a few selected woods have been highly prized and sought after. For example, teak (*Tectona grandis*, in the Verbenaceae), which grows in the seasonally dry rain forests of India, Burma, Thailand, and Indonesia, was recognized in early years as one of the world's strongest and most durable timbers and had especial value in shipbuilding (Fig. 22-10*A*). Only a few good, mature teak trees grow in each square mile of forest; when a marketable teak tree is found it is "ring-barked" and left to "season" where it stands. The reason for this is that a fresh-felled teak tree is too heavy to float to a sawmill. The tree is felled a year later, when elephants haul the logs to a river where they are floated to a sawmill.

Figure 22-9 Three of the most widely used temperate-zone hardwoods. *A*. Hard maple *(Acer saccharum),* used in bowling pins and flooring for bowling alleys (or any floor with heavy usage) and for dance floors; its uniform texture and hardness result in resistance to abrasion. The Romans used it for spears and lances. It is valued for furniture and boat building. *B*. Black walnut *(Juglans nigra),* favored for fine furniture and interior paneling because of the beauty of the grain and its good machining properties. *C*. White oak *(Quercus alba),* good for barrels because the wood is resilient, durable, and impermeable to liquids. Its strength and durability have made it a standard building timber for ships and for strong furniture such as tables and chairs.

Other well-known tropical woods include mahogany, rosewood, ebony, and balsa, only to mention a few. The first Spanish explorers that reached the West Indies soon discovered the red-brown timber we know as mahogany (*Swietenia mahogani,* in the Meliaceae). As early as 1514 it was being used for shipbuilding and by 1715 it was widely used in Europe for high-class furniture. As large trees were cut out in the West Indies, loggers turned to Central America for Honduras mahogany (*Swietenia macrophylla*) and thence to allied species in South American rain forests. Mahogany is one of the largest trees in tropical American rain forests, usually exceeding 33 m in height and 13 m in circumference above large basal buttresses (Fig. 22-10*B*). The value and scarcity of the American mahoganies have led to utilization of related genera in the Meliaceae (e.g., *Khaya ivorensis* and *Etandrophragma cylindrica,* giant forest trees from West Africa). Because the color and quality of wood from *Swietenia* is so highly valued, not only other genera in the Meliaceae, but *Afzelia* in the Leguminosae. *Aucomea* in the Burseraceae, and others are called mahogany.

Rosewood is a name used for several leguminous genera including *Dalbergia* and *Pterocarpus*. The Brazilian rosewood (*Dalbergia nigra*) with its beautiful red-brown wood and network of black streaks has long been a leading timber for luxury woodwork. Many choice pieces have been made during the 350 years that this rare wood has been prized in commerce. Today scarcity and cost limit the regular use of rosewood in solid form to small objects, but it is widely used as a veneer.

Ebony (*Diospyros* spp.) is the symbol of jet-black wood, but most of the hundred-odd ebonies that grow in tropical forests have heartwood that is not wholly black. Instead they show bold, irregular stripes of bright brown, grey, or greenish black on a deep black background. This pattern gives striking decorative effects and leads to their widespread use as surface veneers. Ebony is a relatively small tree growing in the lower stories of tropical rain forests in India, Ceylon, Malaysia, Indonesia,

(A) (B)

Figure 22-10 *A*. Teak *(Tectona grandis)* is among the most important of tropical Asiatic timbers, here shown at Madras in southern India. The man is standing by a ring-barked tree. *B*. A mahogany tree *(Swietenia mahogani)* in the rain forest in Belize; this is one of the world's most valued timber trees.

and Africa. It has long been featured in trade from the East Indies to the Western world and was even commonly used in Roman times.

C. MAN-MADE AND RECONSTRUCTED WOOD

The so-called "man-made woods" are created by cutting logs into components of different shapes and sizes and then reconstructing them by gluing the pieces together.

To make *plywood* large logs are peeled to yield *veneers*. Each log is spun in a huge, powered lathe (Fig. 22-11); as it turns, a large knife cuts into its surface like a great pencil sharpener and skims off a thin, continuous layer of timber. This sheet is called a *rotary cut veneer*. The next stage in forming plywood is to glue three or more (always an odd number) veneers together, with alternate sheets having the grain at right angles to one another. This produces a light, flexible material. Because the wood is strongest along the grain, the alternation of ply produces a stronger plank than one of comparable strong, solid wood. Each sheet of veneer reinforces the adjacent sheet. The center ply of 3-ply is made twice the thickness of its adjacent plies so that quantitatively as much grain runs in one direction as another. In most cases, the center ply or plies consist of cheaper "core" wood. The more expensive, figured "face" veneers are glued to the outside. The fine cabinet woods such as ma-

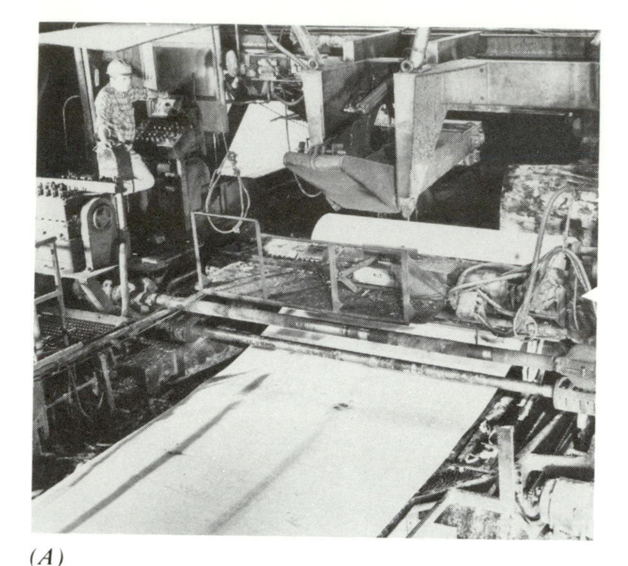

(A)

Figure 22-11 Manner of preparing plywood. *A*. Sheet of veneer being peeled from a slowly revolving log. *B*. Diagram showing relation of grain in plywood to annual growth rings of the log.

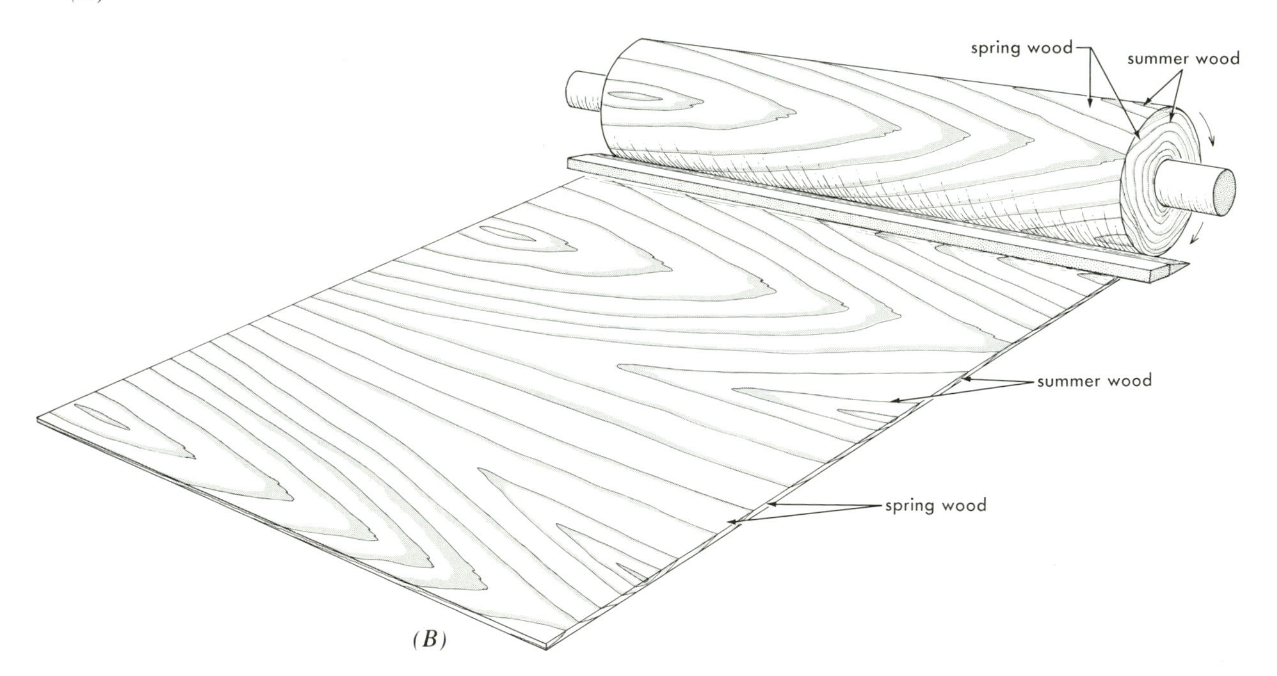

(B)

hogany, teak, and walnut are used for very thin veneers, which are glued to a strong, solid core of the cheaper materials. Both reasonableness of price and beauty in furniture depend upon the use of veneers. The greatest craftsmen in furniture design have always been masters of veneering. It is not possible with solid wood to obtain the striking figure and matched design produced by the veneer knife cutting logs of selected grain at exactly the most favorable angle. Also solid furniture is not as resistant to change in dimension and cracking or splitting as furniture from a well-made veneer. However, if the furniture is to be carved and if parts are subject to hard use, solid wood is preferred.

Laminated boards or *block boards* are built

up on the same principle as plywoods, but thicker or more rigid timbers are used where greater strength is required. These reconstructed timbers can be made in wider sizes than sawn planks, are less prone to warp or split, and can be finished off with a variety of surfaces—decorative or resistant to wear.

Chipboards or *particle boards* are made by chipping wood into small pieces with machines. The chips are then mixed with a powdered plastic glue called a *binder* and spread out between metal plates. This "sandwich" is placed in a press and heated under pressure, which melts the glue and causes it to bind the chips together.

Large-scale commercial development of reconstructed woods has depended upon glues capable of bonding the wood particles permanently. Inefficient glues resulted in plies' frequently failing or peeling—particularly under moist or alternately moist and dry conditions. Within the past three decades, however, tremendous advances have been made in glues and gluing techniques. Thus plywoods capable of withstanding outdoor exposure, marine use, and indoor stresses have now been developed. In fact, with modern glues plywood is said to be stronger than steel, pound for pound.

d. PULPWOOD AND PAPER

The world demand for paper and allied products is now so great that in technologically advanced countries around 40% of all the forest's output is used for them and only 60% is used as solid timber. But over the world as a whole, after allowing for many other purposes that wood serves—as fuel, fencing, simple construction, and so on, the proportion used for paper may be no more than 25%. In the United States it is estimated that nearly a quarter of a ton of paper and fiberboard are consumed annually per capita.

Although most of our paper today is made from wood, until the nineteenth century most of it was made from other fibers. The ancient Egyptians used the stem pith of "papyrus grass," *Cyperus papyrus,* a sedge that grows wild in most soils of the Nile Valley. Although there are some 3000 species of sedges (family Cyperaceae) and probably 700 species of the genus *Cyperus,* this is the only one that was found useful. The papyrus was made mechanically by beating the wetted pulp out into thin sheets. Papyrus is known from as early as 4400 B.P., and much of what we know about the life of those days has been deciphered from the hieroglyphics inscribed on papyri found in the royal tombs.

The Chinese, beginning in the second century A.D. (i.e., 2½ millennia later than Egyptians), developed the modern concept of paper manufacture (Fig. 22-12*A*). They used the stems of several species of grasses such as bamboo and the inner bark of a kind of mulberry (*Brousonnetia papyrifera,* in the Moraceae). They ground them to a pulp with water and drained the suspension over a fine net, originally made of cloth, so that the fibers were deposited as a felted layer. Different plant materials and, especially different refinements of treatment, produced papers of different quality. They also worked out materials for writing and painting on paper, including the soft brushes and "Chinese ink" that are used to this day.

Actually the art of *writing* was developed before paper became available. The Egyptians used hieroglyphics, each character a kind of sentence or statement on their papyrus, but symbols for individual words and sounds soon followed. The Indians inscribed their religious texts in the Pali or Sanskrit languages on palm leaves, by pressing with a sharp wooden point so that the pressed part turned brown. In Europe writers discovered that tannins from oak galls produced a blue-black precipitate when their water supply was rich in iron. This blackish suspension made permanent marks on parchment (made from inner layers of sheepskin or calfskin) and thus was born one of our most useful servants—ink. In the last two decades or so this original ink has been superseded by synthetic blue and black dyes.

By the middle of the nineteenth century practical techniques were developed for con-

(A)

(B)

Figure 22-12 *A*. Papermaking in China in 1634. A vatman plunges a mold into a vat of fibers and water. A sheet of paper forms when he withdraws the mold, sieves out the water, and shakes the mold so that the fibers interlock. *B*. Manufacture of paper today; the "wet" end of a Fourdrinier paper machine. A thin suspension of pulp flows into a moving, wire screen from the head box in the background. The water drains through the screen, leaving behind a mat of fibers in the form of a sheet of paper.

verting wood into pulp. The problem in using wood for paper is to remove as much as possible of the undesired lignin. Not only is lignin brown and turns darker brown with time, but its polymer structure does not give strength to the paper, any more than it does to textile fibers. The lignin content of woods varies from 22% in hickory to 33% in cedar; spruce, which is a major raw material for paper, has about 28% (calculated on the dry weight). In 1852, the first practical attempt was made to remove all or part of this by heating wood chips with alkali under pressure. In America this procedure was changed in 1866 to use calcium or sodium bisulfite instead of alkali, and a few years later this process was developed into the first "sulfite pulp" method in Sweden and Germany. Technical difficulties centering around the high pressure, the acidity of the sulfite, and the evil-smelling by-products prevented its widespread adoption for a further 20 years or so.

Meanwhile the alkali method was still

being developed, and with the addition of sodium sulfate, which became reduced to sulfide during the recovery process, the yield and the quality of the pulp were both improved. This led to the "sulfate pulp" method, which now rivals the sulfite procedure. Both processes depend upon the conversion of the phenolic constituents of the lignin polymer into breakdown products containing sulfonic acid or sulfide groups, which are soluble in the hot extracts and thus leave behind the cellulose. The sulfate method leaves 5 to 8% lignin and the resulting so-called kraft paper is used mainly as wrapping paper and cartons; it is also a raw material for making rayon. By-products of both methods include formic and acetic acids. Because the sulfate method is alkaline, the resins in pine and spruce woods are hydrolyzed and thus a wider variety of softwoods can be used. The terpenes thus liberated are recovered by distillation.

As seen in Figure 22-12, the fibers, suspended in water, are deposited on a metal screem through which the water drains off. The sheet of crude paper is then lifted off manually or mechanically and dried. It is soft and porous ("blotting paper") and is then hardened ("calendered") by being rolled again, this time under high pressure. Some idea of the scale on which paper is made is given by the size of the screen in Figure 22-12*B*. For better writing or printing surface, as for use in books, paper is further surfaced by coating with resins or silicates. The extremely glossy papers, such as those used for printing photographs in this book, have received additional coatings.

The solution of lignin, "lignosulfonic acid," from the sulfite process, containing the accompanying breakdown products, is difficult to find a use for. Some is converted to vanillin, for use as a vanilla substitute in flavoring, but most is discharged. This has created a considerable problem with pollution of streams in the great forested areas.

The above processes have made wood the most economical source for paper pulp, and today 90% of all paper is made from it. A few special kinds of papers, such as those used for cigarettes and banknotes, are still made from flax fiber. However, there is no other fiber as abundant, as relatively easily available, and as economical as wood fiber.

Tree species used for pulping vary widely around the world. In the southeastern United States the yellow pines are commonly used; in the northern U.S. coniferous forests and in Canada and northern Europe spruces, firs, pines, and hemlocks are available; in western North America, it is hemlocks, firs, Douglas fir, larches, and some pines. Also in the northern forests aspen and birches are used. In Australia, Chile, and Spain *Eucalyptus* as well as herbaceous fibers are used. South Africa depends primarily upon introduced eucalypts and pines. In Latin America sugarcane bagasse is often used instead of wood, although the use of tropical hardwoods and conifers for pulp is increasing. "Parana pine" (*Araucaria*) from southern South America supplies much of the Brazilian needs, and use of the Amazonian hardwoods is developing. China apparently depends more upon bamboo, rice, straw, cotton stalks, and bagasse, and in India bamboo also is grown as an important pulping species. Because some tree sources are becoming too scarce or too costly to meet the worldwide demands, research is being directed toward annual or high-yielding shrubs that could substitute as a source of pulp. In the United States many species such as the leguminous sunn hemp (*Crotalaria juncea*—Chapter 21) are being studied as possible future sources of paper.

e. RAYON AND PLASTICS

Rayon was the first of the commercial synthetic fibers and, despite considerable competition from petrochemically derived fibers, is still used for tire cord and in some textiles. About 4% of the North American wood pulp, and over 10% of Japanese wood pulp, are used for dissolved cellulose materials including rayon and other such products as cellophane, photographic film, and acetate plastics. Cotton linters also contribute to the raw materials for rayon. The principle behind production of rayon is old, but it was not until 1855 that a Swiss chemist patented a process for transforming nitrocellulose into fine threads. By 1911 rayon was considered a substitute for silk (often being called "artificial silk"), and by 1923 world production of rayon had equalled that of silk.

The principal method of making cellulose into rayon is by dissolving it in alkali with carbon bisulfide, which converts the cellulose into an orange salt called sodium xanthate; this is squeezed through fine holes into a weakly acid solution that reprecipitates the cellulose, this time as a more lustrous, semitransparent thread (Fig. 22-13). Rayon was the forerunner of the whole series of artificial fibers we have today; not all of these are derived from cellulose; nylon, for example, is a long-chain peptide.

Dissolved cellulose also brought in the use of plastics. The plastics industry came into being in 1868 when cellulose nitrate was combined with natural camphor to produce "celluloid." Then nitrated cellulose was combined with castor oil and pigments to make material that could be formed into artificial leathers. Subsequently combination of cellulose with other substances has yielded many different kinds of plastics.

Figure 22-13 In producing rayon a solution of cellulose is drawn through the holes in the spinneret and is immediately solidified into threads by special chemicals.

5. HOW BARK IS USED

a. CORK

We do not use the bark from trees as much as we do the wood. One outstanding exception is commercial cork, which comes primarily from the bark of *Quercus suber,* the cork oak, a subtropical evergreen tree native to the western Mediterranean. This oak is used most extensively as a source of the commercial product in Portugal, Spain, and northwestern Africa. Cork was known to early Greek and Roman civilizations, where it was used as a seal for jars and casks and even as a float for soldiers crossing rivers.

The cork oak tree grows 15 to 20 m tall and may have a circumference of 3 m or even greater (Fig. 22-14). The exceptionally thick bark adapts it well to the subtropical climate and desiccating winds of the western Mediterranean region.

In most other trees, artificial separation of the bark occurs at the vascular cambium, where any all-round cut will kill the tree. In the cork oak, however, if the separation is made in the summer the break occurs at the cork cambium. This leaves the inner bark (*phellogen*) and vascular cambium unharmed; moreover, the cork cambium continues to be active and produces a fresh protective "cork" or *phellum.* Thus it is

Figure 22-14 Cutting the outer bark from cork oak tree *(Quercus suber)* in Portugal.

possible to harvest cork from the same tree every 10 to 12 years throughout the oak's life span of a century or more.

Rough and fissured bark called virgin cork comprises the first harvest; this material is suitable only for such purposes as insulation, floats for fishing nets, and coarser forms of flooring. Subsequent harvests provide the smooth, fine-textured, high-quality mature cork that is used for bottle stoppers and high-grade tile floors. The harvesting is done with long curved knives that split off large curved slabs of cork up to an inch thick from the trunk and also cut sleeves of bark from the larger branches (Fig. 22-14). A very large tree may yield as much as a ton of cork. Stripped bark is stacked for several weeks, boiled to soften it (also removing tannins and other chemicals), dried, trimmed, graded, and then baled. It thus comes to market usually in flat sheets several square decimeters in size.

Because of the lenticels or "breathing pores" in the bark, to make an effective bottle stopper the cork must be cut so that the lenticels run across the neck of the bottle and not up and down. If you look at any bottle cork you will find the pores with dark brown margins so oriented. In this manner water vapor or volatile alcohol is stopped by the main body of impermeable tissue that extends from top to bottom of the cork. Yet because this tissue is made up of air-filled cells with waterproof walls (i.e., coated with suberin), it is possible to compress the cork and ensure a tight fit all around. The compressed, trapped air insures a tight fit that persists indefinitely until the cork is withdrawn and automatically expands. The presence of enclosed air also accounts for cork's low density, which enables it to be used for floats and life jackets, and also accounts for its low heat conductivity, which makes it an excellent insulator against either heat or cold. Cork lasts almost indefinitely; it is chemically inert, not giving odor or flavor to substances with which it comes into contact, as well as not deteriorating with the contact. Because of these properties, natural cork still has a world market not taken over by any synthetic product.

b. OTHER BARK PRODUCTS

The inner layers of bark of certain trees are rich in chemicals, especially the *tannins*. For example, leather is prepared by steeping the hides of animals in a solution of tannins. These compounds combine with the protein of animal skin, *collagen;* the leather product is much tougher and more resistant than the untreated skin. Usually the washed skin is prepared by treating with lime and sodium sulfide to loosen the hairs and then soaked in a strong tannin extract from the bark. Several species of oak and the edible chestnut are used in Europe, hemlock (*Tsuga*) and tanbark oak (*Lithocarpus*) in America, and *Terminalia* in the tropics. The wattle (*Acacia mearnsii*) is similary used in Australia and South Africa. Tannin is often replaced today by sodium bichromate, which has a similar action. "Oak-tanned" and "chrome-tanned" shoe leathers are almost universal in western countries.

Other barks contain useful chemicals such as quinine from *Cinchona ledgeriana* (Chapter 24) and cinnamon from *Cinnamomum zeylanicum* (Chapter 26). Also, pulverized bark from trees of many kinds is used as a mulch or rooting medium in horticulture.

B. PLANTS GROWN FOR THEIR NITROGEN COMPOUNDS

CHAPTER 23

The Legumes As Food Plants

1. THE PECULIARITIES OF THE PAPILIONOID LEGUMES

The legumes are one of the three largest plant families. Of the three subfamilies described in Chapter 13 that of the Papilionoideae is the most important as the source of food, both to us and to animals. The papilionoid legumes owe their name to the fancied resemblance of the flower to a butterfly (Latin *papilio*). There are five petals arranged as in Figure 23-1; a relatively large *standard,* two smaller *wings,* and two petals joined lengthwise at the base to form the *keel.* Partly within the keel are the 10 stamens, either all joined at their base or 9 joined and 1 free. Between these is the ovary, small at first but growing very rapidly out into a long pod (Fig. 23-2).

When "ripe," the pod splits open down one (or sometimes both) of its sides to release the seeds (Fig. 23-2). These are rather large, with fleshy or hard cotyledons and no endosperm. Some of them germinate with extreme slowness, due to the tight, hard seed coat (*testa*), whose mechanical properties prevent swelling and imbibition of water. This behavior is still more extreme in the subfamily Mimosoideae.

Following germination the growth of the seedling can follow one of two different patterns. In the pea, the shoot or *epicotyl* (= "above the cotyledons") grows out directly from the cotyledons, which therefore stay below ground, *hypogeal* (= "below ground") Fig. 23-3*A*). More commonly, the tissue between and below the cotyledons elongates into a long stem, the *hypocotyl* (= "below the cotyledons"), carrying the cotyledons up above the ground. The true epicotyl then develops from the node that bears the cotyledons (Fig. 23-3*B*). This *epigeal* (= "above ground") development is the more widespread and indeed is the commonest pattern of seedling growth in other dicotyledonous plants.

As food plants the Papilionoideae are second only to the grasses in importance. Because of the symbiotic nitrogen fixation in their nodules (Chapter 9), the plants are normally well supplied with nitrogen, even in poor soils. Correspondingly their seeds have a high protein content, which is the main reason for their worldwide use as food. Especially in primitive agriculture and in the developing countries, where the rest of the diet is mainly tubers or cereal grains, the high-nitrogen legumes are of great value. Some of the cultivated species are rich in oil also. They have a number of minor disadvantages: the plants are generally thin-leaved, transpiring strongly, and therefore need a good water supply. (There are some papilionoid plants of desert or arid soils, but they are not among those used for food.) Because of the iron and molybdenum in the nitrogen-fixing system and the cobalt needed for the growth of the *Rhizobia,* their mineral nutrition requires careful watching, especially in alkaline soils, where these metals may be converted to insoluble

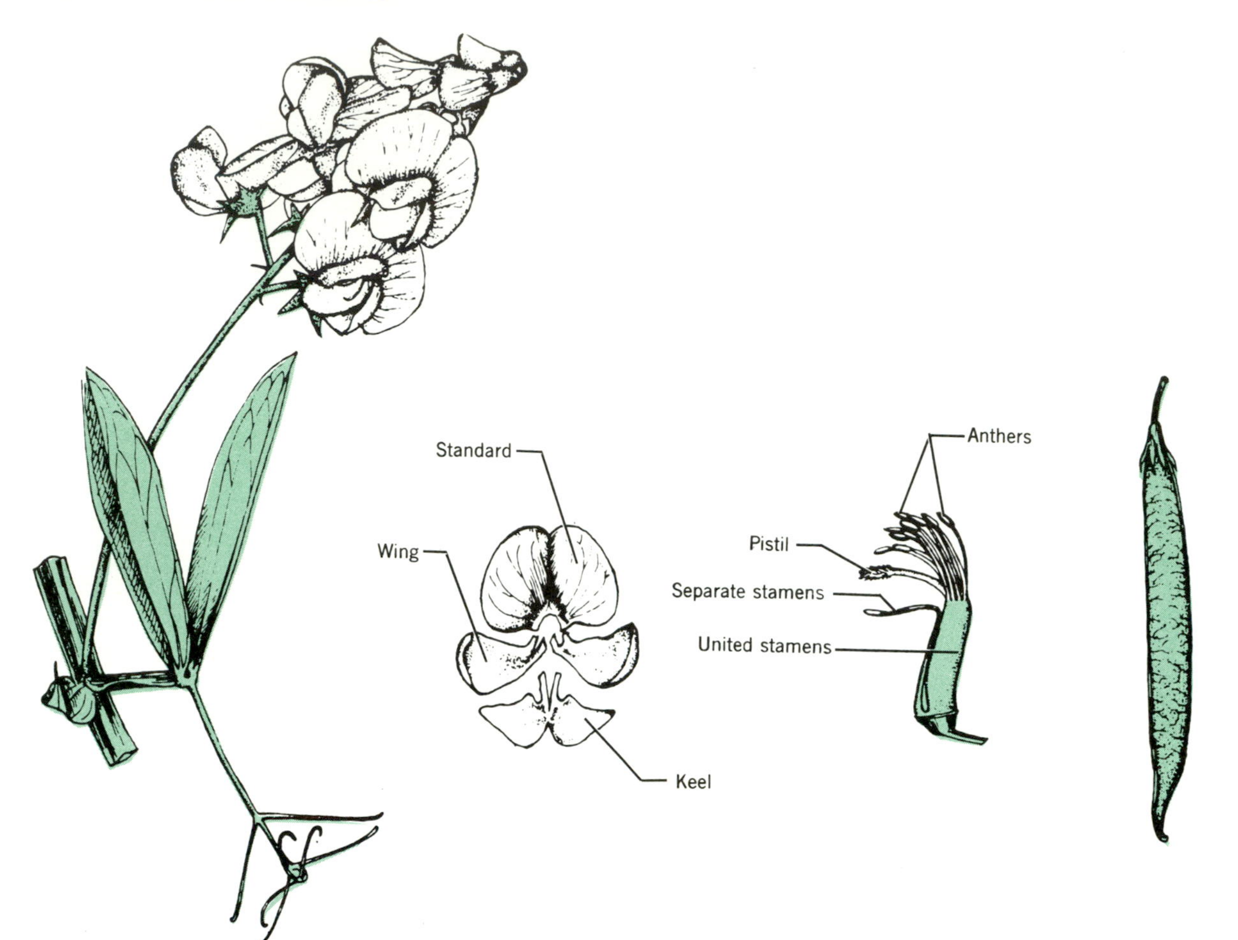

Figure 23-1 Top, leaves, flowers, and fruits of a typical Papilionoid. Bottom, flowers of the wild pea *(Lathyrus silvestris)*.

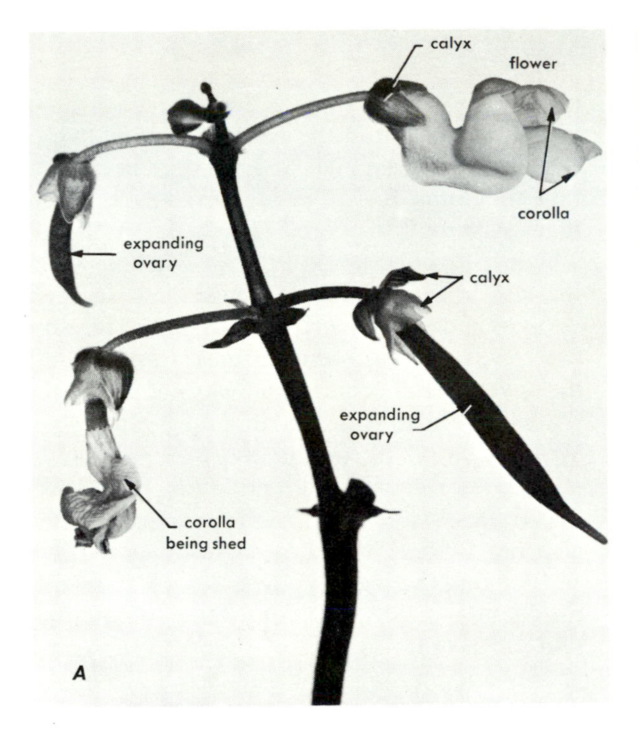

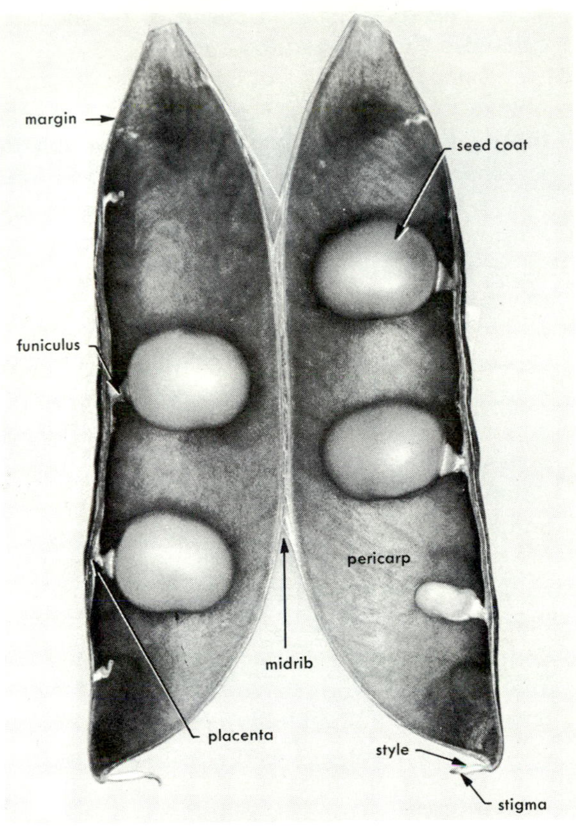

Figure 23-2 *A*. Stages of development of the bean pod. *B*. A pea pod, opened to show attachment of seeds, which are mounted on the margins or sutures of the pod.

(*B*)

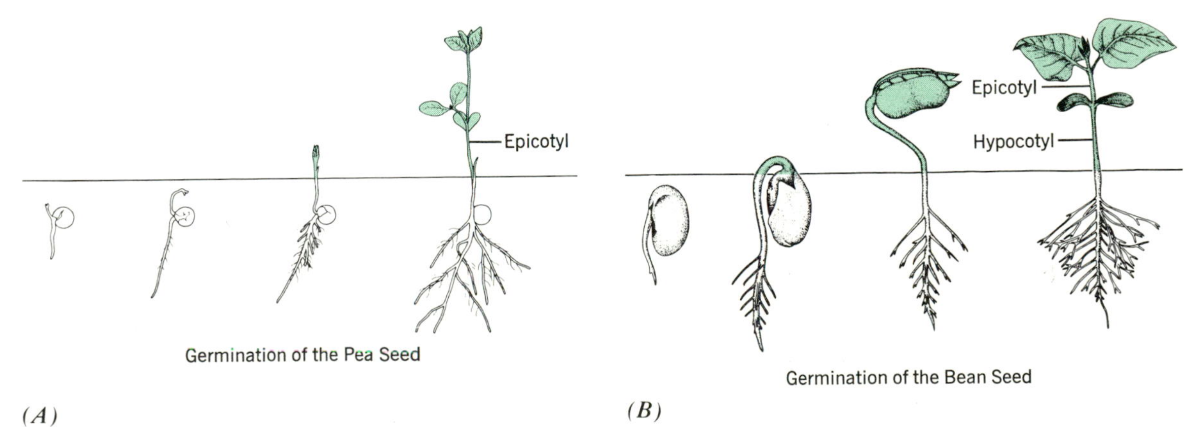

Figure 23-3 *A*. Hypogeal germination of the pea. *B*. Epigeal germination of the bean.

salts. The oxygen requirement of the nodules calls also for well-aerated soils.

Since the proteins of many of the seeds are not very well digested by humans, they need thorough cooking. Some legume seeds contain an inhibitor of proteolytic enzymes, which also does not help the digestion. A few contain some poisonous glycosides. None of these disadvantages has interfered with the steady increase in the cultivation of legumes. The family contains more medicinal plants than any other—especially plants used in the traditional medicine of tropical countries. It includes four of the major crop plants of the world (soybeans, beans, peas, and peanuts) along with numerous minor crops and a group of important forage and pasture plants.

Until now the legumes have not been the subject of nearly as much intensive breeding as the cereals. Polyploidy, for instance, has played no part in their development, and almost all are diploids. In view of the world demand for protein they will undoubtedly receive much more attention in the near future.

2. THE SOYBEAN

Glycine max is by all odds the richest source of vegetable protein in the world. It is a native of east and southeast Asia and has been in cultivation in China for at least 3000 years. From there it spread to Korea and thence to Japan, which it reached about 300 A.D., a time when the present Japanese language had scarcely evolved from its Korean ancestor. The soybean was not introduced into Europe until the seventeenth century and was not much cultivated in North America until the 1930s. Since World War II its use in the U.S.A. has expanded enormously and it now comprises about one-sixth of the total U.S. crop acreage. It is also the United States' second largest export, Japan being the biggest purchaser. Of the world production the U.S.A. contributes better than two-thirds (Fig. 23-4*B*), and most of the rest is grown and consumed in mainland China. In recent years China has even purchased soybean products from the United States.

Wherever soybeans are newly introduced,

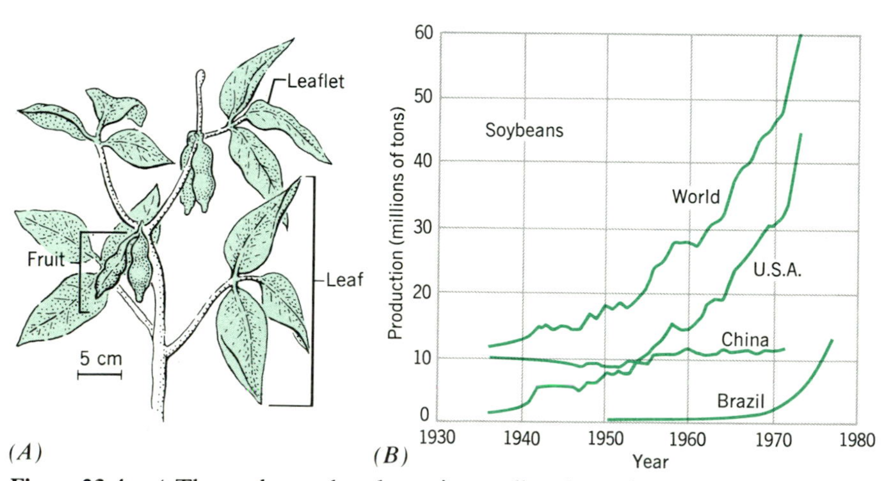

Figure 23-4 *A*. The soybean plant bears its small pods on the main axis. Each pod normally holds only two seeds. *B*. The plot of world production of soybeans shows that the tremendous increase since World War II is almost wholly due to that of the United States, though production in Brazil is increasingly rapidly.

their nodule bacteria, *Rhizobium japonicum,* have to be applied to the soil, because soya does not nodulate readily with other *Rhizobium* species. This is usually done by sprinkling the seed with a liquid culture just before planting.

The cultivated varieties bear larger numbers of pods (see Fig. 23-4*A*), but as there are only two or three seeds per pod the yield is not very high. A field of corn, though its seeds are relatively low in protein, can yield at least as much protein as a field of soybeans. Breeders have so far made little progress in increasing the yield, largely because the genetic material from which to breed is so limited; all the currently grown cultivars in the west stem from six original collections in the Orient. Expeditions have recently gone to China to search for more.

Soybeans have the highest nitrogen content of any agricultural seed, containing up to 44% protein. For comparison, wheat has around 11%. However, the digestibility of the fresh bean is rather low. For this reason much of the consumption is as flour, which is used as filler for meat preparations or made into "textured protein," "imitation bacon," and so on. But the flour has a "beany" flavor that limits the amount that can be incorporated into meats to about 20%. The Japanese surmount these problems by crushing the beans in water and adding salt to precipitate the crude protein as "bean curd" or "Tofu"; this can be eaten with a variety of foods as protein supplement. In addition, the ground beans are fermented with a fungus, *Aspergillus oryzae,* to produce soy sauce, which is rich in glutamic acid, or they can be fermented along with rice to produce a cake called "miso" or "tempe." In addition to their high protein content, the seeds contain about 20% of a most useful oil (Chapter 27).

3. THE BEAN GROUP

The name "bean" is given to members of several genera, of which the genus *Phaseolus* is the most important (Fig. 23-5). It is a genus almost as varied as the genus *Brassica,* the cabbages and mustards (Chapter 12). Most of the *Phaseolus* species come either from tropical America or from Asia, and correspondingly thrive best in areas with a hot summer. The Asiatic species tend to have small seeds, the American species large ones. Five major species are in cultivation.

The kidney or haricot bean (*P. vulgaris*) comes in many cultivars, some dwarf, some climbing; some white-flowered, some purple; some with green pods, some with yellow ("wax beans"), or striped with red. Some are grown for the mature seeds; others are harvested when the long pods are edible. The latter, "snap beans," yield up to 8 tons of the pods per acre (19000 kg/ha). The seeds contain 20 to 25% protein – much less than the soybean, but they are more readily digestible.

It is not clear how these beans came to be called "French beans," for *Phaseolus* is one of the most ancient crop plants of the New World; they were cultivated in the Tehuacán Valley of Mexico before 6000 B.P. They were brought to Europe by the colonizing Spanish and Portuguese in the sixteenth century and have become one of the most valuable of all the gifts to the Old World from the New.

The lima bean (Fig. 23-5), *P. lunatus,* is so-called because it is native to Peru. It was grown there about as early as *P. vulgaris,* but the main center of cultivation now is California. The large, flat seeds of the variety *macrocarpus* are eaten while still green. The dry seeds are called butter beans in England. Some cultivars contain small amounts of the toxic glycoside phaseolunatin, which on digestion releases hydrogen cyanide; however, this comes out on boiling.

The scarlet runner, *P. coccineus,* has large, bright, scarlet flowers and is the species most grown in Europe, because it is the most tolerant of cool summers. It is grown more often for the green (or sometimes violet) pods than for the mature seeds.

The mung bean, *P. aureus,* also called

Figure 23-5 The scarlet runner *(Phaseolus coccineus)* and the lima bean *(Phaseolus lunatus)*. (From Bianchini et al.)

green or golden gram, is Indian or Chinese in origin. It is best known in the West in the form of "bean sprouts," which are the four-day-old etiolated seedlings. The plants are somewhat dwarf and grow well only in warm seasons.

Black gram or urd, *P. mungo,* has been cultivated in India since ancient times, mainly in the south. Both pods and seeds are small, and are eaten boiled or sometimes roasted dry. The seeds are also ground into flour. With the emigration of Indians into Africa and the Caribbean the black gram has become more widespread.

Beans of lesser importance include *Phaseolus* (or *Adzukia*) *angularis,* the adzuki bean, grown in Japan and China; *P. acutifolius,* called the tepary bean in Mexico and the Moth bean or Mat bean in India and Burma. Both these species produce lesser yields per acre than the five major species above.

"Beans" also occur in other genera. Four of these are of some importance: *Vicia faba,* the horsebean, Windsor bean, or broad bean, is from the Mediterranean area. Some 4000 years ago these beans were cultivated by the Swiss lake dwellers and later by the Romans, who appreciated their effect on the soil (though they could not know to what it was due). They are the only beans of European origin and have now become popular in the Orient and in Brazil. A disease called fabism in Southern Italy, which is brought on by eating too many of them, has been traced to small amounts of β-cyanopropionic acid present in the mature beans. In other parts of Europe, where they are regularly eaten in moderate quantities, the disease does not occur.

The lubia, hyacinth, or bomavist bean, *Dolichos lablab,* is eaten in Malaysia and North Africa, but in this case the toxic glycoside is present in larger quantity and is only removed by very thorough boiling. This plant and its relative, *D. uniflorus,* occasionally eaten in south India, are also grown as forage plants in some areas of the tropics.

The jack bean, *Canavalia ensiformis,* dates back about 5000 years in Mexico, where it is probably native. Both the immature beans and the young pods are edible. Its relative, the sword bean, *C. gladiata,* resembles it but is Asiatic in origin. This is by no means the only example (see Chapter 21, section 2) where two species of the same genus have originated in opposite parts of the world.

4. PEAS

The pea (*Pisum sativum*—Fig. 23-6) is of western Asian origin and has been cultivated in Europe for 2 or perhaps 3 millennia. Peas do not appear in Egyptian inscriptions, but they were served at the Roman banquets. About 600 A.D. the pea reached Africa and only in medieval times did it become popular in western Europe. The pea achieved immortal fame to modern biologists as the experimental plant on which Mendel founded the modern science of genetics in 1866 (Chapter 12).

As a crop, peas are subdivided into two varieties, var. *arvense,* the field pea, with purple flowers and greyish, starchy seeds, and *hortense,* the garden pea, with white flowers and more sugary seeds, especially when immature. Both favor more temperate climates than the beans. The pods contain a tough lining that makes them inedible, but the Chinese grow a cultivar ("snow peas") with a soft, edible pod. Dry peas contain about 20% protein, and hence they served as a staple, nitrogenous food for Europeans before the discovery of America brought in the *Phaseolus* beans. Since the equilibrium sugar $\rightleftharpoons$ starch shifts toward starch as the peas mature, the unripe peas are sweeter and considered more of a delicacy, especially in France.

5. PEANUTS OR GROUNDNUTS

Peanuts or Groundnuts, *Arachis hypogaea* (Fig. 23-7), are South American in origin, but the earliest remains that archeologists have found date only from 2800 B.P. In the sixteenth century the Portuguese brought them to East Africa and the Spaniards to the Philippines.

Figure 23-6 Peas *(Pisum sativum)*, chick-peas *(Cicer arietinum)*, and lentils *(Lens esculentum)*. (From Bianchini et al.)

They are now grown throughout the tropics, but since they require 120 cm of rainfall they do not succeed in drier climates. A British attempt to plant them on a massive scale in East Africa after World War II ended in partial failure because of the unreliability of the rainfall. India, China, West Africa, and the southeastern United States are the main producing areas.

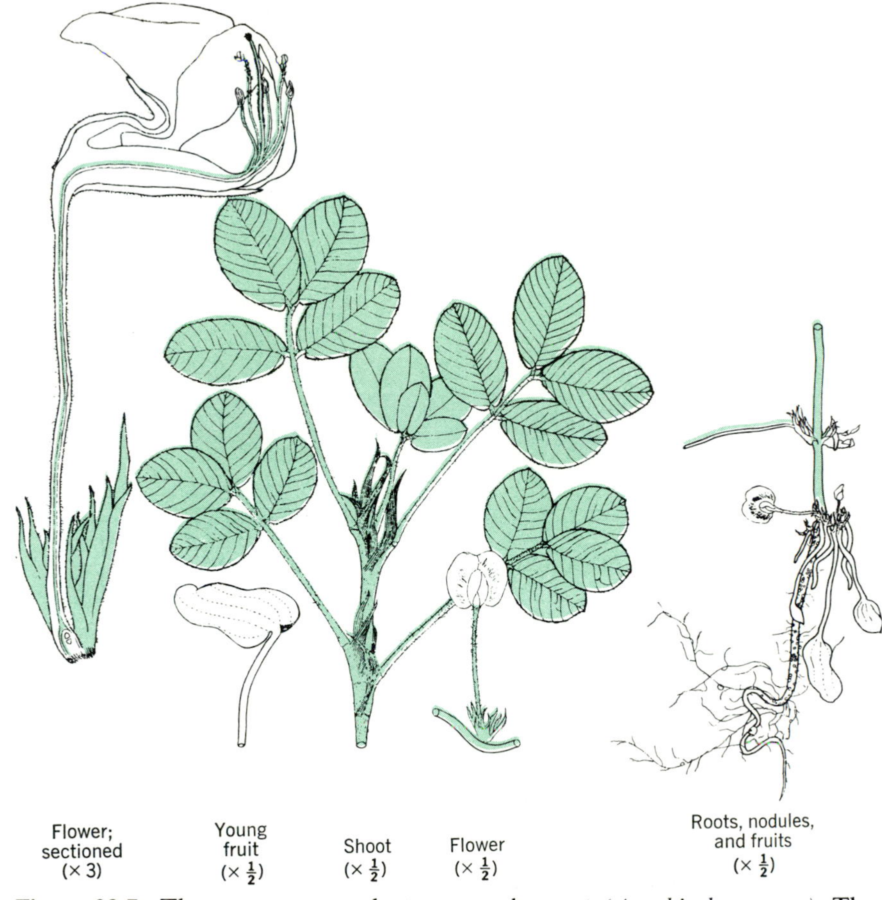

Figure 23-7 The peanut, groundnut, or monkey-nut *(Arachis hypogaea)*. The flowers are yellow, and the pinnate leaves have four leaflets.

The fruits of the peanut are quite exceptional in developing only underground (*hypogaea*= "under earth"). There are both erect and creeping cultivars, which bear their flowers in the axils of the pinnate leaves, and after fertilization a meristem at the base of the ovary gives rise to a long peduncle, which becomes positively geotropic like a root and hence grows downward to the earth (Fig. 23-7). The tip of the ovary hardens, and the whole pod is pushed into the earth to a depth of 3-6 cm. In the erect forms such fruits are borne only at the lower nodes, those formed higher up being unable to reach the ground and thus withering away. Why this happens is not understood. The growing pod becomes fibrous in structure, and, unlike most legumes, it does not split lengthwise. In the small-seeded Valencia type the pod contains up to five seeds, but in the Spanish and Virginia cultivars it has only one or two seeds.

The "nut" (i.e., the seed) contains about 30% protein and can be eaten raw or roasted, or ground into peanut butter or incorporated into candy. In America a good part of that produced (in the southern states) is fed to hogs. Besides protein the peanut contains up to 45% oil (Chapter 27), and its value is about as great for the oil as for the protein.

6. OTHER LEGUMES WITH EDIBLE SEEDS

The lentil, *Lens esculenta* or *Lens culinaris,* is from southeast Asia, where it is still a major food. It is popular in Mediterranean countries too. The small seeds, about 25% protein, break on cooking into the two cotyledonary halves, like split peas (Fig. 23-6). The Latin name refers to the convex-lens shape of these single cotyledons. There are red- and brown-seeded cultivars, and the red varieties readily break up further on cooking into a mush. Of more limited use is the chick-pea or gram, *Cicer arietinum* (Fig. 23-6), from western Asia but now grown mainly in India and South America. Its pods bear only one or two seeds (17% protein), and the yields per acre are well below those of other legumes. The cowpea or black-eyed pea, *Vigna sinensis,* is, in spite of its name (for *sinensis* means from China), probably native to central Africa; it spread east to India and north to the Mediterranean some 3000 years ago. The plant (Fig. 23-8), with its dwarf or climbing cultivars and its trifoliate leaves, looks much like the common bean, except that the pods are narrower and can be a foot long or more. There are large- and small-seeded types. The seed

Figure 23-8 The cowpea or black-eyed pea *(Vigna sinensis),* the principal food legume of Africa, and the asparagus bean or yard-long bean *(Vigna sesquipedalis).* (From Bianchini, et al.)

Figure 23-9 The winged goa bean *(Psophocarpus tetragonolobus)*, with pale blue flowers, four-winged pods, and leaves like those of *Phaseolus*. At right below is a cross section of a pod.

(23% protein) can be eaten fresh or cooked and is popular in the southern United States, but it is often grown more as a minor forage plant than for its seeds. Two very strange-looking pods are those of the Goa bean, *Psophocarpus tetragonolobus*, and the winged pea, *Lotus tetragonolobus*, which have four lengthwise outgrowths or "wings" down the sides of the pods, making them almost square in cross section (Fig. 23-9). Goa beans are eaten in Java, Burma, and New Guinea, and since the plant produces edible tubers as well as seeds it is being considered as a candidate for much wider introduction. The *Rhizobia* that form nodules on both these species appear to be the same as those for *Vigna sinensis;* thus this Rhizobium group has very low specificity for its host plant.

7. FORAGE PLANTS

The wide distribution of herbaceous legumes, their adaptability to a variety of sufficiently moist climates, and their relative independence of nitrogen sources make them ideal for forage and pasture plants. Since the other principal forage plants are grasses, whose growth is very dependent on applied nitrogen, the combination of a grass and a legume makes for the most favorable growth and also for the best mixed diet for the animals. In the Near East where agriculture began, grasses and legumes occurred together in the oak woodlands, and grass–legume mixtures have been found the best pastures ever since. The following are the forage legumes that have attained major status in worldwide agriculture.

a. ALFALFA OR LUCERNE

Some 30 million acres, about one-half of the world's crop of alfalfa (*Medicago sativa*), is grown in the U.S.A. In Europe, where it is known as lucerne, the somewhat cooler climates have limited its distribution. The plant is probably a native of Iran or Iraq; it was known in Greece 2500 years ago and came to the Americas with the Spaniards. For all its wide use, it does require a deep, well-aerated soil (since the nodules need oxygen) and a good supply of phosphate, potassium, and calcium. The warm-weather origin and the need for aeration combine to make it highly successful in the desert wherever irrigation is available. As a result such farmers can cut a crop of alfalfa six to eight times a year. Not content with this, breeders are crossing existing cultivars to obtain hybrid vigor (Chapter 16) and so to increase the yields still further. The perennial habit minimizes the time spent on plowing and seeding. Alfalfa hay is an excellent feed, higher in nitrogen than grass hay.

b. THE CLOVER GROUP

The clovers (*Trifolium* spp.) historically initiated the practice of crop rotation in medieval Europe, for a crop of clover not only yielded good hay but "improved the fertility of the soil" (Chapter 9). As the plants senesce many of the small nodules slough off, and the subsequent ploughing adds the remainder to the soil, so that the amino acids that remain in the nodule tissue, as well as the nitrogen in the rest of the plant, are added to the soil. A typical crop rotation was of three years: wheat, potatoes, clover. Another advantage of rotation was that clover (unlike alfalfa) does not do well in successive crops because its roots are susceptible to fungal attack. The leaves normally contain 20 to 30% of their dry weight as protein, compared to 12 to 15% in most forage grasses.

The small papilionaceous flowers of clover are borne in oval or globular heads. The leaf has three leaflets, as indicated by the Latin name. Of the widely used species, *Trifolium pratense* (red-flowered biennial) and *T. incarnatum* (crimson-flowered annual), which grow up to 1 m in height, are natives of southwest Asia, while *T. repens* (white-flowered), a smaller and semi-creeping plant, is of European origin. This species is the one that suffers most from repeated planting. *Trifolium hybridum* (pink-flowered), known as alsike clover, and *T. alexandrinum* (white- to yellowish-flowered), are intermediate in height; the former is grown in northern fields where the growing season is shorter.

Trifolium subterraneum (mauve-flowered), a creeper with runners largely underground, is grown as a forage plant in western Australia and has several cultivars. However, it offers a good example of the kinds of hazards one can encounter in introducing a new pasture plant. Sheep grazing on one of the cultivars developed infertility and reproductive problems, which were traced to the presence of genistein, a substance that has estrogen (animal female hormone) activity. Many plants contain such substances but generally in amounts too small to exert any physiological effect.

C. KUDZU

Kudzu (*Pueraria phaseoloides* or *P. thunbergiana*, Fig. 23-10) is really a climber but can be grown as a prostrate plant in meadows of the tropics. Its large, trifoliate leaves and blue to purple flowers resemble those of *Phaseolus* or *Dolichos*. Some regard it as a tropical equivalent of alfalfa, but its nitrogen fixation is considerably less. It is estimated that a crop of kudzu adds 60 kg of nitrogen on a hectare of meadow, while good alfalfa can add as much as 400 kg. It spreads rapidly and in parts of the southern United States is becoming a serious weed. Like other herbaceous legumes, kudzu has a rather high water requirement.

Figure 23-10 Kudzu *(Pueraria phaseoloides),* covering an old meadow and spreading on to trees, in the southeastern United States.

CHAPTER 24

Medicinal and Drug Plants

1. MEDICINE, DRUGS, AND RELIGION

Knowledge of the nerve-exciting properties of some plants, and the poisonous properties of others, goes far back in time—perhaps almost as far as that of food plants. Much of the earliest knowledge was not written down but was doubtless kept confidentially by priests and others as a source of power. The medicine man and witch doctor, with his own private collection of herbs and plant extracts, is still with us today in Africa, Malaysia, and elsewhere—fish are still caught by sprinkling certain flowers on the water in central Africa, and religions are based on the hallucinations produced by eating sacred fungi.

Thus the Rig Veda, India's ancient and sacred book of hymns, written down about 1100 B.P., devotes many of its hymns to the virtues of the divine Soma, which probably is the intoxicating mushroom *Amanita muscaria,* the "fly agaric." To this day that mushroom is so much prized among tribes in northern Asia that *one* amanita will buy a reindeer. The intoxicating principle (see section 9) is largely excreted unchanged in the urine, so that the urine is often drunk afterward to obtain a second intoxication. In Mexico about 25 years ago an ancient religion, based on the hallucinations produced by another mushroom, *Psilocybe* sp., was brought to light. Stone images of mushrooms, of religious significance, date from 3000 B.P. Among similarly influential higher plants, the hallucinations induced by the cactus peyote or peyotl have been the center of another religion among North and South American natives for centuries, while the imaginative dreams inspired by opium have formed part of Chinese social custom for perhaps thousands of years. Great numbers of the plant extracts extolled for various illnesses in the medieval herbals, and often dating back to Galen (Chapter 28) were sedatives or stimulants for the nervous system. Thus in this subject myth, religious experience, history, sociology, medicine, and modern science converge.

2. THE ALKALOIDS IN GENERAL

Most of the effects described above are due to alkaloids (one or two exceptions will be noted below). The alkaloids are chemically a rather miscellaneous group of nitrogen compounds, most of which bear a nitrogen atom in a ring with carbon atoms. These nitrogenous compounds can be considered as ammonia, NH_3, with one or more of the H atoms replaced by an organic radical. As such they share an important property with ammonia; namely, they are *alkaline* in solution. In the plant they commonly occur in combination with an organic acid, forming a more or less neutral salt.

The biological quality that the alkaloids have in common is that of acting on the nervous system of animals, either on the nerve endings or on the brain. Unlike most of the vitamins, which function in both animals and plants, the alkaloids seem to have no particular effect on the plants that produce them—no doubt because plants have no detectable nervous system.

The alkaloids are by-products of primary plant metabolism, being derived from naturally occurring organic acids or amino acids (Fig. 6-6). In this chapter we take up a few examples from the more than 5000 alkaloids known, while the small group of alkaloids related to purines will be treated in the following chapter. Some introduction of organic chemical formulae is essential to understand how the compounds are formed in the plant.

A selection of the most important alkaloids and the plants that produce them is given in Table 24-1.

3. POISON HEMLOCK: CONIINE

The leaves of poison hemlock, *Conium maculatum,* contain the simplest of alkaloids, coniine.

TABLE 24-1

Some of the Principal Alkaloids of Flowering Plants

Alkaloid	*Plant*	*Common Name*	*Family*
Atropine	*Atropa belladonna*	Deadly nightshade	Solanaceae
Coniine	*Conium maculatum*	Poison hemlock	Apiaceae (Umbelliferae)
Cocaine	*Erythroxylon coca*		Erythroxylaceae
Digitalin group	*Digitalis purpurea*	Foxglove	Scrophulariaceae
Mezcaline	*Lophophora williamsii*	Peyotl	Cactaceae
Morphine	*Papaver somniferum*	Opium poppy	Papaveraceae
Nicotine—also anabasine and nornicotine	*Nicotiana tabacum* and *N. rustica.*	Tobacco	Solanaceae
Quinine	*Cinchona ledgeriana*	Fever-bark tree	Rubiaceae
Quinidine—also cinchonine	*Cinchona succirubra*	Fever-bark tree	Rubiaceae
Reserpine	*Rauwolfia serpentina*	Snakeroot	Apocynaceae
Solanin	*Solanum tuberosum*	Potato	Solanaceae
Steroid alkaloids (90 compounds)	*Solanum* spp.		Solanaceae
Strophanthin and relatives	*Antearis toxicaria, Strophanthus gratus,* and *S. hispidus*	Upas tree	Urticaceae Apocynaceae
Strychnine—also brucine	*Strychnos nux-vomica*		Loganiaceae (Apocynaceae)
Tubocurarine	*Strychnos* spp.	Curare	Loganiaceae (Apocynaceae)

The plant bears no relation to the forest tree *Tsuga,* which is also called hemlock. In leaf and flower form it resembles the common Queen Anne's lace or cow's parsnip (Fig. 24-1), but the word *maculatum* refers to the tiny red spots on its stems and petioles. The Greeks knew its poisonous quality, and, when Socrates was judged to be an enemy of the people in 399 B.C., he was condemned to die by drinking hemlock extract. Its action is hinted at in Keats' Ode to a Nightingale:"

My heart aches, and a drowsy numbness fills
my sense
As though of hemlock I had drunk.

The poisonous principle is *coniine,* $C_8H_{17}N$. Its role as a by-product of metabolism has been brought out by feeding suspected precursors labeled with ^{14}C to *Conium* plants. If the 8-carbon acid, octanoic acid (an intermediate in fat formation) is synthesized with ^{14}C in the molecule and then injected into the stem of

Figure 24-1 Poison hemlock *(Conium maculatum).* Left, flowering shoot, seedling, and taproot. The finely divided pinnate leaves are typical of the Umbelliferae (compare those of parsley or carrot). (From Muenscher, 1951) Right, the characteristic red- or purple-spotted stem.

Conium, the coniine subsequently extracted contains much radioactivity; the pathway thus probably goes rather directly, via the amide:

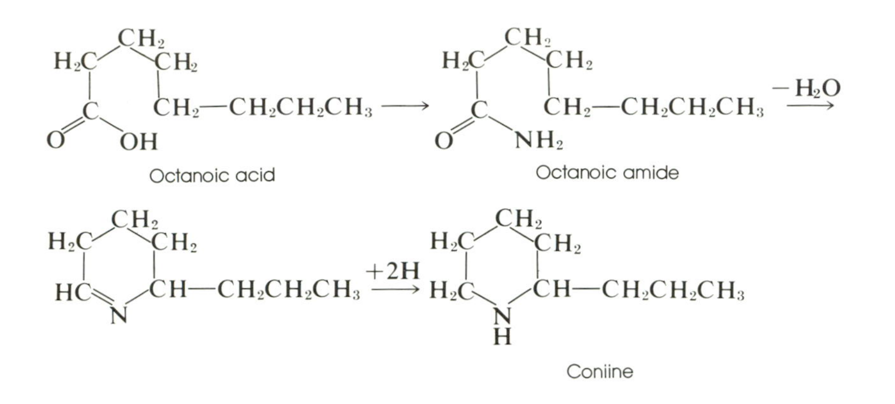

The octanoic acid would doubtless be made by the condensation of four molecules of acetyl CoA, as in the formation of other long-chain acids (see section 6 of Chapter 6).

4. THE PIPE OF PEACE: NICOTINE

The case of nicotine is more revealing as to plant metabolism, and the alkaloid is more widespread, being found in at least six families, and in several genera within the Solanaceae.

Early travelers to the New World found the Caribs smoking the "pipe of peace," and the seeds of tobacco (named for the Caribbean island of Tobago) were brought to Philip II of Spain in 1558. Because the French ambassador distributed them, and his name was Jean Nicot, the plant was given the Latin name *Nicotiana* by Linnaeus. *Nicotiana tabacum* is the species mainly used; but *N. rustica* is also used to some extent, and several other species are grown for their flowers. Its culture for smoking was started by Governor Lane of Virginia and popularized by Sir Walter Raleigh, who planted both tobacco and potatoes on his estate in Ireland in 1586. The custom must have spread rapidly, for King James I (responsible for ordering the current English translation of the Bible) wrote a "Counterblaste to Tobacco" in 1604. Nevertheless it is odd that there is no smoking in Shakespeare's plays, which were written between 1590 and 1613.

Although the pleasantly sedative properties of tobacco have thus been known for nearly 400 years, it was only in the 1950s that the relation between smoking and lung cancer was discovered. The cancer is due to the tar, which is produced by most kinds of leaves on burning, and not to the nicotine—though nicotine does have some undesirable but minor secondary effects of its own. Other, less tasty, leaves are occasionally used for smoking too, but the tumor-inducing tar would be about the same in all burned leaves.

Tobacco growing was started in Virginia about 1612, and in 1660 the production in Britain was made illegal so as not to interfere with the exports from the American colonies. The culture in North America soon spread all over the eastern states from Connecticut to the Carolinas. Tobacco was even used as money for a while in Virginia. The United States now produces about a quarter of the world's supply; other countries producing lesser amounts include China, Brazil, Indonesia, Turkey, Rhodesia, Greece, Russia, and the Philippines.

The tall tobacco plants are commonly de-

Figure 24-2 Leaf and flower of tobacco *(Nicotiana tabacum).* (From Purseglove, 1968)

Figure 24-3 Taking tobacco leaves, threaded on sticks, to the curing shed.

capitated in the summer to make the more basal leaves grow larger, and than harvested at the end of the season when the leaves (Fig. 24-2) are senescing (i.e., yellowing). In some procedures, especially for dark pipe tobaccos, they are stacked together to ferment, which causes browning and converts most of the sugar and starch to organic acids and carbon dioxide. For the lighter cigarette types the leaves are threaded on sticks and the resulting "hands" (Fig. 24-3) are simply dried in large barns, heated either by little fires ("fire-cured") or by metal pipes carrying hot air from stoves outside ("flue-cured"). Most of the flavor and aroma is due not to alkaloids but to other constituents of the leaf, to sugars or sorbitol added during the curing or, in the case of cigar leaf, partly to ammonia set free from the amino acids during senescence. However, since nicotine, like coniine, is a liquid, it does volatize on heating. Thus tobacco yields its drug on smoking, while most drug plants have to be eaten.

Nicotiana tabacum, which has 24 chromosome pairs, is apparently a tetraploid of a hybrid between *N. sylvestris* and another South American species, both having 12 chromosome pairs. The hybrids were probably nearly all sterile and only the chance doubling of the chromosome number led to the self-fertile *N. tabacum. Nicotiana rustica,* the other species with high nicotine content, also has 24 chromosomes. A similar process has been brought about artificially at the next higher level of ploidy, a triploid having been made fertile by chromosome doubling. Thus polyploidy proves of importance in alkaloid plants as well as in cereals (Chapter 16), and in many other plants.

Nicotiana offers a prime example of the localization of alkaloid production within the plant. If isolated young leaves are cultured in nutrient solution they form no nicotine. Scientists grafted tobacco shoots on to tomato roots (the tomato also belongs to the Solanaceae); the shoots grew but their leaves contained almost no nicotine. The reciprocal graft, tomato shoots on tobacco roots, also grew well and produced the typical red fruits but these were bitter, containing nicotine. It follows that *nicotine is synthesized in the roots*. Isolated roots, maintained in nutrient solution, have therefore been used to study the mode of nicotine formation. Several amino acids when fed to the roots led to increased nicotine content, but the B-vitamin *nicotinic acid* gave the best yields. Nicotine contains two rings, those of pyridine and of pyrrolidine. The carbon atoms of nicotinic acid appear in the pyridine ring. The carbon atoms of the amino acids lysine or *ornithine* appear in the pyrrolidine ring. An additional methyl group is supplied by an amino acid, methionine. The whole pathway is thus:

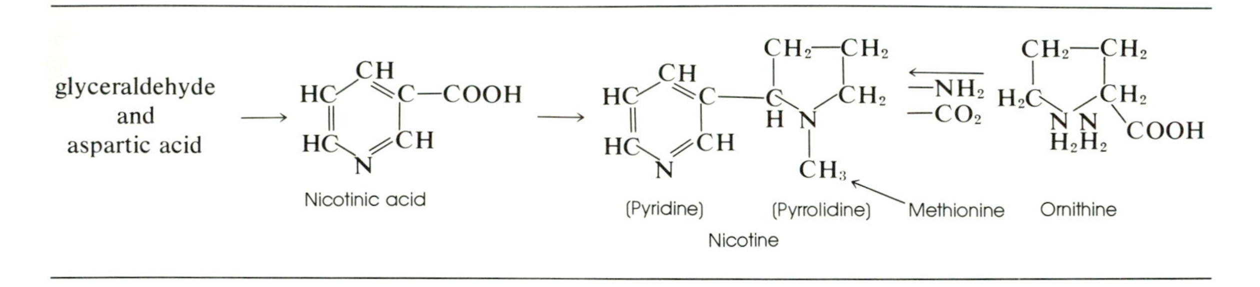

Another species, *N. glauca,* contains a similar alkaloid, anabasine, in which the pyrrolidine ring is replaced by a second pyridine ring, derived from lysine. There is usually some anabasine in ordinary tobacco, too.

5. PAINKILLERS AND SLEEP-BEARERS

a. ATROPINE AND THE LOCAL ANESTHETICS

The alkaloid atropine is found in the leaves and fruit of *Atropa belladonna* (also a member of the Solanaceae), known in Britain, where it is native, as "deadly nightshade," and known to the herbalists as *Solanum lethale* (see Fig. 24-4). The name *belladonna* ("beautiful lady" in Italian) derives from its use by ladies of fashion to dilate the pupils of the eye.

Atropine (structure 3, shown on page 432) and others of the belladonna group of alkaloids are usually stimulatory in low doses and poisonous in larger amounts (Fig. 24-5); most alkaloids behave in the same way, but the stimulatory range is sometimes wide and sometimes very narrow.

The alkaloids of this group are derivatives of *tropane* (structure 1) which in space would have the form shown as 2. Related compounds are found in the pasture weed *Datura stramonium* (Fig. 24-4*C* and 24-4*D*), one of several plants called locoweed (because the leaves and fruits are so rich in alkaloids that cattle that have eaten the leaves are seen to stagger). Other compounds of the same group are found in *Hyoscyamus niger,* henbane (Fig. 24-4*B*), and in *Mandragora* (called the "mandrake" because its branched tubers are thought to resemble a man). All these plants are members of the Solanaceae, or potato family. The most famous tropane alkaloid is cocaine, obtained from the leaves of *Erythroxylon coca,* a member of a small family related to the Rutaceae (citrus group). Cocaine (structure 4, p. 432) is a stimulant that acts by injection or by sniffing. Sherlock Holmes, the detective, was described as injecting it occasionally. Peruvian mountaineers derive stimulation by chewing the coca leaves and as a result are said to be able to work all day without food. However, they ingest only a fraction of the dose that causes addiction.

(A)

(B)

(C)

(D)

Figure 24-4 Four major alkaloid-forming plants of the Solanaceae. *A*. Deadly nightshade *(Atropa belladonna)*. *B*. Henbane *(Hyoscyamus niger)*. *C*. *Datura meteloides*. *D*. Fruit of Locoweed *(Datura stramonium)*.

On injection into the skin or muscle, however, cocaine acts as a *local anesthetic*. This property led to a study of the structures of these and some other compounds with local anesthetic properties, and to the synthesis, on the same general model, of more potent but non-habit-forming compounds like novocain and stovaine, both now widely used in dentistry.

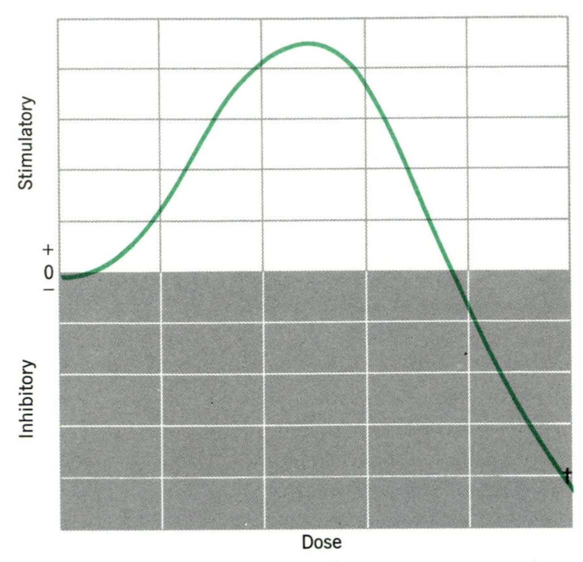

Figure 24-5 Typical curve of response versus dose of an alkaloid; with increasing dose the stimulation becomes an inhibition and toxicity (shown by a dagger) rapidly appears.

It appears that the common property of these local anesthetics is the methylated nitrogen atom at one end of the molecule, separated by a short carbon chain from the benzoyl group at the other. The explanation of this comes from the discovery that when nerve endings of mammals are stimulated a small amount of a special substance is formed that diffuses across from the nerve to activate the muscle end-plate, or, within a nerve junction, to activate another nerve. This substance is *acetylcholine,* structure 5. Probably the local anesthetics act by competing with this substance; in essence they place a benzoyl group (C_6H_5CO), or a close relative of it, where an acetyl group (CH_3CO) should be. This phenomenon of structural analogy gives a broad basis for the action of many drugs. The fact that rabbits can eat belladonna leaves without harm is due to their possessing an enzyme that hydrolyzes the critically important O-CO- grouping.

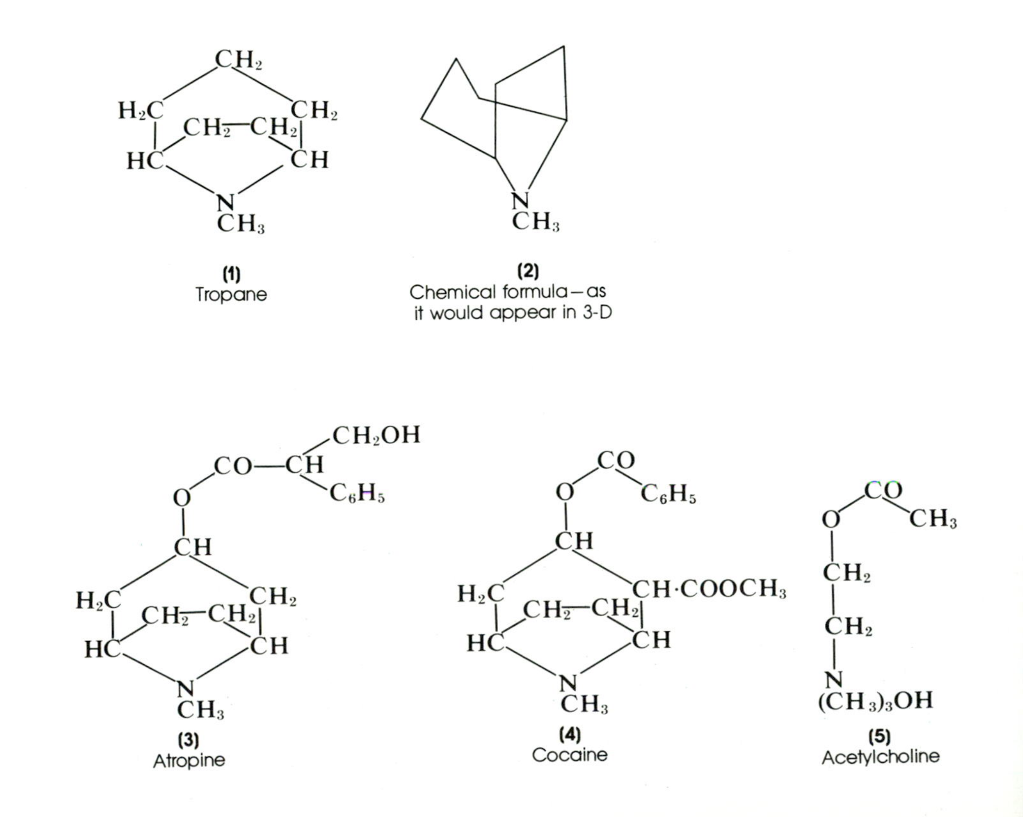

b. OPIUM AND ITS ALKALOIDS

The opium poppy, *Papaver somniferum* of the Papaveraceae, has been known for at least 4000 years, and opium smoking (the morphine volatilizes in the smoke) is an especially ancient custom in China. *Somni-ferum* means in Latin "sleep-bearer." The effects are classically described in De Quincey's "Confessions of an English Opium Eater" (1821); they inspired Coleridge to write the poem "Kubla Khan," which was based on an opium dream.

Raw opium, a mixture of alkaloids of which morphine, thebaine, and codeine are the major constituents, is exuded as a brown gum when the ripening poppy capsule is wounded. Traditionally a spiral cut or scratch is made in the capsule (Fig. 24-6) and the opium scraped off some days later—usually early in the morning, since the morphine content decreases during the day.

Biochemically, morphine seems to be formed in the plant from the condensation and rearrangement of two molecules of the amino acid tyrosine. It can be readily acetylated, chemically, to produce diacetylmorphine or *heroin*. Whereas the dominant action of morphine (named for *Morpheus,* the god of sleep) itself is sedative and sleep inducing, the dominant action of heroin is as a stimulant to the brain, causing euphoria and a sense of internal excitement. Both readily lead to *addiction,* a strange alteration in some brain function in which an adaptation to the drug takes place so that an overpowering need for another dose is felt. With heroin, addiction may even occur after the second dose. The resulting desperate need for the drug, which is expensive, leads to much crime, and accordingly efforts have been made, especially by the United Nations, to stop poppy growing, though a small amount is needed for medicine. Turkey has indeed largely stopped

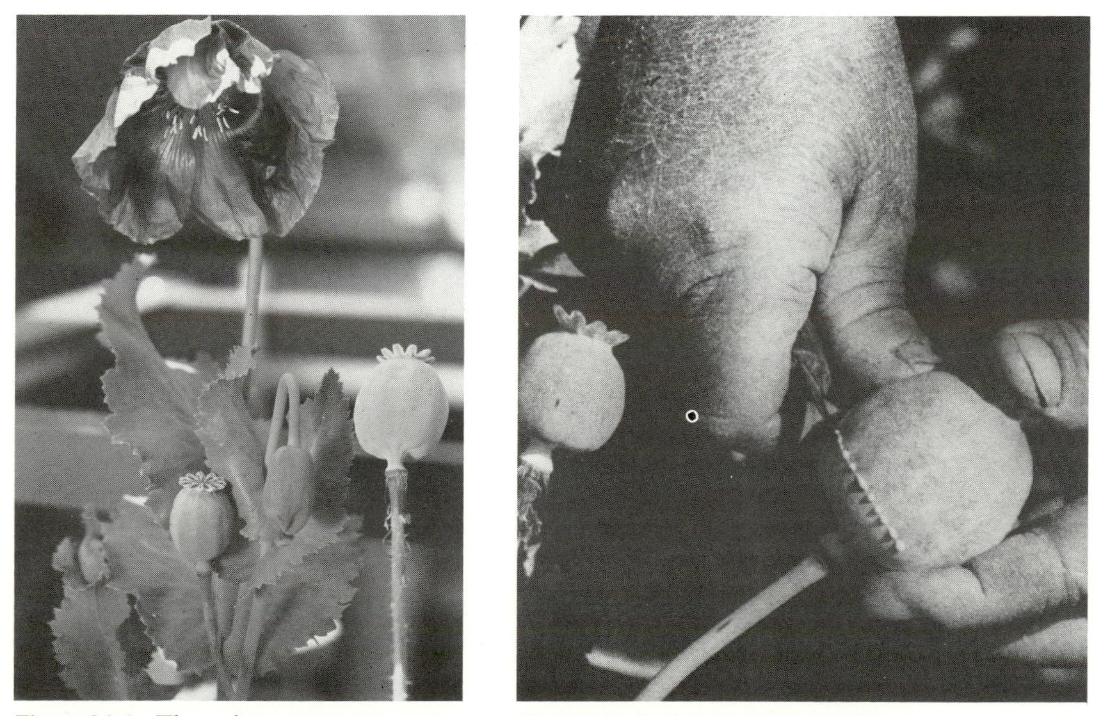

Figure 24-6 The opium poppy *(Papaver somniferum).* Left, the attractive purple flowers, with leaves and young fruit. Right, latex exuding from a fresh incision on a ripe capsule.

opium growing, but the poppy now is mainly grown in the mountainous region between Burma, Laos, and Thailand; the laboratories where acetylation surreptitiously takes place are said to be centered in the south of France; thus it is very much an international problem.

The red-flowered *Erythrina,* a legume often grown as a shade tree for coffee (Fig. 25-3) has very similar alkaloids in its poisonous seeds. It is a surprising fact that the much milder sense of euphoria and stimulation conferred by smoking the leaves of *Cannabis sativus* (marijuana, pot, hashish or bhang) is due to a substance that contains no nitrogen and is therefore not an alkaloid at all, namely tetrahydrocannabinol.

6. QUININE AND THE CONQUEST OF MALARIA

Malaria (the name literally means "bad air") is the world's most widespread disease. It was originally thought to be due to the damp and smelly air arising from swamps, but it is actually caused by a protozoon, *Plasmodium vivax,* which is inoculated into mammals by the bites of mosquitoes. Until recently there were 800 million cases a year worldwide, with 3 million deaths; its victims included Alexander the Great and Albrecht Dürer, and its prevalence in armies defeated many military campaigns.

About 1630 Spanish Jesuits in Lima, Peru, learned that extracts of the bark of a local tree (Quina) could cure malaria. Success with a distinguished patient, Countess Cinchon, resulted in naming the tree *Cinchona.* Many attempts to bring the extract into medicine were made, but there were a number of difficulties: (1) the trees grow in rather inaccessible areas high up (1500–2000 m) on the eastern slope of the Andes, (2) there are several very similar species, the majority of which contain only small amounts of alkaloids, (3) to take the bark for extraction kills the tree. For these reasons several British and Dutch expeditions to bring back seed or seedlings from 1851 onwards were essentially unsuccessful.

Finally Charles Ledger, an Englishman living on the shores of Lake Titicaca in Bolivia, with a servant who knew the good alkaloid-rich species, collected seeds and in 1865 sent some to London and to Java. The Dutch planter in Java, K. W. van Gorkom, planted the seedlings carefully segregated from other plantations so that no crossing would occur; the bark of the resulting trees contained 10 to 13% quinine. Their roots were so weak that they were propagated by grafting, and the resulting plantations now yield about 90% of the world's supply. Trees from Java (Fig. 24-7) have now been planted in Central America.

This species, *Cinchona ledgeriana,* is rather a weak and slow-growing form; the seedlings are very subject to damping-off diseases. It is harvested at about 10 years of age when the trunk is 8 to 10 cm thick; the bark is slit off and extracted with solvents. The quinine base occurs mainly as a salt of an organic acid.

Besides the quinine there are several closely related alkaloids present, but these are not as toxic to the *plasmodium* as is quinine. The stock for the grafting is another *Cinchona* species, *C. succirubra* (or *C. pubescens*), which is sturdier and more disease resistant; but its bark contains only around 5% alkaloids.

Although quinine has a complex formula it was one of the first alkaloids ever to be isolated in pure crystalline form (Pelletier and Caventou, 1820)*. The complexity of the structure of quinine delayed its chemical synthesis for a century and a quarter, until 1944. No sooner had the synthesis been completed than (ironically) synthetic compounds known as quinacrine, paludrine, and others, chemically almost unrelated to quinine, came gradually to replace

* A marble monument to Pelletier and Caventou is on the Boulevard St. Michel in Paris. Such permanent recognition of the efforts of plant scientists is certainly rare!

Figure 24-7 *Cinchona ledgeriana,* the quinine-producing tree. Left, flowering stem and leaves. Right, Cinchona bark being dried.

quinine as antimalarials. Furthermore, after 1950 the battle against malaria shifted to a battle against mosquitoes in the tropics. Spraying houses and barns with DDT to kill mosquitoes cut down malaria deaths by more than two-thirds, to less than 1 million per year worldwide.

7. BRAIN STIMULANTS: THE INDOLE ALKALOIDS

Many brain stimulants are derived from indole (structure 6) and especially from its ethylamine derivative, tryptamine (structure 7) (see p. 436). There are more than 100 of them and they perhaps owe their action on the brain and nervous system to their resemblance to serotonin (structure 8), which is one of the natural activators of the brain; they inhibit the enzyme that oxidizes away the $-NH_2$ group of serotonin. They occur mainly in the Apocynaceae, Leguminosae, Rubiaceae, and Euphorbiaceae and comprise about a quarter of all known alkaloids. Indeed, in a study of a tropical forest in Colombia, indole alkaloids were found in more than half of the species examined. Tryptamine and serotonin are present in bananas and their skins, while at least one tropical plant actually contains 500 ppm of serotonin. Important indole alkaloids include reserpine, from the roots of *Rauwolfia serpentina* (an Indian bush with twisted roots, traditionally used as a remedy for snake bite, but now used as a brain depressant for schizophrenic patients); strychnine, from the seeds of *Strychnos nux-vomica,* which causes painful muscular contractions and distortions, usually fatal; and mezcaline, which is related to tryptamine. This compound is responsible for the hallucinogenic properties of the tiny cactus *Lophophora williamsii,* found in the Sonoran desert and known in Mexico as *peyotl.* Mezcaline (structure 9) is accompanied in the dried plants by about seven other alkaloids. In this compound, as also in muscarine, the nitrogen atom is not in a ring, but it is classed with the alkaloids nevertheless.

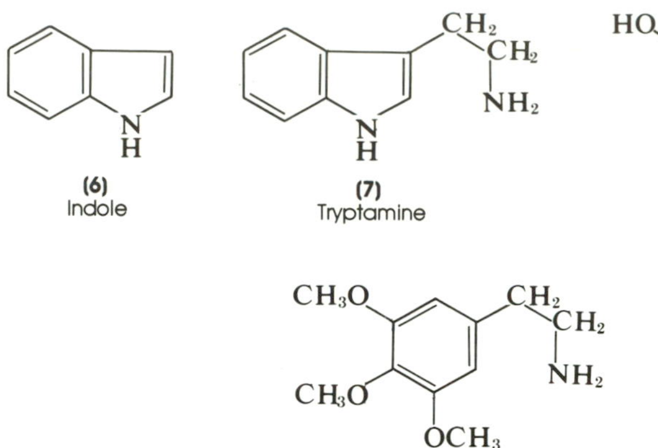

8. HEART STIMULANTS: THE STEROIDAL DRUGS

The foxgloves, *Digitalis purpurea* and *D. lanata,* of the Scrophulariaceae, contain drugs of quite another type, combining a steroid with a sugar. The steroids consist of four fused carbon rings, not containing nitrogen; hence they are not alkaloids. The male and female sex hormones (androgens and estrogens) are steroids. Digitoxin, the main drug in the foxglove, stimulates the heart muscle, prolonging the period of relaxation in each heartbeat; it does not affect the period of contraction. As a result the heart gets a better blood supply and preparations of digitalis are most valuable to those with "cardiac insufficiency," that is, limited vascular supply to the heart.

The handsome plant, *Digitalis purpurea* (Fig. 24-8), with a meter high flowering scape and purple to white bell-shaped flowers, is common all over western Europe and has become naturalized in North America too; its preparations have been in use for centuries. *Digitalis lanata* (the woolly foxglove) is limited to southeast Europe, especially Greece. The drugs, which are similar in both species, are obtained either from the seeds or from the leaves; each yields a different mixture of steroids. Both types are highly toxic in larger doses (as in Fig. 24-5).

Figure 24-8 Foxglove *(Digitalis purpurea),* from the classic work of William Withering, 1785, on the medical uses of extracts of foxglove leaves and seeds.

Similar compounds with similar effects on the heart are present in the flowers and seeds of the common lily-of-the-valley (*Convallaria majalis*). In medieval times this plant was much used in medicine, but it was dropped in the eighteenth century.

9. MEDICINALS, DRUGS, AND POISONS FROM NONFLOWERING PLANTS

The immense variety of nonflowering plants ranges all the way from trees and shrubs to tiny fungi, algae, and bacteria. As might be expected, their output of substances acting on the nervous systems of animals is also immensely varied (Table 24-2). We will take up a few examples chosen on botanical rather than biochemical grounds.

Cycasin is a chemically simple carcinogenic substance occurring in the large, fleshy seeds of a cycad (*Cycas revoluta*) on the island of Guam (Fig. 24-9). The natives soak the fruits in water for 10 days, then make them into flour for use as an emergency food. If unwashed, the flour causes tumors in rats within five to six months. In *Cycas* the active compound, methyl-azoxy-methanol, occurs as a glucoside; in *Encephalartos,* another cycad whose fruits are used for flour in Kenya, it occurs as a compound with glucose and xylose. The tumors are probably due to the azoxy (NO) group. The seeds of some other cycads, including *Cycas circinalis,* contain a simple compound that causes convulsions. Again, however, it is washed out before the seeds are used for food.

Mushrooms contain two very potent drugs. One, *muscarine,* is formed in the orange-colored toadstool *Amanita muscaria* and other *Amanita* species (Fig. 24-10). Some of the amanitas are so poisonous that a single bite can kill a child, but *A. muscaria* in moderate doses has hallucinogenic and sedative effects. The relatively small molecule of muscarine is quickly absorbed, but other active compounds, present

TABLE 24-2

Some Alkaloids and Other Medicinals from Fungi and Bacteria

Alkaloid	*Organism*	*Common Name*	*Group*[a]
Ergotinine, lysergic acid	*Claviceps purpurea*	Ergot	Ascomycetes
Muscarine	*Amanita muscaria*	Fly agaric	Basidiomycetes
Psilocybin	*Psilocybe* spp.		Basidiomycetes
Antibiotic	*Organism*		*Group*
Penicillin	*Penicillium notatum, P. chrysogenum,* and others	Penicillin	Ascomycetes or fungi imperfecti
Cephalosporin	*Cephalosporium* spp.	Cephalosporins	Fungi imperfecti
Streptomycin	*Streptomyces griseus*	Streptomycin	Actinomycetes
Tetracyclins	*Streptomyces aureofaciens* and *Str. rimosus*	Aureomycin and Terramycin	Actinomycetes
Neomycin	*Streptomyces fradiae*	Neomycin	Actinomycetes

[a] See Chapter 29, section 2.

Figure 24-9 *Cycas revoluta,* the cycad whose ground-up seeds cause cancer if not thoroughly washed. The cones are borne in the center, but are not produced for several years.

in smaller amounts, probably participate. Another is *psilocybine,* found in subtropical mushrooms, *Psilocybe* spp. In the religious observances surrounding its use in Mexico, two small mushrooms are eaten and the subject then sees brilliantly colored hallucinations for several hours. The compound is an indole alkaloid containing a phosphate group, the only phospho-indole known in the plant kingdom.

Ergot is an infection of the seed heads of rye with a fungus, *Claviceps purpurea.* The infected grains, which swell and turn purple, are notorious for causing muscular contractions, and the extract is accordingly used sometimes to promote the uterine contractions needed for childbirth. An ergot infection of the rye fields in ancient Greece is believed to have been the cause of the plague of Athens; swelling and gangrene of the arms and legs, probably due to extreme contraction of the blood vessels, were described. Several alkaloids, some of which also occur in the seeds of morning glory (*Ipomoea violacea*), are involved. Although ergotism is now rare, due to improved care in farming and seed selection, occasional outbreaks still occur; there was one in 1951.

Figure 24-10 The divine Soma, or fly agaric *(Amanita muscaria).*

10. ANTIBIOTICS

Finally we must add the immensely important but quite different group of compounds that arrest the growth of bacteria—the *antibiotics.* A few of these are produced by fungi, notably the penicillins from *Penicillium notatum* and *P. chrysogenum,* and the related cephalosporins from *Cephalosporium* spp. Most, however, are from members of the Actinomycetes, which are really closer to the bacteria than to the fungi, but form long thin threads somewhat like fungus hyphae (Fig. 24-11). They include streptomycin, chloramphenicol, the tetracyclins, and a hundred or so more. In most cases other than the penicillins their nitrogen atoms are not contained in the rings but are in side-chains (see section 8). No doubt they rank (with DDT) among the world's greatest lifesavers, and between them they have brought many human bacterial diseases like pneumonia, tuberculosis, gonorrhea, and typhoid under control. Virus diseases, however, remain unconquered.

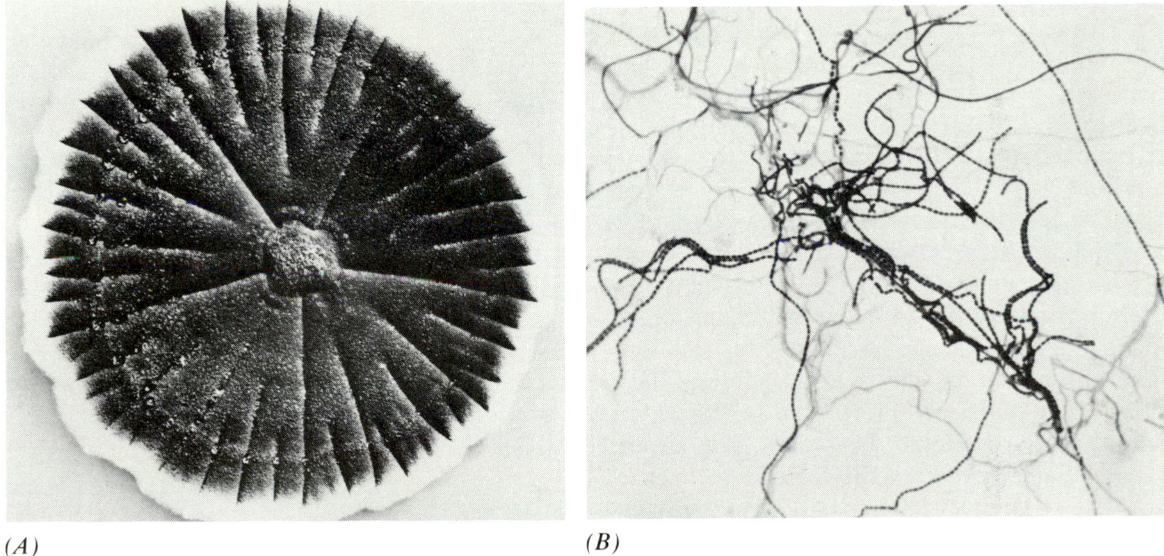

(A) *(B)*

Figure 24-11 *A.* Colony on agar of *Penicillium chrysogenum,* producer of penicillin, *B.* Hyphae of *Streptomyces venezuelae,* producer of chloramphenicol; magnified 1500 ×.

CHAPTER 25

Caffeine Beverages

The need to satisfy thirst is just as great as the need to satisfy hunger. Although water has been, and probably will continue to be, our prime quencher for thirst, we have tended to seek additives that make water either more healthy (by killing microbes), more tasty (by interesting flavors), or more zestful (by stimulants). Fermented drinks appear early in history, because of the prevalence of yeasts and their natural fermentation of grapes and grain, producing wine and beer respectively. Although wine and beer are among the oldest and most cherished of our beverages, tea and coffee, which contain the alkaloid caffeine, have been even more commonly used. Other frequently used caffeine beverages are cocoa, maté, cola, and guaraná. These caffeine beverages are sufficiently important in world markets for several tropical countries to base their economy on the export of coffee or cocoa. As we shall see also, shops where coffee, tea, or cocoa were drunk provided meeting places for discussion of political ideas which had enormous consequence in world affairs.

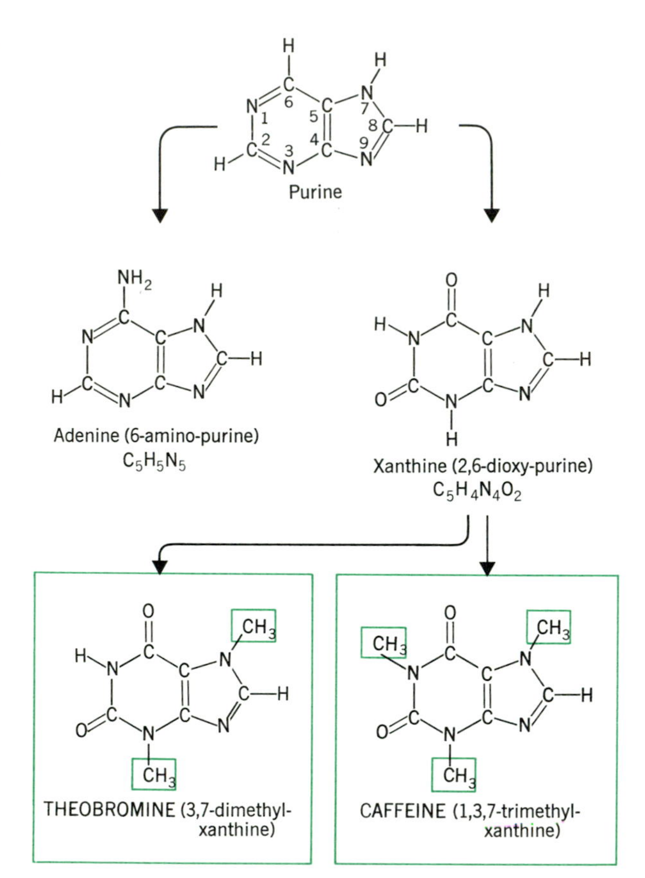

Figure 25-1 Derivation of caffeine and theobromine alkaloids from xanthine, a member of the purine group.

1. THE CAFFEINE ALKALOIDS

Although alkaloids generally are synthesized by numerous taxa scattered through the plant kingdom, the caffeine alkaloids are thought to be more common in higher than in lower plants. They belong to the purine group (Fig. 25-1)—as does adenine (a component of the all-important nucleotide ATP and of DNA and RNA). The closest relative to the caffeine group is the oxygen-containing purine, xanthine. Both caffeine and theobromine possess the same ring structure as xanthine but differ in being methylated. Caffeine and theobromine differ from each other only in that caffeine has one more methyl group attached to the ring than theobromine.

2. BEVERAGE CROPS SUPPORTING TROPICAL ECONOMIES

a. COFFEE

Ninety percent of the world's coffee is produced from *Coffea arabica*, a member of the tropical family Rubiaceae. The native home of the coffee plant is in forests at elevations of from 1500 to 2000 m in the Ethiopian mountains. Few indigenous Africans drink coffee; instead they use the berry as a masticatory. The first drink was probably alcoholic, having been derived from fermentation of the pulp of the coffee berry. Although it is still prepared this way in Arabia, where it was taken in the sixth century, it was in Arabia that coffee was first roasted, in the middle fifteenth century, as we know it today. However, in earlier times in Arabia, the powdered seed was made up in a ball with butter and carried on a desert journey for a stimulant and sustenance. Mocha, the city in Yemen, whose name a type of coffee still bears, remained for two centuries the center of coffee trade, an Arabian monopoly until coffee growing became widespread in the late 1700s.

The practice of drinking coffee spread from the Near East to Cairo (1510) and Constantinople (1550), thence to Venice (1616), and finally to England in 1650 and Holland in 1690. It was in England that social effects from coffee drinking were apparently most felt. The first coffeehouse was opened in the university city of Oxford in 1650 and was followed soon by many in London. The number of coffeehouses increased as they became an important part of the daily life of the better-educated Londoner. They were the center for intellectual discussion and relaxation, and soon each profession and representatives of all kinds of political and religious opinions had their own headquarters in particular coffeehouses. In Miles' Coffeehouse the first ballot box was introduced and used. Insurance of ships and their cargoes started at Lloyd's coffeehouse and is still centered there. When Charles II came to power (1660), Parliament was rarely called and there was little contact between government and the people. The press was regulated as were other informative publications. Consequently the 3000 coffeehouses in England became news centers that kept the public informed and served as public meeting places. When in 1675 Charles II tried to suppress the coffeehouses, the public outcry was so great that the King had to revoke his proclamation.

Coffee was first carried to India, Ceylon, and the East Indies during the sixteenth and seventeenth centuries and from there via Amsterdam to the West Indies and South America (Fig. 25-2*A*). The ultimate source of most of the trees in the New World coffee plantations was a single tree in the botanical garden in Amsterdam. This is another striking example (as we shall see with rubber) of the role played by botanical gardens in the development of agricultural commodities in the tropics. Today one-half of the world's coffee comes from Brazil, with Colombia another large South American producer. Coffee is also an important crop in the Central American countries of Costa Rica, Mexico, Guatemala, and Salvador. In Africa it is a significant crop in Angola, Tanzania, Uganda, Kenya, and the Ivory Coast.

Since the coffee tree is native as an understory tree in upland forests, its cultivation does best in equatorial latitudes at about 1500 to 2000 m altitude. It cannot tolerate freezing, needs high annual rainfall (190 cm), but also requires a dry period following the rains for development of the flower buds. The bushes produce within 3 years and usually last about 40 years. Coffee also requires a fertile soil; thus it does particularly well in the misty cloud forests on volcanic soils such as are found near São Paulo, Brazil, and San José, Costa Rica.

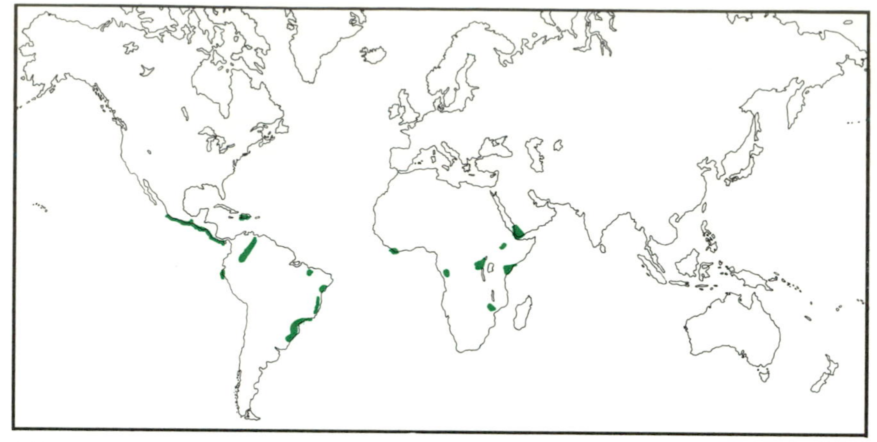

(A) Coffee

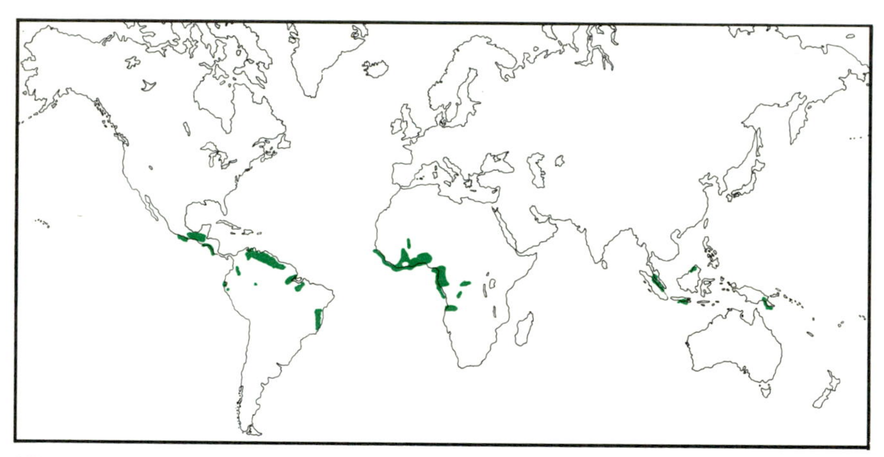

(B) Cocoa

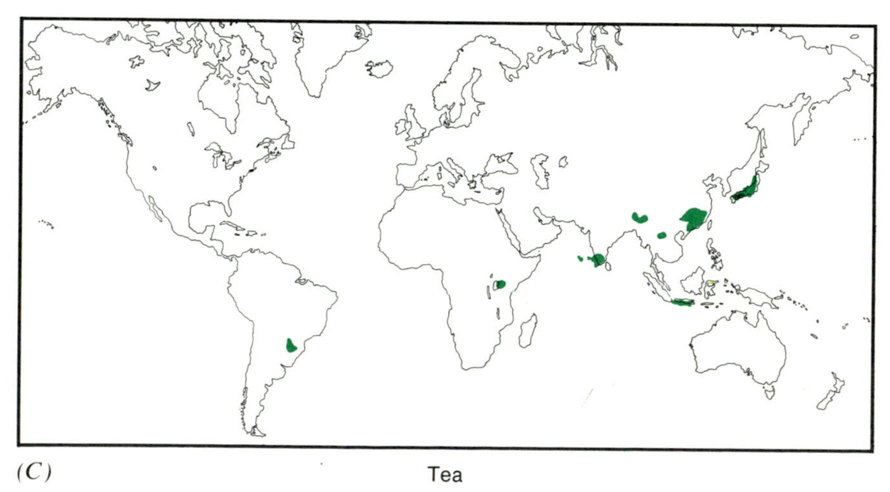

(C) Tea

Figure 25-2 Maps showing the areas where the three major caffeine beverage plants are grown: (*A*, Coffee. *B*. Cocoa. *C*. Tea.) (From Edlin, 1967)

The value of growing coffee in the shade is still controversial. Some coffee growers claim the highest quality coffee comes from shaded conditions, but most of the world's coffee is now grown without shade. Under favorable climatic conditions and intensive cultivation (use of fertilizers and with weed and insect control), as coffee is grown in Brazil, Hawaii, and Kenya, high yields are obtained in the open (Fig. 25-3*A*). In shaded conditions, however, the trees have a longer productive life, and the shade prevents formation of laterites from latosol soils (Fig. 25-3*B*). Where water and soil fertility are limiting, the shade reduces transpiration and evaporation, while the leaf fall provides mulch and nutrients. Also the overstory trees generally are legumes with nodules that fix nitrogen (Chapter 23). One commonly used legume, *Erythrina*, is called "mother of coffee" in Java. Furthermore, the variety of the shade trees tends to provide an environment that reduces insect and weed pests and disease.

Coffea arabica is subject to a serious fungal disease, the leaf rust, *Hemileia vastatrix*. This disease is most serious at low altitude and open habitats, and where trees are debilitated by overbearing. It appears as orange or red pustules (hence the name "rust") on the coffee leaves; the fine spores rapidly cause secondary infections and the many resulting pustules, especially under the stomata, push up the epidermal layer and tear open the leaf. In consequence, the leaves wither and dry up and the plants soon die. The disease seems to have

(A)

(B)

Figure 25-3 *A*. Vast plantations of coffee are grown in the open as shown here in São Paulo, Brazil, where high productivity is maintained through use of fertilizers, pesticides, and herbicides. *B*. Smaller plantations, often in mountainous areas such as this one in Costa Rica, often have overstory trees of bananas and leguminous, nitrogen-fixing trees *(Erythrina)* shown here in background.

originated on wild coffee in Ethiopia, where its effects were not very marked. But when it reached Ceylon (now Sri Lanka) in 1869, the planters there, not understanding its seriousness, neglected it until the whole coffee-planting industry collapsed in a few years. In consequence coffee was replaced by tea, and since England had depended on Ceylon for most of her coffee she soon became, even more than before, a tea-drinking country. To this day Ceylon teas are the most popular types in Britain. A later outcome of the *Hemileia* epidemic was a scientific study of the disease, which was the first plant disease study ever undertaken. A few years ago, *Hemileia* was found to have invaded Brazil, but the invasion has been largely stopped by the use of a combination of two fungicides, a dithiocarbamate and a copper compound. Needless to say, the search for resistant races and also for hybrids has been accelerated.

Coffea robusta is much more resistant than *Coffea arabica* to the *Hemileia* and has accordingly been planted by the colonists in Kenya, where there is now large-scale production; a number of *robusta x arabica* hybrids having moderate resistance have been developed too. However, *C. robusta* is considered less aromatic than *C. arabica*.

Coffee flowers (Fig. 25-4) appear in flushes three or four times a year and are pollinated by insects. When the fleshy fruit ripens to a deep red, it is usually handpicked. It is processed in two ways. In the older dry process used in the Near East, the berries are immediately dried in open areas and the pericarp is mechanically rubbed off. In the wet process, the fruits are thrown into pools where the bad ones float and are easily separated from the good ones. A pulping machine then removes the fleshy pericarp exposing an inner "parchment" that holds two grey-green seeds together (Fig. 24-4). The seeds are then "fermented" and agitated in open trays or large tanks covered with water, to remove the last remains of the pulp, and next spread in the sun to dry. Subsequently, the parchment is rubbed off causing the seeds to separate from each other; they are polished and packed for export.

In either case, the coffee "bean" (seed) is then ready for roasting, which develops the flavors. The roasting caramelizes the glucose and dextrins, which greatly influences the flavor. Coffees from different geographic localities, however, have characteristic flavors that may result from the essential oils, loosely called "caffeol." Caffeine, responsible for the stimulatory effect, occurs only in amounts of 1 to 2% of the dry weight. Caffeine-free coffees are made by solvent extraction, and for this process *C. robusta* is preferred to *C. arabica*.

Various grains such as wheat, rye, and barley when roasted and ground have somewhat the same odor and taste as coffee. By use of these, one can prepare a coffee substitute at low price. Other ingredients in coffee substitutes may be acorns, oak bark, peas, bran, peanuts, or sawdust! In Germany during World War I roasted lupin seeds were used. Chicory root, from *Cichorium intybus,* a composite resembling lettuce, is sometimes roasted and mixed with coffee, as in France and New Orleans.

b. CACAO: "FOOD OF THE GODS"

Chocolate is a seed crop of ancient cultivation in Central America where the Indians believe it to be of divine origin. Their views led Linnaeus to name it *Theobroma*—"food of gods." The specific name is *cacao,* derived from a Mayan dialect where the plant was known as cacahoatl and the drink prepared from the seed was chocoatl. Cacao is the common name for the plant; "cocoa" and chocolate in English-speaking countries are names used for the products. Cacao is a member of the tropical family Sterculiaceae, to which *Cola* also belongs. It probably reached the Mayan and Aztec civilizations from the Brazilian Amazon region. From there it was taken to the Pacific slopes of the Colombian Andes and finally into Central America. The Aztecs so highly

Figure 25-4 *A*. Coffee *(Coffea arabica)* flowers and fruit. *B*. Fruit in section and seeds or "beans" of *C. arabica*. *C*. Beans of *C. Coffea robusta*. (From Masefield et al., 1969)

valued cacao that they used the bean as currency. Montezuma had storehouses of it that he had received for taxes. Cacao was taken to Europe as a curiosity by Columbus, but the Spanish (although they changed its Indian name to *chocolatl* to make it more easily pronounced by Europeans) did not appreciate the Indian method of preparation. In this method, the roasted cacao bean was ground with corn (to absorb the fat) and then boiled in water to which capsicum peppers were added. Although the Indians also substituted vanilla and honey for the peppers, the fat content was still too high for the taste of the Spanish. However, when the Dutch began to grow cacao in Curaçao, C. J. Van Houten in Holland developed a process to remove the fat (cocoa butter), making the beverage more flavorful to Europeans. Cocoa beverage became so popular in France and England that chocolate houses appeared alongside coffeehouses.

The Spaniards introduced the cacao tree into the Philippines in the seventeenth century, and the Dutch carried it to Ceylon and Indonesia (Fig. 25-2). All of these are producing areas today. In South America the growing of cacao was confined to Venezuela by Spanish edict, until their influence waned at the end of the eighteenth century. Because of its Portuguese rather than Spanish heritage, Brazil at the end of the nineteenth century was the largest exporter of cacao in the world. Brazil did not, however, develop large-scale production until after 1888 when slavery was abolished. Cacao was grown by slaves freed from sugar plantations, who were given small plots of land. Until 1900, 81% of the world's cacao came from the New World, principally Brazil. By 1951, however, 60% came from West Africa (35% from Ghana) as a result of the humanitarian efforts of a prominent English Quaker chocolate manufacturer (George Cadbury), who supported a peasant, small-holding production similar to that in Brazil following abolition of slavery. The results were even more successful than those in Brazil, but now most of Brazil's production comes from larger land holdings. In fact, cacao was sufficiently successful in Africa to provide the economic basis that enabled the Gold Coast to be freed as a British colony and become a member of the Commonwealth (Ghana) ahead of many of her neighboring African states. In Ghana the marketing of cacao is done by the government, which uses the difference between the fixed price paid to the farmers and the sales revenue to build roads, schools, and hospitals.

The cacao tree grows natively along river banks in the lowest understory of the Amazonian rain forest. It is essentially an equatorial plant, and, in contrast to coffee, does best where there is no dry season. Like coffee it is frequently cultivated under shade of various leguminous trees and bananas. Although yields can be increased by reducing the shade (since light is the limiting factor for photosynthesis in this case), it must be accompanied by increased fertilization, stepped-up control measures against insects and disease, and irrigation if the water-holding capacity of the soil is poor. Thus the relative values of increased productivity of open sites must be weighed against those of lower soil fertility, possible moisture deficiency, and increased incidence of disease and insects. Presently, in most areas it is more economical to keep the shade, even though there is higher yield in the sun.

Cacao is subject to numerous diseases, one of which is "black pod," a fungal parasite (*Phytophthora palmivora*) that results in black spots in the fruit, leading to rotting. This fungus belongs to the same genus as the fungus that causes potato blight (Fig. 29-4*B*). Also the witches'-broom disease, caused by the fungus *Marasmius perniciosus*, may cause up to 50% mortality in plantations of cacao in its native Amazonian habitat. Fortunately this disease does not occur in Africa and coastal Brazil, which account for 97% of the world's cocoa production today. Swollen shoot is another serious disease, particularly in Ghana. This situation of a tropical plant's meeting with a disease not present in its native land, after being transported to another part of the world, is a reversal of that usually encountered. In other words, the crop plant removed from its area of origin commonly does better, at least under plantation cropping, by escaping the diseases that have evolved along with it.

The flower of cacao grows out of the trunk of the tree, a behavior called *cauliflory* ("flowering from stem"), which is a common feature among tropical plants (Fig. 25-5). The sight of fruits hanging from the trunk of the tree is surprising for most people living in temperate areas. It has been suggested that this type of flowering, which is common in lower-strata trees in the rain forest, occurs to attract moth and butterfly pollinators that move in these strata. When the fruit develops it is carefully handpicked so that another flower will develop from the "cushion." The seeds are then freed from the pulp and put into vats for fermentation (Fig. 25-6*A*). The combined effects of the resulting alcohol and acetic acid, and relatively high temperatures, kill the cotyledons. Enzymes attack the anthocyanins in the red seed turning it to the characteristic deep brown color. Astringency is lost, and the aroma and flavor gradually develop. Then the seed is ready for drying, often in the sun (Fig. 25-6*B*). Further processing, after arrival at a chocolate factory, consists of roasting and pressing the kernels to extract the oils and fats ("cocoa butter"), which make up about one-half of the materials in the seed. Other constituents are *theobromine* (Fig. 25-1), carbohydrate, and protein. Theobromine is extracted from the discarded fruit husks and methylated to produce caffeine for medicinal use. To make cocoa powder about one-half of the fat is removed in a filter press, and then the hard cake is pulverized. In making chocolate candies producers add back about 16% of the cocoa butter to make the mixture malleable for casting into bars. The cocoa butter is solid at room temperature but

(*A*)

(*B*)

(*C*)

Figure 25-5 *A*. Flowers of cacao growing from trunk of plant (cauliflory). *B*. Cacao flower greatly magnified (member of family Sterculiaceae). *C*. Fruits, which develop from these small flowers, clinging to trunk of the small understory tree. This highly productive plant is from plantations of the National Cocoa Institute in Bahia, Brazil.

(A)

(B)

Figure 25-6 Preparation of the cacao seed for sending to the chocolate factory. *A*. Seeds are freed from the pulp of the fruit and put into vats for fermentation where the combined effects of the alcohol, acetic acid, and relatively high temperatures lead to the development of the chocolate aroma and flavor. *B*. Following fermentation, the seeds are dried, often in the sun as shown here in Bahia, Brazil.

melts at body temperature (as every child knows). Sugar, milk solids (for milk chocolate, a product devised by the Swiss), and flavoring are added.

3. TEA: THE UNIVERSAL STIMULANT

Tea is drunk by at least one-half of the world's population. The art of growing tea is at least 4000 years old in the Orient, where it was first used as a medicinal. Later, tea drinking there became a social custom, and from China it spread over the world. A soldier attached to the East India Company found tea bushes growing wild near the China–India border and brought it to the attention of the company, which then developed tea planting in India. Of course, many plants around the world are either collected or grown to be brewed as "teas"—usually because the leaves contain either phenolic or terpenoid components, sometimes accompanied by caffeine alkaloids. What generally is called "tea" (without any reference to the kind of plant from which it is derived) is produced from *Camellia sinensis* (Theaceae). The flower, which we usually do not see in the package of leaves, looks like a simplified flower of the ornamental *Camellia* (Fig. 25-7).

The exact origin of *Camellia sinensis* is not known, although the narrow-leaved form may have originated in southwestern China and the wider-leaved form in northeastern India, Burma, or Thailand. The English word "tea," originally pronounced "tay," comes from the Chinese "te." It was first introduced to Europe in 1610 by the Dutch East India Company and into England in 1664. Although the Dutch are still great tea drinkers, England is now the greatest tea importer. As noted above, tea displaced coffee as the favorite drink in England following the destruction of the coffee in Ceylon (now Sri Lanka) by disease and its replacement by tea plantations (Chapter 29). Tea grows in a wider range of climates than either coffee or cocoa (i.e., an equable moist, subtropical or warm temperate environment), but it requires at least 150 cm of well-distributed rains. China is the greatest tea producer and consumer; most tea there is grown on small plantations for local use. In Japan, India, and Sri Lanka, tea is grown on large plantations for both home consumption and export (Fig. 25-8*A*).

In the plantations, selected leaf clones are started by rooting cuttings in nurseries. These cuttings are transplanted and left to grow unpruned until the plant reaches about 1 m in

Figure 25-7 Flowers of tea *(Camellia sinensis)*. Note the simplified flower of the familiar garden *Camellia.*

height. Then the plant is pruned back each year into flat-topped bushes for easy "plucking," which consists of removing the terminal bud and two or three leaves immediately below it (Fig. 25-8*B* and *C*). In other words, as some brands of tea advertise, the "tender tea" leaves are picked. Plucking these terminal buds stimulates growth of lateral buds (see Chapter 10), which allows harvesting every two weeks. After about 10 years the plants are cut back to the ground and suckers allowed to replace the old bush. Then another cycle of plucking begins after 4 to 5 years of growth. In some areas the plants are grown in partial shade of leguminous trees, as are coffee and cocoa, but more commonly they are grown in the open.

After harvesting, the leaves are treated to produce either of two basic kinds of tea—black or green. The most important constituents in tea leaves are the polyphenolic compounds (about 25%) and caffeine (2.5 to 4.5%). Commercial caffeine, in fact, is obtained from damaged tea and tea wastes. There are traces of essential oils ("theol") which provide some of the aroma, as well as fluoride and riboflavin. The polyphenolics are tanninlike compounds, derivatives of gallic acid and catechin, and occur in the vacuole of the cell. Considerable variation occurs in the polyphenolic content of the leaves depending upon (1) age of the leaves (more polyphenols just following a "flush" of growth), (2) season (higher during the dry season), (3) shade (less with slow growth), and (4) the particular race or strain.

The aroma and flavor of black tea, which constitutes about 75% of exported teas, is due to the oxidation of the polyphenolic compounds as well as to the theol. To accomplish this oxidation, the fresh leaves are withered for 24 hours in well-ventilated indoor racks. The changes that take place here are not well understood but the phenolic compounds are known to be more easily extracted thereafter. The withered leaves are then rolled, breaking the cells and enhancing the enzyme (phenol oxidase) action, in oxidizing the catechins to quinones. The leaves are then spread in cool, moist rooms where the brown-black color and pleasing aroma develop. They are then dried to reduce moisture content to 3% to discourage growth of microorganisms. Black tea is graded by the size of the leaf, being sifted through progressively smaller sieves. The grades have nothing to do with quality or flavor; they are merely a matter of commercial convenience. The first three siftings winnow out the "leaf grades": souchong (the largest leaf), pekoe, and orange pekoe. Subsequent siftings yield the "broken grades," which are important in making tea bags, since they yield up their color and flavor faster than the leaf grades.

For green tea the enzymes responsible for oxidation of the polyphenolic compounds are killed by either steaming the leaves in drums or immediate partial drying in hot pans. Because the oxidation of the polyphenols is prevented, the leaves remain green and there is no "typical" tea aroma (i.e., that of black tea, which is more familiar to people in the Occident). Green tea is consumed primarily in China and Japan, being frequently considered "insipid" by other people.

(A)

(B)

(C)

Figure 25-8 *A*. Large tea plantation in Sri Lanka, one of the world's greatest producers of tea; tea is often picked by hand and the tea plants are pruned to flat-topped bushes for easy "plucking". *B*. Tea being harvested in Indonesia. *C*. The tea harvest comes mainly from the "tender" young growth comprising the terminal bud and two or three leaves immediately below it.

There are also semioxidized types of tea such as Formosa oolong or jasmine (contains jasmine flowers), both of which undergo preliminary sun drying.

Beyond types and grades there are blends made from one or more types and grades. For example, Earl Grey is a black tea blend enhanced with oil of bergamot, a citrus-rind oil often used in perfumery. Darjeeling is usually a blend of highly flavored light, mellow teas from the Darjeeling region of India. Lapsang souchong comes mostly from Taiwan; its distinctly smoky flavor and aroma come from the curing process. Russian tea is sometimes a blend of all-black teas or of black and green teas.

4. OTHER CAFFEINE BEVERAGES OF LOCAL INTEREST

a. YERBA MATÉ

Yerba maté is a universal drink for millions of South Americans, actually only following tea, coffee, and cocoa in consumption (Fig. 25-9). It is made from the leaves of *Ilex paraguariensis* (the holly family, Aquifoliaceae), which grows as a small evergreen tree or shrub in the open forests of southern Brazil, Paraguay, Uruguay, and Argentina. The leaves were used by the Guarani Indians of Paraguay before South America was settled. It can readily be grown from seed, with the first crop ready in a year, although the best yields are obtained from older plants.

Figure 25-9 A yerba maté *(Ilex paraguariensis)* tree growing in open Parana pine forests in southern Brazil. The slender shape of the tree results from repeated cutting of branches to obtain leaves for roasting to make tea.

The small, leafy branches are cut and toasted over a fire. They may then be finally dried in ovens. The "double toast" is especially appreciated by North Americans. The leaves contain some polyphenols and caffeine but far less of both than tea. Hence, it is less astringent and also less stimulating than tea.

b. COLA

Although most of the flavoring and stimulants added to modern cola drinks are now derived from other sources, seeds from the cola trees (*Cola nitida* – Sterculiaceae) native to the west African countries of Ghana and Nigeria were initially used and are responsible for the name for today's popular soft drinks. The cola seed (Fig. 25-10*A*) is extremely rich in caffeine with traces of theobromine and a glucoside, kolanin, which is a heart stimulant. In West Africa the cola "nut" is chewed or may be boiled or pulverized to make a beverage that inhibits fatigue and forestalls hunger. There also it is important in social and religious life.

c. GUARANÁ

Guaraná is the "cola" of Brazil. It comes from the seed of the Amazonian trailing shrub or woody climber *Paullinia cupana,* in the tropical family Sapindaceae (Fig. 25-10*B*). Local people in the Amazonian region grind and pulverize the seeds with water, mix this powder with cassava flour, and mold it into a paste. Dried in smoke, it lasts for years. It may be grated and added to water (which may be carbonated, making the popular soft drink of Brazil). The caffeine contained in one-half teaspoon in a cup of water is equivalent to at least three cups of strong coffee. It also contains some polyphenols and essential oils that give it a bittersweet flavor. It is now being exported to other countries.

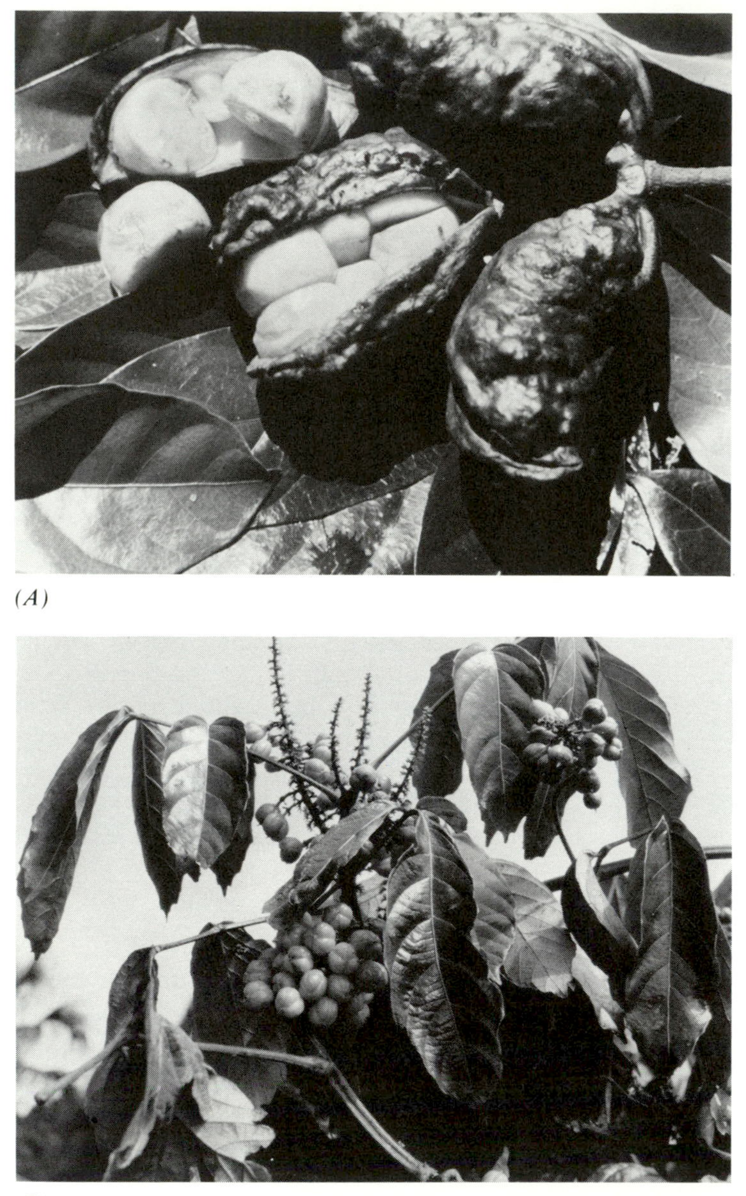

(A)

(B)

Figure 25-10 *A*. Fruit of the cola tree *(Cola nitida)* opened to show the "cola nut" (seed). *B*. *Guaraná* plant *(Paullinia cupana)*, a woody climber from Amazonia with inflorescence showing fruit. The seeds of both of these plants are particularly rich sources of caffeine.

C. PLANTS GROWN FOR THEIR TERPENES AND FATS

CHAPTER 26

Terpene Derivatives: Spices, Perfumes, Resins, and Rubber

The terpenes are the largest and most structurally varied group of products of secondary plant metabolism. An outstanding feature of plants is their ability to produce an almost endless variation on a single chemical theme, and this is beautifully exemplified by the hundreds of terpene derivatives ("terpenoids") derived from a simple C_5 "isoprene building block."

1. BIOSYNTHESIS OF TERPENOIDS

Terpenoids, like the fatty acids (precursors to fats and oils), derive their carbon atoms from fresh acetyl coenzyme A. This in turn is produced from pyruvate, which is only a "few steps" from the product of photosynthesis (Fig. 26-1). Through condensation of three acetyl coenzyme A molecules, the branched 6-carbon mevalonic acid is produced. Mevalonic acid is then phosphorylated through three reactions with a final decarboxylation to form an "activated isoprene" (C_5) unit, *isopentenyl pryophosphate* (IPP). Two molecules of IPP, one of which has been isomerized to a slightly different structure, then condense to form the 10-carbon compound geranyl pyrophosphate (abbreviated geranyl-pp), from which the numerous C_{10} monoterpenes are derived by losing

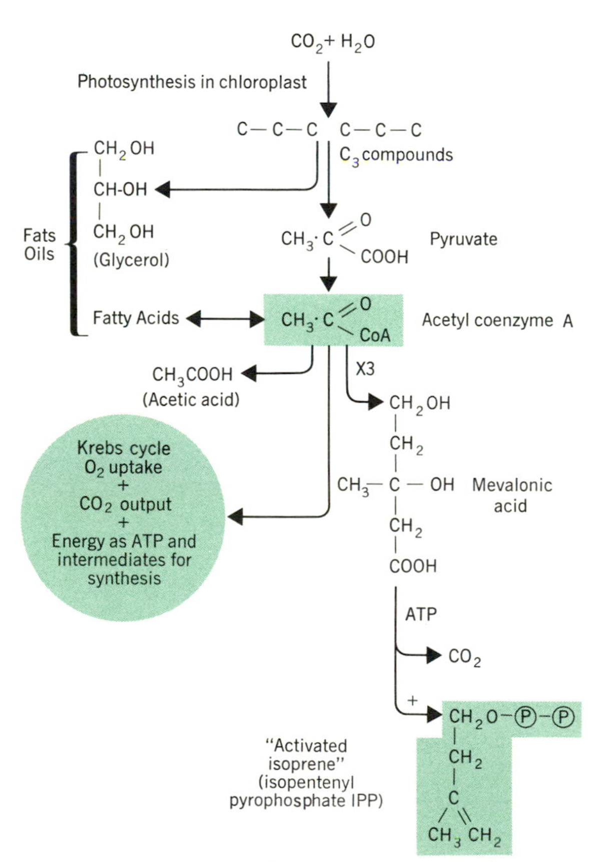

Figure 26-1 Source of carbon for synthesis of the terpenes, and fats and oils.

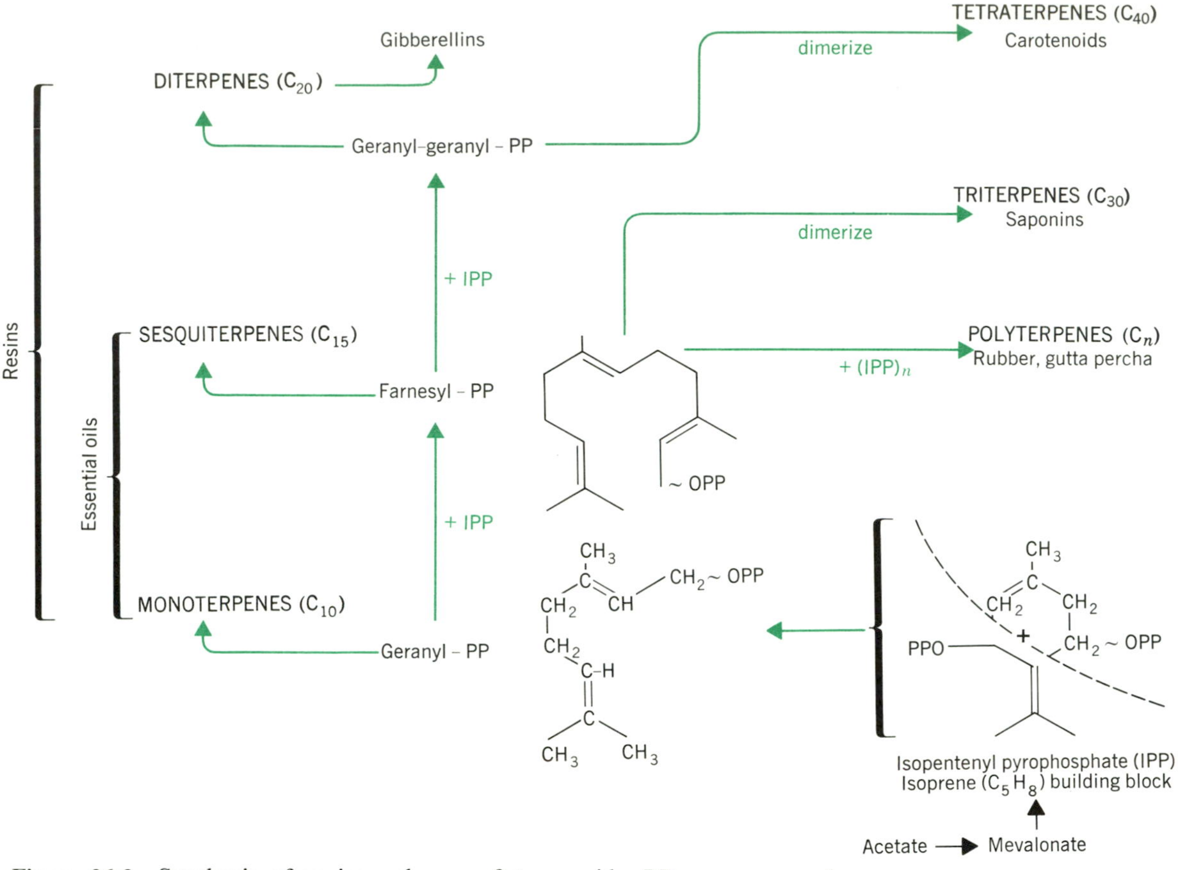

Figure 26-2 Synthesis of various classes of terpenoids. PP means pyrophosphate.

the pyrophosphate group (Fig. 26-2). With the combination of geranyl-PP and another IPP molecule the C_{15} farnesyl-PP, precursor to all sesquiterpenes, is produced. An addition of another IPP unit to farnesyl-PP, or the dimerization of geranyl-PP, forms the precursor (geranyl–geranyl-PP) to the C_{20} diterpenes. Thus each additional IPP adds a 5-carbon unit as shown in Fig. 26-2. Present experimental evidence indicates that mono-, sesqui-, and diterpenes and their derivatives are derived by polymerization of IPP, but the kind of control of this condensation remains unknown. In some cases environmental factors appear to modify these conversions.

The further polymerization of IPP might be expected to give rise to the triterpene (C_{30}) series, but there is no indication that a succession of head-to-tail condensations can give rise to more than four residues (C_{20}). Therefore, dimerization of farnesyl-PP is needed to produce C_{30} triterpenes, and repeated polymerization of IPP units results in polyterpenes; a similar dimerization of geranyl-PP leads to C_{40} tetraterpenes.

These units may be in chains, or cyclized into one or more rings and then reduced, or oxidized into a large variety of derivatives such as alcohols, ketones, aldehydes, and acids (Fig. 26-3). There are almost infinite ways in which

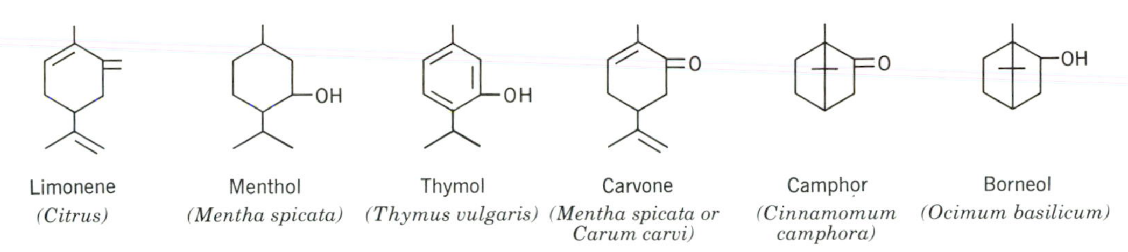

Figure 26-3 Some common monoterpenes. Limonene is a hydrocarbon; Oxygenated compounds include the alcohols (menthol, thymol and borneol) and the ketones (carvone and camphor).

the basic structures may be slightly altered, producing compounds with different physical and chemical properties.

2. THE ROLE OF TERPENOIDS IN PLANTS

Biochemists have traditionally held the view that essential oils were waste products that are formed slowly and irreversibly. Recent evidence, however, has shown dynamic metabolism in perppermint, basil, rose, and geranium. In leaves of peppermint exposed to labeled $^{14}CO_2$, the monoterpenes incorporated the ^{14}C and then lost a large part of the label without any corresponding change in the total amount of monoterpenes present. The essential oils in peppermint are a complex mixture of structurally related monoterpenes, and these labeling studies have shown that one compound has been converted to another. It has also been suggested that these monoterpenes may be used as substrates for energy metabolism when more suitable stored substrates (presumably starch or sugars) become depleted within the secretory cells.

No biochemical functions have yet been reported for monoterpenes. The sesquiterpenoids abscisic acid and xanthoxin have plant regulatory functions (Chapter 10). Although there are few obvious physiological functions of others in the vast array of mono- and sesquiterpenes, it is beginning to be shown how ecologically important these penetrating odors and flavors are in attracting, repelling, or inhibiting the growth of a variety of organisms. Among the diterpenoid (C_{20}) compounds one case has great physiological importance: the gibberellins; these are a group of related compounds having a C_{19} or a C_{20} skeleton derived from the common diterpene kaurene (Chapter 10). Phytol, a diterpenoid, is a constituent of the chlorophyll molecule. Complex mixtures of mono-, sesqui-, and di-, (and sometimes tri-) terpenoids, useful to many plants ecologically as a defense against insects and microorganism attacks, are called *resins*.

It should be pointed out that in general parlance "resin" is often used for *any* "sticky exudate" which is insoluble in water but soluble in organic solvents, and which hardens upon exposure to air. Some plants exude sticky phenolic compounds to the surface of leaves which subsequently harden—and thus they too are often called "resins."

The best known triterpenoid (C_{30}) compounds are the sterols, which are present in many plants. The sterols have been implicated in the structure of animal membranes, and they are known to be important for biological activity (both vegetative and reproductive) in fungi.

Carotenoids, C_{40} terpenoids, are found associated with chlorophylls in all organelles in which photosynthesis takes place and, as discussed in Chapter 5, are thought to be independently important in this key process. Some carotenoids appear as yellow and orange pigments in many organs of the plant, which would also provide cues for both pollinating and dispersal agents.

The polyterpenoids include rubber and gutta-percha, in which the isoprenoid units are linked together in long unbranched chains. They are usually excreted as a component of latex and are not known to have any metabolic roles. However, like resins they may act as insecticides, fungicides, or bactericides.

3. OUR USES OF ESSENTIAL OILS

The lower (C_{10} and C_{15}) volatile mono- and sesquiterpenes have been known and isolated (although seldom as pure compounds) from plants for a variety of human purposes since pre-Christian times. The volatile fragrance, or essence, which made them evident and at the same time readily obtainable by simple distillation of leaves, flowers, and wood, resulted in their being called "essential oils." A large number of plant essences are known and used for flavoring of food (herbs and spices) and as perfumes, medicinals, and other products. Although many essential oils are terpenoid compounds, there are other "essences" that are either phenolic compounds or a mixture of phenolic compounds with terpenoids.

All cultures have used essential oils, and our eating habits are largely governed by the flavor of the food we select. Although three of the five senses are involved in flavor appreciation, smell is the most important sense in savoring the "essences." Generally no single compound is totally responsible for the flavor of any particular plant material, although very important single ones can be isolated. For example, vanillin is sufficiently characteristic of vanilla that artificial vanilla is made of this predominant compound. However, extract from the vanilla bean tastes slightly different from artificial vanilla because of the small amounts of other essential oils and compounds present. In some plants two predominant compounds are needed; for example, thymol and methyl anthranilate in correct proportions give the characteristic aroma of mandarin oranges. On the other hand, in some cases such as chocolate and coffee, years of painstaking work have resulted in identification of over 700 compounds but no single one is responsible for the flavor.

Despite many attempts to correlate chemical structure with flavor sensation, no guiding principles have yet emerged. Compounds with widely divergent molecular structure are similar in flavor, and the converse is also true.

Although we have appeared to use the essential oils more for pleasure than need, they may have served more fundamental roles than we generally give them credit for. For example, essential oils from certain herbs are known to be germicidal (e.g., oregano, cinnamon, clove, camphor, and garlic). Mustard and clove oils have been used as fungicides. Also certain components function more effectively in combination with others. For example, thymol is fungicidal but its effects may be prolonged by adding terpineol and pinene.

Essential oils occur in most organs of the plant but are used most commonly from leaves, flowers, and fruit. We discuss only a few of the more important ones used commonly today as well as those that were sought after in the spice trade through several centuries and thus influenced the course of history.

4. ESSENTIAL OILS USED FOR FLAVOR

a. THE HERBS

The most common plants whose essential oils we use as "herbs" come primarily from the temperate-zone mint and parsley families (Lamiaceae and Apiaceae; Table 26-1). Before tropical spices became available at affordable prices, Europeans used a variety of herbs to

TABLE 26-1

Herb Plants

Mentha piperita (leaf)	Peppermint	Lamiaceae or Labiatae (Mint family)
M. spicata (leaf)	Spearmint	
Satureja hortensis (leaf)	Savory	
Marjorama hortensis (leaf)	Marjoram	
Ocimum basilicum (leaf)	Basil	
Rosmarinus officinalis (leaf)	Rosemary	
Thymus vulgaris (leaf)	Thyme	
Salvia officinalis (leaf)	Sage	
Petroselinum crispum (leaf)	Parsley	Apiaceae or Umbelliferae (Parsley family)
Anthriscus cerefolium (leaf)	Chervil	
Apium graveolens (seed)	Celery	
Coriandrum sativum (seed and leaf)	Coriander	
Carum carvi (seed)	Caraway	
Anethum graveolens (seed)	Dill	
Cuminum cyminum (seed)	Cumin	
Pimpinella anisum (seed)	Anise	
Artemisia dracunculus (leaf)	Tarragon	Asteraceae or Compositae (Sunflower family)

flavor an otherwise dull diet. The Roman recipes of Apicius give great emphasis to herbs. In the Middle Ages herb gardens were tended by monks, who carried the herbs from one country to another. Noblemen grew herbs near their castles, and even the peasants had their own small gardens. Although gourmet cooks today often grow their own herbs, many are grown by modern large-scale agricultural techniques.

In the mint family, essential oils are formed generally in specialized epidermal glands or hairs in the leaves. The oils occur between the cell wall and cuticle of the epidermal hair, where the slightest breakage of the cuticle permits volatilization. Thus many of these herbs are used as dried, crushed leaves, or the essential oil is obtained by steam distillation. In the parsley family the oils occur in epithelial cells lining small pockets or canals in the leaves (e.g., parsley and chervil). The oils appear in the seed of caraway, dill, and anise (Table 26-1).

Table 26-1 lists some of the most important members of the mint family used as herbs in flavoring food. Many of these are illustrated in Figure 26-4. Interestingly, most of these species (excepting the mints, *Mentha,* themselves) grow natively in Mediterranean chaparral (shrub) vegetation. For this reason, they can be grown as a crop in other Mediterranean-type climates such as California. Also it is in plants growing in these semiarid types of habitats that essential oils have been shown to inhibit germination of the seeds of neighboring plants (*allelopathy*).

Peppermint (*Mentha piperita*) and spearmint (*Mentha spicata*) are the two most widely used mints (Fig. 26-4). Both species are familiar home garden plants but their greatest use is for flavoring and medicinals. Distillation of oil of peppermint was known to the Egyptian pharaohs only a few centuries after Christ, but there was little commercial demand for peppermint until the twentieth century. Menthol (Fig. 26-3), the most abundant constituent of peppermint oil, is widely used in lotions, antiseptics, dentifrices, and cigarettes. Commercial menthol, however, is mainly derived from Japanese mint (*M. arvensis*), whereas that used for flavoring comes primarily from *M. piperita* grown either in the Midwest or the west coast of the United States. Mint became an important agricultural crop in the mucklands of southern Michigan and northern Indiana at the turn of this century. However, *Verticillium* wilt discouraged its cultivation there and now the greatest production in America is in the Columbia Basin of Oregon and Washington where it is less subject to this fungal disease. Although commercially grown peppermint derives from clones of *M. piperita* from Surrey, England, and thus has the same genetic background, the flavor and aroma differ between these regions (and from other localities where they are grown). Work done under controlled conditions has shown that diurnal differences in temperature have an effect due to alteration in the balance of daytime photosynthesis and nighttime utilization of photosynthate. Since at night the photosynthate is used for respiration, on warm nights the respiration rate is accelerated, resulting in oxidation of the monoterpenes. Diurnal fluctuations in composition are known for other herb plants such as sage, and some perfumed flowers such as jasmine produce their essential oils only at night.

Other herbs in the mint family (see Fig. 26-4) commonly used for culinary purposes are basil (*Ocimum basilicum*), marjoram (*Marjorama hortensis*), rosemary (*Rosmarinus officinalis*), thyme (*Thymus vulgaris*), savory (*Satureja hortensis*), and sage (*Salvia officinalis*). Many of these herbs have a number of monoterpenoids in common such as borneol or thymol (Fig. 26-3), but the distinctive flavor is due to different quantities of these or unique combinations with other oils. In some cases one particular component predominates, such as borneol in rosemary and carvacrol in savory.

The same situation with regard to terpenoid components exists in the various members of the parsley family used for seasoning. Although many of the more common terpenoids appear as constituents, each herb is usually characterized by a predominant component. For example, cumin (*Cuminum cyminum*) is characterized by cumaldehyde, anise (*Pimpinella anisum*) by anethole, coriander (*Coriandrum sativum*) by coriandrol (= d-linaloöl), and celery seed (*Apium graveolens*) by a combination of d-carvone and smaller amounts of d-limonene (Fig. 26-3).

b. THE SPICES

The difference between "herbs" and "spices" is not absolutely clear-cut, but gener-

(A)

(B)

(C)

(D)

(E)

(F)

Figure 26-4 Some of the most commonly used herbs: *A*. Peppermint. *B*. Sweet basil. *C*. Rosemary. *D*. Thyme. *E*. Tarragon. *F*. Dill. *A-D* are members of the mint family; *E*, the sunflower family and *F*, the parsley family. Compare with Table 26-1.

ally herbs are herbaceous temperate-zone plants whereas spices are derived from tropical trees. There are more exceptions to the rule in the case of spices than herbs. For example, bay is a temperate tree and ginger is a tropical herb. Some of the most important spices prized throughout time and greedily sought after in the spice trade contain essential oils, but the most characteristic component may be either a volatile phenolic compound or an alkaloid.

The Spice Trade and Its Place in History. It is difficult to overestimate the role that spices have played in world history. There are few parallels for sheer romance, even to the search for gold and gems. Over the years several countries tried to monopolize the trade of these predominantly tropical commodities to European civilization. The first monopoly, established by the Arabians, was based upon keeping the Far Eastern source of these tropical treasures secret. During the Greek period spices came from India to Greek-controlled ports along inland trade routes. Interest soared higher during the heyday of the Roman Empire, when the Romans hit upon the advantage of sea routes using the "tricks" of monsoonal winds i.e., sailing ships' being carried by east-to-west prevailing winds from April to September and from west to east from October to March.

While Alexandria, from 641 to 1046, was in the hands of the Arabs, spices again became scarce in Europe due to greater costs than during the Greek and Roman periods. Until the time of the Crusades, which began near the close of the eleventh century, herbs from local gardens were the primary source of flavoring for food. The Crusades reestablished ties between Europe and the large Arab region surrounding the Holy Land, and thus the interest in spices. The great Venetian and Genoese commercial centers provided a dual monopoly in interplay with the Arabians. The Arabians still maintained secrecy regarding the source of some of the important spices, and these secrets were kept in the shipping and marketing to Europe by the Venetian and Genoese merchants. The following examples give some idea of how expensive spices were in the Late Middle Ages in Europe: "A pound of saffron was worth a horse, a pound of ginger a sheep, a pound of nutmeg seven fat oxen and two pounds of mace a cow." By the middle of the thirteenth century, the two Polo brothers (and the son of one) made their famous journey to the Orient to investigate the distribution of the plants producing spices. Finally it was the son, Marco Polo, who wrote a book while in prison about finding a sea route from Europe to the Orient to circumvent the existing monopolies (Fig. 26-5).

The Portuguese were leaders in this exploration for spices. They started exploring the west coast of Africa; Diaz sailed around the Cape of Good Hope and then Vasco da Gama skirted Africa to get to India in 1497 to 1498 (Fig. 26-6). Soon the Portuguese dominated the sea routes to Borneo and the Molluccas (Spice Islands), and they then set up another stranglehold situation, curtailing trade by native islanders so that no spices were to leave the East Indies except on a Portuguese ship.

Meanwhile, the Spanish were in competition with the Portuguese, and Queen Isabella sent Columbus to find a route to India. Instead, in 1492, he discovered America and the West Indies (Fig. 26-6).

The establishment of the Portuguese monopoly was soon challenged by the British, then the Dutch, and finally the French. In 1577, Sir Francis Drake reached the East Indies by going westward around the tip of South America (Fig. 26-6). England consequently in the following century gained virtual control of India and established the powerful East India Company. During the same period the Dutch had broken Portuguese power in the Spice Islands and neighboring regions producing pepper, cinnamon, cloves, mace, and nutmeg. In an effort to maintain control of these regions, they set up a Dutch East India Company, patterned after the British East India Company, which established an even tighter monopoly over the source of the spices than that of the Portuguese.

(A)

(B)

Figure 26-5 *A*. Marco Polo dictating his memoirs from a prison cell in Genoa, 1298. His accounts of the spices and riches of the Orient stimulated the great age of exploration. *B*. This 15th-century miniature from the Livre des Merveilles depicts the black pepper harvest Marco Polo has decribed in his travels near the Malabar Coast.

The British and French decided to break the Dutch control. They succeeded by smuggling out important spice plants, which were taken to grow in some British and French tropical colonies outside the Far East. Thus today many spices from the East Indies are of even greater commercial importance in the West Indies (where the British took them) or on Madagascar or the Seychelles (where the French took them) than in the native homelands.

There are so many interesting spices that we can only discuss a few that are important in history and still so popular that we now think of them as commonplace.

Ginger and Turmeric. One of the first Oriental spices to become known in Europe was ginger (*Zingiber officinale*) in the Zingiberaceae (a small family akin to the lilies) and only third in importance (behind pepper and cinnamon)

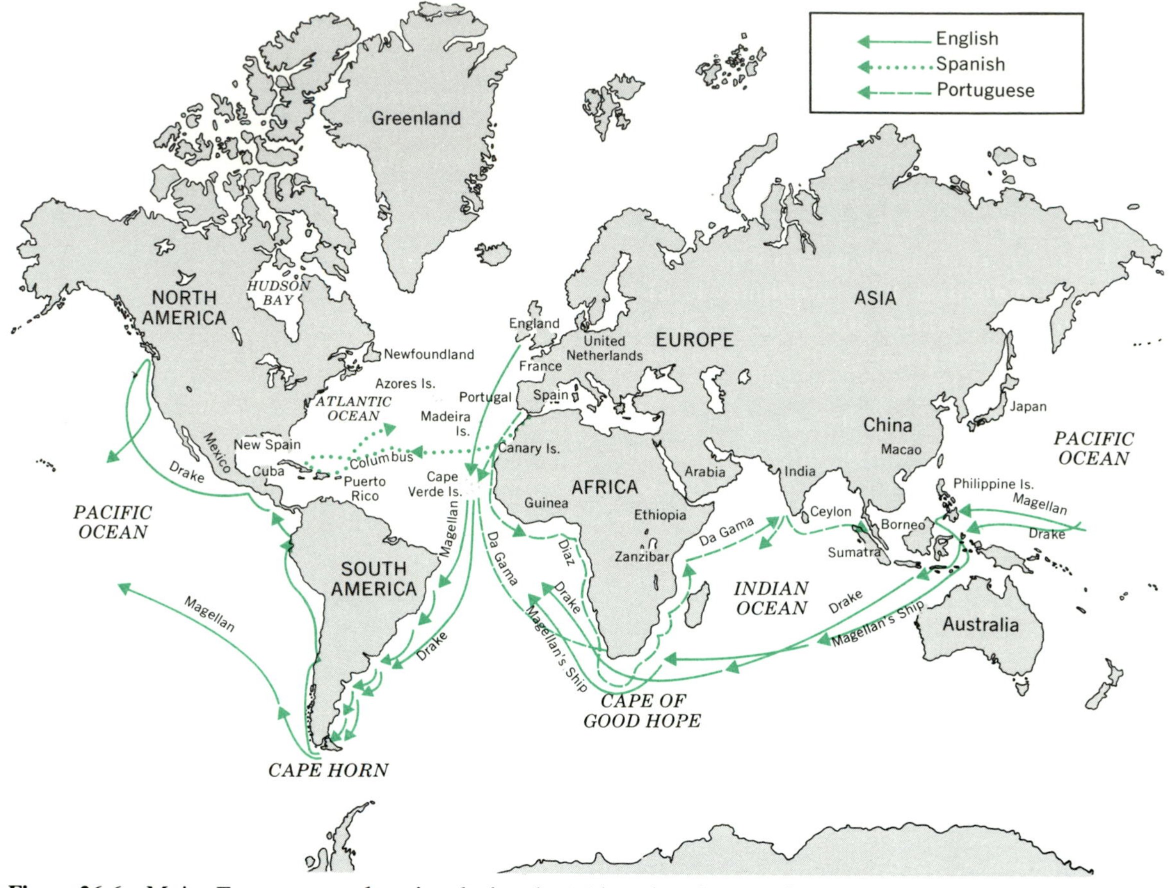

Figure 26-6 Major European exploration during the 15th and 16th centuries, partially in search for spices. Columbus set out from Spain for "the Indies" in 1492 and discovered the West Indies. The Portuguese voyages around Africa were led by Diaz (1486) and da Gama (1497-98). English explorers Magellan (1519-1522) and Drake (1579-80) circumnavigated South America (and the latter also western North America) before heading across the Pacific to the Spice Islands and back via the western coast of Africa.

for centuries. An erect monocot herb indigenous to southern Asia, it was long cultivated by the ancient Chinese and Hindus. Today it is common in most tropical countries but principal exporters are India, Jamaica, and Nigeria. The Latin name "Zingiber" is derived from Sanskrit "singabra," meaning "shaped like a horn"; the shape of the rhizomes does resemble an antler or horn. The tangy, penetrating flavor is due to "ginger oil," which is extracted from the rhizomes. It contains various monoterpenoids, for example zingiberol (producing the characteristic odor) and others. The essential oil is extracted from ground, dried rhizomes to produce ginger ale and ginger beer. Also the green rhizome is crystallized by boiling it in a sugar solution.

Another member of the ginger family, turmeric (*Curcuma longa*), may be used more than any other spice in the Orient because it is the main ingredient of curry powders in India and other eastern countries. It is a wild perennial there but generally is cultivated as an annual crop. Like ginger, it grows from rhizomes,

(A) *(B)*

Figure 26-7 *A*. Black pepper *(Piper nigrum)* vine fruiting in a plantation today in the Amazon Basin. *B*. Peeling bark from young sprout of cinnamon tree in India.

which are dried and ground to give a yellow powder with a strong, musky taste. It also is used as a yellow dye. The main constituent is turmerone, but a number of other common monoterpenoids are present.

Black and White Pepper. The world's most important spice, and one of the earliest items of commerce between the Orient and Europe, is pepper—black and white. Both of these are seeds obtained from berries on a vine of *Piper nigrum* (Piperaceae) (Figs. 26-5*A* and 26-7*B*). Pepper was so valuable in the Middle Ages that rents, dowries, and taxes were often paid with it. Part of the ransom demanded for the city of Rome by Haric the Goth (around 400 A.D.) was 3000 lb of pepper.

The native distribution of *Piper nigrum* is southern India, Ceylon, and Malaysia, but now it is planted in many tropical areas and has become an important commercial crop in the Amazon basin. Peppers with different flavors are known to occur from different geographical areas. Different environments may result in production of a quantitatively different composition of the essential oils, either due to temporal changes or to the development of chemical races in different habitats. The aroma of pepper is due to "oil of pepper," again a complex of monoterpenoids such as pinene, phellandrenes and caryophyllene. The pungent taste is due to a resin, chavicin, and the burning aftertaste to an isomer of it, piperine.

The difference between black and white peppers results from the mode of preparation of the berry (then called the "peppercorn"). Black pepper is picked green and dried until it is shriveled. The black hull is left on and thus when the berry is ground the black fruit wall and white seed parts are mixed. White pepper is made from the fully ripened yellow-red berry, which is soaked in flowing water for about a week. Then the husk or fruit wall is macerated and removed, leaving the white seed.

Capsicum or Red Pepper. The Spaniards arrived in the New World hunting for black pepper but came back in 1493 with multicolored, variously shaped pods of *Capsicum annuum* (Solanaceae), which had been cultivated for several thousand years by the tropical American Indians (Fig. 26-8). Just as the West Indies were named because Columbus was searching for India, so capsicum was called "red pepper" because the Spaniards were hoping to find "pepper." By 1650 *Capsicum annuum*, which is the source of paprika, had spread throughout southern Europe and the African and Asiatic tropics. The compound that controls the flavor is capsaicin, an alkaloid that occurs in the placental region (i.e., where seeds are attached to spongy parts of the fruit wall). Its presence is controlled by one gene. Flavors are modified in different climates and soils. In Hungary the milder forms are used for paprika. In Spain the large green "bell" peppers have

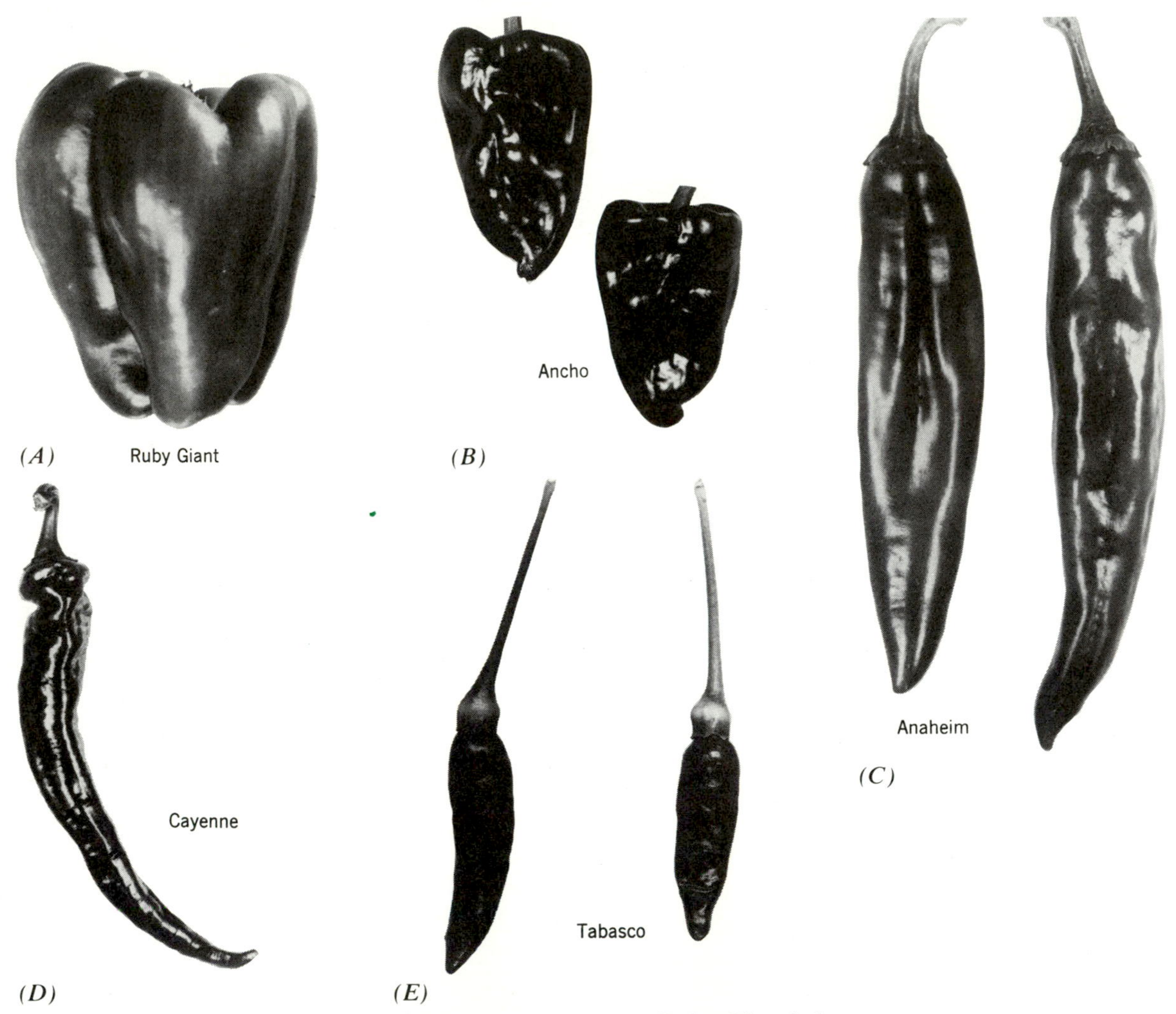

Figure 26-8 Some common types of *Capsicum* peppers. *A*. Ruby Giant belongs to the bell pepper group of *C. annuum*, which is a larger, thick-fleshed mild variety and favorite for salads and for stuffing. *B*. The bland Ancho for chili powder. *C*. The mild Anaheim for paprika. *D*. The Cayenne are hot, pungent peppers used for Cayenne or crushed red pepper. A-D all are varieties of *C. annuum*. *E*. Tabasco, another hot pepper, is a variety of *C. frutescens*.

lost their pungency because capsaicin is absent since both recessive alleles are present in this cultivar. Most of the *chiles* or tabasco types of "hot" peppers are from *C. frutescens,* although in the case of both *C. annuum* and *C. frutescens* the development of many cultivars has led to a confused taxonomy (Fig. 26-8).

Capsicum peppers have very high concentrations of ascorbic acid (vitamin C)—actually, pound for pound, higher than from citrus fruit. In fact, the Hungarian Albert Szent-Györgyi, who won a Nobel prize in 1937 for isolating vitamin C, used paprika peppers and thereby discovered one of the richest sources of it.

Chili peppers are very popular flavoring for foods, particularly in the New World tropics. They not only decrease the monotony of a starchy diet but also increase the flow of saliva, which in turn may stimulate a lagging appetite in hot climates. Additionally, chili pepper may raise the body temperature; this leads to perspiration, which has a cooling effect.

Figure 26-9 Gathering cinnamon *(Cinnamonum zylanicum)* in the 16th century.

Cinnamon. Cinnamon joins black pepper in being one of the two oldest known spices (Fig. 26-9). Cinnamon or "sweet wood" is frequently mentioned in the Bible and has a long history of use during religious rites. Before Christianity it was used to counteract the stench following the offering of burned flesh.

There are two kinds of cinnamon, both derived from species of *Cinnamomum* (Lauraceae). "True" cinnamon comes from the bark of *Cinnamomum zeylanicum,* which occurs natively in southern India and Sri Lanka. Cassia cinnamon, which is what is commonly bought as "cinnamon" in the United States, is from the bark of *C. cassia,* whose home is in South Vietnam and the eastern Himalayas. There is considerable difference in aroma and flavor, the true cinnamon being more delicate, and cassia being stronger, even though the predominant compound in both is a volatile phenolic, cinnamic aldehyde. In Great Britain only the spice taken from *C. zeylanicum* can be called cinnamon whereas that from either species is authorized by the U.S. Food and Drug law.

The spice is obtained in the same manner for both species. Because *C. cassia* is a more slow-growing tree than *C. zeylanicum* it takes a few more years to obtain a harvest of Cassia cinnamon. Although the mature trees of *C. zeylanicum* can attain heights of 10 to about 15 m, in cultivation, for the spice they are cut back two or three years after planting to induce formation of lateral shoots, which form a coppiced bush. With a curved knife the bark is peeled off the shoots in strips. They are harvested during the rainy season when the cambium is active, making it easier to pull off the bark. New shoots will again be ready for harvesting in another two to three years. After the bark has been peeled off, it is left to "ferment" for 24 hours. Then the outer, corky bark is scraped off, leaving the inner bark of phloem, which curls when dried to form the familiar "quill" (Fig. 26-7*B*).

Nutmeg and Mace. Nutmeg and mace, which are both derived from the fruit of *Myristica fragrans* (Myristicaceae), are two important spices with a long history. They were well known to Arabs and Turks during the sixth century and became appreciated in Europe by the twelfth century, attaining particular popularity in England during the fourteenth century. *M. fragrans* is an evergreen tree native to the Moluccas and other islands of the East Indian archipelago but was taken for cultivation by the English, to the West Indies, where both nutmeg and mace are commonly produced today. The East and West Indian grown fruits differ considerably in the piquancy, again probably due to environmental factors influencing the composition of the essential oils.

The fleshy yellow-brown fruit, resembling an apricot (Fig. 26-10), splits in half when ripe, exposing a scarlet, netlike *aril* (an accessory seed covering that is an outgrowth of the ovary), which is the source of mace. Nutmeg is derived from the seed itself. Both spices contain a large number of monoterpenoids; they also contain the highly toxic substance, myristicin, which when taken in excessive amounts can cause fatty degeneration of the liver cells. Although nutmeg and mace contain generally the same components, the quantitative composition is different enough that nutmeg tends to be sweeter and have a more delicate aroma than mace.

Clove. Clove comes from a tree native to the Spice Islands or Moluccas (*Eugenia caryophyllata* in the myrtle family, Myrtaceae, to which *Eucalyptus* belongs). Now the biggest production comes from trees grown on Madagascar and Zanzibar where the French took the plants to break the Dutch monopoly. Today it is used primarily as a food flavoring, although oil of clove has long been utilized as a medicinal.

The word "clove" is derived from the French *clou* meaning "nail," which refers to the dried, unexpanded, nail-shaped flower bud (Fig. 26-11). The buds are picked just before the blossom opens out. Once the flower bud opens, its value as a spice is lost, because a change in the composition of the volatile oils occurs. Eugenol, again a volatile phenolic compound, comprises 70 to 90% of the oils. Other components include monoterpenoids such as limonene, carvone, pinene, and eucalyptol.

Vanilla. Vanilla is the world's most popular flavoring for sweetened food. It comes from a tropical vine orchid (*Vanilla planifolia* and *V. fragrans*) and, other than red peppers, is the only major spice plant that was native to the New World. It occurs in the rain forests of southeastern Mexico, Central America, the West Indies, and northern South America. It is also the only valuable commercial product (excepting the important use as ornamentals) from more than 35,000 species in the family Orchidaceae.

The flavoring is obtained from dried, cured vanilla orchid fruits (Fig. 26-12). The pods or "beans," as they are called, have no flavor when they are picked but the fragrant aroma develops in the curing process, which was known to the Aztecs. Indeed, the beans were sufficiently prized, like cacao "beans," that they were used as a medium of exchange by the Aztecs. For more than three centuries after the arrival of Cortez, Mexico had a monopoly on vanilla, partially because of failure to get the plant to reproduce in the Old World. When it was finally discovered that the specific bee that pollinates the orchid did not occur out of the native range of the plant, techniques were worked out for artificial fertilization. Then vanilla plants were established by the French in Madagascar, the Seychelles, and Zanzibar where much of the world's supply is produced in plantations today.

The vine is grown from cuttings planted close to a supporting tree or stake (Fig. 26-12). The first crop can be harvested in 3 years and the vine is usually abandoned after 10 to 12 years. A healthy vine may bear 100 blossoms, of which only about one-half are usually selected for pollination. Flowering extends over a two-month period but an individual flower only lasts for a day. The beans mature from four to nine months after fertilization.

After harvesting, the time-consuming curing process begins. The beans are first dipped

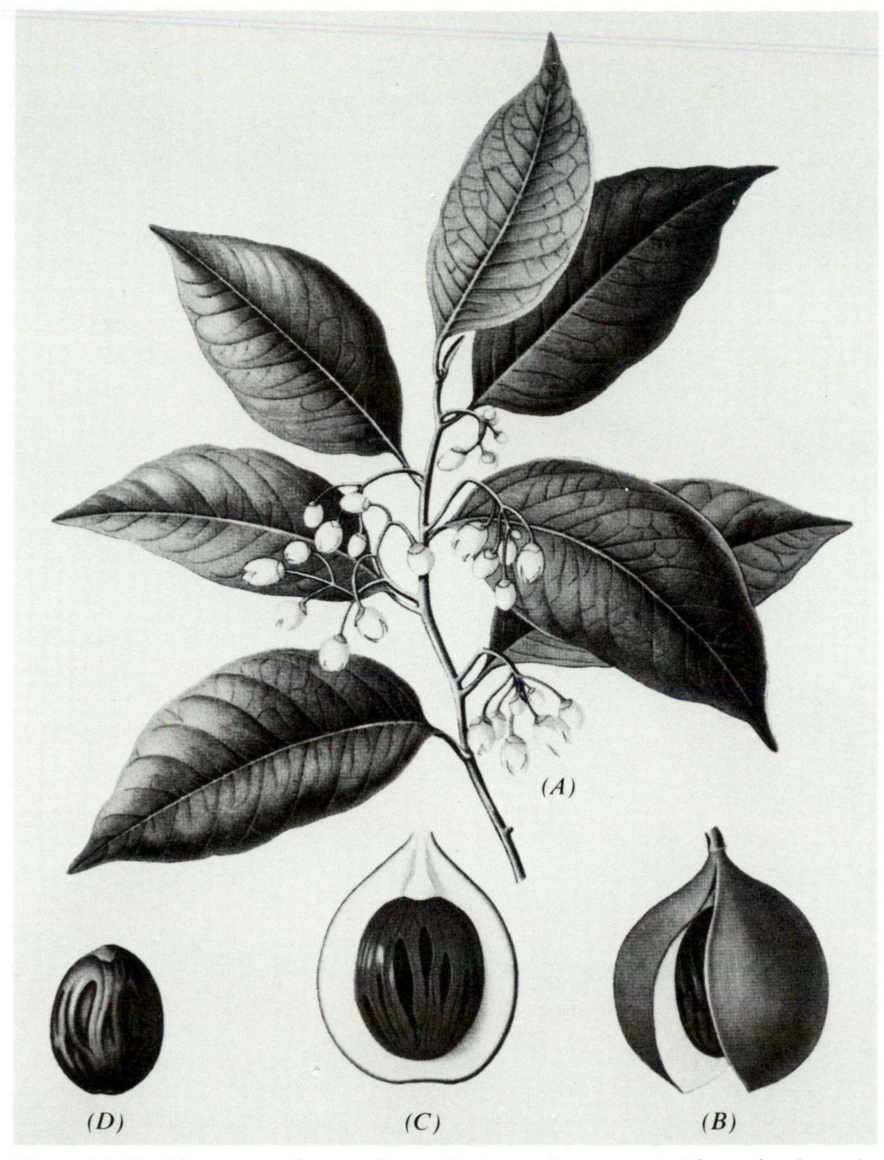

Figure 26-10 Nutmeg and mace from *Myristica fragrans*. *A*. Flowering branch. *B*. Fruit with fleshy mesocarp. *C*. Cross section of fruit showing seed surrounded by an aril, which is source of mace. *D*. Seed, which is source of nutmeg. (From Rosengarten, 1969)

in hot (57° C) water and then successively sweated and dried for five to six months. This process consists of spreading them in the hot sun during the day and folding them in blankets during the night. As the sweating proceeds, a white crystalline phenolic compound, vanillin, is formed. The vanillin, plus other compounds, is then extracted with 35% alcohol. Because this curing process is time-consuming, it increases the cost of natural vanilla. Most "vanilla extract" is synthetic, as there are various inexpensive processes for producing vanillin

Figure 26-11 *A*. Flowering branch of *Eugenia caryophyllata* with unexpanded, nail-shaped flower bud (shown enlarged in *B*), which becomes a clove when dried. (From Rosengarten, 1969)

from eugenol (which is easily obtained from lignin, a residue from paper pulp) and coal tar. Also coumarin (Chapter 6) can serve as a cheap substitute. Although vanillin provides the characteristic aroma and taste of vanilla, the accompanying small amount of terpenoids in the bean adds a "touch" that enables gourmet cooks to distinguish between the real and the artificial.

(A)

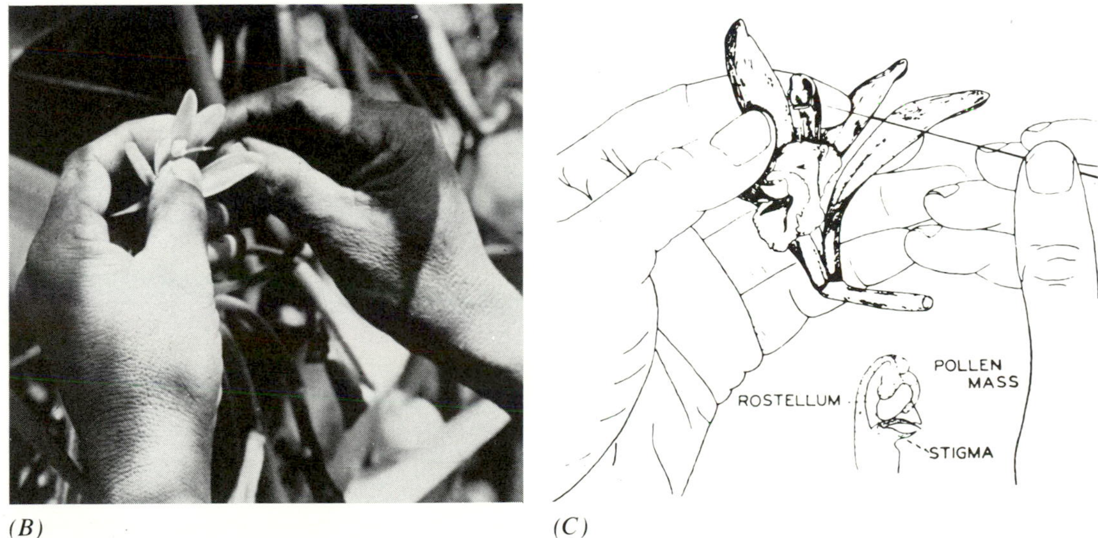

(B) *(C)*

Figure 26-12 *A*. Row of vanilla *(Vanilla planifolia)* vines growing in the Malagasy Republic; the vines are planted against saplings. *B* and *C*. Hand pollination of vanilla flowers. To obtain commercial production of vanilla beans, it is necessary to pollinate the flowers by hand because the bees that usually do this pollinating occur only in the native habitat of the vanilla orchid. The worker lifts the fleshy lip of the rostellum so that the pollen grains from the anther may be smeared with the left thumb upon the stigma. (Drawing from Rosengarten, 1969)

5. THE PERFUMES

In addition to use of essential oils for seasoning, the art of perfumery is as ancient as recorded history. In early days, perfumes were used to mask offensive odors and even may have acted as antiseptics. They have been used as offerings to gods in religious ceremonies in the pre-Christian world and have frequently appealed to the vanity of the idle rich. For example, in Greece a different perfume was used for each part of the body. Use of perfumes, however, declined during the Middle Ages in Europe until the Crusaders brought back information from the Holy Land to France about unusual perfumes and secrets of perfumery. In the twelfth century the perfume industry was created in France by royal charter, and to this day southern France is the center of this industry. It is here that master blenders combine many fragrances into unique creations whose recipes are kept secret. In modern perfumes, pure essential oils are infrequently used. Instead, they are extended with synthetic compounds and, in the case of inexpensive perfumes, with cheaper oils. Fixatives are added to retard volatilization of the essences and to equalize volatilization of numerous components in the perfume so that none predominates. These compounds may be animal oils (such as musk or ambergris) or fatty oils and balsams from plants. Many of the final washwaters in the distillation of perfumes are later used for their residual fragrance in the production of colognes and lotions.

The most important flowers used for perfume production are orange, rose, and jasmine, although violet, geranium, and many others are also used extensively. The perfumery essences are most frequently obtained from flowers, whose nectar glands secrete essential oils that may be attractive to pollinating insects. These floral oils are often extracted by absorption into cold fat, a process called *enfleurage;* the perfume is subsequently extracted from the fat with alcohol. Enfleurage is generally used for delicate essences when harsher distillation techniques might have deleterious effects. It is the chief means of extracting floral essences in the great perfume centers of southern France. Here large acreages of roses, jasmine, geranium, and other flowers are grown for the perfume industry.

Rose oil (or attar of roses) is one of the most prized and most expensive perfume oils. It is primarily obtained from *Rosa damascena,* which is a particularly important crop in Bulgaria and southern France (Fig. 26-13). In the "Valley of Roses", east of Sophia, Bulgaria, rose oil production is the source of employment for several thousand people. The roses bloom in late spring, at which time they are collected for distillation of essential oil (Bulgaria) or solvent extraction or enfleurage (France). Flowers are generally harvested in early morning when the rose is in late bud stage. The principle constituent is citronellol (45 to 65%) with smaller quantities of gerianol, nerol, linaloöl, and others. Because pure rose oil is so expensive ($\frac{1}{2}$ g is obtained from 1000 g of flowers) it is generally extended with less expensive oils.

Another important perfume oil is obtained from several species of *Jasminum* (olive family, Oleaceae), especially *J. grandiflorum,* native to southern Asia. In southern France thousands of kg of this species are harvested, since it requires approximately 50,000 flowers to make 500g of oil. Maximum oil is obtained from flowers picked near daybreak, but because of practical problems of harvesting at this time, buds are often picked in the evening before they open.

Violet essence is obtained mostly from *Viola odorata* var. *semperflorens* (Violaceae). Violet-growing centers are in the Provence region of France. Because greatest fragrance comes from violets that are partially shaded, they are often protected by hedgerows of olive and orange trees. Like jasmine, fragrance is greatest when the blossoms are picked at night or early in the morning. One-half ton of the flowers is needed to yield a pound of essential oil concentrate. The principal constituent is ionone, which also is obtainable from citral, thus providing a basis for synthetic violet perfume.

(A)

(C)

(B)

Figure 26-13 *A*. Roses in cultivation near Sofia, Bulgaria, where they are grown exclusively for their essential oil. *B*. Rose petals being turned in a drying shed at a factory in southern France. 1000g (1 kg) of these petals normally yields about $1\frac{1}{2}$g of attar of roses. *C*. Close view of the rose flower cultivated for perfume.

Perfume oils from various portions of the orange plant have distinctive odors and thus have specific names, such as oil of neroli (from blossoms), petit grain (from leaves and twigs), and bergamot (from fruit rinds). Orange blossom oil comes primarily from *Citrus aurantium* (the bitter orange), grown in Italy, Spain, and Portugal for the fruit, and from the flowers in Provence, France, where the total blossom crop may exceed 2000 tons annually. Most of the oil is obtained by distillation and consists primarily of linaloöl. Orange oil is used mainly as a component in synthetic perfumes and colognes. Bergamot comes from a small orange, *C. bergamia*.

Geranium oil is extracted from the foliage of several species of cultivated "geranium" (Pelaragonium, Geraniaceae) grown for this purpose primarily in northern Africa and Southern France. About 1000 g. of foliage is required to yield 1g. of oil. The principle constituent of geranium oil is geraniol, an important "behind the scenes" terpenoid, because it is used as extender or substitute for more expensive fragrances for a variety of products such as soap.

6. THE RESINS

Resins used in commerce are primarily complex mixtures of terpenoids, containing a volatile fraction of mono- and sesquiterpenes and a nonvolatile fraction of diterpenes and sometimes triterpenes. Resins are synthesized in the epithelial cells lining pockets or canals and then are secreted into these internal cavities. Synthesis generally occurs in all organs of the plant with either a different suite of terpenoids or different quantitative composition of a similar suite in each organ. The composition appears to be genetically controlled and thus little influenced by environmental conditions. Also, genetically controlled variation in terpenoid composition occurs within widely distributed species; that is, it constitutes a chemical form of ecotypic differentiation. Terpenoid resins are synthesized in copious amounts by approximately 10% of the plant families, about two-thirds of these being lowland tropical or subtropical angiosperm trees (including such families as Fabaceae, Burseraceae, and Dipterocarpaceae). All genera in coniferous families, which now occur primarily in temperate regions, synthesize resins, but only those in the Pinaceae and Araucariaceae families produce appreciable quantities.

Resin can fossilize (then it is called *amber*) and thus provides an opportunity to study changes in its chemistry over millions of years of time. Coal scientists have pointed out that resins are highly resistant to chemical change—even to those occurring in the formation of lignites and low-rank bituminous coals. Preservation of resin through long periods of time cannot be attributed only to this resistance to oxidation, but also to the capacity to withstand decomposition by soil microbiota.

The first evidence for resin production occurs in early coniferous trees growing in the coal swamps about 300 million years ago (Chapter 2). The evidence of copious production through the remainder of geologic time to the present day appears correlated with tropical or subtropical conditions. Why should trees have put considerable energy into the production of large quantities of "secondary products" through millions of years? There is evidence that these complex mixtures of terpenoids may serve as defense against the high diversity and intensity of selective pressures exerted by insect and pathogen populations in moist lowland tropical ecosystems. The volatile fractions are either toxic or deterrent to both insect and vertebrate herbivores as well as inhibitory to the growth of fungi. The action may be due to a single compound or to a combination of compounds and concentrations. Variation in composition of the constituents comprising resins also controls the degree of fluidity and rapidity of hardening (when they are exposed on the surface of the plant), which also determines the resin's capacity either to engulf an attacking organism or to heal a wound rapidly that might be subject to attack by microorganisms.

Natural resins have recently been of less economic importance since the development of phenol-, formaldehyde-, and glycerin-based synthetic substitutes; however, natural resins still are used for turpentine, incense, and varnishes, and as mixtures in pharmaceuticals. Some developing tropical countries are now beginning to use natural products again as a major forest resource rather than depend upon substitutes from the petrochemical industry.

a. AMBER

Since the first records of Neolithic man in Europe, approximately 5000 years ago, amber has been used as a barter item (Fig. 26-14). Amber has been cherished not only for its beauty but also because it has been thought to heal many types of illness and also to protect against evil. These ideas have evolved independently in most primitive societies that have found amber. In northern Europe amber was associated with sun worship, which may partially explain the belief in its protective properties. Amber trade routes crossed Europe from the Baltic Sea area, where it was found in great

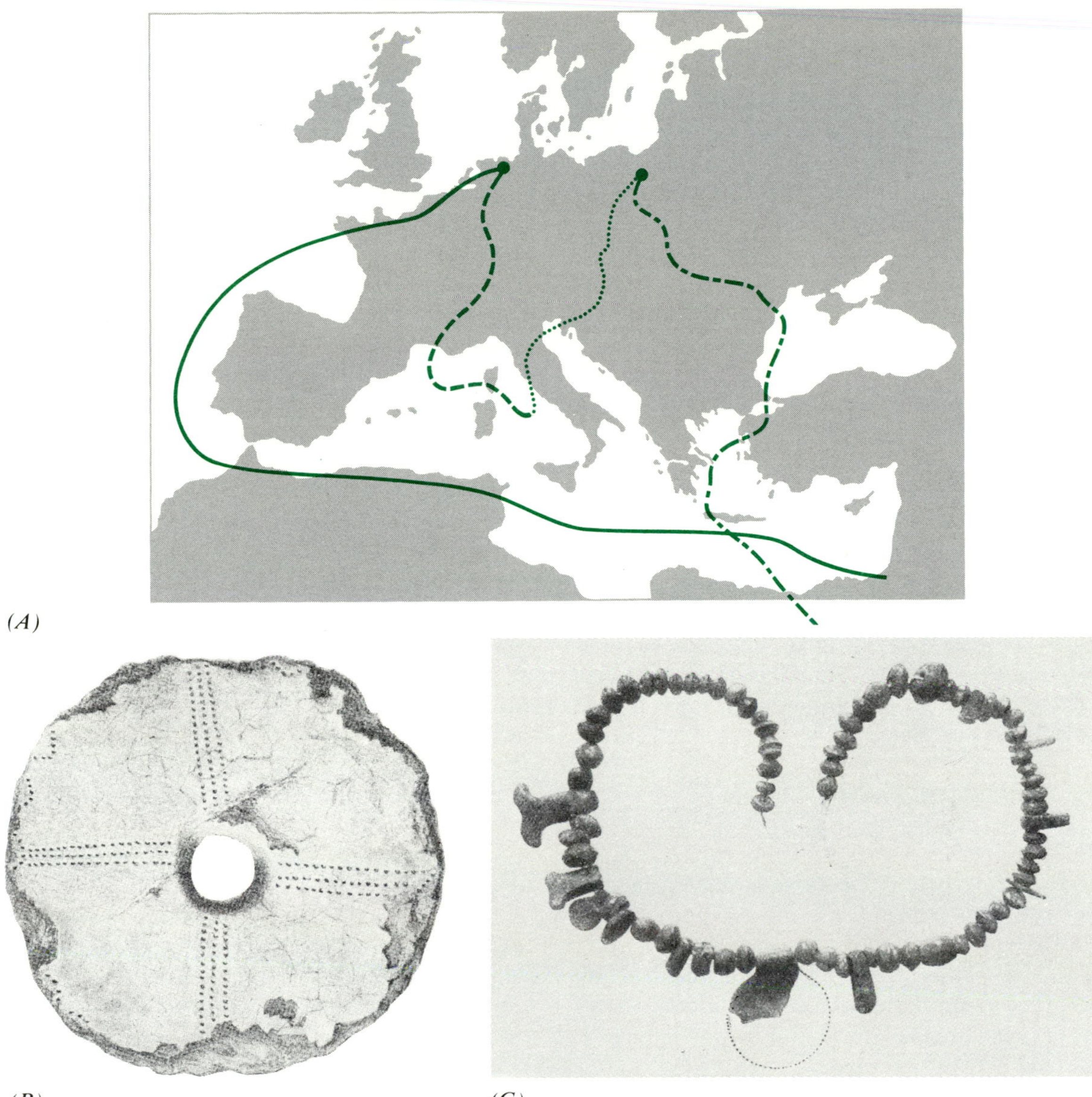

(A)

(B) *(C)*

Figure 26-14 *A*. Map showing trade routes of amber from vast deposits along Baltic and North Seas southward across Europe: Phoenician route (solid line); Greek (long and short dashes); and Roman routes (short dashes, dots). In Neolithic and Roman times the trade route to the Jutland areas went primarily along the Rhine and Rhone river valleys. The Phoenicians went by sea to Jutland. Both the Greek route and another Roman one were directed toward the extensive deposit on the Samland Peninsula (From Bachofen-Echt, 1949). *B*. Amber pendant, ornament from the Stone Age (4000 B.P.) from East Prussia. The cross and ray design has religious meaning: the cult of the sun's wheel. *C*. Amber necklace from western Estonia with pendants (ca. 300 A.D.).

abundance from 5000 to 2000 B.P., to the Adriatic and Black Seas in the Bronze and Iron Ages through to the Greek and Roman periods. Together with tin, it has been considered one of the chief items that led the Romans to penetrate the Gallic regions to the west and north of the Mediterranean. At first amber was the property of the finder but in the fourteenth century the Rittenorder (Order of Knights) made themselves the owners of all amber, and in 1466 anyone withholding it was punished by hanging. The Order promoted the formation of guilds of amberturners, erected warehouses, and conducted sales over a large part of Europe. Extensive commercial operations began in 1837 and even today the largest commercial production and sale of amber comes from large mines on the Baltic shore. Many tons of amber have been excavated by steam dredging or quarrying operations in open pits with subsequent washing by electrical machinery. At one time amber was sorted into about 200 grades, distinguished partly by size and partly by color, to be used primarily for jewelry, smoking items, and varnishes. Amber from Burma and Mexico is sold principally for jewelry.

b. TURPENTINE AND ROSIN

Turpentine consists of the volatile mono- and sesquiterpenoid fractions of resin primarily derived from several species of *Pinus* (pine). *Picea* (spruce), *Abies* (fir), and *Larix* (larch). The nonvolatile primarily diterpenoid fraction is called *rosin*. The largest production comes from those species occurring in warm temperate and subtropical areas.

Although resin is produced in all parts of the pine tree, most is obtained from the trunk by tapping or from the excavated stumps and roots by distillation. The tapping procedure is determined by a vertical resin canal system (Fig. 26-15) that also has horizontal connections in both the wood and bark. Thus horizontal cuts intersect the flow of the resin. In the early days of the U.S. turpentine industry, the chipper made a deep hollow at the base of the tree and successive cuts through the cambium into the wood (Fig. 26-16). This technique greatly decreased the longevity of the tree. Today a shallow wound is made just through the cambium to obtain the resin from both wood and bark in the cambial zone where synthesis is most active (Fig. 26-16). Sulfuric acid is systematically applied to keep the resin from oxidizing as it is exposed to the atmosphere and thus plugging flow from the canals.

Long-term genetic studies have shown that the yield of resin in certain pines, such as *P. elliottii* (slash pine) and *P. caribaea* (Caribbean pine), is under strong genetic control, some geographic races yielding considerably more resin than others. Populations with the highest yield of resin often have the best quality of timber. Thus many fast-growing trees that produce both large quantities of resin and good quality wood have been planted in abandoned fields (where pines flourish normally as a part of secondary succession) in the southern United States for a dual crop (Fig. 26-16). The trees are grown for approximately 10 years, then tapped for resin for the next 10 years, and subsequently harvested for wood at 25 to 30 years. Although, because of increasing labor costs, this practice is decreasing in the United States, this kind of multiple use for resin-producing trees is being advocated for many developing countries.

The settlement of North America was partially due to England's desire to rid herself of dependence on Scandinavian sources of resin, since the "pitch" (as resin was called) was used to calk ships and waterproof rigging. In fact, the industry to produce these ship-related commodities became known as "naval stores." The American colonial settlers, particularly in Virginia and North Carolina, were charged naval stores products as payment for their charters. In those days, naval stores were mainly produced by setting logs on fire in a pit that was covered with soil. The pitch or resin collected in the bottom of the pit along with dirt and was called "tar." Because the workers got tar on their shoes, they were called "tarheel-

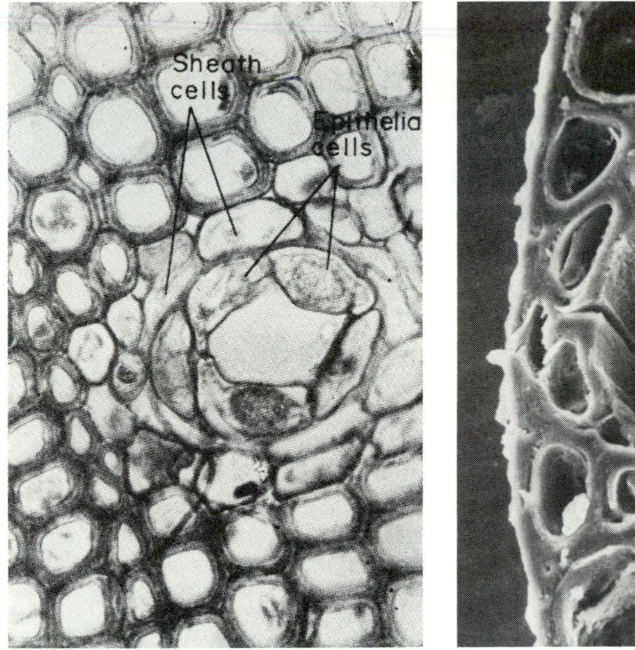

(*A*) (*B*)

Figure 26-15 *A*. Cross section of a vertical resin canal in secondary xylem of *Pinus halepensis* (×480). Epithelial and sheath cells surround each canal. *B*. Horizontal canal within a xylem ray of Douglas fir (*Pseudotsuga menziesii*) (×2500).

ers," which became the nickname for the people of North Carolina. In the 1830s stills used in the Scotch liquor industry were imported for turpentine production, thus providing the derivation of "spirits of turpentine." Worldwide, about one-third of present-day turpentine is obtained from tapping trees, another one-third from distillation of old pine stumps and roots, and another one-third from the sulfate paper processing of coniferous pulp (Chapter 22). In the United States, however, more than one-half of the production comes as a by-product of paper processing. Although at one time coniferous oleoresins were used primarily for their physical-chemical properties in a variety of products such as paint thinners, varnishes, waterproofing, and paper sizing, today they are increasingly being used as chemical raw materials for production of polyterpene resins, insecticides, and many other products. There is even research directed now toward the incorporation of turpentine into automobile fuels. In fact, leguminous trees (*Copaifera*), that produce large amounts, primarily of sesquiterpenes, are being studied as a source of diesel fuel for tropical countries which are petroleum-poor.

C. COPALS

"Copals" are a group of resins that form particularly hard varnishes. All of the leguminous copals are derived from tropical trees in the subfamily Caesalpinioideae. The resin is produced principally in pockets that increase in size by breakdown of the wall of the secretory cells to form large cavities in which the resin can collect. As these large quantities of resin exude from the trunk due to fissures developing from rapid growth or physical injury (Fig. 26-17*B* and 17*D*), so much material falls to the ground that tapping of copal trees has not been necessary to sustain the copal industry. In fact, the quality of the copals for hard varnishes

(A)

(B)

Figure 26-16 *A*. Pine trees grown in plantations in southern U.S. The trees are first tapped for resin then later cut for timber. Trees can both yield high quantities of resin and produce excellent wood. *B*. Different types of cutting used in tapping pine trees in the southern United States for collection of resin. This process is called "boxing" and the cut is often referred to as a "cat face." In the early days of the naval stores industry the wood was cut into deeply (using large hatchets and mallets), but now cuts are just made into the zone of differentiating xylem to prevent the resin from hardening in the air.

seems to improve with burial in the ground. The most important leguminous copals are from *Hymenaea* (Brazilian and Zanzibar copal, Figs. 14-9 and 26-17) and *Guibourtia* (Congo copal) from equatorial rain forests, where enormous quantities are produced. The other major source is from the coniferous family Araucariaceae, including two species of *Agathis* (*A. australis* from New Zealand, and *A. alba* from the Philippines and Indonesia). The copal from *A. australis* ("kauri pine") is famous for the role it played in the early economic development of New Zealand. The large tree is a monarch of the forests and held in esteem similar to the redwood in the United States. Large lumps (up to 100 pounds) were found in layers over extensive areas of the North Island where it was "mined." This copal (considered one of the most valuable hard resins for varnish and linoleum) provided the major export industry, and the lore associated with the "gum diggers" is similar to that of the lumberjack in the western United States. It required a great physical fitness and daring to go up in a bo'sun chair to bring resin lumps down from notches of branches as well as to dig the material from the soil and marshes.

d. DAMMARS

"Dammars" are other commercially important resins obtained from numerous genera, such as *Hopea, Shorea* and *Balanocarpus* in the large tropical family Dipterocarpaceae. In contrast to copals, they dissolve easily in turpentine and thus are used as "spirit varnishes" for decorative work and wallpaper; they are also used for caulking and for incense. Sometimes the trunks of the trees are tapped for the fresh resin, whereas in other cases large masses of natural exudates are collected.

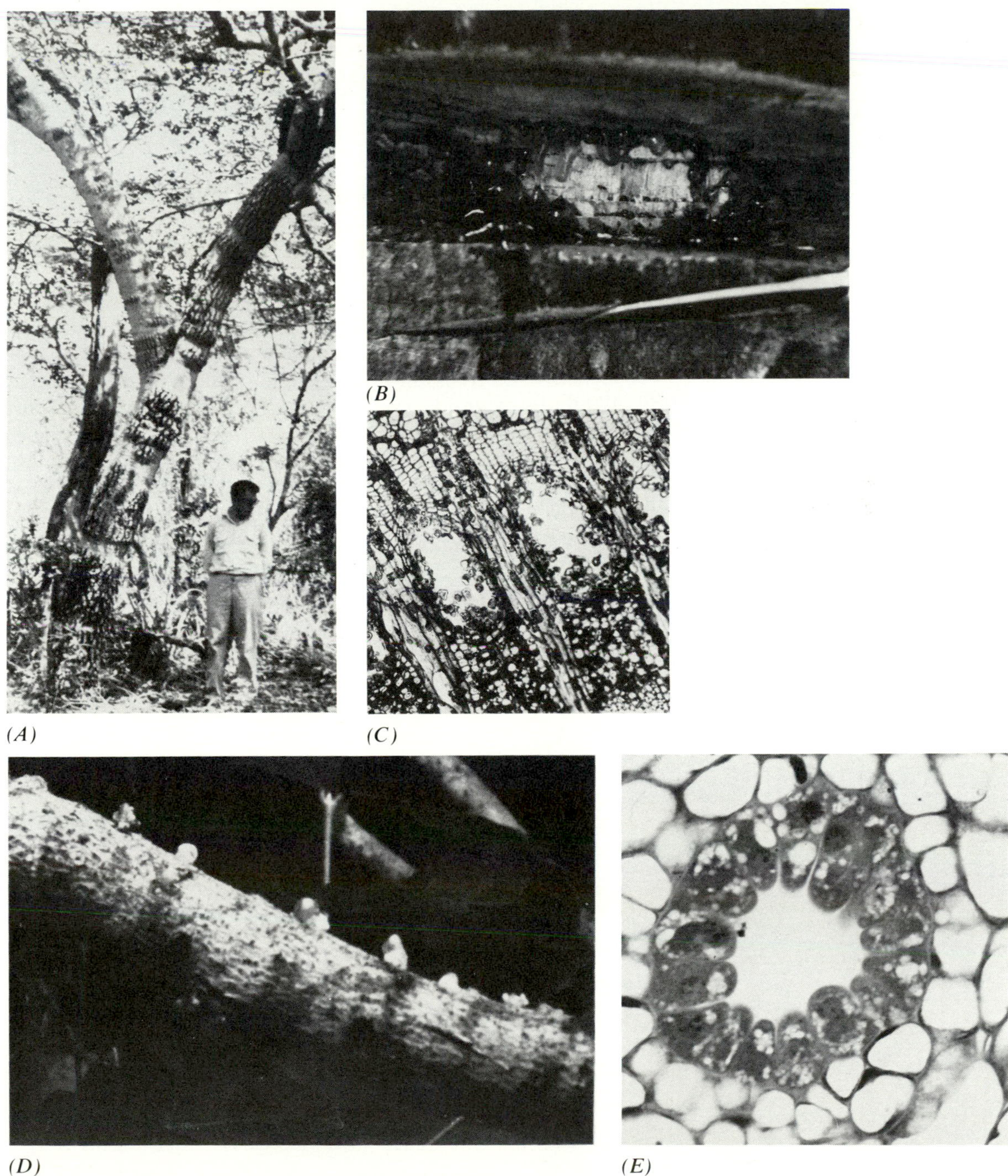

Figure 26-17 Resin production in *Hymenaea courbaril*. *A*. Small tree from Guerrero, Mexico, with rugose patches of vertical fissures from which resin issues. *B*. Cut in trunk of mature tree showing exudation of resin. *C*. Cross-section showing lysigenous pockets developed in cambial zone of mature tree. *D*. Branch showing exuded balls of resin. *E*. Resin pocket in pith of petiole from year-old plant showing densely cytoplasmic secreting cells.

e. INCENSE

The best-known incenses are resins produced by members of the tropical family Burseraceae. In Mexico any burseraceous resin may be called either "copalli" or "incienso." Commercially, the two most important incenses are myrrh and frankincense. Myrrh, from several species of *Commiphora*, is collected primarily in Somaliland; some also comes from Arabia and India. The resin is produced in ducts in the bark and often exudes naturally forming irregular, rounded "teardrops"; it also is tapped commercially.

Frankincense is derived from about 12 species of *Boswellia*, small, shrubby trees growing in northeastern Africa and southern Arabia. It was one of the most important items of commerce when Phoenicians dominated the trade in this area (Fig. 26-18). The resin is synthesized and exuded from ducts in the bark during the wet season and the yellow teardrops are scraped from the bark during the dry season. It is still the essential constituent of incense used today by the Roman Catholic and Greek churches.

Figure 26-18 Sixteenth-century illustration portraying collection of resin oozing from bark of frankincense tree (*Boswellia* in the Burseraceae) from southern Arabia.

7. LATEX AND ITS PRODUCTS

a. RUBBER AND GUTTA-PERCHA

Rubber and gutta-percha are polyterpenes, composed of 50 to 6000 isoprene units, that occur predominantly in the latex of tropical trees and to lesser extent in temperate-zone shrubs and herbs. Latex is a milky (as the Latin word "lac" indicates) or clear-colored juice made up of a complex of materials such as terpenoids (mono-, sesqui-, di-, and tri-, and polyterpenoids), proteins, carbohydrates, tannins, and minerals. Rubber, which can be extracted from the latex by such solvents as benzene, petroleum ether, or chloroform, may constitute up to 60% of the total weight of the latex and as much as 85% of the total solids. It also may represent as much as 20% of the dry weight of the plant. Rubber and gutta are produced by about 2000 species of plants, most commonly belonging to the families Euphorbiaceae, Moraceae, Apocynaceae, Asclepidaceae, and Asteraceae. The most important chemical distinction between rubber and gutta is the mode of linkage of the isoprene units. Rubber has all its isoprene

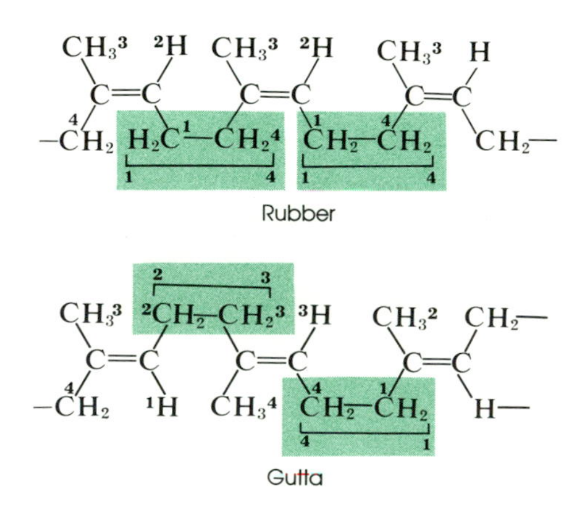

units linked 1–4, as shown. This accounts for its elastic properties and its tendency to crystallize when stretched. In gutta the linkages alternate between 1–4 and 2–3; this gives it a more compact, symmetrical structure that allows its deposition in microcrystals and makes it brittle. The proportion of shorter molecules is greater in rubber from the latex of young plants.

The latex is produced in vessels or special cells (called *laticifers*) quite different from the internal pockets and canals in which resins are produced or the externally located glands and hairs in which the essential oils are secreted. There are two basic types of laticifers: (1) single-celled, nonarticulated and (2) compound, articulated (Fig. 26-19). The first type is called a *laticiferous cell* because it originates from a single cell; by continuing growth it develops tubelike structures that are often branched but do not fuse with other cells. The other type, called a *laticiferous vessel,* originates from a longitudinal chain of cells in which the walls separating the individual cells become perforated or completely resorbed. Thus a tube, resembling xylem vessels in origin, is formed. Vessels generally occur in the outer periphery of secondary phloem tissue, with alternating layers of sieve tubes and parenchyma and laticifers (Fig. 26-19).

The possible function of latex within the plant early fascinated plant anatomists and physiologists. They thought that it might be analogous to the circulatory system in animals and they referred to the secretory cells as "living sap vessels" (*lebenssaftgefässe*). Metabolically it was once thought to be significant as a means of storing "food" (carbohydrates, protein, minerals). However, experiments indicate that these "food materials" in the latex are not readily mobilized to form carbohydrates when the plant is starved. On the other hand, the vessels certainly are adapted as reposi-

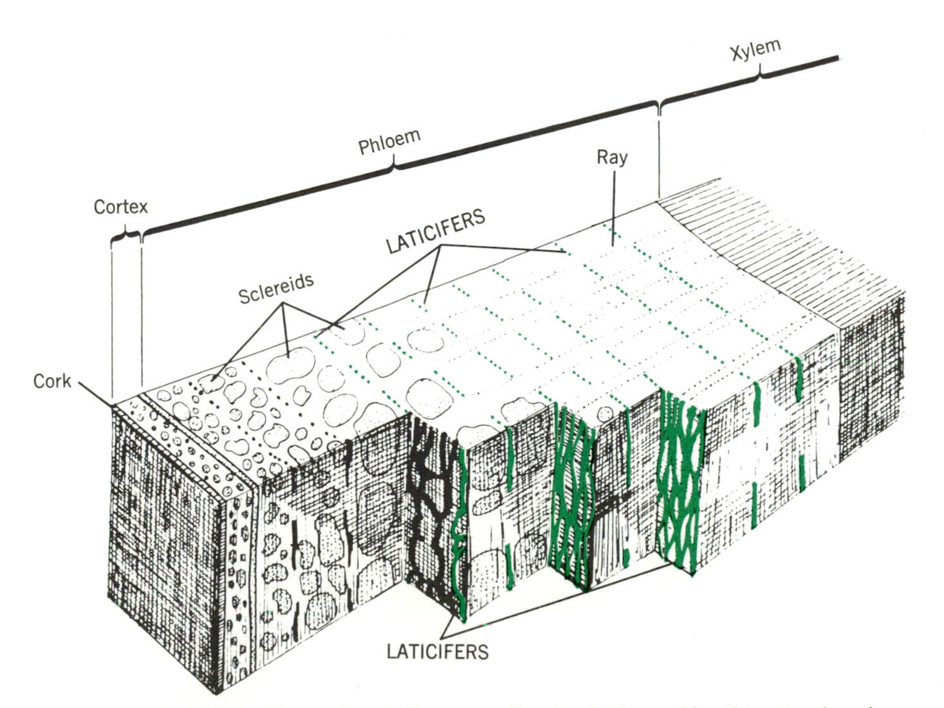

Figure 26-19 Three-dimensional diagram of bark of *Hevea brasiliensis,* showing arrangement of laticiferous vessels in the secondary phloem. (From Fahn, 1972)

tories. The highly polymerized terpenes like rubber probably do not pass across the cell walls from the cell in which they have been synthesized. Yet a variety of materials in the latex, such as the terpenoids, may be fungicidal and/or insecticidal.

b. RUBBER FROM HEVEA

The main source of natural rubber (90% of the world's production) is *Hevea brasiliensis* (Euphorbiaceae), which is native to the Amazon basin. In fact, the boundaries of the Amazonian rain forest are sometimes delimited by the distribution of *H. brasiliensis*. Rubber is produced in this plant by anastomosing vessels in the secondary phloem, which makes it relatively easy to obtain the rubber by regular tapping. The vessels often anastomose 30° from vertical and thus the cuts are made in spirals (Fig. 26-20). When the laticifer is opened, pressure from surrounding cells causes exudation.

Figure 26-20 Rubber tapper collects latex on a plantation in Malaya. Note that the cut is made in spirals because the laticifers anastomose about 30° from vertical.

Latex flow increases with successive tapping, which, of course, is advantageous for commercial collecting. This rapid synthesis following tapping suggests that precursors of rubber must be constantly and quickly available. Rate of flow varies seasonally, the greatest yield occurring during the wet months and the lowest during the dry ones. This relationship appears related to leaf drop during the dry season with consequent reduction in photosynthesis.

After the rubber-bearing latex has been collected from the trees it is generally "coagulated" by either of two ways. In the Amazonian region, where collection is still primarily from native trees scattered through the forest, the latex is poured over a paddle, which then is held over a smoking fire (Fig. 26-21). If the latex is collected in a plantation, it is usually coagulated either by acetic or formic acid. On the other hand, if the material is to be shipped in liquid form, ammonium hydroxide is added as an anticoagulant.

c. HISTORY OF OUR USE OF RUBBER

The collection of rubber from wild trees in contrast to that from cultivated trees grown in vast plantations leads us to the fascinating history of the discovery of rubber by civilized man and related biological problems in the development of its usage. For intrigue and romance, the early history of the rubber industry is a story rivaled only by that of the spice trade.

Europe's first knowledge of rubber appears to have come from balls that bounced, which Columbus brought back from Haiti in 1493. Although bottles were also made of this rubber (from latex from *Castilla* in the fig family, Moraceae), the material was primarily considered a curiosity.

The chemist Sir Joseph Priestley, the discoverer of oxygen, is credited with coining the name "rubber" because he said that "the substance is excellently adapted to rub out marks of a lead pencil." Despite interest, however, in the rare combination of properties such as flexibility, unbreakability, elasticity, and waterproof-

Figure 26-21 After the rubber is collected from native trees in Amazonia, the latex is poured over a paddle, which is held over a smoking fire to coagulate it and thus prepare it for shipment to a factory for processing.

ing, rubber continued to be a rare curiosity. This was partially due to the lack of ability to use the raw material directly. Since latex coagulated in air, articles had to be manufactured from it in the distant tropical forest where the trees grew. An example of the frustrating difficulties were the "gum shoes" made from *Hevea* rubber that were brought back from the Amazon in 1823 to 1824 to New England by Yankee seamen. The rubber shoes were ideal for keeping feet dry and so intrigued the seamen, but the shoes were moulded to fit the Amazonian Indians, whose feet were much smaller than those of the Yankees. Additionally, the shoes in New England cracked in the cold snow and melted when they were put next to the stove to either warm or dry them out. However, about the same time that the Yankees were frustrated about this, Charles Macintosh in Great Britain made a significant breakthrough in using rubber to make raincoats. Macintosh dissolved rubber in naptha and put the mixture on to cloth, where the naptha evaporated, leaving the rubber to waterproof the cloth. Raincoats to this day in Britain are often called "mackintoshes."

The most significant advancement, however, with regard to expanding the use of rubber came with Charles Goodyear's discovery of vulcanization in 1839. In this process the physical characteristics are stabilized by heating rubber with sulfur. One percent of the isoprene monomer becomes cross-linked with bisulfide bridges to form a molecular net. Vulcanization led to a chain reaction of events, starting with the development of tires, which led to a rubber boom in 1853 to 1855. Rubber during this time had to be supplied from wild trees, principally from the Amazon basin, resulting in the creation of local "rubber barons," who accumulated fantastic wealth for a short time. This was epitomized in the town of Manaus in the central Amazonian region, which was a collection center for rubber. But there simply were not sufficient trees to be tapped by crude collecting and preparation techniques to meet the increasing world demand. The only other fair-sized source was *Landolphia,* a vine in the Apocynaceae, which grew in equatorial Africa. The rubber search was so intense that vines were cut to the ground, thus destroying the "goose that laid the golden egg." In fact, the exploitation of African people, who were severely taxed for quotas of *Landolphia* rubber (if not met, women were seized and men put into chains), led to a world scandal.

The consternation over limited supplies of natural rubber led to an exciting episode that completely transformed the rubber industry from using wild trees to using plantation trees, and shifted the maximum production from the Amazonian region to southeastern Asia. These events were initiated by Sir Joseph Hooker, Director of The Royal Botanic Garden at Kew (the world's largest herbarium and botanical garden in London), who had the inspiration of trying to grow *Hevea* in plantations in India and Burma and thus to help the agricultural and economic development of the British Empire. Kew Gardens generally has played an important role in plant exploration as well as introduction of crop plants into new areas (Chapter 28).

A major obstacle was how to get the very short-lived seed from vigorous populations in the Amazon basin to London. In 1872 Hooker had read a book entitled *Rough Notes on a Journey through the Wilderness* by a little-known English adventurer, Henry Alexander Wickham (Fig. 26-22*A*). He thought that this young man might be able to recognize various strains of *Hevea* and be sufficiently ingenious to get seeds from the Amazon to London. Hooker chose his person wisely. Not only did Wickham select what later turned out to be high-yielding plants, but he did find a means of getting 70,000 seeds aboard a ship in the central Amazon basin and then got them to London in such rapid time that 3000 of these short-lived seeds germinated at Kew Gardens. Subsequently, seedlings were shipped to Sri Lanka and the Singapore Bo-

(A)

(B)

Figure 26-22 *A*. Henry Alexander Wickham, the English adventurer who succeeded in getting seeds of *Hevea brasiliensis* from trees in the Amazon Basin to the Royal Botanic Gardens in Kew, England, where 3000 of them germinated and *B*. these seedlings were then sent in a "Wardian case" from the Royal Botanic Gardens to Sri Lanka (then Ceylon) and the Singapore Botanical Gardens from where clonal cuttings were sent to Malaysia, Java, and Sumatra.

tanical Gardens. From there clonal cuttings were sent to Malaysia, Java, and Sumatra (Fig. 26-22*B*). The first commercial rubber from these plantations (11,000 tons) was produced in 1910, and within three years the production rivaled the total collection of rubber from wild species. Over the years since that time an enormous amount of research has gone into plantation rubber; in fact, it is one of the few tropical tree crops that has received as much attention as temperate herbaceous agricultural crops. Wickham's high-yielding trunk is double grafted – first on to a special root stock and second with a particularly vigorous canopy on top. Fortunately for the Asian plantations, the native diseases of *Hevea brasiliensis* were not transported with the seeds; thus it is possible to grow the trees in extensive monocultures. On the other hand, it has as yet been impossible to develop plantations of *H. brasiliensis* in its native home in the Amazon because of the virulent leaf blight *Dothidiella* (Chapter 29). Even when improved stock from Asia was reintroduced into the New World (following many attempts to breed for resistance, yet retain high-yielding characteristics), this fungal disease has wiped out plantations.

Most of the world's natural rubber therefore comes from the Asian plantations, and during World War II there was revived scientific interest in rubber-producing plants when the Japanese controlled the Far East. A search was made for the most promising rubber-producing plants, but in 1945 chemists succeeded in synthesizing rubberlike polymers from styrene and butadiene (available principally from petroleum and coal), which like isoprene, are hydrocarbons. They found that these were capable of copolymerization to form elastomers similar to but not identical with natural rubber. Often today rubber is a mixture from natural and artificial sources.

One other plant in addition to *Hevea* has produced sufficient amounts of rubber to be used commercially. This is the guayule shrub (*Parthenium argentatum*) in the sunflower family, Asteraceae, which is native to upland desert plateaus of northern Mexico and the southwestern United States (Fig. 26-23). Guayule (a Spanish corruption of an Aztec word) produces rubber chemically and physically identical to *Hevea*. In fact, in 1910 guayule provided 50% of the natural rubber in the United States and 10% of that consumed in the world. But reckless exploitation of the wild stands, the Mexican Revolution, and the 1929 depression combined to destroy the industry. During World War II, the United States spent $30 million on the Emergency Rubber Project to develop guayule as a domestic source of rubber again. The crop was grown on high-quality farmland in the Salinas valley of California, but after the war the farmers wanted their land (this is one of the most productive vegetable crop areas in the world). With the renewed availability of cheap rubber from Asia and the new production of synthetic polyisoprene (butadiene) rubber, the government withdrew its funding and ordered the 11,000 acres of guayule fields burned. Furthermore, it was erroneously believed that man-made elastomers would make natural rubber obsolete.

The economic situation has now changed considerably since the end of World War II. The price of both natural and synthetic rubber is steadily increasing. It also appears that *Hevea brasiliensis* may have reached limits for genetic improvement of yield, and the trees are in constant peril from the leaf blight (*Dothidiella*) (Chapter 29) that destroyed the attempted plantations in its native Amazonian home. Furthermore, southeast Asia is politically unsettled. Therefore, there is interest in developing an American source of natural rubber, for its elasticity, resilience, tackiness, and resistance to heat are unmatched by any synthetic rubbers now available. Conventional automobile tires, for example, contain about 20% natural rubber; radial tires, which now dominate the U.S. market, contain as much as 40%. Large tires on aircraft, tractors, and earth-moving vehicles are almost entirely made of natural rubber. Guayule rubber could serve for all of these uses.

In Mexico the government in 1976 set up a plant to process rubber in areas where adult guayule plants grow wild, and by 1979 they were processing 300,000 tons of shrubs annually and producing 30,000 tons of rubber. Since few wild stands of guayule occur in the United States, cultivation will be necessary, and this will require breeding guayule strains that contain uniformly high quantities of rubber. It has also been found that rubber yield can be at least doubled by spraying the shrubs with certain chemicals that have been shown to increase the carotenoids in citrus fruits. In this research, efforts probably will be made for collaboration between the United States and Mexico so that these countries will no longer be dependent upon the tropical rain forest *Hevea* plant.

d. GUTTA-PERCHA AND CHICLÉ

Of the gutta-producing latexes, the traditional *balata* comes from *Palaguium gutta* (Sapotaceae), which grows naturally in India and the South Pacific. It was used by the natives as a moulding material but the British colonists found it had excellent insulating properties. The trees used to be felled to obtain the latex, but now they are tapped.

Chiclé is from the sapodilla tree, *Achras zapota* (also in the Sapotaceae). In Central America this latex provided the original chewing gum. The trees were tapped by itinerant collectors ("chicleros") in the West Indies, southern Mexico, and Guatemala. A series of zigzag gashes was made from the top to the base of the tree (Fig. 26-24). Sixty pounds of chiclé could be collected from a single tree, but then the tree had to be left for a period of time to recover from the tapping. Although chewing gum today is made from a mixture of chiclé and synthetics, there is renewed interest in attempts to collect larger amounts of the natural material.

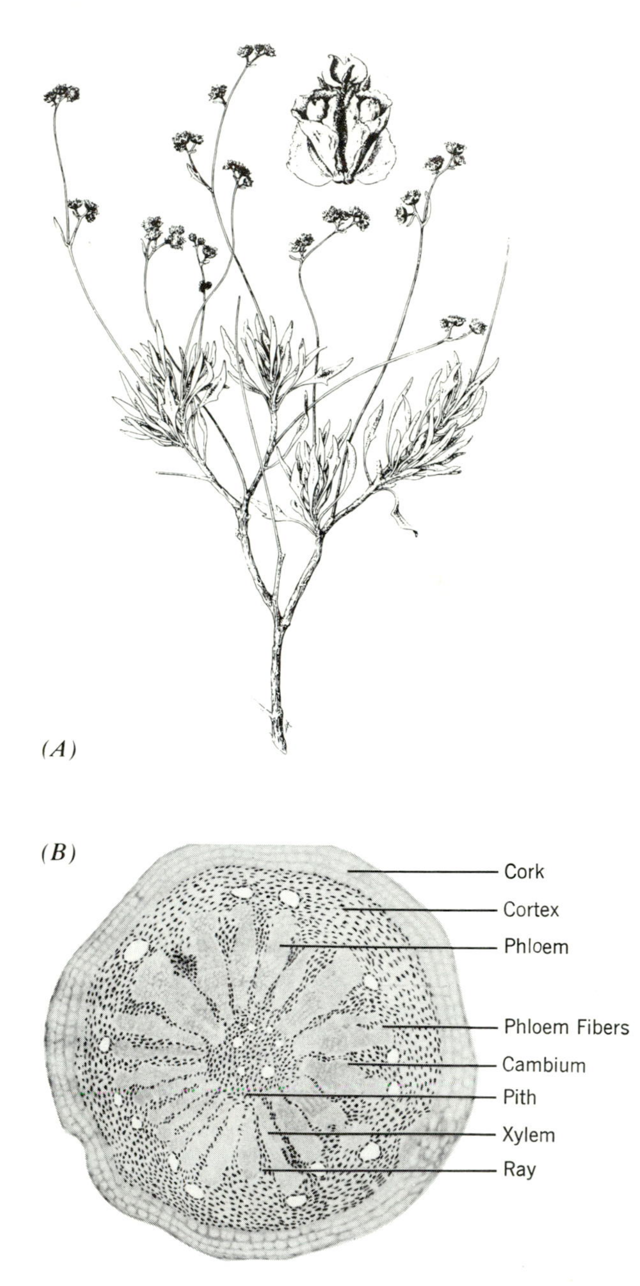

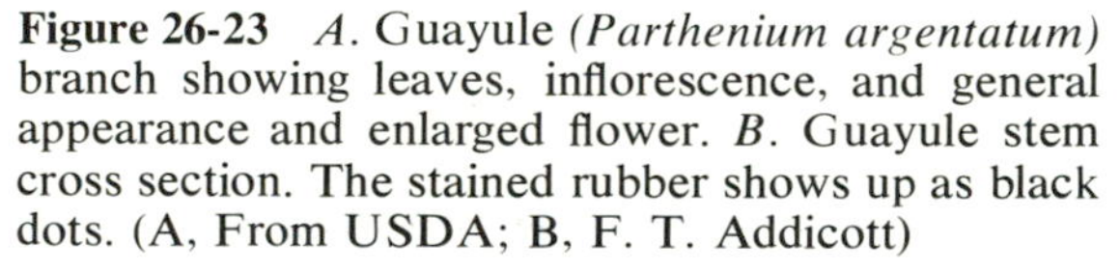

Figure 26-23 *A*. Guayule *(Parthenium argentatum)* branch showing leaves, inflorescence, and general appearance and enlarged flower. *B*. Guayule stem cross section. The stained rubber shows up as black dots. (A, From USDA; B, F. T. Addicott)

(A)

Figure 26-24 *A*. "Cichlero" cutting zigzag gashes in *Achras sapota* to obtain chiclé latex. *B*. Chiclé latex flowing in the gashes.

(B)

CHAPTER 27

Vegetable Oils, Fats, and Waxes

Vegetable oils, fats, and waxes, together termed *lipids,* have been used in much greater amounts than terpenoid products, because of the variety of industrial uses for them. But despite their considerable importance, their history of usage is less spectacular than that of the spices and rubber. Oils and fats particularly seem to have been taken for granted, perhaps because they occur in relative abundance in many plant species and there is a certain degree of interchangeability among the different kinds (i.e., few have truly unique characteristics making any one especially sought after). Thus there have not been expeditions sent out in search of them and imperialist monopolies have not been set up to corner the market on them, as was true in the spice trade. In fact, only a few important oils and waxes come from wild plant sources today, these principally from tropical palms. Most of the oils and fats used for margarine, cooking and salad oils, soaps, paints, and other familiar products come from plants grown as crops in the temperate zone. Those most extensively used for fuels, lubricants, and even to some extent in the manufacture of soaps and detergents have been largely replaced by petrochemical substitutes. However, because these types of oils are essential to the functioning of a modern industrial world, it may well be expected that economic value of plant oils will increase as nonrenewable mineral resources become exhausted. In this respect these oil, fat, and wax plants are just as basic as food plants.

1. DISTRIBUTION OF OILS, FATS, AND WAXES IN THE PLANT

Oils and fats occur in small droplets within the plant cells and are typically most abundant in seeds. The fleshy exocarps of many pome and drupe fruits (such as plums, apples, pears, peaches) contain very low levels of lipids (Chapter 18). However, the fat content of the mature fruit of the oil palm is seldom less than 35% of the fresh tissue weight and often as high as 70%. Other fruits with oil-or fat-rich pericarps are the olive and avocado; in most cases the fatty acid compositions of these pericarp oils or fats are appreciably different from those of the corresponding seed.

The utility of oils and fats to the plant is somewhat obscure, although those in the seed constitute an abundant reservoir of stored energy for the early growth of the seedling. Oils and fats contain about twice as many calories per gram as either carbohydrates or proteins. As a rule seeds rich in oil are also rich in protein, a valuable relationship in choosing and developing food crops.

Wax occurs in almost all vascular plants as an important constituent of *cutin,* which acts as a protective coating on the epidermis of leaves, stems, and fruits, reducing desiccation, oxidation, heavy abrasion, or pest attack. It is also a constituent of *suberin,* the waterproofing material of the walls of cork cells (Chapter 22). At present, leaves are almost the only important sources of vegetable waxes. The cuticle is overlaid with an epicuticular wax (Fig. 27-1), which is the commercial product; this may constitute 4% of the green leaf or 15% of the dry leaf. The one important seed is that of the Jojoba bush, *Simmondsia,* which contains large quantities of a quality wax.

2. CHEMICAL MAKEUP OF OILS, FATS, AND WAXES

Oils, fats, and waxes are all alcoholic esters of fatty acids. Oils and fats are formed by synthesis of fatty acids from carbohydrates (Chapter 6), followed by a combination of these fatty acids through enzymatic action with glycerol to form the esters – triglycerides (Fig. 27-2). Fatty acids are long hydrocarbon chains that carry a terminal carboxyl group, giving them the characteristic of a weak acid. The glycerol forms a link with the carboxyl groups by removal of a molecule of water, thus serving as a binder or carrier for fatty acids. As shown in Figure 27-2, the three fatty acid molecules that become linked to a given glycerol molecule are not necessarily the same kind. The ester linkages can be hydrolyzed by the enzyme lipase; food reserve fat can thus eventually be converted back

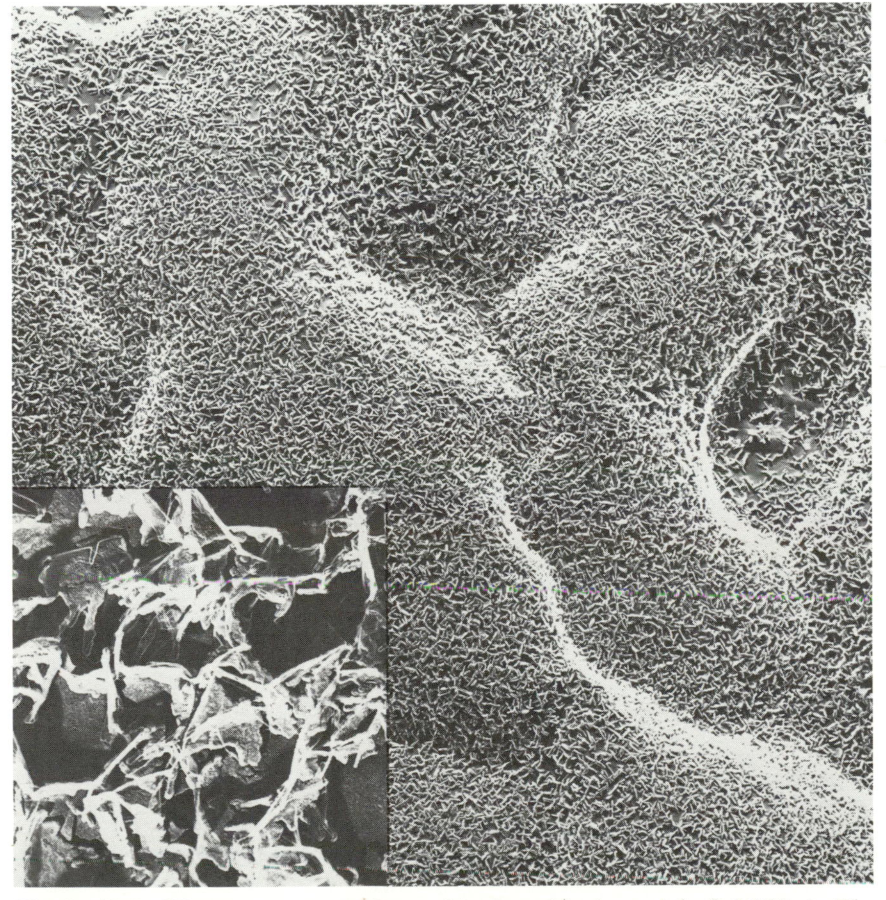

Figure 27-1 Wax on upper surface of juniper *(Juniperus)* leaf (4000×). The outlines of the epidermal cells can be seen as white areas. Below, high magnification (40,000×) of the wax.

Figure 27-2 An oil or fat molecule consists of three fatty acids joined to a glycerol molecule. These bonds are formed by the removal of a molecule of water, as was the case with polysaccharides. The figure shows three of the C_{18} acids combined in one molecule.

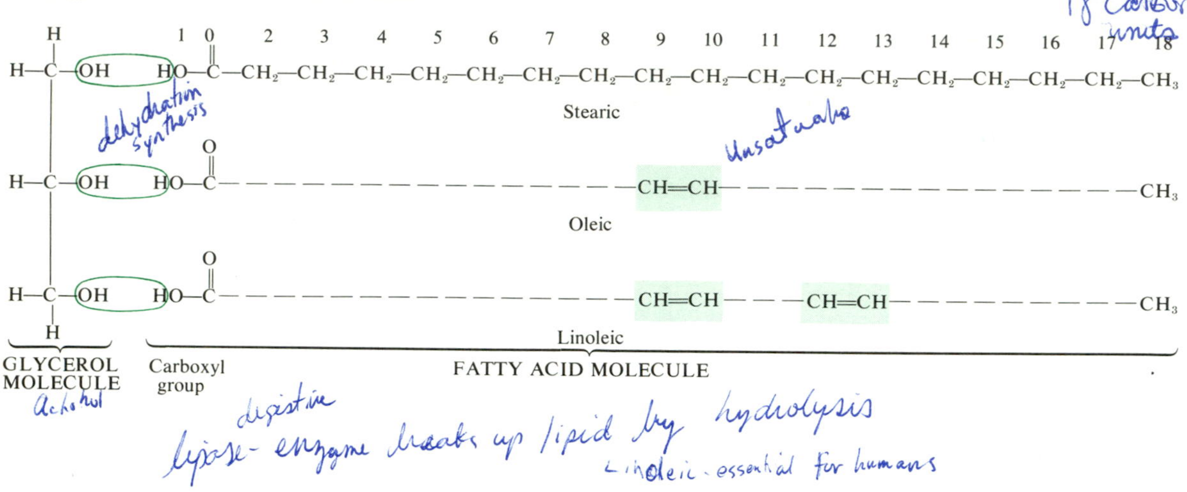

to glycerol and fatty acid. It may be pointed out, however, that linoleic and linolenic acids (Table 27-1) cannot be synthesized by mammals and must be obtained from plant sources; they are therefore called *essential fatty acids*.

Triglycerides are colorless and essentially tasteless. Therefore, the color or taste of fats and oils is due to small amounts of sterols, lecithins, vitamins, and so on. Most fats and oils are unstable and become changed when stored for relatively long periods, particularly at high temperatures and in presence of air, giving rise to aldehydes and ketones that cause the rancid taste.

Waxes are formed by the union of long-chain alcohols (instead of glycerol) with fatty

TABLE 27-1

Some Naturally Occurring Fatty Acids

World Distribution (in percent)	*Saturation*	*Common Name*	*Symbol*[a]	*Melting Point (°C)*
4	Saturated	Lauric	12 : 0	44.2
2		Myristic	14 : 0	53.9
11		Palmitic	16 : 0	63.1
4		Stearic	18 : 0	69.6
24	Mono-unsaturated	Oleic	18 : 1 (9¢)	13.4
34	Polyunsaturated	Linoleic	18 : 2 (9¢, 12¢)	−5
5		Linolenic	18 : 3 (9¢, 12¢, 15¢)	−11

[a]The first figure indicates the number of carbon atoms in the fatty acid chain, and the second gives the number of the unsaturated bond(s); the numbers in parenthesis give their locations. For example, 9¢ means between carbon atoms 9 and 10, etc.

acid molecules. Vegetable waxes contain a wide range of organic compounds and may differ chemically in different organs and with age, season, and local conditions.

3. FATTY ACIDS: SATURATED AND UNSATURATED

Fatty acids occur in straight, even-numbered chains (as a result of the mechanism of their biosynthesis from the two-carbon precursor acetate via acetyl coenzyme A). The product of this synthesis is an acid which, if every carbon atom in the chain bears two hydrogen atoms, is thus *saturated* with hydrogen (Fig. 27-2 and Table 27-1). The general formula for fatty acids of the saturated series is C_nH_{2n+1} COOH (lauric, stearic, myristic, and palmitic in Table 27-1). Removal of hydrogen from specific adjacent carbon atoms in the chain produces double bonds, and the acid is thus *unsaturated* in terms of hydrogen. In most unsaturated fatty acids in plants there is a double bond between carbon atoms 9 and 10; additional double bonds usually occur between C_{10} and the methyl-terminal end of the chain (Table 27-1). In fatty acids containing two or more double bonds, the double bonds are always separated by one of these methylene groups. The double bonds of nearly all of the naturally occurring unsaturated fatty acids have the two hydrogen atoms on the same side, i.e. H H, while H is the more

—C=C —C=C—
H

stable form; thus, for example, oleic acid may be converted on heating with certain catalysts into its isomer elaidic acid, which has a much higher melting point.

The great bulk of plant fatty acids are unsaturated. The seven acids listed in Table 27-1 alone accounted for 94% of the world's commercial vegetable fats in 1969. In general these acids are widely distributed in the lipids throughout the parts of all plants, but often palmitic, oleic, and linoleic predominate.

4. PHYSICAL PROPERTIES

Lipids are water-insoluble organic substances extractable from the cells by organic solvents such as chloroform, ether, and benzene. Vegetable fats differ only slightly from oils in having fatty acid constituents that are more or less solid rather than liquid at ordinary temperatures. Thus a fat in an arctic climate could be an oil in a tropical one. Unsaturated fatty acids have lower melting points than saturated fatty acids (Table 27-1); most oils rich in unsaturated fatty acids are liquids down to −5° C or lower. Because most of the plant fats and oils are composed primarily of unsaturated fatty acids, for simplicity we shall henceforth refer to them only as oils.

The degree of saturation of the constituent fatty acids is an important characteristic of oils. The more double bonds—note sequence in the C_{18} series from saturated stearic through progressively unsaturated oleic (one double bond), linoleic (two), and linolenic (three)—the more likely is the oil to oxidize as a waterproof film (Table 27-1). Oils are often classified for their use according to their ability to oxidize into this film, that is, as nondrying, semidrying, or drying oils. *Nondrying oils* are largely glycerides of saturated and oleic acids with little or no linoleic or linolenic acid present. They remain fluid for prolonged periods upon exposure to air. They occur particularly in tropical plants such as palm, peanut, olive, and castor bean. *Drying oils* are high in glycerides of the most unsaturated types, particularly linoleic and linolenic, but low in oleic compounds. These oils absorb oxygen readily from the air to form an elastic, waterproof film and occur primarily in temperate plants—linseed, soybean, and safflower. The drying oils are most important commercially and are used in such products as paints and varnishes. The *semidrying oils* are intermediate between the two other groups and thus dry slowly and at elevated temperatures; they include such oils as cottonseed, sunflower, sesame, croton, and corn. Both nondrying and semidrying oils are used commercially for soaps, detergents, and other industrial new materials.

The degree of saturation is also significant with regard to vegetable oils as foods, principally as cooking fats and oils and margarines. Recently concern about excessive levels of saturated fatty acids in the diet has increased since it has become known that these are important, along with cholesterol, in producing deposits that lead to hardening of the arteries. Table 27-2 lists the degree of saturation of some common plant fats and compares the saturation with those of common animal fats. It can thus be understood why vegetable oils, such as safflower, corn, and soybean, carry a premium in the manufacture of margarine. Almost all edible oil can be converted to cooking fat or margarine by elevating them to a higher degree of saturation through hydrogenation. This can be done by introducing hydrogen under pressure into heated oils in the presence of a catalyst. Margarine is made in this manner, with the addition of small amounts of milk products, vitamins, and other materials. Margarine is nutritionally equivalent to butter, can have a similar taste to it, and also is lower in saturated fats.

5. MAJOR OIL-PRODUCING PLANTS

Although plant oils are frequently categorized by the relative degrees to which they dry to form a hard film, for our purposes it seems more useful to group them as to their primary use either for food or for industrial purposes.

TABLE 27-2

Fatty Acids in Some Common Foods

Product	*Ratio of Polyunsaturated to Saturated*	*Fatty Acids as Percentage of Total Fat*		
		Polyunsaturated	*Mono-unsaturated*	*Saturated*
Walnuts	10.0	72	17	7
Safflower oil	9.0	72	15	8
Corn oil	4.8	53	32	11
Soybean oil	3.9	59	20	15
Fish (salmon)	3.5	53	25	15
Cottonseed oil	2.0	50	21	25
Peanut oil	1.6	29	47	18
Chicken fat	0.8	26	38	32
Olive oil	0.7	8	76	11
Margarines	0.5	13	57	26
Egg yolk	0.4	12	49	32
Butter	0.07	4	35	55
Beef fat	0.06	3	44	48
Chocolate	0.04	2	37	56
Coconut oil	–	Trace	7	86

Adapted from A. L. Elder and D. M. Rathmann, *Economic Botany* 16, 1962.

a. OILS USED PRIMARILY FOR FOOD

Olive Oil. The most important oil-producing plant in the world, both in history and modern commerce, is the olive (*Olea europea,* in the Oleaceae). Today truly wild olive trees are hard to find; apparently the tree once grew wild on the border of the Persian desert, where it was brought into cultivation about 7000 B.P. The tending of the tree and use of the oil were widely practiced by ancient Assyrians, Egyptians, Greeks, and Romans, and olives are frequently mentioned in the Bible (Fig. 27-3). The Greeks and Romans also transported olive trees to Spain and North Africa. Since they are adapted to a Mediterranean climate with dry summers, in modern times olive trees have been planted in California, Mexico, South Africa, and Australia, where they now constitute an important crop.

The olive tree generally is raised from cuttings and it grows into a spreading tree that endures for centuries. Buds are sometimes grafted on to wild stock in the Mediterranean area. Orchards are generally small, local enterprises, particularly in Spain, Italy, and Greece.

The fruit, which contains 14 to 42% oil, ripens slowly during the summer and starts to fall in the autumn, when it is tediously harvested by hand (Fig. 27-3). Most of the crop is taken immediately to presses where it is crushed to force out the oil. The good oil constitutes the main cooking oil in the Mediterranean region, where it often takes the place of butter. Oil of low grade, such as that from damaged olives, is used in making soaps, as a lubricant,

Figure 27-3 Olive-harvesting methods (left, from a classical Greek vase of about 400 B.C.; right, on a Greek plantation today) have remained much the same for more than 2000 years.

and as a fuel in oil lamps. The press cake is used as a fertilizer in orchards or as cattle feed.

When olives are to be eaten, they are carefully picked at the "strawberry-color" stage for green olives and at the "black" stage for ripe olives. The processing of olives neutralizes the bitter glucoside, oleurapein, present in fresh olives. The black olives are stored for several months in a weak salt solution, where lactic fermentation (Chapter 20) takes place, producing the particular flavor associated with "ripe" olives, and then treated with sodium hydroxide, washed, and stored in a dilute brine. "Green" olives are treated similarly except that lactic fermentation is omitted.

Rape Oil. Because the olive tree does not thrive in the cold, moist climate of northern Europe, northern people early sought a local substitute. They discovered the seed of rape or colza (*Brassica campestris* var. *oleifera*), which now is grown in both Europe and North America as an annual croplike grain. The seeds are threshed out and pressed for the oil; the seed cake is used to feed pigs or cattle.

Corn Oil. Most of the grain crops have little oil in their seed. Corn (*Zea mays*) is an exception in that the embryo yields 50% oil—mainly consisting of oleic and linoleic acids (Table 27-2). It thus has much the same properties as olive oil and is used principally as salad or cooking oil. Although difficult to extract, it is a sufficiently valuable oil for some farmers (particularly in the United States) to grow their entire crop for corn oil. The residue from oil extraction is rich in carbohydrates and protein and used for stock feed. Nonetheless, in terms of the enormous world production of corn, only a small portion goes into the making of corn oil.

Soybean and Peanut Oils. Soybeans (*Glycine max*) and peanuts (*Arachis hypogaea*) are both legumes and thus have been discussed in Chapter 23. Soybean oil now outranks cottonseed as the chief vegetable oil produced in the United States. The soybean contains 13 to 25% oil, 30 to 50% protein, and 14 to 25% carbohydrate. After solvent extraction of the oil, the meal is nearly 50% protein and is used primarily as stock feed. The oil is extensively used in the Far East as food and in the West as salad oil and for the manufacture of margarine and shortening.

The peanut or groundnut contains 36 to 45% oil shelled. The oil is used mainly in edible products such as salad or cooking oils and to a lesser extent for soaps and some industrial purposes. It is high in oleic acid and has moderate amounts of linolenic acid (Table 27-2). In the United States only 15 to 25% of the crop is usually expressed, depending upon the relative demand for peanuts in peanut butter (in the United States frequently more than 50% of the crop is ground for peanut butter) and candy or as salted or roasted peanuts. Like that of the soybean, the press cake of peanut, rich in protein and containing 6 to 8% oil, is excellent livestock feed.

Cottonseed Oil. Prior to the American Civil War and invention of the cottonseed oil mill, the seed from the cotton plant (*Gossypium* spp.) was usually discarded. However, today this seed, a by-product of the cotton fiber industry, is often as valuable as the fiber. The kernels of the seed are about 35% oil, mainly linoleic, oleic, and palmitic (Table 27-2), and 10% protein. Eighty percent of cottonseed is used in production of stock feeds and fertilizers. The United States is the greatest utilizer of cottonseed oil. The protein is used not only for cattle feed but to fortify human foods in regions of protein deficiency (e.g., Incaparina for Latin America).

Safflower and Sunflower Oils. Two members of the sunflower family (Asteraceae or Compositae) are important sources of oils. The flowers of safflower (*Carthamus tinctorius*) have been used as a source of yellow dyes and as a substitute for saffron since ancient times, but only recently have the seeds become prominent as a source of unsaturated oil. The plant is known now only from cultivation but may have originated from several species occurring in northeastern Africa and Asia Minor. Safflower was first cultivated in Egypt and somewhat later in China. It was introduced into Mexico by the Spanish explorers. Perhaps because of its mixed heredity the chemistry of the oils is quite variable, so that considerable effort has been put into breeding plants for specific oil composition and high yields. It has become a favorite candidate for a new crop for drier climates in North America because it is both drought and salt tolerant. The crop does well where barley is grown; however, safflower seed is twice as valuable as barley grain and, if yields are comparable, a far more remunerative crop.

The sunflower (*Helianthus annuus*), an American native, has been cultivated by the American Indians since pre-Columbian times for edible seed (Fig. 27-4). The wild ancestor is no longer known. The species is relatively drought tolerant and thus fits into rotation schemes in small grain-farming areas. Sunflower oil is a major crop of the Soviet Union; in fact, more than half of the sunflower oil of the world is produced there. The oil is mostly used for edible purposes such as shortening and margarines. It is very similar to safflower oil and more stable than soybean oil.

Sesame Oil. The seed of *Sesame indicum* (Pedaliaceae) contains as much as 55% oil, consisting primarily of oleic and linoleic acids. The oil is extracted by expressing and used for salad and cooking oil and the seed is used as a garnish for bakery products. Most of the production comes from areas in China and India where agriculture is not mechanized. Harvesting is often done by hand by hitting the capsules with a stick. Efforts are now being made to develop nonshattering cultivars suitable for mechanized production in the United States, because it has been predicted that sesame seed can yield more per acre than any other annual oil crop under mechanized agriculture.

b. OILS USED FOR BOTH FOOD AND INDUSTRIAL PURPOSES

Palm Oil. Derived from a number of members of the palm family (Palmaceae), the

Figure 27-4 Field of sunflowers grown for their seed.

most important sources are the coconut (*Cocos nucifera*—see Fig. 18-9), the African oil palm (*Elaeis guineensis*—Fig. 27-5), the babassú, and other South American palms (species of *Orbigyna, Attalea,* and *Astrocaryum*). Palms are the single most important plant family in the tropical economy, being used for multifarious purposes; oil is only one of the many products derived from them.

High-yielding palms have been cultivated in Java by the Dutch, and many small farmers in India have palm plantations, but despite their very great utility, palms are not widely cultivated elsewhere, except as ornamentals in landscaped areas. Instead, materials are collected from abundantly distributed wild plants. Most palm fruits used for oil are often simply picked up from the ground where they have fallen. The fruits of babassú palms (the almost ubiquitous *Orbigyna speciosa*), for example, are gathered by local people in Brazil as a major source of sustenance.

A British company, Unilever, has recently made use of tissue culture techniques to produce high-yielding oil palms. It was noticed in plantations that a particular tree (perhaps one in

Figure 27-5 The African oil palm *(Elaeis guineensis)*. A. The plant, B. The fruiting spadex C. Details of fruits and seed. This is one of the world's most important sources of edible and soap-making oil, which is obtained from the fibrous-pulpy mesocarp and from the seed or "kernel". Palm oil before refining is a yellow-orange color due to the pigment carotene, from which the human body can form vitamin A; this gives palm oil a special value in the diet. (From Masefield, 1969)

10,000) produced an excellent balance between saturated and unsaturated fats, grew rapidly, and was resistant to disease. Selective breeding is difficult for oil palms in that (a) they take 15 to 20 years to reproduce, and (b) they are self-sterile; that is, it is not possible to pollinate a superior tree with its own pollen. Also the usual methods of asexual reproduction do not work with oil palm. The production of young plants by tissue culture has been successful, since all of the resulting plants (excluding mutations) are exactly like the parent.

Palm oils may be extracted from the mesocarp of the fruit (pulp oils) or from the endosperm in the seed (kernel oil). Kernel oils generally are high in lauric acid (Table 27-2) and are most commonly used in making soaps. Pulp oils vary; in some cases they resemble kernel oil, whereas others are similar to olive oil. Various oils from palms, especially from the African oil palm, are used in lubricants, margarines, and cosmetics.

Figure 27-6 The castor bean plant *(Ricinus communis)*, an example of the family Euphorbiacaeae, to which the rubber plant, manihot, and poinsettias also belong. The plant often flourishes in disturbed areas such as shown here in a clearing of forest in the Central Amazon Basin.

C. OILS USED PRIMARILY FOR INDUSTRIAL PURPOSES

Castor Bean Oil. One of the most important industrial oils, it is obtained from the seeds of *Ricinus communis* (Euphorbiaceae). The castor bean plant probably was indigenous to Africa but is found both wild and cultivated throughout the world (Fig. 27-6). In temperate areas, it is planted frequently as a summer ornamental. In the past, harvest of castor beans has been chiefly by hand in cheap-labor areas such as Brazil, India, Thailand, and Mexico. However, dwarf strains with capsules that do not shatter readily have been bred, so that castor beans are now machine harvestable.

Oil content of the seed ranges from 35 to 55% and is removed primarily by pressing the seed; the remains are then extracted with solvents. The press cake contains the toxic protein ricin and therefore is unsuitable for stock feed. Castor oil was used in lamps by the ancient Egyptians but today is used most commonly as a lubricant and in the manufacture of a large number of products such as soaps, synthetic rubber, imitation leather, linoleum, oilcloth, and plastics. It is also an ingredient of nylon. Through chemical alteration it can often serve other purposes such as substituting for drying oils, as a plasticizer, and in medicinals.

Tung Oil. Produced from several species of *Aleurites* (Euphorbiaceae), the fruits contain two to five heavy-shelled seeds with a kernel that contains 65% oil. *Aleurites fordii* and *A. montanus* from China produce some of the world's best drying oils. *Aleuritis fordii* was introduced to the New World and now is grown on plantations along the Gulf Coast of the United States. In China it has long been used in medicinals, but its ability to dry rapidly into a tough coating has now made it particularly valued for paints and varnishes.

Linseed Oil. Another excellent drying oil, it is used to form the base of many paints and varnishes. When mixed with powdered cork (or plastic) and spread on a fiber base such as jute, it dries into a hard, waterproof linoleum. The oil is obtained from the small seeds (Fig. 27-7) of

Figure 27-7 Linseed oil, an industrial drying oil, is extracted from the seeds of flax plants shown growing with abundant seed here.

the flax plant (*Linum usitatissimum*), which is also cultivated widely for its linen fiber (Chapter 21). Generally the farmer who is concentrating on the oil crop uses varieties that have been bred to yield much seed but little fiber. Linseed contains as much as 43% oil. The United States and Canada are the chief producers; Argentina and India are also important.

6. MAJOR WAX-PRODUCING PLANTS

Unlike oils, waxes are solid at atmospheric temperatures, although they melt readily when heated. This means they can be used for polishes on surfaces to give a hard, smooth, attractive finish. Because they are highly repellant to water they also are used for waterproofing leather, textiles, and wood.

Vegetable waxes are little used compared to synthetic and petroleum waxes (paraffin) because preference for one wax over another is largely a question of the quantity obtainable. The quantity of wax available from most plants is relatively small and the plants that do yield large amounts generally grow in inhospitable, arid environments. Thus competition from cheaper sources has largely limited use of vegetable waxes to a few special types such as the comparatively expensive but unequaled carnauba wax and the potential substitution of jojoba "oil" (liquid wax) for sperm oil.

a. CARNAUBA WAX

The slow-growing carnauba palm (*Copernicia cerifera*) occurs commonly in the dry, hot savanna areas in northeastern Brazil (Fig. 27-8). It is here that the carnauba, named "the tree of life" by the traveling scientist Alexander von Humboldt (Chapter 14), supports a local economy. At one time the big, fan-shaped leaves of the wild carnauba trees were cut indiscriminately, but more recently the harvest has been restricted by law so as not to destroy the species. Only a certain number of leaves are allowed to be removed during a given year, that is, two cuttings of 10 to 25 leaves each during the dry season (when the wax content is highest). Only recently the trees are being grown in plantations. Following harvest the leaves are shredded, allowed to wither, and then beaten until the wax falls off. The wax is subsequently melted and molded into blocks for export.

Carnauba wax is primarily myricyl cerotate, never chemically synthesized and unique in its hardness, high melting point, and ability to take a high polish. Most of the crude wax goes to the United States where it is used in polishers, lubricants, floor and automobile waxes, carbon paper, phonograph records, plastics, films, and many other products.

b. JOJOBA "OIL"

This so-called "oil" chemically is a liquid wax from jojoba (*Simmondsia chinensis*, Buxaceae), which grows wild over large areas of the Sonoran desert in Arizona, California, and

(A)

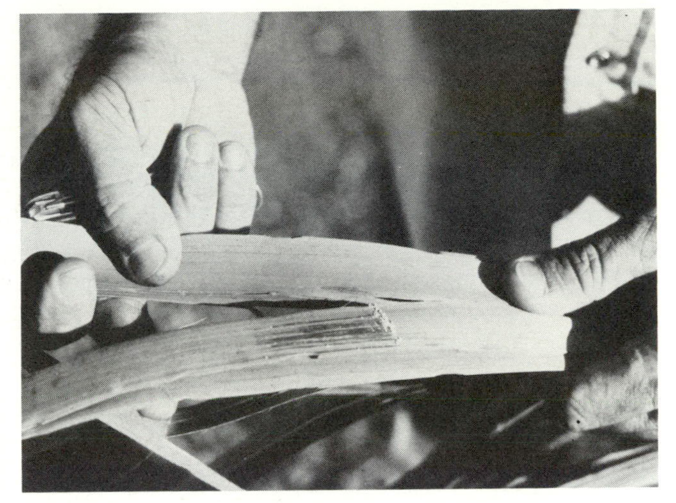

(B)

Figure 27-8 A. Carnauba palms *(Copernicia cerifera)* from Ceará in arid northeastern Brazil where they form extensive stands in periodically wet valleys. The native stands are so large that cultivation of these palms is not common as yet. B. The leaves yield an exceptionally high-quality wax for shoe, furniture, and floor polish. The leaves are cut from September through December with knives at the end of a long pole. The wax occurs in particularly thick layers on the underside of the leaves are shown here, where it has been scraped with a knife.

Mexico. The story of the name of this plant illustrates the problems that plant taxonomists encounter and how the International Code of Nomenclature operates. In 1822 the first specimen to come to the attention of a taxonomist (H. F. Link) was considered to be a kind of box (*Buxus*). Erroneously thinking the specimen had come from China, he named it *Buxus chinensis*. In 1844 T. Nuttall collected the plant from southern California. He recognized it as a genus new to science, and named it in honor of the naturalist T. W. Simmonds, choosing an appropriate specific name *Simmondsia californica*. Now the International Code of Botanical Nomenclature holds that the first species name given must be used, even if, as in this case, it is misleading. So the correct name is *Simmondsia chinesis*. Also, although some taxonomists still think *Simmondsia* belongs in the box family (Buxaceae), others consider it to be a member of a unique family (Simmondsiaceae).

Jojoba seeds contain 50% liquid, unsaturated wax. Waxes of this type are difficult to synthesize commercially, and the only other natural source is the endangered sperm whale producing a wax called *spermaceti*). Because jojoba wax is liquid, it is useful for the manufacture of extreme high-pressure lubricants. Without this wax, the power transmission in many modern heavy-duty vehicles would fail. Jojoba oil is less variable than sperm whale oil and can substitute for all of the uses at present made of sperm whale oil.

The jojoba plant, although usually a low shrub, may attain heights of 3 m and forms a relatively dense cover in the desert that can be browsed by livestock (Fig. 27-9*A*). It is long-lived and can grow without irrigation in areas with rainfall less than 24 cm. Mature plants can yield up to 12 lb of seed (dry weight) and appear to be resistant to serious diseases and pest damage (Fig. 27-9*B*).

Considerable interest has been shown recently in developing a jojoba agriculture in the Sonoran desert, both because it can increase the productivity of arid lands not suitable for conventional crops and because it can substitute for sperm oil, thus conserving the whales (Fig. 27-9*C*). Additionally, it can substitute for beeswax and carnauba wax.

(*A*)

(*B*)

(*C*)

Figure 27-9 A. Jojoba shrubs and large cacti are the dominant plants in some areas near Phoenix, Arizona. Here the lush shrubs illustrate the capacity of the plants to grow well in such desert areas. B. Jojoba seeds ready for harvesting on a native plant. C. Five-year old jojoba plants in cultivation. The plants are tied up to increase population density and production potential per hectare.

PART 5

SOME WORLDWIDE PERSPECTIVES

CHAPTER 28

The Place of Gardens in World Cultural History

1. THE GARDEN OF EDEN AND ITS PHILOSOPHICAL COUNTERPARTS

"And out of the ground made the Lord God to grow every tree that is pleasant to the sight and good for food. . . ." So runs the biblical legend of the Garden of Eden. It embodies what for mankind had long been the concept of a garden—an idyllic place where everything that is good flourishes. The full interpretation of the Adam-and-Eve story is difficult and obscure, but one thing about the Garden is especially characteristic, that in the Garden stood "the tree of the knowledge of good and evil." Strolling in the garden, this seems to mean, is associated with the ability for profound and clarifying thought. Thus a garden is conceived of not only as an idyllic and peaceful spot but as a place potentially stimulating to the mind.

Perhaps the idea of the tree whose fruit confers the knowledge of good and evil is paralleled in the practice of primitive medicine men; even now in South America these men imbibe an extract of a drug plant (*Banisteriopsis*) that is said to make them able to diagnose diseases and foretell their probable outcome. Eve's mythological "apple" may indeed have been derived from ancient knowledge of some plant that yields a harman-like alkaloid with effects on the brain like those discussed in Chapter 24.

The thought-provoking concept of the garden was found again among the early Greeks. The first actual gardens we know about in classical Greece were associated with the *Gymnasia,* the Greeks' sports grounds, of which there were at least three in the suburbs of Athens. Each had a public garden attached to it. These gymnasia were not merely running tracks but, true to the concept of the garden, became the centers of schools of philosophy. The most famous was the Academy,* in which Plato taught his followers.

Epicurus, another famous philosopher, also taught in a garden. His followers were called "they of the garden" and his successor was referred to as "king of the garden." In the Lyceum, a third Athens gymnasium and garden, Aristotle (384 to 322 B.C.) taught, a little later than Plato; his school is called the Peripatetics because he walked in the garden as he taught. Also, adjacent to the Lyceum was yet another garden, that of Theophrastus (370 to 285 B.C.), who was Aristotle's pupil and became his successor. Theophrastus, as noted in

[1] It was of the Academy that Aristophanes wrote:
"But you will below to the Academe go; and under the olive contend
With your chaplet of reed, in a contest of speed, with some excellent rival and friend;
All fragrant with woodbine and peaceful content, and the leaf which the lime blossoms fling
When the plane whispers love to the elm in the grove in the beautiful season of spring."
(From The Clouds, translated by Rogers).

Chapters 3 and 13, used his garden for the study of plants and wrote two multivolume treatises on them. Not only was he thus the first real botanist, but he started a different type of garden tradition, that of the botanic garden.

The garden tradition of modern Japan has, curiously enough, much in common with the Greek ideas, not because the Japanese garden is a place for teaching, but rather because it is, in its conception, a place to stroll in, a place for contemplation and for peace. Buddhist temples are typically surrounded by gardens and most of the royal palaces also have extensive gardens. In the Japanese garden water plays a prominent role, with clipped bushes and carefully pruned conifers around it, together with rocks and an occasional stone lantern (see Fig. 28-1). The philosophical context is at its strongest and strangest in the well-known monastery garden of Ryoanji, where a large field of small white stones, raked into long parallel lines, is broken by five rocks of different sizes. This "garden" without plants is intended for meditation—the extreme of horticultural abstraction.

In a modified spirit, less philosophical and more pattern-oriented, are the formal gardens of France and Italy. These are especially notable in the French chateaus (Fig. 28-2). Here the clipped individual bushes of Japan are substituted by clipped hedges or "vegetable embroidery," and the purpose is purely decorative rather than to provoke thought.

2. BOTANICAL AND MEDICINAL GARDENS OF THE EARLY AND MIDDLE AGES

The second type of garden could be considered primarily utilitarian—a garden intended specifically for the cultivation and study of individual plants. And because plants, or plant extracts, provided the whole basis of early medicine, such gardens were medical as well as botanical. This tradition has a parallel origin in the secret knowledge of the priest, who was also a medicine man and who not only knew where to find the curative herbs in the wild (as do African medicine men today) but grew some of them for his own use in a private enclosure.

It was several centuries after Theophrastus that Dioscorides, in his *Materia Medica* (about 50 A.D.), described some 600 medicinal plants as well as medical preparations of animal origin. As noted in Chapter 13, this was the first book to include drawings of the plants (400 of them in full-page size) and its Latin translation remained influential for over 1000 years.

Possibly even earlier than the Greek and Roman gardens was the beginning of medical botany in China, for there is a tradition that the Emperor Chen-Ning (see rice, section 5 in Chapter 16) wrote a treatise on the medicinal plants about 4700 B.P. It was probably about 24 A.D. that the Chinese began to make tea, not only from *Camellia sinensis,* but also from other plants, and these infusions were used as medicines. The drug ephedrine was also known, as an extract from *Ephedra* spp., and was used to relieve asthma at that time. Chinese use of medicinal plants persists to the present day. Indian traditional medicine (the Ayurvedic system) is also largely based on curative plants. In the West, unfortunately, medical students are no longer taught botany of flowering plants; instead they learn only something (a minimum) about the fungi and actinomycetes that produce the all-important antibiotics.

The botanical gardens came into prominence in Europe only in the sixteenth century. A botanical garden at Cordoba in Spain is said to have been laid out by the Moorish Caliph about 780 A.D., but the first botanical garden of which we have reliable records was at Padua University in 1533. At that time many famous professors combined appointments in medicine and botany; for instance, Caspar Bauhin, after whom the legume *Bauhinia* is named, was professor of anatomy and botany, and physician to the city of Basel. Indeed, one of the small botanical gardens in London is still called the Chelsea "physic garden."

Figure 28-1 Representative Japanese gardens in the vicinity of Kyoto showing the importance of meticulously shaped bushes, the placement of stone ornaments, and the role of water.

Other botanic gardens followed closely after Padua, gradually shifting emphasis to the plants themselves rather than their medicinal uses. By the end of the eighteenth century there were 1600 botanical gardens in Europe, and even one in America (New York).

It was at that time that the great herbals were written. These were botany books, describing plants in terms of their medicinal qualities (see Chapter 13). The earlier herbals, of the thirteenth to fifteenth centuries, had been mixtures of fact and fancy; they included pictures

Figure 28-2 The formal gardens at Chateau Villandry in the Loire district of France. The low hedges forming the "vegetable embroidery" are clipped box *(Buxus sempervirens)*.

of fabulous plants whose fruits were marine animals (!) along with ordinary plants that we can recognize today. But in the sixteenth century this mythology fell away, giving place to careful drawings and descriptions of plants evidently observed growing in the garden as well as those collected from the wild. John Gerard's *Herbal* (1597) contains the first drawing of a potato plant (Fig. 28-3). Andrea Cesalpino, after whom one of the three subfamilies of legumes is named, in 1583 wrote 16 books describing both mature plants and seedlings, which shows that he too grew many plants in his garden. The two herbals of John Parkinson (1629 and 1640) are notable for his attempts at a systematic arrangement of the plants into classes. Some of these are: "Sweet-smelling Purging plants; Venemous, Sleepy and Hurtful plants and their Counterpoysons: Vulnerary or Wound Herbes; Hot and Sharpe Biting plants," and so on. To some extent these classes do coincide with the Linnean families; for instance, the first mentioned class includes mint, thyme, savory, catnip, sage, and basil, which are all members of the mint family (Lamiaceae or Labiatae) a family rich in sweet-smelling terpenoids (Chapter 26). However, he also includes plants that have nothing in common with the Labiatae but their sweet smell. Evidently the coming of Linnaeus, with his structurally based taxonomy, did literally bring botanical order out of medicinal chaos. Even in the latest herbal, that of Nicholas Culpeper (1653), there is no notion of how flowers reproduce. Culpeper describes the stamens as "threads in the middle" and says the petals of some composites "turn into down and fly in the wind."

By the late seventeenth century botanic gardens had become more and more places for scientific research. Sexuality in plants was discovered in the botanical garden at Tübingen by Rudolf Camerarius, Professor of Physic (medicine). He noted that the fruits of a mulberry tree that was isolated in the garden were seedless. He also found that when he isolated one type of flowers on castor bean (*Ricinus communis*) or on corn (*Zea mays*), by removing the staminate flowers, these plants also set no seeds. He thus attributed seed initiation to "that powder which is the most subtle part of the plant," formed in the stamens (i.e., the pollen), and recognized that the style and ovary are the female organs.

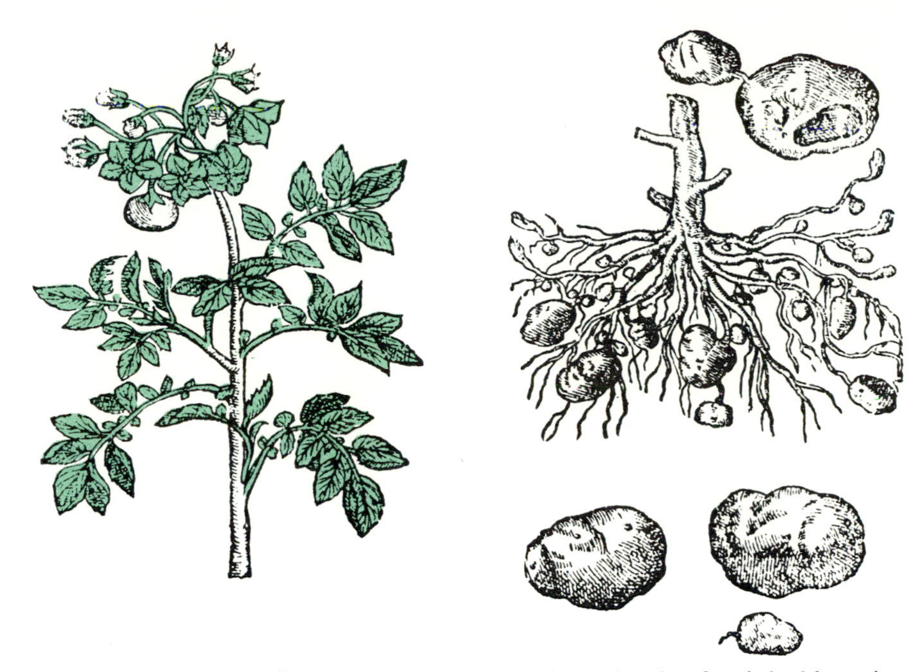

Figure 28-3 The first drawings of the potato plant, shortly after it had been introduced into Europe. (From Gerard's Herbal, *The Historie of Plants,* 1636.)

Johan Gleditsch, director of the Berlin garden, followed up by fertilizing a female palm growing alone in the garden with pollen from a male tree growing at Leipzig (cf. also Fig. 18-12).

The rapid growth of botanical gardens from that time on was partially due to the growing exploration in many parts of the world. Plants were not only being brought back from afar, but botanical gardens were developing new horticultural forms of them. Thus in 1789, *Chrysanthemum indicum,* the ancestor of our cultivated chrysanthemums, was introduced in London. Kew Gardens, near London, has been the most influential of all; it began as a private royal garden early in the eighteenth century and only in 1840 was taken over as a national institution. Kew's function as a plant introduction center is exemplified in the history of rubber (Chapter 26). Also in the last two centuries it was often found that plants in the gardens failed to reproduce, or perennials simply died out, so that to maintain a record of plants that had been grown, as well as those that had been collected in nature, it was necessary to preserve them. This is done by drying the plant, or parts of it, between absorbent sheets, under pressure. The resulting *herbarium specimens,* with details of when, where, and by whom collected and identified, make up a collection that requires little storage space and is permanently available for study. Some of the world's largest reference *herbaria* are in the botanical gardens at Kew, Berlin, New York, Paris, Munich, Harvard and St. Louis (Missouri). These herbaria and gardens have become major centers for systematic studies. At Kew, Bentham and Hooker put together from 1862 to 1896 the *Genera of Plants,* a description of all the world's plant genera. At Berlin Engler and Prantl compiled, from 1897 to 1936, the great *Syllabus of Plant Families,* a still more comprehensive organization of the plant kingdom, which is still commonly used for arranging specimens in herbaria worldwide (Chapter 13).

In the twentieth century increasing numbers of botanical gardens are being developed in tropical countries including Java, Brazil, Venezuela, and Ghana, often in association with universities, so that scientific research can go hand in hand with public parks.

3. GARDENS FOR RAISING FLOWERS, FRUITS, AND VEGETABLES

A third traditional function for gardens represents what the word "garden" mostly means to us today. This function has apparently existed from early times, too, though it has not been so prominent in history. Before 3000 B.P. the Assyrian kings are said to have brought vines and fruit trees from other lands for their gardens. From the inscriptions in Egyptian tombs we know that the ancient Egyptians raised flowers to make funeral wreaths, and there are hieroglyphic pictures of onions, figs, dates, beans, and lettuce, which were probably grown there in home gardens. The Romans (about 150 B.C. to 350 A.D.) raised 16 different vegetables; many of the Roman houses excavated at Pompeii and Herculaneum, which have been preserved in volcanic ash by the eruption of Vesuvius in 79 A.D., were built around enclosed gardens for fruit and flowers (Fig. 28-4). This type of house, enclosing a garden on three sides, is being revived by some modern architects, and it embodies something of the first type of garden, the "pleasaunce" for strolling and for thought. The conception of enclosure, or of privacy, has always been close to the garden idea and is often shown by the use of high brick or stone walls, on the warm, sunny side of which fruit trees or exotic vines can be trained.

China, with its rich natural flora, has been a great center of gardens. When the last Ice Age covered most of Europe, eastern China remained more or less temperate, so that its natural flora is said to contain more flowering plants than all the rest of the temperate Eastern Hemisphere combined. Perhaps as a result, flowers have always been prominent in Chinese art and culture. A number of the European flowers and trees originated in China and came to Europe via Persia and the Mediterranean countries. Some of these betray their travels in their

Figure 28-4 One of the enclosed Roman gardens, the garden of the "House of Silver Nights," at Pompeii, A.D. 79 (restored).

names; for example, the peach and apricot, whose origins are Chinese, are *Prunus persica* and *Prunus armeniaca* respectively.

Indeed the garden of fruits and flowers that is also a "pleasaunce" was prominent in the Middle East in medieval times. At the time of Genghis Khan's *siege* in 1220, the city of Samarkand in Persia is said to have had around it for "several scores of miles, orchards, groves, flower gardens and aqueducts." Coleridge's description of the pleasure-dome that Kubla Khan built in Xanadu (Persia) speaks of "gardens bright with sinuous rills, Where blossomed many an incense-bearing tree."*

From medieval times onward in Europe every squire or landowner had a fruit and vegetable garden, and so did the great institutions. As we saw in Chapter 11, Gregor Mendel, the abbot of the monastery at Brno in Bohemia, laid the foundations of the science of genetics with the peas in the monastery garden there – though this was in the nineteenth century. The royal gardeners in England and France several times went on expeditions to the Mediterranean or the Near East to bring back rare plants for the gardens of the palace. *Tradescantia virginica,* a popular blue-flowered monocot, is named for John Tradescant (head gardener to King Charles I) who found it on a trip through Virginia.

Another development of the same period was to grow orange and lemon trees and oleanders (*Nerium oleander*) in large tubs, so that they could be brought indoors in winter. This custom may have been started by Emperor Frederick II (1194 to 1250), whose gardens at the Nuremberg castle were stocked with orange and lemon trees, which of course could not endure the frosts of central Germany outdoors. The custom spread to England and France in the sixteenth century. The need to bring these trees indoors led to the building in many large houses in Britain and France of an "Orange Court" or an "Orangerie," a spacious gallery with windows facing south. The English "conservatory" of many modest Victorian homes has a similar origin, though usually intended to "conserve" tender houseplants. It often has a partially glassed-in roof and thus led to the modern all-glass greenhouse, the first of which was built in Germany in 1619. The most modern experimental greenhouse, containing a number of different artificial climates, is that at the St. Louis botanical garden (Fig. 28-5).

Of course there were also many hardy fruit trees in western Europe, and these were either native in England or were introduced there, especially in the fifteenth and sixteenth centuries. As early as 1440 an English manuscript gives a list of 78 plants suitable for cultivation, many being herbs for kitchen or medicinal purposes. By the sixteenth century farm and cottage gardens everywhere were growing vegetables and herbs for seasoning.

With the spread of gardens there slowly developed the trade of nurseryman. The fondness of Romans for roses (see section 5) led to the appearance of professional rose-growers as early as the second century A.D., though the main development was in France, Belgium, Holland, and England in the sixteenth and seventeenth centuries. Long before that, and as early as 1250, the possessors of large gardens were selling their surplus produce to the public. But it was the rise of professional seedsmen, orchardists, and nurserymen that was mainly responsible for the rise of the science of horticulture.

4. MODERN HORTICULTURE

Not only do town dwellers, particularly in temperate zones, take pleasure in raising fruit, flowers, and vegetables in home gardens, but large-scale horticulturists have built an extensive industry on supplying such needs. In England a "Gardeners" Company and a "Fruiterers" Company – which we would now call wholesale

* Coleridge's botany betrayed him there, for most "incense-bearing trees" are plants of arid or semiarid soils (Chapter 26).

Figure 28-5 The "Climatron" in St. Louis Botanical Garden, Missouri, U.S.A. Separate areas within this large greenhouse have separate temperature and humidity controls, so that plants from a wide variety of climates can be grown, for example, Amazonian forest, mist forest (with a waterfall), Mediterranean, temperate, and dry tropics. The public can watch research being done also.

nurserymen and orchardists—were chartered in 1605. In the last two centuries large areas of flower growing, (e.g., the bulb fields of Holland, or the carnation fields of eastern Spain) have been developed outdoors, and still greater quantities are raised in greenhouses. The southern counties of England, the region of Russia around Leningrad, and the northeastern states of the United States are notable for their vast acreage of heated greenhouses. The French make more use of low glass frames a foot or two above the ground ("cold frames") and use animal manure, which in its decay warms the soil. Besides flowers, the supply of seedlings and young plants for home gardens is also a major business. One typical catalog of an American nurseryman lists over 300 *genera,* many with three to six species. This great variety is the result of the continued use of a relatively few botanical procedures, most of which were widely adopted only in the last century or so but have ancient origins. Some of these methods have been described in Chapter 12, so we give only a short summary here.

a. SELECTION

Probably the earliest method of plant improvement was selection for seed size. In general this method is not very effective, because variation in the size of seeds is often due as much to fertilizer, water supply, and freedom from disease as to heredity. But selection for size or growth rate of plant, for size, color, and earliness of flowers, for sweetness or succulence of fruits, and for resistance to disease has been of prime importance. Whenever one grows a few hundred seedlings to maturity there are almost always a small number that differ from the average in such useful characters as these. They may be true mutants or may merely represent a favorable combination of genes, but their seeds can be the starting point of an inbreeding program designed to increase the frequency of this useful character, or else of a hybridization program designed to introduce the new character into the line already being grown.

A good deal of selection is also done directly from the wild, to provide mature woody plants for landscaping; these usually require less care than introduced plants.

b. GRAFTING

The essence of successful grafting is to bring the tissue of the *scion* (the new upper part) into contact with the cambium of the stock, so that as the two grow together vascular bundles will differentiate and thus supply the scion with xylem and phloem sap. The principle is simple but the practice is tricky.

To graft a bud, a T-shaped cut, a little bigger than the proposed bud, is made in the bark of a branch of the tree or bush that is to be the stock; the corners of bark are lifted up and the bud, with a thin slice of the cortex of its own stem attached, is inserted (see Fig. 28-6). The base of the bud piece must be cut at the right slope so that it will rest flat against the cambial region of the stock. The graft is usually covered with soft wax ("grafting wax") to keep from drying out and tied tightly with bast or tape. After a few days the part of the branch above the graft is cut away so that there will be no source of auxin above the new bud to inhibit its growth.

To graft a twig, a vigorous shoot about the diameter of a pencil is selected, its base is cut into a long V and inserted into a cleft made near the outermost xylem of the stock, which may or may not be decapitated (Fig. 28-7*A* and 7*B*). If the stock and scion are of the same size the cleft can be located centrally or the stock can be cut off at an angle, the scion cut with the same slope, and the two tied together and waxed.

Still another procedure is to have both stock and scion (about the same age) remain on their own roots, bring a branch of each together, make a shallow slice through the cortex of each, just down to the xylem (Fig. 28-7*C*), and tie them tightly together. The branch of the stock above this "approach graft" is cut off a little later.

c. VEGETATIVE PROPAGATION

As an alternative to grafting, some new varieties can be propagated by the planting of root segments that bear a bud. When the plant has several stems from the base, it can be propagated simply by dividing the base or crown into parts, each having a stem and some of the roots. Plants forming rhizomes, like the iris, or runners, like the strawberry, are of course very easily divided, and the same is true for tubers, each bud of which can become a new plant, and for bulbs, which form secondary bulbils from the base (see Chapter 19).

With trees and shrubs, which usually take some years to flower, vegetative propagation is usual. This can be done by the rooting of cuttings (Chapter 10), or by grafting.

d. POLYPLOIDY

The role of natural polyploids in the history of cultivated plants has been noted several times in this book, most prominently in Chapter 12 and in the discussion of wheat (Chapter 16), cotton (Chapter 21), and tobacco (Chapter 24). In the last half-century or so many garden plants have been improved in this way, using colchicine to induce polyploidy. Some zinnias are even 16-ploid. Tetraploids appear in every nurseryman's listing of vegetables and flowers.

e. CELL AND TISSUE CULTURE

The growth of whole plants from single cells and the possibilities of hybridization of protoplasts were taken up in Chapter 11. Cell cultures have been particularly used for propagation with orchids, because their tiny seeds require complex nutrient media, kept under sterile conditions, and take around five years to reach the flowering stage. In Morel's method, perfected in Paris, the base of the plant is cut up

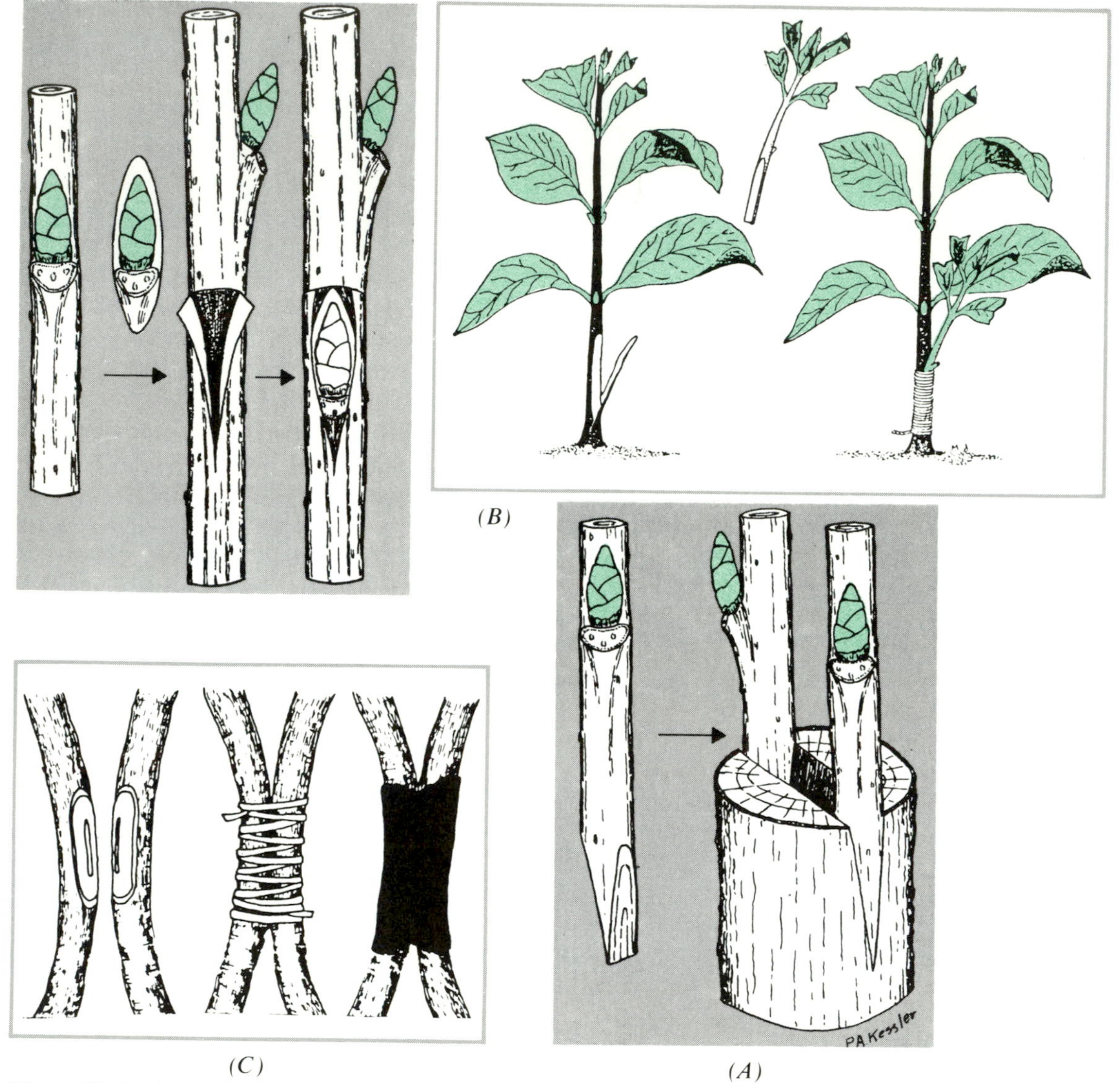

Figure 28-6 An example of the technique of budding: the T-bud or shield bud.
Figure 28-7 Two of the main techniques for grafting. *A*. Cutting the scion and inserting it into the stock. *B*. an actual example of medicinal importance; grafting *Cinchona ledgeriana* on stock of *C. succirubra;* the upper part of the *succirubra* stock will be cut away later. *C*. the approach graft, with both plants on their own roots; one will be cut away after the graft has taken.

into small fragments that are cultured in a liquid nutrient medium. Hundreds of tiny plants are thus produced and, unlike seedlings, all are genetically *identical*. Using this method, orchids are now being propagated in very large numbers, as seen in Fig. 28-8. In another method, the tips of the orchid leaves are cut off and cultured in a similar nutrient, and new plantlets develop from these. The method is now being applied to the propagation of new

Figure 28-8 Raising *Cattleya* orchids wholesale in a modern commercial greenhouse in San Francisco.

Figure 28-9 Propagating *Narcissus* cultivars by tissue culture, at the John Innes Institute in England. The basal segment of a bulb-leaf rests on a nutrient medium containing all minerals, sucrose, vitamins, and hormones. By manipulating the concentrations of auxin and cytokinin (Chapter 10), one can obtain: (top), swelling of the leaf base with the direct formation of a few shoots, or (bottom), a mass of callus from which grow numerous small plantlets, which can then be detached and grown into separate plants.

bulbs and other slow-growing horticultural plants (Fig. 28-9). On a medium high in auxin the cultures generally produce roots; on a high cytokinin medium they form shoots. Thus varying the medium for different species can lead readily to whole plants.

5. FLOWERS AS ROYAL AND RELIGIOUS SYMBOLS

History is full of the representation of gods, rulers, nations, and even emotions by flowers. A few of these still survive, as in the U.S. "state flowers." However, not many citizens know what their state flower is, so the symbolic importance is not great. In Victorian days the "language of flowers" had a great vogue, and selected bouquets could be used to send quite specific messages. The vogue was short-lived, but the old-time flowers were so deeply imbedded in history and custom that they were widely known and long established. Here are a few examples.

a. THE FLEUR-DE-LYS

The *fleur-de-lys* of European heraldry is a diagrammatic or decorative version of the iris flower (Fig. 28-10). The origin of the fleur-de-lys as a symbol is controversial, but it seems likely that the stiff, tall stem of an iris in flower led to its use as a ceremonial scepter by the kings of some early European tribes. The most prominent of such kings was Clovis, king of the Franks (466 to 511 A.D.) who was converted to Christianity in 496; he adopted three fleur-de-lys as his shield, and they are said to represent the Trinity. When, following the Norman Conquest, the kings of England governed part of France, they carried fleur-de-lys on their shields too (Fig. 28-10).

Figure 28-10 Flowers as symbols in heraldry. Above left, medieval shield of France (fleurs-de-lys); right, medieval shield of England (three lions). In center the coat of arms of King Edward III, combining both symbols. Below, the coats of arms of two British colleges, with Tudor roses (Winchester College, Oxford) and both roses and a fleur-de-lys (King's College, Cambridge).

b. THE ROSE

The rose has passed through many modifications in the centuries during which it has been cultivated. In its simplest form, like the present wild roses, it was growing 35 million years ago. The similar dog rose, *Rosa canina,* is common in English hedgerows today and is used as grafting stock for many of the modern cultivars. *Rosa rubiginosa,* the sweetbrier, is also native to Britain, but English colonists in Australia and New Zealand took its seeds with them and in some parts of those countries it has become a nuisance. Other colonists brought the sweetbrier to Virginia, where it has given its name to a women's college.

The prominence of roses in medieval English gardens led to their becoming symbols of rival claimants to the throne. The red rose, *R. gallica,* and the white, *R. alba,* were adopted by the houses of Lancaster and York, respectively, and the prolonged fighting between them has ever since been known as the Wars of the Roses. When Henry Tudor finally took the throne in 1485, the rose became a symbol of the Tudor dynasty (Fig. 28-10).

Double roses are also of very long pedigree. The Romans made excursions to Paestum (near Naples) to see the double roses in flower, just as the Japanese today make spring excursions to see the cherry blossoms. The Summer Damask rose, with about thirty petals, and what is believed to be its hybrid with *R. alba,* both have intense fragrance. This last rose, *R. centifolia,* the "hundred-petaled" or "cabbage rose," is fully double; that is, all its stamens are modified into petals, so that it can set no seed. One form of the 30-petaled rose has pink and white streaks and thus symbolizes the uniting of the warring houses of York and Lancaster. This cultivar, 'York and Lancaster,' still exists.

The rose is also the symbol of fragrance; "A rose is sweeter in the budde than full-bloom," and "Sweet spring, full of sweet days and roses," sang the Elizabethan poets. The Roman custom of wearing wreaths of roses at banquets was to perfume the air (probably none too sweet by the time a large crowd had eaten and drunk their fill). Indeed, roses have long been at the center of the perfume industry (see Chapter 26).

c. THE CHRYSANTHEMUM

The chrysanthemum in its single, daisylike form has also been known from early times. The name literally means yellow flower, but the compact disc surrounded by golden rays obviously suggests the sun. The single yellow-flowered *Chrysanthemum indicum* is (despite its species name) a native of China, but was brought to Japan very early. As a result, the chrysanthemum became the emblem of the Emperors of Japan, and its heraldic equivalent, the rising sun, was the sole device on Japanese flags until a generation ago. The ceramic tiles roofing all Imperial Palaces in Japan are moulded in the form of chrysanthemums (Fig. 28-11).

The chrysanthemum's symbolism derives from the Shinto religion, according to which the emperors are descended in unbroken line from a personage who is essentially the *sun-goddess*. Although this doctrine of divine descent was disavowed by Emperor Hirohito after World War II, the religious ceremonies at the national shrine at Ise, apparently based on a residue of the sun-goddess idea, are still very active.

There is a curious botanical parallel between the histories of medieval Japan and medieval Britain, for in Japan from 1336 to 1392 two emperors held courts, one in Kyoto and the other at Yamato. The continued fighting between the rival groups has been called "The Wars of the Chrysanthemums," like the English Wars of the Roses.

d. THE LILY

The lilies, especially the pure white, fragrant *Lilium candidum,* are the symbols of purity. The Greeks claimed that the white lily sprang from the milk of the goddess of plenty, Hera. In medieval Christian painting, the Madonna often holds lilies, or angels offer lilies to her (Fig. 28-12). The color, the fragrance, the relatively large flowers, and the ease of cultivation (the bulbs are perennial) all combined to make the lily one of the commonest garden flowers of antiquity.

Since *Lilium* is a large genus, and many of the 60 or so species are white, the relations between botany and legend are not clear. The "lilies of the field," whose robes exceeded those of "Solomon in all his glory" (Matthew 6 : 28) were probably the red species *L. chalcedonicum,* but they may actually have been anemones. The bulbs of many *Lilium* species are mild-tasting and are sometimes eaten in the Orient.

Figure 28-11 Tiles on the royal Japanese palace at Kyoto. Every tile, large and small, bears the chrysantheum design.

e. THE LOTUS

The lotus (genus *Nelumbo*), not related to the small trifoliate legume *Lotus*) is a philosophical as well as a religious symbol. It is an aquatic plant whose seed germinates in the black mud of lake bottoms, to produce large leaves and an impressive flower. It therefore symbolizes the essential divinity of humans arising superior to their lower natures. In Buddhism, and especially in Hinduism the lotus is a divine symbol, and its simplified form appears often in Indian and East Asian religious art. Gods are often represented on a lotus throne (Fig. 28-13). The lotus bloom floating on the water also symbolizes the world floating on the ocean.

The sacred lotus, *N. nucifera,* is in the Nymphaeaceae, a small primitive family of dicots. Its seeds and tubers are edible, and in China the tubers are extracted for their starch. The large leaves are sometimes used to wrap up meat, which is then steamed. There are authentic records of lotus seeds, buried in the mud of a lake that has been dried for several centuries, being able to germinate normally. Carbon-14 dating even makes one of them 1000 years old. The related blue and white "water lilies" (they are not lilies) of the Nile, *Nymphea stellata* and *N. lotus,* whose roots are also edible, were sacred to the ancient Egyptians, probably for reasons similar to those above.

Figure 28-12 Leonardo da Vinci's picture of the Annunciation, in the Uffizi Gallery in Florence. This is a part of the picture, showing the angel bearing white lilies for the Madonna.

f. THE MOUTAN

Lastly the tree peony, *Paeonia suffruticosa,* is especially prized in China for its large flowers. Known as moutan, it was brought into cultivation in the Imperial Gardens in the seventh or eighth century A.D.; it symbolizes wealth and honor and has been prominent in Chinese painting for centuries (Fig. 28-14). In the eleventh century new varieties were developed by skilled gardeners, or brought in from the wild, and were propagated by grafting. When the moutan was in bloom the flowers were used to decorate houses and temples.

Figure 28-13 The six Jizo, protectors of the six paths of the soul, in Japanese Buddhism. Each figure stands on a realistically modeled lotus flower.

Figure 28-14 The Moutan, or white peony. Detail of a Japanese screen from about 1600. The design was done in colors on gold paper.

CHAPTER 29

The Global Impact of Plant Diseases and Pests

It is a peculiarity of plant tissue that it provides a good medium for the growth of microorganisms, particularly fungi. Animal tissues, in contrast, tend to be more susceptible to bacteria. The initial stages of the growth of fungi on the outside surface of plants seem to be due to the exudation of traces of organic materials, especially on leaf surfaces; these promote the first stages of germination of fungal spores. The film of moisture that is so often present on leaves in moist climates obviously favors spore germination too. There are comparable exudations from root tips. For example, an exudate from legume roots contains specific substances called *lectins* that attract the *Rhizobia* that form nodules, and an exudate from corn and sunflower roots attracts the roots of the parasitic *Striga* plants. Many of these host–parasite relations, involving some elaborate systems of attraction and repulsion, are only now beginning to be understood, mainly because of the advent of analytical methods of very high sensitivity.

1. THE COLD WAR BETWEEN HOST AND PARASITE

Many parasitic fungi possess enzymes that attack and hydrolyze components of the plant cell wall, especially the pectins and those polysaccharides of glucose and galactose that are linked 1–3, rather than exhibiting the very stable 1–4 linkage as in cellulose (Chapter 21). The fungus *Fusarium* secretes the enzyme *cutinase,* which attacks the cuticle of leaves. A few fungi do attack cellulose itself, and these include the serious pest *Myrothecium verrucaria* (a Phycomycete), which caused such rapid damage to ropes, tents, clothing, and other textiles in the Pacific Campaign of 1942 to 1945 that a special research program had to be organized to combat it. It is not, however, important as a cause of plant disease.

Most plants are not wholly passive in the face of such enzymatic attacks. Among crop plants, breeders have selected over the years numerous varieties resistant to specific diseases. Such resistance is usually conferred by a single gene and its mechanism may be of several kinds; the commonest is that the cells around the site of infection simply die. Because the fungus does not grow well in dead tissue, the infection is thus walled off. This so-called *hypersensitive reaction* is due to excessive sensitivity of the host cells to the enzymes excreted by the fungus. Another kind of resistance, acting in the opposite sense, is that seen in the Anthracnose disease of beans, in which the whole plant is not killed but black spots are caused by limited growth of the fungus. Here the host, *Phaseolus vulgaris,* perhaps stimulated by the wounding that the fungus inflicts, excretes two substances that inhibit those fungal enzymes that hydrolyze the plant cell wall; in addition, it excretes a more aggressive enzyme that attacks the cell walls of the fungus. However, the fungus retaliates in its turn by excreting inhibitors for this latter host enzyme. Such a "battle of the enzymes" may very well be taking place in other plant diseases that have not been studied in such detail.

Another phase of the war between host and parasite is the production by the parasite of substances damaging or toxic to the host. For example, in some wilt diseases, or the Dutch elm disease, the fungus forms quantities of polysaccharide that clog the xylem passages, so that the leaves become deprived of water and thus wilt. In two other cases there are highly specific toxins. In a bacterial leaf-spot disease of tobacco called "wildfire" from its rapid spread, the spot of dead infected tissue on the leaf becomes surrounded with a brown halo due to the diffusion of the toxin into adjacent cells. This toxin was found to be a derivative of the ordinary amino acid methionine; it probably acts by competing with methionine in the normal functions of the leaf cells. A second leaf spot that produces a spreading toxic effect is the "eyespot" disease of sugarcane. This toxin, excreted by the attacking fungus, is a protein, and it binds to a protein in the leaf cell membrane (Fig. 7-2), thus disrupting the functioning of that membrane. In resistant races of the sugarcane the membrane protein is modified so that it does not bind the toxin.

The host plant is often stimulated to fight back with the same types of chemical weapons (called in general *phytoalexins*). Thus when spores of a *Helminthosporium* species germinate on alfalfa leaves, the leaves are stimulated to produce a toxin that inhibits growth of the fungus germ-tube. Carrots similarly form a toxin when infected with the fungus *Ceratocystis fimbriata,* and thus many carrot cultivars can resist the infection. Sweet potatoes (*Ipomea*) when infected with the black rot fungus produce up to 1% of their weight of a terpenoid, ipomeamarone, which inhibits fungus growth. Pea and bean pods produce *pisatin* and *phaseollin,* respectively, when at-

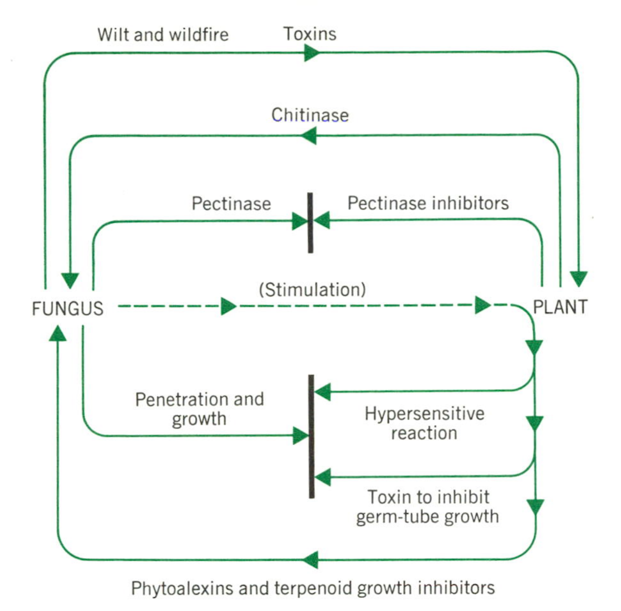

Figure 29-1 Attack and defense; the weapons of war between host and parasite.

tacked by *Monilia* fungi. The highly infective strains of the fungus are those that call forth the least pisatin production, so the pea plant does not "protect itself" against them. Unfortunately neither phaseollin nor the synthetic inhibitors modeled on it are really powerful fungicides. Similarly the late blight of potatoes, described below and in Chapter 19, initiates the production of some phenolic toxins, but unfortunately their toxicity is too weak to prevent the infection. Many of these interactions are summarized in Figure 29-1.

Some parasites do little harm to the host tissue. affecting outward appearances only, such as the scab diseases of apples and potatoes. But the majority of fungus attacks are indeed damaging. Up to a quarter of the world's food crops are probably lost to fungi, and a further portion is eaten by insects and nematodes.

2. THE TYPES OF FUNGI

The fungi constitute a kingdom of their own, with two divisions: the slime molds or Myxomycota and the true fungi or Eumycota. None of the fungi is photosynthetic, and all live either on dead organic matter (*saprophytes*) or on living organisms (*parasites*). The first group also includes, besides slime molds, two small Classes with which we are not here concerned, the Chytridiomycetes and the Oömycetes or water molds. The true fungi are divided into three Classes, each of which comprises hundreds or thousands of species. Class (1) is the Zygomycetes, in which sexual fusion occurs between two similar branch hyphae. These with the two small Classes above have often been grouped together under the name Phycomycetes, the common character being that their plant body consists of fine filaments (*hyphae*) without cross walls, so that nuclei are simply scattered at intervals throughout the cytoplasm (Fig. 29-2). The other two classes are (2) the Ascomycetes, which reproduce sexually by forming a long specialized cell, called an *ascus*, containing eight haploid spores (see Fig. 29-5*B*); and (3) Basidiomycetes, which spend most of their life cycle as cells containing two nuclei each (a "dikaryon"); these fuse only in one special cell, which then forms four haploid spores—growing on stalks out of the corners of this special cell, the *basidium* (Fig. 29-2*C*).

In the Myxomycota or Myxomycetes (slime molds), the spores germinate to small amebae ("swarm cells"), which later come together in a multinucleate mass of cytoplasm, the *plasmodium*—some for feeding, others only to form their reproductive spores (Fig. 29-3).

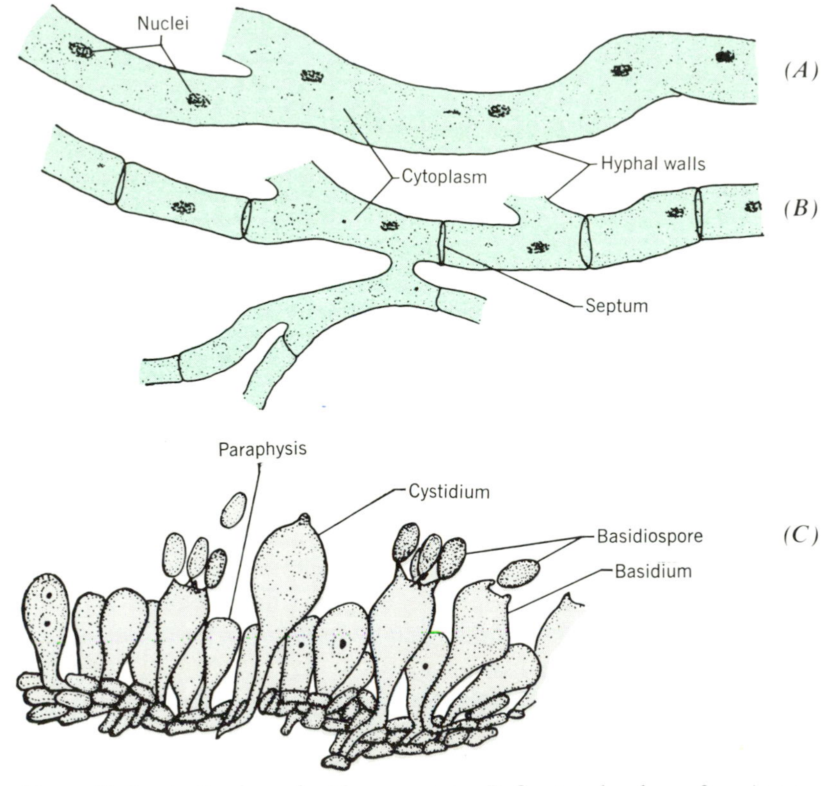

Figure 29-2 *A*. Hyphae of a Phycomycete. *B*. Septate hyphae of an Ascomycete. *C*. The reproductive tissue (hymenium) of a Basidiomycete. The basidia, bearing basidiospores, are separated by sterile outgrowths (paraphyses and cystidia). (From Alexopoulos, 1952)

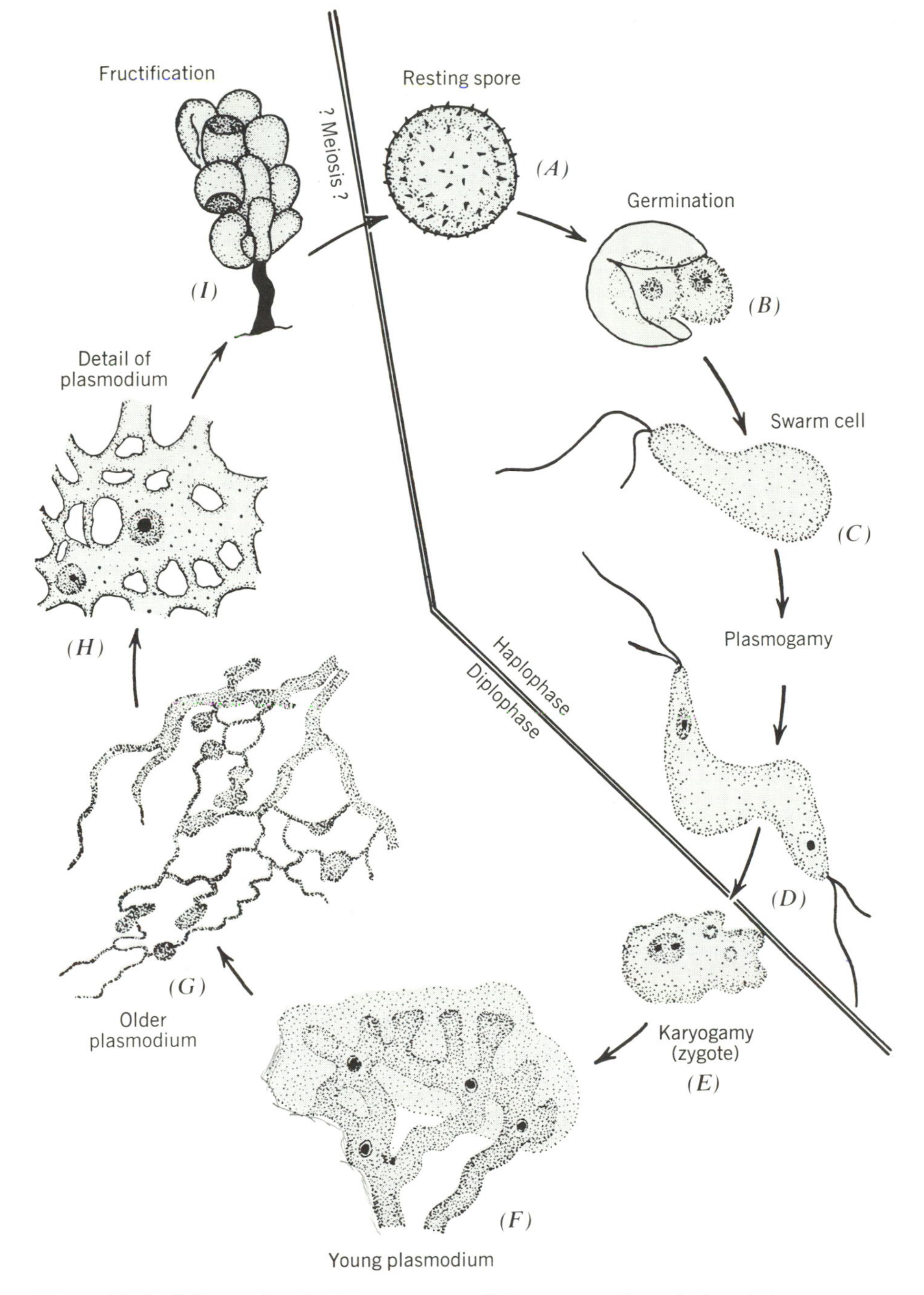

Figure 29-3 Life cycle of a Myxomycete, *Physarum polycephalum*. The spore *(A)* germinates *(B)* to an ameboid cell, which feeds for a while and then fuses with others *(C, D, E)* nuclei fuse too, to form a plasmodium *(F, G, H)*, from which a sporangium *(I)* eventually arises. The spores are probably haploid. (From Alexopoulos, 1952)

There are disease-producing parasites in each group. Of the innumerable fungus diseases of plants about 20 have been well studied, and of these we will take up just 10 (and some of their relatives) that are of major importance and exemplify different aspects of the host–parasite relationship.

3. DISEASES DUE TO PHYCOMYCETES

a. THE DOWNY MILDEW OF GRAPE

In the region near Bordeaux, where the townsfolk stroll through the vineyards, the grape growers resorted to spraying the vines with copper sulfate solution to discourage the children from eating the fruit. In 1882 Professor P. A. Millardet of Bordeaux University noticed that the sprayed plants had no downy mildew, while the other parts of the vineyards were badly infected. He added lime to the solution to precipitate the copper and fix it on the leaves—and thus invented Bordeaux mixture, which has remained one of the most effective fungicides for nearly a century. Thousands of white-washed cottages in France show a blue-green stain where the vines growing against the wall have been sprayed over and over again with Bordeaux mixture.

The hyphae of downy mildew, *Plasmopara viticola,* grow in intercellular spaces in the leaf and fruit tissue of *Vitis vinifera* (grape), forming little swollen branches (*haustoria*) that enter the cells and withdraw nutrients from them. As a result the leaves turn prematurely brown, and the grapes shrink to become "mummy-berries" (Fig. 29-4*A*). To form spores, special hyphal branches (*sporangiophores*) develop and protrude into the air through the stomata, bearing their sporangia (Fig. 29-4*B*). These constitute the visible "down" on the leaf surface.

Many similar downy mildews infect both vegetables and flower crops: blue mold of tobacco seedlings (*Peronospora tabacina*) and the mildews of cucurbits (*Pseudoperonospora*), of lettuce (*Bremia*), and of grasses (*Sclerospora*). They are susceptible not only to Bordeaux mixture but also to the newer Fermate fungicides.

b. LATE BLIGHT OF POTATOES

Phytophthora infestans reproduces asexually by means of motile zoöspores, which germinate to produce hyphae on the surface of a wet leaf of *Solanum tuberosum* (potato). They enter the leaf either through stomata or by hydrolyzing cell walls of the epidermis. Like the downy mildew, their sporangiophores come out as "down" through the stomata (Fig. 29-4*B*), and the spores become motile as soon as they are shed, to repeat the infection on another leaf. But it is when they infect the tubers that the worst damage is caused, for the mycelium sends haustoria into the cells, hydrolyzing and withdrawing their contents; the tubers dry up, turn brown and become secondarily infected with other fungi and bacteria, turning quickly into a stinking black slime.

Since the potato is grown by planting parts of tubers that bear a bud, a single variety may be cultivated over vast areas, and all can be thought of as really parts of the same individual plant. Hence, once the infection begins it has a host "plant" of indefinite size to grow on and can be controlled only by very drastic measures. This is the hazard of large-scale vegetative reproduction. When the disease hit the potato fields of Ireland in 1845, and again in 1846 and later years, it was able to destroy the crop entirely, causing widespread starvation (Chapter 19). Breeders have since grown potatoes for true seed, however, and have thus developed resistant strains (Fig. 29-4*C*). The disease also attacks tomatoes, which are also members of the Solanaceae.

Some 34 other species of *Phytophthora* cause other diseases of roots and tubers. *Phytophthora cinnamomi* (first found on cinnamon trees in Sumatra) attacks many fruit trees and shrubs. Believed to be a native of Malaysia, it has spread worldwide in the last 50 years and is

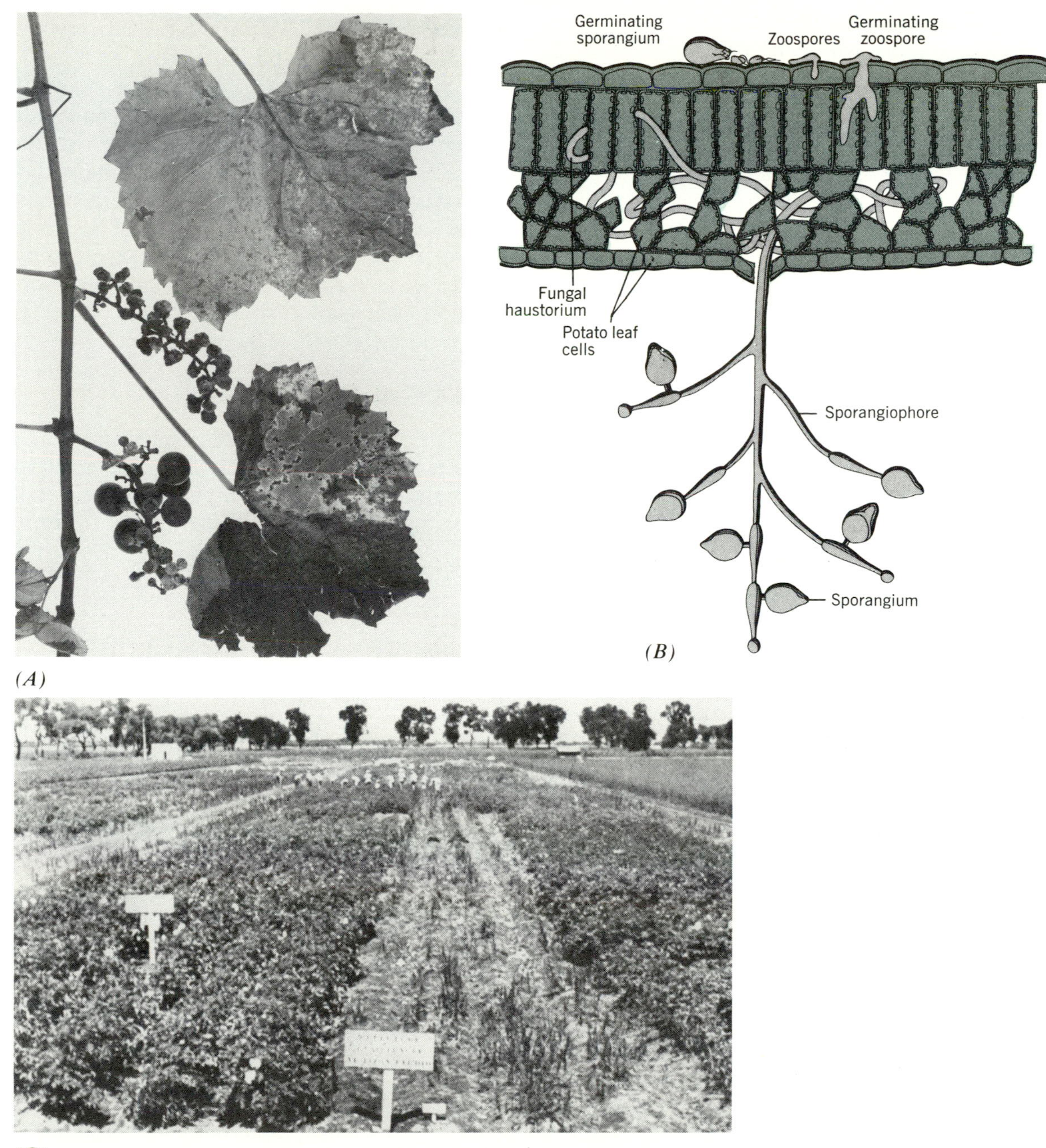

Figure 29-4 *A*. Downy mildew of the grape, with infected leaves and fruit. *B*. Late blight of potato, *Phytophthora infestans*. Zoospores germinating on the leaf and penetrating the epidermis. Sporangiophore growing from hyphae in the leaf and protruding out of a stoma. From the sporangia zoospores will emerge, to germinate again on the leaf as shown at the top. *C*. Testing for resistance to late blight. In the center is a susceptible cultivar, and on either side resistant cultivars. (B, from Raven, Evert, Curtis, 1976)

now doing great damage to *Eucalyptus* forests in Australia, as well as to avocado plantations in Central and South America. The wild species of avocado seem to owe their resistance to the presence in their root cells of a natural fungicide.

4. A DISEASE CAUSED BY A MYXOMYCETE

Clubroot of cabbage, *Plasmodiophora brassicae,* attacks all cultivated forms of *Brassica oleracea,* including cabbage, cauliflower, broccoli, kale, and brussels sprouts. It also attacks smaller crucifers such as mustard and radish.

Its tiny motile zöospores enter the root hairs, much as do the *Rhizobia* of legume nodules (Chapter 9), and form a multinucleate mass, the plasmodium (Fig. 29-3), which spreads from cell to cell through the root. Presently the plasmodium divides up into sporangia, which give rise again to motile zöospores outside the root. Small holes are dissolved in the cell wall for entrance and exit, as well as for movement through the root. It may also divide up into large numbers of resting spores, each with one nucleus, that infect the surrounding soil. Thus once *P. brassicae* has established itself, crucifers cannot be grown again in that location for many years.

The disease does not quite kill the host, but the root cells swell up to about five times their normal size, being full of the plasmodia, and as a result the roots become swollen and "clubbed." Since the crucifers contain a biochemical precursor of auxin, the swelling is probably due to liberation of excess free auxin in the cells. It can be imitated by injecting auxin. Because the swollen roots fail to take up water and nutrient, the plants wilt and starve.

Fortunately the spores germinate poorly at alkaline pH, so that liming the soil is fairly effective as a countermeasure. Methyl bromide is also used in the soil of the seed beds.

5. DISEASES CAUSED BY ASCOMYCETES

Typically the ascomycetes produce an ascus containing eight haploid spores in line (Fig. 29-5), but some of them are "imperfect" and reproduce only by asexual *conidia.* Altogether they cause some 21 plant diseases. One of the most famous is ergot, *Claviceps purpurea,* spread by insects on rye plants and infecting the grains (section 9 of Chapter 24). Four others of special interest are the following.

a. DUTCH ELM DISEASE

The very serious Dutch elm disease was first seen in Holland in 1921 and is believed to have entered America on elm logs imported for making veneer. The fungus, *Ceratostomella*

(A)

Figure 29-5 Apple scab, *Venturia inaequalis. A.* Scabs on leaves and fruit. *B.* The sexual cycle and the (shorter) asexual reinfecting cycle. The germtube from the ascospore penetrates the cuticle and multiplies in the intercellular space below. Note the eight bicellular ascospores formed following fertilization in the sexual cycle (compare Schizosaccharomyces in Fig. 20-2). (From Agrios, 1969)

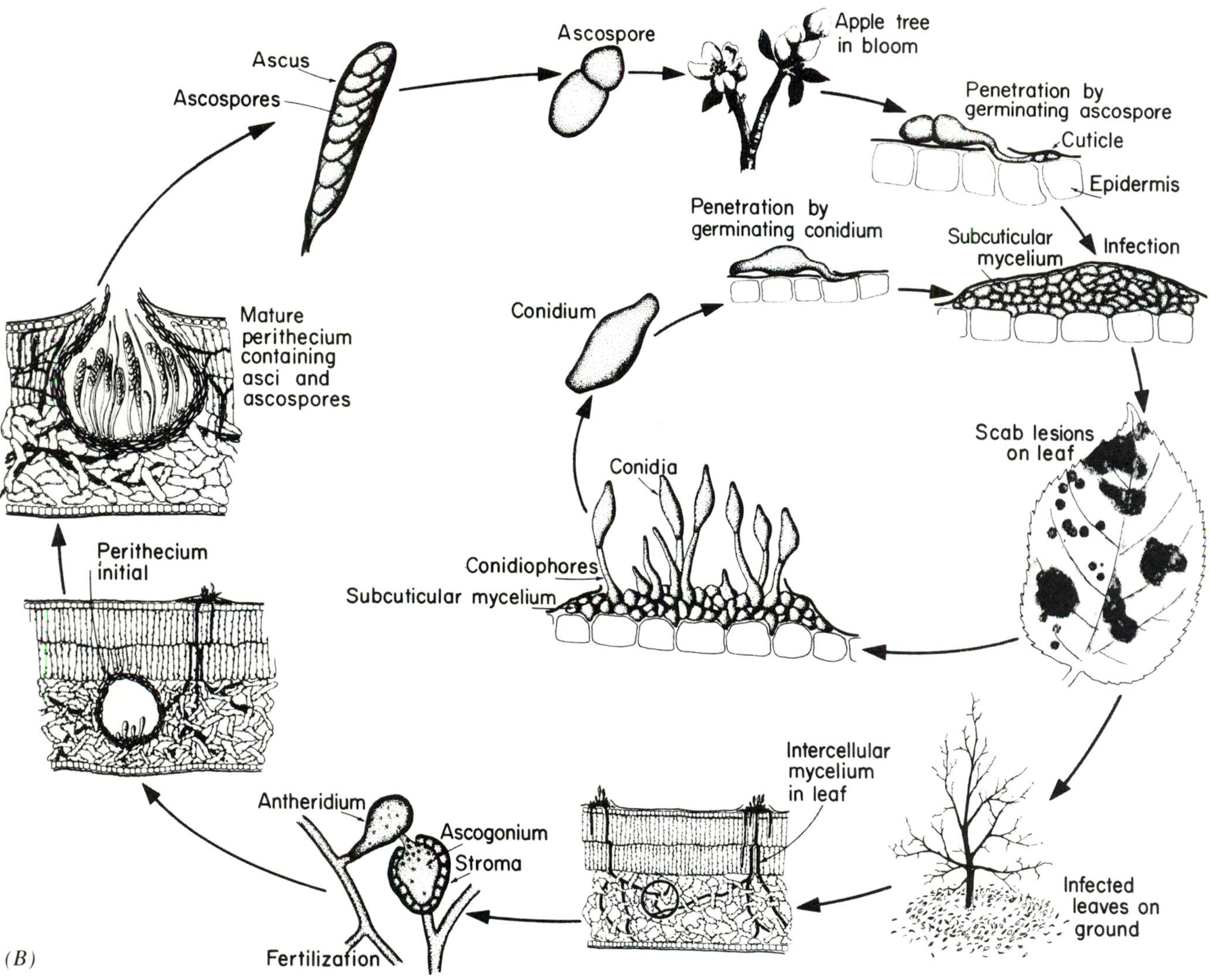
Ascus
Ascospores
Ascospore
Apple tree
in bloom
Penetration by
germinating ascospore
Cuticle
Epidermis
Penetration by
germinating conidium
Subcuticular
mycelium
Infection
Conidium
Mature
perithecium
containing
asci and
ascospores
Scab lesions
on leaf
Conidia
Conidiophores
Subcuticular mycelium
Perithecium
initial
Intercellular
mycelium
in leaf
Antheridium
Ascogonium
Stroma
Infected
leaves on
ground
Fertilization
(B)

Figure 29-5B

ulmi, is carried by the elm beetle, which burrows into the bark and sapwood of American elm (*Ulmus americana*) trees, carrying the spores of the fungus with it. The spores germinate in the xylem sap, and the hyphae grow into the xylem vessels, producing gums that block the water stream and cause the leaves to wilt. The hyphae then spread rapidly all through the affected twigs, forming spores, which further spread the infection.

The control is to spray to kill the beetle, but since DDT has been banned and no other spray is as effective, all the elms in the eastern half of the United States may be lost and the disease is spreading north and west. The English elm, *U. procera*, is now also affected. Perhaps some new and effective insecticide will save the remainder. The Chinese and Japanese elms are resistant and some work on their introduction is going on; it is hoped to introduce resistance by hybridizing one of these with *Ulmus americana*.

Figure 29-6 American chestnut tree killed by the chestnut blight, *Endothia parasitica*.

b. CHESTNUT BLIGHT

Endothia parasitica or chestnut blight has totally destroyed the American chestnut tree (*Castanea dentata*), which was once a major constituent of the forests of eastern North America. The disease was first seen in New York in 1904. Then it spread rapidly into New Jersey and Connecticut. Although control methods were recommended in 1909 they were not adopted; in 1912 the state of Pennsylvania held a conference on control but it was already too late. By 1930 most of the chestnuts as far west as Ohio and as far south as North Carolina were dead (Fig. 29-6). Today only a very few extremely isolated specimens remain.

In 1913 the fungus was found in China on the oriental species *C. mollissima*, but it did little harm there. As with the elms, Chinese and Japanese species seem to be generally resistant and as a result some have been introduced locally and been hybridized with the American species. They produce good (though small) edible nuts, and some of the hybrids are resistant to *Endothia*, but they are not suitable as forest trees.

Although the organism kills the stem cortex, it does not infect the roots, so that sprouts continue to grow from the base; thus young apparently healthy chestnut trees can often be seen in the North American woods, but eventually they all become reinfected. As far as is known the *Endothia* is still fully virulent 75 years after its first appearance.

c. APPLE SCAB

When the spores of *Venturia inaequalis* (apple scab) germinate on the leaves of apple (*Pyrus malus*) they produce a haploid mycelium just under the cuticle. The pushing up of the conidia from this mycelium lifts the cuticle off as a "scab" (Fig. 29-5*A*), and the conidia are discharged to infect other leaves and later also the skin of the fruit. When the leaves fall on the ground at the end of the season the haploid hyphae fuse and the zygote produces a large number of asci contained within a flask-shaped *perithecium* (Fig. 29-5*B*), which overwinters on the dead leaf. Each ascus contains the characteristic eight ascospores. In the early spring the perithecia break open and the ascospores are liberated into the air. (The species name *inaequalis* is given because these ascospores each contain two slightly unequal cells). An asexual phase, located within the scab, leads to formation of conidia that germinate directly on the leaf and thus spread the infection locally.

The control of this disease depends on weather prediction, combined with microscopic observation of the perithecia. When the temperature is warm enough, and just before rain is expected, the perithecia will be ready to discharge. Then the trees and the fallen leaves are sprayed with fungicide; the spraying is repeated for about three weeks thereafter.

d. COFFEE LEAF SPOT

Though not a major disease, *Omphalia* (*Mycena*) *flavida* (coffee leaf spot) on *Coffea arabica* is of special interest because of its basis in known plant hormone relations. When the conidia germinate, the germ-tube enters the leaf and grows into a mycelium that produces a black spot a few millimeters across on the coffee leaf. In itself this is not serious, but if some of the spots are near the base the leaf falls off, which means, of course, serious loss. Since many other spot diseases do not cause leaf fall, it seemed likely that the organism was destroying auxin in the leaf—a process which, as we saw in Chapter 10, would cause abscission. By culturing the organism on media containing auxin it was found that indeed it does produce a powerful enzyme that oxidizes indoleacetic acid to inactive products. The leaf fall is thus due to the cessation of the flow of auxin out of the leaf, resulting in the activation of the abscission layer of cells at the base of the petiole.

6. DISEASES DUE TO BASIDIOMYCETES

Basidiomycetes have growth and reproductive cycles of several degrees of complexity, but they all involve a brief haploid stage and then a longer growth stage in which the *two* haploid nuclei are carried side by side in each cell without fusing. Such cells are *dikaryons*. Finally, at a particular stage, the nuclei do fuse and the zygote produces (sometimes via an intermediate stage) four haploid spores borne on little stalks called basidia (Fig. 29-2*C*). Many of the larger fungi such as mushrooms belong to this group, but most are saprophytes, growing on organic matter in the soil, and do not infect living plants. A number of pathogens of major importance belong to the order Uredinales, the rusts. The most important of these is the stem rust of wheat.

a. WHEAT STEM RUST

Puccinia graminis tritici (wheat stem rust) is probably the one disease that holds the most danger for the world's food. It mutates from time to time so that wheat varieties that resist attack by existing strains of rust fall victim to the new mutants. Thus the world's wheat breeders have to be constantly developing new varieties so as to have new resistant genes ready for introduction as soon as a fresh outbreak is recorded.

The complicated life cycle of the rust (Fig. 29-7) was worked out by many patient investigators. There are no less than five kinds of spores. The disease is initiated by tiny, light binucleate *aeciospores* (dikaryons) that germinate on the moisture film on the wheat leaf or stem, enter through stomata, and grow into a mass of hyphae that forces off the epidermis

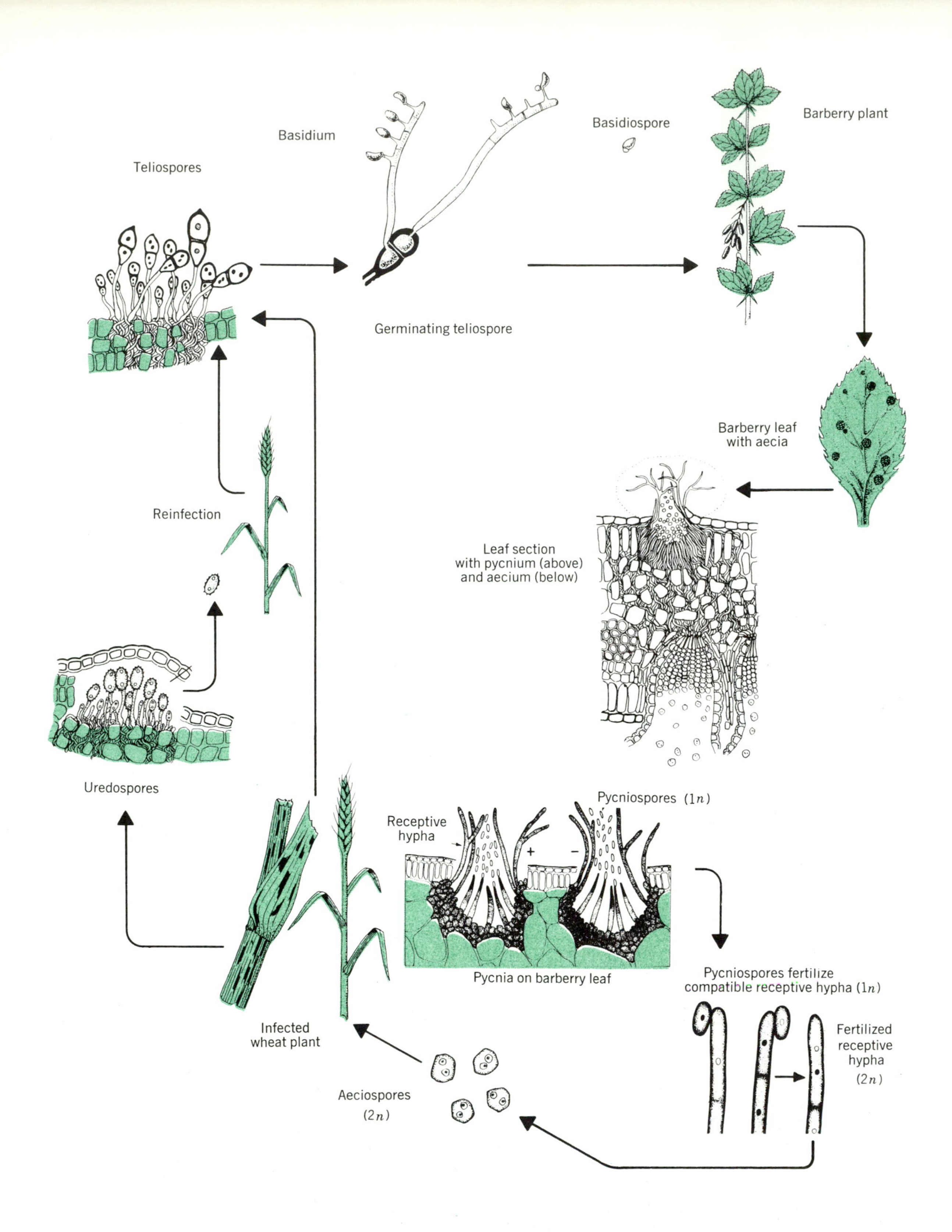
Teliospores
Basidium
Germinating teliospore
Basidiospore
Barberry plant
Barberry leaf
with aecia
Leaf section
with pycnium (above)
and aecium (below)
Pycniospores (1n)
Receptive
hypha
+
−
Pycnia on barberry leaf
Pycniospores fertilize
compatible receptive hypha (1n)
Fertilized
receptive
hypha
(2n)
Aeciospores
(2n)
Infected
wheat plant
Uredospores
Reinfection

(compare apple scab). The result is a blister from which several hundred thousand reddish *uredospores* blow off. These reinfect other plants throughout the remainder of the season. But it is at the end of the season that the life cycle becomes really complex (see Fig. 29-7).

As the weather cools, black overwintering *teliospores* are formed; in these the two nuclei of the dikaryon fuse, and the resulting zygote produces two one-celled spores. Next spring (a cold spell is needed) these germinate to form a minute diploid mycelium that gives rise by meiosis to four haploid *basidiospores,* borne on short stalks. These cannot infect wheat, but infect the common barberry, *Berberis vulgaris.* On the leaves of this plant the spores germinate, penetrate into the tissue, and form a flask-shaped reproductive structure, the *pycnidium.* Inside this some of the hyphae break up to produce *pycniospores,* while others, the so-called *sensitive hyphae,* become fertilized by pycniospores blowing in from a nearby pycnidium. Automatic crossing between the offspring of individual spores is thus achieved. From the fusion, binucleate cells (dikaryons) are formed that produce binucleate hyphae and hence a new structure, the *aecium.* This in turn forms binucleate spores (aeciospores), and it is these that infect the wheat stem.

The rust damages but does not kill its host; in and around the blisters the chlorophyll is bleached and the fungus consumes much of the plant's photosynthate (Fig. 29-8*A*). Grains are formed but they are small and wrinkled (Fig. 29-8*B*) and the yield is only a fraction of the normal.

There are two phases to the control. The first step has been to remove all barberry bushes from the vicinity of wheat fields. This has been done over a long period of years both in the United States and Canada. The second and more continuing step was to develop rust-resistant varieties. For his work in developing rust-resistant dwarf wheats in Mexico, and thus helping to make Mexico self-sufficient in wheat (even, for a while, an exporting country), Norman Borlaug was recently awarded the Nobel Peace Prize.

b. OTHER RUST DISEASES

A comparable case of a pathogenic fungus requiring two alternate hosts is that of the blister rust (*Cronartium ribicola*) on the white pine, *Pinus strobus.* This organism requires wild currant (*Ribes* spp.) as its second host, and foresters in the northeast American states correspondingly root out currant or gooseberry bushes wherever they occur within half a mile of white pine lots or woods.

Flax rust, *Melampsora lini* on *Linum usitatissimum,* is a third member of the Uredinales. It does much of its damage by inserting a haustorium into the leaf mesophyll cells and thus withdrawing part of the cell contents (Fig. 29-9*A*). A fourth member, *Marasmius,* produces the curious distortions of cacao called witches'-broom (Fig. 29-9*B*). Lastly, the coffee rust, which had a great influence on the world's tea and coffee culture, is discussed in Chapter 25.

7. DISEASES DUE TO BACTERIA

The bacteria are the smallest of all plants, commonly 1 to 4 μ in length. A gram of garden soil contains about 30 million bacteria of assorted genera and species. A spadeful of good soil can thus harbor as many bacteria as there are human beings in the whole world. Of the few bacterial infections of plants, three hold special botanical interest.

Figure 29-7 The elaborate life cycle of wheat stem rust, *Puccinia graminis tritici,* with five different kinds of spores. The basidiospores only infect the Barberry; the aeciospores and uredospores infect wheat. (Modified from Greulach & Adams, 1976)

(*A*)

(*B*)

Figure 29-8 *A*. Appearance of wheat stem rust in the field, with uredia and telia on the stems. *B*. Wheat grains from healthy (top) and infected (bottom) plants.

a. FIRE BLIGHT

Fire blight of apple and pear trees was historically one of the very first diseases, plant or animal, to be traced to bacteria. The organism is *Erwinia amylovora,* which is related to the celebrated organism of the human intestines, *Escherichia coli.* Closely related "soft-rot" bacteria attack grocery store vegetables to produce brownish, squashy areas in which the polysaccharides constituting the cell walls have been hydrolyzed and all turgor lost. Fire blight still makes it difficult to raise pears in some parts of North America.

b. FASCIATION

Fasciation of young plants of pea, sweet pea, chrysanthemum, and other species is, like coffee leaf spot, one of the few diseases with a known hormonal basis. Infection by the bacteria (*Corynebacterium fascians*) produces multiple buds at the nodes of the young plants (Fig. 29-10), and the identical symptoms can be produced by injecting pure cytokinins into the stem at the nodes. Cultured on a nutrient medium that contains no adenine or other purines, the bacteria have been shown to produce a cytokinin.

c. CROWN GALL

Crown gall, caused by *Agrobacterium tumefaciens,* results when these bacteria enter a wound; many plants in a variety of families can be infected. The cells divide and enlarge to produce a gall or tumor (Fig. 29-11 and Chapter 10). Unlike legume nodules, however, the tumor tissue soon becomes free of bacteria and

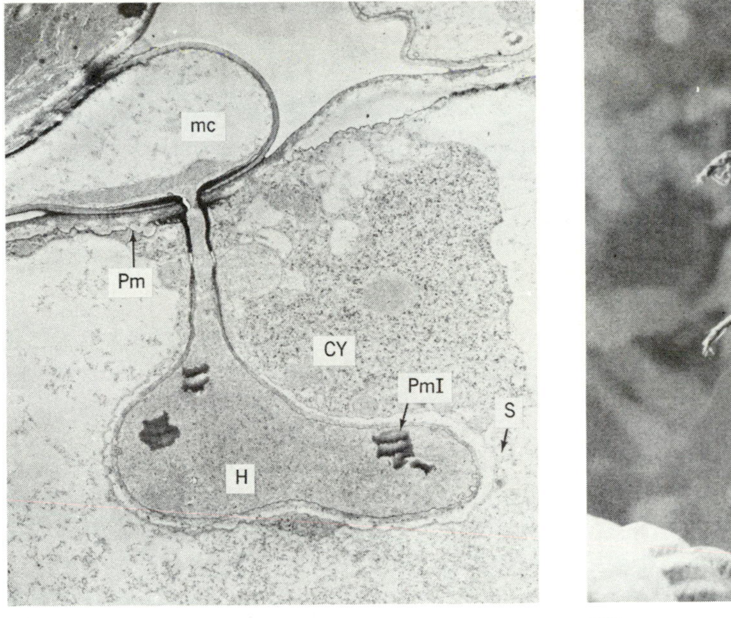

(A) *(B)*

Figure 29-9 *A*. The haustorium, H, of flax rust growing out from a mother cell, mc, in the intercellular spaces of a flax leaf, into a mesophyll cell. Pm, the plasma membrane of the infected cell, and PmI, that of the invaginated cell, whose cytoplasm, CY, is separated from the haustorium wall by the layer S. *B*. Witches' broom of cacao caused by *Marasmius perniciosus*. This organism destroys auxin like the coffee leaf spot of section 5d.

Figure 29-10 Pea seedlings infected by a single prick with a needle carrying the fasciation bacterium. The three plants at left grew in darkness; those at right in the light.

the plant cells have now themselves become tumorous, that is, "transformed" by the influence of the bacteria. Apparently part of the DNA of the bacteria has become incorporated in the nucleus of the plant cell. The resulting cells produce more auxin and more cytokinin than normal tissue, and this increased synthesis probably causes the gall. However, application of hormones alone will not cause the transformation to tumor-forming cells in the absence of the bacteria. Fortunately the disease is seldom lethal, though it is disfiguring. In some plants, resistance to the formation of the bacterial tumors is conferred by the cell walls, which apparently bind the bacteria. Cell walls from the tumors do not possess this binding property.

Figure 29-11 Paris daisy (*Aster* spp.), inoculated on the stem with a culture of the crown gall bacterium forming a primary tumor and, later, secondary tumors on leaf and petiole. This picture is from Erwin Smith's classic study in 1912.

8. DISEASES DUE TO VIRUSES AND MYCOPLASMAS

A *virus* is not an organism but a combination of nucleic acid and protein. On entering a host cell it becomes attached to the hereditary material and causes the host to produce more virus instead of more host nucleic acid and protein. The nucleic acid may contain DNA or RNA; if the latter, a "reverse transcriptase" must be present that "transcribes" the RNA back to DNA. *Mycoplasmas* are more like true microorganisms but of very small size and apparently without rigid cell walls.

The altered metabolism of the plant frequently causes yellowing of the infected tissue. So-called "yellows" of many plants are caused by viruses, injected into the cells by insects. *Barley yellow dwarf,* which attacks all grasses, is spread by aphids, which feed in the phloem by means of a long stylet (Fig. 17-1) and thus introduce virus directly into the conducting tissue. *Aster yellows* and *sugar beet curly-top* are similarly spread by leafhoppers and are caused by mycoplasmas. These diseases can thus be combated by fighting the insects with suitable insecticides.

Tobacco mosaic, "TMV," was historically the first virus to be recognized; M. W. Beijerinck in 1898 thought it might be a *liquid organism,* because it passed through fine-pored porcelain filters that held back bacteria. It was also the first virus to be crystallized (1936). It contains an RNA surrounded by protein molecules. TMV is spread by mechanical wounding, even by so slight a wound as a broken hair on the leaf. Not limited to tobacco, it infects about 150 genera of plants in many families. The only control is through sanitation—removing diseased plants and fallen leaves, and avoiding contact with other tobacco products.

Swollen shoot of Cacao is a continuing menace to cocoa production, especially in West Africa, where it is said to have killed 100 million trees (see Chapter 25). The disease affects the leaves in many ways and distorts the cells of the stem cortex, which swell up, interfering with the transport system, and later die off. Like many other plant viruses, it is carried by aphids.

9. DISEASES DUE TO HIGHER PLANTS

A few of the flowering plants are parasitic, some such as mistletoe (*Viscum*) and dodder (*Cuscuta*) growing on the shoots of woody plants, but those that grow on the roots of their hosts are the more serious. Three genera are members of the Scrophulariaceae:

Striga asiatica and *S. lutea* grow on the roots of corn and of numerous grasses. The plant grows to about a meter high, green with small red flowers in the leaf axils. It is a major pest of corn in South Africa and occasionally appears in the United States. Its minute seeds, up to 500,000 per plant, can lie dormant for years in the soil but are caused to germinate by the substance strigol (the lactone of an organic acid), which is excreted in minute amounts by the corn roots. It has been calculated that each cell of the tiny seeds requires only two to three *molecules* of strigol to activate it.

The infection, known as "witchweed," has been difficult to combat, but fortunately the roots of sunflower and cowpea, which are not susceptible to the parasite, nevertheless excrete strigol or a related compound. This property is now being used to cause the germination of all the dormant seeds so that the young *Striga* plants can then be killed by herbicides.

Alectra spp. grow similarly on the roots of legumes, and *Orobanche minor,* "broom rape," grows on roots of red clover but is not nearly as serious as *Striga* or *Alectra*.

In all these cases the seedling produces a few roots that enter the host roots and there form a haustorium, withdrawing nutrients from the host plant. From the point of entry additional lateral roots are stimulated to develop from the host. Similar swelling and lateral root formation can be induced by extracts of *Alectra* seedlings. The interaction noted at the opening of this chapter thus operates with the higher plant parasites just as it does with the fungi.

10. THE ATTACK OF PLANTS BY NEMATODES

These organisms belong to the class Nemathelminthes. They are primitive eucaryotes; several thousand species are known, and most of them are root parasites. They live mainly in the top 15 cm. or so of soil, attracted to the rhizosphere by the excretions from the roots. The females actually attach to roots and lay egg masses that hatch to produce larvae around $\frac{1}{3}$ mm long; in some species the hatching is stimulated by secretions from the roots. The larvae undergo several molts, and the second stage is usually the one that enters the roots. The mouth has an extrusible stylet that punctures the root epidermis and often also the internal cell walls (Fig. 29-12). Some grow to 3 mm long.

The host specificity of nematodes varies widely; the soybean root nematode attacks a wide variety of legumes, and an Indian race of the nematode *Radopholus* attacks roots of 23 genera, including palms, sugarcane, and a number of legume crops. The root-knot nematode, *Meloidogyne,* is known to have attacked 2000 species, including all cultivated plants. In the latter case the infected roots swell to a gall-like structure in which giant cells are formed around the parasite's head, and cell walls break down, probably due to extrusion of one or more glucanases (Fig. 29-12).

The continuous sucking of cellular contents by nematodes of all types weakens the plant rather than killing it, and infected plants survive in a debilitated state for a long period. Nematode punctures often allow the entry of pathogenic fungi too; wilt and "foot rot" diseases are spread in this way.

Nematodes can be controlled by pesticides that volatilize in the soil. Ethylene dibromide and dichlorpropane are fairly effective, but the soil must be left undisturbed after treatment for two to three weeks. To some extent they are

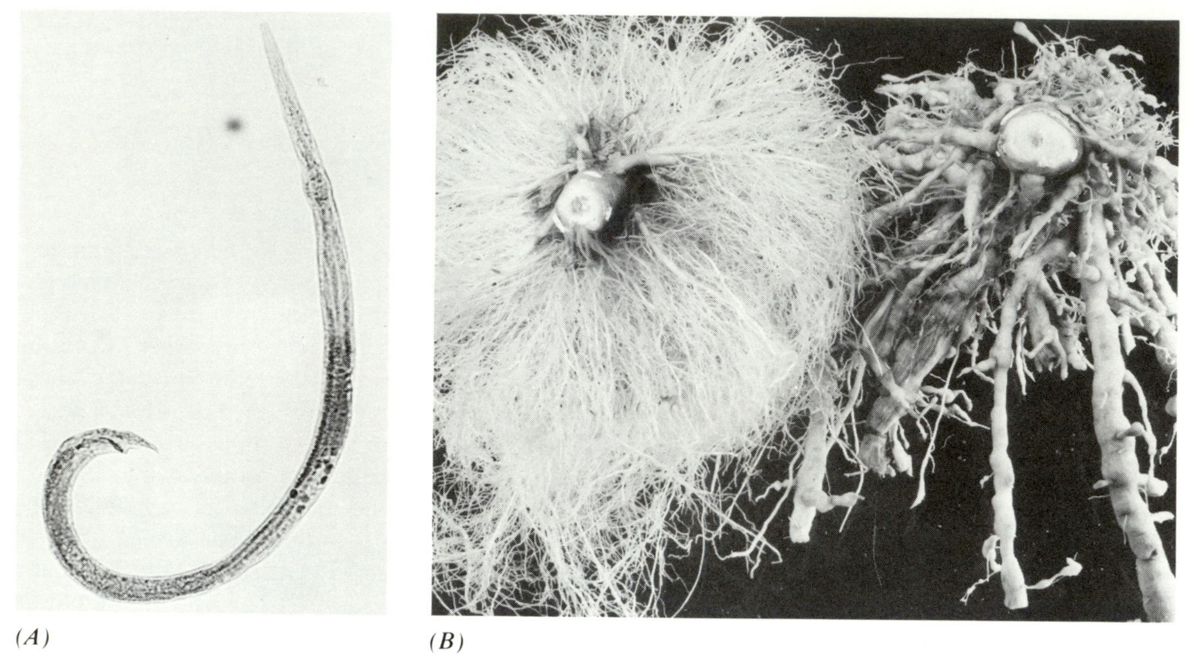

Figure 29-12 *A*. A mature female of the nematode *Paratylenchus*, a common plant parasite. *B*. The effect of the root-knot nematode *Meloidogyne*. At left a healthy tomato root, at right an infected one.

also controlled by fungi, which parasitize them. Some of the large nematodes and large amebae eat the smaller species.

11. THE ATTACK OF PLANTS BY INSECTS

Space does not allow a detailed treatment of the many genera of parasitic insects. From the plant's point of view the attacks can be simply viewed as belonging to three general types:

1. Feeding on leaves by caterpillars, larvae, and some beetles (Fig. 29-13) or on fruits, as by the peach curculio or the cotton boll weevil.
2. Formation of galls, probably by the extrusion of auxins or cytokinins (e.g., gall flies, or the spruce bud-gall aphid, Fig. 29-14).
3. Sucking of the sweet juice of the phloem, especially by aphids.

A historical example of this last type was the "phylloxera" of grapes (caused by an aphid on the roots), which was accidentally introduced from America to Europe in 1878. It nearly ruined the wine industry in France; the vines gradually deteriorated. A complication was that the punctures the aphids made in the root facilitated attack by fungi, especially the many Fusariums that grow in roots. Fortunately although the wine grape, *Vitis vinifera,* is susceptible, the "fox grape" from America, *Vitis labrusca,* and some other wild grapes, are resistant. Thus most of the wine grapes of Europe are now grafted on to American root stocks.

Protection by insecticides requires careful attention to the timing, usually with the aim of getting the insecticide on the plant at just the time of hatching. Plants often have hairs as protective devices: dense hairs, which make it difficult for larvae to move about, stiff, pointed hairs, which can puncture the moving larvae, and hooked hairs, which entangle and hold

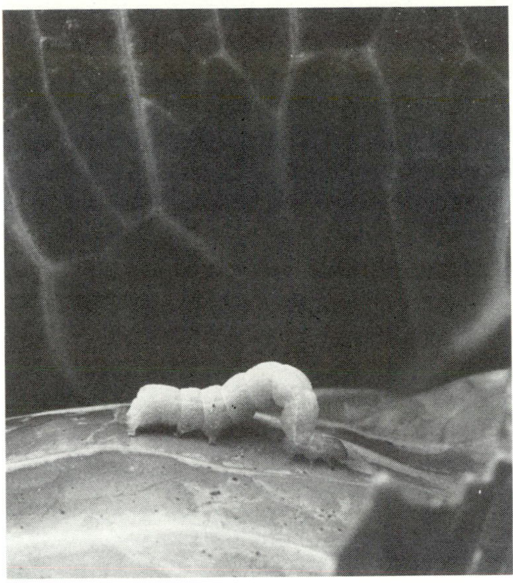

(*A*)

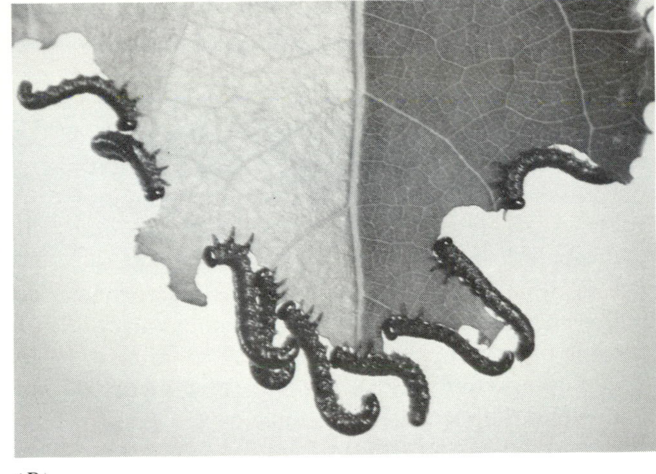

(*B*)

Figure 29-13 *A*. Caterpillar of *Danaus* feeding on a leaf of milkweed, *Asclepias*. *B*. Eastern tent caterpillar feeding on a wild cherry leaf.

(*A*) (*B*)

Figure 29-14. *A*. Leaf galls (caused by mites) on a leaf of alder (*Alnus*). *B*. Stem galls (caused by pupa of a gall-fly) on golden-rod (*Solidago*).

them. Probably all plant hairs function in this way to some extent.

Increasing experimental evidence is indicating the roles played by phenolic compounds such as tannins, quinones, and some terpenoids in deterring insects. Tannins inhibit proteases and other digestive enzymes in the insects' gut. The terpenoid gossypol, produced in the subepidermal glands of cotton, is toxic to larvae of bollworm and tobacco budworm, major cotton pests, but it may be a stimulant to feeding of another pest, the boll weevil. Thus a single compound may be toxic to some insects and stimulating to others. In some cases, the insect may go even further and utilize the compound in its own defense. Some of the problems with the widespread use of insecticides are taken up in Chapter 30.

An example of the extent and variety of insect diseases is furnished by the case of greatest importance to human beings, namely the campaign of the insect world against the wheat plant. Table 29-1, though extensive, is probably incomplete. Some of these pests are of major importance; others not.

12. OUR RETALIATION TO THE DISEASES OF CROP PLANTS

Both fungi and insects center their principal attacks on leaves. The wilt diseases caused by the fungus *Fusarium,* known as root rots, the clubroot of *Brassica, Phylloxera* of grapes, and many nematode diseases are exceptions, and in addition many leaf-inhabiting fungi overwinter in the soil. For this reason the quarantine systems of many of the developed nations are particularly strict with regard to the importation of soil on introduced plants. However, by and large, it is the leaves that represent the most vulnerable area, and hence leaf spraying is the major weapon in the farmer's armamentarium (Fig. 29-15). Against downy mildew of grapes Bordeaux mixture is well established, as we

TABLE 29-1

The Major Insect Pests of Wheat

Insect	*Areas Where Most Active*	*Importance*
Hessian fly (larva)	Canada, North Africa, West Asia, Britain, and New Zealand	The major destroyer; causes losses up to $150 million a year
Winter jointworm	Most wheat-growing areas	Probably second only to the Hessian fly
Wheat stem maggot (larva)	North America (including Mexico)	Destroys up to 2% of North American crop
Cinch bug, green aphid, grain aphids (2 species)	Mainly in Midwest of U.S.A.	Can have up to 20 generations per year
Wheat stem sawfly	Canada and U.S.A.	The worst pest in the Canadian prairies
European and black grain sawflies	Western Europe	
Pale western cutworm and army cutworm	Midwest of U.S.A. and Canada	
Stinkbugs (several species)	Worldwide except South America	
Grasshoppers (five species)	Western U.S.A. and Canada	
Locusts (four species)	Africa, S. Asia, and S.W. Asia	Can wipe out acres of young plants in 24 hours

saw above, but many more powerful fungicides have been developed in the last three decades. Zinc or copper complexes with organic sulfur compounds, such as Fermate, are among the most widely used fungicides, while nicotine, DDT, pyrethrum extract, and certain organic phosphate derivatives such as Parathion and Malathion are among the most effective insecticides. In some wheat-growing areas, Parathion has increased the yields by as much as 10 times.

13. DISEASE PROBLEMS POSED BY PLANT INTRODUCTION

When a new crop plant, or even a plant product, is introduced into a part of the world where it has not previously been growing, it is of course subject to inspection by quarantine officers, who must be well trained in plant pathology. Otherwise there are several possible consequences:

1. The new plant may bring with it a disease that finds in the new habitat another, highly susceptible host. The Dutch elm disease was only a moderate parasite on the European elm but when brought to the United States it proved a most potent pathogen on the American elm. The chestnut blight has a similar history.
2. The plant may escape from a pathogen that normally infects it and thus be more successful in the new area than in the area where it was native. The rubber tree was heavily infected with leaf spot in its native Brazil but was free of that disease in the

Figure 29-15 Dusting potatoes with insecticide in the United States.

Old World because it was brought in by seed. Until the recent outbreak of *Hemilea* in Brazil the same was true for coffee in the New World, and also for the potato for its first two centuries in Europe.

3. The introduced plant may encounter a new pathogen, not present in its native area. This may have been the case for the potato in 1841 and perhaps for the swollen shoot disease of cacao in West Africa more recently.

Of course diseases may spread to new areas by means other than the introduction of infected plants. Of the known plant pathogens the rusts are the only fungi that are *obligate* parasites: other organisms can therefore migrate on a variety of different materials. The rust migrates in spite of its obligate parasitism because there are vast continuous areas of wheat growing. In North America wheat is grown in almost every county from Northern Mexico through Texas to Saskatchewan in Canada. Thus the aeciospores need only to blow a dozen or so miles to find new hosts.

Plants are also subject to nutritional diseases, including excess salinity and the mineral deficiencies discussed in Chapter 8. When one considers the number of enemies that plants have to contend with, it seems remarkable that so many thrive.

CHAPTER 30

The World's Food Supply

1. HUMAN NUTRITION AND MALNUTRITION

In considering the world's food supply, the basic uses of human food must be set out. Firstly, food is required for energy in the form of ATP. All muscular movement (including heartbeat) and most biochemical syntheses require ATP (or other phosphates like UTP or GTP). Since we are not photosynthetic creatures, we have to produce all our ATP through oxidation processes. Weight for weight, fats yield the most ATP, but in ordinary diets carbohydrates are the principal energy source (Eskimos offer an exception: their fatty diet contains minimal carbohydrates). A mole* of glucose (180 g) when fully oxidized to carbon dioxide and water yields about 36 moles of ATP.

Secondly, food is required for tissue building, both for the new tissues of growing young people and the continual replacement of hydrolyzed tissues that goes on at all ages. Since muscle, skin, hair, and nails are all principally proteins (16% nitrogen), and since all cells contain nuclei (19% nitrogen), it is this aspect that requires the nitrogen-containing constituents of the diet.

With a few exceptions, fats, carbohydrates, and proteins can be synthesized from the components of human foods. The exceptions comprise two classes of substances that we cannot synthesize and that must therefore always be supplied in food:

1. Of the twenty *amino acids* present in all proteins, there are nine we cannot synthesize, listed in Table 30-1. However, cysteine and methionine can be formed one from another, so that there are only eight amino acids that *must* be eaten. Since, at least in bacteria, leucine, isoleucine, valine, and threonine all lie on the same general synthetic pathway, it appears that this biochemical pathway is missing in human metabolism. The numbers in parentheses in Table 30-1 are the approximate number of grams of each amino acid needed per day for an adult. These needs introduce a complication in judging the suitability of foods that appear on p. 540.

TABLE 30-1

Amino Acids Essential in Human Diets

Phenylalanine (1.1)
Tryptophan (0.25)
Methionine (1.1)
Cysteine[a]
Threonine (0.5)
Leucine (1.1)
Isoleucine (0.7)
Valine (0.8)
Lysine[b] (0.8)

[a] Not needed if the amount of methionine is adequate.
[b] Can be decreased if the amounts of other amino acids are greater than minimal.

* A mole is the molecular weight in grams: $C_6H_{12}O_6 = 72 + 12 + 96 = 180$.

2. There are smaller amounts of special organic compounds that we, along with most mammals, cannot synthesize. These so-called *vitamins* have miscellaneous uses, and except for vitamin C (of which the amount required is controversial) they are needed only in milligram or even microgram quantities. With the exception of vitamins D and B_{12} all these vitamins are present in edible higher plants. Small amounts of the vitamins D are produced by sunlight acting on ergosterol present in the skin, and are also found in traces in a few plants; larger amounts are in the milk of humans and cattle. The vitamins, their roles, and symptoms of their deficiency are listed in Table 30-2.

The ways in which the human body uses the major plant foods are summarized in this simplified diagram:

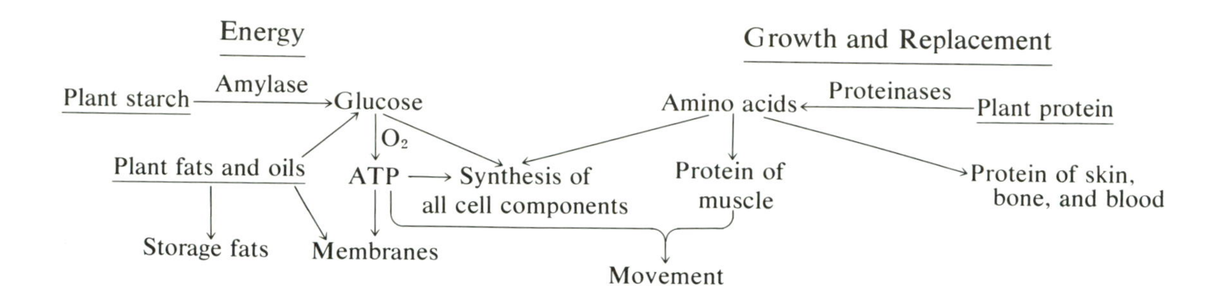

TABLE 30-2

The Principal Nutritional Diseases of Humans

Food Component	*Function*	*Deficiency Disease*
Protein	Body-building	Kwashiorkor
Protein and carbohydrate		Marasmus (essentially starvation)
Vitamin A	Membranes, vision	Night blindness
Thiamine (vitamin B_1)	Coenyzmes for metabolism	Beriberi, polyneuritis
Niacin (vitamin B_3)		Pellagra in humans, black-tongue in dogs
Biotin		Dermatitis, and others
Cobalamin (vitamin B_{12})	Hemoglobin formation	Pernicious anemia
Ascorbic acid (vitamin C)	Membranes, metabolism	Scurvy
Calciferol (vitamins D)	Bone deposition	Rickets, osteoporosis
Tocopherol (vitamin E)	Fertility	Resorption of the fetus, in animals only
Phytomenadione (vitamin K)	Clotting of blood	Failure of blood to clot
Calcium	Bone deposition	Rickets
Iron or cobalt	Hemoglobin formation	Nutritional anemia
Iodine	Thyroid activity	Goiter

The arrows all indicate classes of reactions requiring enzymes and therefore, except for the hydrolysis of starch and proteins, requiring the coenzymes formed from vitamins of the B group.

In addition to the organic foods, nearly the same minerals are required for humans as for plants (Chapter 8), but with some modifications:

1. Much larger amounts of calcium and phosphate are required, because of the large mass of bone (and teeth) to be formed and to be maintained against the slow loss involved in metabolism.
2. Relatively large amounts of iron are needed for the blood hemoglobin as well as for a very similar protein, myoglobin, in muscle.
3. Small amounts of iodine, not needed by plants, are required for the thyroid gland, which produces a derivative of tyrosine containing three iodine atoms; this thyroid hormone controls the rate of metabolism.
4. Vitamin B_{12} contains cobalt and vitamin E contains selenium, but these are needed only in traces.

The above simplified diagram holds also for all mammals, although some can also utilize cellulose, the compound in largest amount in the plant body. Cellulose is attacked, not by the animals themselves, but by anaerobic bacteria resident in the rumen of cattle, sheep, water buffaloes, and camels, or in the small intestine of horses and pigs. The cellulose is fermented to digestible organic acids, with smaller amounts lost as methane and carbon dioxide (Chapter 20).

When food components are missing or insufficient, various *deficiency diseases* result. These are the typical diseases of *malnutrition* (Table 30-2). Often they result from a diet too limited in the number of products included; scurvy was a disease of sailors who, before the invention of steamboats, traveled for months without any source of fresh food; pellagra occurs when the diet is mainly corn, and nutritional anemia when the diet is mainly milk.

2. PLANTS FROM THE NUTRITIONAL VIEWPOINT

As we have seen in previous chapters all plant materials are mixed foods, containing carbohydrates, proteins, usually some fats, some minerals, and at least a selection of vitamins. Let us consider some of the major sources.

a. CARBOHYDRATES

The highest content of carbohydrates is in the edible tubers and roots: manihot, taro, yams, sweet potatoes, and white potatoes. Although these are living tissues and therefore contain *some* protein, the amounts are very small, commonly not over 2 to 3%. A tuber diet unsupplemented with protein is apt to lead, especially in Africa, to the nutritional disease of kwashiorkor. The traditional British "meat and potatoes," the Japanese rice and fish, or the Latin American "corn and beans" are evidently good combinations.

Next in order are the cereal grains. As we saw in Chapter 16, these mostly contain 11 to 14% protein; triticale has 17%, corn has 9%, and rice even less. Although poor in protein, therefore, a diet centering around cereals is acceptable, if with modest supplements. Indeed, most of the human race lives on such diets. Because in the small grains most of the vitamins B and E are in the embryo and aleurone and not in the endosperm, part at least of the whole grain should be included. Corn is very low in calcium and hence the Mexican and Central American practice of cooking corn for tortillas with lime water has good justification. In the rice-eating countries of the Orient a protein supplement is usual; fish, being nearly pure protein, is the ideal supplement. In India gram or other legumes fill a similar role. Polished rice, with embryo and aleurone removed, is almost devoid of thiamine, and indeed thiamine was discovered from this fact. Christian Eijkman, a Dutch medical man in Java, found that the then common disease of beriberi could be cured by feeding the patients rice polishings. Later an extract of rice polishings, in the hands of two skillful Dutch chemists, yielded crystalline thiamine.

b. PROTEINS

The plant foods rich in protein are mainly the leguminous seeds: peas, beans and gram contain about 25% protein, peanuts and bean pods somewhat less, and soybeans much more (38 to 40%). However, the complication with plant proteins concerns the *balance* of amino acids. The eight amino acids essential in our diets are needed in rather definite ratios (Table 30-1). The "best-quality" proteins are therefore those with the same amino acid balance as in our tissues (provided they are digestible). From this viewpoint (only!) the soundest mode of life is cannibalism. The meat of animals comes next closest to the balance of amino acids in human tissues. The proteins of cereals and legumes tend to be low in lysine, methionine, and tryptophan, though rice is adequate in methionine, and the new high-lysine corn (Chapter 16) has two-thirds as much lysine as beef (Figure 30-1). High-lysine mutants have also been developed in both sorghum and barley. Beans are relatively high in lysine, sesame is high in methionine, and both wheat and sesame are equal to beef in tryptophan. The millions of people in India who are vegetarian for religious reasons make extensive use of such supplements. Their total food intake tends to be low by Western standards, but it is often nutritionally balanced.

c. FATS AND OILS

The typical oilseeds – corn, sesame, peanut, sunflower, safflower, coconut, and oil palm – yield most of the edible plant fat. The olive is

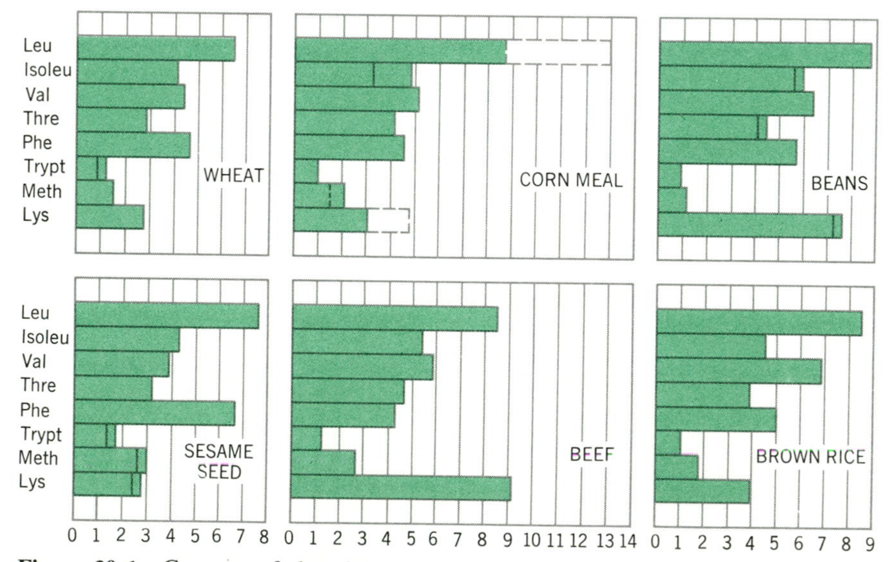

Figure 30-1 Content of the eight essential Amino Acids of Table 30-1 in five plant proteins and in (lean) beef. The data are grams of amino acid per 100 grams of protein. The double lines at the tops of a few columns show values of special importance. In corn meal, the values in dashed lines are those of opaque-2, 60% higher in lysine and correspondingly lower in leucine, isoleucine and methionine.

outstanding as a non-seed source. Except for coconut oil the plant oils are richer in unsaturated fatty acids than animal fats (Table 27-2). The unsaturated fatty acids (liberated from the oils by hydrolysis) are more easily oxidized in the animal or human body, and they promote the oxidation of saturated fats too. Plant foods thus have a dietary advantage over animal fats.

d. THE NUTRITIVE VALUE OF LEAVES

Unlike seeds, the leaves of most plants have rather uniform protein content, averaging around 24%. They are good sources of most vitamins, iron, and usually calcium. Unfortunately the large amount of thick-walled conducting tissue in the leaf veins make most of them indigestible for us. The leaves soft enough for human consumption are mainly limited to the genus *Brassica* (cabbage group), plus lettuce, spinach, sorrel, watercress, and New Zealand spinach. Cattle and other ruminants, able to digest cellulose, can use a far wider range of leaves, including the fibrous grasses and legumes of typical pastures (Chapter 16) and many low shrubs.

The quality of pasture thus indirectly affects the world's protein supply via the domesticated ruminant animals. The annual protein consumption in the form of meat or milk is 15.5 kg per head in the developed countries, 5.8 in the planned economy countries, and 3.3 in the developing countries. The low figure for the latter means not only that plant foods have to supply a high percentage of their protein, but also that many of these people survive on a very low protein diet.

The nutritive quality of pastures is usually much higher when the plants are growing than when they are mature or senescing. The protein content of grasses may fall from over 20% to about 3% in late autumn. For this reason, it is often desirable to feed animals on silage, hay, turnips, grain, or in the tropics cassava, during the dry season. Urea can be converted to protein by cattle, and hence the feeding of synthetic urea, mixed with other materials, is increasing, especially in northern countries. It must be remembered too that in some arid areas the soil and rainfall will not support crops edible by humans, but can give forage for animals; in fact this is often the only productive way to use the land.

3. FACTORS LIMITING THE YIELD OF FOOD

In Chapter 5 we saw how light intensity, CO_2 concentration, and temperature interact in controlling the rates of photosynthesis, and thus of the yield of food. As far as CO_2 is concerned, its concentration is seldom controllable, though under field conditions CO_2 concentration may become appreciably lowered between plants of a dense stand in most climates; therefore a windy day can be beneficial in renewing the air between the plants of a dense crop. At high temperatures and high light intensities, C_4 plants like corn and sugarcane fix CO_2 more effectively than C_3 plants and thus form carbohydrates faster. Factors other than CO_2 require a more extensive review from the viewpoint of yield.

a. LIGHT

Light *intensity* and light *duration* exert quite different influences. In northerly climates, where the sun is relatively low in the sky, the light *intensity* may barely reach the saturation level in spring and fall, especially in Britain, Scandinavia, or northern Russia, where cloudy weather is prevalent. The yields of the typical cool-climate crops, like oats, barley, apples, sugar beets, and *Brassica* plants, are thus, even with the best farming methods, only moderate. In the tropics, on the other hand, the sun is never less than 23° from the vertical at noon, so that average light intensity is much higher. *If* other conditions were favorable, tropical crops would therefore always outyield temperate ones. That the other conditions are *not* commonly favorable is the essence of the food problem for developing countries.

Light *duration* exerts its influence mainly on the time of flowering (see Chapter 10); some plants flower only on long days, some only on short days, and some are neutral. Since in the north, the growing season coincides with the long days, the major cereals of temperate climates are mostly long-day plants. Beans of various kinds, and corn, are day-neutral. Since short-day plants only flower in the spring or fall they find little use as food plants in temperate zones. A few root crops, like chicory and horseradish, are perennials. The *Brassica* group, although believed to be of middle Eastern origin, has long been selected for cool climates and requires long, cool growing seasons; for this reason they are of principal importance in northern and western Europe and northern China. The herbaceous legumes nodulate best on long days.

In the tropics most day lengths vary only between 11 and 13 hours, so that long-day plants seldom make suitable crops unless grown for their vegetative parts. There are a few tropical crops, however, such as okra and some *indica* cultivars of rice, whose day length requirements are so precise that even within a few degrees of the equator (and hence a few minutes change of day length) their flowering is strictly seasonal. The warm winter temperatures in lowland India and southeast Asia make it possible to grow winter-flowering (i.e., short-day) cultivars of rice and even (in semiarid areas) of barley. In tropical areas with heavy rainfall, indeed, winter crops are often best.

b. TEMPERATURE

The average temperature and the extremes of temperature interact with other factors in several ways. In general, growth rate increases with temperature even in plants of cool climates; oats, for instance, have their optimum growth temperature, after the seedling stage, at 30° C (= 86° F), which would be an unusually warm temperature in Scotland or Scandinavia. But high temperature usually causes high transpiration and hence leads to closing of the stomata, which in turn decreases photosynthesis. Also, when the stomata are closed the leaves have lost their cooling system and the leaf temperature soon exceeds the air temperature.

As a practical matter this temperature differential between air and leaves can be used to forecast agricultural yields. In wheat, the amount of the differential measured in the mid-afternoon, and the number of days on which it occurs, have been extensively studied in this regard. If the leaf is hotter than the air this constitutes stress, and the sum of mid-afternoon temperature differentials multiplied by the number of days during which there was this differential $(T_{\text{leaf}} - T_{\text{air}}) \times N_{\text{days}}$ = "Stress-Degree days." The wheat yield in a given area is found to be inversely proportional to the total of "Stress-Degree days" in the season.

Low temperatures, besides slowing growth, have a quite different effect, similar to that of day length, in that some plants require an exposure to low temperature in order to flower subsequently; most cultivars of apples and peaches behave in this way. Winter cultivars of wheat, oats, and barley require moderate freezing for ten or more days (though they can be killed by very severe winters). Biennial plants of all sorts require a cold spell for a while, though often as in parsley, temperatures well above freezing will suffice.

Erratic fluctuations of temperature can be a serious limiting factor. An early frost in the fall will cut short the growing season; late frost in the spring will kill the buds of fruit trees. A recent cold spell in Brazil decreased the coffee crop by a third. But long-term trends present another problem, difficult to evaluate. The average temperature of the whole world is continuously monitored and calculated at Mauna Loa observatory in the Hawaiian islands. It steadily increased from about 1890 to reach a peak in 1943 and then dropped until 1972, losing about 0.5° C in those 29 years. The trend in the past few years is unclear. Some have predicted that the increasing burning of fossil fuels, by raising the carbon dioxide content of the atmosphere, would decrease the reflected heat and thus *raise* world temperatures; others have claimed that increasing amounts of dust and smoke particles in the air would *lower* world temperatures and even that a new Ice Age may be at hand. It is more likely that these long, slow trends in world temperature rest mainly on astronomical causes. Although human activities may slightly modify these changes, we do not understand them as yet.

c. WATER SUPPLY

When rainfall is limiting, crop yields may be determined by the extent, and the timing, of irrigation. In much of California, for instance, with no more than 39 cm of rainfall, vegetables must be irrigated, but in western Europe or the eastern United States, with 75 to 110 cm of rain, irrigation is not needed. The productivity of many areas in Asia and Africa could be greatly increased by irrigation, but to install irrigation systems is expensive—around 1 million dollars a hectare.

The long-term effectiveness of irrigation is interrelated with the salt content of the soil. Enough water must be supplied not only to maintain transpiration (hence to insure stomatal opening) and to compensate for direct evaporation from the soil surface, but also to prevent the gradual accumulation of salts. These problems were taken up in Chapter 7.

An obvious aspect of water supply, of course, is its great variability. When world food reserves are low, a drought or a flood in a productive area can have worldwide repercussions. The great drought of the 1930s in the U.S. Midwest not only produced "dust-bowl" conditions there with much loss of topsoil in the wind, but it decreased crop yields by up to 50%. The smaller drought in 1974 lowered the U.S. cereal yield by some 10%. The six-year drought in northwest Africa resulted in widespread starvation of humans and cattle, and migration of nomadic peoples out of the area. The Soviet Union's dry years in 1972 and 1973 led to their making such huge purchases of American and Canadian wheat, corn, and soybeans (almost 1 billion dollars' worth) that the world reserves reached a dangerously low point. Their still more enormous purchases (25 million tons) in 1979 fortunately coincided with bumper crops in North America.

Flooding can be equally destructive, for soil erosion due to torrential rains can wash away millions of tons not only of the topsoil itself but of the nitrogen and phosphorus it contains. (The shallow, still water of flooded rice fields is of course another matter). Hillside farms for grain should be planted along the contours, in areas subject to heavy rain (see Chapter 31). Curiously, vineyards are usually planted "up and down" on hillsides, because grape roots require good aeration. In some parts of Switzerland it is necessary to carry the soil back from the bottom to the top of the vineyard every spring because of its erosion. Heavy mulching can greatly reduce erosion of soil on a hillside.

Salinity, often a feature of coastal or desert lands, as well as of poor irrigation, is sometimes a limiting factor. Saline soils are in general incompatible with effective food production. An exception is the sugar beet, which actually shows increased growth with moderate salinity; barley is exceptionally salt tolerant, and one cultivar has even yielded a small crop when irrigated with undiluted seawater (Chapter 8).

d. FERTILIZER

Although all 11 mineral elements are needed (so far as known) by all plants, the "limiting factor" rule often becomes important for specific crops, and careful farmers have their soils analyzed to determine which minerals are likely to be in limiting quantities. As a rough guide, leaf crops require high nitrogen, fruit and tuber crops high phosphorus, legume and grain crops a good balance. Chalky soils need additional iron and manganese. There are many special considerations.

An increased use of fertilizers is one of the greatest needs in developing countries. In

1979 all the developing countries together should have imported 9 million tons of fertilizer, but lacked the funds to buy about half of that, even with gifts from many countries totaling over $15 million. (Fertilizer costs around $200/ton). It is far more economical to ship a given weight of fertilizer to such countries than to ship some 20 times that weight of food material (since roughly 1 lb of fertilizer produces about 20 lb of food crops). Synthetic nitrogen fertilizers, in the form mostly of ammonia or urea, have become a necessity in the last quarter-century, but unfortunately the factories that produce them take three to four years to build, and the high temperatures needed in the catalytic process call for considerable fuel usage. We return to the fertilizer problem below.

e. PESTS AND PLANT DISEASES

At least one-third of the world's crops are lost to pests, some before and some after harvest. Thus if all pests could be overcome, the world's food supply would be assured for the next quarter-century at least. The annual loss worldwide is almost equal to Canada's entire grain production. What can be done?

Microbial Diseases. *Microbial diseases* can be combated by:

1. Simple fungicides like Bordeaux mixture or zinc fermate (Chapter 29). These can be sprayed on if the disease is expected, but the more effective procedure is to dust the seed with fungicide just before planting. Almost all agricultural seeds are so treated in North America and Europe.
2. Antibiotics, such as streptomycin, which prevents the growth of blue mold, bacterial fire blight of pears, and other plant diseases. Griseofulvin arrests growth of the Dutch elm disease fungus, but it is too costly for widespread use.
3. Cultural methods, such as not planting annual crops in the same location in successive years. This was probably a subsidiary reason for the practice of rotation of crops, which lasted from the early Middle Ages until the nineteenth century, and for some crops is still used. Also polycultures are sometimes important, particularly in the tropics, as we saw for such crops as coffee and cacao (Chapter 25).
4. Breeding for resistance, usually by introducing genes from other species, as with fragments of *Aegilops* chromosomes brought into wheat (Chapter 16). Among other examples, the rust-resistant wheat cultivars produced by U.S. and Canadian geneticists have fed much of the world's inhabitants for some 40 years, and the mosaic-resistant cultivars of sugarcane produced by the Dutch cane breeders in Java have done a similar service for users of sugar.

Insect Attack. *Insect attack* can be combated by general poisons like lead arsenate or kerosene, which, sprayed on the leaves, are toxic to the chewing insects. Since these can hardly be used on food plants, treatments directed to the specific physiology and pathology of insects are more appropriate. The situation is complicated by the fact that some insects merely carry viruses or bacteria from plant to plant, so that the attack may not be immediately obvious. Several different approaches have been taken.

Toxic Organic Chemicals.

There are several major kinds of toxic organic chemicals (*insecticides*). Contrary to popular opinion, chemicals have been used against insects for centuries. At first there were natural products, obtained from plants. Pyrethrum, the powdered flowers of two chrysanthemum species, was brought to Europe by Marco Polo in the thirteenth century. Pyrethrum flowers are still a major crop in Kenya. Rotenone, extracted from the roots of two genera of tropical legumes (*Derris* and *Lonchocarpus*), and nicotine, extracted from tobacco leaves, came into use as leaf sprays in the last century and are still in use.

A major breakthrough came with the introduction of synthetic chemicals, beginning with DDT, the insecticidal properties of which were discovered during World War II. It saved thousands of soldiers from the lice that carry typhus fever, and for 25 years has been the major weapon against malaria-carrying mosquitoes; however, at least 20 mosquito species have now developed resistance to it. As an insecticide on plants DDT has to be applied in oil solution, because it is so insoluble in water, and the oil may have secondary effects. Also its long persistence in soil led to its being restricted to cotton and citrus crops (where it was most effective) in 1970.

The use of DDT, a benzene derivative with five chlorine atoms, led to the study of other chlorinated hydrocarbons, including Aldrin, Dieldrin, and BHC. Although these include several chemical and physiological types, they all have the weakness of being relatively nonspecific in their action, thus killing other insects, including some predators that parasitize the pest species. They cannot be sprayed at blossom time because of toxicity to bees. In the treatment of cotton, some of the parasites turned out to be more susceptible than the pest bollworms themselves.

A peculiarity of the chlorinated hydrocarbons is their high solubility in lipids. This property, indeed, is what allows them to enter the nerve cells of insects. When ingested by animals the compound therefore becomes sequestered in the fatty tissues, from which it is metabolized only very slowly. A predator eating insects or small animals in a stream containing these compounds thus accumulates the pesticide in its own fatty tissues. Those at the top of the food chain would be expected to accumulate the most (although this concept has been doubted). In any case, these compounds are toxic to fish, and perhaps to some birds, though this is controversial. As to humans, a scientific study of the workers in the Los Angeles DDT factory, who have ingested in the course of their work some 400 times the amount to which ordinary people would be maximally exposed, showed that DDT at least is harmless to adult humans.

A quite different type of chemical insecticide is that of the organophosphates, a large group of organic compounds containing phosphorus, often with sulfur as well. Like DDT, these compounds act on the nerve–muscle junctions of insects, and some of them are taken up so readily by the plant that their level in the phloem is high enough to act systemically against sap-sucking insects like aphids and spider mites. Again, their action is relatively nonspecific, and also they are often quite toxic to animals and humans.

Recently a new type of compound has been developed. It is based on inhibiting the biosynthesis of chitin, the external covering of the insect body. One such compound, called diflubenzuron, is highly effective at low concentrations. Since chitin is not present in the skin of mammals its action should be limited to insects, but as yet it has not been sanctioned for food crops.

Methods Based on Insect Biology

A problem with insecticides is that, when any toxic agent is applied to a statistically varied population, those few individuals having resistance, or producing offspring with mutations toward resistance, have an immense selective advantage and multiply rapidly, so that once resistance has appeared the sensitive population is soon replaced by a resistant one. We noted an analog of this problem in regard to wheat rust in Chapter 29; there, fortunately, genes for rust resistance could be bred into the wheat, but the insect resistance comes from natural mutations, not from the breeder. From the farmer's viewpoint, resistance calls at first for increased dosage of the pesticide, but soon it becomes clear, as it has now with the pink bollworm of cotton, that a change of strategy is needed. Here, therefore, *integrated pest management* is being substituted. In this method the insect population is regularly monitored and action against them is taken only when the population has become great enough to threaten major economic loss. Thereby the conditions for selecting resistant mutants are minimized, and at the same time the development of insects parasitic on the pest (see below) is favored. The type of pesticide or other action can be changed during the season. Integrated pest management (IPM) is a flexible, multidimensional approach utilizing a range of chemical, biological, cultural, and mechanical techniques. The basic premise is that no *single* control method is likely to be successful because of the rapid selective power of insects and because of the many variables, related to location, season, cropping patterns, and individual pests. The approach is being used increasingly on many types of crops other than cotton, such as alfalfa and apples (which receive more pesticides per acre than any other U.S. crop). It is becoming one of the most active areas of research in pest management and is also being expanded to include pathogens and weeds.

Insect attractants and *hormones* now promise to play a most important part on the biological side. Studies of insect physiology and behavior have brought to light the tiny amounts of volatile compounds (*pheromones*) produced by female insects to attract the males. A number of these have been identified and synthesized. Each species appears to use a different compound, sometimes even a specific mixture of isomers of the same compound. The effects are incredibly delicate; 0.04 mg of the gypsy moth pheromone can attract 100 males in 5 minutes, some over a distance of 3.5 k. These compounds therefore make ideal trapping agents, for they act only on the one species that constitutes the pest. Research on just how best to use the pheromones is still under way, but the first large-scale use, against mosquitoes, looks very promising. The resistance problem is automatically bypassed, for a mutation causing insensitivity to the natural attractant would of itself prove lethal, since it would prevent the insects from mating. A similar new lead is provided by synthetic insect hormones, one of which prevents the emergence of the adult form from the larva.

Control by predators has been effective in a few cases. The Japanese beetle, which devours the leaves of grapevines and roses (among others), can be controlled by a spore-forming bacillus that multiplies in the insect's blood. The walnut aphid, which sucks the phloem sap of walnut trees, is controlled now in the United States by a parasite imported either from France (for cool areas) or from Iran (for hotter areas). The tomato horn worm, a very large caterpillar, and the grape leaf hopper are being partly controlled by small wasps that lay their eggs in the larvae. However, it has not proved easy to get an introduced parasite established in a new area. In Hawaii, for instance, 133 species of parasites and predators had been brought in by 1925 but none of them had become established there.

We can hope from all these developments that human ingenuity will steadily prove a match for the great variety and adaptability of the insect world, but research will be continually called for.

Attack by nematodes is less easy to deal with. Since their choice of host plants is moderately specific, some defense can result from continual crop rotation. Some members of the Compositae form compounds in their metabolism that are nematocidal. The recently discovered powerful nematocide, dibromochloropropane, has now unfortunately been found to affect humans, and its use has been temporarily suspended. This field is still wide open for research.

f. WEEDS

Weeds offer special problems (Chapter 12). They absorb water and nutrients from the soil; tall or bushy weeds may even take light off the crop. Pulling weeds by hand is still the method of choice in countries where labor is abundant, but not elsewhere. Simple hoeing is effective for row crops like corn, cotton, potatoes, or sugarcane, and for tree crops, but cannot be used for small grains because they are planted too close together. Some crops (such as sugarcane) grow tall enough and rapidly enough to "shade out" most weeds. But poppies in the wheat fields of western Europe, and mustard in the wheat and barley fields of Britain, reach flowering soon enough to mature their seeds before the crop is harvested. Other "aggressive" weeds have developed both growth rate and general form very similar to those of the crop in which they grow (Chapter 12). Some of these situations can be met by the use of *herbicides*.

The old "weed-killers" like borax, sulfuric acid, or sodium arsenite, used along railroad tracks or on gravel paths, were quite nonspecific in their action. They have already been largely replaced by the discovery that many plants are killed by *auxin* if it is applied in quantities very much greater than those normally present. Indoleacetic acid itself is decomposed rapidly, by enzymes and by bacteria, but some of the synthetic auxins are fairly resistant to decomposition: of these, "2,4-D" (2,4-dichlorophenoxyacetic acid) is the most used. It has little effect on monocotyledons and can therefore be used to free fields of cereal from such dicotyledonous weeds as those just mentioned. Use of 2,4-D on cereal fields in Britain is estimated to have increased grain yields by 30%. Even on row crops it probably causes less damage than hoeing, which tends to break the surface roots.

Discovery of the effectiveness of 2,4-D stimulated research on a host of other substances affecting plants. "Dalapon" (2,2-dichloropropionic acid) was found to have properties the opposite of 2,4-D, namely to kill grasses, with little effect (at the right dosages) on dicotyledons. A compound like 2,4-D but with one more chlorine atom ("2,4,5-T") is effective on woody plants that resist 2,4-D; however, some of the older preparations contained an impurity highly toxic to animals. One group of herbicides can be applied to the soil before the crop is sown, to exert "pre-emergence" treatment; other groups are specially designated for particular crops or weed types. The list is increasing continuously.

In one instance biological control of a weed has been successful; a "prickly pear" (*Opuntia* sp.) native to the New World was introduced into Australia and became a pernicious weed there. A small wasp, *Cactoblastis*, which lays its eggs on the young shoots so that the larvae eat the buds, controls it very well. Some weeds are partly controlled by weevils (small beetles) that bore into the seeds, but up to now biological controls have played only minor parts in large-scale agriculture. In at least one case with insects, resistance has developed to the biological control much like that to the chemical controls.

4. FOOD PROCESSING, STORAGE, AND TRANSPORTATION

The role of these three steps in bringing food to market is not ordinarily part of a book on botany, but they play a major part in the world's food supply and they are certainly factors limiting the food that reaches the consumer. Transportation from farm to market is often a limiting factor in developing countries where roads are few. As small farms in these countries begin to produce food in quantity for the cities, the building of all-weather roads may prove as important as supplying fertilizer or agricultural knowledge.

a. PROCESSING FOOD FOR EATING

Cooking makes foods more digestible and for many foods is essential. It destroys some bacterial toxins, as well as living bacteria that can cause intestinal disease; a drawback is that it can also destroy ascorbic acid, which is readily oxidized at 100°, especially in slightly alkaline solutions. The other vitamins are rather heat-stable.

Extraction with water can remove toxic or bitter substances, as with manihot (Chapter 19), cycad fruits (Chapter 24), or acorns. Olives are extracted with alkali to remove a bitter glycoside. Most of these are ancient processes, and some were initiated to obtain the extract for medicinal purposes.

Milling separates grain into fractions for different uses. Wheat, for instance, is traditionally milled into bran, germ (the embryos), middlings (coarse particles of endosperm), and flour (fine particles of endosperm). The middlings are then further ground into flour. The loss of thiamine in rice aleurone and in wheat germ has been mentioned above. Not only other cereals, but also soybeans, potatoes, and other foods are milled into flour. Milling itself does not change the nutritive value.

b. PROCESSING FOOD FOR STORAGE

The prime problem in food storage is protection against bacteria and fungi. These can grow over a wide temperature range from −1° to about 50°C, but require moisture. Preservation can thus be by drying, freezing, or heating.

Removal of water is the simplest method. Dry beans, nuts, and dry grain or flour keep almost indefinitely, except for attack by weevils, small beetles of the very large family Curculionidae, which apparently generate their own moisture by oxidation of the food. Water can also be removed osmotically by adding salt; salted beans were very much used in early colonial days, and salt pork was traditionally taken on long sea voyages.

Low temperature slows but rarely prevents bacterial growth. A few bacteria will grow slowly at 0° F, so that for long-term storage, temperatures below freezing are essential. Many microorganisms survive the cold and develop at once on warming. Typhoid bacteria have been found in ice cream after a year's storage. The expansion of ice crystals in plant tissue usually disrupts the cells and makes the tissue even more susceptible to microbial growth on warming. A low-cost cooling system for the tropics would be a major advance.

High temperature, like cooking, destroys microorganisms. Their vegetative cells are generally killed after 10 minutes at 65° or 16 seconds at 72°C ("pasteurization") but spores require temperatures above 100°, that is, steam under pressure. Dry heat acts more slowly. The larger the bacterial population the greater is the chance that a few cells will survive, and hence the longer the heat treatment needed.

Storage of whole fruits is a special case, since most fruits do not freeze well. The fruits are usually picked unripe and ripened in the storage chamber. As the fruits mature they produce ethylene (Chapter 10), which in turn hastens their ripening and affects other fruits too. High carbon dioxide pressure (around 20% of atmospheric) antagonizes the ethylene and this "gas storage" has been extensively used for apples. Reduced atmospheric pressure ("hypobaric storage"), on the other hand, draws the ethylene out of the tissue; it takes out some of the more volatile flavoring materials but is now being developed into a practical process.

c. THE PRODUCTION OF MEAT

Conversion of grain to meat by feeding of cattle, hogs, or poultry can be thought of as a form of processing. Much of the starch in the grains is oxidized for energy and some of the protein is also lost, resulting in a net feed conversion rate ranging from about 10% in beef to about 35% in chickens. While some have considered this a waste of potential food (because of the lowered efficiency in using the second level of a trophic food chain) the point to note here is that this type of "processing" confers a needed elasticity on the agricultural market. For in years like 1977 when there is a record grain crop and world reserves are high, land can be taken out of wheat and put into pasture, thus increasing the meat supply. Such alternative functions are made use of in Australia, New Zealand, and the United States at least; Argentina's beef is mainly from permanent pasture. In this balance, as with many other agricultural products, the determining factors are economic rather than biological.

5. THE WORLD POPULATION AND ITS GROWTH

Food supply has to be measured against food consumption, and in the past the growth of food supply has tended to parallel the growth of population. Put crudely, as more food became available, more babies survived. At the time described in Chapter 3, when agriculture began, the entire human population of the world is estimated to have been perhaps under several million; it is now 5000 times as much! That food and population have increased in parallel is to be expected, since for nine-tenths of the time since agriculture began, each family or group has raised (or gathered, or hunted) its own food. Only in the last few centuries has the nonfarming population become a large fraction of the total. In central Africa, for example, nearly all families still raise their own food, but in western Europe and the United States about 5% of the population now raise the food for the other 95%. But this ratio can only be sustained with sophisticated farming methods, abundant fertilizer, and a favorable climate. It seems unlikely that it can go much lower. In the United States, Australia, Argentina, and the Soviet Union, the amount of agricultural land is high relative to the population. Table 30-3 sets out suggestive facts (rounded data) on the ratio of land to people in six major countries. The figures show that Japan and Germany, with relatively dense populations, manage to make up for the shortage of tillable land by generous use of fertilizers. Brazil and the United States are intermediate. India, with almost three times the population of the United States and less than half the amount of land per head, is using not *more* fertilizer but far *less*. Since other developing countries yield similar figures, there is evidently room for great improvement in this factor. Of course the difference is offset in part by the much greater use of hand labor, which though producing far less crop *per person*, favors higher crop yields *per unit area*.

The heart of the problem rests in the fact that the rates of growth of the human population of individual countries and areas are now no longer parallel to the growth of agricultural productivity. Population growth can be described in various terms; "crude birth rates," that is, the number of *live births per year per*

TABLE 30-3

Approximate Ratios of Land to People in Six Major Countries

Country	*Population Total (millions)*	*Percent Urban*	*Cultivated Land (hectares per head × 100)*	*Amount of Fertilizer Used (kilograms per hectare)*
Japan	100	70	7	238
Germany	70	75	19.5	162
U.S.A.	210	70	93	55
Brazil	107	60	35	50[a]
U.S.S.R.	225	50	104	–
India	550	19	33	2.5

[a]60% of this is used in São Paulo state, the center of intensive agriculture.

thousand of total population, are probably the simplest measure. However, they are not the same as the actual rates at which the population increases because individuals survive to different ages in different parts of the world. Figure 30-2 presents crude birth rates for a representative series of countries, both developing and developed. Unfortunately the data are none too reliable, since in many countries census-taking is incomplete, and hence some of the figures in the upper half of the Figure are probably too low.

One can see at a glance that the major food-exporting countries fall into the lower group. There are a few slight exceptions; Thailand exports some rice, and Mexico was for a time a cereal-exporting country, until the growth of its population caught up with the improvements in corn and wheat culture described in Chapter 16.

The countries in the upper group, however, represent a much larger total population than those down below. India has almost 600 million, China about 950 million (as far as known); birth rate data are very uncertain for these countries. Latin America has at least 280 million. In all, the developing countries thus comprise about three-quarters of the world's population, and it is this group that is growing most rapidly. For example, India adds about 11 million people, half the population of California, *every year*. It is to this group, therefore, that we have to look *either* for major increases in the rate of food production *or* for major decreases in the birthrate.

A more critical measure of population growth is the "rate of natural increase," which is defined as the difference between the annual birthrate and the death rate, both expressed per thousand of the population. For most western countries with good medical attention the death rate averages about 14.3 per 1000. It can be seen from the table, therefore, that except for immigration the countries at the low end of the lower group are barely maintaining themselves. The high rates of natural increase are in part due to the decreases in the death rate that

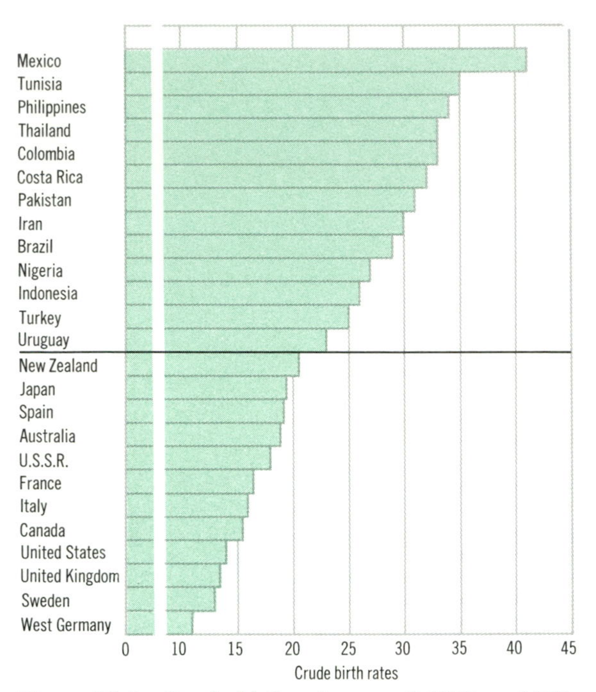

Figure 30-2 Crude birth rates, as of 1972 or 1973, from selected countries. The data are live births per year per thousand of total population. The average value for Latin America as a whole is about 30. The values for USSR and Germany have greatly decreased in the last 40 years, and that for the United States has decreased since the data were provided.

have taken place in the last 50 years or so, due to improvements in medical care, better sanitation, control of malaria, and other factors. Until very recently, no similar effort has been directed towards decreasing the birth rate. That problem is also social and cultural, since the tradition of large families in many countries is hard to change. In Mexico and India the *average* family size is about 7. Desire for smaller families and postponement of marriage thus rank along with available contraceptives and free abortion as major factors.

6. THE OUTLOOK

The balance between population and food supply will obviously be a close one. World food production will have to go on increasing for many years if mass starvation is to be avoided.

Is this possible? In the light of the preceding sections we set out here the factors, favorable and unfavorable, that contribute to helping us answer this all-important question.

a. FACTORS GENERALLY UNFAVORABLE

First of all, it is clear that no major enlargement of the world's agricultural lands can be expected. The only land not yet sown to food crops is likely to be either of low productivity or else to be already in use for important non-food crops like timber or cotton. In the 1960s the Soviet Union brought a sizable area in central Asia (Kazakhstan) into cultivation, but it has such low average rainfall that its yields are likely to be very sensitive to weather variation. Just now considerable areas of forest in the tropics, especially the Amazon valley, are being cut down for use as farm land, but the soil is low in nutrients and subject to the formation of laterites (see Chapter 15). On the contrary, some land now under cultivation in North America should probably be taken out for a few years, since the continual plowing and cropping is leading to serious soil erosion. The U.S. Soil Conservation Service considers that the erosion problem has become much worse recently (see Chapter 31). Thus the balance between population and food is not likely to be swung much by putting additional areas into food production. Any idea, therefore, that "biomass" can be raised for fuel on the scarce agricultural land is obviously unrealistic.

A second generally unfavorable aspect is that of the conflicting needs of town and country populations. In India and Pakistan, and also in West Africa, the prices of food (mainly cereals) have been deliberately kept low in order to help the townspeople, with the result that the farmer cannot earn enough to buy equipment and fertilizer; thus the standard of farming remains primitive. Superimposed on this is the steady drift of people from rural areas to the towns and cities. Havana, Caracas, and Rio de Janeiro are examples of cities becoming surrounded with vast areas of shantytowns, where poor rural families hope to find better living conditions and higher-paid urban kinds of work. The separation of Pakistan from India, and later of Bangladesh and Pakistan, and the advent of repressive regimes in southeast Asia have led to floods of refugees into the cities near the borders, such as Calcutta and Karachi, and into Malaysia, and these add to the already high food demand of those urban areas. The resulting "urban sprawl," as in other growing cities too, swallows up good farm land.

The high price of oil set by the agreements among the oil-producing countries has raised the costs of mechanized agriculture everywhere, and in particular makes it more difficult for the developing countries to improve their farming efficiency. The fuel spent in transporting food from farm to merchant is also involved here. Draft animals (horses, oxen, water buffaloes, camels, llamas) are alternatives on small farms, but they have to be fed; the United States used to devote *several million* acres of agricultural land to raising feed for farm horses.

In the major food-producing countries the utilization of practically all available land for agriculture, with accompanying heavy use of fertilizers, leads to the drainage of nutrients into streams and lakes, causing the excessive growth of aquatic plants. Most of these die in the low temperature and low light intensity of winter, and their decay pollutes these bodies of water. Unfortunately the continued world demand for food will make it difficult for this factor to be much improved, though more careful monitoring of farm runoff, and less waste of fertilizer in some areas, will certainly help.

Fertilizer supply will be increasingly important as tropical countries develop the practice of multiple cropping, that is, successive crops on the same land within one season. In China, where hand labor is exceedingly plentiful, it is customary to transplant whole crops from the nursery to the field, and in the vicinity of large cities as many as six crops are raised one after the other, by rapidly transplanting half-grown plants immediately after harvest of the preceding crop. In most of southeastern

China, raising three crops per year (at least one being rice) is now basic farm practice. The Ganges valley presents another climate and soil where three crops a year could readily be grown, but they will need fertilizer.

b. FACTORS GENERALLY FAVORABLE

To offset the discouraging numbers in Figure 30-2, there has been a gradual but real decrease in birth rate in a number of developing and semideveloped nations, including Taiwan, Tunisia, Costa Rica, Egypt, and Chile. If these decreases, which average about one birth per thousand of population per year, continue, rates comparable with those in the lower group of Figure 30-2 should be reached by the turn of the century. It is claimed that both the spread of education, and the move of people to the cities, tend to lower the birth rate somewhat. Increase in per capita income has the same tendency. There is also a strong move in China, India, and Eastern Europe for official bodies to bring information and assistance on family planning to the poor, both urban and rural, and this should soon be affecting the statistics in these large populations. Indeed many East European countries have already shown steady decreases in their crude birth rates over the last 20 years. The annual growth rate of mankind as a whole was 1.9% in 1970 and is now down to 1.7%.

The unlikelihood of adding much to the area under cultivation has thus far been offset by small but steady increases in the productivity of the existing land. Even in India grain production has doubled in the last 25 years. The increases have been better still in Java, where the new rice cultivars have been well received, fertilizer is being used, and villagers are moving back out of the towns. That such increases are modest compared with what is possible is exemplified by the fact that rice can be produced on experiment stations with a yield of 10,000 kilograms per hectare, while the actual world average is only 2200. The maximum yields of wheat, corn, and legumes on well-run modernized farms are similarly around four times the world averages, and even small subsistence farms could raise their yields considerably.

To these increases in yield achievable by the best farm practices and optimum use of fertilizer must be added the increases due to scientific developments. Hybrid corn raised the corn yields in the United States by some 50% with little or no increase in land, water, or fertilizer. The new dwarf cultivars of rice (Chapter 16) can increase yields under primitive conditions by 25 to 30%, and with adequate fertilizer by 300%. The introduction of semi-dwarf wheat caused marked but less spectacular increases, especially in areas of low rainfall like northern Mexico and west Texas. The new types of insecticide based on insect hormones and attractants should also, in due time, increase yields of some crops.

The world's food reserves have recovered rapidly from their dangerously low levels of a few years ago. Since in the years 1976 to 1979 the United States experienced record or near-record yields of both wheat and corn, and Soviet harvests were (until 1979) normal, the emergency is at least temporarily over. This does not, of course, offer any long-term basis for relaxing efforts to keep crop production at maximum levels, particularly in the United States, which provides most of the world's export cereals (even including rice!) and soybeans (Fig. 30-3).

c. THE BALANCE

Apart from temporary crises due to unfavorable weather or unpredicted pests, the long-term outlook seems to be one of bare and slow improvement. Overall food production in the developed countries is increasing at about 2.8% per year and has continued to do this for many years. Even in the developing countries the overall average increases at over 2.0% per year; in Java it is 4%. In India the amount of grain produced *per capita,* which was 142 kg in 1950, rose to 162 kg in 1974; thus, it kept just ahead of the growth of population. Worldwide, the U.N. projections, though only to 1985, show a

Figure 30-3 Helping to feed the world. Large-scale harvesting of wheat in Pennsylvania. This type of immensely productive mechanized farming makes the United States and Canada the world's leading wheat exporters.

world population increase of 2% per year and a food production increase of 2.7% per year. In Latin America and non-Communist Asia the population is expected to increase 0.2 to 0.4% faster than the food supply, but in the rest of the world it will increase 1.0 to 1.9% more slowly than the food supply. The onset of government-supported family planning has yet to show its effect in most areas, and this effect may well increase the favorable side of the balance. Lowering of the birth rate is the world's A-1 urgency. Meanwhile, however, the United Nations estimates that 460 million people (more than twice the population of the United States) are underfed.

After all, the most elementary responsibility of a government should surely be to see that its people are fed. That many governing bodies have overlooked this minimal requirement is perhaps partly because it is hard to assess the intricate relations between the results of scientific and technical endeavor in different fields. Thus the industrial revolution of the nineteenth century in Britain would probably not have been possible without the simultaneous development of intensive agriculture. Similarly, the embarrassingly high rate of increase of tropical populations has resulted in part from the impact of modern medicine on tropical diseases and thus on the death rate. The gradual spread of education is having multiple effects, from greater expectations for consumer goods on the one hand to better farming practices and smaller family size on the other. The results of such opposing tendencies are hard to predict. It must also be noted that failure to ensure sufficient supplies of fuel energy for the productive agriculture of 10 years hence may bring the world's food reserves down again. One need not argue as to *which* source of energy to use; we shall need coal, oil, conservation, and nuclear and solar energy; all five.

To keep the balance just favorable will need cooperation not only from governments, but from every citizen, especially from those like the readers of this book, who have enough scientific background to appreciate the complexities.

CHAPTER 31

Modification and Preservation of Ecosystems: Optimal Land Usage

To understand and measure the changes that humans have made to natural ecosystems, it is helpful to view our place in Nature. Our flexibility as organisms is often emphasized, that is, our freedom from those evolutionary traits that would sharply delimit our activities or choice of habitats, as is true with most other animals. What is often overlooked, however, is that we exist by virtue of plants and animals that have become highly evolved and specialized during the last 2 to 3 billion years. As discussed throughout this book, angiosperms, and a mammalian fauna dependent upon them, are the most obvious of the specialized plants and animals from which we have primarily obtained our sustenance. Our existence would have been inconceivable with only the organisms of the remote geological past, although they are still important to us as decomposers and as members of food chains. Also the persistence of numerous species of gymnosperms, along with many kinds of woody angiosperms, have furnished facilities without which we would have been severely handicapped. Indeed our presence seems to have depended upon trees, since our ancestors, who lived in trees, were obliged to develop the free shoulder articulation, grasping hands, and stereoscopic vision that serve us so well.

To modern urbanized humans the organization of ecosystems, which cycle materials and energy, regulate the moisture economy, and make possible the formation of soil, is less obvious in importance to our survival than specialized kinds of plants and animals. Any component species within an ecosystem is considered to survive by virtue of its niche; by occupying a niche, it also assumes a role in relation to its surroundings. For further survival, it is necessary that the role not be a disruptive one. Organic systems or processes are expressions of thermodynamic laws tending to approach a condition of minimal stress or steady state. Warnings have been sounded by many who study these complex interactions that we need more scientific knowledge to meet emergencies arising from rapidly moving human impact on these ecosystems. It is also necessary to formulate long-range perspectives so that the promise of their use may be weighed against caution about their destruction.

1. HISTORICAL VIEW OF OUR IMPACT ON ECOSYSTEMS

When *Homo sapiens* first evolved he probably affected natural ecosystems no more than most other animals (and probably not as much as elephants and hippopotami, which were part of his environment). On the other hand, environmental impacts on early humans were immense. Natural catastrophes, such as floods, earthquakes, and volcanic activities, were more devastating than they are today; in fact, their effects have resulted in a long history of our struggle to "conquer the environment." Additionally, our human mark on the world's ecosystems *remained* insignificant as long as populations were small, distribution of people was localized, and technologies were simple. Let us now look at the manner and intensity in which humans have modified the plants in natural ecosystems through use of fire, agriculture, and industrialization.

a. FIRE

Fire was one of the first technologies that increased our mobility. It enabled us to move into and be able to exist in cold climates. Evidence for our use of fire collected from natural sources occurs in cold Eurasia 300,000 to 350,000 years ago and 40,000 to 70,000 years ago in Africa (Chapter 3). Fire also greatly aided in the processing of food through cooking and by 38,000 B.P. fire-making was a part of our cultural heritage.

Fire as a tool to clear the forest for agriculture and grazing brought three kinds of effects on ecosystems: (1) widespread ones, affecting sizable areas of forest and grassland, (2) inherently repetitive ones, hitting the same area frequently, and (3) highly selective ones, exterminating some species and encouraging others. However, fire is a natural part of the environment, playing a significant role in natural cycles under certain climatic conditions (Fig. 31-1). For example, in California with its dry Mediterranean climate throughout the summer, the shrubby chaparral ecosystem is rejuvenated by fire on a 15- to 25-year cycle, and the giant *Sequoia* forests in the Sierra Nevada are maintained in a steady-state condition by fire on essentially a 100-year cycle. These natural fire cycles are much longer than many of those created by human beings today. In African savannas, for example, the land may be burned at least twice a year. This has been a constant theme of human activities—extreme acceleration of processes compared with those of natural evolution. However, compromise is possible for management practice, which we discuss later in this chapter.

Figure 31-1 In 10 U.S. national parks, fires are being managed so that they play their natural role in the ecosystems. This fire in Kings Canyon National Park (1973) is typical of the natural burns, which are closely observed but not suppressed unless they become dangerous to life or property, in the parks in the Sierra Nevada Mountains of California.

b. AGRICULTURE

Although advances in hunting technology affected the structure of natural ecosystems, the greatest early impact on them came with the advent of agriculture. Hunter–gatherers still were very much a part of nature, competing with other organisms for the food supply, but cultivators started to modify vegetation to suit human needs. They interfered with the normal flow of energy in the biosphere and diverted it to the products they could eat. As we have discussed throughout this book, humans have decided which plants grew where, protected useful plants from disease, and even altered the course of plant evolution by bringing into being variations within species that would not have survived without human care. A series of interrelated events then and now continues to lead to either destruction or modification of natural ecosystems. Massive quantitative increases in food supply as a result of cultivation of plants have resulted in population increases. The necessity of tending crops had made some populations sedentary. Successful production of food in the hands of relatively few people promotes the division of labor and increases development of technology and cultural advances. Thus with both the increases in numbers of people and the concentration of their activities in urban areas, there has been substantial change in their relation to both natural and humanly created ecosystems.

Although there are many reasons why civilizations have developed and flourished, one condition is essential: there must always be a sufficient production of the necessities of life—particularly food. Some of our first civilizations were built on irrigated agriculture. This was primarily because the irrigated lands remained productive much longer than did the lands where rainfall furnished the water for crops. A secondary reason was probably that farm production was more dependable on irrigated lands, where drought was not likely to be catastrophic.

Most historians agree that the first civilizations were developed in the valleys of the Nile, Tigris and Euphrates, and the Indus (Chapter 3). These valleys shared three common features: (1) the soil was fertile, (2) rainfall was scant, but (3) water supply was dependable for crops because irrigation was used. Fertile soil and dependable water supply enabled farmers to produce a large surplus of food and ensured continuity of food supply for many generations. Civilization spread from these irrigated valleys to other areas, partially resulting from population pressures. Then a series of migrations ended the era of purely local impact on ecosystems.

Most of the area surrounding the Mediterranean Sea, except Egypt, presents a striking example of human impact on natural ecosystems as urban areas developed. At one time nearly every section of the Mediterranean supported a progressive and virile civilization: Crete, Syria, Lebanon, Tunisia, Algeria, Sicily, Greece and Turkey. Yet generally these countries today are not prosperous. In fact, this is one of the main regions of which it has been said that "civilized man has left a desert in his foot prints."

As civilization spread from valleys of the Nile and the Tigris and Euphrates to shores and islands of the Mediterranean, civilized humans came to cultivate different kinds of land. They had learned over a period of 2000 years or more how to practice permanent agriculture on irrigated land and had used shifting agriculture on nonirrigated land for 3000 years. But apparently never before had they attempted to practice *permanent* agriculture on sloping land with rainfall as the primary water supply.

The farmers on the island of Crete were probably the first to build a civilization based on rain-fed agriculture. The story of agriculture and its impact on the land by the Phoenicians in Lebanon is similar to that of the Minoans on the island of Crete, although the time, place, and details vary.

The Phoenicians, who borrowed most of their culture (including agricultural practices) from Mesopotamia, Egypt, and Crete, probably settled the country known today as Lebanon between 4500 and 4000 B.P. Their homeland consisted of a relatively narrow strip of coastal plain and a foothill zone that rose steeply into the Lebanon Mountains. The coastal plain and foothills had fertile soil and were covered with grasses and forests of the famed "Cedars of Lebanon" when these people settled there (Fig. 31-2*A*). There was also sufficient rain to give adequate yields of grains, grapes, olives, and other crops grown by these people. Early in

(A)

(B)

Figure 31-2 *A*. Remains of the Cedars of Lebanon. There are only four small protected groves now in Lebanon where there once were beautiful stands covering thousands of hectares. The eroded hills in the background are now unproductive. This is an example of extreme mismanagement of land and transfer of inappropriate agricultural techniques from one environment to another. *B*. An English man-made landscape balancing utilization of the land for agriculture, woodlots, or patches of forest. Villages also are often endowed with gardens and street trees.

their history the Phoenicians found that their timber was an easily marketed product to the people of the treeless plains of Egypt and Mesopotamia. The narrow coastal and foothills area became inadequate to provide food for the expanding population, and cultivated fields began to creep up the cleared slopes. In addition the Phoenicians traditionally were a nomadic tribe that herded goats where the deforested areas were not cultivated. The goats took over, preventing the recovery of forests that could protect soil, prevent floods, and provide a continuing economic asset. Deforestation, cultivation based on the techniques successful under irrigation along river valleys, and activities of "scavenger" goats brought on erosion, which turned Lebanon into a "well-rained-on desert."

Like Lebanon, Palestine, "the Promised Land," which 3000 years ago was "flowing with milk and honey," has been so devastated by erosion that soils were washed from fully half the area of the hill lands. However, modern Israel has restored much of the land and stopped ongoing destruction. Techniques of soil conservation, water control, and forestry are being applied to the remainder of the soils; importantly also, goats are being restrained! Millions of trees have been planted; fires and grazing are controlled. This limited land, however, did not recover on its own, but by massive infusion of manpower, machinery, technology, capital, and furious will to survive.

These glaring examples of mismanagement of land 4000 years ago are unfortunately being repeated in some areas (e.g., San Salvador and Brazil) today, despite active soil conservation research and education throughout the world—particularly in the developed nations. During the last 2000 years it is estimated that about one-third of the topsoil in croplands of North America has been lost. Much of this was lost in the Midwest during the dust bowl days (1934 to 1938); immediately following that period various methods of soil conservation, such as contour plowing, crop rotation, application of livestock manure, planting of cover crops, and "nontillage" or "minimum tillage" techniques were extensively employed. However, even though soil is lost to erosion each year it is also being formed continuously. The rate of soil formation (Chapter 15) is difficult to measure, but under ideal management it may be formed at the rate of 1 in. (2.5 cm) in about 30 years. Under normal agricultural conditions the rate is nearer 2.5 cm every 100 years, (i.e., about 1.5 tons of topsoil formed per acre per year). Under natural conditions soil formation averages a rate of 2.5 cm in 300 to 1000 years. All of these estimates, of course, are averages over different climates, bedrock types, and natural ecosystem types. Thus, in well-managed soils, there is the opportunity to form soil at a rate even greater than that lost to erosion. However, the *average* annual loss of topsoil under all types of agricultural practices in the United States has been around 12 tons per acre. Perhaps more alarming is that the erosion problem in the developing countries is estimated to be twice as severe as in the United States.

Soil erosion adversely affects agricultural crop productivity because of (1) selective removal of plant nutrients and organic matter by wind and water, (2) poor structure, (3) gross removal of topsoil, and (4) increased runoff, reducing water availability to crops and causing flood damage to other crops. These losses, which affect crop productivity, may be temporarily offset by heavy use of fertilizer and irrigation. But use of fertilizer and irrigation is demanding in terms of energy requirements, and both loss of topsoil and energy resources are long-range problems that should be kept in mind today. The present economic pressures to put aside soil conservation methods known to be effective need to be carefully evaluated in terms of the "trade-off" of somewhat higher levels of productivity.

Agricultural practices as we think of them today have their roots in European Renaissance times (1400 to 1600). The voyages of the great explorers and the birth of scientific experimentation took place during this cultural revival in Western Europe. The voyages of Columbus, Vasco da Gama, and others resulted in exchange of crops, domesticated animals, and farming practices between different continents. For example, when Columbus arrived in America, he found the Indians using techniques different from those in Europe. The Indians practiced intercropping of corn and legumes and planted the seeds individually in rows, whereas the Europeans practiced crop rotation —growing one crop one year and another one the next year on the same field—and spread seed by broadcasting them over the land.

c. INDUSTRIALIZATION

The Scientific–Industrial Revolution that occurred between 1700 and 1750 was responsible for another burst of population in the Western World and its consequent environmental effects. In Great Britain, for example, most of the magnificent English oaks (*Quercus robur*) were cut to build ships in the sixteenth century; then essentially the remainder of the forests were cut in the eighteenth century to make charcoal to smelt iron. Following this era was a period when large areas were laid waste to mine coal. However, as Great Britain is an example in which natural ecosystems were destroyed with technological advance, it also is a heartening story of gradual substitution of man-made landscapes for the natural ones (Fig. 31-2*B*). This transformation was the result of early and persistent restoration efforts aided by a benign climate. Recovery of the land was much easier here than in the Mediterranean area because the gentle rainfall and coolness help create the "greenness" of the vegetation for which the British Isles are so famous. Also the real effort toward restoration began in 1895 in the reign of Queen Victoria with a National Planning Program. In 1930, as many as 250,000 voluntary school and college students worked on collecting data for a 12-volume work *Land of Britain,* and this program helped to produce well-educated citizens aware of the relative values in preservation and modification of ecosystems. Thus there has been considerable reforestation and protection of good agricultural land. Today forestry in Scotland has reached such levels of success that good agricultural land can justifiably be used for production of wood. Moreover, the pastures and forests of Britain, France, and Germany (as well as other European countries) fill admirably the role of stabilizers of the landscape.

Certainly, on an evolutionary scale no other organism has had such impact on ecosystems in such a short period of time (essentially 10,000 years) as human beings have had. This rapidity, however, emphasizes that only now, somewhat in retrospect, are human beings in the western technological societies philosophically seeing the value of considering themselves as members of the ecosystems in which they live. This shift in perspective is leading to recognition that, while some natural ecosystems are utilized, others need to be protected and preserved and decisions made on the amount of modification that is acceptable.

2. LAND USAGE MODEL FOR TODAY BASED ON ECOSYSTEM CHARACTERISTICS

Only 30% of the biosphere is land; of that 20% is unusable (i.e., areas of desert or perpetual snow that essentially support no vegetation). The remaining 104 million square km of biologically productive land may be divided into three types.

1. Unconverted natural ecosystems—primarily northern coniferous and tropical forests
   26×10^6 sq km
2. Converted or degraded ecosystems—managed forests or grazing land
   67.3×10^6 sq km
3. Artificial ecosystems—primarily agriculture
   10.4×10^6 sq km

These figures indicate that the explosive growth of population and industrialization today has made land-use problems acute except in far northern forests and until recently tropical forests. Now the tropical forests are also disappearing at a rapid rate (Chapter 14). We have long competed with other forms of life for space. Increasingly we are becoming our own competitors—a situation intensified by the diversity of our interests that affect land use (residence, business, agriculture, recreation, waste disposal industry, and transportation. The effect has been confusing, as can be observed in many metropolitan areas.

Allocation of space by planning is not a new idea. As noted above it has been developed in Great Britain for a long time and in other parts of western Europe as well. However, continual problems exist. For example, in Holland, where planning is extremely advanced, they are facing the crisis of their population's saturating the land. In the United States, resistance to planning results from our still having some margin of safety in unused resources and by many people's conviction that an economy can continue to expand almost without limits. To face these kinds of problems, an American scientist, Eugene Odum, has pointed out the relevance of ecosystem development theory to land planning via a minimodel (Fig. 31-3, Table 31-1) in which the total landscape is compartmentalized into *productive* (young or growth type), *protective* (mature or steady-state type), and *compromise* (multiple-purpose or intermediate type) environments that can be linked to the *urban-industrial environment*. He reasoned that by analyzing the flow of energy and materials, and organisms (including humans), between the compartments, limits could be determined that ultimately must be set for each compartment in order to maintain regional and global balances. Furthermore, it would provide means of assessing energy drains on ecosystems by harvest, pollution, and results of other human activities. We shall use this ecologically based model as an integrative approach in discussing the decisions facing humans today regarding modification and preservation of world ecosystems.

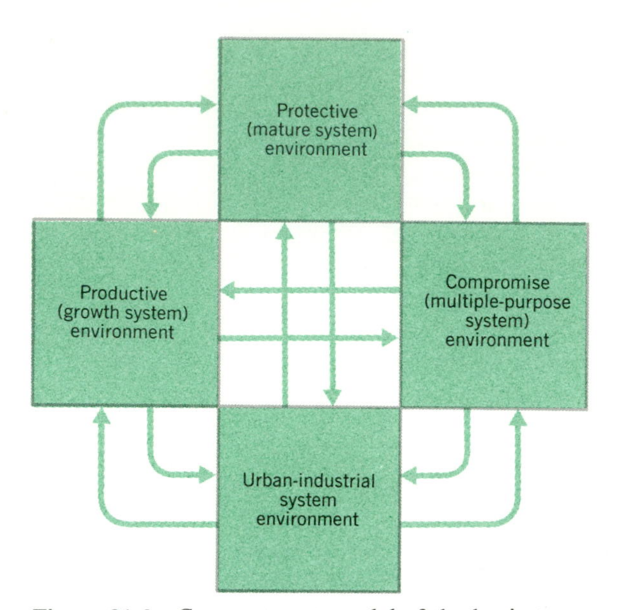

Figure 31-3 Compartment model of the basic types of environments required by humans, partitioned according to ecosystem development and life cycle resource criteria (From E. Odum, 1969).

TABLE 31-1

Contrasting Characteristics of Young and Mature Ecosystems

Young	*Mature*
Production	Protection
Growth	Stability
Quantity	Quality

a. PRODUCTIVE ENVIRONMENT (YOUNG OR GROWTH SYSTEM)

Humans have generally favored productive systems to satisfy their needs. For example, the goal of agriculture or tree farming is to achieve high rates of production of readily harvestable products with little biomass left to accumulate on the landscape. In other words, the goal is maximum yield and high production/biomass efficiency (Chapter 14). On the other hand, the evolutionary strategy, as seen in the outcome of the successional process, is directed toward the reverse efficiency—a high biomass/production ratio. Humans have generally been preoccupied with obtaining as much "production" from the landscape as possible by developing and maintaining early stages of succession in ecosystems. In temperate-zone agriculture (as well as in some places in the tropics) stability is maintained by considerable input of "cultural" energy and materials both on and off the farm. Fertilizers provide nutrients and irrigation provides water where needed. Because the crops here are primarily planted as monocultures, chemical pesticides and herbicides generally are used to protect against predators and epidemic diseases (Chapters 29 and 30). Also machine labor requires an enormous energy input. Production of raw agricultural commodities in the United States ranks third in energy consumption after the steel and petroleum-refining industries. About equal amounts of fuel energy (including petroleum and electricity) are used on and off the farm. Among the purchased inputs, the manufacture of fertilizers uses the most energy (28%) and the production of pesticides the least (0.6%). Consumption of petroleum products on the farm for plowing, harvesting, and other uses accounts for 51% of the energy used by U.S. agriculture. We now know rather precisely how much "cultural" energy it requires to produce many U.S. food crops. These energy accounts range from 1 to 30×10^3 calories per acre per year among the 24 principal cereal, fruit, and vegetable crops (Fig. 31-4). Oats, corn, soybeans, and wheat use the least cultural energy; deciduous fruits and sugar are only moderately energy intensive, with citrus fruits using a bit more energy because of specialized requirements for irrigation and frost and pest control. Peanuts, rice, and potatoes are midway on the scale with annual vegetables and fruits the most energy intensive. These latter crops need considerable fertilizer and pesticides and grow best under frequent irrigation. In addition the energy used in processing and marketing, geared to consumer preferences for uniformity and quality, can far exceed that needed to produce the crop.

Needless to say, modern civilization would be inconceivable if the energy now required by agriculture had to come from human muscles instead of from gasoline and electricity. But it is a fact, nevertheless, that if fossil fuels are to remain the most important source of power, the sheer size of the world population will make it impossible to continue for long the energy deficit spending on which agriculture depends in prosperous industrialized countries. Also there would be no hope of extending these practices to the developing countries, which constitute the largest part of the world. No matter how the situation is rationalized, present scientific agriculture is only possible as long as cheap sources of energy are available. As world supplies of petroleum dwindle, the modern farmer, like the modern technologist, will become increasingly ineffective unless energy derived from other sources can be supplied in large amounts at low cost, or alternatives to reducing energy inputs are developed. Alternatives such as rotation of crops and use of green manure might in some cases reduce the energy demand of chemical fertilizers and pesticides. Even in the United States energy expenditure could be reduced by substituting some man-power currently displaced by mechanization. Thus, the future of land management is intimately bound to sources of energy, as are all other aspects of our present human life.

To attempt to meet these problems in tropical developing countries such as Central America and Africa, emphasis at research stations is being put on increasing yield in subsistence-type agriculture. There it is now recognized that some successful labor-intensive agriculture should be maintained along with the energy-intensive agriculture needed to feed the ever-burgeoning populations that are increas-

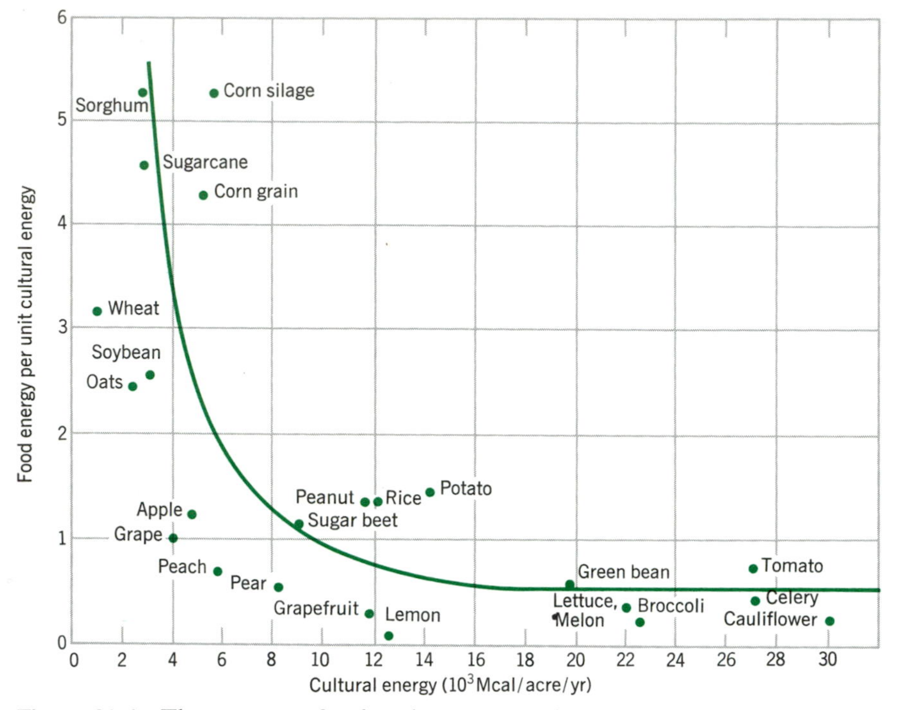

Figure 31-4 The amount of cultural energy used to produce 24 grain, forage, fruit, and vegetable crops and the amount of food energy in calories that each plant yields per unit of cultural energy invested. The ratio is the measure of food energy efficiency. Remember that 1 acre = 0.405 ha. (From Heichel, 1976)

ingly migrating into urban areas. This trend toward strengthening the success of the small farmer is epitomized by the concept "small is beautiful" emphasized by the British economist E. F. Schumacher. Furthermore, tropical agricultural aid programs throughout the world are now investigating means of improving small-scale agricultural practices.

Additional problems have arisen in the tropics where transfer of temperate-zone agricultural techniques may be inappropriate. For example, we discussed the value of polycultures versus monocultures in growing of coffee and cacao (Chapter 25). Diversified farming or polycultures may result in somewhat lower yields than monocultures, but there is considerable saving on fertilizers and even irrigation water in seasonally dry areas. More importantly, however, less application of chemical pesticides may be needed, because the diversity of kinds of plants, simulating natural conditions, may be a general deterrent to epidemic outbreaks of insects and pathogens. Furthermore, the cover provided by polycultures protects latosol soils from developing into laterites (Chapter 15). The use of polycultures, particularly in rain forest sites, however, does not deny the value of some intensively cropped monocultures under other conditions as indicated previously.

b. COMPROMISE ENVIRONMENT (INTERMEDIATE OR MULTIPURPOSE)

The differences between clear-cutting and selective cutting techniques in harvesting timber present an excellent comparison of productive versus compromise or multiple-use systems. Clear-cutting of timber and subsequent creation of single-tree plantations provides a controlled yield such as that in annual crop plants. In some cases, such as Douglas fir (Chapter 22), the trees do not regenerate rapidly in the shade of selectively cut forests. With proper management clear-cutting of such species can both provide high yields and protect watersheds (Fig. 31-5*A*). On the other hand, clear-cutting practices may have adverse effects on watersheds and be aesthetically displeasing. Thus selective harvesting of forests offers an alternative compromise system (Fig. 31-5*B*). In fact, many foresters find that this practice offers the "best of both worlds" in that there is acceptable sustained yield, yet it provides watershed protection, is aesthetically pleasing, and leaves forests for various types of recreation. Thus, too, it provides a multiple-use environment.

Because ecosystems have complex budgets of inputs and outputs of energy and minerals, and because we lack precise information about these relationships and internal functions that maintain ecosystems, it is often difficult to assess human impact such as deforestation and agriculture on the biosphere. As a result neither foresters, agriculturists, nor land planners can assess on either a short or long-term basis the full range of consequences a particular activity may have.

In deforestation, as previously noted, there is a removal of the vegetative cover, which may lead to significant erosion of the soil and underlying rock, particularly where the slopes are steep and the rainfall high. In studies of the relationship between various types of vegetative cover, land use, precipitation, and erosion in small watersheds in the southern United States, it is clear that vegetation plays a dominant role in reducing sediment yields. The Hubbard Brook Ecosystem is one of the few long-term experiments, initiated some 15 years ago, to measure mineral budgets in both undisturbed and disturbed sites within an ecosystem in the same watershed in the White Mountains of New Hampshire. The experimental forest is dominated by northern hardwoods underlain by a uniform bedrock. In determining the mineral budget, the investigators estimated the input by analyzing the mineral content of the rain and snow and calculated the output by constructing concrete weirs that channeled the water flowing out of selected areas. They discovered that the natural forest was extremely efficient in conserving minerals. Annual imput–output budgets for inorganic nitrogen and phosphorus show a net gain with potassium near balance, even though large amounts of these nutrients are mobile. Other experiments at Hubbard Brook have included: (1) complete deforestation with the cut timber left, neither roads nor skid trails constructed, and herbicides not applied to prevent regrowth of vegetation, (2) clear-cut watershed, with skid trails constructed, merchantable timber removed, and no herbicides used, and (3) strip-cut watershed, in which timber was harvested in strips 25 m wide alternating with uncut strips 50 m wide, the strips roughly paralleling the elevation contours.

In the deforested watershed, annual runoff increased by more than 30% after the cutting, and mineral concentrations increased dramatically in streams about five months after cutting for all major ions except NH_4^+ and SO_4. Concentrations of dissolved substance reached a peak during the second year. Nitrate–nitrogen losses were most extreme; these resulted from accelerated decomposition of organic matter on the forest floor coupled with increased nitrification. Table 31-2 shows the net loss or gain (precipitation minus stream loss) of nutrients in kilograms per hectare during the second full water year. Generally the pattern of nutrient loss from the clear-cut experiment (W101) paralleled that of the deforested (W2) watershed but was not as large (particularly with regard to nitrate and to lesser extent with calcium and potassium). In the strip-cut watershed, losses of nitrate–nitrogen and potassium were below those of the deforested or clear-cut watersheds. Of special interest is the presumed biological effect of the uncut vegetation on reducing nitrate concentrations in stream water. These results have

(A)

(B)

Figure 31-5 *A*. Patches of clearcut lodgepole pine *(Pinus contorta)* in Montana. Since the clearcut patches are small (8-12 hectares), effects on watershed are kept to a minimum. *B*. An example of selective cutting. Under this method of management, mature timber is removed every few years along with crooked, defective, and oher less desirable trees. Also the areas may be used for recreation.

suggested to the Hubbard Brook investigators that strip-cutting may be a reasonable management alternative for timber harvest in northern hardwoods.

Within the Cayuga Lake forested watershed in central New York, export of inorganic nitrogen (NO_3^- and NH_4^+ in stream water was directly related to the extent of agricultural ac-

TABLE 31-2

Net Loss or Gain (Precipitation Minus Stream Loss) of Nutrients (in kg/ha) during Second Full Water Year (1 June-31 May) after Cutting in the Hubbard Brook Experimental Forest

Nutrient	*[a]Undisturbed*	*Strip-cut (W4)*	*Clear-cut (W101)*	*Deforested (W2)*
Ca	−10.2	−25.2	−21.2	− 90.4
Mg	− 2.3	− 4.2	− 4.6	− 17.9
K	− 0.8	− 4.0	− 7.4	− 35.8
Na	− 5.2	−10.4	− 6.1	− 17.3
H	+ 0.9	+ 1.0	+ 1.0	+ 0.5
NH_4-N	+ 2.0	+ 1.8	+ 1.9	+ 1.9
NO_3-N	+ 0.7	− 8.0	−10.4	−142
P	+ 0.1[b]	—	—	+ 0.08[b]
SO_4-S	− 4.3	− 6.1	− 0.3	0
Cl	+ 1.7	− 4.4	+ 5.6	− 3.7
SiO_2-Si	−16	−31	−24	− 32
Total	33.4	81.7	65.5	336.6

[a] 9-yr average for all undisturbed watersheds. (Likens et al., 1974).
[b] If particulate matter losses were included, total P for undisturbed would = +0.087; for (W2) = −0.104. (From G. E. Likens and F. H. Borman, Proc. First International Congress of Ecology, 1974)

tivity in the various tributary watersheds. In contrast, the phosphorus outputs were more closely related to the percentage of watersheds in urban usage. At Lake Mendota, Wisconsin, the primary source of phosphorus input for the lake was sewage whereas the source of nitrogen was land runoff. All evidence now indicates that human activity in a particular watershed (e.g., logging, agriculture, road building, mining, or urbanization) tends to alter, remove, or overwhelm the steady-state capabilities of natural terrestrial ecosystems. As a result, drastically increased amounts or proportions of dissolved substances and particulate matter are exported from the watershed and may accelerate the rate of eutrophication in aquatic ecosystems. These studies allow us to make relative judgments regarding specific land-use practices—and to decide whether to move from a strictly productive environment to a compromise one. Also these studies have alerted us to the necessity of monitoring aquatic systems to know where certain chemicals reach dangerous levels for us as well as for other organisms.

Despite the general rule that mature ecosystems tend to recycle minerals tightly, there are cases where, even without human intervention, Nature fails to complete the recycling process. The accumulation of peat, coal, oil, guano, and other deposits of organic origin are examples of this situation. These materials have become chemically stabilized after being only partially decomposed. Paradoxically, we help somewhat in the completion of the cycle when we burn peat, coal, and oil.

In addition to deliberately managed compromise systems, a more-or-less regular, natural disturbance can maintain an ecosystem at some intermediate point in the successional sequence, resulting in a compromise between youth and maturity. Thus, we can deliberately use fire either to maintain a steady state or to set back succession to some desired point. In the southeastern United States, light fires of moderate frequency can maintain a pine forest against the development of older successional stages, which are less desirable economically. The fire-controlled forest yields less wood than a tree farm (i.e., young trees, all about the same age, planted in rows and harvested on a short rotation schedule), but it provides greater protection for the landscape, wood often of higher quality, and a home for the variety of game and birds that could not survive in a tree farm. Some fire-maintained ecosystems then are an example of a compromise between "production simplicity" and "protection diversity."

c. URBAN–INDUSTRIAL ENVIRONMENT

As the efficiency of agricultural systems has increased, fewer individuals have been needed for food production and there has been expansion of urban and industrial complexes. By 1830 the increased agricultural output was leading to greater urbanization and more people employed in industries (Chapter 30). By the turn of this century, Great Britain was the only nation essentially urbanized; however, by 1965, urbanization was a world trend. In the United States now 53% of the people are concentrated in 213 urban areas.

One of the most critical problems is that the urban-industrial complexes have expanded on to some of the best agricultural land, that is, on prime soils. It has been shown throughout world history that cropland is twice as likely as non-cropland to be urbanized because cities generally arose around agricultural centers. Since 1945 in the United States the total loss of good cropland to urbanization, highways, and other special uses is 18 million hectares, an area the size of Nebraska. The U.S. automobile system requires a minimum area of 260,000 hectares just for parking. California, the number-one state in terms of agricultural income (double that of any other state) and diversity of products grown exemplifies this problem (Fig. 31-6). Despite Classes I and II agricultural soils comprising only 6% of the area of California, 80% of the Los Angeles area, and 70% of the San Jose area is built on these prime soils. Rapid loss of the best agricultural soils is now being recognized, so that measures can be taken to try to counter the results of urban population pressures to sprawl across the world's food baskets.

Industrialization and urban sprawl has also had consequential effects on native, ornamental and agricultural plants due to air pollution (Fig. 31-6). Nitrogen oxides, produced by automobile exhaust, undergo changes in the atmosphere that lead to production of ozone (O_3). With increase in concentration of O_3, damage has occurred in a wide variety of plants. Forests in southern California and certain crops, particularly those in which the leaves are eaten, have been seriously damaged (Fig. 31-7). Also sulphur dioxide, produced by ore-smelting processes and by burning of low-grade coal, may have injurious effects on plants; also, depending on environmental conditions, sulphur dioxide may be converted to sulfuric acid. The acidity of rain and snow has increased significantly in industrial regions such as the northeastern United States during recent years. Not only do these rains have immediate effects on some plants, but they also are suspected of lowering soil pH, causing excessive mineral leaching with long-range effects on ecosystems. Furthermore, ozone and sulphur dioxide often occur together. When this happens, they synergistically react in ways that are not yet understood, causing damage at concentrations considerably below levels that would result in damage to plants if either were present alone.

d. PROTECTIVE ENVIRONMENT (MATURE OR STEADY-STATE SYSTEM)

Until we can determine more precisely how far we may safely go in expanding intensive agriculture and urban sprawl at the ex-

Figure 31-6 Urban sprawl in Los Angeles over some of the highest quality agricultural soil in California. Note also in the background the pollution that has had significant effects on natural, agricultural, and ornamental plants.

(A) *(B)*

Figure 31-7 Effects of air pollution on ponderosa pine in San Bernardino National Forest near Los Angeles, California. *A*. Healthy trees, *B*. Ten years later these pines are dying due to photochemical smog.

pense of remaining natural ecosystems (existing now in the United States primarily as government land held as wilderness areas, parks, and wildlife refuges), it is good insurance to hold as much land inviolate as is possible, for several reasons. First is the scientific need to understand some of the complexity of ecosystem structure and function. As indicated in the studies of mineral budgets in forests, years are required to obtain the kind of data about the function of ecosystems so necessary for our evaluation of critical land-use problems. Secondly, some relatively undisturbed ecosystems should be maintained to protect genetic reservoirs for future breeding programs. Genetic resources of crops are being threatened as old, indigenous materials are replaced by recently developed cultivars. Although replacement is essential for significant increases in productivity, the genetic resources being lost may also be essential for future breeding work and continued improvement of the plants. World collections of some of the major crop plants are substantial (e.g., corn and rice) but others (e.g., cassava and sweet potatoes) are seriously inadequate. No collection as yet is complete and plant breeders warn that genetic vulnerability remains a genuine hazard. Thirdly, wilderness areas need to be protected for recreation (Fig. 31-8). As Rene Dubos has eloquently put it, these are important to provide "mysteries transcending daily life and to recapture direct awareness of cosmic forces from which we have emerged." However, the heavy use of wilderness areas in the Western World is making it difficult to manage these finite, dwindling, and nonsubstitutable resources. Since they comprise less than 3% of the contiguous United States, recently the U.S. Forest Service has pointed out the necessity for thinking about "biocentricity" (i.e., maintenance of natural energy and material flows within wilderness

Figure 31-8 Hiking in a protected environment – Grand Teton National Park, Wyoming.

ecosystems) in contrast to "anthropocentricity" (making human use the primary objective) in managing wilderness areas. All of these reasons suggest that preservation of natural areas "is not a peripheral luxury for society but a capital investment from which we expect to draw interest" as stated by Odum.

A major unknown, in setting aside mature systems, is the *size* necessary for maintenance of the proper functioning of the ecosystem. This question has recently become acute regarding the Redwood National Park in California. Are flooding and fire important in maintaining the ecosystem and, if so, how are they introduced so as to simulate natural cycles? Is it possible to clear-cut redwood forests immediately adjacent to the park areas and not disrupt the undisturbed area? These are problems for which we still do not have definitive answers, but in the case of the redwoods, we only have a few small areas left untouched by disruptive human activity. Setting aside reserves in tropical areas such as the Amazon Basin poses additional problems. We know even less about the nature of tropical rain forests than forests of temperate zones, particularly because of the tremendous biotic diversity and complexity of interactions among plants and animals. Remaining areas of undisturbed rain forests are being so rapidly developed in an attempt to accommodate exploding populations that it has been predicted by numerous tropical biologists that the majority of mature rain forests will have been destroyed by the turn of the century (Chapter 14). Determination of the minimum effective size of rain forest is particularly difficult because tree species in rain forests typically occur at densities of 0.3 per hectare *or less,* indicating that large areas are necessary to ensure continuation of viable plant populations as well as associated animals. Tropical scientists have suggested that in the Amazon basin any preserve should be 2000 km^2 at a minimum and preferably 10,000 km^2 if ecosystem structure is to be effectively preserved. Ideally a preserve would include a watershed of an entire river system.

e. IMPLEMENTATION OF AN INTEGRATED APPROACH TO LAND USE

Although some ancient civilizations despoiled their environments and similar processes are going on today, this does not have to be inevitable. We can create environments and manage them so that they can be ecologically stable, economically profitable, and aesthetically pleasing. Many riches of Nature have been brought to light only in regions that have "humanized"—agricultural lands, gardens, and parks created and maintained by human toil. The most pleasant and probably the safest landscape in which to live is one of diversity—crops, natural forests and other vegetation, lakes, roadsides, marshes, seashores and waste places, in other words, a mixture of communities of different ecological character. As individuals we more or less instinctively surround our houses with trees, shrubs, and grasses in gardens or parks, at the same time as other people attempt to obtain high yields from the corn or wheat field or orchard. Society needs to explore ways to deal with the landscape as a whole so that technological skills will not run too far ahead of our understanding of the impact of change. The soil conservation movement in America is an excellent example of a program dedicated to the consideration of the whole farm or watershed as an ecological unit. However, soil conservation organizations have remained too exclusively farm-oriented and have not yet risen to the challenge of the urban-rural landscape. More attention may have to be given to ecologically based models, such as we have discussed here, that attempt to predict necessary global balances between productive, protective, and multiple-use environments and their links to urban and industrial environments. Implementation, however, will require procedures for zoning the landscape and the complexities of decisions by industry and government as to where and how to set restrictions. As with the critical issues involving global food supply (Chapter 30), a balanced perspective toward use of the world's land will also probably demand action by both the scientist and the informed citizen.

Suggested Readings

CHAPTER 2

Barghoorn, E. S. 1971. "The Oldest Fossils." *Scientific American* **224**:30-42. (The first evidence that bacteria lived 3 billion years ago.)

Delevoryas, T. 1966. *Plant Diversification*. Holt, Rinehart and Winston, New York (Useful short summary of major trends in plant diversification through geologic time.)

Foster, A. S. and Gifford, B. M. Jr. 1974. *Comparative Morphology of Vascular Plants*. 2nd ed. W. H. Freeman and Company, San Francisco. (A well-presented general account of the comparative morphology of vascular plants containing excellent summary data.)

Fox, Sidney and Dose, Klaus, 1977. *Molecular Evolution and the Origin of Life*. Maral Dakkar, Inc., New York. (An interesting current viewpoint.)

Rhodes, F. H. T. 1962. *The Evolution of Life*. Penguin Books, Harmondsworth, England. (The first 6 chapters present, very readably, the material of Chapter 2, and much more besides.)

Stebbins, G. L. 1974. *Flowering Plants, Evolution Above the Species Level*. The Belknap Press of Harvard University Press. (A fascinating treatment of the origin and subsequent diversification of the angiosperms by one of the great theorticians of plant evolution.)

CHAPTER 3

Harlan, J. R. 1971. "Agricultural origins; centers and non-centers." Science 174; 468-474. (Discussion of how agriculture seems to have developed in centers but also differed over vast areas in various parts of the world.)

Hutchinson, Joseph, Clark, J. G. G., Jope, E. M. and Riley R. (Organizers of Symposium for the Royal Society and the British Academy). 1977. *The Early History of Agriculture*. Phil. Trans. R. Soc. London B 275:1-23. Oxford University Press. (An outstanding collection of relatively recent papers from a symposium initiated by the Department of Archeology at Cambridge University.)

Simmonds, N. W. 1976 *Evolution of Crop Plants*. Longman, New York. (A relatively up-to-date compendium of information regarding the evolution of the most important crop plants with good maps and bibliography for each crop.)

Streuver, Stuart (ed.) 1971. *Prehistoric Agriculture*. American Museum Science books in Anthropology. The Natural History Press, Garden City New York. (A collection of interesting articles on prehistoric agriculture by leading authorities in the field.)

CHAPTER 4

Esau, K. 1977. *Anatomy of Seed Plants*. John Wiley & Sons, Inc., New York. (A complete text.)

Fahn, A. 1974. *Plant Anatomy,* Pergamon Press, Oxford and New York, 511 pp. (A complete text.)

O'Brien, T. P., and M. E. McCully. 1970. *Plant Structure and Development,* MacMillan, New York, Toronto, Collier-MacMillan. (Outstanding for its beautiful micrographs, and notable for the combination of structure and function.)

CHAPTER 5

Bassham, J. A. 1968. "The Path of Carbon in Photosynthesis." *In, Bio-Organic Chemistry, Readings from Scientific American,* Chapter 24, W. H. Freeman and Company, San Francisco. (Very clear and useful presentation, written when the subject was "hot".)

Leopold, A. C., and P. E. Kriedemann. 1975. *Plant Growth and Development*. Chapter I (covers also material of Chapter 7). McGraw Hill, Inc., New York.

Zelitsch, I. 1972. *Photosynthesis, Photorespiration and Plant Productivity*. Academic Press, New York. Chapters 3 and 4. (Full coverage including chloroplast structure and the relations between photosynthesis and metabolism.)

CHAPTER 6

Hall, J. L., J. T. Flowers, and R. M. Roberts. 1974. *Plant Cell Structure and Metabolism*. Longman Group Limited, London (425 pp.). (Presents plant tissues at the electron-microscope level, and particularly strong on the biochemical side.)

Lehninger, A. L. 1968. "Energy Transformation in the Cell." *In, Bio-Organic Chemistry, Readings from Scientific American,* Chapter 25. W. H. Freeman and Company, San Francisco.

Steward, F. C. 1964. *Plants at Work*. Chapter 7, "The Release of Energy."

Ribereau-Gayon, P. 1972. *Plant Phenolics*. Hefner Publishing Company, New York (250 pp.). (A most readable treatment of flavanols, anthocyanins, tannins, etc., which are very lightly discussed in this book.)

CHAPTERS 7 AND 8

Dainty, J. 1969. "The Water Relations of Plants." Chapter 12 in *Physiology of Plant Growth and Development*. M. B. Wilkins, (ed.). McGraw Hill, Inc., London.

Dixon, H. H. 1914. *Transpiration and the Ascent of Sap in Plants*. (The classic on water movement in plants. Chapter IV is the most important part.)

Epstein, E. 1973. "Roots." *Scientific American* **228:** 48-55. (Succinct account by an expert.)

Leopold and Kriedemann (above).

Revelle, R. 1971. "Water." Chapter 6 *in, Man and the Ecosphere, Readings from Scientific American,* W. H. Freeman and Company, San Francisco.

Richardson, M. 1968. Translocation in Plants. Chapters 1, 2, and 3. *In, Studies in Biology,* No. 10, St. Martin's Press, New York. (A neat, short account of both xylem and phloem movement, including both structure and function.)

CHAPTER 9

Delwiche, C. C. 1970. "The Nitrogen Cycle." Chapter 7 *in, The Biosphere, Readings from the Scientific American,* W. H. Freeman and Company, San Francisco. (Brief but inclusive.)

Thimann, K. V. 1963. "The Fixation of Nitrogen." Chapter 9 in *The Life of Bacteria*. 2nd Edition, MacMillan, Inc., New York. (Chapter 10, Nitrification, denitrification and the Nitrogen Cycle, in the same, supplements the material further.)

CHAPTER 10

Galston, A. W., P. J. Davis, and F. L. Salter. 1980. *The Life of the Green Plant.* Chapters 3 and 9 through 12. The Prentice Hall, Inc., Englewood Cliffs, N.J. (An extensive treatment of growth and development and their hormonal control.)

Leopold, A. C., and P. E. Kriedemann. *Plant Growth and Development.* McGraw Hill, New York. 2nd Edition. (The chapters following Chapter I deal with translocation, the growth process, the influence of environmental factors, etc.)

Mayer, A. M., and A. Poljakoff-Mayber. 1975. *The Germination of Seeds.* 2nd Edition. Pergamon Press, New York and Oxford. (A book of professional type, with extensive bibliography, but very readable and suggestive of further experiments.)

Street, H. E., and H. Opik. 1970. *The Physiology of Flowering Plants: Their Growth and Development.* Edward Arnold Publishers, London; American Elsevier Co., New York (263 pp.). (Chapters 2, 7, 8, 9, and 10 treat of growth and development, with minimum biochemistry; other chapters cover mineral nutrition and water relations, with some differences from Chapters 7 and 8 here.)

Thimann, K. V. 1977. *Hormones in the Whole Life of Plants.* Univ. of Massachusetts Press, Amherst, Mass. (Centers on the hormones and the processes they control, in seeds, flowers, fruits, roots and whole plants.)

Wareing, P. F., and I. D. J. Phillips. 1978. *The Control of Growth and Differentiation in Plants.* 2nd Edition. Pergamon Press, New York and Oxford. (Chapters 4, 5, 7, and 9 are especially pertinent; a clear discussion, somewhat stressing the theoretical basis.)

CHAPTER 11

Ayala, F. J. (ed.) 1976. *Molecular Evolution.* Sinauer Associates, Inc. Sunderland, Mass. (An interesting series of papers by outstanding authorities in the field.)

Crow, James. 1966. *Genetic Notes.* Burgess Publishing Company, Minneapolis, Min. (A clear, concise summary of genetics valuable to students.)

Darwin, Charles. 1962. *The Voyage of the Beagle.* Natural History Library, Doubleday and Company, Inc., Garden City, N.Y. (A reissue of Darwin's chronicle of the expedition that led him eventually to his theory of evolution. This is the best introduction to Darwin and his ideas.)

Levine, Louis. 1973. *Biology of the Gene.* The C. V. Mosby Company, St. Louis, Mo. (One of the very best short genetics books.)

Peters, James A. (ed.) 1960. *Classic Papers in Genetics.* Prentice-Hall, Inc., Englewood Cliffs, N.J. (A collection of papers by most of the researchers responsible for the most significant developments in genetics.)

Stebbins, C. L. 1966. *The Process of Organic Evolution.* Prentice-Hall, Inc., Englewood Cliffs, N.J. (A brief review of organic evolution by one of the most active workers in the field.)

CHAPTER 12

Anderson, E. 1967. *Plants, Man and Life.* University of California Press, Berkeley and Los Angeles. (Enjoyable reading that excites thinking about plant genetics.)

Brewbaker, J. L. 1964. *Agricultural Genetics.* Prentice-Hall, Inc., Englewood Cliffs, N.J. (A relatively short text on practical genetics problems, especially good for polyploidy.)

Clausen, Jens. 1951. *Stages in the Evolution of Plant Species.* Cornell University Press, Ithaca, N.Y. (A useful summary of trends in the studies of ecotypic differentiation by one of the pioneer workers.)

Frankel, O., and E. Bennett (eds.) 1970. *Genetic Resources in Plants—Their Exploration and Conservation.* F. A. Davis Company, Philadelphia, Pa. (Presentation of present concern for loss of genetic variability for breeding of crop plants and general relationship to conservation practice.)

Hutchinson, J. (ed.) 1965. *Crop Plant Evolution.* Cambridge University Press, England. (Series of interesting but technical articles.)

Grant, Verne. 1971. *Plant Speciation.* Columbia University Press, New York. (Excellent introduction to most aspects of plant evolution.)

CHAPTER 13

Cronquist, A. 1968. *The Evolution and Classification of Flowering Plants.* Houghton Mifflin Company, Boston, Mass. (Presentation of one of the most widely accepted current classification systems for angiosperms.)

Heslop-Harrison, J. 1967. *New Concepts in Flowering Plant Taxonomy.* Harvard University Press, Cambridge, Mass. (A useful and readable discussion of concepts, although somewhat out of date.)

Heywood, V. H. 1976. *Plant Taxonomy.* 2nd Edition. The Institute of Studies in Biology No. 5. Edward Arnold Publishers, London. (A brief introduction to major concepts of taxonomy.)

Raven, P. H., B. Berlin, and D. Breedlove. 1971. "The Origins of Taxonomy," *Science* **174**:1210-1233. (A review of historical development of taxonomy with suggestions as to the directions in which it may develop most meaningfully.)

CHAPTER 14

Billings, W. D. 1966. *Plants, Man and the Ecosystems.* Wadsworth Publishers, Inc., Belmont, Calif. (Excellent summary of plant ecology approached from the viewpoint of a physiological ecologist.)

Borman, F. H., and G. E. Likens. 1967. "Nutrient Cycling," *Science* **153**:424-449. (Good general discussion, emphasizing watershed models.)

Phillipson, J. 1966. *Ecological Energetics.* Edward Arnold Publishers, London. (A readable "thumbnail" account of principles.)

Whittaker, R. H. 1975. *Communities and Ecosystems.* 2nd Edition. MacMillan, Inc., New York. (One of the best available overviews of ecological principles concerned with the community and ecosystem level of organization.)

Woodwell, G. M. 1971. "Toxic Substances and Ecological Cycles." Chapter 12 *in Man and the Ecosphere, Readings from Scientific American.* W. H. Freeman and Company, San Francisco.

CHAPTER 15

Brady, N. C. 1974. *The Nature and Properties of Soils.* 8th Edition. MacMillan, Inc., New York. (The most comprehensive introductory text on soil science.)

Bridges, E. M. 1970. *World Soils.* Cambridge University Press, England. (A simplified presentation of world soils with excellent color photographs of profile types.)

Dale, T., and V. G. Carter. 1953. *Topsoil and Civilization.* University of Oklahoma Press. Norman. (A fascinating account of the role soils have played in the rise and fall of civilization.)

Kellogg, C. E. 1970. "Soil." Chapter 11 in *Plant Agriculture, Readings from Scientific American.* W. H. Freeman and Company, San Francisco. (Excellent readable overview of soil formation and major soil types.)

CHAPTER 16

Curtis, B. C., and Johnston, D. R. 1970. "Hybrid Wheat." Chapter 16 in *Food and Agriculture, Readings from the Scientific American.* W. H. Freeman and Company, San Francisco.

Heiser, C. B. 1973. *Seed to Civilization.* W. H. Freeman and Company, San Francisco. (Chapter 5, Grasses, the Staff of Life, is the most pertinent, but agricultural and economic plants are treated throughout the book.)

Mangelsdorf, P. C. 1970. "Hybrid Corn." Chapter 15 in *Food and Agriculture, Readings from the*

Scientific American. W. H. Freeman and Company, San Francisco.

Mangelsdorf, P. C. 1974. *Corn, Its Origin, Evolution and Improvement*. Harvard University Press, Cambridge, Mass. (Very complete in all respects; especially strong on corn evolution and the associated controversies.)

Zohary, D., and H. Kuckuk. 1970. "Wild Wheats and Primitive Wheats." Chapters 19 and 20 in *Genetic Resources in Plants—Their Exploration and Conservation*. O. H. Frankel and E. Bennett, (eds.) F. A. Davis Company, Philadelphia.

CHAPTER 18

Biale, J. B. 1970. "The Ripening of Fruit." Chapter 9 *in Plant Agriculture, Readings from Scientific American*. W. H. Freeman and Company, San Francisco.

Edlin, H. L. 1967. *Man and Plants*. Aldus Books, London. Chapter 8, "Soft-Stemmed Seed and Fruit Crops"; Chapter 9, "Orchards and Vineyard." (A readable discussion of many important fruits and beautiful color photographs.)

Miller, E. V. 1954. "The Natural Origin of Some Popular Varieties of Fruits," *Economic Botany* **8**:337-348.

CHAPTER 19

Correll, D. S. 1962. *The Potato and Its Wild Relatives*. Stechert-Hafner. Pennsauken, N.J. (An extensive classic systematic, genetic and geographic study.)

Coursey, D. G. 1967. *Yams*. Tropical Agriculture Series, Longman Green and Co. London. (A good discussion of agricultural aspects of the yam.)

Heiser, C. B. 1969. *Nightshades*. W. H. Freeman and Company, San Francisco. Chapter 2, "Earth's Apples." (A delightful discussion of potatoes.)

CHAPTER 20

Alexopoulos, C. J., *Introductory Mycology*. 2nd Edition. John Wiley & Sons, Inc., New York.

Phaff, H. J., and E. Mrak. 1977. *The Yeasts*. 2nd Edition. (A thorough yet most readable treatment of all the yeast general.)

Thimann, K. V. 1963. *The Life of Bacteria*. 2nd Edition. Macmillan Co., New York Chapter 11, "The Alcoholic Fermentation." (The first part is general and historical, the last part technical.)

Wagner, P. 1974. "Wines, Grapevines and Climate." *Scientific American offprints in Agriculture*, No. 1298. W. H. Freeman and Company, San Francisco and New York.

CHAPTER 21

Barber, G. A., A. D. Elbine, and W. Z. Hassid. 1964. "Synthesis of Cellulose." *Journal of Biological Chemistry* **239**:4056-4058. (An important paper that pioneers work on the mode of formation of cellulose in plants.)

Kirby, R. H. 1963. *Vegetable Fibres*. Leonard Hill, London (463 pp.). (A generous supply of botanical and agricultural details, not out of date in any important way.)

CHAPTER 22

Cooks, C. B. 1948. "Cork and Cork Products," *Economic Botany* **2**:293-302. (An interesting account of harvesting of the bark and its many uses as cork.)

Edlin, H. L. 1969. *What Wood is That?* The Viking Press, New York. (A manual of wood identification, with 40 actual wood samples, amply illustrated, and text with many interesting facts.)

Hidy, R. W. et al. 1963. *Timber and Men*. MacMillan, Inc., New York. (A good general discussion of the utilization of trees.)

Meyers, N. 1980. *Conversion of Tropical Moist Forests*. National Academy of Sciences, Washington, D.C. (Clear statement of current need for research to balance the utilization and conservation of one of the world's great resources which is being rapidly depleted.)

CHAPTER 23

Duke, J. A. 1981. *Handbook of Legumes of World Economic Importance*. Plenum Press, New York. 344 large pages, 63 contributors. (A valuable and complete reference book, with notes on uses, ecology, harvesting, etc., for each species.)

Heiser, C. B. 1973. *Seed to Civilization*. W. H. Freeman and Company, San Francisco. Chapter 6, "Poor Man's Meat, the Legumes."

CHAPTER 24

Hansen, H. A. 1978. *The Witch's Garden*. The Unity Press., Santa Cruz, 128 pp. (Entertaining essays on poisonous and medicinal plants as used in classical and mediaeval times.)

Kreig, M. B. 1964. *Green Medicine*. Rand McNally and Company, Chicago. (A popular-style paperback, but quite inclusive in coverage.)

Schultes, R. E., and A. Hofmann. 1973. *The Botany and Chemistry of Hallucinogens*. (A notable example of successful collaboration between a botanist and a chemist; the level is basically professional.)

Tippo, O., and W. L. Stern. 1977. *Humanistic Botany*. W. W. Norton and Company, Inc., New York. Chapter 9, "Medicinal Plants." (A lighthearted presentation, in 45 pages.)

Wheelwright, E. G. 1974. *Medicinal Plants and Their History*. Dover Publishing Company, New York. (Primarily historical.)

CHAPTER 25

Eden, T. 1958. *Tea*. Longman, Green and Company, New York. (Excellent compilation.)

Porter, R. H. 1950. "Maté – South American or Paraguay Tea," *Economic Botany* **4**:37-51. (Interesting information on local products.)

Urquhart, D. H. 1961. *Cocoa*. Second Edition. Longman, Green and Company, London. (A full account of history, botany, cultivation and marketing of cocoa.)

CHAPTER 26

Dodge, B. S. 1959. *Plants That Changed the World*. Little, Brown and Company, Boston, Mass. Chapter 6, "The Weeping Tree." (An account of the exciting manner in which *Hevea* seeds were "stolen" from the Amazon and sent to Asia.)

Langenheim, J. H. 1969. "Amber: A Botanical Inquiry," *Science* **163**:1157-1169. (A summary of resins through time and possible evolutionary significance.)

Masselman, George. 1967. *The Money Tree (the Spice Trade)*. McGraw-Hill, Inc., New York. (An enjoyable popularization, emphasizing the historical importance of the spice trade.)

Parry, J. W. 1969. *Spices:* Vol. I, *The Story of Spices;* Vol. II, *Morphology, Histology, Chemistry*. Chemical Publishing Company, New York. (A valuable source of information otherwise difficult to find.)

Polhamus, L. G. 1962. *Rubber*. World Crop Series; Inter-science Publishers, New York. (An excellent compendium of information on all aspects of rubber.)

Rosengarten, Frederic. 1969. *The Book of Spices*. Livingston Publishing Company, Philadelphia, Pa. (A fine compendium of botanical and culinary information about spices.)

CHAPTER 27

Beevers, H. 1961. *Respiratory Metabolism in Plants*. Harper & Row, Publishers, Inc., New York. Chapter 12, "Relationships to Fat Matobolism." (Concise, clear discussion of fat metabolism.)

Goodwin, T. W., and E. J. Mercer. 1972. *Introduction to Plant Biochemistry*. Chapter 8, "Lipid Metabolism." (An up-to-date detailed treatment of the biochemical reactions constituting lipid metabolism in plants.)

Hilditch, T. P., and P. N. Williams. 1964. *The Chemical Constitution of Natural Fats*. 4th Edition. John Wiley & Sons, Inc. (A compendium of detailed information on chemistry and distribution of fats and fatty acids.)

CHAPTER 28

Agrios, G. N. 1969. *Plant Pathology,* Academic Press, Inc., New York. (A complete treatise on all aspects of plant diseases.)

Large, E. C. 1946. *The Advance of the Fungi*. Jonathan Cape, London. (An early semipopular account of fungus diseases; Chapter I, "Late Blight of Potato," is among the best of its kind, especially in bringing out the ramifications of a major plant disease.)

National Research Council, 1968. *Plant Disease, Development and Control*. Washington, D.C., National Academy of Sciences. (A 190-page book, agriculturally oriented, and very useful as a major reference.)

Stapley, J. H., and F. C. H. Gayner. 1969. *World Crop Protection*. Vol. 1, *Pests and Diseases*. Liffe Books, London; Chemical Rubber Company, Cleveland, Ohio. (Brief accounts of the major diseases of all the world's crops, and how they are being combated.)

Wilson, C., W. E. Loomis, and T. A. Steeves. 1971. *Botany,* 5th Edition. Holt, Rinehart & Winston, New York. (Pp. 566-582 cover the smuts and rusts, fingi imperfecti, and plant disease control. The treatment is brief but reasonably up to date.)

CHAPTER 29

Li, H. L. 1959. *Garden Flowers of China*. Ronald Press, New York (240 pp.). (Interesting discussion of Moutan, Lotus, etc.)

Webber, R. 1968. *The Early Horticulturists*. A. M. Kelley, Publishers, New York. (In the guise of biographies, a lively history of gardens and horticulture in Britain from the sixteenth to the early twentieth century.)

CHAPTER 30

Heiser, C. B. Seed to Civilization. The Story of Food. 2nd. Ed., 1981. San Francisco, W. H. Freeman. (A lively semipopular treatment of the world's food, with particularly well balanced discussions of meat and cereals.)

Hendricks, S. B. 1969. "Food from the Land," Chapter 4 in *Resources and Man,* prepared by National Research Council. W. H. Freeman and Company, San Francisco. (An authoritative statement on the world food outlook, somewhat out of date but well balanced.)

Scientific American (Board of Editors). 1976. *Food and Agriculture*. W. H. Freeman and Company, San Francisco. (12 essays, especially good on the agriculture of developing countries.

CHAPTER 31

Arvill, Robert. 1970. *Man and Environment*. Penguin Books, Harmondsworth, England. (A good general statement.)

Fosberg, F. R. 1973. "Temperate zone influence on tropical forest land use: a plea for sanity." in *Tropical Forest Ecosystems in Africa and South America: A Comparative Review*. B. J. Meggers, E. S. Ayensu and W. D. Duckworth (eds.). Smithsonian Institution Press, Washington, D.C. (A succinct comment from an authority on the subject).

Gregor, H. F. 1971. "Urban pressures on California Land." In *California: Its People, Its Prospects,* R. W. Durrenberger (ed.), National Press Books, Palo Alto.

Hinrichs, N. (ed.), 1971. *Population, Environment and People*. McGraw-Hill Book Co., New York. (Statements by delegates to the First National Congress on Optimum Population and Environment held in Chicago, 1970; perhaps unique in representing broad spectrum of American interests and organizations.)

Leopold, A. 1949. *A Sand County Almanac,* Oxford University Press, England. (A classic philosophical statement of the need of humans to reevaluate the use of land.)

Stamp, L. D. 1955. *Man and the Land*. Collins, London. (This excellently documented study of the British Isles is exemplary.)

Metric Equivalents

The metric system is the most common system of weights and measures, being used in almost all countries outside English-speaking North America, and is used universally in scientific research. Furthermore, it is a simpler system because all conversions from one unit to another are based on powers of 10. The basic units are the centimeter, gram and second.

	Metric	U.S.
Length:	1 meter (m) = 39.37 inches = 1.094 yards 1 kilometer (km) = 0.621 miles 1 decimeter (10 cm) = 3.94 inches 1 centimeter (cm) (10 mm) = 0.3937 inches 1 millimeter (mm) = 0.001 meter 1 micrometer or 1 micron (μ) = 10^{-6} meter 1 nanometer (nm) or 1 millimicron (mμ) = 10^{-9} meter 1 Angstrom (A.U.) = 10^{-10} meter	1 yard (36 inches) = 0.914 meter 1 mile (1760 yards, 5380 feet) = 1.61 kilometer 1 foot (12 inches) = 30.48 centimeters 1 inch = 2.54 centimeters
Area:	1 hectare (ha) (10,000 sq. meters) = 2.471 acres 1 square meter (m^2) = 1.196 square yards or 1550 square inches	1 acre = 0.405 hectare 1 square yard = 0.836 square meter 1 square inch = 6.45 square centimeters
Weight:	1 gram (g) = 0.0353 ounces 1 kilogram (kg) = 2.206 pounds 1 metric ton (1000 kg) (sometimes writen tonne) = 2206 pounds	1 ounce = 28.3 grams 1 pound avoirdupois (lb) = 453 grams 1 long ton (2240 pounds) = 1015 kilograms
Volume:	1 liter (1) (1000 cubic cm) = 33.8 fluid ounces = 1.057 quarts = 0.2642 gallons U.S. 1 cubic meter (1000 liters) = 35.3 cubic feet 1 milliliter (ml) = 0.379 fluid ounce	1 fluid ounce = 29.57 milliliters 1 quart = 946.3 milliliters 1 gallon U.S. (gal) = 3.785 liters 1 cubic foot = 28.3 liters
Yield:	1 ton (metric) per hectare = 893 pounds per acre	1 ton (long) per acre = 412 kilograms per hectare (kg/ha)

Glossary

Abbreviation	Meaning
A.S.	Anglo-Saxon
dim.	diminutive
F.	French
Gr.	Greek
It.	Italian
L.	Latin
M.E.	Medieval English
M.L.	Medieval Latin
N.L.	New Latin
O.E.	Old English
R.	Russian
Sp.	Spanish

A. Abbreviation for an *angstrom,* 0.0001 of a micron. There are 10,000,000 angstroms in a millimeter, and 10 in a nanometer.

Abscisic acid. A hormone inducing dormancy, stomatal closure, growth inhibition, and other responses in plants. Abbreviated *ABA* or *AbA*.

Abscission zone (L. abscissus, cut off). Zone of thin-walled cells extending across the base of a petiole, the breakdown of whose cell walls separates the leaf or fruit from the stem.

Absorption spectrum. A graph showing the ability of a substance to absorb light of various wavelengths.

Achene. Simple, dry, one-seeded indehiscent fruit, with seed attached to ovary wall at one point only.

Acid. A substance that can donate a hydrogen ion (H^+).

Actinomorphic (Gr. aktis, ray + morphe, form). Said of flowers of a radiating or star pattern, capable of bisection into similar halves in two or more planes.

Action spectrum. A graph showing the relative effectiveness of different wavelengths of light in causing a physiological response.

Active transport. The energy-expending process by which a cell moves a substance across the plasma membrane, often from a lower concentration to a higher concentration, against a diffusion gradient.

Adaptation. (L. ad, to + aptare, to fit). Adjustment of an organism to the environment.

Adenine. A purine base, $C_5H_5N_5$, present in nucleic acids and nucleotides, as well as in ATP and ADP.

ADP. Adenosine diphosphate.

Adsorption (L. ad, to + sorbere, to suck in). The concentration of molecules or ions at a surface or an interface.

Adventitious (L. adventicius, not properly belonging to). Referring to a structure arising from an unusual place, e.g., buds at other places than leaf axils, roots growing from stems or leaves.

Aeciospore (Gr. aikia, injury + spore). One of the dikaryotic asexual spores of rust fungi.

Aecium (plural *aecia*) (Gr. aikia, injury). In rust, a reproductive body that produces aeciospores.

Aerobe (gr. aer, air + bios, life). An organism that uses molecular oxygen in its respiratory process.

Agar (Malay agaragar). A gel-forming substance obtained mainly from certain species of red algae.

Aggregate fruit (L. ad, to + gregare, to collect; to

bring together). A fruit developing from the several separate carpels of a single flower; e.g., a strawberry; compare with *multiple fruit*.

Aleurone layer (Gr. aleuron, flour). The outermost cell layer of the endosperm of wheat and other grains.

Alga (plural, *algae*) (L. alga, seaweed). A member of the large group of aquatic plants containing chlorophyll and thus able to carry out photosynthesis.

Alkali. See *Base*.

Alkaloid. A nitrogen-containing organic compound (*alkaline* in solution) that stimulates or depresses the nervous or muscular systems of animals.

Alleles (Gr. allelon, of one another). Variant forms of the same gene.

Allopolyploid (Gr. allos, different + poly, many + ploos, fold + eidos, form). A polyploid in which one or more sets of chromosomes come from different species or widely different strains.

Alpine (L. Alpes, the Alps Mountains). Meadowlike vegetation at high elevation, above tree line.

Alternate. Referring to bud or leaf arrangement in which there is one bud or one leaf at a node.

Alternation of generations. The alternation of haploid (gametophytic) and diploid (sporophytic) phases in the life cycle of many organisms; the phases (generations) may be morphologically similar or distinct, depending on the organism.

Amino acid. An organic compound containing both an amino group ($-NH_2$) and an acid group ($-COOH$).

Ammonification (Ammon, Egyptian sun god, near whose temple ammonium salts were first prepared from camel dung, + L. facere, to make). Decomposition of amino acids to ammonia.

Amylase. An enzyme that hydrolyzes starch.

Amyloplast (L. amylum, starch + plastos, formed). Cytoplasmic organelle specialized to store starch. Abundant in roots and in storage organs such as tubers.

Anaerobe (Gr. an, not + aer, air + bios, life). An organism able to live in the absence of free oxygen, or in greatly reduced concentrations of free oxygen.

Anaphase (Gr. ana, up + phais, appearance). That stage in mitosis in which half chromosomes or sister chromatids move to opposite poles of the cell.

Anatomy. The study of form and structure.

Androecium (Gr. andros, man + oikos, house). The aggregate of stamens in the flower of a seed plant.

Angiosperm (Gr. angeion, a vessel + sperma from speirein, to sow, hence a seed or germ). Literally a seed borne in a vessel; thus a group of plants whose seeds are borne within a matured ovary.

Anion. Any negative ion.

Annual (L. annus, a year). A plant that completes its life cycle within one year and then dies.

Annual ring. Ring of xylem in wood, which indicates the annual increment of growth in trees growing in temperate regions.

Annular vessels (L. annularis, related to a ring). Vessels with rings of secondary wall material.

Anther (M. L. anthera, from Gr. anthos, flower). The pollen-bearing portion of a stamen.

Anthocyanin (Gr. anthos, a flower + kyanos, dark blue). A blue, purple, or red vacuolar pigment.

Antibiotic. A substance or extract that inhibits the growth of bacteria or fungi.

Anticlinal cell division. Cell division in which the newly formed cell wall is perpendicular to the axis of the organ surface (cf. Periclinal).

Antipodal (Gr. anti, opposite + pous, foot). Referring to cells at the end of the embryo sac opposite that of the egg apparatus.

Apex (L. apex, a tip, point, or extremity). The tip, point, or angular summit of anything; the tip of a leaf; that portion of a root or shoot containing apical and primary meristems.

Apical dominance. The inhibition of lateral buds or meristems by the apical meristem.

Apical meristem. A mass of meristematic cells at the very tip of a shoot or root.

Arboretum (L. arbor, tree, also arboretum, a place grown with trees). A place, usually outdoors, set aside for the display of living plants, including herbs and shrubs as well as trees; contrasts with an herbarium, which displays dead, preserved remains of plants.

Aril (M. L. arillus, a wrapper for a seed). An accessory seed covering formed by an outgrowth at the base of the ovule.

Ascus (plural, *asci*) (Gr. askos, a bag). A specialized cell, characteristic of the Ascomycetes, in which two haploid nuclei fuse, immediately after which meiosis and mitosis produce eight ascospores still contained within the ascus.

-ase. A chemical suffix indicating an enzyme.

Asexual (Gr. a, not + L. sexualis, sexual). Any type of reproduction not involving the union of gametes or meiosis.

Assimilation (L. assimilare, to make like). The transformation of food into cell material.

Atom (Gr. atomos, indivisible). A unit of matter, consisting of a dense, central nucleus surrounded by negatively charged electrons that are in constant motion. The nucleus consists of positively charged protons and uncharged neutrons.

ATP. Adenosine triphosphate, a high-energy organic phosphate of great importance in energy transfer in cellular reactions.

Autopolyploid (Gr. autos, same, self + Gr. poly, many + eidos, form). A polyploid in which the chromosomes all come from the same species, usually resulting from the doubling of chromosomes of the same individuals.

Autotrophic (Gr. auto, self + trophein, to nourish with food). Pertaining to an organism that is able to manufacture its own food.

Axil (L. axilla, armpit). The upper angle between a petiole of a leaf and the stem from which it grows.

Axillary bud. A bud formed in the axil of a leaf.

Auxin (Gr. auxein, to increase). A hormone that regulates many aspects of plant growth and development, especially cell elongation.

Backcross. The crossing of a hybrid with one of its parents or with a genetically equivalent organism; a cross between an individual whose genes are to be tested and one that is homozygous for all recessive genes involved in the experiment.

Banner (M. L. bandum, a standard). Large, broad, and conspicuous petal of legume type of flower.

Bark (Swedish bark, rind). The external group of tissues, from the cambium outward, of a woody stem or root.

Base. A substance that can accept a proton (H^+) or one that dissociates to form OH^- ions.

Basidiospore (M. L. basidium, a little pedestal + spore). See *Basidium*.

Basidium (plural, *basidia*) A reproductive cell of the Basidiomycetes in which nuclei fuse and meiosis occurs, to produce usually four haploid basidiospores. It may be a special club-shaped cell, a short filamentous cell, or a short four-celled filament.

Berry. A simple fleshy fruit, the ovary wall fleshy and including one or more carpels and seeds.

Biennial (L. biennium, a period of two years). A plant that requires two years to complete its life cycle. Flowering is normally delayed until the second year.

Binomial (L. binominis, two names). Two-named; in biology each species is generally indicated by two names, the genus to which it belongs and its own species name.

Bioassay (Gr. bios, life + L. exagere, to weigh or test). To test for the presence or quantity of a substance by using an organism's response to treatment with that substance.

Biotin. A vitamin of the B complex.

Bordered pit. A pit in a tracheid or vessel member having a distinct rim of the cell wall overarching the pit.

Bract (L. bractea, a thin plate of precious metal). A thin modified leaf, from the axil of which arises a flower or an inflorescence.

Bud. An undeveloped shoot, largely meristematic tissue, generally protected by modified scale-leaves. Also a swelling on a yeast cell that will become a new yeast cell when released.

Bud scale. A modified outer leaf of a bud.

Bulb (Gr. bolbos, an onion). A short, flattened, or disk-shaped underground stem, with many fleshy scale-leaves filled with stored food.

Bundle sheath. The sheath of parenchyma cells that

surrounds the vascular bundle of leaves, sometimes called border parenchyma.

C_3 cycle. The Calvin-Benson cycle of photosynthesis, in which the first stable products after CO_2 fixation are three-carbon molecules.

C_4 cycle. The Hatch-Slack cycle of photosynthesis, in which the first stable products after CO_2 fixation are four-carbon molecules.

Callose (L. callum, thick skin + ose, a suffix indicating a carbohydrate). An amorphous polysaccharide deposited around pores in sieve tube members.

Callus (L. callum, thick skin). A mass of thin-walled cells, usually developed as the result of wounding or in tissue cultures.

Calorie (L. calor, heat). The amount of heat needed to raise the temperature of 1 g of water 1° C (usually from 14.5 to 15.5° C), also called gram-calorie; 1000 calories = 1 kilocalorie.

Calyx (Gr. kalyx, a husk, cup). Sepals collectively; the outermost flower whorl.

Cambium (L. cambium, one of the alimentary fluids of the human body supposed to carry nutrients). A layer, usually one or two cells thick, of persistently meristematic tissue, that divides to give rise to secondary tissues, resulting in growth in diameter.

Capillaries (L. capillus, hair). Very small spaces, or fine bores in a tube.

Capsule (L. capsula, dim. of capsa, a case). A simple, dry, dehiscent fruit, with two or more carpels.

Carbohydrate (from carbon + hydrate, containing water). Compounds with the general formula $C_x(H_2O)_y$ or $C_xH_{2y}O_y$ (also x can equal y).

Carbon fixation. The attachment of CO_2 to an organic compound to produce an organic acid.

Carotene. A reddish-orange plastid pigment of formula $C_{40}H_{56}$; named from the carrot to which it gives the color.

Carotenoids. A class of fat-soluble compounds that includes carotenes and xanthophylls; most of them are yellow, orange, or red.

Carpel (Gr. karpos, fruit). A floral leaf bearing ovules along its margins.

Caryopsis (Gr. karyon, a nut + opsis, appearance). A simple, dry, one-seeded, indehiscent fruit, with pericarp firmly united all around to the seed coat.

Casparian strip. Suberized strip that impregnates the radial and transverse walls of endodermal cells.

Catalyst (Gr. katelyein, to dissolve). A substance that accelerates a chemical reaction but that is not used up in the reaction.

Cation (Gr. kata, downward + ion, going). Any positive ion.

Cation exchange (Gr. kata, downward). The replacement of one positive ion (cation) by another, as on a negatively charged clay particle.

Catkin (literally a kitten, first used in 1578 to describe the inflorescence of the pussy willow). A type of inflorescence, really a spike, generally bearing only flowers of one sex, which eventually falls from the plant entire.

Cell (L. cella, small room). The smallest unit of material in the organism that is capable of self-reproduction. It is surrounded by a plasma membrane and contains a store of DNA together with a metabolic system.

Cell wall. A layer of material, chiefly cellulose and other polymers, that is laid down outside the plasma membrane of most plant cells, protecting the protoplast and limiting its expansion.

Cellulose (cell + ose, a suffix indicating a carbohydrate). A polysaccharide occurring in the cell walls of the majority of plants; it is composed of hundreds of glucose molecules linked together in a definite pattern.

Cenozoic (Gr. kainos, recent + zoe, life). The geologic era extending from 65 million years ago to the present.

Centromere (L. centrum, center + Gr. meros, part). Specialized part of chromosomes where spindle fibers attach and to which two chromatids connect. Two *kinetochores,* one on each chromatid, compose one centromere.

Chemotropism (chemo + Gr. tropos, a turning). Influence of a chemical substance on the direction of growth.

Chernozem (R. cherny, black + zem, earth). A soil characteristic of some grassland vegetation in warm areas with moderate rainfall; dark in color because of a high content of organic matter; a *mollisol.*

Chiasma (Gr. chiasma, two lines placed crosswise). The cross formed by breaking, during prophase 1 of meiosis, of two nonsister chromatids of homologous chromosomes and the rejoining of the broken ends of different chromatids.

Chitin (Gr. chiton, a coat of mail). A polymer in which the monomer unit is the modified sugar N-acetyl glucosamine; it is the principal stiffening material in the cell walls of most fungi and in the exoskeletons of insects and crustaceans.

Chlorophyll (Gr. chloros, green + phyllon, leaf). The green pigment found in the chloroplast, acting to absorb light energy in photosynthesis.

Chloroplast (Gr. chloros, green + plastos, formed). Specialized cytoplasmic body, containing chlorophyll and carotenoids, in which photosynthesis takes place.

Chlorosis (Gr. chloros, green + osis, diseased state). Failure of chlorophyll development, because of a nutritional disturbance or because of an infection.

Chromatid. The half-chromosome during prophase and metaphase of mitosis, and between prophase 1 and anaphase II of meiosis.

Chromatin (Gr. chroma, color). Substance in the nucleus that readily takes artificial staining; being that portion that bears the determiners of hereditary characters; made up of DNA and protein.

Chromosome (Gr. chroma, color + soma, body). A condensed mass of chromatin, visible during cell division.

Citric acid cycle. A system of reactions that contributes to the catabolic breakdown of acids in respiration and that provides building materials for a number of important anabolic pathways. Also called the *Krebs cycle* and the *tricarboxylic acid (TCA) cycle*.

Class (L. classis, one of the six divisions of Roman people). A taxonomic group below the division level but above the order level.

Clay. Soil particles less than 2 microns ($2\mu m$) in diameter, composed mainly of aluminum (Al), oxygen (O), and silicon (Si).

Climax community (Gr. klima, ladder). Stable community in a successional series that is more or less in equilibrium with existing environmental conditions.

Clone (Gr. klon, a twig or slip). The aggregate of individual organisms produced asexually from one sexually produced individual.

Coal Age. The Carboniferous period, beginning 345 million years ago and ending 280 million years ago.

Coalescence (L. coalescere, to grow together). A condition in which there is union of separate parts of any one whorl of flower parts; synonyms are *connation* and *cohesion*.

Coenocyte (Gr. koinos, shared in common + kytos, a vessel). A plant filament whose protoplasm is continuous and multinucleate and without any division by walls into separate protoplasts.

Coenzyme. A substance, usually nonprotein and of low molecular weight, necessary for the action of some enzymes.

Coleoptile (Gr. koleos, sheath + ptilon, feather). The tubular first leaf in germination of grasses, which sheaths the succeeding leaves.

Collenchyma (Gr. kolla, glue + -enchyma, a suffix, derived from parenchyma and denoting a type of cell or tissue). A stem tissue composed of cells that fit rather closely together and with walls thickened at the angles of the cells.

Colloid (Gr. kolla, glue + eidos, form). Referring to matter composed of particles, ranging in size from 0.1 to 0.0001 micron, dispersed in some medium, as clay particles in soil.

Colony (L. colonia, a settlement). A growth form characterized by a group of closely associated, but poorly differentiated, cells; sometimes filaments can be associated together in a colony (as in Nostoc), but more typically it consists of unicells.

Community (L. communitas, a fellowship). All of the populations within a given habitat; usually the populations are thought of as being interdependent.

Companion cells. Cells associated with sieve tubes.

Complete flower. A flower having four whorls of floral leaves; sepals, petals, stamens, and carpels.

Compound leaf. A leaf whose blade is divided into several distinct leaflets.

Cone (Gr. konos, a pine cone). A fruiting structure composed of modified leaves or branches, which bear sporangia on their surface; usually arranged in a spiral order, as in the pine cone.

Conduction (L. conducere, to bring together). Act of moving or conveying water through the xylem.

Conidia (singular, *conidium*) (Gr. konis, dust). Asexual reproductive cells of fungi, arising by the cutting off of terminal or lateral cells of special hyphae, or by being pushed out from a flask-shaped cell.

Conifer (cone + L. ferre, to carry). A cone-bearing tree.

Conjugation (L. conjugatus, united). Process of sexual reproduction involving the fusion of isogametes or of specialized cell extensions.

Conservative (L. conservare, to keep). Said of a taxonomic trait whose expression is not much modified by the external environment; a trait that is constant unless its genetic base is changed.

Convergent evolution. The process in which successive progeny, originally of quite distinct parents, gradually come to appear more and more alike.

Copal (Sp. Copal, Mexican copalli, incense). A hard, lustrous resin yielded by various tropical trees, used for varnishes and other materials where a durable surface is desired.

Cork. An external, secondary tissue impermeable to water and gases, produced by the cork cambium.

Corm (Gr. kormos, a trunk). A short, solid, vertical, enlarged underground stem in which food is stored.

Corolla (L. corolla, dim. of corona, a wreath, crown). Petals, collectively; usually the conspicuously colored flower whorl.

Cortex (L. cortex, bark). The primary tissue of a stem or root that is bounded externally by the epidermis. Internally it is bounded in the stem by the phloem and in the root by the pericycle.

Cotyledon (Gr. kotyledon, a cup-shaped hollow). Seed leaf; two, generally storing food, in dicotyledons; one in monocotyledons.

Covalent bond. A binding force that holds two atoms together by sharing a pair of electrons.

Cristae (L. crista, a crest). Crests or ridges forming the infoldings of the inner mitochondrial membrane.

Crop rotation. The practice of growing different crops in regular succession to aid in the control of diseases, to increase soil fertility, and to decrease soil erosion.

Cross-pollination. The transfer of pollen from a stamen to the stigma of a flower on another plant (except in clones).

Crossing-over. The exchange of corresponding segments between chromatids of homologous chromosomes.

Cultivar. A uniform group of cultivated plants obtained by breeding or selection, and propagated as a pure line.

Cuticle (L. cuticula, dim. of cutis, the skin). The waxy layer on the outer wall of epidermal cells.

Cutin (L. cutis, the skin). The waxy substance of the cuticle.

Cyme (Gr. kyma, a wave, a swelling). An elongate determinate flower cluster, i.e., the first flower to open and to lengthen its pedicel is the central or apical one; development then follows from this downwards. Cf. *raceme*.

Cytochrome (Gr. kytos, a receptacle or cell + chroma, color). A group of electron-transport proteins serving as carriers in mitochondrial oxidations and in photosynthetic electron transport.

Cytokinesis (Gr. kytos (above) + kinesis, motion). Division of cytoplasmic constituents at cell division.

Cytokinin. A class of hormones that participate in controlling cell division and other developmental processes in plants.

Cytology (Gr. kytos, (above) + logos, word, speech, discourse). The science of dealing with the cell.

Cytoplasm (Gr. kytos + plasma, form). All the protoplasm of a cell outside the nucleus.

Cytosine. A pyrimidine base, $C_4H_5N_3O$, found in DNA and RNA.

Dammar (Malay, damar, resin). An oleoresin from a variety of tropical trees, primarily from Southeast Asia.

Deciduous (L. deciduus, falling). Referring to trees and shrubs that lose their leaves in the fall.

Decomposer (L. de, meaning undo + componere, to put together). An organism that obtains food by breaking down dead organic matter into simpler molecules.

Dehiscent (L. dehiscere, to split open). Opening spontaneously when ripe, splitting into definite parts.

Deletion (L. deletus, destroyed, wiped out). Used here to designate an area, or region, lacking from a chromosome.

Denitrification (L. de, to denote an act undone + nitrum, nitro, a combining form indicating nitrogen + facere, to make). Conversion of nitrates into nitrites, into gaseous oxides of nitrogen, or into free nitrogen.

Deoxyribose nucleic acid (DNA). Hereditary material whose characteristic sugar is deoxyribose, $C_5H_{10}O_4$. The DNA molecule carries hereditary information.

Detritus (L. detritus, worn away). Particulate organic matter released in the decomposition of dead organisms or parts of organisms (such as plant litter).

Diatom (Gr. diatomos, cut in two). Member of a group of golden brown algae with siliceous cell walls fitting together much as do the halves of a pill box.

Dicotyledon (Gr. dis, twice + kotyledon, a cup-shaped hollow). A plant whose embryo has two cotyledons.

Dictyosome (Gr. diktyon, a net + soma, body). One of the component parts of the Golgi apparatus; in plant cells a complex of flattened double lamellae.

Differentiation (L. differe, to carry different ways). Developmental change of a cell leading to the presence of features that equip the cell for performing specialized functions.

Diffuse porous. Wood with an equal and random distribution of large xylem vessel members throughout the growth season.

Diffusion (L. diffusus, spread out). The movement of molecules from a region of higher concentration to a region of lower concentration.

Dikaryon (Gr. di, two + karyon, nut). A hypha or mycelium in which each cell contains two haploid nuclei, the two usually derived from different parent organisms.

Dioecious (Gr. di, two + oikos, house). Unisexual; having the male and female elements in different individuals.

Diploid (Gr. diploos, double + eidos, form). Having a double set of chromosomes per cell; usually a sporophyte generation.

Distillation. A process in which a volatile constituent of a mixture is driven off by heating; it is usually but not necessarily condensed again in another part of the apparatus and thus recovered.

Division. A major portion of the plant kingdom; equivalent to phylum.

DNA. Deoxyribonucleic acid.

Dominant (L. dominari, to rule). Referring, in ecology, to species of a community that receive the full force of the macroenvironment; usually the most abundant of such species.

Dormant (L. dormire, to sleep). Being in a state of reduced physiological activity such as occurs in seeds, buds, etc.

Dorsiventral (L. dorsum, the back + venter, the belly). Having upper and lower surfaces distinctly different, as many leaves do.

Double bond. A covalent bond formed by four electrons.

Double fertilization. The fusion of the egg and sperm (resulting in a $2n$ fertilized egg, the zygote) and the simultaneous fusion of the second male gamete with the polar nuclei (resulting in a $3n$ primary endosperm nucleus which then forms endosperm); a unique characteristic of all angiosperms.

Drupe (L. drupa, an overripe olive). A simple, fleshy fruit, derived from a single carpel, usually one seeded, in which the exocarp is thin, the mesocarp fleshy, and the endocarp stony.

Early wood. That portion of an annual ring formed during spring, characterized by large cells and thin walls, also called *spring wood*.

Ecology (Gr. oikos, home + logos, discourse). The study of life in relation to environment.

Ecosystem (Gr. oikos, house + symstanai, to place together). An inclusive term for a living community and all the factors of its nonliving environment.

Ecotype (Gr. oikos, house + typos, the mark of a

blow). Genetic variant within a species that is adapted to a particular environment yet remains interfertile with all other members of the species.

Edaphic (Gr. edaphos, soil). Pertaining to soil conditions that influence plant growth.

Egg (A.S. aeg, egg). A female gamete.

Electron (L. electrum, amber, from which electricity was first produced by friction). An elementary particle of matter bearing a unit of negative electrical charge. Low in mass and in constant rapid motion, electrons repel one another and are attracted to the positively charged atomic nucleus. A pair of electrons rotating about two nuclei constitutes a chemical bond.

Electron transport chain. A membrane-bound system that controls the flow of electrons from reduced to oxidized compounds, so that some of the energy carried by the electrons is used to form ATP. The chain consists of several compounds (carriers) that alternately accept and donate electrons. Found in mitochondria and chloroplasts.

Electron microscope. A microscope that uses a beam of electrons rather than light to produce a magnified image.

Element (L. elementa, the first principles). In modern chemistry, a substance that cannot be divided or reduced by any known chemical means to a simpler substance.

Embryo (Gr. en, in + brein, to swell). A young sporophytic plant, while still retained in the gametophyte or in the seed.

Embryo sac. The female gametophyte of the angiosperms; generally a seven-celled structure; the seven cells are two synergids, one egg cell, three antipodal cells (each with a single haploid nucleus), and one endosperm mother cell with two haploid nuclei.

Endocarp (Gr. endon, within + karpos, fruit). The inner layer of the fruit wall (pericarp).

Endodermis (Gr. endon, within + derma, skin). The layer of living cells with characteristically thickened walls and no intercellular spaces, which surrounds the vascular tissue in nearly all roots and certain stems and leaves.

Endogenous (Gr. endon, within + gens, race). Produced within the cell or organism.

Endoplasmic reticulum (Gr. endon, within + plasma, anything formed; L. reticulum a small net). A system of membrane-bound cisternae found in the cytoplasm.

Endosperm (Gr. endon, within + sperma, seed). The nutritive tissue formed within the embryo sac of seed plants; it is often consumed as the seed matures, but remains in the seeds of cereals.

Enzyme (Gr. en, in + zyme, yeast). A protein that acts as a catalyst to speed chemical reactions.

Epicotyl (Gr. epi, upon + kotyledon, a cup-shaped hollow). The upper portion of the axis of an embryo or seedling, above the cotyledons.

Epidermis (Gr. epi, upon + derma, skin). A superficial layer of cells occurring on all parts of the primary plant body, except the root cap.

Epigyny (Gr. epi, upon + gyne, woman). The arrangement of floral parts in which the ovary is embedded in the receptacle so that the other parts appear to arise from the top of the ovary.

Epiphyte (L. epi, upon + phyton, a plant). A plant that grows upon another plant, but is not parasitic.

Epistatic (N. L. fr. Gr., epistasis, a stopping). Used to describe a gene whose action determines whether or not the effects of another gene will occur.

ER. Endoplasmic reticulum.

Ethylene. C_2H_4, a gaseous hormone that participates in the control of many developmental processes in plants.

Etiolation (F. etiolier, to blanch). A condition involving increased stem elongation, poor leaf development, and lack of chlorophyll, found in plants growing in the absence, or in a greatly reduced amount, of light.

Eucaryote (L. eu, true + karyon, a nut, referring in modern biology to the nucleus). Any organism characterized by having cellular organelles, including a nucleus, bounded by membranes.

Eutrophication (Gr. eu, true + trophein, to nourish). Pollution of bodies of water resulting from natural, geological, or biological processes such as siltation or encroachment of vegetation or accumulation of detritus.

Evapotranspiration (L. evaporare, e, out of + vapor, vapor + F. transpirer, to perspire). The process of water loss in vapor form from a unit surface of land directly and from leaf surfaces.

Exine (Gr. ex, outside). Outer coat of pollen.

Exocarp (Gr. ex, without, outside + karpos, fruit). Outermost layer of fruit wall (pericarp).

Exogenous (Gr. ex, outside, + genos, race, kind). Produced outside of, originating from, or due to external causes.

Family (L. familia, family). In plant taxonomy, a group of genera; families are grouped in orders.

Fascicular cambium. Cambium within vascular bundles.

Fats. Compounds of glycerol with simple organic acids. The proportion of oxygen to carbon is much less in fats than it is in carbohydrates. Fats in the liquid state are called oils.

Fermentation (L. fermentum, a drink made from fermented barley, beer). Enzymatic breakdown of organic compounds by a process that does not involve molecular oxygen.

Ferredoxin. An electron-transferring protein containing iron, involved in photosynthesis and in nitrogen fixation.

Fertilization (L. fertilis, capable of producing fruit). The union of egg and sperm.

Fertilizer. A material, organic or inorganic, that supplies one or more of the elements that promote the growth of plants.

Fiber (L. fibra, a fiber or filament). An elongated, tapering, thick-walled strengthening cell occurring in various parts of plant bodies.

Field capacity. The amount of water retained in a soil (generally expressed as percent by weight) after larger capillary spaces have been drained by gravity.

Filament (L. filum, a thread). The stalk of a stamen bearing the anther at its tip.

Flora (L. floris, a flower). An enumeration of all the species that grow in a region; also the collective term for all the species that grow in a region.

Floret (F. fleurette, a dim. of fleur, flower). One of the small flowers that make up the inflorescence of the composites (called a head) or the spike of the grasses.

Follicle (L. folliculus, dim. of follis, bag). A simple, dry, dehiscent fruit, with one carpel, splitting along one suture.

Food chain. Food web, the path along which energy is transferred within a community from producers to consumers to decomposers.

Fossil (L. fossa, a ditch). Any impressions, impregnated remains, or other trace of an animal or plant of past geological ages that has been preserved in the earth's crust.

Fruit (L. fructus, that which is enjoyed, hence product of the soil, trees, cattle, etc.). A matured ovary; in some seed plants other parts of the flower may be included; also applied, as *fruiting body,* to reproductive structures of some fungi.

Fucoxanthin (Gr. phykos, seaweed + xanthos, yellowish brown). The brown pigment of the brown algae.

Fungus (plural *fungi*) (L. fungus, a mushroom). A eukaryotic organism that lacks chlorophyll, has cell walls that contain chitin, and that reproduces by means of spores.

Fusiform initials (L. fusus, spindle + form). Meristematic cells in the vascular cambium that develop into xylem and phloem cells comprising an axial system.

Gamete (Gr. gametes, a husband, gamete, a wife). A protoplast that fuses with another protoplast to form the zygote in the process of sexual reproduction.

Gametophyte (gamete + Gr. phyton, a plant). The gamete-producing plant.

Gene (Gr. genos, race, offspring). A group of base pairs in the DNA molecule that determines one or more hereditary characters.

Gene recombination. The appearance of gene combinations in the progeny different from the combinations present in the parents.

Generation (L. genus, birth, race, kind). Any phase of a life cycle characterized by a particular chromosome number, as the gametophyte generation and the sporophyte generation.

Genetic code. The three-symbol system of base-pair sequences in DNA; referred to as a code because it determines the amino acid sequence in the enzymes and other proteins synthesized by the organism.

Genetics (Gr. genesis, origin). The science of heredity.

Genotype (gene + type). The assemblage of genes in an organism (cf. phenotype).

Genus (plural *genera*) (Gr. genos, race, stock). A group of related species.

Geotropism (Gr. ge, earth + tropos, turning). A growth curvature induced by gravity.

Germination (L. germinare, to sprout). The beginning of growth of a seed, spore, or other once-dormant structure.

Gibberellins. A class of hormones that participate in controlling many developmental processes in plants.

Glucose (Gr. glykys, sweet + ose, a suffix indicating a carbohydrate). A common hexose sugar, $C_6H_{12}O_6$.

Glume (L. gluma, husk). An outer and lowermost bract of a grass spikelet.

Glycogen (Gr. glykys, sweet + genes, born). A polysaccharide built of glucose, akin to starch, serving as a food reserve in animals and fungi and some prokaryotes.

Glycolysis (Gr. glykys, sweet + lysis, a loosening). Decomposition of sugar compounds without involving free oxygen; early steps of respiration.

Golgi body (Italian cytologist Camillo Golgi 1844-1926, who first described the organelle). In plant cells a series of flattened plates, more properly called *dictyosomes*.

Grafting. A union of different individuals in which a portion, the scion, is inserted on a root or stem, the stock.

Grana (singular *granum*) (L. granum, a seed). Structures within chloroplasts, seen as a series of parallel lamellae with the electron microscope.

Ground meristem (Gr. meristos, divisible). A primary meristem that gives rise to cortex and pith.

Ground tissues. Those primary tissues (parenchyma, collenchyma, and sclerenchyma) that provide such basic functions as storage, support, and secretion.

Guanine. A purine base, $C_5H_5N_5O$, present in both DNA and RNA.

Guttation (L. gutta, drop, exudation of drops). Exudation of water from plants, in liquid form.

Gynoecium (Gr. gyne, woman + oikos, house). The aggregates of carpels in the flower of a seed plant.

Gymnosperm (Gr. gymnos, naked + sperma, seed, sperm). The common name for a group of divisions, including the Conifers, which bear exposed seeds, in contrast to the flowering plants that bear the seeds enclosed in a fruit (mature ovary); cf. Angiosperm.

Haploid (Gr. haplos, single + eidos, form). Having a single complete set of chromosomes, or referring to an individual or generation containing such a single set of chromosomes per cell; usually a gametophyte generation.

Hardwood. General name for the angiospermous trees (e.g., oak, birch, maple, mahogany) or for the wood from these.

Haustorium (plural, *haustoria*) (M. L. *haustrum,* a pump). A projection that acts as a penetrating and absorbing organ.

Head. An inflorescence, typical of the composite family, in which flowers are sessile and grouped closely on a receptacle.

Heartwood. Wood in the center of old secondary stems that is plugged with resins and tyloses and is not active.

Helix (Gr. helix, anything twisted). The spiral form.

Hemicellulose. A class of polysaccharides of the cell wall, less resistant to enzymes than cellulose.

Herb (L. herba, grass, green blades). A seed plant that does not develop woody tissues, hence herbaceous.

Herbal (L. herba, grass). A book that contains the names and descriptions of plants, especially those that are thought to have medicinal uses.

Herbarium (L. herba, grass). A collection of dried and pressed plant specimens.

Herbicide (L. herba, grass + cidere, to kill). A chemical used to kill plants; some are chemically related to auxin.

Heredity (L. hereditas, being an heir). The transmission of morphological and physiological characters of parents to their offspring.

Heterocyst (Gr. heteros, different + kystis, a bag). An enlarged colorless cell that occurs in the filaments of certain blue-green algae; associated with nitrogen fixation.

Heterogametes (Gr. heteros, different + gamete). Gametes dissimilar from each other in size and behavior, like egg and sperm.

Heterosis (Gr. heterosis, alteration). Hybrid vigor, the superiority shown by a hybrid over either of its parents in any measurable character.

Heterospory (Gr. heteros, different + spore). The condition of producing microspores and megaspores.

Heterotrophic (Gr. heteros, different + trophein, to feed). Referring to a plant obtaining organic food from outside sources.

Heterozygous (Gr. heteros, different + zygon, yoke). Having two different alleles for a given gene in the diploid cell.

Hexose (Gr. hexa, six + -ose, suffix indicating carbohydrate). One of the many carbohydrates with six carbon atoms ($C_6H_{12}O_6$).

Higher vascular plant. Common name for extant seed-producing plants; hence, now includes only gymnosperm and angiosperm divisions although in the past other groups did produce seeds.

Hilum (L. hilum, a trifle). The scar on a seed, which marks the place where the seed broke from the stalk.

Homologous chromosomes (Gr. homos, the same + logos, relation). Members of a chromosome pair; they may be heterozygous or homozygous.

Homospory (Gr. homos, one and the same + spore). The condition of producing one sort of spore only.

Homozygous (Gr. homos, one and the same + zygon, yoke). Having two identical alleles for a given gene in the diploid cell.

Hormone (Gr. hormaein, to excite). A compound that is normally produced by a plant and whose function is to act as a signal in controlling development.

Horticulture. The art or science of culturing gardens for fruit, flowers, or vegetables—often used specially for the cultivation of fruit trees.

Humidity, relative (L. humidus, moist). The ratio of the weight of water vapor in a given quantity of air to the total weight of water vapor that quantity of air is capable of holding at the temperature in question, expressed as percent.

Humus (L. humus, the ground). Decomposing organic matter in the soil.

Hybrid (L. hybrida, offspring of a tame sow and a wild boar, a mongrel). The offspring formed by mating two plants or animals that differ genetically.

Hydathode (Gr. hydro, water + O.E. thoden, stem or thyddan, to thrust). A structure, usually on leaves, which releases liquid water during guttation.

Hydrolysis (Gr. hydro, water + lysis, loosening). Reaction of a compound with water, resulting in decomposition into less complex compounds.

Hypanthium (L. hypo, under + Gr. anthos, flower). Fusion of calyx and corolla part way up their length to form a cup, as in many members of the rose family.

Hypha (plural *hyphae*) (Gr. hyphe, a web). A slender, elongated, threadlike cell or filament of cells of a fungus.

Hypocotyl (Gr. hypo, under + kotyledon, a cup-shaped hollow). That portion of an embryo or seedling between the cotyledons and the radicle or young root.

Hypogyny (Gr. hypo, under + gyne, female). A condition in which the receptacle is convex or conical, and the flower parts are situated one above another in the following order, beginning with the lowest: sepals, petals, stamens, carpels.

IAA. Indoleacetic acid, a major plant hormone.

Imbibition (L. imbibere, to drink). The absorption of liquids or vapors into the ultramicroscopic spaces or pores, especially in seeds.

Imperfect flower. A flower lacking either stems or pistils.

Imperfect fungi. Fungi reproducing only by asexual means.

Incomplete flower. A flower lacking one or more of the four kinds of flower parts.

Indehiscent (L. in, not + dehiscere, to divide). Not opening by valves or along regular lines.

Inferior ovary. An ovary more or less (sometimes completely) attached to the calyx and corolla.

Inflorescence (L. inflorescere, to begin to bloom). A flower cluster.

Integuments (L. integumentum, covering). Cell layers around the ovule that develop into the seed coat.

Interphase (L. inter, between + phase). Period between mitotic divisions, consists of G1, pre-DNA synthesis phase; S, DNA synthesis; and G2, post-DNA synthesis phase.

Intercalary (L. intercalare, to insert). Descriptive of meristematic tissue or growth not restricted to the apex of an organ, i.e., growth at nodes.

Intercellular (L. inter, between + cells). Lying between cells.

Interfascicular cambium (L. inter, between + fasciculus, small bundle). Cambium that develops between vascular bundles.

Internode (L. inter, between + nodus, a knot). The region of a stem between two successive nodes.

Intine (L. intus, within). The innermost coat of a pollen grain.

Intracellular (L. intra, within + cell). Lying within the cell.

Introgressive hybridization (L. intro, to the inside + gressus, stepped + hybrida, halfbreed). Backcrossing between hybrids and the original parental stock.

Involucre (L. involucrum, a wrapper). A whorl or rosette of bracts surrounding an inflorescence.

Ion. An atom or molecule that has a net negative or positive electrical charge because the number of electrons does not equal the number of protons.

Irregular flower. A flower in which one or more members of at least one whorl are of different form from other members of the same whorl; zygomorphic flower.

Isodiametric (Gr. isos, equal + diameter). Having diameters equal in all directions, as a ball or a cube.

Isogametes (Gr. isos, equal + gametes). Gametes similar in size and behavior.

Isolating barriers. Any barrier to the crossing (hybridization) of plants; includes biological, physiological, and ecological categories.

Isomers (Gr. isos, equal + meros, part). Two or more compounds having the same molecular formula but different internal structure; for example, glucose and fructose, both having the formula $C_6H_{12}O_6$.

Karyogamy (Gr. karyon, nut + gamos, marriage). The fusion of two nuclei.

Keel (A.S. ceol, ship). A structure of the legume flower, made up of two petals loosely united along their edges.

Kinetochore (Gr. kinein, to move + chorein, to move apart). Specialized portion of a chromosome that marks the point of spindle fiber attachment.

Krebs cycle. See *Citric acid cycle*.

Lamella (plural, *lamellae*) (dim. of L. lamina, a thin plate). A cellular membrane, e.g., those seen in chloroplasts and those separating cell walls from one another.

Late wood. That portion of an annual ring that is formed during summer and fall, characterized by small diameter cells with thick walls, also called *summer wood*.

Lateral bud. A bud that grows out of the side of a stem.

Laterite (L. later, a brick). A soil characteristic of rain forest vegetation; color is red from oxidized iron in the A horizon; in the oxisol soil order.

Latex (L. latex, juice). A milky or clear-colored juice made up of a complex of materials such as terpenes, proteins, carbohydrates, phenolics, etc.

Leach (O.E. leccan, to moisten). To extract a soluble or movable substance (as ions or clay particles or organic fragments by causing water to filter down through a material (as a soil horizon).

Leaflet. Separate blade of a compound leaf.

Leaf primordium (L. primordium, a beginning). A lateral outgrowth from the apical meristem, which will become a leaf.

Legume (L. legumen, any leguminous plant, particularly bean). A simple, dry dehiscent fruit with one carpel, usually splitting along two sutures.

Lemma (Gr. lemma, a husk). Lower bract that subtends a grass flower, but above the glumes.

Lenticel (M. L. lenticella, a small lens). An opening in the bark that permits the passage of gas.

Leucoplast (Gr. leukos, white + plastos, formed). A colorless plastid.

Liana (F. liane from lier, to bind). A vine or climbing plant.

Lichen (Gr. leichen, thallus plants growing on rocks and trees). A composite plant consisting of a fungus living symbiotically with an alga.

Lignification (L. lignum, wood + facere, to make). Impregnation of a cell wall with lignin.

Lignin (L. lignum, wood). A polymer of phenolic substances impregnating the cellulose framework of certain plant cell walls.

Ligule (L. ligula, dim. of lingua, tongue). In grass leaves, an outgrowth from the upper and inner side of the leaf blade where it joins the sheath.

Linkage. The grouping of genes on the same chromosome.

Linked characters. Characters of a plant or animal controlled by genes grouped together on the same chromosome.

Lipase (Gr. lipos, fat + -ase, suffix indicating an enzyme). An enzyme that breaks fats into glycerol and fatty acids.

Lipids (Gr. lipos, fat + L. ides, suffix meaning son of; now used in sense of having the quality of). Compounds that are insoluble in water and soluble in fat solvents.

Liter. 1000 cubic centimeters.

Loam (O. E. lam or Old Teutonic lai, to be sticky, clayey). A soil having 30 to 50% sand, 30 to 40% silt, and 10 to 25% clay.

Lobed leaf (Gr. lobos, lower part of the ear). A leaf divided by clefts or sinuses.

Locule (L. loculus, dim. of locus, a place). A cavity of the ovary in which ovules occur.

Lodicules (L. lodicula, a small coverlet). Two scale-like structures that lie at the base of the ovary of a grass flower.

Lower vascular plant. Common name for extant vascular plants that do not produce seeds (even though in the past some of their members did produce seeds); the Psilophyta, Lycophyta, Sphenophyta, and Pterophyta divisions.

Lumen (L. lumen, light, an opening for light). The cavity of the cell within cell walls.

Lysis (Gr. lysis, a loosening). A process of disintegration or cell destruction.

Macronutrient (Gr. makros, large + L. nutrire, to nourish). An essential element required by plants in relatively large quantities.

Megaphyll (Gr. megas, great + phyllon, leaf). A leaf whose trace is marked with a gap in the stem's vascular system.

Megasporangium (Gr. megas, large + sporangium). Sporangium that bears megaspores.

Megaspore (Gr. megas, large + spore). The haploid spore of vascular plants, which gives rise to a female gametophyte (compare microspore).

Megasporophyll (Gr. megas, large + spore + Gr. phyllon, leaf). A leaf bearing megasporangia.

Meiosis (Gr. meioun, to make smaller). Two special cell divisions occurring in the life cycle of every sexually reproducing plant and animal, halving the chromosome number and effecting a segregation of genetic determiners.

Membrane (L. membrana, skin covering the separate members of the body). A limiting protoplasmic surface, consisting of protein and lipid, which surrounds cells and cellular organelles.

Meristem (Gr. meristos, divisible). Undifferentiated tissue, the cells of which are capable of active cell division and differentiation into specialized tissue.

Mesocarp (Gr. mesos, middle + karpos, fruit). The middle layer of a fruit wall (pericarp).

Mesophyll (Gr. mesos, middle + phyllon, leaf). Parenchyma tissue of leaf between epidermal layers.

Mesophyte (Gr. mesos, middle + phyton, plant). A plant that normally grows in moist habitats.

Mesozoic (Gr. mesos, middle + zoe, life). A geologic era beginning 225 million years ago and ending 65 million years ago.

Metabolism (Gr. metabole, change). The overall set of chemical reactions occurring in an organism or cell.

Metabolite (Gr. metabole, change + ites, one of a group). A chemical that is a normal cell constituent capable of being metabolized.

Metaphase (Gr. meta, after + physis, appearance). The stage of mitosis during which the chromosomes, or at least the kinetochores, lie in the central plane of the spindle.

Metaxylem. Last formed primary xylem.

Microbody (Gr. mikros, small + body). A cellular organelle, always bounded by a single membrane, frequently spherical, from 20 to 60 nm in diameter, containing a number of enzymes.

Microfibrils (Gr. mikros, small + fibrils, dim. of fiber). Very small fibers of the cell wall.

Micrometer (Gr. mikros, small + metron, measure). One millionth (10^{-6}) of a meter, or 0.001 millimeter; also called a *micron,* and abbreviated μm.

Micronutrient (Gr. mikros, small + L. nutrire, to nourish). An essential element required by plants in relatively small quantities.

Microphyll (Gr. mikros, small + phyllon, leaf). A leaf whose trace is not marked with a gap in the stem's vascular system; microphylls are thought to represent epidermal outgrowths.

Micropyle (Gr. mikros, small + pulon, orifice, gate). A pore leading from the outer surface of the ovule between the edges of the integuments down to the surface of the nucellus.

Microsporangium (plural, microsporangia) (Gr. mikros, little + sporangium). A sporangium that bears microspores.

Microspores (Gr. mikros, small + spore). A spore that, in vascular plants, gives rise to a male gametophyte.

Microsporophyll (Gr. mikros, little + spore + Gr. phyllon, leaf). A leaf bearing microsporangia.

Microtubule (Gr. mikros, small + tubule, dim. of tube). A tubule 25 nm in diameter and of indefinite length, occurring in the cytoplasm of many types of cells.

Middle lamella (L. lamella, a thin plate or scale). Thin layer separating two adjacent protoplasts and remaining as a distinct cementing layer between adjacent cell walls.

Mitochondrion (plural, *mitochondria*) (Gr. mitos, thread + chondrion, a grain). A membrane-bounded organelle responsible for intracellular respiration.

Mitosis (plural *mitoses*) (Gr. mitos, a thread). Nuclear division involving duplication of the chromosomes and their separation into two daughter nuclei; the chromosome number remains constant. Compare with *meiosis.*

Molecule (F. mole + cule, a dim.; literally, a little mass). A unit of matter, the smallest portion of an element or a compound that retains chemical identity with the substance in mass containing two or more atoms. Some organic molecules may contain hundreds of atoms.

Monocotyledon (Gr. monos, solitary + kotyledon, a cup-shaped hollow). A plant whose embryo has one cotyledon.

Monoecious (Gr. monos, solitary + oikos, house). Having the male and female reproductive organs in separate structures, but borne on the same individual.

Monophyletic (Gr. mono, single + phyle, tribe). Said of organisms having a common (but sometimes quite ancient) ancestor.

Morphology (Gr. morphe, form + logos, discourse). The study of form and its development.

Multiple fruit. A cluster of mature ovaries produced by a cluster of flowers, as a pineapple.

Mutation (L. mutare, to change). A sudden, heritable change appearing as the result of a change in genes or chromosomes.

Mycelium (Gr. mykes, fungus). The mass of hyphae forming the body of a fungus.

Mycology (Gr. mykes, fungus + logos, discourse). The branch of botany dealing with fungi.

Mycorrhiza (Gr. mikes, fungus + rhiza, root). A symbiotic association between a fungus and the root of a higher plant.

NAD. Nicotinamide adenine dinucleotide, a coenzyme capable of being reduced.

NADH. Reduced NAD.

NADP. Nicotinamide adenine dinucleotide phosphate, a coenzyme capable of being reduced.

NADPH. Reduced NADP.

Nanometer (Gr. nanos, dwarf). One millionth (10^{-16}) of a millimeter, equals 10 angstroms; abbreviated nm.

Natural classification. A classification scheme based on the phylogenetic nature of the organisms classified; contrasts with an artificial classification, which separates organisms on the basis of convenient traits, but fails to show the evolutionary relationships among the organisms.

Natural selection. The effect of the environment on channeling the genetic variation of organisms down particular pathways.

Net productivity. The arithmetic difference between energy produced in photosynthesis and energy lost in respiration.

Net venation. Veins of leaf blade visible to the unaided eye, branching frequently and joining again, forming a network.

Niche (It. nicchia, a recess in a wall). The role played by a particular organism in its ecosystem.

Nitrification (L. nitrum, nitro, indicating nitrogen + facere, to make). Oxidation of ammonium salts to nitrates through the activities of specialized bacteria.

Nitrogen fixation. The process of reducing N_2 gas to ammonia and incorporating it into the cell; accomplished only by certain prokaryotes.

nm. See *nanometer*.

Node (L. nodus, a knot). Slightly enlarged portion of the stem where leaves and buds arise, and where branches originate.

Nodule. An outgrowth, usually on roots, caused by microorganisms.

Nucellus (L. nucella, a small nut). Tissue composing the chief part of the young ovule, in which the embryo sac develops; megasporangium.

Nucleic acid. A polymeric molecule consisting of subunits called nucleotides, linked together in a chain.

Nucleolus (L. nucleolus, a small nucleus). A darkly staining body in the nucleus.

Nucleosides. Components of nucleic acid consisting of a base and a sugar; in DNA, the sugar is deoxyribose, and in RNA, ribose. In both DNA and RNA three of the bases are adenine, guanine, and cytosine; the fourth base in DNA is thymine, and in RNA it is uracil.

Nucleotide. A nucleoside to which a phosphate unit is attached.

Nucleus (L. nucleus, kernel of a nut). A membrane-bounded organelle that contains most of the DNA in eukaryotic cells; it also contains the nucleolus.

Numerical taxonomy. A field of taxonomy that places equal weight on all of the characters used to distinguish between different taxa.

Nut (L. nux, nut). A dry, indehiscent, hard, one-seeded fruit, generally produced from a compound ovary.

Order. A taxonomic category below class and above family.

Organ (L. organum, an intrument or engine of any kind). A part or member of a plant body adapted for a particular function.

Organelle. A membrane-bound specialized region within a cell such as the mitochondrion or dictyosome.

-ose. A chemical suffix indicating a carbohydrate.

Osmosis (Gr. osmos, a pushing). Diffusion of a solvent through a differentially permeable membrane.

Ovary (L. ovum, egg). Enlarged basal portion of the pistil, which becomes the fruit.

Ovule (L. ovulum, dim. of ovum, egg). A rudimentary seed, containing, before fertilization, the female gametophyte, with egg cell, all being surrounded by the nucellus and one or two integuments.

Oxidation. The removal of electrons or hydrogen, or the addition of oxygen to a compound.

Palea (L. palea, chaff). The upper one of the two bracts that subtend a grass flower.

Paleoecology (Gr. palaios, ancient). A field of ecology that reconstructs past vegetation and climate from fossil evidence.

Paleozoic (Gr. palaios, ancient + zoe, life). A geologic era beginning 570 million years ago and ending 225 million years ago.

Palisade parenchyma. Elongated cells, containing many chloroplasts, found just beneath the upper epidermis of leaves.

Palmately veined (L. palma, palm of the hand). Descriptive of a leaf blade with several principal veins spreading out from the upper end of the petiole.

Panicle (L. panicula, a tuft). A branched inflorescence that bears a racemose flower cluster on each branch.

Pappus (L. pappus, woolly, hairy seed or fruit of certain plants). Scales or bristles representing a reduced calyx in composite flowers.

Parallel venation. Type of venation in which veins of a leaf blade that are clearly visible to the unaided eye are parallel to each other.

Parasite (Gr. parasitos, one who eats at the table of another). An organism deriving its food from the living body of another plant or animal.

Parenchyma (Gr. parenchein, an ancient Greek medical term meaning to pour beside, based on the belief that the liver and other internal organs were formed by blood diffusing through the blood vessels and coagulating, thus designating ground tissue). A tissue composed of cells that usually are not differentiated and have thin walls; site of most essential processes such as photosynthesis, secretion, and storage.

Parent material. The original rock or depositional matter from which the soil of a region has been formed.

Parthenocarpy (Gr. parthenos, virgin + karpos, fruit). The development of fruit without fertilization.

Parthenogenesis (Gr. parthenos, virgin + genesis, origin). The development of a gamete into a new individual without fertilization.

Pathogen (Gr. pathos, suffering + genesis, beginning). An organism that causes a disease.

Pathology (Gr. pathos, suffering + logos, discourse). The study of diseases, their effects on plants or animals, and their treatment.

Peat (M. E. pete, of Celtic origin, a piece of turf used as fuel). A mass of semicarbonized vegetable tissue, such as *Sphagnum,* formed by partial decomposition in water.

Pectin (Gr. pektos, congealed). A class of polymers of the cell wall that are built chiefly of partially oxidized sugars.

Pedicel (L. pediculus, a little foot). Stalk or stem of the individual flower of an inflorescence.

Peduncle (L. pedunculus, a late form of pediculus, a little foot). Stalk or stem of a flower that is borne singly; or the main stem of an infloresence.

Pentose (Gr. pente, five + ose, indicating carbohydrate). A five-carbon sugar, $C_5H_{10}O_5$.

Peptide. Two or more amino acids linked by peptide bonds.

Perennial (L. perennis, lasting the whole year through). A plant that lives an indefinite number of years.

Perfect flower. A flower having both stamens and pistils.

Perianth (Gr. peri, around, about + anthos, flower). The petals and sepals taken together.

Pericarp (Gr. peri + karpos, fruit). The fruit wall, developed from the ovary wall.

Periclinal cell division. Cell division in which the newly formed cell wall is parallel to the axis of the organ, or to its surface.

Pericycle (Gr. peri + kyklos, circle). Tissue, generally of a root, bounded externally by the endodermis and internally by the phloem.

Periderm (Gr. peri + Gr. derma, skin). Protective tissue that replaces the epidermis after secondary growth is initiated. Consists of cork, cork cambium, and phelloderm.

Perigyny (Gr. peri + gyne, female). A flower structure in which the sepals, petals and stamens are attached *around* the ovary. Cf. epigyny and hypogyny.

Permafrost (L. permanere, to remain + A. S. freosan, to freeze). Soil that is permanently frozen; usually found some distance below a surface layer that thaws during warm weather.

Petiole (L. petiolus, a little foot or leg). Stalk of a leaf.

Petrochemicals. Those organic chemicals derived from petroleum.

PGA. 3-Phosphoglyceric acid, a three-carbon compound formed as an intermediate in the fermentation of sugars, and also the first stable product of carbon fixation in the C_3 cycle of photosynthesis.

Phelloderm (Gr. phellos, cork + derma, skin). A layer of cells formed in the stems of some plants from the inner cells of the cork cambium.

Phellogen (Gr. phellos, cork + genesis, origin). Cork cambium, a cambium giving rise externally to cork and, in some plants, internally to phelloderm.

Phenotype (Gr. phanein, to show + type). The bodily characteristics of an organism.

Phloem (Gr. phloos, bark). Food-conducting tissue, consisting of sieve tubes or sieve cells, companion cells, parenchyma, and fibers.

Phosphorylation. A reaction in which a phosphate group is added to a compound, e.g., the formation of hexose phosphate from a hexose and inorganic phosphate.

Photon. A quantum of light; the energy of a photon is proportional to its frequency: $E = h\nu$ where E is energy, h is Planck's constant (6.62×10^{-27} erg-second), and ν is the frequency.

Photoperiod (Gr. photos, light + period). The length of day or period of daily illumination of a plant.

Photophosphorylation. A reaction in which light energy is converted into chemical energy in the form of ATP, produced from ADP and inorganic phosphate.

Photoreceptor (Gr. photos, light + L. receptor, a receiver). A light-absorbing molecule involved in converting light into some metabolic (chemical energy) form, e.g., chlorophyll and phytochrome.

Photosynthesis (Gr. photos, light + syn, together + tithenai, to place). A process in which light energy is used to drive the formation of organic compounds.

Phototropism (Gr. photos, light + Gr. tropos, a turning). Influence of light on the direction of plant growth.

Phycocyanin (Gr. phykos, seaweed + kyanos, blue). A blue pigment that participates in the photosynthesis of blue-green algae.

Phycoerythrin (Gr. phykos, seaweed + erythros, red). A red pigment that participates in the photosynthesis of red algae.

Phylogeny (Gr. phylon, race or tribe + genesis, beginning). The evolution of a group of related individuals.

Phylum (Gr. phylon, race or tribe). A primary division of the animal or plant kingdom.

Physiology (Gr. physis, nature + logos, discourse). The science of the functions and activities of living organisms.

Phytochrome. A pigment found in green plants controlling development in response to certain light stimuli.

Phytoplankton (Gr. phyton, a plant + planktos, wandering). Free-floating plants, collectively.

Pigment. A substance that absorbs visible light and hence appears colored.

Pinnately veined (L. pinna, a feather). Descriptive of a leaf blade with single midrib from which smaller veins branch off, somewhat like the divisions of a feather.

Pioneer community. The first stage of a succession.

Pistil (L. pistillum, a pestle). Central organ of the flower, typically consisting of ovary, style, and stigma.

Pistillate flower. A flower having pistils but no stamens.

Pit. A thin area of a secondary cell wall.

Pith. The parenchymatous tissue occupying the central portion of a stem.

Placenta (plural *placentae*) (L. placenta, a cake). The tissue within the ovary to which the ovules are attached.

Placentation (L. placenta, a cake + -tion, state of). Manner in which the placentae are distributed in the ovary.

Plasmalemma (Gr. plasma, form + lemma, a husk of a fruit). A synonym for *plasma membrane*.

Plasma membrane. The membrane that separates the living protoplast from the external environment; found in all cells.

Plasmodesma (plural, plasmodesmata) (Gr. plasma, form + desmos, a bond, a band). Fine protoplasmic thread passing through the wall that separates two protoplasts.

Plasmodium (Gr. plasma, form + mod. L. odium, something of the nature of). In Myxomycetes, a coenocytic mass of protoplasm, with no surrounding wall.

Plasmolysis (Gr. plasma, form + lysis, a loosening). The separation of the cytoplasm from the cell wall due to removal of water from the protoplast.

Plastid (Gr. plastis, a builder). A class of organelles, including the chloroplast, also chromoplasts, containing other pigments, and amyloplasts, containing starch.

Plastoquinone. A quinone, one of a group of compounds involved in the transport of electrons during photosynthesis in chloroplasts.

Pleiotropism. The capacity of a gene to affect a number of different characteristics.

Plumule (L. plumula, a small feather). The first bud of an embryo or that portion of the young shoot above the cotyledons.

Plywood. A building material consisting of two or more thin sheets or layers of wood glued together.

Podzol (R. pod, under + zola, ashes). A soil characteristic of taiga vegetation; color of the A horizon is gray because of excessive leaching; in the spodosol soil order.

Pollen (L. pollen, fine dust). A collective term for pollen grains.

Pollen grain. A microspore containing mature or immature microgametophyte (male gametophyte).

Pollen tube. The parasitic, colorless male gametophyte of seed plants, which grows through the pistil of an angiosperm.

Pollination. The transfer of pollen from a stamen or staminate cone to a stigma or ovulate cone.

Pollutant (L. polluere, to dirty + lutum, mud). A substance produced by human activity and introduced to the environment; it may be gas, liquid, or solid, easily broken down or long lasting.

Polymer. A molecule that is made by coupling together many small molecules (monomers) that are similar to one another.

Polymerization. The chemical union of monomers such as glucose, nucleotides, or monoterpenes to form polymers such as starch, nucleic acids, or polyterpenes, respectively.

Polynomial (Gr. polys, many + L. nomen, name). A scientific name for an organism composed of more than two words; compare *binomial*.

Polynucleotides (Gr. polys, many). Long-chain molecules composed of units (monomers) called nucleotides; nucleic acid is a polynucleotide.

Polyphyletic (Gr. polys, many + phyle, tribe). Referring to organisms that did not have an ancestor in common.

Polyploid (Gr. polys, many + ploos, fold + eidos, form). A plant or tissue with more than two complete sets of chromosomes per cell.

Polyribosome. An aggregation of ribosomes; frequently called simply polysome.

Polysaccharides (Gr. polys, many + sakcharon, sugar). Polymeric molecules such as starch and cellulose, consisting of numbers of simple sugar molecules linked together.

Pome (Gr. pomme, apple). A simple fleshy fruit, the outer portion of which is formed by the floral parts that surround the ovary.

Pore spaces. Spaces between soil particles that may be filled with air or water and into which root hairs may penetrate.

Prairie (L. pratum, meadow). Grassland vegetation, with trees essentially absent; considered to have more rainfall than does the steppe.

Predation (L. predatio, plundering). A form of biological interaction in which one organism is destroyed by another; parasitism is a form of predation.

Primary meristems. Meristems of the shoot or root tip giving rise to the tissues of the primary plant body.

Primary tissues. Those tissues, epidermis, xylem, phloem, and ground tissues, that are formed from the primary meristems.

Primitive (L. primus, first). Refers to a taxonomic trait thought to have evolved early in time.

Primordium (L. primus, first + ordiri, to begin to undertake). The earliest initial stage of any organ or part of an organ.

Procambium (Gr. pro, before + cambium). A primary meristem that gives rise to primary vascular tissue and, in most woody plants, to the vascular cambium.

Producer (L. producere, to draw forward). An organism that produces organic matter of itself and for other organisms (consumers and decomposers) by photosynthesis.

Prokaryotes (L. pro, before + karyon, a nut, referring in modern biology to the nucleus). The bacteria and blue-green algae that do not have the DNA separated from the cytoplasm by an envelope.

Prophase (Gr. pro, before + phasis, appearance). An early stage in nuclear division, characterized by coiling of the chromosomes and formation of the mitotic spindle.

Protease (protein + -ase, a suffix indicating an enzyme). An enzyme breaking down a protein.

Protein (Gr. proteios, holding first place). A polymeric molecule made up of amino acids linked together in a chain by peptide bonds.

Proteolytic. Causing the hydrolysis of proteins (proteolysis) to their constituent amino acids.

Proterozoic (Gr. protero, before in time + zoe, life). The earliest geologic era, beginning about 4.8 billion years ago and ending 570 million years ago; also called the Precambrian era.

Protoderm (Gr. protos, first + derma, skin). A primary meristem that gives rise to epidermis.

Proton (Gr. protos, first). The nucleus of a hydrogen atom is a single positively charged particle, the proton; the nuclei of all other elements consist of protons and neutrons; the mass of a proton is 1.67×10^{-24} g.

Protoplast (Gr. protoplastos, formed first). The organized living unit of a cell.

Protoxylem. First formed primary xylem.

Purines. A group of compounds in which carbon and nitrogen atoms form a double ring structure, one ring with six atoms, the other with five atoms; includes adenine and guanine, as well as caffeine.

Pyrimidines. A group of compounds in which carbon and nitrogen atoms form a 6-membered ring; includes three constituents of nucleic acids—cytosine, thymine, and uracil.

Quantum (L. quantum, how much). The elemental unit of energy.

Raceme (L. racemus, a bunch of grapes). An elongate flower cluster whose development is indeterminate, i.e., the opening of flowers and the elongation of their pedicels takes place in order from the base up.

Rachis (Gr. rhachis, a backbone). The main axis of a spike; in compound leaves, the extension of the petiole that in an entire leaf would form a midrib.

Radicle (L. radix, root). That portion of the plant embryo that develops into the seedling root.

Ray system (L. radius, a beam or ray). System of cells (usually parenchyma) in wood that are oriented perpendicular to the long axis of the stem. They are formed from ray initials of the vascular cambium.

Receptacle (L. receptaculum, a reservoir). The enlarged end of the pedicel or peduncle to which other flower parts are attached.

Recessive. Describing a gene whose phenotypic expression is masked by a dominant allele, making the resulting heterozygotes phenotypically indistinguishable from dominant homozygotes.

Recombination (L. re, again + combinatus, joined). The mixing of genotypes that results from sexual reproduction.

Reduction (L. reductio, a bringing back). A chemical reaction involving the removal of oxygen from, or the addition of hydrogen or an electron to, a substance.

Regular flower. A flower whose corolla is made up of similarly shaped petals equally spaced and radiating from the center of the flower.

Reproductive isolation. The separation of populations in time or space so that genetic flow between them is cut off.

Resin. A plant exudate (mixture of either terpenoids or flavonoids) that is insoluble in water but soluble in organic solvents, and hardens on exposure.

Resin duct. Pockets, glands or canals; resin is secreted from epithelial cells lining the duct which may occur in any plant organ.

Respiration (L. re, again + spirare, to breathe). In the cell, the catabolic process by which sugars and other fuels are oxidized and broken down, with some of the energy captured in the form of ATP.

Rhizoid (Gr. rhiza, root + eidos, form). A cellular filament that performs the function of a root.

Rhizome (Gr. rhiza, root). An elongated, underground, horizontal stem.

Ribose. One of the pentose sugars, present in RNA.

Ribosomes (ribo, from RNA + Gr. somatos, body). Small particles 10 to 20 nm in diameter, containing RNA and protein; active in protein synthesis.

Ring porous. Wood with large xylem vessel members mostly in the early wood; compare with *diffuse porous*.

RNA. Ribonucleic acid.

Root cap. A mass of living cells covering and protecting the apical meristem of a root; site of perception of gravity in geotropism.

Root hairs. Long fine cells growing out from the epidermis of a root, and providing a means to increase the absorptive surface of the root.

Root pressure. Pressure in the xylem arising as a result of osmosis in the root.

Rosette. A shoot with a very short stem, composed of several unelongated internodes with fully expanded leaves.

Runner. A stem that grows horizontally along the ground surface.

Sand. Soil particles between 50 and 2000 μm in diameter.

Sap. (1) A name applied to the fluid content of the xylem or sieve elements of the phloem; (2) the fluid contents of the vacuole are called cell sap.

Saprophyte (Gr. sapros, rotten + phyton, a plant). An organism deriving its food from the dead body or the nonliving products of another plant or animal.

Sapwood. Peripheral wood that actively transports xylem sap.

Savanna (Sp. sabana, a large plain). Vegetation of scattered trees in a grassland matrix.

Sclerenchyma (Gr. skleros, hard + -echyma, a suffix denoting tissue). A strengthened tissue composed of cells with heavily lignified cell walls.

Scutellum (L. scutella, a dim. of scutum, shield). The single cotyledon of a grass embryo.

Secondary tissues. Those tissues, xylem, phloem, and periderm, that form from secondary meristems.

Secretory structures. Specialized plant structures such as nectaries, glands, and canals that secrete secondary plant substances.

Seed (A. S. sed, anything which may be sown). Popularly as originally used, anything that may be sown; i.e., "seed" potatoes, "seeds" of corn, sunflower, etc.; botanically, a seed is the matured ovule without accessory parts.

Seed coat. A hardened, outer layer of the seed, derived from the integument(s) of the ovule, often regulating germination.

Seed plant. Common name for members of the gymnosperm and angiosperm groups, and for extinct members of other groups that produced seeds.

Senescence. The sequence of physiological changes that leads ultimately to the death of an organ or an organism.

Sepals (M. L., sepalum, a covering). Whorl of sterile, leaflike structures that usually enclose the other flower parts.

Septate (L. septum, fence). Divided by crosswalls into cells or compartments.

Septum (L. septum, fence). A dividing wall or partition; frequently a crosswall in a fungal or algal filament.

Sessile (L. sessilis, low, dwarf, from sedere, to sit). Sitting, referring to a leaf lacking a petiole or a flower, or a fruit lacking a pedicel.

Sexual reproduction. Reproduction that requires meiosis, formation of two types of gametes, and fertilization for a complete life cycle.

Sheath. Part of the leaf that wraps around the stem, as in grasses.

Sibling species. Species morphologically nearly identical but incapable of producing fertile hybrids.

Sieve cell. A long and slender sieve element without a companion cell, with relatively unspecialized sieve areas, and with tapering end walls that lack sieve plates; found in gymnosperms and ferns.

Sieve plate. The perforated wall area in a sieve tube through which pass strands connecting sieve tube protoplasts.

Sieve tube. A long cellular tube specialized for the conduction of food materials.

Sieve tube members. An enucleate phloem cell primarily responsible for photosynthate transport; separated from other sieve tube members by sieve plates.

Silique (L. siliqua, pod). The fruit characteristic of Cruciferae (Brassicaceae); two-celled, the valves splitting from the bottom and leaving the placentae with a partition stretched between.

Silt. Soil particles between 2 and 50 μm in diameter.

Simple pit. Pit not surrounded by an overarching border; in contrast to bordered pit.

Softwood. General name for the coniferous trees (gymnosperms) or for the wood made from them.

Soil (L. solum, soil, solid). The uppermost stratum of the earth's crust, which has been modified by weathering and organic activity into (typically) three horizons: an upper A horizon that is leached; a middle B horizon in which the leached material accumulates; and a lower C horizon that is unweathered parent material.

Soil texture. Refers to the amounts of sand, silt, and clay in a soil, as a sandy loam, loam, or clay texture.

Species (L. species, appearance, form, kind). A group of individuals usually interbreeding and having many characteristics in common.

Sperm (Gr. sperma, the generative substance or seed of a male animal). A male gamete.

Spermatophyte (Gr. sperma, seed + phyton, plant). A seed plant.

Spike (L. spica, an ear of grain). An inflorescence in which the main axis is elongated and the flowers are sessile.

Spikelet (L. spica, an ear of grain + dim. ending -let). The unit of inflorescence in grasses; a small spike.

Spindle (A. S. spinel, an instrument employed in spinning thread by hand). Referring in mitosis and meiosis to the spindle-shaped intracellular aggregate of microtubules involved in chromosome movement.

Sporangium (spore + Gr. angeion, a vessel). Spore case.

Spore (Gr. spora, seed). A reproductive cell that develops into a plant without union with other cells.

Sporophyte (spore + Gr. phyton, a plant). In alternation of generations, the plant in which meiosis occurs and which thus produces haploid spores.

Spring wood. See *Early wood.*

Stamen (L. stamen, the vertical warp in a loom). Flower structure made up of an anther (pollen-bearing portion) and a filament.

Staminate flower. A flower having stamens but no pistil.

Starch (M. E. sterchen, to stiffen). A polysaccharide composed of glucose; the chief food storage material of many plants.

Stele (Gr. stele, a post). The central cylinder, inside the cortex, of roots and stems of vascular plants.

Sterile. (1) Completely uninfected by microorganisms, or (2) (of a soil) unable to support growth of plants.

Steroids. A group of substances, natural or synthesized, containing four fused (nonaromatic) rings of carbon atoms and a variety of substituents or side-chains.

Stigma (L. stigma, a prick, a spot, a mark). Receptive portion of the style to which pollen adheres.

Stipule (L. stipula, dim. of stipes, a stock or trunk). A leaflike structure growing out from either side of the leaf base.

Stolon (L. stolo, a shoot). A stem that grows horizontally along the ground surface.

Stoma (plural, *stomata*) (Gr. stoma, mouth). Epidermal structure on stems and leaves composed of two guard cells plus the small pore between them, through which gases pass.

Stroma (Gr. stroma, a bed or covering). A mass of protecting vegetative filaments; the background substance of chloroplasts, probably the location of the carbon cycle of photosynthesis.

Style (Gr. stylos, a column). The slender column of tissue that arises from the top of the ovary and through which the pollen tube grows down from the stigma.

Suberin (L. suber, the cork oak). A waxy material present in cell walls of cork tissue and endodermis.

Substrate. A molecule that engages in a reaction catalyzed by an enzyme.

Succession (L. successio, a coming into the place of another). A sequence of changes of the species that inhabit an area, from an initial pioneer community to a final climax community.

Sucrose. Table sugar, a disaccharide, $C_{12}H_{22}O_{11}$, made of a molecule of glucose linked to a molecule of fructose.

Superior ovary. An ovary completely separate and free from the calyx.

Symbiosis (Gr. syn, with + bios, life). An association

of two different kinds of living organisms with mutual benefit.

Sympetaly (Gr. syn, with + petalon, leaf). A condition in which petals are united.

Syncarpy (Gr. syn, with + karpos, fruit). A condition in which carpels are united.

Synergids (Gr. synergos, toiling together). The two nuclei at one end of the embryo sac, which, with the third (the egg), constitute the egg apparatus.

Synsepaly (Gr. syn, with + sepals). A condition in which sepals are united.

Taiga (Teleut taiga, rocky mountainous terrain). A broad northern belt of vegetation dominated by conifers; also a similar belt in mountains just below alpine vegetation.

Taproot. The primary root of a plant formed in direct continuation with the root tip or radicle of the embryo; a tapering main root from which arise smaller, lateral roots.

Taxon (plural, *taxa*) (Gr. taxis, order). A general term for any taxonomic rank, from subspecies to division.

Taxonomy (Gr. taxis, order + nomos, law). Systematic botany; the science dealing with the describing, naming, and classifying of plants.

Telophase (Gr. telos, completion + phase). The last stage of mitosis, in which daughter nuclei are reorganized from the separated chromosomes.

Tendril (L. tendere, to stretch out, to extend). A slender coiling organ that aids in the support of stems.

Terpene (Gr. terpen, L. terpentin, turpentine). Hydrocarbons derived from an isoprene (C_5H_8) building block, containing two (as in some essential oils) to hundreds of these units (as in rubber) joined together; also *terpenoid.*

Testa (L. testa, shell). The outer coat of the seed.

Tetrad (Gr. tetradeion, a set of four). A group of four, usually referring to the haploid spores immediately after meiosis.

Tetraploid (Gr. tetra, four + ploos, fold + eidos, form). Having four sets of chromosomes per nucleus.

Thallus (Gr. thallos, a sprout). Plant body without true roots, stems or leaves.

Thymidine. A nucleoside incorporated in DNA, but not in RNA.

Thymine. Methyl uracil, $C_5H_6N_2O_2$, a pyrimidine occurring in DNA, but not in RNA.

Tiller (O. E. telga, a branch). A grass stem arising from a lateral bud at a basal node; tillering is the process of tiller formation.

Tissue. A group of cells that perform a collective function.

Tonoplast (Gr. tonos, stretching, tension + plastos, molded, formed). The cytoplasmic membrane bordering the vacuole; formerly thought to regulate the pressure exerted by the cell sap.

Tracheid (Gr. tracheia, windpipe). An elongated, tapering xylem cell, with lignified pitted walls, adapted for conduction and support.

Tracheophytes (Gr. tracheia, windpipe + phyton, plant). Vascular plants.

Translocation (L. trans, across + locare, to place). The transfer of food materials or products of metabolism.

Transpiration (F. transpirer, to perspire). The giving off of water vapor from the surface of leaves.

Triose (Gr. treis, three + -ose, suffix indicating a carbohydrate). Any three-carbon sugar, $C_3H_6O_3$.

Tritium. A hydrogen atom, the nucleus of which contains one proton and two neutrons; it is written as ^3H; the more common hydrogen nucleus consists only of a proton.

Trophic level. A step in the movement of energy through an ecosystem.

Tropism (Gr. trope, a turning). An orientation of the direction of growth in an organ, guided by an external stimulus such as light or gravity.

Tuber (L. tuber, a bump, swelling). A much enlarged, short, fleshy underground stem tip.

Tundra (Lapp tundar, hill). Meadowlike vegetation at low elevation in cold regions that does not experience a single month with average daily maximum temperature above 50° F.

Turgid (L. turgidus, swollen, inflated). Swollen, distended; referring to a cell that is firm due to water uptake.

Turgor pressure (L. turgor, a swelling). The pressure within the cell resulting from the absorption of water into the vacuole and the imbibition of water by the protoplasm.

Umbel (L. umbella, a sunshade). An inflorescence, the individual pedicels of which all arise from the apex of the peduncle or flower stem.

Uracil. A pyrimidine, $C_4H_4N_2O_2$, found in RNA but not in DNA.

Uredospore (L. uredo, a blight + spore). A red, one-celled summer spore in the life cycle of the rust fungi.

Vacuole (L. dim of vacuus, empty). A watery solution forming a portion of the protoplast distinct from the protoplasm.

Vascular (L. vasculum, a small vessel). Referring to any plant tissue or region consisting of or giving rise to conducting tissue, e.g., bundle, cambium, ray.

Vascular bundle. A strand of tissue containing primary xylem and primary phloem (and procambium if present) and sometimes enclosed by a bundle sheath of parenchyma or fibers.

Vascular cambium. Cambium giving rise to secondary phloem and secondary xylem.

Vascular plant (L. vasculum, small vessel). The common name for any plant that has xylem and phloem.

Vegetable. Used as an adjective, for the plant kingdom; as a noun, any plant or plant organ that is edible (colloquial).

Vegetation (L. vegetare, to quicken). The plant cover that clothes a region; it is formed of the species that make up the flora, and is characterized by the abundance and form (tree, shrub, herb, evergreen, deciduous plant, etc.) of certain of them.

Veneer. A thin sheet of wood, usually with attractive grain, used to cover ordinary, less decorative wood.

Vernalization (L. vernalis, belonging to spring + izare, to make). The promotion of flowering by applying periods of extended low temperature; seeds, bulbs, or entire plants may be so treated.

Vessel. A series of xylem elements whose function it is to conduct water and mineral nutrients.

Vessel element. A xylem cell derived from the vascular cambium or procambium; a portion of a vessel.

Virus. A combination of nucleic acid (DNA or RNA) and protein that can infect either procaryotic or eukaryotic organisms, modifying their metabolism or behavior. Some viruses can even transfer their nucleic acid to the nucleus of the host.

Vitamins (L. vita, life + amine). Naturally occurring organic substances, necessary in small amounts for the normal metabolism of plants and animals.

Water potential. The difference between the activity of water molecules in pure distilled water at standard temperature and pressure, and the activity of water molecules in any other system; the activity of these water molecules may be greater (positive) or less (negative) than the activity of the water molecules under standard conditions.

Weathering. Physical and chemical change in parent material that leads to soil formation.

Weed (A. S. woed, used at least since 888 in its present meaning). Generally a herbaceous plant or shrub not valued for its use or beauty, growing where unwanted, and regarded as using ground or hindering the growth of more desirable plants.

Whorl. A circle of three or more flower parts, or of leaves.

Wood (M. E., wode, wude, a tree). Secondary xylem.

Xanthophyll (Gr. xanthos, yellow, brown + phyllon, leaf). A yellow chloroplast pigment.

Xerophyte (Gr. xeros, dry + phyton, plant). A plant that normally grows in dry habitats.

Xylem (Gr. xylon, wood). A plant tissue consisting of tracheids, vessel elements, parenchyma cells, and fibers; wood.

Zygomorphic (Gr. zygon, a yoke + morphe, form). Having bilateral symmetry; said of organisms, or a flower, capable of being divided into two symmetrical halves only by a single longitudinal plane passing through the axis (cf. Actinomorphic).

Zygote (Gr. zygon, a yoke). A protoplast resulting from the fusion of gametes.

Credits

PART I OPENER Courtesy of The Metropolitan Museum of Art

CHAPTER 1

(Insets) From W. A. Jensen and F. B. Salisbury, *Botany: An Ecological Approach.* Reprinted by permission of Wadsworth Publishing Co., Inc., Belmont, Calif., 1972.

CHAPTER 2

Figure 2-1 (A) From J. W. Schopf, E. S. Barghoorn, M. D. Mason and R. O. Gordon, *Science,* **149**:1366, 1965; (B) and (C) From E. S. Barghoorn and S. A. Tyler, *Science,* **147**:566, 1965.

Figure 2-3 Courtesy J. W. Schopf, University of California at Los Angeles.

Figure 2-4 Courtesy J. W. Costerton et al., *J. Bact,* **94**:1764-1777, 1967.

Figure 2-5 Courtesy Myron C. Ledbetter, Brookhaven National Laboratory and Keith Porter, University of Colorado, Boulder, Colo.

Figure 2-6 Adapted from W. A. Jensen and F. B. Salisbury, *Botany: An Ecological Approach.* Reprinted by permission of Wadsworth Publishing Co., Inc., Belmont, Calif., 1972.

Figure 2-7 From H. N. Andrews, *Studies in Paleobotany,* John Wiley & Sons, Inc., New York, 1961.

Figures 2-9, 2-10, 2-13, and 2-14 From M. Neushul, *Botany,* Hamilton Publishing Co., Santa Barbara, Calif., 1974.

Figures 2-8 and 2-12 (A) and (B) Adapted from P. H. Raven, R. E. Evert, and H. Curtis, *Biology of Plants,* 2nd ed., Worth Publishers, Inc., New York, 1976, pp. 126 and 301. (C) Courtesy J. M. Pettitt.

Figure 2-16 Adapted from E. O. Wilson, T. Eisner, W. L. Briggs, R. E. Dickerson, R. L. Metzenberg, R. D. O'Brien, M. Susman, and W. E. Boggs, *Life on Earth,* Sinauer Inc., Publishers, Stamford, Conn., 1973.

CHAPTER 3

Figure 3-1 Adapted from E. O. Wilson, T. Eisner, W. R. Briggs, R. E. Dickerson, R. L. Metzenberg, R. D. O'Brien, M. Susman and W. E. Boggs, *Life on Earth,* Sinauer, Inc., Publishers, Stamford, Conn., 1973.

Figure 3-2 Drawing by Zdenêk Burian. From J. Angusta and Z. Burian, *Prehistoric Man,* Artia Publishers, Prague, Czechoslovakia.

Figure 3-3 From K. W. Butzer, "Environment, Culture and Human Evolution," *American Scientist,* **64,** 1976.

Figure 3-4 (Top) From J. R. Harlan, "Agricultural Origins: Centers and Noncenters," *Science,* **174,** 1971. Copyright © 1971 American Association for the Advancement of Science.

Figure 3-5 (A), (B), and (C) Courtesy of W. J. Braidwood. From "The Prehistoric Project," The Oriental Institute, The University of Chicago, Chicago, Ill.

Fig. 3-6 Adapted from J. Iversen, "Forest Clearance in the Stone Age," Copyright © March 1954 by Scientific American, Inc. All rights reserved.

Figure 3-7 Adapted from M. D. Coe, "The Chinampas of Mexico." Copyright © July 1964 by Scientific American, Inc. All rights reserved.

Figure 3-8 From J. R. Harlan, "The Plants and Animals That Nourish Man." Copyright © September 1976 by Scientific American, Inc. All rights reserved.

PART II OPENER Jerome Wexler/Photo Researchers.

CHAPTER 4

Figure 4-2 (B) Photo by Neil Hallam & Terence O'Brien.

Figure 4-3 (Diagrams) From T. E. Weier, C. R. Stocking and M. G. Barbour, *Botany,* 5th Edition, John Wiley & Sons, 1974; (Photos) Courtesy Andres S. Bajer.

Figure 4-4 (B) Courtesy J. Heslop-Harrison.

Figure 4-5 From B. A. Meylan and B. G. Butterfield, *Three Dimensional Structure of Wood,* Chapman & Hall, Ltd, London.

Figures 4-6 and 4-23 From W. A. Jensen and F. B. Salisbury, *Botany: An Ecological Approach.* Reprinted by permission of Wadsworth Publishing Co., Inc, Belmont, Calif., 1972.

Figure 4-6 (E) From T. Kerr and I. W. Bailey, "The Cambium and Its Derivative Tissues, Number X. Structure. Optical Properties and Chemical Composition of the So-called Middle Lamella," *Journal of the Arnold Arboretum,* 15, 1934.

Figure 4-7 From Katherine Esau, *Plants, Viruses and Insects,* Harvard University Press, Cambridge, Mass., 1961.

Figures 4-8 (A) and 4-9 From John Troughton and Lesley A. Donaldson, *Probing Plant Structure,* McGraw-Hill, Inc., New York, 1972. Courtesy John Troughton.

Figure 4-10 (A) and (B) From Katherine Esau, *Plants, Viruses and Insects.* Harvard University Press, Cambridge, Mass. 1961. (C) From Sorokin and Thimann, *Protoplasma,* **59**:345, (1964).

Figure 4-11 (A) and (B) Carolina Biological Supply Company; (D) K. Esau, *Anatomy of Seed Plants,* John Wiley & Sons, Inc., New York, 1977.

Figure 4-11 (C), 4-13 (A, left), and 4-15 From C. L. Wilson, W. E. Loomis, and T. A. Steeves, *Botany,* 5th ed. Copyright © 1952, 1957, 1962, 1967, and 1971 by Holt, Rinehart & Winston, Inc., New York. Reprinted by permission of Holt, Rinehart & Winston.

Figure 4-12 (A), (B), (E) and (F) Grant Heilman. (C) and (I) John O. Sumner/National Audubon Society Collection. (D) and (G) From T. E. Weier, C. R. Stocking and M. G. Barbour, *Botany: An introduction to plant biology* 5th ed., John Wiley & Sons, Inc., New York, 1974. (H) Eric J. Hosking/National Audubon Society Collection. (J) and (K) Runk/Schoenberger, Grant Heilman.

Figure 4-13 (A) (Center) From L. Jost, *Vorlesungen über Pflanzenphysiologie,* Gustav Fisher Verlag, Jena 1913; (B) Courtesy Alan Haney, Warren Wilson College, Swannanoa, N. C.

Figures 4-14, 4-17, 4-18, 4-20 and 4-21. From T. E. Weier, C. R. Stocking, and M. G. Barbour, *Botany: An Introduction to Plant Biology,* 5th ed. John Wiley & Sons, Inc., New York, 1974.

Figure 4-16 Courtesy Howard F. Towner, Department of Biology, Loyola Marymount University, Los Angeles.

Figure 4-24 (A) From O'Brien and M. McCully, *Plant Structure and Development,* Macmillan Publishing Company, Inc., New York, 1969. (B) Courtesy Myron C. Ledbetter, Brookhaven National Laboratory.

Figure 4-25 From W. A. Jensen, "Fertilization in Flowering Plants," *Bioscience,* **23,** 1973.

CHAPTER 5

Figure 5-1 From W. H. Hoover, "The Dependence of CO_2 Assimilation in a Higher Plant on the Wavelength of the Radiation," *Smithsonian Miscellaneous Collection,* **95:** (21), 1937.

Figure 5-2 Courtesy S. C. Holt, University of Massachusetts, Amherst.

Figure 5-3 From R. Emerson and C. M. Lewis, "The Photosynthetic Efficiency of Phycocyanin in Chroöcocus, and the Problem of Carotenoid Participation in Photosynthesis," *Journal of General Physiology,* **25**:579–595, 1942.

Figure 5-5 Courtesy Howard Clark Douglas and Edwin S. Boatman, University of Washington, Seattle

Figure 5-7 From D. I. Arnon, "Proton Transport in Photooxidation of Water: A New Perspective on Photosynthesis." *Proceedings of the National Academy of Sciences,* **78**:2942–2946, 1981.

Figure 5-8 (A) Courtesy Dr. M. D. Hatch, CSIRO, Canberra, ACT, Australia. From M. D. Hatch, in *CO_2 Metabolism and Plant Productivity,* eds., R. H. Burris and C. C. Black, University Park Press, Baltimore, Md., 1976. (B) Hawaii Visitors Bureau.

Figure 5-10 Hugh Spencer/National Audubon Society Collection-Photo Researchers.

Figure 5-11 Adapted from O. Björkman and P. Holmgren, "Adaptability of the Photosynthetic Apparatus to Light Intensity in Ecotypes from Shaded and Exposed Habitats," *Physiol. Plant,* **16,** 1963.

CHAPTER 6

Figure 6-3 Courtesy David Ringo, University of California, Santa Cruz.

Figure 6-4 From F. C. Gregory, I. Spear, and K. V. Thimann. "The Interrelation Between CO_2 Metabolism and Photoperiodism in Kalanchoe," *Plant Physiology,* **29**:220–229, 1956.

CHAPTER 7

Figure 7-2 From J. Heslop-Harrison, *Cellular Recognition Systems in Plants,* Edward Arnold, London, 1978. (Scheme first proposed by Singer and Nicholson, 1972.)

Figure 7-5 (Left) From N. A. Maximov, *Plant Physiology,* eds., R. B. Harvey and A. E. Murneek, McGraw-Hill, Inc., New York, 1938. (Right) From T. L. Rost, M. G. Barbour, R. M. Thornton, T. E. Weier, and C. A. Stocking, *Botany: A Brief Introduction to Plant Biology,* John Wiley & Sons, Inc., New York, 1979.

Figure 7-6 From V. A. Greulach and J. E. Adams, *Plants: An Introduction to Modern Botany,* 3rd ed. John Wiley & Sons, New York, 1976.

Figure 7-7 (Above) From D. T. McDougall, *Studies in Tree Growth by the Dendrographic Method,* Carnegie Institution of Washington, Publication No. 462, 1936. (Below) From T. Kozlowski and C. H. Winget, "Diurnal and Seasonal Variation in Radii of Tree Stems," *Ecology,* **45**:149–155, 1964. Copyright © 1964, The Ecological Society of America.

Figure 7-8 From T. L. Rost, M. G. Barbour, R. M. Thorton, E. T. Weier, and C. R. Stocking, *Botany: A Brief Introduction to Plant Biology,* John Wiley & Sons, Inc., New York, 1979.

Figure 7-9 Courtesy John Troughton. From John Troughton and Lesley A. Donaldson, *Probing Plant Structure,* McGraw-Hill, Inc., New York, 1972.

Figure 7-11 Modified from J. V. G. Loftfield, *The Behavior of Stomata,* Carnegie Institution of Washington, Publication No. 34, 1921.

Figure 7-12 From T. A. Mansfield and R. J. Jones, "Effects of Abscisic Acid on Potassium Uptake and Starch Content of Stomatal Guard Cells," *Planta,* **101**:147–158, 1971. Courtesy T. A. Mansfield, University of Lancaster, England.

Figure 7-13 Courtesy Dr. T. A. Steeves, C. L. Wilson, W. E. Loomis, T. A. Steeves, *Botany,* 5th ed., Holt, Rinehart & Winston, Inc., New York, 1971.

Figure 7-14 From K. A. Grossenbacher, "Autonomic Cycle of Rate of Exudation of Plants," *American Journal of Botany,* **26**:107–109, 1939.

Figure 7-15 (A) Bureau of Reclamation, United States Department of the Interior. (B) and (C) Grant Heilman.

CHAPTER 8

Figure 8-1 From V. I. Palladin, *Plant Phyiology* (translated by B. E. Livingston), P. Blakiston's Sons., Philadelphia, Pa., 1926.

Figures 8-2, 8-3, and 8-4 From D. R. Hoagland and D. I. Arnon, "The Water Culture Method for Growing Plants Without Soil," University of California Agricultural Experiment Station Circular 347, 1938.

Figure 8-5 (A) and (B) From D. R. Hoagland, "Inorganic Plant Nutrition," *Chronica Botanica,* Waltham, Mass., 1944.

Figure 8-6 From D. R. Hoagland and T. C. Broyer, "General Nature of the Process of Salt Accumulation by Roots with Description of Experimental Methods," *Plant Physiology,* **11**:471–507, 1936.

CHAPTER 9

Figure 9-1 From B. Frank, *Ueber die Pilzsymbiose der Leguminosen,* Paul Parey, Berlin, 1891.

Figure 9-3 (A) and (B) Courtesy P. J. Dart, United Nations, New York.

Figure 9-4 Courtesy W. J. Brill, University of Wisconsin, Madison.

Figure 9-5 From B. Bowes and D. Callahan, *Bulletin of the Harvard Forest,* 1975–1976. Photo courtesy J. G. Torrey, Harvard University, Cambridge Mass.

CHAPTER 10

Figure 10-1 Grant Heilman.

Figure 10-3 From F. W. Went and K. V. Thimann, *Phytohormones,* Macmillan Publishing Company, Inc., New York, 1937.

Figure 10-4 (A) From H. Ikuma and K. V. Thimann, "Action of Gibberellic Acid on Lettuce Seed Germination," *Plant Physiology,* **35**:553–560, 1960. (B) From S. S. Chen and K. V. Thimann, "Studies on the Germination of Light-Inhibited Seeds of *Phacelia tanacetifolia,*" *Israel Journal of Botany,* **13**:57–73, 1964.

Figure 10-5 (Left) From P. W. Brian, G. W. Elson, H. G. Hemming, and M. Radley, *Journal of the Science of Food and Agriculture,* **5**:602–612, 1954. (Right) Photo courtesy of Dr. Fausto Lona, Istituto Botanico, Parma, Italy.

Figure 10-6 (A) From K. V. Thimann, "Growth and Growth Hormones in Plants," *American Journal of Botany,* **44**:49–55, 1957.

Figure 10-7 From A. V. Chadwick and S. P. Burg, "An Explanation of the Inhibition of Root Growth Caused by Indole-Acetic Acid," *Plant Physiology,* **43**:415–420, 1967.

Figure 10-8 From B. Phinney, "Growth Responses of Single-Gene Dwarf Mutants to Gibberellic Acid." *Proceedings National Academy of Sciences,* **42**:185–189, 1955.

Figure 10-9 From M. Wickson and K. V. Thimann, "The Antagonism of Auxin and Kinetin in Apical Dominance," *Physiologia Plantarum,* **11**:62, 1958.

Figure 10-10 From F. Skoog and C. O. Miller, "Chemical Regulation of Growth and Organ Formation in Plant Tissues Cultured *in Vitro,*" *Symposia of the Society for Experimental Biology,* **11**:118–131, 1957. Cambridge University Press, London, and Academic Press, Inc., New York.

Figure 10-11 (A) From H. A. A. van der Lek, *Over de Wortelvorming van Houtige Stekken.* (Dissertation, Utrecht). H. Veenman en Zonen, Wageningen, Holland, 1925. (B) From P. C. Reano, *Philippine Agriculturist,* **29**:87–99, 1940. (C) From A. I. Hitchcock and P. W. Zimmerman, *Florists' Exchange,* 1938.

Figure 10-12 From G. Melchers, "The Physiology of Flower Initiation." Lectures given at Imperial College, London; published by Max Planck Gesellschaft, Tübingen, 1952.

Figure 10-13 From J. Nitsch and L. Somogyi, *Annales de la Société Nationale de Horticulture,* **16**:466–470, 1958.

Figure 10-14 From H. A. Borthwick, Agricultural Research Service, Plant Industry Station, U. S. Department of Agriculture.

Figure 10-15 Modified from K. Hamner, "Hormones and Photoperiodism," *Cold Spring Harbor Symposia on Quantitative Biology,* **10**:49–59, 1942.

Figure 10-16 From C. Martin and K. V. Thimann, "The Role of Protein Synthesis in the Senescence of Leaves. I. The Formation of Protease," *Plant Physiology,* **49**:64–71, 1972.

Figure 10-17 Courtesy R. H. Wetimore, Harvard University, Cambridge, Mass.

Figure 10-18 From J. P. Nitsch, "Phytohormones et biologie fruitière," *Fruits* (France), **8**:91–97, 1953.

Figure 10-19 From K. V. Thimann, "Toward an Endocrinology of Higher Plants," *Recent Progress in Hormone Research,* **21**:579–596, 1965. Copyright © 1965, by Academic Press, Inc., New York.

CHAPTER 11

Figure 11-1 (A) and (B) The Bettmann Archive.

Figure 11-2 (A) The Bettmann Archive.

Figure 11-7 (A) From F. J. Ayala, "Molecular Genetics Evolution," *Molecular Evolution* (F. J. Ayala, ed.), Sinauer Associates, Inc., Publishers, Sunderland, Mass, 1976. (B) From T. E. Weier, C. R. Stocking, and M. G. Barbour, *Botany: An Introduction to Plant Biology,* 5th ed., John Wiley & Sons, Inc., New York, 1974.

Figure 11-8 From G. L. Stebbins, *Vistas in Botany* (W. Turril, ed.). Copyright © 1959, Pergamon Press, Ltd., New York.

Figure 11-9 Adapted from P. H. Raven, R. E. Evert, and H. Curtis, *Biology of Plants,* 2nd ed., Worth Publishers, Inc., New York, 1976, p. 141.

Figure 11-10 (A) and (B) Courtesy Charles Heiser, Department of Plant Science, Indiana University. Bloomington, Ind. (C), (D), and (E). From T. E. Weier, C. R. Stocking, and M. G. Barbour, *Botany: An Introduction to Plant Biology,* 5th ed., John Wiley & Sons, Inc., New York, 1974.

PART III OPENER Pierre Berger/National Audubon Society Collection-Photo Researchers.

CHAPTER 12

Figure 12-1 and legend from C. L. Stebbins. *Processes of Organic Evolution,* 1966, pp. 42-43. Reprinted by permission of Prentice-Hall, Inc., Englewood Cliffs, N.J.

Figure 12-3 From S. Carlquist, *Island Life,* Natural History Press, Garden City, N.Y., 1965.

Figure 12-4 Courtesy J. Antonovics, Duke University, Durham, N. C.

Figure 12-6 Adapted from J. Clausen, *Stages in the Evolution of Plant Species,* Copyright © 1951 by Cornell University, Ithaca, N. Y. Used by the permission of the publisher, Cornell University Press.

Figure 12-8 From D. Cavagnaro, "Circus of Quercus," *Pacific Discovery,* **27**, 1974.

Figure 12-9 From James F. Shepard, Dennis Bidney, and Elias Shahin, "Potato Protoplasts in Crop Improvement," *Science,* Volume 208, April 4, 1980.

Figure 12-10 Adapted from W. A. Jensen and F. B. Salisbury, *Botany: An Ecological Approach.* Reprinted by permission of Wadsworth Publishing Co., Inc., Belmont, Calif., 1972.

Figure 12-11 W. Atlee Burpee Seed Company.

Figure 12-12 Courtesy Calvin O. Qualset, Department of Agronomy and Range Science, University of California, Davis. Photo by Jack Clark, University of California.

Figure 12-13 Courtesy S. C. H. Barrett, University of Toronto, Toronto, Canada.

CHAPTER 13

Figure 13-1 (A) and (B) Courtesy the Hunt Institute for Botanical Documentation, Carnegie-Mellon University, Pittsburgh, Pa.

Figure 13-2 Culver Pictures.

Figure 13-2 (B) Adapted from modification by Lyman Benson from Bessey's original chart. Original publication: C. E. Bessey, *Annals of the Missouri Botanical Garden,* **2**, 1915.

Figure 13-3 From T. L. Rost, M. C. Barbour, R. M. Thornton, T. E. Weier, and C. R. Stocking, *Botany: A Brief Introduction to Plant Biology,* John Wiley & Sons, Inc., New York, 1979.

Figures 13-4 and 13-6 Adapted from G. H. M. Lawrence, *Taxonomy of Vascular Plants.* Copyright © 1951 by Macmillan Publishing Co., Inc., New York.

Figures 13-5 and 13-10 (B) From L. H. Bailey, *Hortus Third,* Copyright © by Cornell University, L. H. Bailey Hortorium, 1976.

Figures 13-7 and 13-13 From C. L. Wilson, W. E. Loomis, and T. A. Steeves, *Botany,* 5th ed. Copyright © 1952, 1957, 1962, 1967, 1971 by Holt, Rinehart & Winston, Inc. Reprinted by permission of Holt Rinehart & Winston.

Figures 13-8, 13-10 (A) and 13-11 From T. E. Weier, C. R. Stocking, and M. G. Barbour, *Botany: An Introduction to Plant Biology,* 5th ed., John Wiley & Sons, Inc., New York, 1974.

CHAPTER 14

Figure 14-1 From W. B. Clapham, Jr., *Natural Ecosystems,* Figure 6-16, p. 211, Macmillan Publishing Co., Inc., New York, 1973.

Figure 14-2 From T. L. Rost, M. G. Barbour, R. M. Thornton, T. E. Weier, and C. R. Stocking, *Botany: A Brief Introduction to Plant Biology,* John Wiley & Sons, Inc., 1979. (Original from P. W. Richards, *Tropical Rain Forest,* Cambridge University Press.)

Figure 14-3 (A) Courtesy Miller Redwood Company. (B) Copyright © Keith Gunnar/National Audubon Society

Collection-Photo Researchers. (C) Copyright © Ned Haines/Rapho-Photo Researchers.

Figure 14-4 (A) Adapted from R. H. Whittaker, "Plant Populations and the Bases of Plant Indication," *Angewandte Pflanzensoziologie,* 1954. (B) Adapted from R. H. Waring and J. Major, "Some Vegetation of the California Coastal Redwood Region in Relation to Gradients of Moisture, Nutrients, Light and Temperature," *Ecological Monographs,* **34,** 1964 by permission of Duke University Press, Durham, N. C.

Figure 14-5 Adapted from R. H. Whittaker, *Communities and Ecosystems,* 2nd ed., Figure 49, p. 164, Macmillan Publishing Co., Inc., New York 1975. Copyright © 1975 R. H. Whittaker.

Figure 14-7 From W. M. Laetsch, *Plants Basic Concepts in Botany,* Figure 1-14. Copyright © 1979 by Little, Brown & Company, Boston, Mass.

Figure 14-8 Adapted from H. L. Edlin, *Man and Plants,* Aldus Books, London, 1967.

Figure 14-9 Jean H. Langenheim.

Figure 14-10 Copyright © Leonard Lee Rue III 1971/ Photo Researchers.

Figure 14-11 U. S. Department of Agriculture.

Figure 14-12 Grant Heilman.

Figure 14-13 U.S. Forest Service.

Figure 14-14 Fred Bruemmer.

Figure 14-15 Copyright © Charlie Ott/National Audubon Society Collection-Photo Researchers.

Figures 14-8 and 14-20 From R. L. Smith, *Ecology and Field Biology,* 3rd ed. Figure 6-5, p. 163 and Figure 6-2, p. 156. Copyright © 1980 by R. L. Smith. Reprinted by permission of Harper & Row, Publishers, Inc., New York.

Figure 14-21 (E), (F), and (G) Larry West/Bruce Coleman.

Figured 14-22 (B), (C), (D), and (E) Jack Dermid.

CHAPTER 15

Figure 15-1 From V. A. Greulach and J. E. Adams, *Plants: An Introduction to Modern Botany,* 3rd ed., John Wiley & Sons, Inc., New York, 1976.

Figure 15-2 Adapted from N. C. Brady, *The Nature and Property of Soils,* 8th ed., Figure 7-20, p. 190. Copyright © 1974, Macmillan Publishing Co., Inc., New York.

Figure 15-3 From N. C. Brady, *The Nature and Property of Soils,* 8th ed., Figure 7-23, p. 196. Copyright © 1974, Macmillan Publishing Co., Inc., New York.

Figure 15-4 (A) Courtesy Bruce F. Bohor, Geologist, Denver Geological Survey.

Figure 15-4 (B) From N. C. Brady, *The Nature and Property of Soils,* 8th ed., Figure 4-2, p. 74. Copyright © 1974, Macmillan Publishing Co., Inc., New York.

Figure 15-6 From H. O. Buckman and W. C. Brady, *The Nature and Property of Soils,* 7th ed., Figure 12-1, p. 295. Copyright ©, 1969, Macmillan Publishing Co., Inc., New York.

Figure 15-7 (A) and (B) Courtesy V. S!ankis, Ontario Forest Research Center, Maple, Ontario. (C) Courtesy F. H. Meyer, University of Hannover, Institute for Landscape Management and Nature Conservation, West Germany.

Figure 15-8 Copyright © Luc Bouchage/Rapho Photo Researchers.

PART IV OPENER Grant Heilman.

CHAPTER 16

Figures 16-1, 16-4, 16-13 (Left), and 16-15 From H. L. Edlin, *Man and Plants,* Aldus Books, London, 1967 (Figure 16-1, drawing by Edward Poulton; Figure 16-13 and 16-15, drawings by Jill Mackley).

Figure 16-2 Courtesy Archeological Superintendent of Rome, Museo Nazionale Romano.

Figure 16-3 Adapted from A. S. Boughey, *Man and the Environment.* The Macmillan Company, New York and London, 1971.

Figure 16-5 (Top) Copyright © Almasy. (Bottom) Philippines Press and Publications Division, Department of Tourism, Rizal Park, Manila.

Figure 16-6 Gordon N. Converse (Chief Photographer), *The Christian Science Monitor.*

Figure 16-7 (A) and (B) The International Rice Research Institute (C) From R. F. Chandler, "New Horizons for an Ancient Crop," Symposium on World Food Supply, International Botanical Congress, 1969. Printed by Allis Chalmers Company.

Figure 16-8 Courtesy Dr. Galinat. Original drawing by Walton C. Galinat, University of Massachusetts Suburban Experiment Station.

Figure 16-9 Robert S. Peabody Foundation for Archaeology, Andover, Mass.

Figure 16-10 (G) From H. Garrison Wilks, "Teosinte: The Closest Relative of Maize," The Bussey Institute of Harvard University, 1967.

Figure 16-11 Courtesy of the American Museum of Natural History, New York.

Figure 16-12 From Farmers' Bulletin, No. 1744. U. S. Department of Agriculture, Bureau of Plant Industry, Soils and Agricultural Engineering, Washington, D. C.

Figure 16-13 (Right) Courtesy K. W. Hilu, from K. W. Hilu and J. M. J. De Wet, "Racial Evolution in *Eleusine coracana* ssp. Coracana (Finger Millet)," *American Journal of Botany,* **63**:1311–1318, 1976.

Figure 16-14 Courtesy Rodale Press, Inc., Emmaus, Pa.

CHAPTER 17

Figure 17-1 (Top, left and right) Martin H. Zimmermann (Harvard Forest, Petersham, Mass.) (Bottom) Martin H. Zimmermann, *Science,* **133**:73, 161.

Figures 17-2 (A), 17-4 (Left) and 17-5 (A) and (B) From H. L. Edlin, *Man and Plants*, Aldus Books, London, 1967 (Figure 17-2, drawing by Jill Mackley; Figure 17-5 A and B, drawing by Edward Poulton).

Figure 17-2 (B) From W. G. Freeman and S. E. Chandler, *The World's Commercial Products*, London, 1906.

Figure 17-3 (Top and bottom) Grant Heilman.

Figure 17-4 Courtesy Don Goldman, Heritage Conservation and Recreation Service, U. S. Department of the Interior.

Figure 17-5 (C) Grant Heilman.

Figure 17-6 (A) and 17-7 From G. B. Masefield, S. G. Harrison, M. Wallis, *The Oxford Book of Food Plants*, illustrated by B. E. Nicholson. Copyright © Oxford University Press, 1969.

Figure 17-6 (B) Grant Heilman.

Figure 17-8 Courtesy Tate & Lyle Ltd., Reading, England.

CHAPTER 18

Figure 18-1 (A) Copyright © 1976 Jeanne White/Photo Researchers. (B) Copyright © Karl H. Maslowski/National Audubon Society Collection-Photo Researchers (C) Copyright © Lynwood M. Chace/National Audubon Society Collection-Photo Researchers. (D) Courtesy John N. A. Lott (E) and (F) Copyright © Charles J. Ott/National Audubon Society Collection-Photo Researchers.

Figure 18-5 (Top) From R. J. Weaver et al., "Response of Clusters of Vinifera Grapes to 2, 4-Dichlorophenoxyacetic Acid and Related Compounds," *Hilgardia*, Vol. 31, No. 5, August 1961. (Center) From R. J. Weaver, "Use of Gibberellins In Grape Production." *The Blue Anchor Magazine*, November 1958. (Bottom) From R. J. Weaver et. al, "Effect of Kinins on Fruit Set and Development in *Vitis vinifera*," *Hilgardia*, Vol. 37, No. 7, January 1966.

Figure 18-6 (Top and bottom) From Cheong-Yin Wong, "Induced Parthenocarpy of Watermelon, Cucumber and Peppers by Use of Growth Promoting Substances," *Proceedings of the American Society for Horticultural Science*, Vol. 36, 1938.

Figure 18-7 From J. A. Biale, "The Ripening of Fruit. Copyright © May 1954 by Scientific American, Inc. All rights reserved.

Figures 18-8, 18-9, 18-10 (A), and 18-12 From G. B. Masefield, S. G. Harrison, M. Wallis, *Oxford Book of Food Plants*, illustrated by B. E. Nicholson. Copyright © Oxford University Press, New York, 1969.

Figure 18-10 (B) National Maritime Museum.

Figure 18-11 (A) From R. F. Scagel, R. J. Bandoni, G. E. Rouse, W. B. Schofield, J. R. Stein, and T. M. C. Taylor, *Plant Diversity: An Evolutionary Approach*. Reprinted by permission of Wadsworth Publishing Co., Inc., Belmont, Calif., 1969. (B) From T. L. Rost, M. G. Barbour, R. M. Thornton, T. E. Weier, and C. R. Stocking. *Botany: A Brief Introduction*, John Wiley & Sons, Inc., New York 1979. (C) Adapted from S. B. Hrdy and W. Bennett, "The Fig Connection," *Harvard Magazine*, September-October 1979.

Figure 18-13 Courtesty of the Trustees of the British Museum.

Figures 18-14 and 18-15 Grant Heilman.

CHAPTER 19

Figure 19-1 Adapted from H. L. Edlin, *Man and Plants*, Aldus Books, London, 1967.

Figure 19-2 Adapted from T. E. Weier, C. B. Stocking, and M. G. Barbour, *Botany: An Introduction to Plant Biology*, 5th ed., John Wiley & Sons, Inc., New York, 1974.

Figure 19-4 Adapted from W. W. Robbins, T. E. Weier, and C. B. Stocking, *Botany*, 2nd ed., John Wiley & Sons, Inc., New York, 1957.

Figure 19-5 (D) Grant Heilman.

Figure 19-8 (A) Grant Heilman, (B) By M. C. Lincoln, from C. L. Wilson, W. E. Loomis, and T. A. Steeves, *Botany*, 5th ed. Copyright © 1952, 1957, 1962, and 1971 by Holt, Rinehart & Winston, Inc., New York.

Figure 19-9 (A), (B), and (C) Jean H. Langenheim.

Figure 19-10 (A) Musée de l'Homme, Paris. (B) The Mansell Collection, London.

Figure 19-11 Walter Chandoha.

Figure 19-12 (A) Courtesy Shell Photos, London. (B) Peter Buckley/Photo Researchers.

Figure 19-13 (A) and (B) Jean H. Langenheim.

CHAPTER 20

Figure 20-1 From R. D. de la Rivière, *Pasteur, Extraits de ses Oeuvres*. Gauthiers-Villars, Paris, 1967.

Figure 20-2 (A) The Fleischmann Laboratories, Standard Brands, Inc. (B) From H. J. Phaff, M. W. Miller and E. M. Mrak, *The Life of Yeasts*, 2nd ed., Harvard University Press, Cambridge, Mass., and London, 1978.

Figure 20-3 Jean H. Langenheim.

Figure 20-4 J. Pavlovsky/Rapho-Photo Researchers.

Figure 20-5 Courtesy Wine Institute.

CHAPTER 21

Figure 21-1 (A) and (B) Courtesy A. Frey-Wyssling, Zurich, Switzerland.

Figure 21-2 and 21-7 (A) From I. W. Bailey, "The Walls of Plant Cells," American Association for the Advancement of Science, Publication No. 14, 1934.

Figure 21-3 (A), (B), and (C) From I. W. Bailey, "Aggregations of Microfibrils and Their Orientations in the Secondary Wall of Coniferous Tracheids," *American Journal of Botany*, **44** (5):415–418, 1954.

Figure 21-4. From Raoul d'Harcourt, *Textiles of Ancient Peru* (translated by S. Brown). University of Washington Press, Seattle, 1974.

Figure 21-5 (Left and right) National Cotton Council, Memphis, Tenn.

Figure 21-6 Courtesy James McDonald Stewart, from *American Journal of Botany*, **62**:723–730, 1975.

Figure 21-7 (A) From I. W. Bailey. (B) From R. W. Schery, *P ants for Man*, 2nd ed., Prentice-Hall, Inc., Englewood Cliffs, N. J., 1972. (D) courtesy U. S. Department of Agriculture.

Figure 21-8 (A) and (B) From Stephen J. Zand, *Kapok: A Survey of Its History, Cultivation and Use*, 1941. (C) W. H. Hodge/Peter Arnold.

Figure 21-9 (Left) From H. G. Edlin, *Man and Plants*, Aldus Books, London, 1967.

Figure 21-10 (A), (B), and (C) FAO.

Figure 21-11 J. Ph. Charbonnier, "Realités"/Photo Researchers.

Figure 21-12 UN Photo, issued by FAO.

Figure 21-13 (A) K. V. Thimann. (B) W. H. Hodge/Peter Arnold.

CHAPTER 22

Figures 22-1, 22-2 22-3, 22-5, 22-7, 22-11 (B) From T. E. Weier, C. R. Stocking and M. G. Barbour, *Botany: An Introduction to Plant Biology*, 5th ed., John Wiley & Sons, Inc., New York, 1974.

Figure 22-4 From B. A. Meylan and B. G. Butterfield, *Three Dimensional Structure of Wood*, Chapman & Hall, Ltd., London.

Figure 22-6 From W. W. Robbins, T. E. Weier, and C. B. Stocking, *Botany*, 2nd ed., John Wiley & Sons, Inc., New York, 1957. Original from U. S. Department of Agriculture Misc. Circ. 66.

Figure 22-8 (A) Mary M. Thacher/Photo Researchers. (B) Cover and title page of "An Act for the Preservation of White and the Other Pine Trees Growing in Her Majesty's Colonies," 1710. Courtesy Chronica Botanica.

Figure 22-9 (A), (B), and (C) Courtesy St. Regis Paper Company.

Figure 22-10 (A) and (B) Courtesy Field Museum of Natural History, Chicago.

Figure 22-11 (A) Courtesy American Plywood Association.

Figure 22-12 (A) The British Library, Department of Oriental Manuscripts and Printed Books. (B) Courtesy Westvaco Corporation.

Figure 22-13 Courtesy Viscosuisse.

Figure 22-14 Courtesy Portuguese National Tourist Office.

CHAPTER 23

Figure 23-1 (Top) From G. H. M. Lawrence, *Taxonomy of Vascular Plants*. Copyright © 1951 by Macmillan Publishing Co., Inc. New York. (Photo) Courtesy W. Rauh, University of Heidelberg, Federal Republic of Germany.

Figure 23-2 From T. L. Rost, M. C. Barbour, R. M. Thornton, T. E. Weier, and C. R. Stocking, *Botany: A Brief Introduction to Plant Biology*, John Wiley & Sons, Inc., New York, 1979.

Figure 23-3 From *A Laboratory Manual for General Botany*, 5th ed., by Margaret K. Balbach, Lawrence Bliss, and Harry J. Fuller, copyright © 1977, 1969, 1962, 1956, and 1950. Holt, Rinehart & Winston. Reprinted by permission of Holt, Rinehart & Winston.

Figure 23-4 Adapted from Folke Dovring, "The Soybean." Copyright © February 1974 by Scientific American, Inc. All rights reserved.

Figures 23-5, 23-6, and 23-8 From F. Bianchini, F. Corbetta, and M. Pistoia, *The Complete Book of Fruits and Vegetables*. Copyright Arnoldo Mondadori, 1973. Crown Publishers, Inc. New York, 1976.

Figure 23-7 From J. W. Purseglove, *Tropical Crops—Dicotyledons*, John Wiley & Sons, Inc., New York, 1968.

Figure 23-9 From Board on Science and Technology for International Development Report, "The Winged Bean." National Academy of Sciences, Washington, D. C., 1975.

Figure 23-10 Courtesy Ray Dickens, Auburn University, Auburn, Ala.

CHAPTER 24

Figure 24-1 (Left) Reprinted with permission of Macmillan Publishing Co., Inc., New York, from *Poisonous Plants of the U.S.* by W. C. Muenscher. Copyright © 1951 by Macmillan Publishing Co., Inc. (Right) From J. P. Smith, "California's 'Borgia' Plants," *Fremontia* 1, No. 3, 3–7, 1973. Courtesy James P. Smith, Humboldt State University, Arcata, Calif.

Figure 24-2 From J. W. Purseglove, *Tropical Crops—Dicotyledons*, Vol. 2, John Wiley & Sons, Inc., New York, 1969.

Figure 24-3 Grant Heilman.

Figure 24-4 (A) Copyright © Henry Mayer/National Audubon Society Collection-Photo Researchers. (B) Courtesy W. Rauh, University of Heidelberg, Federal Republic of Germany. (C) Courtesy HCD de Wit, The Netherlands. (D) Grant Heilman.

Figure 24-6 (Left) Courtesy University of California, Lawrence Berkeley Laboratory. (Right) Courtesy A. D. Krikorian, State University of New York at Stonybrook.

Figure 24-7 (Above) From Freeman and Chandler, 1906. (Below) W. H. Hodge/Peter Arnold.

Figure 24-8 From William Withering "An Account of the Foxglove and Some of Its Medical Uses," 1785. Courtesy Aldus Archives, London.

Figure 24-9 W. Rauh, University of Heidelberg, Federal Republic of Germany.

Figure 24-10 From C. J. Alexopou'os, *Introductory Mycology*, John Wiley & Sons, Inc., New York, 1952.

Figure 24-11 (A) Courtesy Chas Pfizer & Company, Inc. (B) Prepared by M. L. Littman, Army Institute of Pathology.

CHAPTER 25

Figure 25-2 From H. L. Edlin, *Man and Plants,* Aldus Books, London, 1967.
Figure 25-3 (A) and (B) Jean H. Langenheim.
Figure 25-4 From G. B. Masefield, S. G. Harrison, and M. Wallis, *The Oxford Book of Food Plants,* Illustrated by B. E. Nicholson. Copyright © Oxford University Press, New York, 1969.
Figure 25-5 (A), (B), and (C) Photo by Isaias Alves, CEPLAC, Itabuna, Brazil.
Figure 25-6 (A) and (B) Photo by Isaias Alves, CEPLAC, Itabuna, Brazil.
Figure 25-7 Photo Library, FAO, photo by W. Williams.
Figure 25-8 (A) Photo Library, FAO, photo by C. Sanchez. (B) Copyright © Paolo Koch/Rapho-Photo Researchers. (C) Courtesy Ceylon Tea Center, London.
Figure 25-9 Jean H. Langenheim.
Figure 25-10 (A) W. H. Hodge/Peter Arnold. (B) Jean H. Langenheim.

CHAPTER 26

Figure 26-4 (A) Walter Chandoha (B), (C), (D), and (F) Grant Heilman. (E) George Whiteley/Photo Researchers.
Figure 26-5 (A) and (B) The Mansell Collection, London.
Figure 26-7 (A) Jean H. Langenheim. (B) K. V. Thimann.
Figure 26-8 (A), (C), (D), and (E) U. S. Department of Agriculture. (B) Courtesy Petoseed Company, Inc., Calif.
Figure 26-9 National Library of Medicine.
Figures 26-10 and 26-11 From Frederic Rosengarten, Jr., *The Book of Spices,* Livingston Publishing Company, Philadelphia, Pa. 1969. Original from Gehring and Neiweiser, Bielefeld, Federal Republic of Germany.
Figure 26-12 (A) and (B) Madagascar Vanilla Growers. (Right) Drawing by T. Beamish, Seychelles, from Frederic Rosengarten, Jr., *The Book of Spices,* Livingston Publishing Company, Philadelphia, Pa., 1969.
Figure 26-13 (A) Courtesy Bulgarian Tourist Office. (B) and (C) Photos by Appollot. Grasse, France.
Figure 26-14 (A) Adapted from Bachofen-Echt, *Der Bernstein und seine Einschlüsse,* Springer-Verlag, Wien, 1949. (B) and (C) From Arnolds Spekke, *The Ancient Amber Routes and the Geographical Discovery of the Eastern Baltic,* M. Goppers, Stockholm, Sweden, 1957.
Figure 26-15 (A) From A. Fahn, *Plant Anatomy,* 2nd ed., Pergamon Press, Ltd. (B) From B. A. Meylan and B. G. Butterfield, *Three Dimensional Structure of Wood,* Chapman & Hall, Ltd.
Figure 26-16 (A) Runk-Schoenberger/Grant Heilman. (B) Jean H. Langenheim.
Figure 26-17 Jean H. Langenheim.
Figure 26-18 National Library of Medicine.
Figure 26-19 From A. Fahn, *Plant Anatomy,* Reprinted with permission of Pergamon Press, Ltd.
Figure 26-20 United Nations.
Figure 26-21 Photo by Ghillean Prance, New York Botanical Garden.
Figure 26-22 (A) and (B) Crown Copyright: reproduced with the permission of the Controller of Her Majesty's Stationery Office and of the Royal Botanic Gardens, Kew.
Figure 26-23 (A) U. S. Department of Agriculture. (B) F. Addicott, Department of Botany, University of California, Davis.
Figure 26-24 (A) Courtesy William Wrigley Jr. Company. (B) W. H. Hodge/Peter Arnold.

CHAPTER 27

Figure 27-1 (Both) B. E. Juniper, School of Botany, University of Oxford.
Figure 27-3 (Left) Aldus Archives, from H. L. Edlin, *Man and Plants,* Aldus Books, London, 1967. (Right) Fritz Henle/Photo Researchers.
Figure 27-4 Grant Heilman.
Figure 27-5 From G. B. Masefield, S. G. Harrison, M. Wallis, *The Oxford Book of Food Plants,* Illustrated by B. E. Nicholson. Copyright © Oxford University Press, New York, 1969.
Figure 27-6 Jean H. Langenheim.
Figure 27-7 Grant Heilman.
Figure 27-8 (A) and (B) Jean H. Langenheim.
Figure 27-9 (A), (B), and (C) Photos by Kelley Dwyer, 1981, Jojoba Commodities Group, North Hollywood, Calif.

PART V OPENER Paolo Koch/Rapho-Photo Researchers.

CHAPTER 28

Figure 28-1 (A), (B) and (C) From G. Mosher, *Kyoto: A Contemplative Guide,* Charles E. Tuttle Company, Inc., Tokyo, Japan, 1964.
Figure 28-2 Gene Heil/Photo Researchers.
Figure 28-3 From John Gerard, *The Historie of Plants,* 1636. Reprinted by Minerva Press, London 1971.
Figure 28-4 Editorial Photocolor Archives.
Figure 28-5 Srenco Photography.
Figures 28-5 and 28-6 From V. A. Greulach and J. E. Adams, *Plants: An Introduction to Modern Botany,* 3rd ed. John Wiley & Sons, Inc., New York, 1976.
Figure 28-8 Copyright © Chuck Ashley/Rapho-Photo Researchers.

Figure 28-9 (Top and bottom) Courtesy G. Hussey, John Innes Institute, Norwich, England.
Figure 28-11 K. V. Thimann.
Figure 28-12 Editorial Photocolor Archives.
Figure 28-13 Musée Guimet and Réunion des Musées Nationaux, France.
Figure 28-14 Courtesy Rinzai-ji, Inc., Los Angeles, Calif., Myoshin-ji-Betsuin, Japan.

CHAPTER 29

Figures 29-2 and 29-3 From C. J. Alexopoulos, *Introductory Mycology,* John Wiley & Sons, Inc., New York, 1952.
Figure 29-4 (A) Photo by Howard H. Lyon, Department of Plant Pathology, Cornell University, Ithaca, N.Y. (B) From P. H. Raven, R. E. Evert, and H. Curtis, *Biology of Plants,* 2nd ed., Worth Publishers, Inc., New York, 1976, p. 218. (C) From E. C. Stakman and J. G. Harrar, *Principles of Plant Pathology,* Ronald Press, New York, 1957.
Figure 29-5 (A) Photo by Howard H. Lyon, Department of Plant Pathology, Cornell University, Ithaca, N.Y.
Figures 29-5 (B), and 29-7 From G. N. Agrios, *Plant Pathology,* Academic Press, Inc., New York, 1969.
Figure 29-6 Connecticut Agricultural Experiment Station, New Haven.
Figure 29-8 (A) and (B) U. S. Department of Agriculture.
Figure 29-9 (A) From Charles E. Bracker, Purdue University. (B) From E. C. Stakman and J. G. Harrar, *Principles of Plant Pathology,* Ronald Press, New York, 1957.
Figure 29-10 K. V. Thimann.
Figure 29-11 From Erwin F. Smith, "The Structure and Development of Crown Gall: A Plant Cancer," U. S. Department of Agriculture, June 29, 1912.
Figure 29-12 (A) and (B) Runk/Schoenberger/Grant Heilman.
Figure 29-12 (A) Dupont. (B) Grant Heilman.
Figures 29-14 (A) and (B) and 29-15 Grant Heilman.

CHAPTER 30

Figure 30-3 Grant Heilman.

CHAPTER 31

Figure 31-1 National Park Service, photo by Bill Jones.
Figure 31-2 (A) Copyright © 1980 Stephanie Dinkins. (B) Noel Habgood FRPS.
Figure 31-4 From E. Odum, "The Strategy of Ecosystem Development," *Science* **164,** 1969. Copyright © 1969, American Association for the Advancement of Science.
Figure 31-5 (A) U. S. Forest Service (B) Charles Lockard–United States Forest Service.
Figure 31-6 Georg Gerster/Rapho-Photo Researchers.
Figure 31-7 (A) and (B) United States Forest Service, San Bernardino, National Forest.
Figure 31-8 George Bellerose/Stock, Boston, Mass.

INDEX OF PERSONAL NAMES

SUBJECT INDEX

This index includes terms, subjects and plant names (both common and scientific). Glossary entries are not included. Bold-face numerals refer to illustrations.

continued from inside front cover

HUMANS

YEARS AGO (THOUSANDS)	DATE	
6×10^3	4000 BC	Origin of agriculture in China
5×10^3	3000 BC	**Rise of Egyptian Civilization**
	2800	Irrigation in Nile Valley: first use of papyrus
	2700	Medicinal plants developed in China
4×10^3	2000 BC	
	1700	Oldest trees now living germinated
3×10^3	1000 BC	Development of spice trade routes

PLANT SCIENCE

YEARS AGO (THOUSANDS)	DATE	
	370-286	Theophrastus, first botanist
	75 BC	Lucretius' ideas of origin of iife
2×10^3	**Birth of Christ**	
	55 AD	**Julius Caesar conquers Gaul and Britain**
	105	Rise of Roman agriculture; selection of seeds. Paper invented in China
		Constantinople ("New Rome") Established
	800 (ca)	Harun al Raschid irrigates the desert around Baghdad. Agriculture in eastern North America
1×10^3	1000 AD	
	1492	**Columbus Discovers America** Period of active exploration for spices
	1500	The great herbals
	1680-1710	Hooke and Leeuwenhoek, microscopy
	1700	**Industrialization Begins**
250	1727	Stephen Hales, sap movement discovered